STUDIES IN CLASSICAL AND QUANTUM NONLINEAR OPTICS

Proceedings of the Fifth International Topsøe Summer School on Nonlinear Optics, Aalborg, Denmark, August 3-8, 1992

STUDIES IN CLASSICAL AND QUANTUM NONLINEAR OPTICS

Edited by

Ole Keller

Nova Science Publishers, Inc.

Art Director: Christopher Concannon
Graphics: Elenor Kallberg and Maria Ester Hawrys
Manuscript Coordinator: Roseann Pena
Book Production: Tammy Sauter, and Michelle Lalo
Circulation: Irene Kwartiroff and Annette Hellinger

Library of Congress Cataloging–in–Publication Data available upon request

ISBN 1-56072-168-5

6080 Jericho Turnpike, Suite 207
Commack, New York 11725
Tele. 516-499-3103 Fax 516-499-3146
E Mail Novasci1@aol.com

Printed in the United States of America

CONTENTS

PREFACE

In the period August 3 - August 8, 1992, the Fifth International Topsøe Summer School of Nonlinear Optics was held at the University of Aalborg. The school was arranged by the Institute of Physics at the University of Aalborg in cooperation with the Danish Research Academy in an attempt to stimulate the education of the young physicists, and to strengthen the international cooperation within the domain of nonlinear optics.

The recognized International Topsøe Summer Schools, sponsored by Haldor Topsøe A/S, are devoted to subjects within the natural sciences and are held annually at one of the Danish universities.

The present Summer School also was the second one in a series of International Aalborg Summer Schools on Nonlinear Optics. The First International Summer School was held in the summer of 1989 at the Institute of Physics.

This book is based on the lectures and contributed papers presented at the school. About 75 experienced and younger scientists from 20 different countries, viz. Bulgaria, China, Denmark, Egypt, England, Estonia, France, Germany, India, Lithuania, Norway, Poland, Romania, Russia, Scotland, Spain, Sweden, Switzerland, USA, and Uzbekistan, participated, and 34 invited lectures were given by 18 distinguished physicists. Each of the invited speakers was asked to write a review article within their respective fields. Besides these articles, the book contains short papers written by many of the remaining participants.

The programme of the summer school emphasized a treatment of basic physical properites related to the nonlinear interaction of light and matter. Nonlinear effects in atomic systems and in condensed matter were discussed, both theoretical and experimental aspects covered, and classical phenomena as well as phenomena requiring a field-quantized description were presented. In a balanced manner themes of current interest in the research literature and general principles of importance for newcomers to the field of nonlinear optics were dealt with.

Specific financial support was provided by Det Obelske Familiefond, Sparekassen Nordjylland, and the University of Aalborg.

Here, I wish to express my thanks to Dr. Haldor Topsøe, who took part in the opening of the school, to the invited speakers for undertaking the burden of writing the review articles and giving the lectures, to the remaining participants without whom the school would never have been held (!), and to coworkers at the Institute of Physics for their outstanding practical contributions in one way or another.

It is my hope that this book, albiet covering only a tiny fraction of the important topics in nonlinear optics, can indicate how charming and exciting this branch of physics, which we all love, is.

Aalborg, August 1993

Ole Keller

Studies in Classical
and Quantum Nonlinear Optics

ISBN 1-56072-168-5

OPTICAL PHASE CONJUGATION AND ITS APPLICATIONS TO RESONANCE FLUORESCENCE

Henk F. Arnoldus
Department of Physics, Mendel Hall, Villanova University,
Villanova, Pennsylvania 19085, USA

Thomas F. George
Departments of Chemistry and Physics,
Washington State University, Pullman Washington 99164-1046, USA

ABSTRACT

Optical phase conjugation via four-wave mixing in a nonlinear crystal is considered. Two strong counterpropagating pump lasers irradiate a medium, and a probe beam is incident under a different angle. The three fields couple through the third-order susceptibility, with the result that a phase-conjugated image of the probe is generated. The frequency of the produced field is shifted, and its amplitude can be expressed as a Fresnel coefficient times the amplitude of the incident wave. These Fresnel reflection coefficients can be evaluated exactly, although their form depends in a complicated way on the various parameters of the system.

As an application, we study the emission of fluorescence by an atom near such a phase conjugator. This requires a quantized version of four-wave mixing. It appears that an atom near a four-wave mixer can excite spontaneously via emission of a photon. In order to satisfy energy conservation, two pump photons are annihilated in this process. Essentially, the two pump photons are emitted by the medium towards the atom, and are absorbed by the atom in two stimulated excitations. The net effect is the emission of a fluorescent photon and a spontaneous excitation from the ground state to the excited state. The spectral distribution of these photons has been evaluated.

1. INTRODUCTION

An electromagnetic plane wave has an electric field with a space ($\vec{r}$) and time (t) dependence given by

$$E(\vec{r},t) = e^{i\vec{k}\bullet\vec{r}-i\omega t} + e^{-i\vec{k}\bullet\vec{r}+i\omega t} \quad . \tag{1}$$

For such a wave, its phase-conjugated replica is by definition a wave with the same time dependence, but with a complex conjugated phase (spatial part) with respect to the original wave. Therefore, the phase conjugated field of (1) is

$$E(\vec{r},t)_{pc} = e^{-i\vec{k}\bullet\vec{r}-i\omega t} + e^{+i\vec{k}\bullet\vec{r}+i\omega t} \quad . \tag{2}$$

Comparison with (1) shows that this field is related to the original field according to

$$E(\vec{r},t)_{pc} = E(\vec{r},-t) \quad , \tag{3}$$

so that the phase-conjugated image of a plane wave is equal to the time-reversed wave. This is again a plane wave, and it counterpropagates the original wave. Therefore, (perfect) phase conjugation is identical to time reversal. Any device that can transform an optical wave front into its time-reversed replica is called a phase conjugator (PC). Obviously, such a device can not exist because it would violate causality. However, there are many optical configurations which can generate a phase-conjugated image with high accuracy. The first experimental demonstration of an optical phase conjugator was given by Zel'dovich *et al* [1]. They used a cell with methane gas, and stimulated Brillouin scattering in the gas effectively turns a field into its phase-conjugated image. A few years later, phase conjugation through four-wave mixing in liquids and solids was realized [2 - 4], and contemporary PC's utilize photorefractive media. The great advantage of these devices is that there is no need for external pump lasers. Although, a practical limitation of photorefractive PC's is the slow response of the medium.

A different topic in contemporary optics is the influence of the presence of media on spectroscopic properties of atoms and molecules. An excited state |e> of a two-level atom has a radiative lifetime of 1/A, with A the Einstein coefficient for spontaneous decay to the ground state |g>. With ω_o the transition frequency and $\vec{\mu}$ the atomic dipole moment operator, the value of A for decay in empty space is given by

$$A = \frac{\omega_o{}^3}{3\pi\varepsilon_o\hbar c^3}|\langle e|\vec{\mu}|g\rangle|^2 \quad . \tag{4}$$

When the atom is close (within a few optical wavelengths) to the surface of a medium, say a dielectric or a mirror, then its lifetime can change dramatically. Since the Einstein coefficient is also equal to the full width at half maximum of the absorption profile of the transition, it follows that the medium also affects the shape (and strength) of this spectral line. In a similar way, the resonance fluorescence spectrum of a laser-driven atom near a medium is modified. Suppose that the surface of the medium is the plane z = 0, that the medium occupies the region z < 0, and that the atom is located at position $\vec{r} = h\vec{e}_z$, with h > 0. Then the Einstein coefficient will become a function of h and depend on the dielectric constant of the medium. In addition, the coefficient A will depend on the orientation of the atomic transition dipole moment matrix element $<e|\vec{\mu}|g>$, because of the loss of spherical symmetry. An atom with a perpendicular dipole moment, with respect to the surface, will decay at a different rate than an atom with its dipole moment parallel to the surface. Decay rates for other orientations can than be expressed in terms of these two special ones. This is due to the remaining cylindrical symmetry of the configuration [5]. More general, in the presence of linear media of any kind and shape, all changes in the optical properties of an atom enter through a modification of the Einstein coefficient. In this paper we shall consider an atom near the surface of a four-wave mixing PC. As it turns out, for such a nonlinear medium the modifications of atomic behavior are quite different. Besides a change in Einstein coefficient, there is also a change in the dynamical properties of the atom. Furthermore, the mechanism of emission of fluorescence near a PC is essentially different than near a linear medium.

2. THE FOUR - WAVE MIXING PHASE CONJUGATOR

We shall consider a phase conjugator which operates through four-wave mixing,

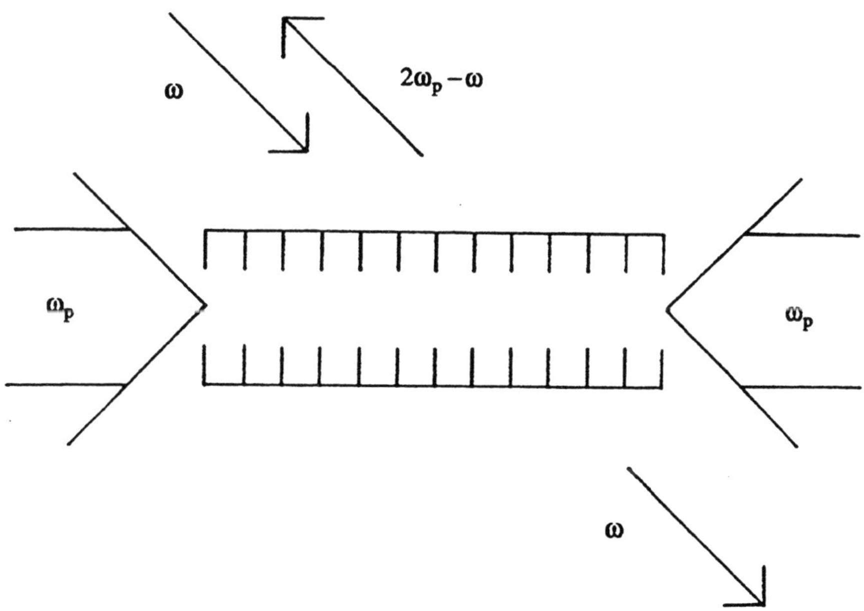

Fig. 1. Schematic setup of a four-wave mixing phase conjugator.

in a configuration as shown schematically in Fig. 1. A layer of nonlinear material with thickness d is irradiated by two strong counterpropagating pump lasers with frequency ω_p. The wave incident from the region z > 0 has frequency ω, wave vector $\vec{k}$, angle of incidence θ_i, amplitude E_o, and polarization vector $\vec{\varepsilon}_\sigma$; e.g., this field is given by

$$\vec{E}(\vec{r},t) = E_o \, Re \, \vec{\varepsilon}_\sigma e^{i(\vec{k}\bullet\vec{r}-\omega t)} \quad . \tag{5}$$

The subscript σ indicates the polarization, which can be surface polarized $(\sigma = s)$ or plane polarized $(\sigma = p)$. The vector $\vec{e}_\sigma$ is perpendicular to the wave vector $\vec{k}$, and the dispersion relation in vacuum gives the restriction $k = \omega / c$. The medium is assumed to be transparent (dielectric constant equal to unity) so that there is no ordinary reflection into the specular direction. The four-wave mixing is brought about by the third-order susceptibility tensor $\vec{\vec{\chi}}^{(3)}$, which couples the incident wave to an almost counterpropagating wave with frequency $2\omega_p - \omega$. The phase-conjugated wave can thus be represented as

$$\vec{E}(\vec{r},t)_{pc} = E_o \, Re \, P_\sigma \vec{\varepsilon}_\sigma' \, e^{i(-\vec{k}'\bullet\vec{r}-(2\omega_p-\omega)t)} \quad . \tag{6}$$

The new wave vector $-\vec{k}'$ is such that $\vec{k}'$ has the same parallel component with respect to the surface of the medium as $\vec{k}$, as follows from the boundary conditions on the electric and magnetic fields at the surface. The magnitude of $\vec{k}'$ is equal to

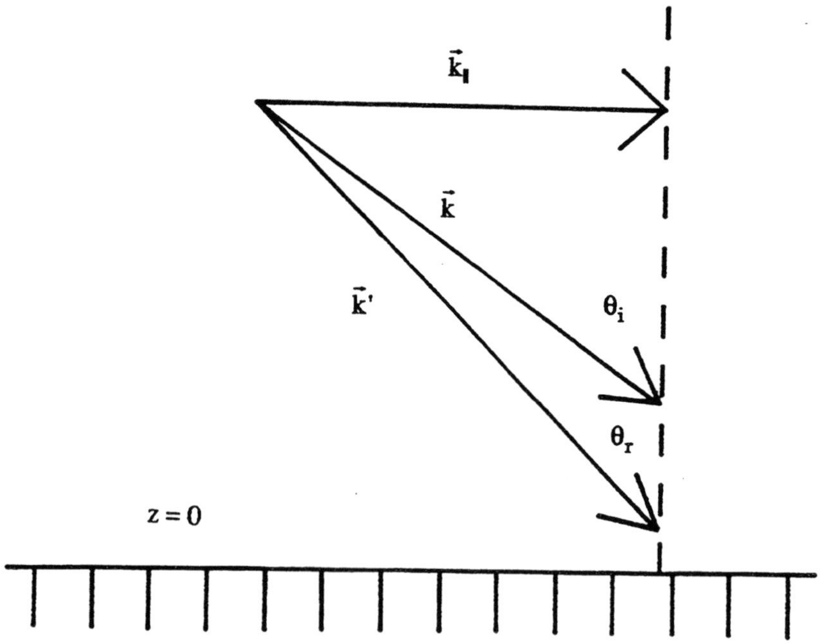

Fig. 2. Diagram for the calculation of the angle of reflection, as given by (7).

$k' = (2\omega_p - \omega)/c$, as follows from the dispersion relation in vacuum. This value of $\vec{k}'$ determines the angle of reflection θ_r, given the incident wave vector $\vec{k}$ and the frequency mismatch. To be specific, the reflection angle is related to the angle of incidence according to

$$\sin(\theta_i) = \left(2\frac{\omega_p}{\omega} - 1\right)\sin(\theta_r) \quad , \tag{7}$$

as follows from the construction in Fig. 2. Notice that the phase-conjugated wave travels opposite to $\vec{k}'$. The polarization of the reflected wave is the same as for the incident wave (either s- or p-polarized), except that $\vec{\varepsilon}_\sigma$ is perpendicular to the vector $\vec{k}'$. Finally, the number P_σ is the complex-valued Fresnel reflection coefficient. This number depends in a complicated way on the parameters in the problem. The derivation of explicit expressions (different for s- and p-polarization) can be found in Ref. 6.

3. FRESNEL COEFFICIENTS

The absolute value of a Fresnel reflection coefficient P_σ is effectively the ratio of the amplitudes of the reflected field and the incident field, just as for linear media. The

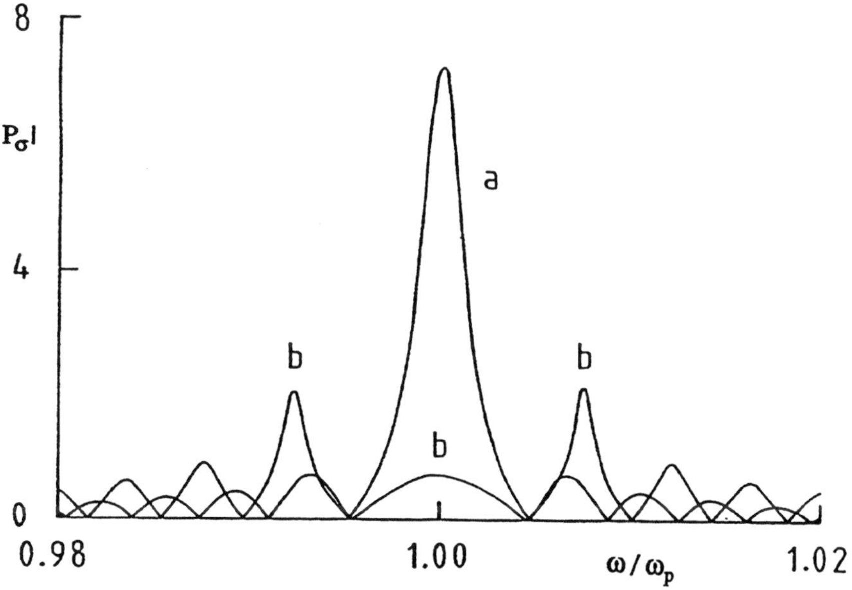

Fig. 3. Absolute value of the Fresnel coefficients as a function of the frequency ω, in units of the pump frequency ω_p. Curve *a* is for s-polarization, and curve *b* is for p-polarization. All other parameters are the same for both curves.

phase, $\arg(P_\sigma)$, represents the phase shift between the two waves. For linear media, this is always zero or π, but for nonlinear media this phase shift can have any value. An important property of the Fresnel coefficients is that they depend on the frequency ω of the incident field. Figure 3 shows the dependence of $|P_\sigma|$ on ω, and as could be expected, there is a sharp resonance at the pump frequency ω_p. It should be noted that this dependence is a geometrical effect, and not the frequency dependence of the susceptibility tensor. This has not been taken into account, and would give rise to an additional frequency dependence. Recall that for linear media there is essentially no frequency dependence for Fresnel coefficients, except through the frequency dependence of the dielectric constant. Another interesting phenomenon is that a Fresnel coefficient can be larger than unity, indicating an amplification of the reflected wave with respect to the incident wave. This is never possible for linear media due to energy conservation. Here, the energy for amplification is supplied by the pump beams. In Fig. 4 the Fresnel coefficient P_s is plotted in the complex plane as a function of the frequency. Point *b* corresponds to the peak at resonance ($\omega = \omega_p$) in Fig. 3, and points *a* and *c* are the two side maxima. At resonance the reflection coefficient is purely imaginary, which corresponds to a phase shift of $\pi/2$. The arrowheads in the figure indicate the direction of increasing frequency.

Another interesting behavior of Fresnel coefficients is their dependence on the

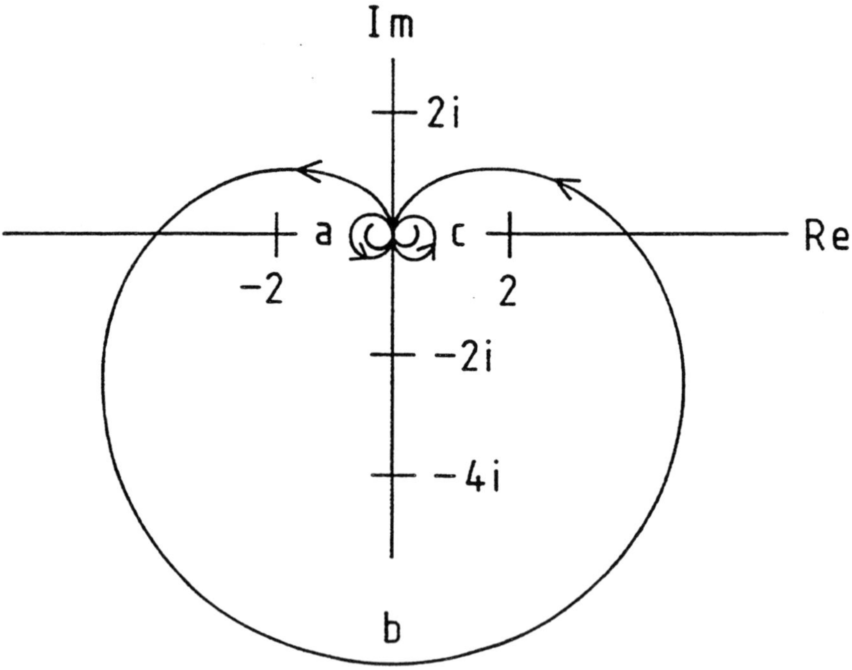

Fig. 4. Polar diagram in the complex plane of the Fresnel coefficient P_s for the same parameters as in Fig. 3. Along the curve, the relative frequency ω/ω_p varies.

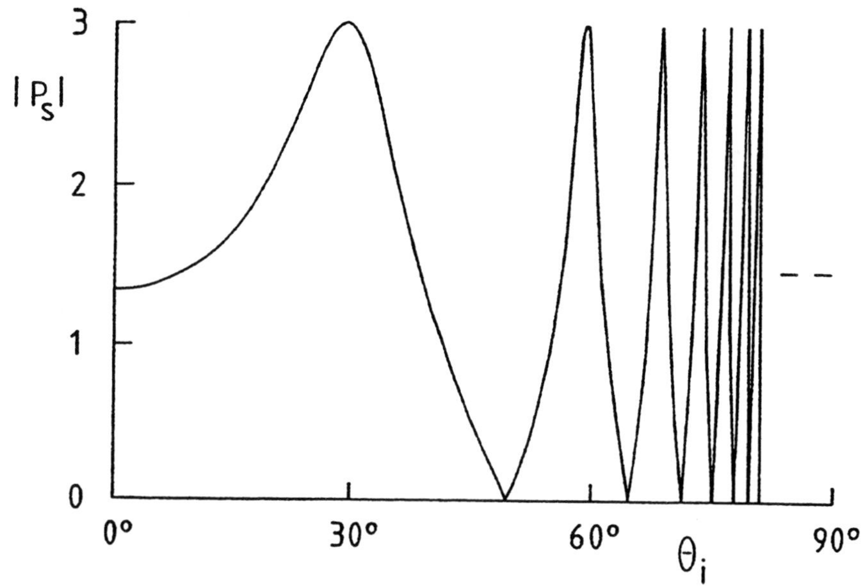

Fig. 5. Fresnel coefficient for s-waves as a function of the angle of incidence.

angle of incidence θ_i. As shown in Fig. 5, there is a strong oscillatory dependence on θ_i. The maximum reflection is not attained for normal incidence, but for some finite angle. Also interesting is that for many angles of incidence the reflectivity is identically zero. At the maxima in between, the reflectivity has approximately the same value at every peak (for this particular set of parameters).

4. ATOMIC RELAXATION

Next we want to consider the behavior of a two-state atom which is located at a normal distance h from the surface of the PC, and on the z-axis. Spontaneous decay of an excited atom is brought about by the interaction between the atomic electric dipole moment $\vec{\mu}$ (a quantum operator) and the electric field of the vacuum at the location of the atom. This electric dipole coupling is governed by the interaction Hamiltonian

$$H_{ar} = -\vec{\mu} \bullet \vec{E}(h\vec{e}_z)_{vacuum} \quad . \tag{8}$$

In empty space, the electric field operator can be represented by the standard plane-wave expansion

$$\vec{E}(\vec{r},t)_{vacuum} = \sum_{\vec{k}\sigma} \sqrt{\frac{\hbar\omega}{2\varepsilon_o V}} a_{\vec{k}\sigma} \vec{\varepsilon}_{\vec{k}\sigma} e^{i(\vec{k}\bullet\vec{r}-\omega t)} + \text{H.c.} \qquad \text{(empty space)} \quad , \tag{9}$$

with V the quantization volume, $\omega = ck$ the dispersion relation in vacuum, $\vec{e}_{\vec{k}\sigma}$ a unit transverse polarization vector for the mode $\vec{k}\sigma$, and $a_{\vec{k}\sigma}$ the annihilation operator for a photon in the mode $\vec{k}\sigma$. The time dependence in (9) represents the Heisenberg picture, and for t = 0 this will be taken to coincide with the Schrödinger picture. This vacuum field can now be considered as an incident field for reflection at the four-wave mixer. There are two contributions to this. Traveling waves with $k_z < 0$ are incident on the medium from the region z > 0, and they give rise to a phase-conjugated reflected wave in z > 0 and a transmitted wave in the region z < d, just as in Fig. 1. On the other hand, waves with $k_z > 0$ enter the medium from the region z < d, and in the region z > 0 they generate transmitted waves, which are not phase-conjugated. In the quantized version of four-wave mixing, as we need it here, a 'phase conjugation' becomes a 'Hermitian conjugation'. This implies that the annihilation part in (9) will be converted into a creation part upon reflection. Explicitly, the field in the region z > 0 becomes

$$\vec{E}(\vec{r},t)_{vacuum} = [\sum_{\vec{k}\sigma,k_z<0} \{E_{\vec{k}\sigma}\vec{\varepsilon}_{\vec{k}\sigma}e^{i(\vec{k}\cdot\vec{r}-\omega t)} + E^{\dagger}_{\vec{k}\sigma}P^{*}_{\vec{k}\sigma}\ \vec{\varepsilon}\,'_{\vec{k}\sigma}\ e^{i(-\vec{k}'\cdot\vec{r}-(2\omega_p-\omega)t)}\} + \sum_{\vec{k}\sigma,k_z>0} E_{\vec{k}\sigma}T_{\vec{k}\sigma}\vec{\varepsilon}_{\vec{k}\sigma}e^{i(\vec{k}\cdot\vec{r}-\omega t)}] + \text{H.c.} \quad . \tag{10}$$

Here we introduced the notation

$$E_{\vec{k}\sigma} = \sqrt{\frac{\hbar\omega}{2\varepsilon_o V}}\, a_{\vec{k}\sigma} \quad , \tag{11}$$

and $T_{\vec{k}\sigma}$ represents the Fresnel transmission coeficient. The field (10), with t set equal to zero, couples to the atomic dipole as given by eq. (8).

The Hamiltonian of free evolution of the two-state atom is

$$H_a = \hbar\omega_e P_e + \hbar\omega_g P_g \quad , \tag{12}$$

in terms of the projectors onto the atomic levels,

$$P_e = |e\rangle\langle e| \quad , \quad P_g = |g\rangle\langle g| \quad . \tag{13}$$

The free evolution of the radiation field is determined by the Hamiltonian

$$H_r = \sum_{\vec{k}\sigma} \hbar\omega(a^{\dagger}_{\vec{k}\sigma}a_{\vec{k}\sigma} + \tfrac{1}{2}) \quad , \tag{14}$$

in self-explanatory notation. Then the equation of motion for the density operator ρ of the atom, the field, and the interaction is given by

$$i\hbar\frac{d\rho}{dt}=[H_a+H_r+H_{ar},\rho] \quad . \tag{15}$$

The effect of the presence of the nonlinear medium is now parametrically included through the Fresnel reflection and transmission coefficients in the expression for the electric field of the vacuum. Finally, the dipole operator $\vec{\mu}$ in the interaction Hamiltonian can be expressed in terms of atomic operators as

$$\vec{\mu}=|e>\vec{\mu}_{eg}<g| + |g>\vec{\mu}^*_{eg}<e| \quad , \tag{16}$$

with $\vec{\mu}_{eg}=<e|\vec{\mu}|g>$ the transition dipole matrix element.

In this section we are interested in the time evolution of the atomic state, rather than the evolution of the entire system. Therefore, we introduce the partial atomic density operator $\rho_a(t)$, defined as

$$\rho_a(t)=\mathrm{Tr}_{\mathrm{field}}\,\rho(t) \quad . \tag{17}$$

This operator represents the atomic state, irrespective the state of the radiation field. It is a standard technique to derive an equation of motion for $\rho_a(t)$ with reservoir theory [7]. The result is

$$i\hbar\frac{d\rho_a}{dt}=[H_a,\rho_a]-i\hbar\Gamma\rho_a \quad . \tag{18}$$

Here, Γ is a Liouville (or super) operator which accounts for the relaxation of the atomic density operator. With Π an arbitrary atomic operator, the effect of Γ is the operation

$$\Gamma\Pi=\tfrac{1}{2}A_e\{P_e\Pi+\Pi P_e-2|g><e|\Pi|e><g|\}$$

$$+\tfrac{1}{2}A_g\{P_g\Pi+\Pi P_g-2|e><g|\Pi|g><e|\} \quad , \tag{19}$$

on Π. The parameters A_e and A_g are defined as

$$A_e=A\{1+\tfrac{1}{2}\bar{P}^2\} \quad , \tag{20}$$

$$A_g=\tfrac{1}{2}A\bar{P}^2 \quad . \tag{21}$$

The parameter $\bar{P}^2$ is the square of the Fresnel reflection coefficient, averaged over the angle of incidence, and including a weight factor [8]. It is assumed that for both s-waves and p-waves this gives the same result, which is a slight approximation.

For an atom in empty space the steady-state density operator is given by $\overline{\rho}_a=|g><g|$, showing that the atom is in the ground state (an overbar indicates a steady-state value). This gives $\overline{n}_e=0$ and $\overline{n}_g=1$ for the steady-state populations of the

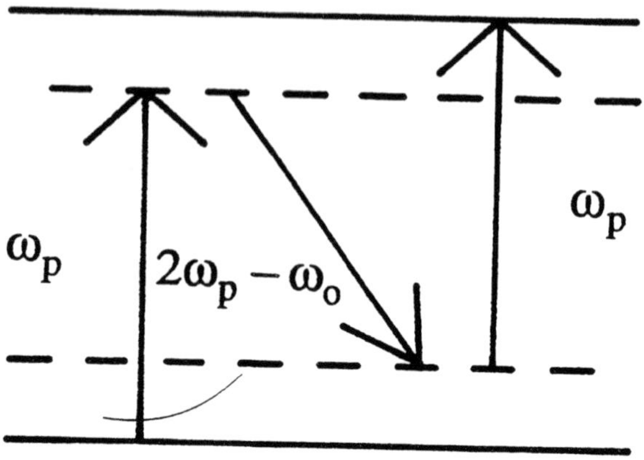

Fig. 6. Three-photon process, leading to spontaneous excitation. In between the absorptions of the two photons with frequency ω_p, a fluorescent photon with frequency $2\omega_p - \omega_o$ is emitted.

levels |e> and |g>, respectively. When we take the matrix elements <e|...|e> and <g|...|g> of (18), we obtain

$$\frac{dn_e}{dt} = -A_e n_e + A_g n_g \quad , \tag{22}$$

$$\frac{dn_g}{dt} = -A_g n_g + A_e n_e \quad , \tag{23}$$

which is a set of rate equations for the populations of the levels. With the condition

$$n_e(t) + n_g(t) = 1 \quad , \tag{24}$$

the set (22), (23) can be solved easily. If we take the steady-state limit of the solution (or simply set the right-hand sides equal to zero), we obtain

$$\bar{n}_e = \frac{A_g}{A_e + A_g} = \frac{\frac{1}{2}\bar{P}^2}{1+\bar{P}^2} \quad , \tag{25}$$

$$\bar{n}_g = \frac{A_e}{A_e + A_g} = \frac{1+\frac{1}{2}\bar{P}^2}{1+\bar{P}^2} \quad . \tag{26}$$

From (25) we see that the population of the excited state is finite for $\vec{P}^2 \neq 0$. This is one of the remarkable properties of an atom near a PC, which is impossible for any environment with linear materials. The responsible process for this spontaneous excitation of the atom is shown in Fig. 6. As we shall see in the next section, the atomic dipole polarizes the medium even when it is in the ground state. In four-wave mixing processes inside the medium, this polarization field is converted into a field that contains real photons. These photons are emitted by the medium, towards the atom. Subsequent absorption of a photon can than lead to excitation. Actually, two photons are absorbed and one is emitted spontaneously, as shown in Fig. 6. This follows from either energy conservation or the results from the next section. After such a three-photon process, the atom is in the excited state and can decay in the usual way. This leaves the atom in the ground state, and consequently a succeeding three-photon event can excite the atom again. These processes continue indefinitely, giving rise to a finite population of the excited level.

5. FLUORESCENCE

The electric field (10) is the Heisenberg-picture representation of the vacuum field near a PC, in the absense of the atom. Its value at t = 0 determines the evolution of the atomic density operator, as shown in the previous section. On the other hand, the coupling to the atomic dipole will modify this field also, since spontaneous decay gives rise to the emission of fluorescent photons. This changes the vacuum field to the new Heisenberg representation, including fluorescence:

$$\vec{E}(\vec{r},t)=[\sum_{\vec{k}\sigma,k_z<0}\{E_{\vec{k}\sigma}(t)\vec{\varepsilon}_{\vec{k}\sigma}e^{i\vec{k}\cdot\vec{r}} + E^{\dagger}_{\vec{k}\sigma}(t)P^{*}_{\vec{k}\sigma}\ \vec{\varepsilon}^{\,\prime}_{\vec{k}\sigma}\ e^{-i(\vec{k}'\cdot\vec{r}+2\omega_p t)}\} + \sum_{\vec{k}\sigma,k_z>0} E_{\vec{k}\sigma}(t)T_{\vec{k}\sigma}\vec{\varepsilon}_{\vec{k}\sigma}e^{i\vec{k}\cdot\vec{r}}] + \text{H.c.} \quad . \tag{27}$$

The time dependence is now accounted for by the operator

$$E_{\vec{k}\sigma}(t)=\sqrt{\frac{\hbar\omega}{2\varepsilon_o V}}\,a_{\vec{k}\sigma}(t) \quad , \tag{28}$$

which in turn is determined by the time dependence of the annihilation operator $a_{\vec{k}\sigma}(t)$. In addition, on the second line of (27) we have the time dependent factor $\exp(-2i\omega_p t)$. This factor represents parametrically the time dependence of the square of the annihilation operator for a pump photon. From this we see that the second line of (27) represents the phase-conjugated field, as it is generated in a four-wave mixing process. The annihilation operator for two pump photons is coupled to the creation operator for a regular photon (either of a vacuum mode or of fluorescence, as shown below). This explains why in Fig. 6 the process has to involve two pump photons and one fluorescent photon, and such that the annihilation of two pump photons goes together with the creation of a fluorescent photon. Energy conservation then requires that the fluorescent photon must have an energy equal to $2\omega_p - \omega_o$, as follows from the diagram. In the next section we shall prove this more directly.

The time dependence of the annihilation operator $a_{\vec{k}\sigma}(t)$ for a photon in the mode $\vec{k}\sigma$ is governed by Heisenberg's equation

$$i\hbar \frac{d}{dt} a_{\vec{k}\sigma}(t) = [a_{\vec{k}\sigma}(t), H_a(t) + H_r(t) + H_{ar}(t)] \quad . \tag{29}$$

Working out the equal-time commutators then yields

$$i\hbar \frac{d}{dt} a_{\vec{k}\sigma}(t) = \hbar\omega a_{\vec{k}\sigma}(t) - \vec{\mu}(t) \bullet [a_{\vec{k}\sigma}(t), \vec{E}(h\vec{e}_z, t)] \quad , \tag{30}$$

which can be integrated as

$$a_{\vec{k}\sigma}(t) = a_{\vec{k}\sigma}(0) e^{-i\omega t} + \frac{i}{\hbar} \int_0^t dt' \, e^{i\omega(t'-t)} \, \vec{\mu}(t') \bullet [a_{\vec{k}\sigma}(t'), \vec{E}(h\vec{e}_z, t')] \quad . \tag{31}$$

Here, $a_{\vec{k}\sigma}(0)$ is the annihilation operator in the Schrödinger picture. When we substitute (31) into (28) and the result into (27), then the first term on the right-hand side of (31) gives precisely the vacuum field from (10). Hence we can write the field as

$$\vec{E}(\vec{r}, t) = \vec{E}(\vec{r}, t)_{vacuum} + \vec{E}(\vec{r}, t)_{fluorescence} \quad , \tag{32}$$

and the fluorescence part corresponds to the second term on the right-hand side of (31). The resulting expression for the fluorescence field is, however, cumbersome.

Observations of fluorescence radiation are made by a photomultiplier in the far field, for which r is much larger than a wavelength. Therefore, we only need the asymptotic value of the electric field, which can be found with the method of stationary phase [9]. After some lengthy calculations we obtain [10]

$$\vec{E}(\vec{r}, t)_{fluorescence} = \frac{\omega_o^2 e^{-i\omega_o \tau}}{4\pi\varepsilon_o c^2 r} \{\vec{M} - (\vec{M} \bullet \hat{r})\hat{r}\} + H.c. \quad . \tag{33}$$

Here, $\hat{r}$ is the unit vector into the direction of observation, τ is defined as

$$\tau = \frac{h}{c} \cos\theta \quad , \tag{34}$$

with θ the direction of observation, and the operator $\vec{M}$ is given by

$$\vec{M} = \vec{\mu}(t)^{(+)} - e^{-2i\omega_p t} P^* \vec{\mu}(t)^{(-)} \quad . \tag{35}$$

We have assumed that the Fresnel coefficient is the same for s- and p-waves for this θ, which is a slight simplification. It is exact for special cases, like the most important one with $\theta = 0$ (photomultiplier on the z-axis, observing perpendiculer to the surface). Finally, the superscripts (+) and (-) indicate the positive and negative frequency part of the Heisenberg-picture dipole operator, respectively. The corresponding Schrödinger operators are (values for t = 0)

$$\vec{\mu}^{(+)} = |g> \vec{\mu}_{eg}^{\,*} < e| \quad , \quad \vec{\mu}^{(-)} = |e> \vec{\mu}_{eg} < g| \quad , \tag{36}$$

which are proportional to the atomic lowering and raising operator, respectively.

6. SPECTRAL DISTRIBUTION

Equation (33) gives formally the electric field of the fluorescence radiation which is emitted spontaneously by the atom near the PC. In this section we derive the spectral distribution of this radiation, and the intensity. Note that without the presence of the PC, the right-hand side of (33) remains identical in form, except that the operator $\vec{M}(t)$ reduces to $\vec{\mu}(t)^{(+)}$. In that case, the radiation field contains no photons (in the steady state, for which the atom is in state |g>). We shall see that in the presence of the four-wave mixer the fluorescence radiation does contain photons, which are emitted at a steady rate. This is never possible near any linear medium.

We shall assume that the radiation is detected by a photomultiplier, operating as a photon counter, at a distance r from the atom. Furthermore we assume, for simplicity, that the field is filtered by a polarizer which transmits the $\vec{e}_d$-component of radiation. The positive frequency part of the filtered fluorescence then is

$$E(t)^{(+)} = \vec{E}(\vec{r},t)^{(+)}_{\text{fluorescence}} \bullet \vec{e}_d^{\,*} \quad , \tag{37}$$

and since $\vec{e}_d$ is perpendicular to $\hat{r}$, this becomes

$$E(t)^{(+)} = \frac{\omega_o^2 e^{-i\omega_o \tau}}{4\pi\varepsilon_o c^2 r} \vec{M}(t) \bullet \vec{e}_d^{\,*} \quad . \tag{38}$$

Then we introduce the raising operator d, which is given in the Schrödinger picture by

$$d = |e><g| \quad , \tag{39}$$

and we take the matrix element $\vec{\mu}_{eg}$ to be real. Then $E(t)^{(+)}$ can be written as

$$E(t)^{(+)} = \gamma\{d^{\dagger}(t) - P^* e^{-2i\omega_p t} d(t)\} \quad , \tag{40}$$

and here all constant factors are combined into the parameter γ.

From the theory of photon detection [11] it follows that the spectral distribution of any field $E(t)^{(+)}$ can be represented as

$$I(\omega)=\frac{\xi}{\pi}Re\int_0^{\infty}d\tau\, e^{i\omega\tau} < E(0)^{(-)}E(\tau)^{(+)} > \quad , \tag{41}$$

where the parameter ξ is an overall factor which accounts for the detecter efficiency, the aperture, etc. The negative-frequency part $E(t)^{(-)}$, as it appears for t = 0 in (41), is the Hermitian conjugate of the positive-frequency part $E(t)^{(+)}$, and the brackets <...> indicate a quantum average. The total intensity of the radiation is given by

$$I_{total} = \int d\omega\, I(\omega) \quad , \tag{42}$$

which yields with (41)

$$I_{total} = \xi < E(0)^{(-)}E(0)^{(+)} > \quad . \tag{43}$$

When we substitute (40) with t replaced by τ into (41), and also the Hermitian conjugate with t set equal to zero, then the expression for $I(\omega)$ splits up into four parts. Each part contains a quantum correlation function between operators d and $d^{\dagger}$. These functions can be evaluated by standard methods, for instance with the Liouville operator formalism. We find:

$$< d(0)d(\tau) > \; = \; < d^{\dagger}(0)d^{\dagger}(\tau) > \; = \; 0 \quad , \tag{44}$$

$$< d(0)d^{\dagger}(\tau) > \; = \; \bar{n}_e\, e^{-i\omega_o\tau-\frac{1}{2}(A_e+A_g)\tau} \quad , \tag{45}$$

$$< d^{\dagger}(0)d(\tau) > \; = \; \bar{n}_g\, e^{i\omega_o\tau-\frac{1}{2}(A_e+A_g)\tau} \quad . \tag{46}$$

After evaluating the integral in (41), we obtain for the spectral distribution of the fluorescence:

$$I(\omega)=\frac{I_e}{\pi}Re\frac{1}{\frac{1}{2}(A_e+A_g)-i(\omega-\omega_o)} + \frac{I_g}{\pi}Re\frac{1}{\frac{1}{2}(A_e+A_g)-i(\omega-(2\omega_p-\omega_o))} \quad . \tag{47}$$

It appears that the spectrum consists of two Lorentzian lines, both of which have a half-width at half maximum equal to $\frac{1}{2}(A_e+A_g)$. The strengths of the lines are I_e and I_g, respectively, with

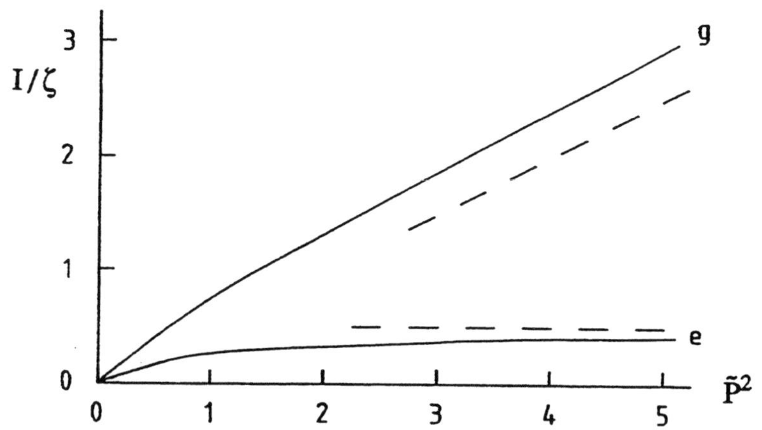

Fig. 7. Dependence of the line strengths on the reflectivity. Curve *e* corresponds to ordinary fluorescence, and curve *g* is the intensity of the line which is due to three-photon events. The dashed lines indicate the asymptotic values.

$$I_e = \zeta \bar{n}_e \quad , \quad I_g = \zeta \bar{n}_g \tilde{P}^2 \quad , \tag{48}$$

in terms of the steady-state level populations from (25) and (26). Here we have set $\zeta = \xi\gamma^2$ for the overall parameter. The line with its strength proportional to the population of |e> is located at $\omega = \omega_o$, and this is obviously ordinary fluorescence during decay from the excited state to the ground state. The second line is centered around $\omega = 2\omega_p - \omega_o$, and its intensity is proportional to the population of the ground state. Therefore, this must be the radiation which is emitted during a three-photon event, as depicted in Fig. 6. With the expressions for $\bar{n}_e$ and $\bar{n}_g$, the inequality

$$I_g \geq 2I_e \tag{49}$$

is easy to verify. This shows that the second line is at least twice as strong as the first, which seems to indicate that three-photon processes occur at least twice as much as ordinary decay. This, however, can not be the case. After a three-photon event the atom is in the excited state, and for the next process to happen the atom has to decay first to the ground state and emit a photon with frequency ω_o. This would imply that the two different processes alternate, which would make the line strengths equal. The explanation for this apparent contradiction is that not all photons necessarily end up in the detector. It can be shown that all three-photon photons propagate towards $z > 0$, whereas all ordinary photons are emitted isotropically. Consequently, half the number of ordinary photons do not travel into the direction $z > 0$, and can therefore never be detected. This explains the factor of 2 in (49). Another interesting fact to notice is that the line strengths only depend on the reflectivity $\tilde{P}^2$, apart from the overall factor ζ. The behavior of the line strengths as a function of $\tilde{P}^2$ is illustrated in Fig. 7. The total intensity is given by

$$I_{total} = I_e + I_g = \tfrac{1}{2}\zeta\bar{P}^2 \frac{3+\bar{P}^2}{1+\bar{P}^2} \quad . \tag{50}$$

7. LASER - DRIVEN ATOM

As a last topic we consider the situation where the atom is irradiated by a strong laser beam which is nearly resonant with the atomic transition frequency ω_o. Let the electric field at the position of the atom be given by

$$\vec{E}(t)_L = Re\, \vec{E}_o e^{-i(\omega_L t + \phi(t))} \quad , \tag{51}$$

where $\phi(t)$ is a random phase which accounts for the laser linewidth. We shall take this phase to be the independent-increment process, which yields a Lorentzian laser profile [12]. Then the atomic electric dipole couples to this external field, just as to the electric field of the vacuum. This gives rise to an interaction Hamiltonian $-\vec{\mu} \bullet \vec{E}(t)_L$, and therefore the equation of motion (18) has to be modified to

$$i\hbar \frac{d\rho_a}{dt} = [H_a - \vec{\mu} \bullet \vec{E}(t)_L \, , \rho_a] - i\hbar\Gamma\rho_a \quad . \tag{52}$$

Essentially, this interaction Hamiltonian should be added to the Hamiltonian in (15), but it can be shown that this simply leads to (52). When writing out the interaction Hamiltonian with the help of (16) for the dipole operator, we see that $-\vec{\mu} \bullet \vec{E}(t)_L$ has four terms. In the rotating-wave approximation we drop the two terms which correspond to non-energy-conserving processes. These are the coupling between the atomic lowering operator and the positive-frequency part (annihilation part) of the laser field, and the coupling between the raising operator and the negative-frequency part. Then the interaction Hamiltonian reduces to

$$-\vec{\mu} \bullet \vec{E}(t)_L = -\tfrac{1}{2}\hbar\Omega \, d \, e^{-i(\omega_L t + \phi(t))} + H.c. \quad , \tag{53}$$

with

$$\Omega = \hbar^{-1}\vec{E}_o \bullet \vec{\mu}_{eg} \tag{54}$$

the Rabi frequency. The time dependence of the Hamiltonian in (52) can be simplified by the following transformation of the density operator:

$$\sigma(t) = e^{-i(\omega_L t + \phi(t))P_g} \rho(t) e^{i(\omega_L t + \phi(t))P_g} \quad . \tag{55}$$

The equation of motion for $\sigma(t)$ then becomes

$$i\frac{d\sigma}{dt}=[H_d+P_g\frac{d\phi}{dt},\sigma]-i\Gamma\sigma \quad , \tag{56}$$

where H_d is the dressed-atom Hamiltonian, defined by

$$H_d=\Delta P_g-\tfrac{1}{2}(\Omega d+\Omega^* d^\dagger) \quad . \tag{57}$$

Here, $\Delta=\omega_L-\omega_o$ is the detuning of the laser frequency from resonance. Then we average the transformed density operator over the stochastic laser phase, which yields [12]

$$i\frac{d}{dt}\langle\sigma\rangle=[H_d,\langle\sigma\rangle]-iW_o\langle\sigma\rangle-i\Gamma\langle\sigma\rangle \quad , \tag{58}$$

where the notation $\langle\sigma\rangle$ indicates the average of σ over the random process. The Liouville operator W_o takes into account the effect of the laser phase, and is given by

$$W_o\Pi=\lambda[P_g,[P_g,\Pi]] \quad , \tag{59}$$

with λ the laser linewidth and Π an arbitrary operator.

The steady-state solution $\overline{\langle\sigma\rangle}$ of (58) is easily found. For the population of the excited state we obtain

$$\overline{n}_e=\frac{\frac{1}{2}|\Omega|^2\eta+A_g(\Delta^2+\eta^2)}{|\Omega|^2\eta+2\overline{A}(\Delta^2+\eta^2)} \quad , \tag{60}$$

and the coherence between the two levels is

$$\overline{\langle e|\sigma|g\rangle}=\frac{\frac{1}{2}\Omega A(i\eta-\Delta)}{|\Omega|^2\eta+2\overline{A}(\Delta^2+\eta^2)} \quad . \tag{61}$$

Here we introduced the parameters

$$\overline{A}=\tfrac{1}{2}(A_e+A_g) \quad , \quad \eta=\lambda+\overline{A} \quad . \tag{62}$$

8. RESONANCE FLUORESCENCE

The intensity of the emitted resonance fluorescence is again given by (43), and the field by (40). Evaluation then gives

$$I_{total}=\zeta(\overline{n}_e+\overline{n}_g\bar{P}^2) \quad , \tag{63}$$

which is formally identical to the sum of the expressions in (48). The difference is, of course, that the population of the upper level is now given by (60) rather than by (25). Working out (63) gives

$$I_{total} = \zeta \frac{\frac{1}{2}|\Omega|^2 \eta(1+\bar{P}^2) + A_g(\Delta^2+\eta^2)(3+\bar{P}^2)}{|\Omega|^2\eta + 2\bar{A}(\Delta^2+\eta^2)} \quad . \tag{64}$$

In the low-intensity limit ($|\Omega|^2 \to 0$) this reduces to (50). Another interesting limit is the case of a very strong driving field, for which we have

$$I_{total} = \tfrac{1}{2}\zeta(1+\bar{P}^2) \quad , \quad |\Omega|^2 \to \infty \quad . \tag{65}$$

For a free atom this saturation limit is ½ (apart from the factor ζ), which is equal to the maximum population of the excited state. Near a PC, however, this upper limit can grow indefinitely.

The spectral distribution of the resonance fluorescence is again formally given by (41), but the evaluation of this expression is now considerably more complicated. The result is given explicitly in Ref. 13.

9. SEPARATED LINES

The dressed-atom Hamiltonian H_d has as its eigenvalue equation

$$H_d|\pm\rangle = \mp\tfrac{1}{2}\hbar\Omega'|\pm\rangle \quad , \tag{66}$$

and the generalized Rabi frequency Ω' is given by

$$\Omega' = \mathrm{sgn}(\Delta)\sqrt{|\Omega|^2+\Delta^2} \quad . \tag{67}$$

This Hamiltonian determines the free evolution of an atom in a monochromatic laser field with frequency ω_L, as follows from (58). The two additional terms then account for the laser linewidth and for relaxation due to spontaneous emission. If the Einstein coefficient A and the laser linewidth λ are small compared to $|\Omega'|$, then the dressed-atom Hamiltonian will dominate the time evolution. In this limit the resonance fluorescence spectrum consists of three lines, which form the so-called Mollow triplet [14]. The central line is the Rayleigh (R) line, which is located at $\omega = \omega_L$. Then, at $\omega = \omega_L - \Omega'$ there is a sideband, which is called the fluorescence (F) line. The reason for this is that in the low-intensity limit this line is positioned at $\omega \approx \omega_L - \Delta = \omega_o$, the atomic resonance. The third line is the three-photon (T) line at $\omega = \omega_L + \Omega'$. For an atom in a monochromatic laser field (and no PC) the spectrum is symmetric around $\omega = \omega_L$. A finite laser linewidth tends to increase the strength of the F-line at the expense of the T-line. The strength of the R-line is independent of the laser linewidth.

Let us write

$$\Lambda = \omega - \omega_L \quad , \quad \delta = \omega_L - \omega_p \quad , \tag{68}$$

for the frequency of a fluorescent photon relative to the frequency of the driving laser, and the frequency mismatch between the driving laser and the pump laser, respectively.

With other notations defined below, the resonance fluorescence spectrum in the limit of separated lines is found to be

$$I(\omega)=\frac{I_{Rc}}{\pi}Re\frac{1}{\lambda-i\Lambda}+\frac{I_{Ri}}{\pi}Re\frac{1}{\lambda+u+v-2w-i\Lambda}$$
$$+\frac{I_F}{\pi}Re\frac{1}{y_F-i(\Lambda+\Omega')}+\frac{I_T}{\pi}Re\frac{1}{y_T-i(\Lambda-\Omega')}$$
$$+\frac{I_{Rc'}}{\pi}Re\frac{1}{\lambda-i(\Lambda+2\delta)}+\frac{I_{Ri'}}{\pi}Re\frac{1}{\lambda+u+v-2w-i(\Lambda+2\delta)}$$
$$+\frac{I_{F'}}{\pi}Re\frac{1}{y_F-i(\Lambda+2\delta-\Omega')}+\frac{I_{T'}}{\pi}Re\frac{1}{y_T-i(\Lambda+2\delta+\Omega')} \quad . \tag{69}$$

Any line of the form

$$\frac{I}{\pi}Re\frac{1}{\alpha-i(\Lambda-\upsilon)} \quad , \tag{70}$$

with $\alpha>0$, is a Lorentzian with intensity I, half-width at half-maximum equal to α, and is positioned at $\Lambda=\upsilon$, or $\omega=\omega_L+\upsilon$. Therefore, the fluorescence spectrum (69) consists of eight Lorentzians.

The first line is the coherent Rayleigh line at the position of the laser frequency and with linewidth equal to the linewidth of the driving laser. In the limit of a vanishing linewidth, this coherent part becomes a delta function. The second line is the incoherent part of the R-line at $\omega=\omega_L$, but it has an additional broadening by u + v - 2w. The parameters u, v, and w are defined as

$$u=A_e g_-^2+A_g g_+^2+2\lambda g_o^2 \quad , \tag{71}$$

$$v=A_e g_+^2+A_g g_-^2+2\lambda g_o^2 \quad , \tag{72}$$

$$w=(4\lambda-\lambda')g_o^2 \quad , \tag{73}$$

in terms of the three other parameters

$$g_\pm=\frac{\Omega'\mp\Delta}{2\Omega'} \quad , \quad g_o=\frac{|\Omega|}{2\Omega'} \quad , \tag{74}$$

where λ' is a second parameter accounting for the laser linewidth. For the special case of the phase-diffusion model, we have $\lambda'=4\lambda$, which gives w = 0. The third line is the F-line at $\Lambda=-\Omega'$, and the fourth line is the T-line at $\Lambda=\Omega'$. These four lines form the famous Mollow triplet, although the widths and the strengths of the lines are modified by the presence of the PC. For the widths of the F- and T-lines we find

$$y_F = \overline{A}(1+2g_o^2)+2\lambda g_o^2+\lambda' g_+^2 \quad , \tag{75}$$

$$y_T = \overline{A}(1+2g_o^2)+2\lambda g_o^2+\lambda' g_-^2 \quad , \tag{76}$$

respectively. These four lines can be considered as an incident field on the surface of the PC, and they are reflected as the phase-conjugate image of the Mollow triplet. The image lines are indicated by a prime. Since an incident wave with frequency ω is reflected with a frequency $2\omega_p-\omega$, the images of the two components of the R-line are expected at $\omega=2\omega_p-\omega_L$, or $\Lambda=-2\delta$. This appears to be the case indeed, as follows from (69). Similarly, the F'- and T'-lines are respectively located at $\Lambda=\Omega'-2\delta$ and $\Lambda=-\Omega'-2\delta$, as could be expected. This implies that the order of the triplet is reversed; for instance, for $\delta=0$ the F'- and the T-line coincide, and the T'-line is located on top of the F-line. It also follows from (69) that the widths of the lines are unaltered upon reflection. From (75) and (76) we obtain

$$y_T - y_F = \frac{\lambda'\Delta}{\Omega'} \geq 0 \quad , \tag{77}$$

showing that the width of the T-line is larger than the width of the F-line, when the laser linewidth and the detuning are nonzero.

For the intensity of the coherent Rayleigh line we find

$$I_{Rc} = \zeta g_o^2 \frac{(u-v)^2}{(u+v)(u+v-2w)} \quad , \tag{78}$$

and for the incoherent component we have

$$I_{Ri} = 2\zeta g_o^2 \frac{u(v-w)+v(u-w)}{(u+v)(u+v-2w)} \quad . \tag{79}$$

The total strength of the R-line then becomes

$$I_{Rc}+I_{Ri} = \zeta \frac{|\Omega|^2}{4(|\Omega|^2+\Delta^2)} \quad , \tag{80}$$

independent of spontaneous decay, laser linewidth and the presence of the phase conjugator. The strengths of the two sidebands are

$$I_F = \zeta g_-^2 \frac{v}{u+v} \quad , \quad I_T = \zeta g_+^2 \frac{u}{u+v} \quad , \tag{81}$$

for which it holds that

$$I_F \geq I_T \quad . \tag{82}$$

For the reflected components of the R-line it holds that

$$I_{Rc'} = \bar{P}^2 I_{Rc} \quad , \quad I_{Ri'} = \bar{P}^2 I_{Ri} \quad , \tag{83}$$

as could be expected. For the sidebands, however, we obtain

$$I_{F'} = \zeta \bar{P}^2 g_-^2 \frac{u}{u+v} \quad , \quad I_{T'} = \zeta \bar{P}^2 g_+^2 \frac{v}{u+v} \quad , \tag{84}$$

and their strengths are not just the phase-conjugate intensity reflection coefficient $\bar{P}^2$ times the strengths of the incident lines. It still holds that the F'-line is stronger than the T'-line, but the intensity is distributed differently over the two sidebands, as compared to the incident Mollow triplet.

ACKNOWLEDGMENTS

This work was supported in part by the National Science Foundation under Grant CHE - 9196214. We also acknowledge the National Science Foundation for a Research Opportunity Award.

REFERENCES

1. B. Ya. Zel'dovich, V. I. Popovichev, V. V. Ragulski, and F. S. Faizullov, *JETP Lett.* 15 (1972) 109.
2. D. M. Bloom and G. C. Bjorklund, *Appl. Phys. Lett.* 31 (1977) 592.
3. P. V. Avizonis, F. A. Hopf, W. D. Bomberger, S. F. Jacobs, A. Tomita, and K. H. Womack, *Appl. Phys. Lett.* 31 (1977) 435.
4. S. M. Jensen and R. W. Hellwarth, *Appl. Phys. Lett.* 32 (1978) 166.
5. H. F. Arnoldus and T. F. George, *Surf. Sci.* 205 (1988) 617.
6. H. F. Arnoldus and T. F. George, *J. Mod. Opt.* 36 (1989) 31.
7. W. H. Louisell, *Quantum Statistical Properties of Radiation*, New York: Wiley, 1973, Ch. 6.
8. H. F. Arnoldus and T. F. George, *Quantum Nonlinear Phenomena* 1 (1992) in press.
9. M. Born and E. Wolf, *Principles of Optics*, Oxford, Pergamon, 1980, p. 752.
10. H. F. Arnoldus and T. F. George, *Phys. Rev.* A 43 ((1991) 6156.
11. J. H. Eberly and K. Wodkiewicz, *J. Opt. Soc. Am.* 67 (1977) 1252.
12. H. F. Arnoldus and G. Nienhuis, *J. Phys. B: At. Mol. Phys.* 16 (1983) 2325.
13. H. F. Arnoldus and T. F. George, *J. Phys. B: At. Mol. Opt. Phys.* 24 (1991) 2653.
14. B. R. Mollow, *Phys. Rev.* 188 (1969) 1969.

Studies in Classical
and Quantum Nonlinear Optics

ISBN 1-56072-168-5

DYNAMICS OF SPATIAL SOLITON COUPLING

A.D. Boardman and K. Xie
Joule Laboratory, Department of Physics,
University of Salford, Salford, M5 4WT, UK

1. INTRODUCTION

It is many years now since the basic physics, concerning the self-trapping, of a powerful beam of light in a nonlinear medium was thoroughly discussed [1]. Some of the imagery evoked in that discussion will be produced here and the crucial question of beam stability, and how to achieve it, experimentally, will be addressed. The latter discussion will be presented in the knowledge that recent stimuli to the field involving some pioneering experiments have appeared [2-5]. The most important involve the use of planar waveguides [2,3].

A beam of light with a rectangular distribution of energy, a radius r and a wavelength λ can propagate through a dielectric, non-magnetic, medium that has a linear refractive index n. It is well-known, however, that the beam will spread out due to diffraction. Indeed, at a given radius r the light (laser) beam can be imagined to act as its own aperture and rays of light will diffract through it, with an approximate diffraction angle $\theta_d = \lambda/2nr$. If the material becomes nonlinear, in a Kerr-like manner, [i.e. third-order only with nonlinearity proportional to intensity], with a nonlinear coefficient α, then the dielectric constant $\epsilon_L = n^2$ of the linear medium will change, by a small amount, to $\epsilon_L + \alpha|\mathbf{E}|^2$, where $\mathbf{E}$ is the electric field carried by the beam. If the medium has a positive α then the refractive index *inrceases* to $n + (\alpha/2n)|E|^2$ and, as shown in Figure 1, it is possible to balance the tendency of the beam to self-diffract with a *nonlinear* tendency to self-focus. At a critical angle θ_c

$$n \sin \frac{\pi}{2} = \left(n + \frac{\alpha}{2n} |\mathbf{E}|^2 \right) \cos \theta_c \tag{1}$$

and, since α is usually rather small, $\theta_c^2 \simeq (\alpha/n^2)\ |\mathbf{E}|^2$. A balance occurs, therefore, at $\theta_c = \theta_d$, where

$$\alpha|\mathbf{E}|^2 = \lambda^2/4r^2 \tag{2}$$

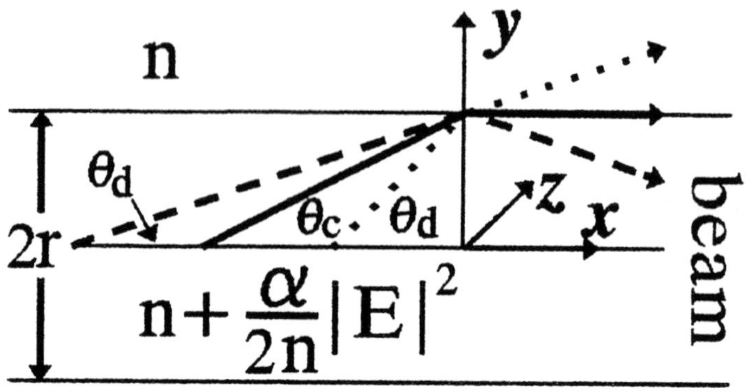

Figure 1. Illustration of balancing diffraction (negative lens effect) with nonlinearity (self-focusing, positive lens effect) $\theta_c < \theta_d$: rays lost; $\theta_c > \theta_d$: confinement.

Consider now a linearly polarised beam carrying an electromagnetic field $\mathbf{E} = (E_x, 0, 0)$, $\mathbf{H} = (0, H_y, 0)$, where **H** is the magnetic field. Since a rectangular distribution of energy is being considered here (extensions to more complicated beams add mathematical difficulty but do not add anything new to the physics) it is necessary only to consider, if the energy flow is required, the simple expression for the time-averaged modulus of the Poynting vector along the z-direction. This is $S = \frac{1}{2} R_e (E_x H_y^*)$, where R_e denotes the real part and the * denotes the complex conjugate. For TE *or* TM waves, propagating along z, in which $E_z \simeq 0$, as it is in many practical cases, then $S = (\epsilon_0/2)$ nc $|E|^2$ watts/m^2, where ϵ_0 is the permittivity of free space and the power flow along z is $P = \int S dx dy$.

Given a beam of radius r the critical power at the balance is [1]

$$P_{cr} = \pi r^2 \left(\frac{\epsilon_0}{2} nc\, |E|^2\right) = \frac{\pi}{8} \frac{\epsilon_0\, nc}{\alpha} \lambda^2 \qquad (3)$$

Note that P_{cr} differs a little from [1] in the use of modern units, the use of the less precise $\lambda/2nr$, rather than $0.61\, \lambda/2nr$, and the definition $(\alpha/2n)|E|^2$, rather than $n_2|E|^2$. The important feature of equation (3) is that, if $P > P_{cr}$, rays are forced towards the axis of the beam in a *self-focusing* action. If $P < P_{cr}$ then *diffraction* takes over. These properties of the medium occur because P_{cr} is fixed, i.e. ϵ_0, n, c, α and λ are fixed parameters of the system and it is necessary for any launched input power P to be precisely equal to P_{cr} for stability to occur. Inevitably, in any real experiment the power launched into a waveguide will be either $P < P_{cr}$ or $P > P_{cr}$, even if only infinitesmally different. In either case, such a fluctuation will lead to instability taking the form of a 'runaway' to diffraction or to self-focusing. This, then, is the problem with self-trapped beams in which diffraction and self-focusing is finely balanced in a bulk (infinite with no boundaries) medium. Mathematically, it means that such beams are, by their nature, unstable in a multi-dimensional system and this has been established, for a longtime, in the literature [6].

The situation described above, although interesting, in principle, would appear to have very little experimental promise because of its instability. A new interest has arisen in this area, however, going under the name of spatial solitons, which will be defined properly later on. This interest arises because some forms of stabilisation have been proposed that are proving to be rather successful. The first stabilisation method [4] involves controlling the perturbations that can lead to instability by a method based upon a modulation due to interference fringes [4,5]. This will not be discussed here since the aim of this chapter is to study *spatial solitons*. The latter can be created within a nonlinear *planar* waveguide and reduces the propagation to a two-dimensional one.

The instability referred to above that arises in a three-dimensional medium has a P_{cr} that is fixed, in the sense that self-adjustment of the beam such as its radius, while focusing, cannot change it. On the other hand, if a beam is launched into a planar waveguide that has interfaces parallel to the x-z plane, then guiding confinement will occur in the y-direction. Suppose that the guide is weakly nonlinear, so that the modal field intensity is unaffected by nonlinearity to first order, then the power needed for diffraction and self-focusing to balance becomes

$$P_{cr} = \frac{\pi r d}{2} \epsilon_0 \, nc \, |E|^2 = \frac{\pi}{8} \frac{\epsilon_0 nc}{\alpha} \frac{d}{r} \lambda^2 \qquad (4)$$

where $2d$ is the thickness of a guiding layer with *linear* refractive index n.

r is now the largest beam dimension and $\pi r d$ is the area of a beam with most of its energy confined an elliptic cross-section, and within the guide. This is shown in Figure 2, which also contains a schematic explanation of the stability. The value of P_{cr} does not now arise from the cancellation of r^2 but has left within it a factor $\left(\frac{1}{r}\right)$ [2-5]. As r changes, due to attempts to self-diffract or self-focus, the system is always forced back to stability. This is the elegant feature of propagation in a waveguide and is creating some excitement in the field [2,3]. Naturally, since open waveguides, made of transparent dielectric materials, are used, real beams carry some part of their energy outside the waveguide. This is because tangential field components are continuous at the

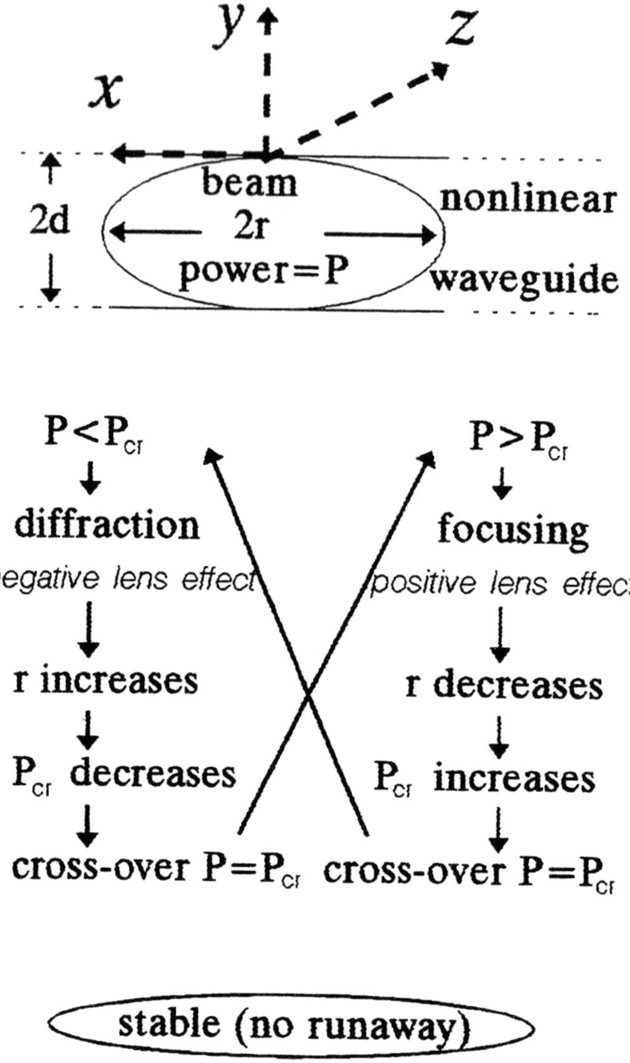

Figure 2. Stable beam propagation in a planar waveguide. Launched powers $P < P_{cr}$ or $P > P_{cr}$ return towards $P = P_{cr}$. Diffraction/self-focusing *balance* in the x-direction; *linear guiding* in y-direction; *propagation* in z-direction.

boundaries but, as shown in Figure 3, most of the energy is carried, as sketched in Figure 2, so that the 'balancing' arguments are correct, even for realistic cases.

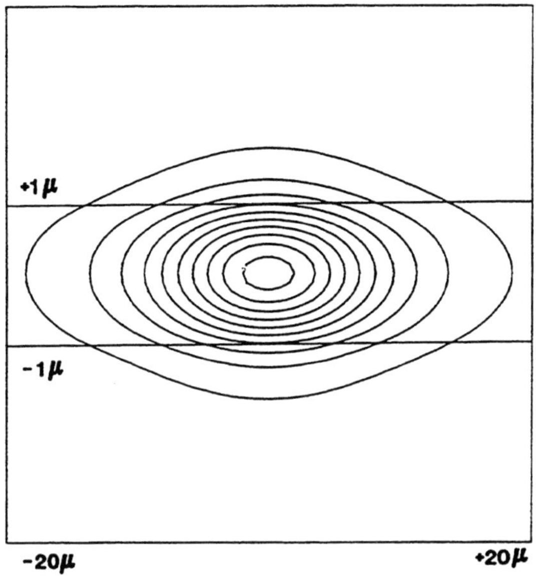

Figure 3. Contour plot of the intensity distribution of a stabilised beam in a planar waveguide. Note that the contours are further apart as the field weakens into the cladding regions.

2. EVOLUTION EQUATION IN A SINGLE WAVEGUIDE

The waveguide is assumed to be both weakly nonlinear, and weakly guiding, in accordance with many experimental situations. Hence, for propagation along the z-axis, waveguide confinement in the y-direction and nonlinear-diffraction balancing in the x-direction, the electric field of the beam has components (E_x, E_y, E_z), where $|E_z| << |E_x|, |E_y|$. The guide is also assumed to support only its fundamental mode, linearly polarised along x or y. Hence, the electric field modulus is

$$E \simeq \begin{cases} E_x/2 + cc & \text{TE mode} \\ E_y/2 + cc & \text{TM mode} \end{cases} \qquad (5)$$

Under this, scalar, approximation the basic wave equation is

$$\nabla^2 E + \frac{\omega^2}{c^2} E + \mu_0 \omega^2 P = 0 \qquad (6)$$

where ω is the angular carrier frequency, μ_0 is the permeability of free space and P is the total polarisation. The most important situation concerns isotropic planar waveguides, for which the material has a centre of symmetry. For such materials

$$\begin{aligned} &\text{LINEAR:} && P^{(1)} = \epsilon_0\, \chi^{(1)}\, E \\ &\text{NONLINEAR:} && P^{(3)} = 3/4\, \epsilon_0\, \chi^{(3)}\, |E|^2\, E \end{aligned} \tag{7}$$

where $\chi^{(1)}$ is a linear susceptibility and $\chi^{(3)}$ is the third-order nonlinear susceptibility [7]. In order to obtain (7), it should be emphasised that the material is assumed to have a centre of symmetry and to be of highest symmetry group [8].

The development of the basic equation needed to describe what happens to a beam in a waveguide proceeds as follows. Let the field $E(x,y,z)$ be defined as

$$E(x,y,z) = \Gamma\, A(y)\, B(x,z) \exp\left(\frac{i\omega}{c}\beta z\right) \tag{8}$$

where

- Γ is a possible normalisation factor
- $A(y)$ is the *linear* modal field
- $B(x,z)$ is a *slowly varying* function of z
- β is the *linear* carrier wavenumber

Γ is selected so as to make the power density equal to $(\epsilon_0 nc/2)\, |B|^2$ i.e.

$$\frac{1}{2}\epsilon_0\, nc\, \Gamma^2\, |B(x,z)|^2 \int A^2(y)dy = \epsilon_0 \frac{nc}{2} |B(x,z)|^2 \tag{9}$$

Hence, $\Gamma^2 = 1/\int A^2 dy$, where the explicit functional dependence of $A(y)$ is left out, for convenience. This omission will take place, quite frequently, for A, and also for other variables. The presence of the nonlinearity perturbs β to $\bar{\beta}$ so that equation (6) breaks up into

$$2i\frac{\omega}{c}\beta\frac{\partial B}{\partial z} + \frac{\partial^2 B}{\partial x^2} + \frac{\omega^2}{c^2}(\bar{\beta}^2 - \beta^2)\, B = 0 \tag{10}$$

$$\frac{\partial^2 A}{\partial y^2} + \frac{\omega^2}{c^2}(\epsilon_L + \Delta\epsilon - \bar{\beta}^2)\, A = 0 \tag{11}$$

where $\Delta\epsilon = \alpha|E|^2$ and $\alpha = \frac{3}{4}\chi^{(3)}$. $\Delta\epsilon$ is, in practice, only a small perturbation so it does not, to a good approximation, affect the modal field A(y). Hence $\overline{\beta}$ can be obtained using a simple first-order perturbation theory. The result is

$$\overline{\beta^2} - \beta^2 = \frac{\int \Delta\epsilon\, A^2\, dy}{\int A^2\, dy} = \alpha \frac{\int A^4\, dy}{(\int A^2\, dy)^2} |B|^2 \tag{12}$$

and the *effective mode width* is $a = \frac{1}{2}(\int A^2 dy)^2/\int A^4 dy$. After making the transformations $\frac{\omega}{c} z \rightarrow 2\beta z$, $\frac{\omega}{c} x \rightarrow x$, $\sqrt{\alpha'/2}\, B \rightarrow \psi$, $\alpha' = \alpha/2a$, the equation for B reduces to the following standard nonlinear Schrödinger equation (NLS)

$$i\frac{\partial\psi}{\partial z} + \frac{\partial^2\psi}{\partial x^2} + 2|\psi|^2\psi = 0 \tag{13}$$

Equation (13) has many very interesting solutions but the ones of interest here are called *spatial solitons*. In fact, it is the lowest order, stationary, solutions, of this type, that are selected here because they not only have the form of stable beams, but also propagate without change of shape.

The lowest order, fundamental, solutions of equation (13) have the analytical form

$$\psi = \eta \operatorname{sech}[\rho(x-x)] \exp[i\,\xi/2\,(x-x) + i\theta] \tag{14}$$

As shown in Figure 4, ρ is the inverse of the width, $\eta = \rho$ is the peak amplitude and $x = x_0 + \xi z$ is the centre of the beam. The interpretation of ξ is that $\tan^{-1}(\xi)$ is the angle that the propagation direction of the beam makes to the z-axis. θ is the phase and is equal to $\theta_0 + (\xi^2/4 + \eta^2)z$, where θ_0 is just some initial value that is of no physical importance.

The validity of the solution given by equation (14) can be most easily checked by simple substitution, rather than any advanced inverse scattering technique. For example, if $Sh = \operatorname{sech}[\rho(x-x_0-\xi z]$, $Th = \tanh[\rho(x-x_0-\xi z)]$ and $Exp = \exp[i\xi/2\,(x-x_0-\xi z) + i\theta_0 + i(\xi^2/4 + \rho^2)z]$, then

- $\frac{\partial\psi}{\partial z} = [\xi\rho^2\, ShTh - i\rho\,(\xi^2/4 - \rho^2)Sh]\, Exp$
- $\frac{\partial^2\psi}{\partial x^2} = [\rho^3\, ShTh^2 - \rho^3 Sh^3 - i\xi\,\rho^2 ShTh - \xi^2/4\,\rho Sh]\, Exp$
- $2|\psi|^2\,\psi = 2\rho^3\, Sh^3\, Exp$

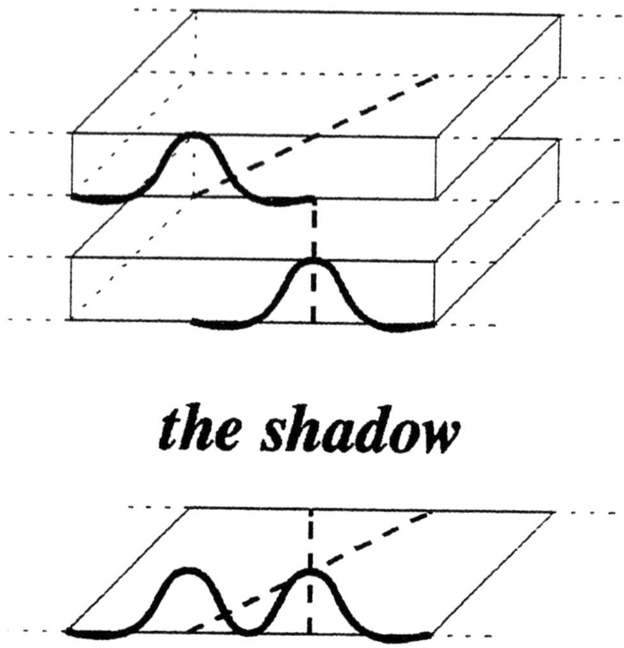

Figure 4. Interacting spatial solitons in coupled, identical, planar waveguides.

and equation (13) is, obviously, satisfied.

Many solutions exist to equation (13), however, that are generated by initial forms $\psi(x,0) = N\rho \text{ sech } (\rho x)$ in which the integer N designates the *soliton* order [6, 7]. In this chapter only $N = 1$ solitons are investigated, for which the beam profile does not change shape with propagation. The $N > 1$ solitons are interesting, but the beam profile does change. In fact, for $N > 1$, $|\psi|^2$ is periodic with a normalised period $\rho^2 z_0 = \pi/4$.

z_0 is a useful length scale of a potential soliton system, even for $N = 1$. For instance, a given spatial soliton beam, may have a width D_0, where the beam amplitude reduces to sech(1) of the peak amplitude. This is given by $D_0 = (\lambda/2\pi)\ \rho^{-1}$, and the soliton period (in real units) is $z_0 = (\beta/\lambda)\ \pi^2 D_0^2$. As hinted at above, the value of z_0 may appear to be irrelevant for constant shape $N = 1$ solitons but it is a good scale against which to judge whether a system will support solitons or not. Typically, for $D_0 \simeq 8.5\mu m$, $\lambda \simeq 0.62\mu m$ and an effective index $\beta \simeq 1.53$, $z_0 \simeq 1.55mm$. Such a value for z_0 encourages the belief that the creation of spatial solitons planar waveguides is a realistic objective. The time-averaged modulus of the Poynting vector is $S = (1/2)\ \epsilon_0$ cn $|E|^2$ Watts/m^2 so that the power density, along the x-direction, is $I = \int Sdy$ $= (2a/nn_2)\ |\psi|^2$ Watts/m. The total power in a fundamental spatial soliton is, therefore,

$$P = \iint Sdxdy = \int Idx = \int \left(\frac{2a}{nn_2}\right) \rho^2 \operatorname{sech}^2(\rho x)$$
$$= \frac{\lambda}{2\pi}\left(\frac{4a\rho}{nn_2}\right) = \left(\frac{\lambda}{2\pi}\right)^2 \frac{4a}{nn_2 D_0} \tag{15}$$

If $a = 1.8\mu m$, $D_0 = 8.5\mu m$, $\lambda = 0.62\mu m$, $n = 1.53$ and $n_2 = 3.4 \times 10^{-16}\ cm^2/W$ then $P \simeq 158kW$, which is experimentally achievable. Indeed, even lower powers are now possible [3]. The final, general, points to make about these two-dimensional stable spatial solitons are

- confinement due to the nonlinearity is in *one transverse dimension* only
- *linear* guiding occurs in the other transverse dimension because of the weak nonlinearity
- propagation occurs along the third dimension.

They can be easily excited in a planar waveguide, if the initial power of the beam exceeds a certain threshold value. More could be said about the basic soliton behaviour, especially in connection with chirp [7], but it is much more interesting to go on to the topic of soliton interaction the foundations of which will be addressed in the next section.

3. COUPLED SCHRÖDINGER EQUATIONS

Figure 4 shows two identical waveguides, each carrying a spatial soliton. These solitons interact with each other and one way to think of this process is to imagine that the upper soliton casts a shadow into the second guide. An analysis of the interaction then proceeds as if the two solitons E_1 and E_2 are, respectively, associated with the two waveguides of Figure 4 and that the beams have identical polarisation and carrier frequency. The scalar wave equations then become

$$j = 1,2: \quad \nabla^2 E_j + \frac{\omega^2}{c^2} E_j + \mu_0\, \omega^2 P_j = 0 \tag{16}$$

where the polarisations P_j are

$$j = 1,2: \quad P_j = \epsilon_0 \chi^{(1)} E_j + \epsilon_0 \chi^{(1)} E_{3-j} + \epsilon_0 \alpha |E_j|^2 E_j \tag{17}$$

Here P_j contains terms corresponding to

- unperturbed linear polarisation
- linear perturbation from a neighbouring guide
- nonlinear self-perturbation

Nonlinear perturbations from a neighbouring guide are ignored for the moment. Such a step is perfectly satisfactory provided that the omitted terms are genuinely second-order compared to the self-perturbation term.

The evolution equations can be derived along similar lines to that of the isolated guide, considered earlier. Given amplitudes $A_1(y)$, $A_2(y)$ in guides 1 and 2, respectively, the normalisation factor Γ is given by $\Gamma^2 = 1/\int A_1^2 dy = 1/\int A_2^2 dy$ and

$$\overline{\beta_j^2} - \beta^2 = \alpha' |B_j|^2 + 2\nu \frac{B_{3-j}}{B_j} \tag{18}$$

where

$$\alpha' = \frac{\alpha \int A_j^4 dy}{(A_j^2 dy)^2}, \quad \nu = \frac{\chi^{(1)}}{2} \frac{\int A_1 A_2 dy}{\int A_1^2 dy} = \frac{\chi^{(1)}}{2} \frac{\int A_1 A_2 dy}{\int A_2^2 dy} \tag{19}$$

After making exactly the same transformations as before, the following *coupled* equations emerge

$$j = 1,2: \; i \frac{\partial \psi_j}{\partial z} + \frac{\partial^2 \psi_j}{\partial x^2} + 2|\psi_j|^2 \psi_j + 2\nu \psi_{3-j} = 0 \tag{20}$$

If we now couple solitons with different polarisation in the *same* waveguide, but with the same frequency, then the electric field is $\mathbf{E} = \frac{1}{2} [E_x \hat{\mathbf{x}} + E_y \hat{\mathbf{y}} + cc]$. The [8,9] coupling between E_x and E_y leads to

$$k,j = x,y; \quad P_j^{(3)} = \epsilon_0 \alpha \left[(|E_j|^2 + |E_k|^2) E_j + f\, (E_j^* E_k - E_j E_k^*) E_k \right] \tag{21}$$

where, for example. $f = \frac{1}{3}$ for electronic distortion nonlinearity.

If the TE, TM wavenumbers, β_x and β_y are associated with the field components E_x and E_y, respectively, then $\nu = 2\beta(\beta_x - \beta_y)$, where β is the average wavenumber $(\beta_x + \beta_y)/2$ and the coupled equations for ψ_x and ψ_y are

$$i\,\frac{\partial \psi_j}{\partial z} + \frac{\partial^2 \psi_j}{\partial x^2} \pm 2\nu\psi_j + 2(|\psi_x|^2 + |\psi_y|^2)\psi_j + 2f\,(\psi_j^*\psi_k - \psi_j^*\psi_k)\,\psi_k = 0, \; (j = x, y) \quad (22)$$

where

- ψ_x : linearly polarised mode in the x-direction
- ψ_y : linearly polarised mode in the y-direction.

The transformations

$$\psi_1 = \frac{\sqrt{1-f}}{2}(\psi_x + i\psi_y) : \text{left-handed elliptically polarised mode}$$

$$\psi_2 = \frac{\sqrt{1-f}}{2}(\psi_x - i\psi_y) : \text{right-handed elliptically polarised mode}$$

lead to

$$j = 1,2: \; i\,\frac{\partial \psi_j}{\partial z} + \frac{\partial^2 \psi_j}{\partial x^2} + 2\nu\psi_{3-j} + 2(|\psi_1|^2 + \mu|\psi_{3-j}|^2)\psi_j = 0 \quad (23)$$

where $\mu = (1+f)/(1-f)$, and the equations describe the coupling of elliptically polarised modes. f is $\frac{1}{3}$ for the electronic nonlinearity and $f = 0$ for thermal nonlinearity.

4. SOLVING THE COUPLED EQUATIONS

On the face of it, equations (23) and (20) are impossible to solve, analytically. Usually it is necessary to resort to substantial numerical computation. There is another way, however, to make analytical progress and it requires the adoption of a variational method. This relies upon the use of what is called a Lagrangian density, makes of Hamilton's principle and introduces *trial* functions. The latter are so-called because they are not exact solutions but are functions that are *tried* through the adjustment of parameters, until some minimum energy state is reached.

4.1 Variational Principle

The basic idea is to construct a quantity L, called a *Lagrange density* [10, 11] which is a function of functions of the basic parameters of the system. For example, in dynamics the parameter may be time and the functions coordinates and velocities. For waveguides containing spatial solitons, the parameters are x and z and L is a function of ψ_1, ψ_2, $\frac{d\psi_1}{dz}$, $\frac{d\psi_2}{dz}$, $\frac{d\psi_1}{dx}$ and $\frac{d\psi_2}{dz}$. This is an elegant method that requires the adjustment of ψ_1 and ψ_2 to ensure that

$$\delta J = \delta \iint L dx dy = 0 \tag{24}$$

Any additional constraints upon the system must be satisfied as well, but the main point is that $\iint L dx dy$ must be stationary and (24) is a generalisation of Hamilton's principle. The latter usually refers to dynamics problems. For example, consider $J = \int_{t_1}^{t_2} L dt$, where the parameter is the time t. If $L = L\left(q, \frac{dq}{dt}\right)$, where q is some variable then δJ (the *variation* in J) can be found from a simple Taylor expansion about the stationary state. This proceeds as follows. Let q *vary* to bQ and $\dot{q}$ *vary* to $b\frac{dQ}{dt}$, where b is a small quantity. If this is done then

$$\delta J = \int_{t_1}^{t_2} \left[L\left(q + bQ, \dot{q} + b\frac{dQ}{dt}, t\right) - L\left(q, \frac{dq}{dt}, t\right) \right] dt$$

$$\simeq b \int_{t_1}^{t_2} \left[\frac{\partial L}{\partial q} - \frac{d}{dt}\left(\frac{\partial L}{\partial \left[\frac{\partial q}{\partial t}\right]} \right) \right] Q \, dt = 0 \tag{25}$$

Hamilton's principle states that $\delta J = 0$ i.e. J is stationary, with respect to small changes in the variables, provided that

$$\frac{\partial L}{\partial q} - \frac{d}{dt}\left(\frac{\partial L}{\partial \left[\frac{\partial q}{\partial t}\right]} \right) = 0 \tag{26}$$

which is the Euler-Lagrange equation. From this example, it can be seen how the coupled equations for ψ_1 and ψ_2 can be processed.

Firstly, the Lagrangian density is

$$L = \sum_{j=1}^{2} \left[\frac{i}{2} \left(\psi_j^* \frac{\partial \psi_j}{\partial z} - \frac{\partial \psi_j^*}{\partial z} \psi_j \right) - \left| \frac{\partial \psi_j}{\partial x} \right|^2 + |\psi_j|^4 \right]$$

$$+ 2\nu \left(\psi_1 \psi_2^* + \psi_1^* \psi_2 \right) + 2\mu |\psi_1 \psi_2|^2 \tag{27}$$

and the Euler-Lagrange equations are

$$\frac{\partial}{\partial z} \frac{\partial L}{\partial \left[\frac{\partial \psi_j^*}{\partial z} \right]} + \frac{\partial}{\partial x} \frac{\partial L}{\partial \left[\frac{\partial \psi_j^*}{\partial x} \right]} - \frac{\partial L}{\partial \psi_j^*} = 0 \tag{28}$$

It is easy to check that equations (28) yield equations (20). The next step is use *trial functions* in (28). These trial functions will now contain z-dependent variables, such as amplitudes and phase, that will satisfy the Euler-Lagrange equations. Hence, the problem is reduced to solving coupled nonlinear equations in the new trial function variables in such a way that $\delta J = 0$. This is a perfectly valid solution of the problem and demonstrates the beauty of the elegance of a method that was invented in the nineteenth century.

The trial functions to be used here are

$$\psi_j = \eta_j \operatorname{sech}[\rho_j (x - x_j)] \exp\left[i\xi_j/2\,(x - x_j) + i\theta_j\right] \tag{29}$$

which are in a form that satisfies the *uncoupled* equations. This time, however, the Euler-Lagrange equations will operate on the trial function variables η_j, ρ_j, ξ_j, x_j and θ_j. Since it is easy to integrate over x, before the variational process begins, this is done first to create a reduced Lagrangian. The application of the Euler-Lagrange equations then leads to ten coupled equations for η_j, ρ_j, x_j, ξ_j and θ_j. The solution of these equations yields the whole evolution picture for the spatial solitons.

Before we go on to a detailed discussion it is worth making a quotation here [10]. *It usually happens that the differential equations for a given phenomenon are worked out first, and only later is the Lagrange function found, from which the differential equations can be obtained. This is not to minimise the importance of finding the Lagrange density L, for it is of considerable utility to find what physical quantity must be minimised to obtain the differential equations for the phenomena, and the form of the variational equations often suggests fruitful analogies and generalisations.* This is precisely what will happen here and, indeed, rather more.

4.2 TE-TM Coupling

Equations (22) are the coupled equations for the interaction between a TE-polarised spatial soliton and a TM-polarised spatial soliton. This is the case where spatial solitons with *different* polarisation are coupled in the *same* frequency. This interaction will serve to illustrate the way in which the Lagrangian analysis is applied before going on to detail, in more physical terms, what the precise outcome of a soliton interaction will be.

The Lagrangian density is [12]

$$L = \frac{i}{2}\left(\psi_x^* \frac{\partial\psi_x}{\partial z} - \frac{\partial\psi_x^*}{\partial z}\psi_x\right) + \frac{i}{2}\left(\psi_y^* \frac{\partial\psi_y}{\partial z} - \frac{\partial\psi_y^*}{\partial z}\psi_y\right) + |\psi_x|^4 + |\psi_y|^4$$

$$- \left|\frac{\partial\psi_x}{\partial x}\right|^2 - \left|\frac{\partial\psi_y}{\partial x}\right|^2 + 2|\psi_x|^2\,|\psi_y|^2 + f\,(\psi_x\psi_y^* - \psi_x^*\psi_y)^2 + 2\nu(|\psi_x|^2 - |\psi_y|^2) \quad (30)$$

Since two widely spaced beams can not have their shape affected by any interaction a useful trial function to use here is the exact solution, in the absence of mutual coupling, with the form

$$\psi_j = \eta_j \,\text{sech}\,[\eta_j\,(x-x_j]\,\exp\left[i\,\frac{\xi_j}{2}(x-x_j) + i\theta_j\right] \quad (31)$$

The parameters η_j, x_j, ξ_j and θ_j, however, will now be treated as *variational* parameters. The reduced Lagrangian is

$$L = \int L dx = \sum_{j=1}^{2}\left[\left(\xi_j\frac{\partial x_j}{\partial z} - \frac{2\partial\theta_j}{\partial z}\right)\eta_j + \frac{2}{3}\eta_j^3 - \frac{\xi_j^2}{2}\eta_j\right] + L_{xy} \quad (32)$$

where

$$L_{xy} = 4\nu(\eta_1-\eta_2) + 2(1-f)\eta_1^2\eta_2^2 \int \text{sech}^2[\eta_1(x-x_1)]\text{sech}^2[\eta_2(x-x_2)]dx$$

$$+ 2\eta_1^2\eta_2^2 f \int \text{sech}^2[\eta_1(x-x_1)]\,\text{sech}^2[\eta_2(x-x_2)]\cos[\xi_2(x-x_2)-\xi_1(x-x_1)+2\theta_2-2\theta_1]dx \quad (33)$$

It is now necessary to calculate the derivative of L with respect to θ_1, θ_2, x_1, x_2, ξ_1, ξ_2, η_1 and η_2. The integration over x reduces the variational problem to $\delta\int L dz = 0$ so that the Euler-Lagrange equations yields

$$\frac{d}{dz}\left(\frac{\partial L}{\partial\left[\frac{d\theta_j}{dz}\right]}\right) - \frac{\partial L}{\partial\theta_j} = 0 \quad (j = 1,2) \tag{34a}$$

$$\frac{d}{dz}\left(\frac{\partial L}{\partial\left[\frac{d\eta_j}{dz}\right]}\right) - \frac{\partial L}{\partial\eta_j} = 0 \quad (j = 1,2) \tag{34b}$$

$$\frac{d}{dz}\left(\frac{\partial L}{\partial\left[\frac{dx_j}{dz}\right]}\right) - \frac{\partial L}{\partial x_j} = 0 \quad (j = 1,2) \tag{34c}$$

$$\frac{d}{dz}\left(\frac{\partial L}{\partial\left[\frac{d\xi_j}{dz}\right]}\right) - \frac{\partial L}{\partial\xi_j} = 0 \quad (j = 1,2) \tag{34d}$$

Equations (34) are a set of eight much simpler nonlinear ordinary differential equations. Note that the word ordinary is included here to show that the resulting equations do not contain any partial derivatives.

The application of equations (34) to equation (32) gives

$$\frac{d\eta_j}{dz} = -\frac{1}{2}\frac{\partial L_{xy}}{\partial\theta_j}\ , \quad \frac{d\theta_j}{dz} = \frac{\xi_j^2}{4} + \eta_j^2 + \frac{1}{2}\frac{\partial L_{xy}}{\partial\eta_j} - \frac{\xi_j}{2\eta_j}\frac{\partial L_{xy}}{\partial\xi_j} \tag{35a, b}$$

$$\frac{d}{dz}(\xi_j\eta_j) = \frac{\partial L_{xy}}{\partial x_j}\ , \quad \frac{dx_j}{\partial z} = \xi_j - \frac{1}{\eta_j}\frac{\partial L_{xy}}{\partial\xi_j} \tag{35c, d}$$

These are the evolution equations for the quantities η_j, x_j, ξ_j and θ_j, introduced through the trial functions (31). For example,

$$\frac{d\eta_j}{dz} = (-1)^j\, 2f\eta_1^2\eta_2^2 \int \operatorname{sech}^2[\eta_1(x-x_1)]\operatorname{sech}^2[\eta_2(x-x_2)]$$

$$\sin[\xi_2(x-x_2)-\xi_1(x-x_1) + 2\theta_2 - 2\theta_1]dx \tag{37}$$

The other expressions in equations (25) are also readily obtainable, but require rather more effort.

The interesting point about equations (35) is that a *particle representation* is immediately suggested, involving quantities equivalent to *mass*, *momentum* and so on. Briefly, it is not difficult to show, from equation (35), that

$$\frac{d}{dz}(\eta_1+\eta_2) = -\frac{1}{2}\left[\frac{\partial L_{xy}}{\partial \theta_1} + \frac{\partial L_{xy}}{\partial \theta_2}\right] = 0 \tag{38}$$

implying *conservation of mass*, while

$$\frac{d}{dz}(\eta_1\xi_1+\eta_2\xi_2) = \frac{\partial L_{xy}}{\partial x_1} + \frac{\partial L_{xy}}{\partial x_2} = 0 \tag{39}$$

implying *conservation of momentum*. Such ideas will be placed on more general basis in the next section.

4.3 Conserved and Normalised Quantities

Given the general trial function of equation (29), where η_j sech $[\rho_j(x-x_j)]$ is used, with $\eta_j \neq \rho_j$, it is possible to make the following definitions and draw the following conclusions from the variational analysis

mass

$$\int \psi_1^2 dx = 2\,\frac{\eta_1^2}{\rho_1}\ ,\quad \int \psi_2^2 dx = 2\,\frac{\eta_2^2}{\rho_2} \tag{40}$$

conservation of mass

$$\frac{d}{dz}\left[\frac{\eta_1^2}{\rho_1} + \frac{\eta_2^2}{\rho_2}\right] = 0 \tag{41}$$

centre of mass

$$\int x|\psi_1|^2\, dx = 2\,\frac{\eta_1^2}{\rho_1}x_1\ ,\quad \int x|\psi_2|^2\, dx = 2\,\frac{\eta_2^2}{\rho_2}x_2 \tag{42}$$

stationary centre of mass

$$\frac{d}{dz}\left[\frac{\eta_1^2}{\rho_1}x_1 + \frac{\eta_2^2}{\rho_2}x_2\right] = 0 \tag{43}$$

conservation of momentum

$$\frac{d}{dz}\left[\frac{\eta_1^2}{\rho_1}\xi_1 + \frac{\eta_2^2}{\rho_2}\xi_2\right] = 0 \tag{44}$$

In the light of these most interesting conclusions, drawn from the applications of the Euler-Lagrange equations to the trial functions (29), the following normalised quantities can also be introduced

mean mass: $$2M = \frac{\eta_1^2}{\rho_1} + \frac{\eta_2^2}{\rho_2} \tag{45}$$

mass difference: $$\frac{\eta_2^2}{\rho_2} - \frac{\eta_1^2}{\rho_1} = 2Mn \tag{46}$$

momentum difference: $$2M\xi = \frac{\eta_2^2}{\rho_2}\xi_2 - \frac{\eta_1^2}{\rho_1}\xi_1 \tag{47}$$

mass centre separation: $$2\Delta = \frac{\eta_2^2}{\rho_2}x_2 - \frac{\eta_1^2}{\rho_1}x_1 \tag{48}$$

phase difference: $$\theta = \theta_2 - \theta_1 \tag{49}$$

These definitions serve as a means to normalise all parameters and variables. Some care needs to be taken, however, since, from the Schrödinger equation, the 'scale' of z and x are quite different. Hence, using M to scale with, it is necessary to make the transformations $M^2 z \rightarrow z$ and $Mx \rightarrow x$. Furthermore, to complete the normalisation the following transformations must also be used: $\rho_j/M \rightarrow \rho_j$, $\xi/M \rightarrow \xi$ and $\nu/M^2 \rightarrow \nu$. Note that μ scales as unity so that $\mu \rightarrow \mu$.

4.4 Particle in a Potential Well

Basically, two main cases could be studied now, in order to lift the calculation out the apparent complexity of the coupled differential equations. These are $\nu/M^2 \ll 1$ or $\nu/M^2 \gg 1$, with the former corresponding to a strong nonlinear interaction, or a weakly birefringent guide. This will be considered here as an example. First, the restriction $\nu/M^2 \ll 1$ removes this factor from the coupled equations. It is then, for the sake of illustration, only necessary to restrict the investigation to beams of equal amplitude (n = 0), whose initial phase (θ_j) are equal, opposite or orthogonal ($\theta = 0, \pm\pi, \pm\pi/2$). If this action is taken then, leaving the question of stability until later,

$$\frac{dn}{dz} = 0 \, , \quad \frac{d\theta}{dz} = 0 \tag{50a}$$

$$\frac{d\Delta}{dz} = \xi + 2f \cos(2\theta) \int x \, \mathrm{sech}^2(x+\Delta)\mathrm{sech}^2(x-\Delta)\sin(2\xi x)dx \tag{50b}$$

$$\frac{d\xi}{dz} = 2 \int [\tanh(x-\Delta)-\tanh(x+\Delta)]\mathrm{sech}^2(x+\Delta)\mathrm{sech}^2(x-\Delta)[1-f+f\cos(2\theta)\cos(2\xi x)]dx \tag{50c}$$

These equations can be developed by using

$$-\frac{d\xi}{dz}\frac{d\Delta}{dz} + \frac{d\Delta}{dz}\frac{d\xi}{dz} = 0 \tag{51}$$

to yield

$$\frac{\xi^2}{2} - 4(1-f)\left[\frac{2\Delta\cosh(2\Delta)}{\sinh^3(2\Delta)} - \frac{1}{\sinh^2(2\Delta)}\right]$$

$$-\frac{4\pi f\cos(2\theta)}{\sinh(\pi\xi)}\left[\frac{\sin(2\xi\Delta)\cosh(2\Delta)}{\sinh^3(2\Delta)} - \frac{\xi\cos(2\xi\Delta)}{\sinh^2(2\Delta)}\right] = C \tag{52}$$

, where C is an integration constant, determined from the initial conditions. For thermal nonlinearity, $f = 0$, so that

$$\frac{\xi^2}{2} - 4\left[\frac{2\Delta\cosh(2\Delta)}{\sinh^3(2\Delta)} - \frac{1}{\sinh^2(2\Delta)}\right] = C \tag{53}$$

Now

$$\frac{dC}{dz} = \frac{\partial C}{\partial \Delta}\cdot\frac{d\Delta}{dz} + \frac{\partial C}{\partial \xi}\cdot\frac{d\xi}{dz} = 0 \tag{54}$$

so that

$$\frac{d\xi}{dz} = -\frac{\partial C}{\partial \Delta} \, , \quad \frac{d\Delta}{dz} = \frac{\partial C}{\partial \xi} \tag{55}$$

Hence, for $f = 0$,

$$\frac{d\Delta}{d\xi} = \xi \tag{56a}$$

$$\frac{d\xi}{dz} = 4\left[\frac{6\cosh(2\Delta)}{\sinh^3(2\Delta)} + \frac{4\Delta}{\sinh^2(2\Delta)} - \frac{12\Delta\cosh^2(2\Delta)}{\sinh^4(2\Delta)}\right] \tag{56b}$$

The substitution of equation (56a) into equation (53) and equation (56b) leads to

$$\frac{1}{2}\left(\frac{d\Delta}{dz}\right)^2 - 4\left[\frac{2\Delta\cosh(2\Delta)}{\sinh^3(2\Delta)} - \frac{1}{\sinh^2(2\Delta)}\right] = C \tag{57}$$

$$\frac{d^2\Delta}{dz^2} = 4\left[\frac{6\cosh(2\Delta)}{\sinh^3(2\Delta)} + \frac{4\Delta}{\sinh^2(2\Delta)} - \frac{12\Delta\cosh^2(2\Delta)}{\sinh^4(2\Delta)}\right] \tag{58}$$

Here 2Δ is the relative distance between beam maxima and z is equivalent to a 'time' coordinate. Hence equation (57) is an energy conservation law with $\frac{1}{2}\left(\frac{d\Delta}{dz}\right)^2$ interpreted as kinetic energy with the potential energy being given by

$$U(\Delta) = -4\left[\frac{2\Delta\cosh(2\Delta)}{\sinh^3(2\Delta)} - \frac{1}{\sinh^2(2\Delta)}\right] \tag{59}$$

The total energy C is then equivalent to the interaction energy ('force') between the two spatial solitons. The potential $U(\Delta)$ is shown in Figure 5 which shows that the beams are always attractive to each other in that particular case. For numerical purposes, the actual potential function $U(\Delta)$ can, in this special case, be fitted by $U'(\Delta) = -\frac{4}{3}\,\mathrm{sech}^2\left(\frac{4}{3}\Delta\right)$. Hence, the interaction force becomes

$$F = \frac{d^2\Delta}{dz^2} = -\frac{\partial U}{\partial\Delta} \simeq -\frac{\partial U'}{\partial\Delta} = -\frac{32}{9}\,\mathrm{sech}^2\left(\frac{4}{3}\Delta\right)\tanh\left(\frac{4}{3}\Delta\right) \tag{60}$$

As the solitons move apart work is done by this force and the soliton velocity, $\frac{d\Delta}{dz}$, will decrease until it reaches zero. At this point, the soliton returns. The 'escape velocity' that permits a soliton to get away (to infinity) from the well is

$$\xi_e = 2\sqrt{2}\left[\frac{2\Delta_0\cosh(2\Delta_0)}{\sinh^3(2\Delta_0)} - \frac{1}{\sinh^2(2\Delta_0)}\right]^{\frac{1}{2}} \simeq \frac{2\sqrt{2}}{\sqrt{3}}\,\mathrm{sech}\left(\frac{4}{3}\Delta_0\right) \tag{61}$$

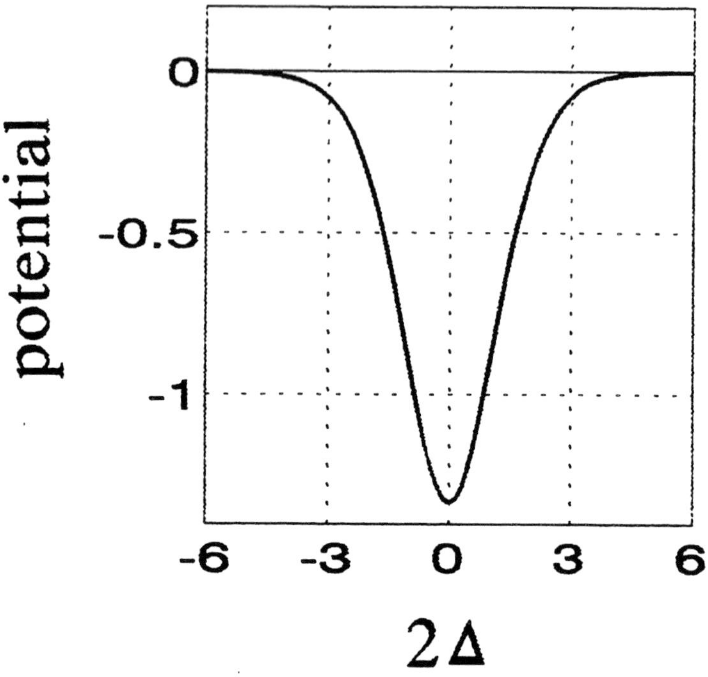

Figure 5. Potential energy as a function of separation for two spatial solitons of equal amplitude in a thermally nonlinear material. The solitons are attracted to each other.

, where Δ_0 is the initial separation distance of the spatial solitons. In terms of beam language, the parameter ξ is related to the beam propagation angle, with respect to the z-axis, by $\phi = \tan^{-1}(\xi_e)$. This means that, if two beams propagate at an angle $2\phi_0 > 2\tan^{-1}(\xi_e)$, they will separate, finally, while if they propagate at an angle $2\phi_0 < 2\tan^{-1}(\xi_e)$ the beams will keep crossing each other, periodically. This feature is sketched in Figure 6, and is simulated numerically in Figure 7.

As shown in Figure 6 the beams, for $2\phi_0 < 2\tan^{-1}(\xi_e)$, after crossing, will achieve a maximum separation Δ_r, before turning back towards each other, where Δ_r is given by

$$-4\left[\frac{2\Delta_r\cosh(2\Delta_r)}{\sinh^3(2\Delta_r)} - \frac{1}{\sinh^2(2\Delta_r)}\right] = C \qquad (62)$$

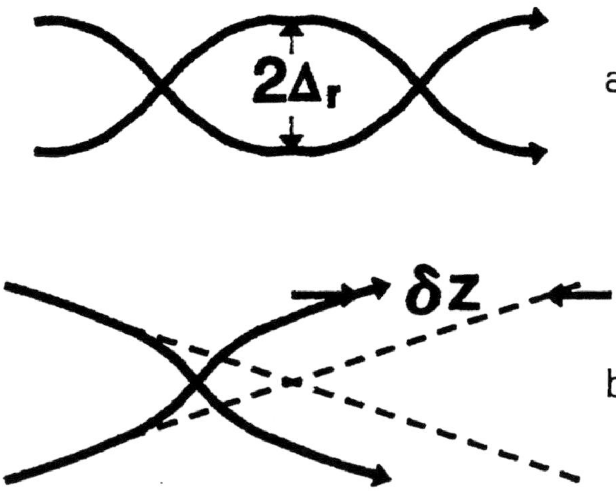

Figure 6. TE-TM interaction of two spatial solitons of equal amplitude crossing each other in a thermally nonlinear medium. (a) relative velocity less than escape velocity (b) relative velocity greater than escape velocity.

or $\Delta_r = 3/4 \cosh^{-1} [1/\sqrt{-3C/4}]$. The soliton paths criss-cross each other, in an oscillatory pattern, with a period equal to

$$Z_p = 4 \int_0^{\Delta_r} \frac{dz}{d\Delta} d\Delta \simeq \sqrt{6} \int_0^{\Delta_r} \frac{d\Delta}{\left[\operatorname{sech}^2\left(\frac{4}{3}\Delta\right) - \operatorname{sech}^2\left(\frac{4}{3}\Delta_r\right)\right]^{\frac{1}{2}}}$$

$$= \frac{3\sqrt{3}}{4\sqrt{2}} \pi \cosh\left(\frac{4}{3}\Delta_r\right) = \frac{3\pi}{2\sqrt{2}}\left(\frac{1}{\sqrt{-C}}\right) \tag{63}$$

For an initial velocity greater than ξ_e the solitons will cross each other and proceed to infinity, with a velocity $\pm \sqrt{2C}$. They will not meet each other again but, as shown in Figure 6, a 'memory' of the interaction appears as a spatial shift

$$\delta z = 2 \int_0^{\infty} \left[\frac{dz}{d\Delta} - \left(\frac{dz}{d\Delta}\right)_{\infty}\right] d\Delta$$

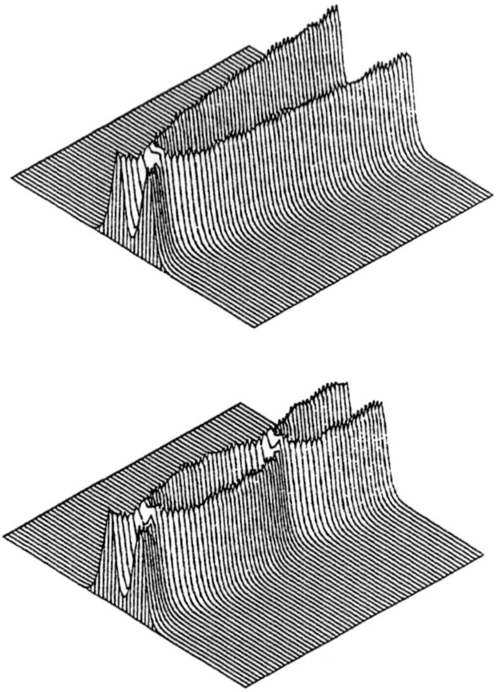

Figure 7. Numerical simulation of two spatial solitons, whose relative velocity is greater or less than the escape velocity, in a thermally nonlinear medium.

$$\simeq \frac{\sqrt{3}}{\sqrt{2}} \int_0^{\infty} \left[\frac{1}{\left[\operatorname{sech}^2\left(\frac{4}{3}\Delta\right) + \frac{3C}{4} \right]^{\frac{1}{2}}} - \frac{1}{\left[\frac{3C}{4}\right]^{\frac{1}{2}}} \right] d\Delta$$

$$= \frac{3}{4\sqrt{2C}} \log\left(\frac{3C}{3C+4}\right) \tag{64}$$

The simulation in Figure 7 shows events in which the solitons separate.

5. AN INVESTIGATION OF STABILITY

There are various other types of interaction that can be imagined for spatial solitons depending upon the configuration. A TE-TM interaction in a single waveguide under thermal nonlinearity and other special conditions has been discussed but now the spatial solitons can be imagined as coupled in a more general way involving two waveguides. The interaction scheme is as follows

A	B
equal amplitude equal phase or opposite phase	superimposed co-propagating

stationary solution

$\Delta = 0, \xi = 0, n = 0, \theta = 0, \pm \pi$

STABILITY?

change in A conditions	change in B conditions

For example, the system can consist, as shown in Figure 8, of two identical solitons in two separate, yet coupled, planar waveguides. In equilibrium (stationary [unperturbed] state) the guiding system is perfectly balanced, with solitons of equal amplitude. Clearly for the stationary state there is no energy exchange between the guides and no relative sideways movement between the two spatial solitons. The question then is whether such a system is stable for various values of ν and $\mu = (1-f)/(1+f)$. In other words, the stationary solutions and stability of the pair coupled equations

$$i \frac{\partial \psi_1}{\partial z} + \frac{\partial^2 \psi_1}{\partial x^2} + 2\nu\psi_2 + 2(|\psi_1|^2 + \mu|\psi_2|^2)\,\psi_1 = 0 \tag{65a}$$

$$i \frac{\partial \psi_2}{\partial z} + \frac{\partial^2 \psi_2}{\partial x^2} + 2\nu\psi_1 + 2(|\psi_1|^2 + \mu|\psi_1|^2)\,\psi_2 = 0 \tag{65b}$$

will be investigated, for two counter-rotating elliptically polarised spatial soliton beams. Consider a change in the B conditions so that the perturbation causes a change in the beam centre (shift in Δ) and beam propagation direction (ξ). It

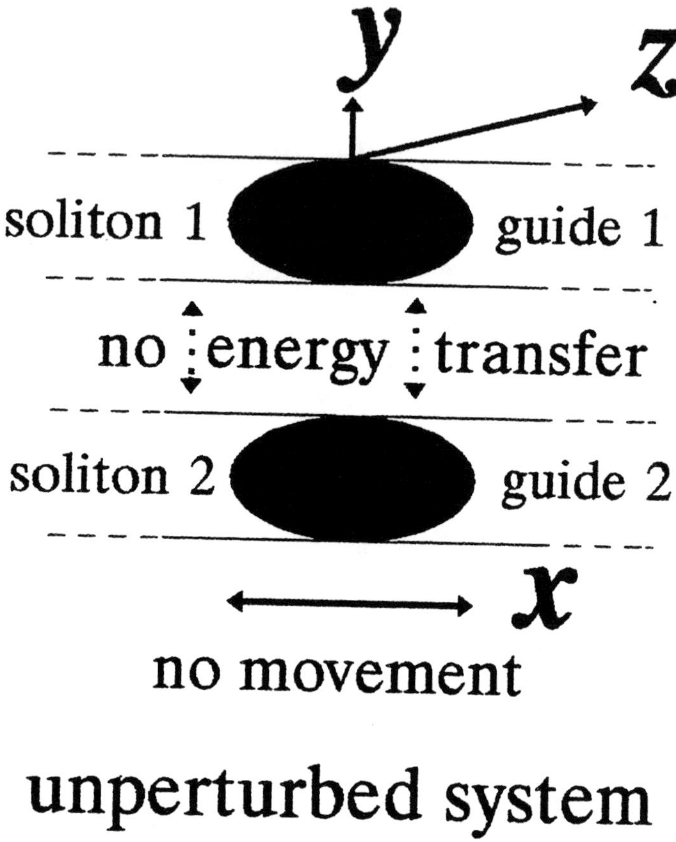

Figure 8. Two spatial solitons in separate waveguides. The solitons are superimposed and co-propagating. The unperturbed system is perfectly balanced with equal amplitudes, no energy transfer between the guides, and no sideways movement of the solitons.

also causes a change in beam width ρ_j but the perturbation is symmetric with respect to the two beams so that $\rho_1 = \rho_2$ (always), $n = 0$ and $\theta = 0, \pm \pi$. Hence, dn/dz and $d\theta/dz = 0$. After making the normalisations previously defined, and after performing a considerable amount of labour the Euler-Lagrange equations give

$$\frac{2}{3}\rho^2 - \frac{4}{3}\rho + \frac{\xi^2}{2} - \frac{4\nu\pi\sin(\Delta\xi)}{\sinh(2\rho\Delta)\sinh\left(\frac{\pi\xi}{2\rho}\right)} - 4\mu\rho\left[\frac{1}{\tanh^2(2\rho\Delta)}-1\right]\left[\frac{2\rho\Delta}{\tanh(2\rho\Delta)}-1\right]$$

$$= -4\nu - \frac{2}{3}(1+\mu)^2 \tag{66}$$

$$\frac{1}{3}(1-\rho) + \mu\left[\frac{1}{\tanh^2(2\rho\Delta)}-1\right]\left[\frac{2\rho\Delta}{\tanh(2\rho\Delta)}-1\right]$$

$$-\frac{\gamma}{\rho}\;\frac{\pi\sin(\Delta\xi)}{\sinh(2\rho\Delta)\sinh\left(\frac{\pi\xi}{2\rho}\right)}\left[\frac{2\rho\Delta\cosh(2\rho\Delta)}{\sinh(2\rho\Delta)} - \frac{\frac{\pi\xi}{2\rho}\cosh\left(\frac{\pi\xi}{2\rho}\right)}{\sinh\left(\frac{\pi\xi}{2\rho}\right)}\right]$$

$$-\mu\,\frac{2\rho\Delta\,\mathrm{sech}^2(2\rho\Delta)}{\tanh^4(2\rho\Delta)}\{2\rho\Delta\,[2+\mathrm{sech}^2(2\rho\Delta)] - 3\tanh(2\rho\Delta)\} = 0 \tag{67}$$

from which we obtain

$$\lim_{\substack{\Delta\to 0\\ \xi\to 0}}\left(\frac{d\xi}{d\Delta}\right)^2 = \lim_{\substack{\Delta\to 0\\ \xi\to 0}}\frac{1}{2}\left(\frac{d^2\xi^2}{d\Delta^2}\right) = -\frac{16\rho^2}{3}\;\frac{\nu + \frac{4}{5}(1+\mu)\mu}{1+\frac{\pi^2}{3}\frac{\nu}{(1+\mu)^2}} \tag{68}$$

If $\lim_{\substack{\Delta\to 0\\ \xi\to 0}} \frac{d^2\xi^2}{d\Delta^2} > 0$ then the system is unstable, similarly for $\left(\frac{d\xi}{d\Delta}\right)^2$, and, therefore, instability is predicted for the conditions

$$\frac{\frac{4}{5}\mu(1+\mu)+\nu}{1+\frac{\pi^2\nu}{3(1+\mu)^2}} < 0 \tag{69}$$

Finally, the two curves

$$\frac{4}{5}\mu(1+\mu)+\nu = 0 \tag{70a}$$

$$1 + \frac{\pi^2 \nu}{3(1+\mu)^2} = 0 \tag{70b}$$

separate regions that are stable or unstable in the (ν,μ) plane, as shown in Figure 9. Such curves have been referred to a stability edges [12]. It is interesting that the Lagrangian analysis presented here leads to different conclusions. It should be noted, however, that the work of [12] contains a linear stability analysis which contrasts rather strongly with the variational, nonlinear, approach.

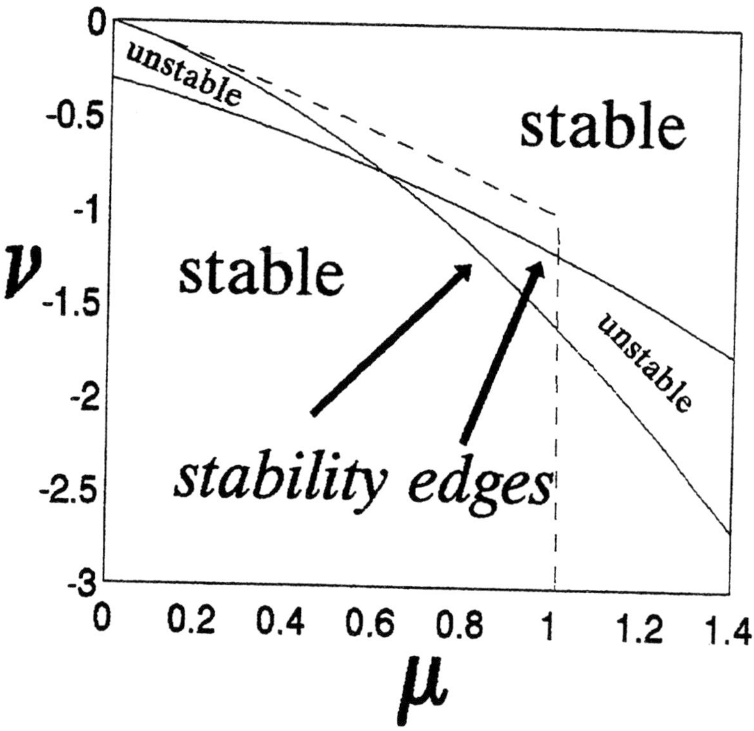

Figure 9 (ν,μ) plane showing stable and unstable regions for type B (symmetric) perturbations. —— Lagrangian theory. ---- denotes a linear stability analysis [12] for which the region contained by ---- and the coordinate axes is unstable.

6. CONCLUSIONS

This chapter contains a detailed account of the theory of spatial solitons in planar optical waveguides. Every effort has been made to present simple physical arguments, backed up by detailed mathematical presentations. The nature of the nonlinear polarisation has been carefully presented and its interpretation in terms of possible soliton interactions has been given. It is emphasised that spatial solitons arise as a consequence of balancing nonlinear self-focusing with diffraction. Indeed, diffraction is regarded as a positive lens effect and self-focusing is regarded as a negative lens effect.

A full derivation of the nonlinear Schrödinger equation is given for a weakly nonlinear waveguide and some of its solutions are discussed. The mathematical work then goes on to consider the interaction of spatial solitons, both through nonlinear polarisation effects in a single waveguide and through the coupling between two waveguides. Several fascinating possibilities are expressed, including one that involves an oscillatory interaction. The dynamics of spatial solitons are developed from Lagrangian variational analysis and it turns out that this leads naturally to a particle representation. In other words, it is shown that it is possible to think of the interacting spatial solitons in terms of such familiar concepts as mass, momentum and so on. Hence, the development of the theory leads to one of the basic characteristics of Lagrangian formulations. This is that, although the basic equations are known in one form, their expression in terms of a Lagrangian and trial functions leads to a strong enhancement of the theory. This is because it is an elegant approach which presents opportunities for using analogies such as the particle representation we have described.

Finally, the topic of spatial solitons is only just beginning and the discovery that these quantities can be readily generated in planar guides opens up the fascinating prospect that soliton dynamics will become commonplace in integrated optics.

ACKNOWLEDGEMENT

Both authors acknowledge financial support from the UK SERC Nonlinear Optics Initiative within a collaborative project on Nonlinear Guided Optical Waves.

REFERENCES

1. S.A. Akhmanov, A.P. Sukhorukov, and R.V. Khokhlov; "Self-focusing and diffraction of light in a nonlinear medium"; *Soviet Physics USPEKHII*; Vol. 93 No. 1 1968; pp.609-636.

2. J.S. Aitchison, Y. Silberberg, A.M. Weiner, D.E. Leaird, M.K. Oliver, and J.L. Jackel; "Spatial optical solitons in planar glass waveguides"; *J. Opt. Soc. Am. B*; Vol. 8 No. 6 1991; pp.1290-1297.

3. J.S. Aitchison, K. Al-Hemyari, C.N. Ironside, R.S. Grant, and W. Sibbett; "Observation of spatial solitons in AlGasAs waveguides"; *Electronics Letters*; Vol. 28 1992; pp.1879-1880.

4. A. Barthelemy, S. Manuef, and F. Froehly; "Propagation soliton et autoconfinement de faisceaux laser par non-linearite de Kerr"; *Opt. Comm.*; Vol. 55 1985; pp.201-206.

5. F. Reynaud, and A. Barthelemy; "Soliton beam propagation in nonlinear Kerr media"; *Guided Wave Nonlinear Optics*, edited by D.B. Ostrowsky, and R. Reinisch; (Dordrecht/Boston/London, NATO ASI Series E, Applied Sciences) 1992; pp.319-340.

6. V.E. Zakharov, and A.M. Rubenchik; "Instability of waveguides and solitons in nonlinear media"; *Sov. Phys. JETP*; Vol. 38 1974; pp.494-500.

7. G.P. Agrawal; *Nonlinear Fiber Optics*; (Boston: Academic Press, Inc.) 1989; pp.51-143.

8. A.D. Boardman, P. Egan, T. Twardowski, and M. Wilkins; *Nonlinear Waves in Solid State Physics*; (New York and London: NATO ASI Series B: Plenum Press); Vol. 247 1990; Edited by A.D. Boardman, M. Bertolotti and T. Twardowski; pp.1-51.

9. A.D. Boardman, and P. Egan, "Third-order nonlinear electromagnetic TE and TM guided waves"; *Nonlinear Surface Electromagnetic Phenomena*; (Amsterdam: North Holland); 1991; Edited by H.E. Ponath and G.I. Stegeman; pp.73-287.

10. P.M. Morse, and H. Fesbach; *Methods of Theoretical Physics*; (New York: McGraw-Hill); 1953; pp.275-347.

11. B.L. Moisewitsch; *Variational Principles*; (London: Interscience Publishers); 1966; (310pp).

12. C. Pare, and M. Florjanczyk; *Phys. Rev. A*; Vol. 41 No. 11 1990; pp.6287-6295.

Studies in Classical
and Quantum Nonlinear Optics

ISBN 1-56072-168-5

NEAR DIPOLE-DIPOLE INTERACTION EFFECTS IN NONLINEAR AND QUANTUM OPTICS

Charles M. Bowden
Weapons Sciences Directorate, AMSMI-RD-WS-ST
Research, Development, and Engineering Center
U. S. Army Missile Command
Redstone Arsenal, Alabama 35898-5248, USA

I. INTRODUCTION

For static electric fields, the well-known Lorentz-Lorenz relation [1] relates the microscopic local static electric field, $\mathbf{E}_L$, which couples with an atomic or molecular dipole moment, to the macroscopic field $\mathbf{E}$ and volume polarization $\mathbf{P}$ in the dielectric medium, and is given as the so-called local field correction as,

$$\mathbf{E}_L = \mathbf{E} + \left[\frac{4\pi}{3} + s\right]\mathbf{P} \tag{1-1}$$

where s is a structure factor equal to zero for cubic or spherical symmetry. A well-known fact in condensed matter physics is that Eq. (1-1), together with Maxwell's equations, leads to the well-known Clausius-Mossotti equation relating the microscopic molecular polarizability α to the macroscopic dielectric function ϵ of the medium,

$$\alpha = \frac{3}{4\pi n}\left[\frac{\epsilon - 1}{\epsilon + 1}\right] \tag{1-2}$$

where n is the volume density. Alternatively, using the constitutive relation,

$$\mathbf{P} = \chi \mathbf{E} \,, \tag{1-3}$$

the macroscopic susceptibility is related to the atomic or molecular polarizability, α ,

$$\chi = \frac{n\,\alpha}{1 - 4\pi/3n\alpha} \,. \tag{1-4}$$

The corresponding problem for time varying, propagating fields is fundamentally different, and nontrivial, since causality conditions and retardation effects must be explicitly taken into account. Previously, it has been shown [2] using the extinction theorem [3] that for a stationary, monochromatic, time-dependent, propagating field in a linear, isotropic, homogeneous medium, the relation (1-1) (with s = 0) is valid and so the Clausius-Mossotti relation [2] then follows. A similar development has been provided for crystalline and amphorous materials by Van Kranendonk and Sipe, as well as by Hynne and Bullough [4]. These works dealt with linear dielectric media, where the macroscopic field **E** and the macroscopic polarization **P** are related by a constitutive relation. This leaves open the question concerning propagating fields in nonlinear media.

The remainder of this review article is organized in the following way. The next section will be used to present a self-consistent development of the Maxwell-Bloch formulation [5] for the electromagnetic field propagation in dense, nonlinear media, having many atoms within a cubic resonance wavelength. It is shown that the resulting Bloch equations in the homogeneously-broadened limit, are identical to the usual optical Bloch equations [5], except that a population excitation dependent shift in the atomic resonance is induced by the near dipole-dipole interaction [6]. This nonlinearity is sufficient to cause a plethora of new and interesting optical phenomena. Section 3 is used to discuss new results derived using the generalized Maxwell-Bloch formulation for dense material. The final section is used for summary and conclusion.

2. MAXWELL-BLOCH FORMULATION FOR DENSE MEDIA

The model consists of a collection of homogeneously-broadened two-level atoms coupled to the electromagnetic field in the dipole and semiclassical approximations. In Appendix A, we show that the system is described by the optical Bloch equations [5], for the level population difference W and the slowly varying, off-diagonal matrix element R_{ab} which is proportional to the slowly varying polarization envelope P,

$$\frac{dW}{dt} = -\gamma_L (W+1) + \frac{\mu}{\hbar}\left[\mathsf{E}_L^{*}\mathsf{R}_{ab} + \text{c.c.}\right],$$ (2-1)

$$\frac{dR_{ab}}{dt} = -\left(i\Delta + \gamma_T\right)\mathsf{R}_{ab} - \frac{\mu}{2\hbar}\mathsf{E}_L W,$$ (2-2)

where $|a\rangle$ represents the upper level and $|b\rangle$ the lower level, separated at zero field strength by the atomic energy level spacing $\hbar\omega_0$. Furthermore, $\Delta = \omega_0 - \omega$ is the detuning of the field central frequency, ω, from atomic resonance, μ is the matrix element of the transition dipole moment, and γ_L and γ_T are the homogeneous relaxation and dephasing rates, respectively. The microscopic slowly-varying electric field envelope E_L is the electric field which drives an atom of polarization amplitude p in the dipole interaction approximation, $H_I = -\mathbf{E}_L \bullet \mathbf{p}$.

The macroscopic electric field amplitude **E** is related to the macroscopic polarization in the medium **P** by Maxwell's wave equation,

$$-\nabla \times \nabla \times \mathbf{E} - \frac{1}{c^2}\frac{\partial^2 \mathbf{E}}{\partial t^2} = \frac{4\pi}{c^2}\frac{\partial^2 \mathbf{P}}{\partial t^2},$$ (2-3)

where c is the speed of light in vacuum. (See Appendix A.) In terms of the slowly-varying amplitudes E and P,

$$\mathbf{E} = \frac{1}{2}\left[\mathcal{E}\ e^{-i\left(\omega t - \mathbf{k}\bullet\mathbf{x}\right)} + \text{c.c.}\right], \tag{2-4}$$

$$\mathbf{P} = \frac{1}{2}\left[\mathcal{P}\ e^{-i\left(\omega t - \mathbf{k}\bullet\mathbf{x}\right)} + \text{c.c.}\right], \tag{2-5}$$

$$\mathcal{P} = 2i\mu n \mathcal{R}_{ab}, \tag{2-6}$$

and $\mathbf{P} = \mathbf{p}n$. Here, Eqs. (2-1) and (2-2) for the atomic variables are coupled to Eq. (2-3) in the field and polarization variables via $\mathcal{R}_{ab}$, provided there exists a relation, $\mathcal{E}_L = \mathcal{E}_L(\mathcal{P})$.

It is to be noted that the microscopic field $\mathcal{E}_L$ in Eqs. (2-1) and (2-2) is not, in general, equivalent to the macroscopic field $\mathcal{E}$ of Maxwell's equation that appears in Eqs. (2-4) and (2-5). Fundamentally at least, $\mathcal{E}_L$, the field that drives an atom, does not contain the atom's self-field, whereas the macroscopic Maxwell field, Eqs. (2-3) and (2-4), contain all local fields present [4]. Neglecting this discrepancy achieves a negligible error in solving Eqs. (2-1) – (2-3) with initial and boundary conditions, provided that the density of atoms is such that there is only a small fraction of an atom within a cubic atomic resonance wavelength, on the average. However, if the conditions are such that there are many atoms within a cubic resonance wavelength, then the error can be considerable in equating $\mathcal{E}_L$ to $\mathcal{E}$.

It has been pointed out by Van Kranendonk and Sipe [4] that the Lorentz-Lorenz relation, Eq. (1-1), arises as a "correction" in the transition from microscopic to macroscopic electrodynamics for linear systems, due to the neglect of the atom's self-field. We shall show that the relation, Eq. (1-1), is satisfied in general and that this results in a self-consistent set of generalized Maxwell-Bloch equations which can lead to a plethora of new and interesting nonlinear optical phenomena.

In an earlier work [7] Bowden and co-workers developed the set of Bloch-like equations for dense media coupled to the Maxwell wave equation in the slowly-varying envelope approximation (SVEA), within the framework of quantum electrodynamics. This was done for a long pencil-like medium, and a set of generalized Maxwell-Bloch equations for a homogeneous medium having many atoms within a cubic resonance wavelength was derived. Later, similar results for large Fresnel number [8] were also calculated. In both approaches the plane wave assumption was invoked to obtain closed-form results. Here, however, we use the method of Bowden and Dowling [6] to consider a similar medium but entirely relax the restriction to plane wave behavior by using an exceedingly simple, compact, and general approach invoking Green's function methods. Our results are then entirely general for dense media and can readily be adopted to inhomogeneous broadening [9]. We shall discuss this aspect later on in this chapter. Our procedure here [6] is to generalize the idea given originally by Lorentz [8] and used by Jackson [1] for static fields; that is, we assume that the local field $\mathbf{E}_L(\mathbf{x},t)$ that drives an atom is the sum of two contributions,

$$\mathbf{E}_L = \mathbf{E} + \mathbf{E}_i \tag{2-7}$$

where $\mathbf{E}$ is the average macroscopic field and $\mathbf{E}_i$ is the internal field at the atom due to polarization of the near dipoles within a cubic wavelength λ^3. This internal field itself can be written as the difference of two fields, namely,

$$\mathbf{E}_i = \mathbf{E}_{near} - \mathbf{E}_P \ , \tag{2-8}$$

where $\mathbf{E}_{near}$ is the actual contribution from dipoles within a cubic region of size d^3 where d is the typical intermolecular spacing. $\mathbf{E}_P$ is a counterterm that takes into account the correction in going from microscopic to macroscopic variables. We are assuming $d \ll \lambda$. Under this assumption, the field $\mathbf{E}_{near}$ can be calculated assuming static conditions since the effects of retardation are completely negligible. Jackson [1] presents the argument of Lorentz [10] to show that

$$\mathbf{E}_{near} = s\,\mathbf{P} \tag{2-9}$$

where s is exactly zero for a cubic lattice and nearly so for other reasonably isotropic materials. We shall take $s \equiv 0$ as our *ansatz*. Then it remains to compute the term $-\mathbf{E}_P$ in Eq. (2-8) that is a correction term which accounts for the error in going from the process of averaging at the scale of dipole spacing d to the intermediate scale l where the macroscopic notion of a polarization **P** holds, but we still have many dipoles in a cubic wavelength. Hence,

$$d << l << \lambda \tag{2-10}$$

is the relevant ordering of length scales. Since the counterterm $-\mathbf{E}_P$ is a slowly-varying macroscopic quantity at the intermediate scale l, we compute it using a manifestly macroscopic approach. For convenience, we assume a harmonic time dependence $e^{\pm i\omega t}$ at the scale $l << \lambda$ where the slowly-varying quantities are effectively constant. (The restriction to a harmonic dependence simplifies the calculation, but it is not required.) If we neglect all multipole fields higher than dipole and assume there is no free charge or current in the medium, then for the charge density ρ and the current density **J** we have

$$\rho = -\nabla \bullet \mathbf{P} \tag{2-11a}$$

$$\mathbf{J} = \frac{\partial \mathbf{P}}{\partial t} = -i\omega\,\mathbf{P} \tag{2-11b}$$

for the negative frequency term of **P**. (We shall explicitly calculate only the negative frequency part of $\mathbf{E}_P$, the positive frequency component follows similarly.) Using the Coulomb gauge electromagnetic Green's function [11] we may write the scalar potential φ and the vector potential **A** as

$$\varphi(\mathbf{x},t) = -4\pi \int_{v \ll \lambda^3} d^3x' \int \frac{d^3k'}{(2\pi)^3} \frac{e^{ik'\bullet(x-x')}}{k'^2} \nabla' \bullet \mathbf{P}(\mathbf{x},t'),$$

(2-12a)

$$\mathbf{A}(\mathbf{x},t) = -4\pi ik \int_{v \ll \lambda^3} d^3x' \int \frac{d^3k'}{(2\pi)^3} \frac{e^{ik'\bullet(\mathbf{x}-\mathbf{x}')}}{k'^2} e^{ik|\mathbf{x}-\mathbf{x}'|} \left(\overleftrightarrow{\mathbf{1}} - \hat{\mathbf{k}}'\hat{\mathbf{k}}'\right) \bullet \mathbf{P}(\mathbf{x}',t)$$

(2-12b)

where $\overleftrightarrow{\mathbf{1}}$ is the unit dyadic and $e^{ik|\mathbf{x}-\mathbf{x}'|} \approx 1$ is a retardation factor that gives a negligible contribution in a small macroscopic volume $v \ll \lambda^3$. In the Coulomb gauge, as is well known, φ changes instantaneously with $\mathbf{P}$ and contains no retardation factor. The longitudinal and transverse electric fields, $\mathbf{E}_P^{\parallel}$ and $\mathbf{E}_P^{\perp}$, respectively, can be obtained in the Coulomb gauge [1] by

$$\mathbf{E}_P^{\parallel} = -\nabla\varphi,$$

(2-13a)

$$\mathbf{E}_P^{\perp} = \frac{i}{k} \nabla \times \nabla \times \mathbf{A},$$

(2-13b)

where Eq. (2-13b) for the transverse field is obtained by combining the harmonic version of Ampère's law, $\mathbf{E} = (i/k)\,\nabla \times \mathbf{B}$, with the relation $\mathbf{B} = \nabla \times \mathbf{A}$. Evaluation of $\mathbf{E}^{\parallel}$ from Eq. (2-13a) using the expression (2-12a) for φ yields in a straightforward fashion

$$\mathbf{E}_P^{\parallel} = -4\pi\mathbf{P}$$

(2-14)

upon integrating by parts and using the representation of the δ function,

$$\delta(\mathbf{x}) = \int \frac{d^3k}{(2\pi)^3}\, e^{i\mathbf{k}\bullet\mathbf{x}} \,. \tag{2-15}$$

Computing the transverse electric field $\mathbf{E}_P^{\perp}$ requires a bit of care in handling the transverse δ function in $\mathbf{x}$ space. With the definition

$$\overleftrightarrow{\partial}^{\perp}(\mathbf{x}) \equiv \int \frac{d^3k}{(2\pi)^3}\, e^{i\mathbf{k}\bullet\mathbf{x}} \left(\overleftrightarrow{\mathbf{1}} - \hat{\mathbf{k}}\hat{\mathbf{k}}\right), \tag{2-16}$$

we may regularize the integral by multiplying through by a factor $\Omega^2/(k^2 + \Omega^2)$, where it is understood that $\Omega \to \infty$ at the end of the calculation. The result is then [12]

$$\overleftrightarrow{\partial}^{\perp}(\mathbf{x}) = \frac{2}{3}\overleftrightarrow{\mathbf{1}}\,\delta(\mathbf{x}) + \left.\frac{\eta_\Omega(\mathbf{x})}{4\pi|\mathbf{x}|^3}\left(3\,\hat{\mathbf{x}}\hat{\mathbf{x}} - \overleftrightarrow{\mathbf{1}}\right)\right|_{|\mathbf{x}| \neq 0} \,, \tag{2-17}$$

where $\eta_\Omega(\mathbf{x})$ is a cutoff dependent regularizer [12] that has the limiting forms,

$$\eta_\Omega(\mathbf{x}) \sim \begin{cases} 1, & \Omega|\mathbf{x}| \gg 1 \qquad \text{(2-18a)} \\ \Omega^3\,|\mathbf{x}|^3\,, & \Omega\,|\mathbf{x}| \ll 1\,, \qquad \text{(2-18b)} \end{cases}$$

that shows $\overleftrightarrow{\partial}^{\perp}(\mathbf{x})$ to be nonsingular at the origin. Combining the expression (2-13b) for the transverse electric field $\mathbf{E}_P^{\perp}$ with the vector potential, Eq. (2-12b), and the above expression (2-17) for the transverse δ function dyadic, yields

$$\mathbf{E}_{\mathbf{P}}^{\perp}(\mathbf{x},t) = \frac{8\pi}{3}\mathbf{P}(\mathbf{x},t) + \int_{V/\{\mathbf{x}\}} d^3x'\, \frac{\eta(\mathbf{R})}{R^3}\left(3\,\hat{\mathbf{R}}\hat{\mathbf{R}} - \overleftrightarrow{\mathbf{1}}\right) \bullet \mathbf{P}(\mathbf{x}',t)\,, \tag{2-19}$$

where $\mathbf{R} \equiv \mathbf{x} - \mathbf{x}'$ and $v/\{\mathbf{x}\}$ is a small spherical region of volume $v \ll \lambda^3$ with the origin at $\mathbf{x}$ removed. The angular integration of the second term in Eq. (2-19) vanishes, giving a net result of $\mathbf{E}_P^{\perp} = (8\pi/3)\mathbf{P}$. The total field at the point $\mathbf{x}$ is the sum of the longitudinal and transverse fields; hence

$$\begin{aligned} \mathbf{E}_P &= \mathbf{E}_P^{||} + \mathbf{E}_P^{\perp} \\ &= -4\pi\,\mathbf{P} + \frac{8\pi}{3}\mathbf{P} \\ &= -\frac{4\pi}{3}\mathbf{P}\,. \end{aligned} \tag{2-20}$$

This is the negative frequency result, but the positive frequency calculation yields precisely the same expression and so the sum of these also obeys Eq. (2-20). Thus, from Eq. (2-8), we have the total local field as

$$\begin{aligned} \mathbf{E}_L &= \mathbf{E} + \mathbf{E}_i \\ &= \mathbf{E} + \mathbf{E}_{near} - \mathbf{E}_P \\ &= \mathbf{E} + 0 - \left[-\frac{4\pi}{3}\mathbf{P}\right] \\ &= \mathbf{E} + \frac{4\pi}{3}\mathbf{P}\,, \end{aligned} \tag{2-21}$$

yielding a similar expression for the slowly varying field,

$$\mathsf{E}_L = \mathsf{E} + \frac{4\pi}{3}\mathsf{P}\,. \tag{2-22}$$

We reiterate that this novel derivation of the result is simple, compact, and makes no use of constitutive relations and hence holds for propagation in nonlinear as well as linear media.

If Eq. (2-22) is used in Eqs. (2-1) and (2-2), to eliminate the microscope local field, $\mathbf{E}_L$, the result is a set of generalized Bloch equations which are self-consistent with the Maxwell wave equation for the macroscopic field, Eq. (2-3),

$$\frac{dW}{dt} = -\gamma_L (W+1) + \frac{\mu}{\hbar}\left[\mathbf{E}^* \mathbf{R}_{ab} + c.c.\right], \tag{2-23}$$

$$\frac{dR_{ab}}{dt} = -\left[\gamma_T + i\left(\Delta + \varepsilon W\right)\right]\mathbf{R}_{ab} - \frac{\mu}{2\hbar}\mathbf{E}\, W, \tag{2-24}$$

where the near dipole-dipole local field correction parameter,

$$\varepsilon = \frac{4\pi}{3\hbar}\mu^2 n \tag{2-25}$$

has units of frequency. These equations are Bloch equations, generalized to media of densities such that there are many atoms within a cubic wavelength. They are identical to the usual optical Bloch equations, except for the first term on the right-hand side of Eq. (2-24) involving the product $W\mathbf{R}_{ab}$ and the constant ε, Eq. (2-25), which expresses the strength of the near dipole-dipole interaction (NDD). This additional term is nonlinear in the atomic variables, varying as $W\mathbf{R}_{ab}$, and constitutes a dynamic shift in the resonance frequency that is population-dependent. This modification also implies a static shift of the isolated atom resonance frequency. This new contribution to the Maxwell-Bloch equations gives rise to an array of novel and interesting physical phenomena in nonlinear and quantum optics. Equations (2-23) and (2-24) for the atomic variables, and (2-3) for the field variables, therefore, constitute a generalized, self-consistent, Bloch-Maxwell formulation for optically dense media.

If we wish to write Eqs. (2-23) and (2-24) for the self-consistent optical Bloch equations in terms of real variables, it is convenient to use the canonical transformation

$$\rho_{ab} \equiv \rho\, e^{-i\left(\omega t+\mathbf{k}\bullet\mathbf{x}+\varphi\right)} \quad , \tag{2-26}$$

where we choose the phase φ such that

$$\mathsf{E} = -\left|\mathsf{E}\right| e^{-i\varphi} \ . \tag{2-27}$$

This Eq. (2-27) implies that we have imparted an explicit phase factor of π to the arbitrary phase of the slowly varying field amplitude E, for the sake of convenience.

The effect of this transformation is to recast Eqs. (2-23) and (2-24) with the identification

$$\mathsf{E} \rightarrow \left|\mathsf{E}\right| , \tag{2-28a}$$

$$\Delta \rightarrow \Delta - \dot{\varphi} \ . \tag{2-29b}$$

In terms of real variables, let us define

$$\mathsf{R}_{ab} \equiv \frac{1}{2}(V+iU) \ , \tag{2-30}$$

where, as before, $\rho \equiv i\,\mathsf{R}_{ab}$. In terms of the real variables V and U, then, we have

$$\dot{U} = -\gamma_T U - \left(\Delta - \dot{\varphi} + \varepsilon W\right)V , \tag{2-30a}$$

$$\dot{V} = -\gamma_T V + \left(\Delta - \dot{\varphi} + \varepsilon W\right)U + \Omega W , \tag{2-30b}$$

$$\dot{W} = -\gamma_L(W+1) - \Omega V, \tag{2-30b}$$

where $\Omega \equiv \mu |E|/\hbar$ is the Rabi frequency, and ε is the local field correction constant as before.

These generalized equations have already been analyzed in connection with intrinsic optical bistability [7], self-phase modulation in self-induced transparency [13,14], linear and nonlinear spectral shifts [15,16], propagational effects in nonlinear media [17], inversion and ultrafast optical switching effects [18,19], statistical effects in superfluorescence and amplified spontaneous emission [20], and systems exhibiting coherence-induced lasing without inversion and the related high indices of refraction without absorption [21].

It is anticipated that many other new and novel physical phenomena will be derived from this new formulation, possibly leading to new and important experimental results and device applications [18,19].

3. NEAR DIPOLE-DIPOLE INTERACTION EFFECTS IN OPTICAL SYSTEMS

Here we review some of the new physical phenomena which have been derived from the coupled set of equations, Eqs. (2-23), (2-24), and (2-3), and some more general implications of Eq. (2-22). In Subsection 3.1 we discuss predictions concerning linear and nonlinear spectral shifts [15] implied by Eqs. (2-23) – (2-24), together with corresponding experimental studies [16]. The implications for intrinsic optical bistability, [7, 22, 23], are presented in Subsection 3.2; self-induced transparency and self-phase modulation [32], [33] are discussed in Subsection 3.3; and novel effects in optical switching and adiabatic inversion [18, 19] are presented in Subsection 3.4. Effects on amplified quantum fluctuations in superfluorescence and amplified spontaneous emission [20] are addressed in Subsection 3.5. Some aspects of the influence of near dipole-dipole (NDD) interactions on induced quantum coherence effects, including lasing without inversion (LWI), field-induced transparency (FIT), and field-induced enhancement of index of refraction

(EIR) [47], are discussed in the Subsection 3.6. Finally, we illustrate some effects of inhomogeneous broadening using a perturbative scheme to generate closed-form Bloch-like equations. We show how the results can be used as a two-level, Bloch-Maxwell-like model for semiconductor lasers.

We restate our fundamental equations again here for clarity, and in the slowly varying phase and envelope approximation (SVEA).

$$\frac{\partial W}{\partial t} = -\gamma_L (W+1) + \frac{\mu}{\hbar}\left[\mathcal{E}\,\mathcal{R}^*_{ab} + \mathcal{E}^*\mathcal{R}_{ab}\right], \tag{3-1}$$

$$\frac{\partial \mathcal{R}_{ab}}{\partial t} = -\left[\gamma_T + i(\Delta+\epsilon W)\right]\mathcal{R}_{ab} - \frac{\mu}{2\hbar}\mathcal{E}\,W, \tag{3-2}$$

$$\epsilon = \frac{4\pi}{3\hbar}\mu^2 n, \tag{3-3}$$

$$\frac{1}{c}\frac{\partial \mathcal{E}}{\partial t} + \frac{\partial \mathcal{E}}{\partial z} = -\frac{4\pi\omega\mu n}{c}\mathcal{R}_{ab}, \tag{3-4}$$

where the last equation is the linearized form of the Maxwell wave equation, Eq. (2-3), in SVEA. In terms of real variables, where we have used Eqs. (2-26), (2-27), and (2-29),

$$\dot{U} = -\gamma_T U - \left(\Delta - \dot{\varphi} + \epsilon W\right)V, \tag{3-5}$$

$$\dot{V} = -\gamma_T V + \left(\Delta - \dot{\varphi} + \epsilon W\right)U + \Omega W, \tag{3-6}$$

$$\dot{W} = -\gamma_L (W+1) - \Omega V, \tag{3-7}$$

$$\frac{\partial \Omega}{\partial t} + c\,\frac{\partial \Omega}{\partial z} = -\,\frac{2\pi\omega\mu^2 n}{\hbar}\,V\,, \tag{3-8}$$

$$\Omega\,\frac{\partial \phi}{\partial t} = \frac{2\pi\omega\mu^2 n}{\hbar}\,U\,. \tag{3-9}$$

The two forms,which are equivalent, will be useful in what follows.

3.1 Linear and Nonlinear Spectral Shifts

Friedberg, Hartmann and Manasseh [15] were the first to analyze spectral shifts associated with NDD. In steady-state, $dW/dt = d(R_{ab}/dt) = 0$; Eqs. (3-1) and (3-2) together with the constitutive relation $p = \chi\varepsilon$, give for the susceptibility,

$$\chi = -\,\frac{i\,\mu^2\,n\,W}{\left[\gamma_T + i\,(\Delta + \epsilon W)\right]}\,, \tag{3-10}$$

with real and imaginary parts determined by

$$\chi = \chi' + i\,\chi''\,, \tag{3-11}$$

$$\chi' = -\,\frac{\mu^2\,n\,W\,(\Delta + \epsilon W)}{\left[\gamma_T^2 + (\Delta + \epsilon W)^2\right]}\,, \tag{3-12}$$

$$\chi'' = -\,\frac{\mu^2\,n\,\gamma_T W}{\left[\gamma_T^2 + (\Delta + \epsilon W)^2\right]}\,. \tag{3-13}$$

For weak fields, W = −1, and Eqs. (3-12) and (3-13) show a shift of the resonance and absorption peak that is proportional to the strength of the NDD, $\in$, Eq. (3-3). It is noted that even for $\Delta = 0$, Eq. (3-12) provides a nonzero contribution to the refractive index due to the term proportional to $\in$ in the numerator. If the field is saturation, then $W \rightarrow 0$ and the NDD has no effect. For intermediate fields, however, the absorptive and dispersive components, Eqs. (3-13) and (3-12), depend nonlinearly on the field strength.

More recently, Boyd, et al. [16], studied linear and nonlinear NDD induced spectral shifts from reflection and surface phase conjugation measurements from dense sodium vapor at a sapphire window. They measured shifts which are larger by several percent than predicted from Eqs. (3-1) – (3-4), using (3-3). They ascribe the discrepancy to collisional effects.

Intensity-dependent spectral shifts, as well as spectral broadening due to NDD have been predicted for pump, weak probe, conditions [23] – [25]. Measurement of the Lorentz shift has also been proposed using parametric interactions in a strongly driven, two-level system [26]. Intensity-dependent shifts were also predicted for reflection from dense vapors at a dielectric interface [27], taking into account the self-reflected wave effects [28] – [29].

3.2 Intrinsic Optical Bistability

In steady-state, Eqs. (3-1) and (3-2) combine to yield an equation of state which is cubic in the relative level population W as a function of the field intensity $|\varepsilon|^2$,

$$(W+1)\left[\gamma_T^2 + (\Delta + \in W)^2\right] = -\frac{\mu^2 \gamma_T}{\hbar^2 \gamma_L} |\varepsilon|^2 W . \tag{3-14}$$

The term in $\in$, the NDD strength, causes (3-14) to be cubic in W, rather than linear, and produces local bistability in the medium determined by the local intensity and a threshold value for $\in$,

$$\in/\gamma_T \gtrsim 3 \text{ for } \Delta = 0 .$$

This same expression was earlier derived from a fully quantum treatment [7]. It is stressed that Eq. (3-14) expresses a local (in space) condition and the response of the medium as a whole is governed by integrating the time-independent field, Eq. (3-4) with appropriate boundary conditions for the input field. This amounts to using Eq. (3-14) in conjunction with the integration of the field equation in the form

$$\frac{d\,|\varepsilon|^2}{dz} = \frac{2\pi\hbar\omega n}{c}\,\frac{\gamma_L}{\gamma_T}\,(W+1) \; . \tag{3-15}$$

Equation (3-15) was integrated by Ben-Aryeh, Bowden, and Englund [30], which resulted in a first-order spacial phase transition in the spacially-dependent complex index of refraction in the material due to the local intrinsic optical bistability (IOB) distribution in the material. Inguva and Bowden calculated the dynamical onset of this first-order phase transition by integration of the temporally and spacially dependent equations [17]. The SVEA results showed remarkable agreement with corresponding results retaining second order in space [22].

The first prediction of conditions for IOB caused by NDD was given by Bowden and Sung [31] as a first-order phase transition in the Dicke model [31], using the mean field approximation. Later, the mean field approximation was relaxed in the work of Hopf, Bowden, and Louisell [22], who introduced the Lorentz-Lorenz correction into the optical Bloch equations and integrated the steady-state equations using second-order in the spacial derivatives. Conditions which preclude IOB in gases have been analyzed by Friedberg, Hartmann, and Manasseh [33].

The effects of quantum fluctuations in IOB were discussed by Ben-Aryeh and Bowden [34], and it was established that squeezing below the shot noise is not intrinsic to IOB. The effects of a squeezed vacuum on IOB, on the other hand, were shown by Singh, Rai, Bowden, and Postan [35] to produce novel and interesting effects. The main result of that work was the establishment of local bistability in the population difference or polarization as a function of relative phase between the incident field and the squeezed vacuum field. Thus, in the presence of a squeezed vacuum, switching may be accomplished by changing the slowly varying phase of the incident field, with the amplitude, or intensity, held constant.

Intrinsic optical bistability in a ring cavity, including transverse effects, was treated by Scalora, Haus, and Bowden [36]. Although chaotic behavior did not occur over a large region of the parameter space due to large absorption, Farey-tree-like structures appeared for large enough detuning from resonance. The periods in the output wave were found to be quite sensitive to the Fresnel number.

It has been shown that ultrashort light pulses passing through a thin film of dense material can exhibit transient, bistable behavior in reflection and transmission on a time scale of superradiation [37]. A requirement is that the excitation field frequency be greater than the atomic resonance frequency. The mechanism leads to a new kind of nonlinear transparency.

3.3 Self-Induced Transparency

The coupled Maxwell-Bloch equations, Eqs. (3-5) – (3-9), under conditions for invariant pulse propagation, including NDD effects, are [38]

$$\dot{U} = -(\Delta - \dot{\varphi} + \in W)\,V \ , \tag{3-16}$$

$$\dot{V} = (\Delta - \dot{\varphi} + \in W)U + \kappa\varepsilon W \ , \tag{3-17}$$

$$\dot{W} = -\kappa\varepsilon V \ , \tag{3-18}$$

$$\dot{\varepsilon} = -(\kappa/m^2)\,V\,, \tag{3-19}$$

$$\left(\dot{\varphi} + \frac{\Delta k}{\delta}\right)\varepsilon = \frac{\kappa}{m^2}\,U\,. \tag{3-20}$$

Here $\kappa = (2\mu/\hbar)$, Δk is a quantity that denotes possible dispersive effects, where k is the wave vector corresponding to the field carrier frequency ω. It is to be noted here that $\bullet \equiv (\partial/\partial\tau)$, where $\tau = t - z/\mathcal{V}$ is the retarded time variable and $\mathcal{V}$ is the velocity of the pulse. In regard to further notation, $\delta \equiv (1/\mathcal{V}) - (1/c)$ where c is the speed of light in vacuum, and $1/m^2 \equiv \pi\rho\hbar\omega/(c\eta_0\delta)$, where ρ is the number density of atoms and η_0 is the linear index of refraction of the medium.

These equations have been first solved analytically by Stroud, Bowden, and Allen [39] for the field amplitude dependence, and then by Bowden, Postan, and Inguva [38] in both amplitude and phase. A first integral of these equations gives a first-order equation for the phase in terms of the amplitude of the field [38],

$$\dot{\varphi} = \frac{c_1}{\varepsilon^2} + \epsilon\,\frac{m^2}{8}\,\varepsilon^2\,, \tag{3-21}$$

where c_1 is a constant of integration. Since the first term, for $c_1 \neq 0$, leads to a singularity in $\dot{\varphi}$ in the wings of the pulse, as $\varepsilon \rightarrow 0$, we take $c_1 = 0$. The second term in Eq. (3-21) causes intensity-dependent chirp in the pulse temporal evolution which depends uniquely upon the NDD, i.e., for $\epsilon \neq 0$.

Full integration of the equations of motion yields expressions for the field amplitude, $\varepsilon(\tau)$, and slowly varying phase, $\varphi(\tau)$,

$$\varepsilon(\tau) = \frac{\varepsilon_o \cos\left(\frac{1}{2}\Phi\right)}{\left[\cosh^2(\lambda\tau) - \sin^2\left(\frac{1}{2}\Phi\right)\right]^{1/2}} , \tag{3-22}$$

$$\varphi(\tau) = \varphi_o + \frac{\epsilon\, m^2 \varepsilon_o^2}{8\lambda} \cot\left(\frac{1}{2}\Phi\right) \arctan\left[\tan\left(\frac{1}{2}\Phi\right) \tanh(\lambda\tau)\right] . \tag{3-23}$$

Here ε_0 is the peak amplitude of the pulse envelope of temporal width λ and Φ lumps the rest of the parameter space [38]. In the absence of NDD, $\epsilon = 0$, and $\Phi = 0$ yielding the usual hyperbolic secant temporal dependence for the amplitude, Eq. (3-22), the McCall-Hahn result [40].

For $\epsilon \neq 0$, however, Eqs. (3-22) – (3-23) yield a solitary pulse which is self-phase modulated, where the envelope departs from that of a hyperbolic secant by the second term in the denominator of Eq. (3-22), and, though a solitary pulse, is not of the Sine-Gordon variety. When amplitude and phase are combined, Eqs. (3-22) and (3-23), and for sufficiently large values of ϵ, single solitary pulse behavior persists for the field tuned near atomic resonance. The effect of the phase is simply to narrow the pulse envelope. This is reasonable, since one observes from Eq. (3-23) that the effect of the phase is to cause a near step-function modulation of the pulse near its peak, which arises due to the hyperbolic tangent temporal functional dependence depicted in Eq. (3-23). If the field frequency, ω, is tuned above atomic resonance, ω_0, however, the self-phase modulation causes single pulse breakup into multiple pulses as a function of off-tuning, with a limiting condition of a continuous pulse train [38]. The effect of self-phase modulation is a phenomenon due entirely to NDD, and is therefore a clear signature of NDD effects in SIT. An area theorem does not exist in the case of SIT with NDD.

3.4 Novel Optical Switch and Adiabatic Inversion

For an ultrashort, externally applied field, Eqs. (3-5) (3-9) reduce to

$$\frac{\partial U}{\partial t'} = -\left(\tilde{\Delta} - \dot{\varphi} + \tilde{\epsilon} W\right) V \tag{3-24}$$

$$\frac{\partial V}{\partial t'} = \left(\tilde{\Delta} - \dot{\varphi} + \tilde{\epsilon} W\right) U + \tilde{\Omega} W \tag{3-25}$$

$$\frac{\partial W}{\partial t'} = -\tilde{\Omega} V \quad , \tag{3-26}$$

where the time variable has been scaled, $t' = t/t_P$, where t_P is a characteristic interaction time associated with the temporal width of the pulse,

$$\tilde{\Delta} = (\omega_a - \omega)\,\tau_P \quad , \quad \tilde{\epsilon} = \frac{4\pi\mu^2}{3\hbar}\rho\,\tau_P \quad \text{and} \quad \tilde{\Omega} = \frac{\mu E L \tau_P}{\hbar}$$

where ρ is the density of atoms and $L = (\eta_o^2 + 2)/3$ and η_o is the background index of refraction. These equations, with $\dot{\varphi} = 0$, were integrated using various pulse shapes [18] for the externally applied field, including hyperbolic secant, Gaussian and Lorentzian. It was found that when the peak Rabi frequency Ω_o equals ϵ, $\Omega_o/\epsilon = 1$, the system is left in an inverted state of passage of the pulse. In fact there is a range of values $(\Omega_o/\epsilon) = 1 \pm \delta$ which leaves the system fully inverted where δ depends upon $\epsilon\,\tau_P$, being larger for shorter pulse widths. As Ω_o/ϵ is increased, the system is left alternatively in the ground or the fully excited state in a periodic repetition.

It is observed that ϵW in Eqs. (3-24) and (3-25) plays the role of an intrinsic self-chirp [19]. Consider the case where the medium is excited at exact resonance by an applied field of the form

$$\tilde{\Omega}(t') = \tilde{\Omega}_o \,\mathrm{sech}\,(t') , \tag{3-27}$$

where $\tilde{\Omega}_0 = (\tilde{\epsilon}^2 + 1)^{1/2}$. The solution to Eqs. (3-24) – (3-26) is

$$U = -\frac{\tilde{\epsilon}}{\sqrt{\tilde{\epsilon}^2 + 1}}\,\mathrm{sech}\,(t'),\ V = -\frac{1}{\sqrt{\tilde{\epsilon}^2 + 1}}\,\mathrm{sech}\,(t')\,,\ W = \tanh\,(t'). \tag{3-28}$$

These are exactly the equations that Allen and Eberly present for inversion by a chirped hyperbolic secant pulse [5], if $-\tilde{\epsilon}W$ is identified with $\partial\phi/\partial t$, as suggested by Eqs. (3-24) and (3-25). This result can therefore be interpreted as inversion due to self-chirping. This phenomenon constitutes a unique optical switch [18] and is explained in terms of the quasi-adiabatic following model as adiabatic inversion due to intrinsic self-chirping [19].

3.5 Quantum Statistics of Superfluorescence and Amplified Spontaneous Emission

Recently, interest in superfluorescence (SF) and amplified spontaneous emission (ASE) has been revived by the very interesting experiments of Boyd, et al. [41]. In these experiments, a transition from SF to ASE was observed in the emission from $KCl{:}O_2^-$ in the solid state. The parameter controlling the transition was the dipole dephasing rate, $\gamma_\perp$, which was varied by controlling the temperature variation (10–30K) of the solid. A crucial parameter in their analysis was the number of atoms cooperating in the emission. Their analysis was based upon the number of ions in the neighborhood of 10^9 cm^{-3}, whereas in reality the number could have been as high as 10^{10} to 10^{13} cm^{-3}. Since the density of atoms participating in the collective emission can be rather high, the role of dipole-dipole interaction among neighboring atoms becomes important.

The equations of motion for the slowly varying operators obtained from a fully quantum mechanical treatment [7], where the atomic operators were averaged over a finite volume on the order of a cubic resonance wavelength, are given here in the retarded time frame,

$$\frac{\partial \sigma_z}{\partial \tau} = -\gamma_{\parallel}\left(\sigma_z + N\right) - \sigma_+ \Omega_R^{(+)} - \Omega_R^{(-)} \sigma_- - 2\sigma_+ F^{(+)} - 2F^{(-)} \sigma_- + G^z(z,\tau)$$

(3-29)

$$\frac{\partial \sigma_+}{\partial \tau} = i(\Delta + i\gamma_{\perp})\sigma_+ - i\epsilon\, \sigma_+ \sigma_z + \frac{1}{2}\Omega_R^{(-)} \sigma_2 + F^{(-)} \sigma_z + G^{(+)}(z,\tau)$$

(3-30)

$$\frac{\partial \Omega_R^{(-)}}{\partial z} = -g\, \sigma_+ \ .$$

(3-31)

Here, $\Omega_R^{(+)}$ / $\Omega_R^{(-)}$ are the positive / negative frequency components of the reaction field and $F^{(+)}$ / $F^{(-)}$ have similar connotations for the free field. With the initial state of the field as the vacuum, the free (vacuum) field satisfies conditions for a Gaussian random process with zero mean,

$$\langle F^{(+)}(\tau)\, F^{(-)}(\tau')\rangle = \frac{1}{N\tau_R}\, \delta(\tau - \tau') \, .$$

(3-32)

The source terms $G^z(z,\tau)$ and $G^+(z,\tau)$ were added to account for collisionally induced fluctuations, and are connected with γ through the fluctuation-dissipation theorem [42],

$$\langle G^{(+)}(z', \tau')G^{-}(z,\tau)\rangle_R = \gamma_{\perp}\delta(z - z')\delta(\tau - \tau') \, ,$$

(3-3

$$\langle G^z(z,\tau)G^{\pm}(z', \tau')\rangle_R = 0 \, ,$$

where $\langle\ \rangle_R$ denotes reservoir average.

The initial and boundary conditions for SF are

$$\langle\sigma^+(z,0)\rangle = 0 \quad , \quad \langle\sigma^z(z,0)\rangle = +1$$
$$\langle\Omega_R^+(z,0)\rangle = 0 \ , \tag{3-34}$$

so that the system is prepared in an initial state of complete inversion and radiates to the vacuum. The linearized regime of quantum initiation can be solved analytically [20]. Solutions which include the nonlinear regime are obtained by numerical integration of the equations of motion and propagation, Eqs. (3-29) – (3-31) using the "stochastic model" Monte Carlo approach [43,44]. Thus, the Monte Carlo method can be compared with analytical results in the linearized regime by comparing the numerical ensemble results with the quantum average. The "stochastic method" numerical ensemble was shown to adequately represent the quantum averages for situations, like SF, where quantum fluctuations and nonlinearity are not simultaneously important [43,44]. It probably gives good results as well in the ASE regime, though in this regime it is not strictly valid. A different set of stochastic equations can be used in this case [43], but are not nearly as convenient for numerical calculations.

Phase waves, which are stochastically induced, macroscopic, sudden shifts in the signal amplitude phase, were discussed originally by Hopf [45] in connection with SF, and later by Hopf and Overman [46] in relation to swept-gain superfluorescence. More recently, phase waves were shown by Englund and Bowden [47,43,44] to cause spontaneously induced solitons in stimulated Raman scattering. This prediction was confirmed within experimental error in the experiments of Carlsten, MacPherson, and Swanson [48].

From numerical ensemble calculations, Rai and Bowden [20] analyzed phase wave distributions, pulse shapes, and delay time pulse height statistics as a function of dephasing rate, from nearly pure SF through completely developed ASE. Parameters used in the calculations were taken from the experimental conditions [41] and the strength of the NDD contribution, $\in$, was estimated from those conditions, $\in/\gamma_{\perp} \approx 1$. The results [20] indicate that phase wave statistics play a significant role in the effects of amplified quantum fluctuations and that NDD effects, i.e., $\in \neq 0$, also plays a significant role [49].

3.6 Quantum Induced Coherences

The contribution of NDD in a system consisting of a dense collection of saturable two-level atoms is to cause a population-difference-dependent shift in the atomic resonance frequency (intrinsic dynamic chirp) as exhibited by Eqs. (3-1) – (3-4). A major contribution is a linear as well as a nonlinear spectral shift in IOB as already discussed in Section 3.1 above. On the other hand, novel quantum coherence effects are manifested in terms of adiabatic inversion and optical switching, also discussed in Section 3.4 above.

A different class of intrinsically induced quantum coherence effects arises in systems characterized by more than two energy levels, i.e., at least two distinct coupled channels for absorption and emission (quantum V or λ system), when NDD effects are introduced via the substitution Eqs. (2-21) or (2-22) for the local field.

It has recently been demonstrated [50] that field-induced absorptionless index of refraction [51,52], as well as inversionless gain [53], can be enhanced more than an order of magnitude due to cross-coupling distinct coupled decay/excitation channels caused by eliminating the local field, i.e., the field that drives an atom, in terms of the macroscopic field and polarization, Eq. (2-21) or Eq. (2-22). This comes about because the introduction of (2-21) causes the equations of motion to become nonlinear in the atomic variables, including products of off-diagonal density matrix elements corresponding to distinct excitation channels.

3.7 Generalized Bloch-Maxwell Formulation for Semiconductor Lasers

A perturbative scheme introduced by Graham and Cho [54] was used by Bowden and Agrawal [55] to obtain a set of generalized Bloch equations which is based upon averaging the Bloch equations, generalized for NDD, over the intraband distributions by the introduction of new dynamical variables [55]. The new equations, to first order in the perturbation are [55]

$$\frac{dW}{dt} = -\gamma_L(W+W_{th}) + i\Omega P^* - i\Omega^* P + \langle\Lambda\rangle \, , \tag{3-35}$$

$$\frac{dP}{dt} = -\gamma_T\left(1+i\Delta\right)P + \gamma_T \, s^2 S - \frac{i}{2} \, \Omega W - i\epsilon \, PW \quad , \tag{3-36}$$

$$\frac{dS}{dt} = -\gamma_T\left(1+i\Delta\right)S + \gamma_T P + \frac{1}{2}\,\Omega U + \epsilon PU \, , \tag{3-37}$$

$$\frac{dU}{dt} = -\gamma_c U - \left(\Omega S^* + S\Omega^*\right) - 2\epsilon\left(P^* S + PS^*\right), \tag{3-38}$$

$$\frac{d\Omega}{dt} = -\kappa\Omega + igP \, . \tag{3-39}$$

Here, s^2 is the perturbation parameter which contains all of the information concerning the joint density of states,

$$s^2 = \frac{\langle\delta^2 p_{ss}\rangle}{\langle p_{ss}\rangle} \tag{3-40}$$

and $\delta = (\omega-\omega_0)/\gamma_T$, where ω_0 is arbitrary and the averages are over the joint density of states [55], and the subscripts in Eq. (3-40) signify evaluation at steady-state. The new variables which are generated in the perturbative scheme are, through first-order

$$S = \frac{\langle i\, \delta\, p \rangle}{s^2} , \tag{3-41}$$

$$U = \frac{\langle \delta\, w \rangle}{s^2} . \tag{3-42}$$

Thus, it is seen that S represents the inhomogeneous broadening for the polarization, whereas U is sensitive to asymmetry in the population distribution. The lower case variables appearing in the averages (3-40) – (3-42) are k-dependent, i.e., $p \equiv p(k)$ in (3-41). The various parameters in Eqs. (3-35) – (3-39) are: Λ is the pumping rate related to the injection current; γ_c is the intraband relaxation; γ_T is the dipole dephasing rate for electronic transitions; γ_L is the population relaxation rate related to interband transitions; ϵ is the strength of the NDD and W_{th} is the thermal equilibrium contribution to the electron population distribution. These equations were shown [55] to reduce, in the rate equation limit, to the usual equation of motion for the carrier distribution, when $\epsilon = 0$,

$$\frac{dN}{dt} = \frac{I_{in}}{qV} - \gamma_L N - \frac{\gamma_L}{(1+s^2)} \frac{|\varepsilon|^2}{I_S} \left(N - N_o\right) \tag{3-43}$$

where $N_o = -\frac{N}{2}(W - W_{th})$, $I_S = \frac{\hbar^2}{\mu^2}\gamma_T\gamma_L$, $N_o = -\frac{N_t}{2} W_{th}$, $I_{in} = qVN_t \frac{\langle \Lambda \rangle}{2}$.

The dynamical instabilities associated with Eqs. (3-35) – (3-39) have been analyzed by Bowden, Singh, and Agrawal [56]. It is found generally that the threshold for instabilities is reduced by the NDD contributions.

4. SUMMARY AND CONCLUSIONS

A generalized set of coupled Maxwell-Bloch equations has been derived, from first principles, Eqs. (3-1) – (3-4), suitable for representing linear and nonlinear electromagnetic field-matter interactions in media of densities such that there are many atoms within a cubic resonance wavelength. The new equations open all of nonlinear and quantum optics to re-examination for new, interesting, and novel physical phenomena. A plethora of new results has already evolved from detailed examination of these equations under various conditions, and some of them have been reviewed briefly in Section 3. It is anticipated that many new and important predictions will soon be forthcoming, which will lead to new and interesting experimental investigations and eventually to many useful new and novel devices.

APPENDIX A

For the purpose of review and the establishment of convention and notation, we derive here the usual Maxwell-Bloch equations for our model of a collection of identical two-level atoms. Let the upper and lower levels of each atom have energies e_a and e_b, respectively, with an energy difference given by $\hbar\omega_0 \equiv e_a - e_b$. The Hamiltonian for an atom interacting with an electric field $\mathbf{E}_L$ is given by

$$H = e_a R_{aa} - e_b R_{bb} - \mathbf{p} \bullet \mathbf{E}_L \ , \tag{A1}$$

where the operator R_{ij} is defined as

$$R_{ij} \equiv |i\rangle\langle j| \ , \tag{A2}$$

and the dipole moment and the field are given by

$$\mathbf{p} = \mu\left(R_{ab} + R_{ba}\right)\hat{\mathbf{r}} \ , \tag{A3}$$

$$\mathbf{E}_L = \left(E_L^+ + E_L^-\right)\hat{\mathbf{r}} \tag{A4}$$

where we have assumed that locally the field and dipole have the same spatial orientation $\hat{\mathbf{r}}$. The positive frequency component of the field E^+ can be written in terms of the slowly-varying component $\mathcal{E}$ via

$$E_L^+ = \frac{1}{2}\,\mathcal{E}_L\, e^{-i\left(\omega t - \mathbf{k}\bullet\mathbf{x}\right)} \quad , \tag{A5}$$

where the negative frequency given by $E^- \equiv (E^+)^*$. In addition, the slowly-varying component, which is complex, can be written in terms of a magnitude and a phase using the convention

$$\mathcal{E}_L = -\left|\mathcal{E}_L\right| e^{-i\varphi} \tag{A6}$$

that implies

$$E_L = -\left|\mathcal{E}_L\right| \cos\left(\omega t + \mathbf{k}\bullet\mathbf{x} + \varphi\right) , \tag{A7}$$

since $E_L = E_L^+ + E_L^-$.

The time evolution of the density matrix ρ obeys the Heisenberg equation

$$i\hbar\dot{\rho} = [H,\rho] \tag{A8}$$

where

$$H \equiv \begin{bmatrix} e_a & -\mu E_L \\ -\mu E_L & e_b \end{bmatrix} \tag{A9}$$

and

$$\rho \equiv \begin{bmatrix} \rho_{aa} & \rho_{ab} \\ \rho_{ba} & \rho_{bb} \end{bmatrix} . \tag{A10}$$

From the matrix equation Eq. (A8) we easily obtain the following three equations for the components of the density matrix,

$$\dot{\rho}_{aa} = \frac{i\mu E_L}{\hbar} \left(\rho_{ba} - \rho_{ab} \right) , \tag{A11a}$$

$$\dot{\rho}_{bb} = \frac{-i\mu E_L}{\hbar} \left(\rho_{ba} - \rho_{ab} \right) , \tag{A11b}$$

$$\dot{\rho}_{ab} = -i\omega_o \rho_{ab} + \frac{i\mu E_L}{\hbar} \left(\rho_{bb} - \rho_{aa} \right) , \tag{A11c}$$

recalling that $\rho_{ab} = \rho^*_{ba}$. Setting $W \equiv \rho_{aa} - \rho_{bb}$ to be the inversion, we may rewrite Eqs. (A11) as

$$\dot{W} = \frac{2i\mu E_L}{\hbar} \left(\rho_{ba} - \rho_{ab} \right) \tag{A12a}$$

$$\dot{\rho}_{ab} = -i\omega_o \rho_{ab} - \frac{i\mu E_L}{\hbar} W . \tag{A12b}$$

Rewriting the electric field E_L and the dipole moment $\mathbf{p}$ in terms of the slowly-varying variables $\mathcal{E}$ and $\mathcal{R}$, we have

$$E_L = \mathbf{E}_L \bullet \hat{\mathbf{r}} = \frac{1}{2} \left[\mathcal{E} \, e^{-i\left(\omega t - \mathbf{k} \bullet \mathbf{x}\right)} + \mathcal{E}^* e^{i\left(\omega t - \mathbf{k} \bullet \mathbf{x}\right)} \right] \tag{A13}$$

and

$$p = \langle \mathbf{p} \bullet \hat{\mathbf{r}} \rangle = \left(\rho_{ab} + \rho_{ba}\right) = \frac{1}{2}\left[\mathsf{R}\, e^{-i\left(\omega_o t - \mathbf{k} \bullet \mathbf{x}\right)} + \mathsf{R}^{*}\, e^{i\left(\omega_o t - \mathbf{k} \bullet \mathbf{x}\right)}\right], \tag{A14}$$

where we have defined the slowly-varying variable R implicitly by

$$\rho_{ab} \equiv \mathsf{R}\; e^{-i\left(\omega_o t - \mathbf{k} \bullet \mathbf{x}\right)}, \tag{A15a}$$

$$\rho_{ba} \equiv \mathsf{R}^{*}\; e^{i\left(\omega_o t - \mathbf{k} \bullet \mathbf{x}\right)}. \tag{A15b}$$

If we now insert these expressions for ρ_{ab} and ρ_{ba} into the Eqs. (A12), we obtain, in the rotating wave approximation, i.e., neglecting terms that vary with a phase of $(\omega_o + \omega)t$ compared with those that vary at $(\omega_o - \omega)t$,

$$\dot{W} = \frac{i\mu}{\hbar}\left(\mathsf{E}_L \mathsf{R}^{*} - \mathsf{E}_L^{*} \mathsf{R}\right) \tag{A16a}$$

$$\dot{\mathsf{R}} = -i\omega_o \mathsf{R} + i\omega \mathsf{R} - \frac{i\mu}{2\hbar}\mathsf{E}_L W. \tag{A16b}$$

Setting $\Delta \equiv \omega_o - \omega$, and adding phenomenological damping and relaxation terms, γ_L and γ_T, and defining $i\mathsf{R}_{ab} \equiv \mathsf{R}$ to conform to the notation of Hopf, Bowden, and Louisell [22], we have

$$\dot{W} = -\gamma_L (W+1) + \frac{\mu}{\hbar}\left(\mathsf{E}_L \mathsf{R}_{ab}^{*} + \mathsf{E}_L^{*} \mathsf{R}_{ab}\right), \tag{A17a}$$

$$\dot{R}_{ab} = -\left(\gamma_T + i\Delta\right) R_{ab} - \frac{\mu}{2\hbar} E_L W \quad . \tag{A17b}$$

Realizing that the field E_L driving the atom is the local field, substitution of Eq. (2-22) into (A17) to eliminate E_L, and using Eq. (2-6), gives us the Bloch equations (2-23) and (2-24), that include the near dipole-dipole term.

The derivation of the Maxwell wave equations, Eq. (2-3), begins with Maxwell's equations in a charge free and current free polarization medium, namely

$$\nabla \bullet \mathbf{D} = 0 , \tag{A18a}$$

$$\nabla \times \mathbf{E} = -\frac{1}{c} \frac{\partial \mathbf{B}}{\partial t} , \tag{A18b}$$

$$\nabla \times \mathbf{B} = \frac{1}{c} \frac{\partial \mathbf{D}}{\partial t} , \tag{A18c}$$

with the substantive relation

$$\mathbf{D} = \mathbf{E} + 4\pi\mathbf{P} . \tag{A19}$$

Combining Eqs. (A18b), (A18c), and Eq. (A19), we obtain

$$\nabla \times \nabla \times \mathbf{E} = -\frac{1}{c} \frac{\partial}{\partial t} \left(\nabla \times \mathbf{B}\right) = -\frac{1}{c^2} \frac{\partial^2 \mathbf{D}}{\partial t^2}$$

$$= -\frac{1}{c^2} \frac{\partial^2}{\partial t^2} \left(\mathbf{E} + 4\pi \mathbf{P}\right) = -\frac{1}{c^2} \frac{\partial^2 \mathbf{E}}{\partial t^2} - \frac{4\pi}{c^2} \frac{\partial^2 \mathbf{P}}{\partial t^2} , \tag{A20}$$

which gives immediately Eq. (2-3), as desired.

REFERENCES

1. J. D. Jackson, (Ed.), *Classical Electrodynamics*, second edition, (New York: John Wiley) 1975, Ch. 4.
2. M. Born and E. Wolf, (Eds.), *Principles of Optics*, (New York: McMillian Co.) 1964, Ch. 2.
3. P. P. Ewald, *Ann. D. Physik*, Vol. 49, 1916, p. 1; C. W. Osen, *Ann. D. Physik*, Vol. 48, 1915, p. 1.
4. J. Van Kranendonk and J. E Sipe, in *Progress in Optics XV*, E. Wolf (Ed.) (Amsterdam: North-Holland) 1977, p. 245; F. Hynne and R. K. Bullough, *Phil. Trans. R. Soc. Lond. A*, Vol. 312, 1984, p. 251; *ibid.*, Vol. 321, 1987, p. 305; *ibid.*, A, Vol. 330, 1990, p. 253.
5. L. Allen and J. H. Eberly, (Eds.), *Optical Resonance and Two-Level Atoms*, (New York: John Wiley) 1975.
6. C. M. Bowden and J. P. Dowling, *Phys. Rev. A*, Vol. 47, 1993, in press.
7. Y. Ben-Aryeh, C. M. Bowden, and J. C. Englund, *Phys. Rev. A*, Vol. 34, 1986, p. 3917.
8. C. M. Bowden, "Near Dipole-Dipole Interaction Effects in Quantum and Nonlinear Optics," in *Proceedings of International Conference on Quantum Optics, Hyderabad, India*, G. S. Agarwal and R. Inguva (Eds.), (Plenum: New York), in press.
9. C. M. Bowden and G. P. Agrawal, "Generalized Bloch-Maxwell formulation for single mode semiconductor lasers," unpublished.
10. H. A. Lorentz, *Wiedem. Ann.*, Vol. 9, 1880, p. 641; *ibid*, Vol. 11, 1881, p. 70.
11. A. O. Barut and J. F. Van Huele, *Phys. Rev. A*, Vol. 32, 1985, p. 3187.
12. C. Cohen-Tannoudji, J. Dupont-Roc, and G. Grynberg, in *Photons and Atoms* (New York: John Wiley and Sons) 1989, Ch. I.
13. C. M. Bowden, A. Postan, and R. Inguva, *J. Opt. Soc. Am.*, Vol. 8, 1991, p. 1081.
14. C. R. Stroud, Jr., C. M. Bowden, and L. Allen, *Opt. Commun.*, Vol. 67, 1988, p. 386.

15. R. Friedberg, S. R. Hartmann, and J. T. Manasseh, *Phys. Rep.*, Vol. 7, 1973, p. 101; *Phys. Rev. A*, Vol. 40, 1989, p. 2446.
16. J. J. Maki, M. S. Malcuit, J. E Sipe, and R. W. Boyd, *Phys. Rev. Lett.*, Vol. 67, 1991, p. 972.
17. R. Inguva and C. M. Bowden, *Phys. Rev. A*, Vol. 41, 1990, p. 1670.
18. M. E. Crenshaw, M. Scalora, and C. M. Bowden, *Phys. Rev. Lett.*, Vol. 68, 1992, p. 911.
19. M. E. Crenshaw and C. M. Bowden, *Phys. Rev. Lett.*, Vol. 69, 1992, p. 3475.
20. J. Rai and C. M. Bowden, *Phys. Rev. A*, Vol. 46, 1992, p. 1522.
21. J. P. Dowling and C. M. Bowden, unpublished.
22. F. A. Hopf, C. M. Bowden, and W. H. Louisell, *Phys. Rev. A*, Vol. 29, 1984, p. 2591.
23. R. Friedberg, S. R. Hartmann, and J. T. Manasseh, *Phys. Rev. A*, Vol. 40, 1989, p. 2446.
24. R. Friedberg, S. R. Hartmann, and J. T. Manasseh, *J. Phys. B: At. Mol. Opt. Phys.*, Vol. 24, 1991, p. 3981.
25. R. Friedberg, S. R. Hartmann, and J. T. Manasseh, *Opt. Commun.*, Vol. 75, 1990, p. 263.
26. R. Friedberg, S. R. Hartmann, and J. T. Manasseh, *Phys. Rev. A*, Vol. 42, 1990, p. 494.
27. R. Friedberg, S. R. Hartmann, and J. T. Manasseh, *Phys. Rev. A*, Vol. 42, 1990, p. 5573.
28. S. R. Hartmann and J. T. Manasseh, *Optics Lett.*, Vol. 16, 1991, p. 1349.
29. L. Rosa-Franco, *Phys. Rev. Lett.*, Vol. 55, 1985, p. 249; *J. Opt. Soc. Am. B.*, Vol. 11, 1987, p. 1878; *Opt. Lett.*, Vol. 15, 1990, p. 1230.
30. Y. Ben-Aryeh, C. M. Bowden, and J. C. Englund, *Opt. Commun.*, Vol. 61, 1987, p. 147.
31. C. M. Bowden and C. C. Sung, *Phys. Rev. A*, Vol. 19, 1979, p. 2392.
32. R. H. Dicke, *Phys. Rev.*, Vol. 93, 1954, p. 99.

33. R. Friedberg, S. R. Hartmann, and J. T. Manasseh, *Phys. Rev. A*, Vol. 39, 1989, p. 3444.

34. Y. Ben-Aryeh and C. M. Bowden, *J. Opt. Soc. Am. B*, Vol. 8, 1992, p. 1168.

35. S. Singh, J. Rai, C. M. Bowden, and A. Postan, *Phys. Rev. A.*, Vol. 45, 1992, p. 5160.

36. M. Scalora, J. W. Haus, and C. M. Bowden, *Phys. Rev. A.*, Vol. 41, 1990, p. 6320.

37. M. G. Benedict, V. A. Malyshev, E. D. Trifonov, and A. I. Zaitsev, *Phys. Rev. A.*, Vol. 43, 1991, p. 3845.

38. C. M. Bowden, A. Postan, and R. Inguva, *J. Opt. Soc. Am. B*, Vol. 8, 1991, p. 1081.

39. C. R. Stroud, Jr., C. M. Bowden, and L. Allen, *Opt. Commun.*, Vol. 67, 1988, p. 387.

40. S. L. McCall and C. S. Hahn, *Phys. Rev.*, Vol. 183, 1969, p. 457.

41. M. S. Malcuitt, J. J. Maki, D. J. Simkin, and R. W. Boyd, *Phys. Rev. Lett.*, Vol. 59, 1987, p. 1189; J. J. Maki, M. S. Malcuit, M. G. Raymer, R. W. Boyd, and P. D. Drummond, *Phys. Rev. A.*, Vol. 40, 1989, p. 5135.

42. M. Lax, *Phys. Rev.*, Vol. 145, 1966, p. 110.

43. J. C. Englund and C. M. Bowden, *Phys. Rev. A.*, Vol. 42, 1990, p. 2870.

44. J. C. Englund and C. M. Bowden, *Phys. Rev. A.*, Vol. 46, 1992, p. 578.

45. F. A. Hopf, *Phys. Rev. A.*, Vol. 20, 1979, p. 2064.

46. F. A. Hopf and E. A. Overman, II, *Phys. Rev. A.*, Vol. 19, 1979, p. 1180.

47. J. C. Englund and C. M. Bowden, *Phys. Rev. Lett.*, Vol. 57, 1986, 2661.

48. D. C. MacPherson, R. C. Swanson, and J. L. Carlsten, *Phys. Rev. A.*, Vol. 40, 1989, p. 6745; *Phys. Rev. A.*, Vol. 39, 1989, p. 6078.

49. J. Rai, R. Inguva, and C. M. Bowden, in *Proceedings of the Sixth Rochester Conference on Coherence and Quantum Optics*, J. H. Eberly, L. Mandel, and E. Wolf (Eds.) (New York: Plenum), 1989, p. 961.

50. J. P. Dowling and C. M. Bowden, "Near Dipole-Dipole Effects in Lasing Without Inversion: An Enhancement of Gain and Absorptionless Index of Refraction," submitted for publication.
51. M. O. Scully, *Phys. Rev. Lett.*, Vol. 67, 1991, p. 1855; M. O. Scully and Shi-Yao Zhu, *Opt. Commun.*, Vol. 87, 1992, p. 134; M. Fleischhauer, C. H. Keitel, M. O. Scully, and Chang Su, *Opt. Commun.*, Vol. 87, 1992, p. 109.
52. M. O. Scully, *Phys. Rev. Lett.*, Vol. 67, 1991, p. 1855; M. Fleischhauer, et al., *Phys. Rev. A*, Vol. 46, 1992, p. 1468.
53. S. Harris, *Phys. Rev. Lett.*, Vol. 62, 1989, p. 1022; K. J. Boller, A. Imamoglu, and S. E. Harris, *ibid.*, Vol. 66, 1991, p. 2593; J. E. Field, K. H. Hahn, and S. E. Harris, *ibid.*, Vol. 67, 1991, p. 3062; O, Kocharovskaya and Ya. I. Khanin, *Pis'ma Zh. Eksp. Teor. Fiz.*, Vol. 48, 1988, p. 581 [*JETP Lett.*, Vol. 48, 1988, p. 630]; O. Kocharovskaya and P. Mandel, *Phys. Rev. A*, Vol. 42, 1990, p. 523; O, Kocharvskaya, P. Mandel, and S. Radeonychev, *Phys. Rev. A*, Vol. 45, 1992, p. 1997; M. O. Scully, S. Y. Zhu, and A. Gavrieliedes, *Phys. Rev. Lett.*, Vol. 62, 1989, p. 2813; A. Lyras, X. Tang, P. Lambropoulos, and J. Zhang, *Phys. Rev. A.*, Vol. 40, 1989, p. 4131; S. Basile and P. Lambropoulos, *Opt. Commun.*, Vol. 78, 1990, p. 163.
54. R. Graham and Y. Cho, *Opt. Commun.*, Vol. 47, 1983, p. 52.
55. C. M. Bowden and G. P. Agrawal, "Generalized Bloch-Maxwell Formulation for Semiconductor Lasers," submitted for publication.
56. C. M. Bowden, S. Singh, and G. P. Agrawal, "Laser Instabilities and Chaos in Dense Media: Local Field Corrections," to be submitted for publication.

Studies in Classical
and Quantum Nonlinear Optics

ISBN 1-56072-168-5

MICROWAVE CAVITY QUANTUM ELECTRODYNAMICS: I, ONE AND MANY RYDBERG ATOMS IN MICROWAVE CAVITIES; AND II, FUNDAMENTAL THEORY OF THE MICROMASTER

R.K. Bullough‡, B.V. Thompson‡, N. Nayak and N.M. Bogoliubov***

‡ Department of Mathematics, UMIST, Manchester M60 1QD, UK
* Now at S.N. Bose National Centre for Basic Sciences, BG-45, Sector-II, Salt Lake City, Calcutta 700091, India.
** Now at Department of Physics, University of Jyväskylä, SF-40100, Finland and Steklov Mathematical Institute, Fontanka 27, St. Petersburg, Russia

1. INTRODUCTION

This short survey of microwave cavity q.e.d. summarizes theoretical work done in Manchester going back 13 or more years. This originally concerned [1-11] a detailed investigation of the Dicke model [12] of N_A 2-level atoms on the same single site interacting with radiation. The model was viewed as one of considerable interest to theoretical quantum optics because it was exactly solvable and, for example, it was possible to evaluate (numerically) all of its correlation functions for a description of the evolution of the photon statistics in collective spontaneous emission in the presence of black-body radiation [6,8-10] or to evaluate them for the N_A atom system driven by coherent radiation [1-7,10,11,13,14]. On the other hand the non-physical assumption that all of the N_A atoms occupied the same site in space seemed to vitiate any claim that predictions of the model could be confirmed by experiments.

The whole situation changed when Haroche and colleagues reported experiments done on Rydberg Na atoms in low-Q microwave cavities which demonstrated [6,8-11] collective giant oscillator formation by N_A atoms in thermal equilibrium - the Ref. [15] - and the details of the evolution of the photon statistics in collective super-radiant emission - the Ref. [16] - observations wholly consistent, and in detail, with the predictions [6,8-11] of the Dicke model driven by incoherent radiation. These observations and their theory have helped to create the relatively new science of *cavity* q.e.d. and this is what these two lectures I, and II, are about.

The work in Manchester going back some 13 years [1-11] has involved S.S. Hassan, Ain Shams University, Egypt; R.R. Puri and S.V. Lawande, BARC, Bombay; G.S. Agarwal, University of Hyderabad, India; J. Timonen, University of Jyväskylä, Finland; G.P. Hildred, SERC PDRA at UMIST until 1985; B.M. Garraway, now at Imperial College, London; and M.R.B. Wahiddin, now at Malaya University, Kualar Lumpar; as well as three of the authors, R.K.Bullough, B.V.Thompson and N.Nayak. Recently N.M. Bogoliubov, Steklov Mathematical Institute, St. Petersburg and SERC VF in Manchester December 1990 - November 1991 brought expertise on quantum group theory which led us to develop alternative q-boson theories of cavity q.e.d., and some of this work is described in the Lecture II. The authorship of this article is what it is because the new work described in it, which is in the Lecture II, is due to these four people.

The work over 13 years surveyed in the two lectures, I and II, was presented to the Summer School in Nonlinear Optics by R.K.Bullough. The work has led to the many theoretical papers [1-11] and to a number of predictions - not all of which could be summarised either at the School or in this report. As explained two of these predictions were strikingly confirmed in the observations on Na reported [15,16] in 1982. Haroche's group also confirmed Purcell's cavity factor η_{cav} in 1983 [17] : η_{cav} appears naturally in the master equation theory [6,8-10] described in Lecture I - see §4 below.

Subsequently in work initiated in a collaboration with G.S. Agarwal during 1985-86 [10,11,18-20] we used new numerical techniques to calculate the fluorescence spectra from single Rydberg atoms in microwave cavities with quality factors Q (logarithmic decrements $Q\omega^{-1}$) covering almost seven orders of magnitude - the currently accessible experimental range, $5 \times 10^3 \lesssim Q \le 10^8$. Here recent experiments on Rydberg ^{85}Rb and ^{87}Rb atoms at the $Q = 10^8$ [21,22] begin [23] to offer further opportunity for *quantitative* comparisons between theory and experiment.

The drive of the two lectures was, however, towards a fundamental theory of the micromaser, and this theory is ultimately presented as a part of the Lecture II. The *quantitative* simulation of the action of the one ^{85}Rb atom micromaser as it is in operation at the Max Planck Institute, Garching, near Munich under the direction of Professor Herbert Walther required new numerical methods to handle the *qualitatively* significant black-body radiation. These techniques were already developed in principle for the calculations of the fluorescence spectra, but two of the authors (BVT and NN) in particular were instrumental in applying these techniques to the micromaser problem. Qualitative to quantitative agreement between theory [11,24,25] and experiment [26,27] here is already achieved in terms of revivals in atomic inversion, cavity field photon number and intensity-intensity correlation, the last indicating periods of photon anti-bunching and sub-Poissonian photon statistics. Sub-Poissonian photon statistics of the cavity field of the micromaser is observed now in [28,29] and is the subject of our papers [23,30,31] and of the Lecture II below. But Lecture II also makes strong predictions about pure Fock state evolution. The three recent papers [23,30,31] on the micromaser cover some of this material in different ways, but the Lecture II, which starts at the §6 of this survey can be read as a research paper in its own right consisting largely of new material comparing closely only with Ref.[31] which is not yet published.

2. MICROWAVE CAVITY Q.E.D.

Microwave cavity q.e.d. concerns Rydberg atoms undergoing microwave transitions in nearly resonant or resonant microwave cavities. Rydberg atoms are typically hydrogen-like atoms of the Group I sequence (Na, K, Rb) excited to states with high principal quantum numbers n from which they can make microwave transitions to states with the same n or a nearby n (e.g. Na, $30S_{\frac{1}{2}} \to 30P_{\frac{1}{2}}$, $\lambda = 2.2$mm; ^{85}Rb, $63P_{3/2} \to 61D_{5/2}$, $\lambda = 13.98$mm). At these frequencies ($\nu \sim 10^{10}- 10^{11}$ Hz) cavity modes are well spaced and only single modes near to a resonance with the atomic transition need to be considered. Atomic sizes $\propto n^2$, so dipole matrix elements $p \propto n^2$, and the coupling between the atoms and the cavity field amplitudes $\mathcal{E}$ in the cavity mode, proportional to $\Omega \equiv p\,\mathcal{E}\hbar^{-1}$, the Rabi frequency, is large. These several features simplify cavity q.e.d. against 'ordinary' q.e.d. - which is a field theory with infinitely many modes - and for example a good model for much of cavity q.e.d. is the Jaynes-Cummings model [32]. This model consists of one 2-level atom coupled to a single electromagnetic field mode in a cavity which is of infinite Q. It is 'quantum integrable' i.e. exactly solvable [34-36], and we give the actual solution in the §§3 and 5 below.

One 2-level atom coupled to one cavity mode has two quantum degrees of freedom [33]. In this the J-C model [32] contrasts with 'ordinary' q.e.d. which has an infinite number of modes and a continuous infinity of degrees of freedom. Nevertheless the J-C model exhibits a fundamental quantum coupling. For one 2-level atom behaves like a single fermion, with two states and occupation numbers which can be labelled 0,1; while one e.m.

field mode is a quantum oscillator with boson occupation numbers 0,1,2,3,... . Thus the one 2-level atom and one e.m. field mode together exemplify a form of the fundamental fermion-boson coupling of 'ordinary' q.e.d. It is a *nonlinear* quantum coupling, and it leads to new fundamental nonlinear quantum physics. This is the fundamental interest of microwave cavity q.e.d.; and its relative simplicity makes it ideal for a joint theoretical-experimental study of the fundamentals of q.e.d.

In practice finite temperatures introduce, to a greater or less degree, important additional effects due to microwave black-body radiation. This breaks the integrability of the J-C model, for example. So that even at temperatures $T = 0.5^{\circ}K$ of the most recently reported experiments on the micromaser [28,29], where the Planck function $\bar{n}_{thermal} = (e^{\beta\hbar\omega_o}-1)^{-1}$, at the atomic transition frequencies ω_o and temperature $T = \beta^{-1}k_B^{-1}$, is $\bar{n}_{thermal} \approx 0.15$ (i.e. where there is about one sixth of one photon on average with an energy $\sim \frac{1}{6} \times 10^{-23}J$), the effect of these black-body photons at the Q's now achieved ($Q = 3\times10^{10}$ [28,29]) is to change *qualitatively* the theory of the micromaser from the corresponding theory at $Q = \infty$. For $T>0$ and $Q<\infty$, the complete integrability of the J-C model is not enough: problems of damping via the black-body radiation and of pumping through the black-body radiation and through the advent of atoms in their inverted states in the case of the micromaser, transform the problem from an exactly solvable quantum system to a far-from-equilibrium nonlinear system for which sophisticated numerical techniques are needed to provide any solution. The study of the *qualitative* changes in the action of the micromaser induced by black-body radiation is the essential content of the Lecture II.

The Lecture I, apart from this introduction §2 to microwave cavity q.e.d., consists of §§ 3, 4 and 5. The §3 analyses the 2-level atom. This simple object has a surprisingly rich mathematical structure expressed through its algebras. From it we show how, in a low-Q cavity, Einstein A- and B- coefficient theory for the interaction of radiation and atoms breaks down, and how N_A 2-level atoms becomes one 'giant quantum oscillator' for large enough N_A. The §4 then compares this many-atom equilibrium collective behaviour with the experimental observations of Haroche and colleagues [15]. Dynamical collective behaviour observed as super-radiant emission of radiation is also compared with the second group of experiments [16] by Haroche and colleagues, and the significance of Purcell's cavity factor η_{cav} on the Einstein A- coefficient Γ_o is explained and compared with a third set of experiments [17]. The §4 which is concerned with master equations in low-Q cavities ($Q \sim 5\times10^3$) then extends that theory to the case of N_A Rydberg atoms in *squeezed* vacua and giant oscillator behaviour in this case is predicted.

The §5 is concerned with calculations and experiments on the fluorescence from single atoms in microwave cavities. The single cavity mode frequency is ω (the atomic frequency is ω_o and $\omega \approx \omega_o$). The e.m. field-atom coupling is g (~50kHz) and the cavity damping factor $\kappa = \frac{1}{2}\omega Q^{-1}$. The range of coupling available to experiment including the $Q = 3\times10^{10}$ reported for the micromaser in [28,29] is now $10^3 \gtrsim \kappa g^{-1} \geq 0.0007647$: the corresponding range of Q is $5 \times 10^3 \lesssim Q \leq 3\times10^{10}$. For the fluorescence spectra described in this §5 we cover $5\times10^3 \leq Q \leq 10^8$: $Q = 10^8$ is the Q-value for the recent experiments [21,22] on ^{85}Rb and ^{87}Rb atoms. The vacuum field Rabi splitting [18,36] is probably the point of most interest in these spectra and this can be seen in the calculations for $\kappa g^{-1} \lesssim 1$

($Q \geq 10^6$). At $\kappa g^{-1} \approx 10^{-2}$ as in the experiments [21,22] the two peaks should be easily resolved. There are intrinsic difficulties (§5) in observing microwave fluorescence spectra, but Haroche's group have been able to use the atoms themselves as their own, very sensitive, detectors of the spectrum. Splitting for $N_A = 3$ is clearly seen [22] and $N_A = 1$ is intended.

There is also a 2-photon cavity q.e.d. Haroche's group successfully demonstrated a 2-photon micromaser using 40S ⟶ 39P, 39P ⟶ 39S transitions in ^{85}Rb at 68.41587 GHz in a Niobium cavity with $Q = 10^8$ [37,38]. The Stark shifts are very small [11]. In this connection we have computed [39-41] collapse and revival in the inversion $\langle S^z \rangle N_A^{-1}$ for both $N_A = 1$ [39,40] and $N_A \geq 2$ [41] atoms in both 2-photon on-resonance and 2-photon off-resonance in high-Q cavities, $Q = \infty$. We have also computed revivals in $g^{(2)}(0) = \{\langle a^{\dagger 2} a^2 \rangle / \langle a^\dagger a \rangle^2\} - 1$ for $N_A = 1$ [39] and $N_A \geq 2$ [40,41] as a function of time t. This quantity is the intensity-intensity correlation as a function of t and describes the photon statistics and antibunching. The atoms are in each case in 2-photon resonance or off-resonance with a $Q = \infty$ cavity containing a coherent state field. When the 2-photon process is *non*-degenerate we find a remarkable recurrence behaviour for the inversion $\langle S^z \rangle$ when $N_A = 1$ [40]. We have also investigated squeezing in the degenerate case and the non-degenerate case [39,40]. We cannot report further on these high-Q results for 2-photon cavity q.e.d. in these two lectures: the reader is referred to [39-41] for more details. We note here in the §4 however that in low-Q cavities giant 2-photon quantum oscillators can be formed as $N_A \to \infty$.

We are not able to report on the *coherently driven* Dicke model in these two lectures either: the reader is referred to [1-7,10,11,13,14] and particularly to [10,11]. In [11] the main analysis concerns the evolution of N_A-atom atomic coherent states (su(2) coherent states) by N_A-Rydberg atoms in low-Q cavities. These states are *squeezed* and this collective squeezing is described in that paper. In the case of the coherently driven Dicke model 'phase transitions' both on- and off-resonance are predicted to occur and this complicates the squeezing. For example on exact resonance where in the absence of any cavity there is still a phase transition we predict [11,42,43] squeezing in on-resonant resonance fluorescence which decreases with N_A (a result which is consistent with the formation of an su(2) coherent state). Off-resonance where there is no phase transition we predict the squeezing in resonance fluorescence increases with N_A [11,42-44]. Ref. [44] also analyses squeezing in super-radiance from an initial collective atomic coherent state. Results for the coherently driven Dicke model in low-Q cavities were also reported in [10] where the main emphasis was on predictions for phase transitions, second order on resonance, first order off resonance, induced by the driving field. The first order transition exploits the two independent detunings between three frequencies, that of the driving field, that of the cavity, and that of the atoms. We give a quantum theory of optical bistability for about 50 Na Rydberg atoms off-resonance with the two detunings in a low-Q cavity. These various and obviously important theoretical predictions for cavity q.e.d. have still to be verified experimentally and we refer the reader to the papers [10,11,42] for further details. We note that the beginnings of a theory of super-radiance into a squeezed vacuum is reported in [45,46] and fluorescence spectra in squeezed vacua are reported in [45]. The effect of a squeezed vacuum on giant oscillator formation by Rydberg atoms is described in [45,46] and in the §4 of these lectures. The effect of a squeezed vacuum on the dielectric properties of a slab of many atoms forming a Fabry-Perot cavity at *optical frequencies* is described in our paper [47] at the Summer School.

3. WHY CAVITY Q.E.D.? THE 2-LEVEL ATOM AND ITS ALGEBRAS

Cavity q.e.d. concerns (as explained) one (N_A=1) or more (N_A>1) *atoms* interacting with radiation in cavities of finite quality factor Q. As explained close to any non-degenerate resonance atoms behave like fermions. The resultant fermi-bose coupling is typical of ordinary quantum electrodynamics where free charges, electrons, positrons, interact with radiation. This is why we talk about *cavity* quantum electrodynamics.

The atom, which close to a non-degenerate resonance acts like a fermion, has become a two-state system, the 2-level atom. Conceptually a 2-level atom has two, and only two, non-degenerate energy levels separated by an energy $\hbar\omega_0$. The two states are $|e\rangle$ (say) and $|g\rangle$ (say) (e is 'excited', g is 'ground'). There is a dipole matrix element p connecting these two levels and with p≠0 transitions of frequency ω_0 take place between them. The $p = e\langle e|x|g\rangle$ is chosen real, and e is the charge on the electron. This electron would be the single 'optical' electron of the Group I sequence Na, K, Rb for example. The simple 2-level structure is surprisingly *rich* in mathematical terms and has proved to be widely applicable to the physics of quantum optics. This wide application is because the interest of quantum optics lies in the non-linearity. This nonlinearity always enters through the atoms (and not through the radiation), and atoms become nonlinear only at or very close to a resonance.

The standard reference to 2-level atom theory is the book 'Optical Resonance and Two-level Atoms' by L. Allen and J.H. Eberly now a Dover publication [85]. An early paper by one of the authors (RKB), hard to get hold of, is [86].

The rich structure of the 2-level atom expresses itself through a number of physically different *algebras*. By an algebra we mean in physical terms a closed set of commutation relations. The orbital angular momentum operators J^x, J^y, J^z satisfy the algebra $[J^x, J^y] = i\hbar J^z$ with two other similar expressions found by cyclic permutation of x, y, z. The algebra is closed in terms of the three independent operators $(J^x, J^y, J^z) = \vec{J}$.

In the quantum mechanics of 2-level atoms we become interested in spin commutation relations $[S^x, S^y] = i\hbar S^z$, etc. By scaling out the $\hbar$ and introducing raising and lowering operators $S^\pm$ we reach the algebra

$$[S^\pm, S^z] = \mp S^z,\ [S^+, S^-] = 2S^z\ . \tag{1}$$

This is the so-called su(2) Lie algebra.

As a *two*-state system the 2-level atom has the two-state spin description $|\uparrow\rangle$ (spin-up), the state $|e\rangle$, and $|\downarrow\rangle$ (spin-down), the state $|g\rangle$. Raising and lowering operators are then (in terms of $|e\rangle$, $|g\rangle$)

$$S^+ = |e\rangle\langle g|,\ S^- = |g\rangle\langle e|$$
$$S^z = \tfrac{1}{2}(|e\rangle\langle e| - |g\rangle\langle g|)\ . \tag{2}$$

These operators satisfy the su(2) spin algebra (1) and form a *two-dimensional representation* of su(2): they operate on the 2-dimensional Hilbert space $|e\rangle$, $|g\rangle$. There are at least four other 2-dimensional, or essentially 2-dimensional, representations of physical interest!

First of all a single 2-level atom is two *constrained* fermions:- suppose $a^\dagger$(a) and $b^\dagger$(b) create (annihilate) a 'particle' in $|\uparrow\rangle = |e\rangle$ ($|\downarrow\rangle = |g\rangle$); $a^\dagger$, a and $b^\dagger$, b are fermion operators so

$$aa^\dagger + a^\dagger a = 1\ ;\quad bb^\dagger + b^\dagger b = 1\ ;$$
$$(a^\dagger)^2 = (a)^2 = 0\ ;\ (b^\dagger)^2 = (b)^2 = 0;\ a^\dagger b + ba^\dagger = 0;\ \text{etc.} \tag{3}$$

Then

$$S^+ = a^\dagger b,\ S^- = b^\dagger a,\ S^z = \tfrac{1}{2}(a^\dagger a - b^\dagger b) \tag{4}$$

is a representation of the su(2) Lie algebra (1). Thus the 2-level atom is *two* fermions. But $a^\dagger a + b^\dagger b = 1$ (for one optical electron in the atom) constrains these two fermions to one effective fermion.

Secondly, a single 2-level atom is two constrained bosons! This is the so-called Jordan-Schwinger representation of su(2) with the additional constraint. For raising (lowering) operators $a^\dagger$(a) and $b^\dagger$(b) put

$$[a, a^\dagger] = 1\ , \quad [b, b^\dagger] = 1$$
$$[a^\dagger, b] = [a, b] = 0\ . \tag{5}$$

Then

$$S^+ = a^\dagger b,\ S^- = b^\dagger a,\ S^z = \tfrac{1}{2}(a^\dagger a - b^\dagger b) \tag{6}$$

is a representation of the su(2) Lie algebra (1). But then again there is the constraint $a^\dagger a + b^\dagger b = 1$ (for one electron) so these two bosons are constrained to one boson which, because of the constraint, is confined to precisely two levels. Thus this boson system is really a two-state system and fermion-like.

In fact one can easily see that the 2-level system is one (and only one) fermion and (alternatively) that it is one and only one constrained boson. The Primakov-Holstein transformations put

$$S^- = (2S-\hat{n})^{\frac{1}{2}}a,\quad S^+ = a^\dagger(2S-\hat{n})^{\frac{1}{2}} = (2S+1-\hat{n})^{\frac{1}{2}}a^\dagger,$$
$$S^z = (\hat{n}-S) \tag{7}$$

with $\hat{n} = a^\dagger a$. Then if $S = \frac{1}{2}$, and $a^\dagger a + aa^\dagger = 1$ with $(a)^2 = (a^\dagger)^2 = 0$ the representation (7) is a representation of the su(2) Lie algebra (1). This is plainly 2-dimensional since $\hat{n}$ has the eigenvalues 0,1 belonging to $|0\rangle$ (or $|g\rangle$) and $|1\rangle$ (or $|e\rangle$) only. But if $S = \frac{1}{2}$, and $[a, a^\dagger] = 1$ one finds (7) becomes a representation of su(2) also and

$$S^+ = a^\dagger(1-\hat{n})^{\frac{1}{2}}\ ,\quad S^- = (1-\hat{n})^{\frac{1}{2}}a\ . \tag{8}$$

So the oscillator algebra (which is called the Heisenberg-Weyl algebra) $[a, a^\dagger] = 1$ with $\hat{n} = a^\dagger a$ is constrained to states $|0\rangle = |g\rangle$ and $|1\rangle = |e\rangle$.

The boson-fermion alternative descriptions of one 2-level atom are a form of 'bose-fermi equivalence'. This equivalence is a feature of the so-called quantum integrable models in one space and one time dimensions [33-35,48,49]. These models, which are field theories, have a large infinity of degrees of freedom. An outstanding example is the quantum sine-Gordon equation (technically to be read in normal order)

$$\frac{\partial^2\phi}{\partial x^2} - \frac{1}{c^2}\frac{\partial^2\phi}{\partial t^2} = \frac{m^2}{c^2}\sin\phi \tag{9}$$

for which the quantum field $\phi(x,t)$ has the equal time commutation relations $[\phi(x,t), \phi^\dagger(x',t)] = \delta(x-x')$ and $x \in \mathbb{R}$ labels the degrees of freedom. This system is bose-fermi equivalent to the massive Thirring model [50] a coupled fermion model.

In cavity q.e.d. we are concerned with relatively few degrees of freedom but the experiments of Haroche and co-workers [15] using Na atoms in low-Q cavities ($Q \sim 5\times10^3$) making microwave transitions between the $30S_{\frac{1}{2}} \leftrightarrow 30P_{\frac{1}{2}}$ states at wavelength $\lambda = 2.2$ mm can be interpreted as displaying some aspects of this bose-fermi equivalence which is already exhibited by one single 2-level atom. We explain all of this in what follows.

A fundamental model in quantum optics has been the Jaynes-Cummings model [32]. This consists of one 2-level atom coupled to a single mode of the field. The field mode is quantised in some box which in effect acts as a cavity of infinite Q. Plainly in practice we shall need to couple the system to a heat bath at finite temperatures $T = \beta^{-1}k_B^{-1}$. Then there is the question of whether Q remains high (say $Q = 3\times10^{10}$ of the maser experiments [28,29]) or Q is low (say $Q = 5\times10^3$). These two extreme cases are extreme cases of the range of Q now available to experiments, and these exhibit qualitatively different physics. This qualitative difference in the physics is a consequence of the effects of black-body radiation.

At $Q = \infty$ the J-C model is a completely integrable quantum system with *two* quantum degrees of freedom. The Hamiltonian can be taken in the form

$$H = \hbar\omega\, a^{\dagger} a + \hbar\omega_0\, S^z + \hbar g(a^{\dagger} S^- + S^+ a) \tag{10}$$

and $g \in \mathbb{R}$ with dimensions of frequency is the coupling constant between matter and field. It is useful to know that in the one ^{85}Rb atom micromaser [26-29], which can be modelled by (10), $g \sim 44$kHz. The free 2-level atom with transition frequency ω_0 has the Hamiltonian $\hbar\omega_0 S^z$; S^z is a spin operator with the representation (2) ; $S^{\pm}$ are the raising and lowering operators of (2). The photons (bosons) have the free-Hamiltonian $\hbar\omega a^{\dagger} a$ ($\omega \approx \omega_0$) and the interaction is the bilinear one $\hbar g(a^{\dagger} S^- + S^+ a)$. Because this is fermion-boson coupling the system is *non-linear* (it becomes linear if the $S^{\pm}$ are unconstrained boson operators).

At this point it is worthwhile recalling where we have come from:- The free non-relativistic electron of charge e, mass m_e and momentum **p** has the coupling with the radiation field $(2m_e)^{-1}(-2\mathbf{p}.\mathbf{A}ec^{-1} + e^2c^{-2}\mathbf{A}^2)$, where **A** (in Gaussian units) is the vector potential and c is the velocity of light *in vacuo*. For an optical electron bound to an atom there is a canonical transformation [51] which, in dipole approximation, can express this interaction as $-\mathbf{d}.\mathcal{E}$ where **d** is the atomic dipole operator and $\mathcal{E}$ is the electric field ($\mathcal{E} = -c^{-1}\partial\mathbf{A}/\partial t$). In our 2-level-atom-single-mode language $\mathbf{d} = \hat{\mathbf{u}}p(S^+ + S^-)$ where $\hat{\mathbf{u}}$ is a unit vector in the direction of $p = e \langle e|x|g\rangle$ (called the x-direction) and $\mathcal{E}$, also along x, is, up to a choice of phase, $i(2\pi\omega/\hbar V)^{\frac{1}{2}}(a-a^{\dagger})$ where V is the volume of the cavity for quantisation (the volume of the chosen microwave cavity). We actually choose $\mathcal{E}$ to be a momentum like canonical variable, so that, after scaling to canonical variables q, Π, $q \equiv (\omega^{-\frac{1}{2}}\hbar^{\frac{1}{2}}/\sqrt{2})(ae^{i\delta}+a^{\dagger}e^{-i\delta})$ and $\Pi \equiv (\omega^{+\frac{1}{2}}\hbar^{\frac{1}{2}}/\sqrt{2})(ae^{i(\delta-\frac{1}{2}\pi)} + a^{\dagger}e^{-i(\delta-\frac{1}{2}\pi)})$ which satisfy $[\Pi,q] = \frac{1}{2}\hbar(e^{-i\frac{1}{2}\pi} - e^{+i\frac{1}{2}\pi}) = -i\hbar$, for any choice of the phase δ. Then this leads finally to the interaction $\hbar g(S^+ + S^-)(a+a^{\dagger})$ where $g = (2\pi\omega/\hbar V)^{\frac{1}{2}}p$ with dimension of frequency. We now make the so-called rotating wave approximation, which is energy conserving, so as to reach $\hbar g(S^+ a + a^{\dagger} S^-)$ and this is the interaction used in (10). Note that this interaction started as fermion-boson coupling, but this fermion-boson coupling re-asserts itself as explained since $S^{\pm}$ are fermion operators in the sense already explained. The important point is that this coupling, like the original fermion-boson coupling, is nonlinear. In the Lecture II, when we look at some new bosons which we call the q-bosons, we in effect ask if there are other non-linearities contained in the interaction between the atom and the field in a microwave cavity, but the answer to this question is still not known.

Returning to the Hamiltonian (10), it is now easy to see that because the interaction is made energy conserving by making the rotating wave approximation the total number operator

$$N = S^z + a^{\dagger} a + \tfrac{1}{2} \tag{11}$$

is an operator constant of the motion i.e. $[H, N] = 0$. A quantum system is completely *quantum* integrable if, when there are M quantum degrees of freedom there are M independent mutually commuting operators, and one of these operators can be taken as the Hamiltonian H [33]. The J-C model (10) has *two* degrees of freedom: one of these is the quantum oscillator for which $[a, a^{\dagger}] = 1$ and from which canonical coordinate operators are $\Pi \equiv (\omega^{\frac{1}{2}}\hbar^{\frac{1}{2}}/\sqrt{2})(a+a^{\dagger})$, $q \equiv i(\omega^{-\frac{1}{2}}\hbar^{\frac{1}{2}}/\sqrt{2})(a - a^{\dagger})$ as was explained. The other degree of freedom is the atom, a spin-$\frac{1}{2}$ system. A canonical *pair* can in effect be identified for this spin system through operators S^+ and S^- given that $\frac{1}{2}(S^+S^- + S^-S^+) + (S^z)^2 = \frac{1}{2}(\frac{1}{2}+1)I$, where I is the unit operator.

The J-C model then has *two* quantum degrees of freedom and *two* mutually commuting operators H, N. It can in fact be integrated through a sequence of 2×2 eigenvalue problems because at fixed N with eigenvalue n (say), and as explained further in §5, the eigenstates $|g\rangle|n\rangle$ and $|e\rangle|n-1\rangle$, each belonging to the eigenvalue n, span a 2-dimensional Hilbert sub-space. H is then diagonalised for fixed N=n on that sub-space. The states $|n\rangle$ are n-photon states.

This then is the completely integrable J-C model.

The experiment of Haroche *et al.* [15] in effect shows how $N_A = 1$ fermion, the 2-level atom, becomes one giant collective boson as the number N_A of 2-level atoms becomes very large. At very high Q this would mean changing $S^{\pm}$, S^z in the J-C Hamiltonian (10) to the *collective* N_A-atom operators

$$S^{\pm} = \sum_{i=1}^{N_A} S_i^{\pm}\ , \quad S^z = \sum_{i=1}^{N_A} S_i^z\ ; \tag{12}$$

for which each $S_i^{\pm}$, S_i^z satisfies the su(2) Lie algebra (1), and operators for different i commiute. Then, in consequence, the collective operators $S^{\pm}$, S^z satisfy the algebra (1) also. However, these collective operators now form an N_A+1 dimensional representation of su(2), acting on a Hilbert space of that dimension. The model (10) with the collective operators (12) is the Tavis-Cummings model [52] and is apparently quantum integrable. We do not pursue this model further in this article however. Instead we want next to examine the physics of the collective operators (12).

Atoms occupy volume in space and can be located at different sites $\mathbf{x}_i$ (say). One could define operator densities

$$S^{\pm}(\mathbf{x}) = \sum_{i=1}^{N_A} S_i^{\pm}\,\delta(\mathbf{x}-\mathbf{x}_i)\ ; \quad S^z(\mathbf{x}) = \sum_{i=1}^{N_A} S_i^z\,\delta(\mathbf{x}-\mathbf{x}_i) \tag{13}$$

and in quantum dielectric theory [2,47,51] for example this must be done. In this case there are *retarded* interactions between each pair of atoms. Evidently in (12) all of the atoms are placed on the same site $\mathbf{x}_i = \mathbf{x}_0$ (say) (and the density factors $\delta(\mathbf{x}-\mathbf{x}_0)$ are discarded). This single-site N_A-atom model is exactly the Dicke model [12]. But since atoms cannot, by the Pauli principle, occupy the same sites $\mathbf{x}_0$ what about the physics of this model? In effect the experiments of Haroche *et al.* [15,16] show that at the microwave wavelengths $\lambda \gtrsim$ mm the Dicke model *is* a very good physical model. The idea is that at these wavelengths retardation effects over *macroscopic* regions ~ mm are negligible: thus cavity sizes must be ~ mm in dimension (as was the case [15,16]).

We need to analyse the Dicke model. The Dicke model has the Hamiltonian

$$H = H_0 \equiv \hbar\omega_0 \sum_{i=1}^{N_A} S_i^z \tag{14}$$

only. It acts on the Hilbert space spanned by the 2^{N_A} states $|f_1, f_2, \ldots, f_{N_A}\rangle \equiv |f_1\rangle|f_2\rangle \ldots |f_{N_A}\rangle$ where $|f_i\rangle$ is either $|e_i\rangle$ or $|g_i\rangle$ for atom i. In spin language $|f_1, \ldots, f_{N_A}\rangle = |\uparrow,\uparrow,\uparrow\ldots,\uparrow\rangle$ for $|f_1\rangle = |e_1\rangle$, $|f_2\rangle = |e_2\rangle, |f_3\rangle = |e_3\rangle, \ldots, |f_{N_A}\rangle = |e_{N_A}\rangle$, etc. In adding spins we know that we can find groups of states with total spin S, S-1,..., 1. Evidently $S = \frac{1}{2}N_A$. The collective Dicke states are then $|S,m\rangle$ where $S = \frac{1}{2}N_A, \frac{1}{2}N_A-1, \ldots, \frac{1}{2}$ (N_A odd) or $S = \frac{1}{2}N_A, \frac{1}{2}N_A-1, \ldots, 1, 0$ (N_A even). The m labels eigenstates of S^z : $S^z|S, m\rangle = m|S, m\rangle$ and $-S \le m \le +S$. The total ground state $|\downarrow, \downarrow, \ldots, \downarrow\rangle$ has $S = \frac{1}{2}N_A$ and $m = -S$; and there are $2S+1 = N_A+1$ states for $S = \frac{1}{2}N_A$, including the ground state and the totally excited state $|\uparrow, \uparrow, \ldots, \uparrow\rangle$. The energies are $-\frac{1}{2}N_A\hbar\omega_0$ (ground state) and $+\frac{1}{2}N_A\hbar\omega_0$ (total excited state). For $S = \frac{1}{2}N_A-1$ there are N_A-1 degenerate sets each with N_A-1 states : the ground state energies of these sets are each $(-\frac{1}{2}N_A+1)\ \hbar\omega_0$. The sequence thereafter is similar with $\frac{1}{2}N_A(N_A-3)$ degenerate sets having $S = \frac{1}{2}N_A-2$ and each with N_A-3 states, then $S = \frac{1}{2}N_A-2$, $S = \frac{1}{2}N_A-3$, $\frac{1}{2}N_A-4, \ldots$ and all the 2^{N_A} states are systematically exhausted this way [12].

The collective dipole operator is proportional to $S^+ + S^-$. Since these operators conserve total spin S (they raise or lower on m) and the "addition of angular momentum" puts $\langle S, m|S', m'\rangle \propto \delta_{ss'}\ \delta_{mm'}$, it follows that no matrix elements connect Dicke states with different S. Thus transitions take place only up and down the 'ladders' of the 2S+1 states with the same S. Thus, if by some means one of the N_A-1 ladders with $S = \frac{1}{2}N_A-1$ or one of the $\frac{1}{2}N_A(N_A-3)$ ladders with $S = \frac{1}{2}N_A-2$, etc. can be entered, excitation is trapped there with a minimum energy which is greater than the total ground state energy $-\frac{1}{2}N_A\hbar\omega_0$.

An obvious example is the $N_A=2$ system. This is the spin S=1 triplet and spin S=0 singlet.

$$
\begin{aligned}
|1,1\rangle &= |e_1\rangle|e_2\rangle \\
|1,0\rangle &= \frac{1}{\sqrt{2}}(|e_1\rangle|g_2\rangle + |g_1\rangle|e_2\rangle) \\
|1,-1\rangle &= |g_1\rangle|g_2\rangle
\end{aligned}
\tag{14a}
$$

and

$$
|0,0\rangle = \frac{1}{\sqrt{2}}\ (|e_1\rangle|g_2\rangle - |g_1\rangle|e_2\rangle), \tag{14b}
$$

a total of $2^2 = 4$ states. If $|0,0\rangle$ can be reached the excitation is trapped there. The reader might like to evaluate the $N_A = 3$ system which has an $S = \frac{3}{2}$ quadruplet and *two* $S = \frac{1}{2}$ doublets (a total of $8 = 2^3$ states).

The N_A+1 states with $S = \frac{1}{2}N_A$ include both the total ground state and the total excited state, each of which can be relatively easily reached in experiments. Thus this 'ladder' with $S = \frac{1}{2}N_A$ for N_A atoms is the physically most significant one. It forms a $2S+1 = N_A+1$ representation of the su(2) Lie algebra (1). The SU(2) *group* is a double cover of the O(3) group. And [53] the O(3) group is the group of the classical pendulum in 3 dimensions (a rigid rod fixed at one end so that the other end can cover the surface of a sphere). We can therefore think of the N_A+1 states with $S = \frac{1}{2}N_A$ forming a 'quantum pendulum'. Semi-classically this pendulum is the Bloch vector whose end point covers the surface of a sphere [53]. This sphere has unit radius but scales to radius N_A for N_A 2-level atoms. Dicke's idea [12] was that the system of N_A atoms all excited in the spin state $|\uparrow,\uparrow,\ldots,\uparrow\rangle$ will exhibit collective spontaneous emission, called 'super-radiance': the quantum pendulum moves from total spin up to total spin down $|\downarrow,\downarrow,\ldots,\downarrow\rangle$, the total ground state, radiating N_A photons of energy $\hbar\omega_0$. But in the neighbourhood of states with equal numbers of atoms excited and in their ground states (namely states $|S,m\rangle$ with $S = \frac{1}{2}N_A$ where $m \simeq 0$) the spontaneous *radiation rate* w goes as the square N_A^2 of the number of atoms -

precisely $w \approx \frac{1}{4}N_A^2\Gamma_0$ where Γ_0 is the 1-atom A-coefficient, $\Gamma_0 = \frac{4p^2}{3}\omega_0^3\hbar^{-1}c^{-3}$. The collective spontaneous emission thus becomes *pulse*-like with a sech^2 profile [10] and with the radiation rate per particle enhanced by the factor $\frac{1}{4}N_A$ at peak. We analyse this super-radiant emission from Rydberg Na atoms in the next section, §4.

The conclusion so far then:- The Dicke operators (12) act on the Hilbert sub-space of Dicke states $|S = \frac{1}{2}N_A, m\rangle$, $-\frac{1}{2}N_A \le m \le +\frac{1}{2}N_A$ and this sub-space forms a nonlinear 'quantum pendulum'.

In thermal equilibrium, i.e. in equilibrium with block-body radiation of frequency $\omega \approx \omega_0$, the quantum pendulum will remain on average in some collective thermally excited state. We calculate this state next.

First of all notice that for the N_A-atom collective operators (12) the Primakov-Holstein transformation (7) for boson operators a, $a^\dagger$ still applies but $S = \frac{1}{2}N_A > \frac{1}{2}$: then $\hat{n} = a^\dagger a$ has allowed values 0, 1, ..., N_A and the oscillator H-W algebra $[a, a^\dagger] = 1$, $\hat{n} = a^\dagger a$ is constrained to the N_A+1 states $|0\rangle$, $|1\rangle$, ..., $|N_A\rangle$ as compared with the two states $|0\rangle$, $|1\rangle$ for $S = \frac{1}{2}$ (see below (8). Note that we do not know of any corresponding *fermion* description for $N_A>1$: there is only (7) with $S = \frac{1}{2}$ and $aa^\dagger + a^\dagger a = 1$ in fermion description).

If for $N_A>1$ we define $\bar{S}^{\mp} = (2S)^{-\frac{1}{2}} S^{\mp}$, then

$$[\bar{S}^{\pm}, S^z] = \mp \bar{S}^{\pm}, \quad [\bar{S}^+, \bar{S}^-] = (S)^{-1}S^z . \tag{15}$$

Evidently

$$\bar{S}^+ = a^\dagger(1-(2S)^{-1}\hat{n})^{\frac{1}{2}}, \quad \bar{S}^- = (1-(2S)^{-1}\hat{n})^{\frac{1}{2}}a,$$
$$S^z = \hat{n} - S, \tag{16}$$

and as $S \to \infty$ (i.e. $\frac{1}{2}N_A \to \infty$) $\bar{S}^+ \to a^\dagger$, $\bar{S}^- \to a$ and $[\bar{S}^+, \bar{S}^-] \to -1$ i.e. becomes $[a, a^\dagger] = +1$. Thus as $S = \frac{1}{2}N_A \to \infty$ the N_A-atom Dicke model becomes a 'giant quantum oscillator'. This operation of taking an infinite dimensional representation of su(2) and forming the algebra of a, $a^\dagger$ and $\hat{n} = a^\dagger a$, the H-W algebra, is known as 'group contraction' [54,55].

We can now guess at the physical situation in thermal equilibrium when $N_A \to \infty$. The giant oscillator ought to be occupied on the average by $\bar{n} = (e^{\hbar\beta\omega_0}-1)^{-1}$ black-body photons : $(e^{\beta\hbar\omega_0}-1)^{-1}$ is the Planck function. This means that the mean number N_+ of excited atoms in thermal equilibrium ought to be $N_+ = \bar{n}$ (in the Dicke state $|\frac{1}{2}N_A, m\rangle$ the number of excited atoms is $m+\frac{1}{2}N_A$ but we want the *average* excitation). Einstein A- and B- coefficient theory suggested a very different result : for one atom a state in the Hilbert space is $c_1|e\rangle + c_2|g\rangle$ and the probability of being in $|e\rangle$ ($|g\rangle$) is $|c_1|^2$ ($|c_2|^2$) with $|c_1|^2 + |c_2|^2 = 1$. Then $|c_1|^2/|c_2|^2 = e^{-\beta\hbar\omega_0}$ in equilibrium where $\beta^{-1} = k_BT$. The mean occupation of the upper state $|e\rangle$ is $|c_1|^2 \equiv N_+$ and in equilibrium $|c_1|^2 + e^{+\beta\hbar\omega_0}|c_1|^2 = 1$ so

$$N_+ = |c_1|^2 = \bar{n}\,(1 + 2\bar{n})^{-1} \tag{17}$$

(after a little algebra) where $\bar{n} = (e^{\beta\hbar\omega_0}-1)^{-1}$. Thus for $N_A > 1$ A- and B-coefficient theory seems to mean

$$N_+ = N_A\,\bar{n}(1+2\bar{n})^{-1}\,. \tag{18}$$

Moreover N_-, the mean number of atoms in the lower state, would be $N_- = N_A(1+\bar{n})(1+2\bar{n})^{-1}$ and $N_+/N_- = e^{-\beta\hbar\omega_0}$ so the distribution remains a Boltzmann distribution for all values of N_A.

Elementary statistical mechanics shows [56] that for $N_A=2$ 2-level atoms, with access to the whole 4-dimensional Hilbert space (14a) with (14b), relation (18) certainly applies. Likewise for $N_A=3$, 4, etc. However for the reduced Hilbert space which forms the quantum pendulum

$$N_+ = \frac{X+2X^2}{1+X+X^2} = \frac{\bar{n}(1+3\bar{n})}{1+3\bar{n}+3\bar{n}^2}\,,\quad N_+ = \frac{X+2X^2+3X^3}{1+X+X^2+X^3} = \frac{\bar{n}(1+4\bar{n}+6\bar{n}^2)}{1+4\bar{n}+6\bar{n}^2+4\bar{n}^3} \tag{19}$$

for $N_A = 2$, and $N_A= 3$ respectively [56]. In these expressions we have introduced $X \equiv \bar{n}(1+\bar{n})^{-1} = e^{-\beta\hbar\omega_0}$. In terms of X, $N_+ = \bar{n}(1+2\bar{n})^{-1}$ for $N_A = 1$ is $X/(1+X)$, while the expected result is $N_+ = \bar{n} = X/(1-X)$ for $N_A = \infty$. We thus expect to see a curve for N_+ against N_A starting at $X/(1+X)$ for $N_A = 1$ and rising *not* as $N_A\{X/(1+X)\}$ for increasing N_A but rising eventually to $X/(1-X)$ simply, as $N_A \to \infty$. The ratios N_+/N_- from (19) are no longer Boltzmann. And, as $N_A \to \infty$ we expect that $N_+/N_- \sim \bar{n}(N_A-\bar{n})^{-1} = [N_A(e^{\beta\hbar\omega_0}-1)-1]^{-1}$ which is of course not Boltzmann either.

In a remarkable early experiment with Rydberg N_A atoms making $30S_{\frac{1}{2}} \leftrightarrow 30P_{\frac{1}{2}}$ transitions in equilibrium with black-body radiation at temperatures T = 300°K and 900°K ($\bar{n}$ = 47 and 137 respectively) in a low-Q cavity ($Q \sim 5 \times 10^3$) Haroche and colleagues [15] were able to see a part, the large N_A part, of just this curve. To see why a *low*-Q cavity should be used we need some master equation theory for the density operators ρ and this follows next.

4. MASTER EQUATIONS : GIANT QUANTUM OSCILLATORS

A fundamental master equation for the density operator ρ for matter and field operators coupled in a Hamiltonian H to a single cavity mode in a very high Q cavity in which the cavity mode is coupled to a heat bath at temperature T is [18]

$$\frac{d\rho}{dt} = \frac{-i}{\hbar}[H,\rho] - \kappa(\bar{n}+1)[a^\dagger a\rho+\rho a^\dagger a - 2a\rho a^\dagger] - \kappa\bar{n}[aa^\dagger\rho+\rho aa^\dagger-2a^\dagger\rho a]\;: \tag{20}$$

$\bar{n} = (e^{\beta\hbar\omega}-1)^{-1}$, the Planck function for the mode of frequency ω and $\kappa \equiv \frac{1}{2}\omega Q^{-1}$. For N_A Rydberg atoms with microwave frequencies ω_0 and $\omega \approx \omega_0$ we can adopt the Dicke model for the atoms (Haroche's experiments confirm this assumption) and the Hamilton H is the Tavis-Cummings Hamiltonian (10) in which $S^\pm$, S^z are the collective operators (12).

Now the point of the *low*-Q cavity:- For low-Q cavities one can use projection techniques [20], or adiabatic elimination of the e.m. field mode operators a, $a^\dagger$ (at low-Q photons escape instaneously from the cavity), to reach at exact resonance $\omega = \omega_0$ of mode and atoms

$$\frac{d\rho_A}{dt} = \frac{-i}{\hbar}[H_0,\rho_A] - g^2\kappa^{-1}(\bar{n}+1)[S^+S^-\rho_A+\rho_A S^+S^- -2S^-\rho_A S^+] - g^2\kappa^{-1}\bar{n}[S^-S^+\rho_A+\rho_A S^-S^+-2\,S^+\rho_A S^-] \tag{21}$$

where ρ_A is the reduced density operator ("density matrix") for the atoms alone: g is the coupling constant in H, eqn.(10) and takes the value

$g^2\kappa^{-1} = \frac{1}{2}\eta_{cav}\Gamma_0$ where Γ_0 is the A-coefficient ($\Gamma_0 = \frac{4}{3}\frac{\omega_0^3}{\hbar c^3}p^2$) and η_{cav} is Purcell's cavity factor [9,57] : $\eta_{cav} = 3\pi c^3/V\omega_0^2\ \kappa$ where V is the cavity volume. Off resonance, $\omega \neq \omega_0$, one also finds there is a radiative level shift in the low-Q cavity [10,20].

Readers interested in alternative routes to the master equation (21) will find, in particular, a fundamental analysis of the numbers $\bar{n}$, $\bar{n}+1$ appearing in it, and which are *free* black-body radiation field numbers, in the Ref. [87] and its references: operator reaction field theory [60,61] is used and an approximate *decorrelation* procedure decorrelating field and matter is needed to reach these free field numbers. Relevant and equally fundamental papers are [88,89] (Ref. [88] also makes reference through its Ref. 11 to a first report of a theory of photon antibunching).

However, the point of (21) and the low-Q cavity is for present purposes that the Hamiltonian, after the projection, is exactly the Dicke model Hamiltonian H_0 of eqn.(10) : $H_0 = \hbar\omega_0 S^z$.

For equilibrium $d\rho_A/dt = 0$ in the master equation (21) and we can solve it exactly (or use the equilibrium statistical mechanics exactly). We find in terms of the parameter $X = \bar{n}(1+\bar{n})^{-1}$ that for arbitrary N_A [6,9-11]

$$N_+ = \frac{X[1-(N_A+1)X^{N_A}+N_A X^{N_A+1}]}{[1-X-X^{N_A+1}+X^{N_A+2}]} \tag{22}$$

and for $N_A = 1$, $N_+ = X/(1+X) = \bar{n}/(1+2\bar{n})$ and for $N_A \to \infty$, $N_+ = X/(1-X) = \bar{n}$ as hazarded in §3. From (22) we see indeed that N_+/N_- follows Boltzmann's distribution only for $N_A = 1$. Ref. [6] proves that the probability of n atoms being excited satisfies *Bose-Einstein* statistics when $N_A \longrightarrow \infty$ - so N_+ is indeed $N_+ = \bar{n}$.

We note that for k-photon transitions, rather than the 1-photon transitions of the 2-level atom, we find [11] exactly the same *formula* for N_+ but $X \equiv (\bar{n})^k(1+\bar{n})^{-k}$ and for $N_A = 1$, $N_+ = X/(1+X) = \bar{n}^k/[(1+\bar{n})^k+\bar{n}^k] \sim \frac{1}{2}$ for large $\bar{n}$ for any k including k = 1. However for $N_A \to \infty$ there is a giant k-photon quantum oscillator and $N_+ \to X/(1-X) \sim k^{-1}\bar{n}$ for large $\bar{n}$.

So far the only experiment is the very beautiful 1-photon one by Haroche and colleagues [15]. He used the $30S_{\frac{1}{2}}$, $M_J = \pm\frac{1}{2} \to 30P_{\frac{1}{2}}$, $M_J = \pm\frac{1}{2}$; $\Delta M_J = \pm 1$ transitions with a 'degeneracy' of two. This would mean that the asymptote of N_+ against N_A as $N_A \to \infty$ is $2\bar{n}$ not $\bar{n}$. There is a complication because an atomic beam of Na atoms in $30S_{\frac{1}{2}}$ enters the cavity and the beam subsequently exits from the cavity with the atoms in either $30S_{\frac{1}{2}}$ or $30P_{\frac{1}{2}}$. The cavity transit time was 2.5μ sec. So the question is whether thermal equilibrium can be reached in < 2.5μ sec. Now the radiation rate is collective and gains the factor $\frac{1}{4}N_A$ at peak. Thus if N_A is large enough the rate will be enhanced to more than $(2.5\mu\ \text{sec})^{-1}$. In practice $2g^2\kappa^{-1} = \eta_{cav}\Gamma_0$, the rate constant in low-Q cavities, satisfies this condition for $N_A \gtrsim 2{,}000$. Results observed by Haroche and colleagues [15] for T=300°K and 900°K are displayed here as the Fig.1

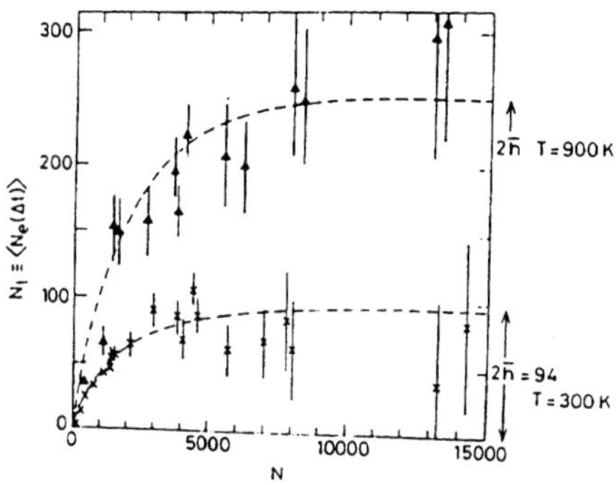

Fig.1 Plot of N_+ against N_A (called N). The asymptotes $2\bar{n}$ for the giant oscillator are clearly evident from the dashed curves.

The dashed curves in Fig.1 are computed *semiclassically* (in [15]) and take into account failure to reach equilibrium for the smaller N_A. Thus the observed curves only support quantum mechanics through the asymptotes $2\bar{n}$. We have ourselves compared the quantum theory and the semiclassical theory numerically taking into account failure to reach equilibrium for small $N_A <$ 2,000. Results are given in the Fig.2 taken from [9]

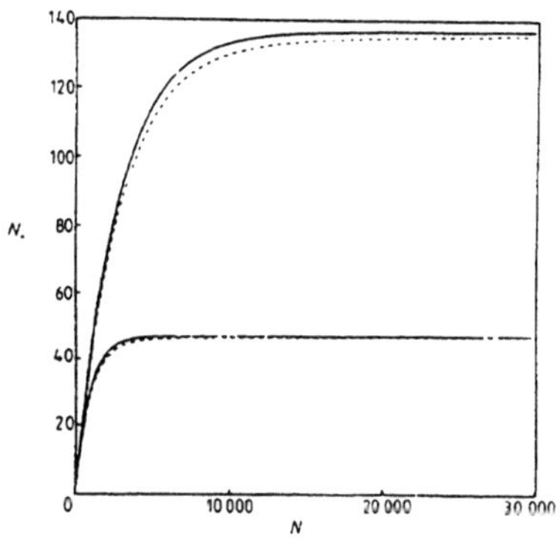

Fig.2 Comparison of semiclassical theory (full curves) and quantum theory (broken curves) for $N_+(t)$ at t = 2.5μ sec for $\bar{n}$ = 47 (T = 300°K) and $\bar{n}$ = 137 (T = 900°K).

The biggest difference in Fig.2 is $\Delta N_+ \sim 4$ for $\bar{n}$ = 137, $N_A \sim 5000$ and failure to reach equilibrium within a time t = 2.5μ sec for the smaller N_A wholly masks the significant difference between the quantum theory and the semiclassical theory ($N_+ = \bar{n}$ and so is never less than $\bar{n}$ in the steady state in the semiclassical theory). Both theories as observed agree well with experiment. But by increasing $g^2\kappa^{-1}$ by (say) a factor of 100 with $N_A < 100$ (conditions achieved in [17]) and with varied transit times < 6μ sec one should easily see an approach to equilibrium below $2\bar{n}$. The low-Q condition is $N_A\, g^2\kappa^{-1} \ll \kappa$ [9] and an observation of the asymptote (<33 for $\bar{n}$ = 47) would be a nice confirmation of the quantum theory. The experimental problem is to increase $g^2\kappa^{-1}$ and keep $\bar{n}$ large.

We turn to the *dynamics* of the master equation (21). One can already *see* the Dicke factor N_A in (21) in so far as H_o goes as N_A but S^+S^- etc in the decay terms go as N_A^2. Notice that, at T = 0, $\bar{n} = 0$ and $g^2\kappa^{-1} = \frac{1}{2}\eta_{cav}\Gamma_o$. In free space $g^2\kappa^{-1} \to \frac{1}{2}\Gamma_o$ and (21) is in the form appropriate for Dicke's [12] theory. We can expect the single point assumption $\mathbf{x}_i = \mathbf{x}_o$ for all the atoms to be valid as before for microwave frequencies. Thus, for cavity sizes ~ mm., N_A Rydberg atoms in a low-Q cavity could be ideal for the study of super-radiance. At higher frequencies (smaller λ) the theory of super-radiance is bedevilled by propagation effects between the atoms [58].

We can put (21) in the 'Dicke representation' in which $P_m(t) \equiv \langle \frac{1}{2}N_A, m \mid \rho_A(t) \mid \frac{1}{2}N_A, m\rangle$ is the probability of finding $N_- = (\frac{1}{2}N_A - m)$ atoms in their ground states. The relevant experiment was performed by Haroche and colleagues using the $29S_{\frac{1}{2}}$ to $28P_{\frac{1}{2}}$ transitions of Na. They interpreted $P_m(t)$ thus defined as the probability of $\frac{1}{2}N_A - m$ photons emitted at time t, not quite the same thing [9]. In this Dicke representation [9]

$$\frac{dP_m}{d\bar{t}} = (1+\bar{n})\nu_{m+1} P_{m+1}(\bar{t}) + \bar{n}\nu_m P_{m-1}(\bar{t}) - [(1+\bar{n})\nu_n + \bar{n}\nu_{n+1}]\, P_m(\bar{t}) \qquad (23)$$

with $\bar{t} \equiv \eta_{cav}\Gamma_o t$, $\nu_m = (\frac{1}{2}N_A + m)(\frac{1}{2}N_A - m + 1)$ and $|m| \le \frac{1}{2}N_A$. Calculation is difficult for large N_A although we confirmed [9] the actual observational results for which N_A = 3200. The Fig.3 shows other numerical results for fewer N_A, namely N_A = 100. In it one can see evolution from $P_m(0) = \delta_{m,\frac{1}{2}N_A}$ to $P_m(\bar{t})$, a monotonically decreasing Bose-Einstein distribution, as $\bar{t} \to \infty$. The statistics is *Poissonian* at intermediate times. Indeed this can be seen explicitly through the Fig.4 which plots the coherence functions $g^{(n)}(0) = (\mathrm{Tr}\, \rho_A(\bar{t})(S^+)^n(S^-)^n)/(I(\bar{t}))^n$ where $I(\bar{t}) \equiv \mathrm{Tr}\, \rho_A(\bar{t})(S^+S^-)$ for n = 2 and

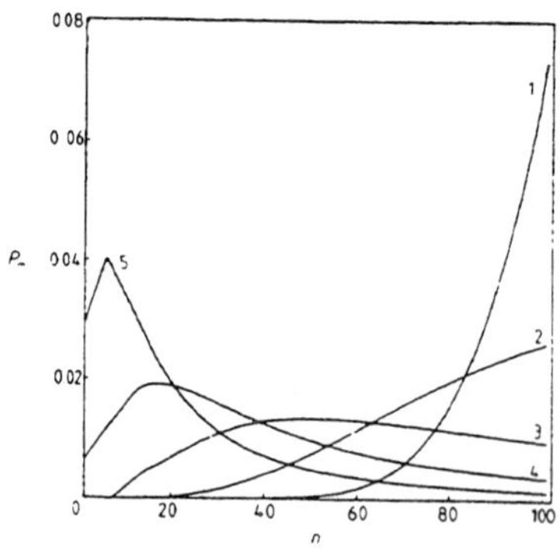

Fig.3 Probability distribution of atoms over the Dicke states, $P_m(\bar{t})$, at $\bar{t}$ = 0.02,0.03,...,0.06 (labelled 1-5) for N_A = 100 initially excited atoms in a black-body cavity field with $\bar{n}$ = 1. P_m is plotted against n ($\equiv m + \frac{1}{2}N_A$) - all atoms excited is n = 100; n = 0 only is all atoms in the ground state.

$n = 3$, as functions of $\bar{t}$. (The notation $g^{(n)}(0)$, which however depends on time $\bar{t}$, refers to delays τ in the coherence functions here set to zero). Notice that $g^{(2)}(0)$ is close to unity at the super-radiant peak, and $g^{(3)}(0)$ is likewise. These results show considerable measures of *coherence* and a Poisson photon statistics at $\bar{t} \approx 0.04$ (although no off-diagonal elements of ρ_A ever develop). A more extended analysis is already given in [10].

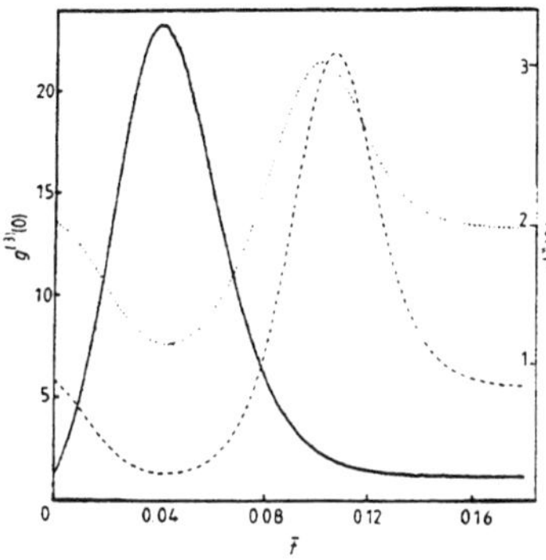

Fig.4 Degree of second-order coherence $g^{(2)}(0)$ (scale on right, asymptote 2) and third order coherence $g_3(0)$ (scale on left, asymptote 6). The hard line is the transient intensity $I(\bar{t})$ with the super-radiant peak at $\bar{t}$ = 0.04.

We complete this §4 by extending (21) to the case of N_A Rydberg atoms in a broad band squeezed vacuum. For the analytical details for deriving (23a), which has been carried out by operator reaction field theory [60] for a broad band squeezed vaccum (without any cavity) the reader is referred to [61]. The result for ρ_A is

$$\begin{aligned}\frac{d\rho_A}{dt} = & -i(\omega_o'+\Delta_N)[S^z,\rho_A] \\ & -\tfrac{1}{2}\Gamma_o(N+1)\,[S^+S^-\rho_A + \rho_A S^+S^- - 2\,S^-\rho_A S^+] \\ & -\tfrac{1}{2}\Gamma_o N[S^-S^+\rho_A + \rho_A S^-S^+ - 2\,S^+\rho_A S^-] \\ & +(\tfrac{1}{2}\Gamma_o M^*-i\Delta_{M^*}^{(-)})e^{2i\omega_p t}\,[S^-S^-\rho_A + \rho_A S^-S^- - 2\,S^-\rho_A S^-] \\ & +(\tfrac{1}{2}\Gamma_o M+i\Delta_M^{(-)})e^{-2i\omega_p t}\,[S^+S^+\rho_A + \rho_A S^+S^+ - 2\,S^+\rho_A S^+]. \end{aligned} \qquad (23a)$$

The important numbers are $N \equiv \langle a^\dagger a\rangle$ for each mode of the squeezed vacuum and $M \equiv \langle a^\dagger a^\dagger\rangle$. One proves $|M|^2 \le N(N+1)$ with equality for minimum uncertainty squeezed vacuum modes. The frequency $2\omega_p$ is that of an optical parametric oscillator supposed to generate the squeezed vacuum. The ω_o' is the vacuum shifted value of ω_o : the Δ_N, $\Delta_M^{(-)}$ are 'renormalisations' (ω_o' gains the radiative shift Δ_N in the squeezed vacuum (which is an excited state of an ordinary vacuum)). Notice that when $M = 0$ (23a) reduces to the Dicke form (21) since $N = \langle a^\dagger a\rangle$ at the vacuum shifted frequency ω_o'.

The Dicke model described in §3 is concerned with the su(2) Lie algebra equation (1) and its N_A+1 dimensional representations. This algebra becomes the Heisenberg-Weyl algebra as $N_A \to \infty$ as explained and the model itself becomes the giant quantum oscillator. One can argue from (23a) that in the squeezed vacuum the mean occupation number of this oscillator, namely N_+, should still be $N_+ = \langle a^\dagger a \rangle = N$. Thus the curve of N_+ against N_A should reach the same asymptote as for black-body radiation with mean number $\bar{n} = \langle a^\dagger a \rangle$. In the Fig.4S we plot N_+ against N_A from $N_A = 2$ to $N_A \to \infty$ through even values. Only even N_A values are taken and this is because we can easily solve (23a) (with the renormalisations omitted) precisely for even values of N_A, for which the equilibrium state is a *pure state*. More details are in [46]. The Fig.4S (S for squeezed!) uses $N = \bar{n} = 47$ and the solid-line (for the squeezed vacuum) compares with the dashed line (for black-body radiation). Both asymptotes as $N_A \to \infty$ are the same at $N = \bar{n} = 47$, but the curves plainly differ substantially for smaller N_A. Unfortunately no experiment has been performed yet to identify the solid-line in the case of the squeezed vacuum. Note that this solid line is no longer monotonically increasing. Apparently [62] this is a feature of the squeezed vacuum - a feature which needs analytical investigation.

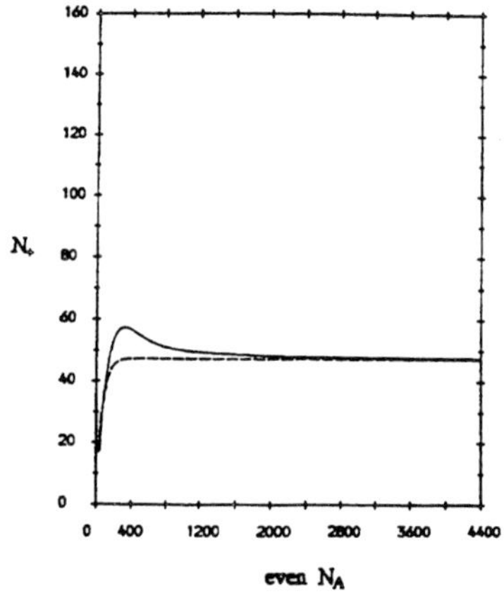

Fig.4S Plot of N_+ in the squeezed vacuum against N_A for $N = 47$ and $|M|^2 = N(N+1)$ calculated from the master equation (23a) in the steady state: renormalisations $\Delta M^{(-)}$ are set to zero and only the pure state solutions for N_A even are considered. For comparison (dashed line) N_+ against N_A for $\bar{n} = 47$ is also shown. The steady state quantum curve compares with the observations (with asymptote $2\bar{n} = 94$) shown in Fig.1.

5. FLUORESCENCE SPECTRA

We have calculated fluorescence spectra for a *single* Rydberg atom in cavities with Q in the range low to high, $5 \times 10^3 \lesssim Q \leq 10^8$, at temperatures in the range $0 < T < 6^\circ K$. There is a fundamental difficulty with such calculations since a continuous spectrum (as calculated) presupposes a continuous run of modes close to the single cavity mode of the model Hamiltonian: this Hamiltonian is the J-C Hamiltonian (10). It might appear that it is sufficient to measure the fluorescence spectrum at right angles to the cavity outside the cavity - so that a continuous run of modes in the vacuum could be used. However, for Q's $\gtrsim 10^6$ or even somewhat smaller, such a physical hole in the cavity, giving access to the vacuum modes outside, dramatically reduces the Q.

This situation was apparently fundamental and moreover the fluorescent photons have very small energies, $\sim 10^{-23}$J and only few such photons would be collected, too weak to be measured.

In a series of still ongoing experiments Haroche and colleagues used first ^{85}Rb atoms and the $39S_{\frac{1}{2}} \leftrightarrow 39P_{3/2}$ transitions at 68.5 GHz, in a cavity with $Q = 10^8$ [21]. They introduced a c.w. microwave probe which could be tuned against the microwave transition frequences and monitored the states of the outcoming atoms. Atoms in the $39S_{\frac{1}{2}}$ state enter the cavity and emerge from the cavity in either $39S_{\frac{1}{2}}$ or $39P_{\frac{1}{2}}$, and the fraction of atoms emerging in each of these states is monitored by ionisation detectors. In this way the atoms act as their own very sensitive detectors! The c.w. probe, tuned across the spectral profile, in effect provides the continuous run of cavity modes not available in the cavity. We have proved [20,23] that the spectrum, which has become the probability of finding an outcoming atom in the upper state, is exactly the spectrum we have calculated in Refs. [18,20].

In these first experiments [21] *vacuum field Rabi splitting* (see below) was observed for $N_A = 30$ and $N_A = 15$ but became obscured for smaller N_A. In further experiments on the $39S_{\frac{1}{2}}$, $F = 3 \rightarrow 39P_{\frac{1}{2}}$ transition in ^{87}Rb, N_A was reduced to $N_A = 3$, while the expected dependence of the vacuum field Rabi splitting on $\sqrt{N_A}$ was also observed [22]. We calculated the spectrum for $N_A = 1$ and the $39S_{\frac{1}{2}} \rightarrow 39P_{\frac{1}{2}}$ transitions of ^{85}Rb in [23] and both Rabi splitting and first side bands are clearly seen. We now use the $N_A = 1$ J-C model to indicate the origin of vacuum field Rabi splitting.

The master equation is the master equation (20) for arbitrary Q, large or small. The Hamiltonian is the J-C Hamiltonian (10). To solve this model for $Q = \infty$ one works with a Hilbert sub-space and a fixed value of $N = a^\dagger a + S^z + \frac{1}{2}$. This subspace is spanned by $|n, e\rangle \equiv |e\rangle|n\rangle$ and $|n+1, g\rangle \equiv |g\rangle|n+1\rangle$ where the $|n\rangle$ are photon number states and N has the common eigenvalue n+1; the number n=0,1,2,... . H has a 2x2 matrix form and its eigenstates prove to be

$$|n,\pm\rangle = \begin{Bmatrix} \cos\phi_n \\ -\sin\phi_n \end{Bmatrix} |n+1, g\rangle + \begin{Bmatrix} \sin\phi_n \\ \cos\phi_n \end{Bmatrix} |n,e\rangle \tag{24}$$

where

$$\tan\phi_n = (2g\sqrt{n+1})(\Omega_{n,\Delta} - \Delta)^{-1}, \qquad \omega_{n,\pm} = \omega(n+\tfrac{1}{2}) \pm \tfrac{1}{2}\Omega_{n,\Delta},$$

$$\Omega^2_{n,\Delta} = 4g^2(n+1) + \Delta^2, \qquad \Delta = \omega_o - \omega. \tag{25}$$

The ground state is $|0,g\rangle$, eigenenergy $-\frac{1}{2}\hbar\omega_o$: there is no vacuum level shift of this ground state energy in the high Q cavity (and atoms vacuum shifted outside the cavity ought to be attracted into it!) At resonance $\Delta = \omega_o - \omega = 0$, $\Omega^2_{n,o} = 4g^2(n+1)$ and the excited eigenenergies are

$$\hbar\omega_{n,\pm} = \omega_o(n+\tfrac{1}{2}) \pm g\sqrt{n+1} \quad ; \; n = 0,1,2,\dots . \tag{26}$$

For n = 0, the first excited states, a pair of degenerate states at energies $+\frac{1}{2}\hbar\omega_o$ are *split* to $\hbar\omega_{o,\pm} = \frac{1}{2}\hbar\omega_o \pm \hbar g$. *This is the vacuum field Rabi splitting.*

The transitions to the ground state have the *two* split frequencies $\omega_o \pm g$, and the analogue of spontaneous emission in the cavity is double peaked. The Figs. 5 and 6 show our computations reported in [18] for this single atom vacuum field splitting. For $g\kappa^{-1} = 2$ (Fig.5) $Q \sim 2 \times 10^6$ and $Q = \frac{1}{2}\omega g^{-1}\, g\kappa^{-1} \sim 10^6\, g\kappa^{-1}$ for Figs 5,6.

Note that the peak widths are determined entirely by the finite temperatures T. The Figs. (7a,b,c) taken from [19] show results at still higher values of Q.

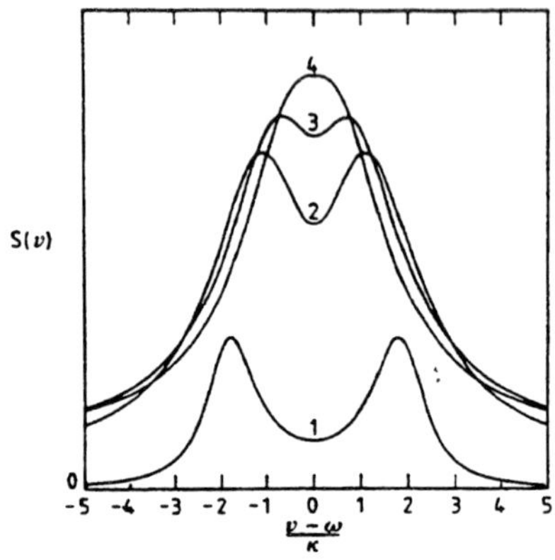

Fig.5 Spectrum of spontaneous emission $S(\nu)$ as a function of $(\nu-\omega)\kappa^{-1}$ for $g\kappa^{-1}$ = 2 and various values of $\bar{n}$= (1) 0.1, (2) 1.0, (3) 2.0, (4) 5.0.

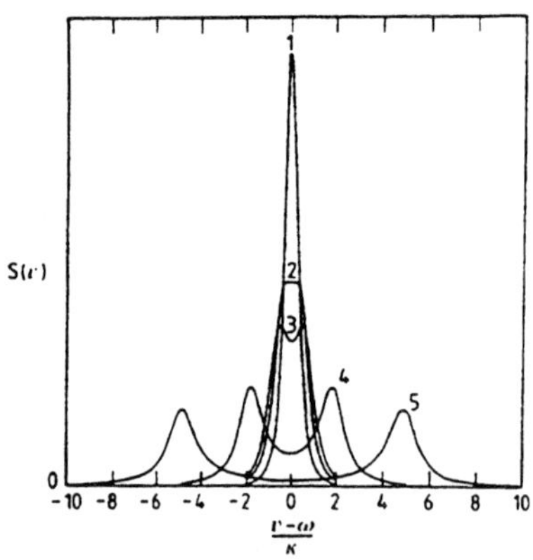

Fig.6 Spectrum of spontaneous emission $S(\nu)$ as a function of $(\nu-\omega)\kappa^{-1}$ for $\bar{n}$ = 0.1 but varying $g\kappa^{-1}$ = (1) 0.5, (2) 0.75, (3) 0.9, (4) 2.0, (5) 5.0.

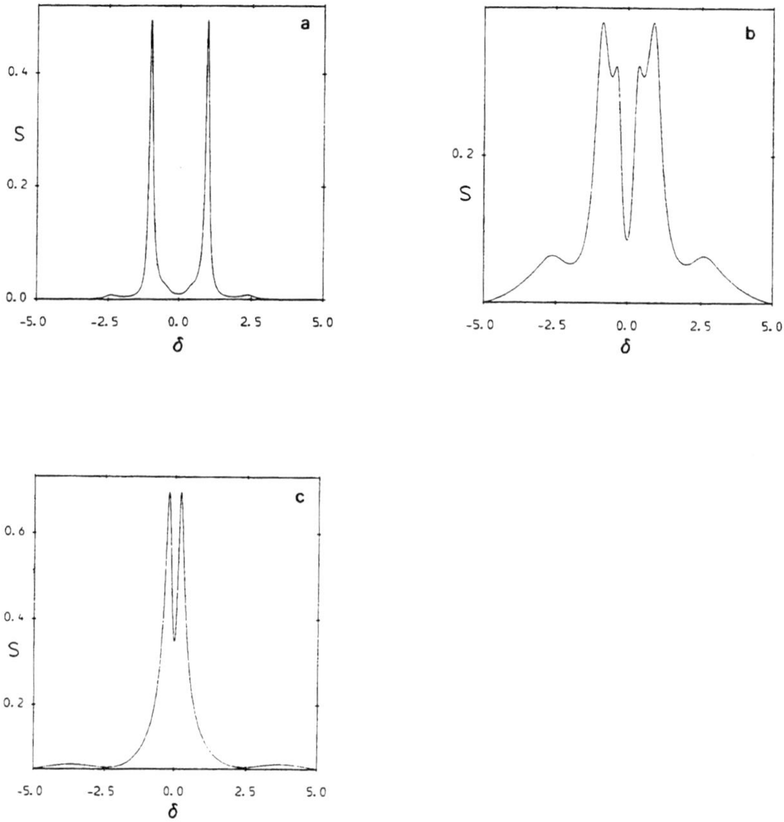

Fig.7 The steady state spectrum in a cavity with finite Q, κg^{-1} = 0.1 and for different cavity temperatures such that $\bar{n}$ = (a) 0.1, (b) 1.0, (c) 5.0.

The Fig.8 is a time averaged spectrum in a perfect cavity Q = ∞ [19] but a coherent field with varying $\bar{n} = \langle a^{\dagger} a\rangle$ is included. The Fig.8 here is case (b) $\bar{n}$ = 1.0 of Fig.1 of [19]. The essential point is that these high Q spectra Figs. 7 and 8 show both vacuum field Rabi splitting and contributions from sets of *four* allowed transitions which at resonance are

$$\nu-\omega = \pm\, g(\sqrt{n+1} + \sqrt{n}),\ \pm\, g(\sqrt{n+1} - \sqrt{n}) \tag{27}$$

where n = 1,2,...

The numerical methods for calculating these spectra Figs. 5-7 are described briefly in [10] and in [18]. They employ matrix continued fraction techniques as described there. Complete detail will be given in [20]. In the meantime similar calculations have been performed by Zoller and colleagues [63]. The problem of observing these spectra which delayed our publication [20] was not addressed in [63]. Here we shall describe these matrix continued fraction techniques a little further in the Lecture II following

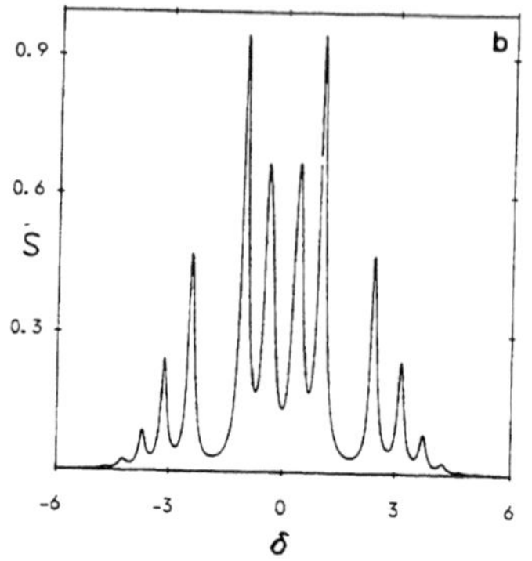

Fig.8 Time averaged spectrum in a perfect cavity $Q = \infty$ as a function of $\delta = (\nu-\omega)\, g^{-1}$: $\bar{n} = 1.0$ (see text). Notice how peaks at $\delta = \pm(\sqrt{n+1} - \sqrt{n})$ eventually fill in the central part about $\delta = 0$ but peaks at $\delta = \pm(\sqrt{n+1} + \sqrt{n})$ create side bands.

which is devoted to the theory of the micromaser. In the case of the micromaser we are no longer concerned with spectra (2-body correlations) but with the evolution of the inversion, the cavity photon number, and the photon statistics, essentially 1-body objects. For these the continued fraction calculations are a little simpler.

6. LECTURE II, FUNDAMENTAL THEORY OF THE MICROMASER - INTRODUCTION

This short introduction summarises the content of this Lecture II. The analysis is developed at some length in the sections, §7, §8 etc. which follow. New results are reported and the whole constitutes a self-contained *paper* reporting these various results.

The micromaser currently in operation at the Max Planck Institute, Garching, Munich uses a stream of velocity selected Rydberg ^{85}Rb atoms which enter a Niobium super-cooled cavity at 0.5°K (or less) in the 'upper' $63P_{3/2}$ state $|e\rangle$ and which can make resonant transitions at 21.456 GHz to the 'lower' $61D_{5/2}$ state $|g\rangle$. Photons dumped in the cavity build up a microwave cavity field $\mathcal{E}$. Relevant *times* are $T_c \sim 0.2$ sec, the cavity damping time $(T_c = (2\kappa)^{-1})$, $t_{int} \approx 35\mu$ sec, the time in which an atom interacts with the cavity field $\mathcal{E}$, and $T_p \approx 7{,}000\ \mu$ sec, the inverse of the 'repetition rate' i.e. the average rate at which atoms enter and leave the cavity. For these values there is, with probability ≈ 1, *at most* one atom in the cavity (for more than 99% of the time the cavity is empty of atoms) and during t_{int} a good model for the system, for which $Q = 3\text{x}\ 10^{10}$ is now achieved [28,29] is the J-C model of Lecture I, with Hamiltonian (10). However, since $Q < \infty$ and $T = 0.5$°K [28,29] these several times together mean that the system conforms to standard Scully-Lamb laser theory except only that the atomic decay is controlled by t_{int} and not by the spontaneous emission time Γ_0^{-1} (the inverse A-coefficient) of Scully-Lamb theory. This has the consequence that the *two* parameters now in the theory admit strictly quantum 'sub-Poissonian' statistics in the cavity field : these fields have variance σ with $0 < \sigma < 1$ whereas a single mode laser well above threshold is in a coherent state with Poisson statistics and $\sigma=1$. The quantum character of $\mathcal{E}$ can exhibit itself in the 'revivals' of the atomic inversion, the photon number and the intensity - intensity correlation (which shows the $\sigma < 1$). An extreme case is the Fock state, a pure n-state $|n\rangle$ of $\mathcal{E}$ which has $\sigma=0$. We investigate experimentally realisable conditions for the creation of such Fock states in the micromaser. Remarkably enough the very few microwave black-body photons (which have Planck function $\bar{n} \approx \frac{1}{6}$ at $0.5^{\circ}K$) play a crucial role in a theory which now becomes very *different* from Scully-Lamb theory.

There is a second theme of the material which follows: this extends the theory of the micromaser from one expressed in terms of ordinary cannonically quantised bosons to a theory given in terms of the recently discovered "q-bosons". It is an open question whether q-bosons and q-quantisation [23,54,55,64] have any physical significance as such. However, the q-boson theory from a pragmatic point of view simply introduces additional non-linearities into a theory described wholly in terms of ordinary bosons [65]. In these terms the question is - Is the J-C Hamiltonian (10) good enough for the micromaser where opportunity to check on fundamental details and assumptions of a quantum theory is now open to experimental investigation? As noted already in §3 the J-C Hamiltonian (10) is taken in rotating wave approximation for example. The q-boson theory adds an infinite set of additional nonlinearities to the J-C Hamiltonian. These are (mathematically) very natural. It might just be that these additional nonlinearities represent real physics for the model and these consequences can be observed in a micromaser experiment. The observable revivals are changed and the J-C eigenspectrum is changed. These changes are calculated in the material following.

7. Q-BOSON AND BOSON CAVITY Q.E.D. : FUNDAMENTAL THEORIES OF THE MICROMASER

The material which now follows reports on two aspects of the action of the one ^{85}Rb atom-one mode micromaser currently in operation [26-29] at the Max Planck Institute, Garching by Munich, Germany. There have been three other reports [23,30,31] but the present material amplifies the two reports [23,30] already published and puts the third report [31] which is still to appear, into context within the more general framework of cavity q.e.d. as described in the Lecture I, §§1-5.

The first, and major, report which is given in the §§10-12 of the following, concerns a new theoretical investigation, numerically based, of the action of the micromaser derived from conventional quantum mechanics. As noted in §6, in the recent experiments [28,29] there is at most one ^{85}Rb atom in the maser cavity during one atomic transit time $t_{int} \sim 35\mu$ sec. But, even at *precise* atomic velocity selection, atoms still enter stochastically with an exponential distribution parameterised by a mean repetition rate T_p^{-1} (say): $T_p \sim 7{,}000\mu$ sec. in the experiments — so the cavity is empty of atoms with probability ≈ 1 for some 99% of the time. The main conclusion of the §§10-12 then supports the conclusion of [23,30,31] namely that if the empty-of-atoms time $t_e \equiv T_p - t_{int}$ is reduced towards the values of t_{int} itself, the system acts more like a maser with infinite cavity Q_o even at the now realised Q and temperature T [28,29] $Q = 3 \times 10^{10}$, $T = 0.5^{o}K$. Thus, in distinction to previous discussions, eg. that by Meystre *et al.* [66], we suggest that it may be possible to create Fock states in the micromaser at this Q and T. However, as T_p is reduced, the theoretical problem becomes complicated by the possibility of more than one atom being in the cavity at the same time - the effects of which on Fock state evolution we are still investigating [23].

The conclusion that Fock states will emerge within single atom theories is based on the numerical results already reported in [23]. These results are amplified in the present paper by further studies *in the current regime of micromaser operation* which is $T_p \approx t_e >> t_{int}$ [28,29]. Even in this regime the conclusion is that, as T_p moves towards t_{int}, even though $T_p >> t_{int}$ always, the variance for the maser cavity field defined here by ([28,29] use Q_a for our symbol σ)

$$\sigma \equiv [\langle(a^\dagger a)^2\rangle - (\langle a^\dagger a\rangle)^2]/\langle a^\dagger a\rangle \tag{28}$$

systematically falls. The natural unit of time is g^{-1} where, as in the material of Lecture I, g is the real valued coupling constant between the atom and the field: g = 44kHz in the experiments [28,29] but we shall not need this number as such in this paper. Thus for gt_{int} = 1.047 and gT_p = 720 with the Q = 3×10^{10} and the mean black-body photon number $\bar{n}_{th}$ = 0.15 at the temperature T = 0.5^oK of the experiments [28,29] we find σ = 0.281 becoming σ = 0.204 for gT_p = 216. The gt_{int} is chosen to *minimise* σ (more or less - see the Fig 15 in §12). The values of σ are lower than those observed [28,29] (the smallest σ observed is $\sigma \approx 0.3$ for gT_p = 308). The fields are sub-Poissonian ($\sigma < 1$) but still differ substantially from the σ=0 required by a cavity field in a pure Fock state (n-state, $|n\rangle$) of n photons. The fall of σ with T_p in this large T_p regime is nevertheless consistent with the numerical evidence of [23] that Fock states, σ=0, will emerge for the much smaller T_p where $gT_p \gtrsim gt_{int}$. The practical message for Fock state creation seems to be reduce T_p *to some extent* - so that the probability of more than one atom in the cavity at the same time still remains small.

The second analysis of this paper again takes up the recent idea [23,54,67] that the micromaser may be used as an example to distinguish between a quantum mechanical system quantised by normal, canonical, quantisation ('conventional quantum mechanics') and one in which quantisation is imposed by the recently discovered 'quantum groups' [33,35,54,55,64,68]. The Jaynes-Cummings model eqn.(10) of one 2-level atom and one quantised cavity mode is fundamental to the theory of the micromaser [69-71]. We show how the model can be given a quantum group quantisation by using the q-deformed bosons ('q-bosons' [33,54,55,64,72-74]) recently described. We show that q-dependent features, depending on the value of the real parameter q of the q-bosons, would appear in the quantum revivals of the micromaser as already suggested [67], and in the second and higher Rabi split eigenenergies. The vacuum field Rabi splitting, defined in §5 and currently being observed [22,23], is not changed. We give the relevant master equation for $N_A \geq 1$ q-deformed 2-level atoms coupled to a single cavity mode coupled to a heat bath : the cavity mode has q-boson excitations to and from the heat bath. We show that an experiment like that of [15] *does not distinguish* between q-quantisation and ordinary canonical quantisation. However, super-radiance experiments like those of [16] could do so. New experiments are needed to determine whether there *is* any physics in the q-quantisation described by the q-bosons.

The material which follows is presented in the following way. In §8 we introduce the cavity q.e.d. of microwave q-bosons via the theory of quantum groups and in §9 we introduce and solve the q-deformed J-C model. In §10 we sketch our methods for computing the effects of black-body radiation on single atoms occupying (microwave) cavities with Q in the range $10^3 \lesssim Q \leq 3\times10^{10}$. In §11 we make a fundamental analysis of the canonically quantised micromaser, introducing the 'trapping states' [66,69-71] and also drawing a comparison with the ordinary laser much as this was done in [70]. The §12 reports sets of numerical results which support the conclusions of [23], all but two of these evaluated in the $T_p \gg t_{int}$ regime. A conclusion on all of the material §§8-12 is given in the final §13. The §14 offers further reading.

The reader who has worked his way through the Lecture I, the §§1-5, and who is interested only in our recent results for the ^{85}Rb atom micromaser may now read the §§ 10,11 and 12, with some back reference. The work on q-quantisation and the q-bosons in §§ 8 and 9 is self-contained and the reader who is interested in the algebras of quantum optics may like to compare the work developed here with that presented in the §3.

8. THE CAVITY QUANTUM ELECTRODYNAMICS OF Q-BOSONS

The Jaynes-Cummings model [32] of one 2-level atom coupled to a single cavity mode and cavity $Q = \infty$ and which has the Hamiltonian (10) is as explained in §3 an example of a "completely integrable" quantum system [33,35]. A quantum integrable system with N quantum degrees of freedom has N independent operator constants $\hat{I}_K$ which commute : $[\hat{I}_K,\hat{I}_\ell] = 0$, and the $\hat{I}_K$ commute with the Hamiltonian $\hat{H}$.

The Jaynes-Cummings model has, as in §3 and we now use a 'hat' to indicate an operator,

$$\hat{H} = \omega_0 S^z + \omega a^\dagger a + g(a^\dagger S^- + S^+ a), \quad \hat{N} = S^z + a^\dagger a + \tfrac{1}{2}\ , \tag{29}$$

and $[\hat{H},\hat{N}] = 0$: the single cavity mode (frequency ω) with bose operators $a, a^\dagger$ and $[a,a^\dagger] = 1$ provides one quantum degree of freedom : the single atom (frequency ω_0), with the spin-$\frac{1}{2}$ operators S^z, $S^\pm$ satisfying the su(2) Lie algebra eqn.(1) namely

$$[S^\pm,S^z] = \mp S^z,\ [S^+,S^-] = 2S^z, \tag{30}$$

but with total spin $S = \frac{1}{2}$, provides a second quantum degree of freedom. As is shown in §§3 with 5 the system can indeed be solved for its energy eigenvalues and eigenstates : the problem of diagonalising an infinite dimensional matrix for $\hat{H}$ is reduced to diagonalising a sequence of 2x2 matrices characterized by the eigenvalues n of $\hat{N}$: $\hat{N}|n,\pm\rangle = n|n,\pm\rangle$, $n = 0,1,2,\ldots$; $\hat{H}|n,\pm\rangle = \{(n-\frac{1}{2})\omega \pm \frac{1}{2}[(\omega-\omega_0)^2 + 4g^2n]^{\frac{1}{2}}\}|n,\pm\rangle$. Note the change $n+1 \to n$ compared with eqns. (24), (25).

The integrable dynamical systems are associated with the Lie algebras $\tilde{g}$ [33,35,54,55,64,68] (we use $\tilde{g}$ here to distinguish from coupling constant g). In the case of classical integrable dynamical systems in one space and one time dimensions these algebras are extended to the 'loop' algebras $\tilde{g}\otimes[\lambda,\lambda^{-1}]$, $\lambda \in \mathbb{C}$ [33,35] (an element of these algebras is an element of $\tilde{g}$ multiplied by a Laurent polynomial in λ). The classical system can be *quantised* by 'deforming' $\tilde{g}$: the route is that $\tilde{g}$ is *deformed* to $\tilde{g}_q$ by a parameter $q = e^t$ and $t \sim \hbar$ e.g. $t = \mu\hbar$, $\mu \in \mathbb{R}$: $\tilde{g}_q$ is a Hopf algebra. Dual to $\tilde{g}_q\otimes[\lambda,\lambda^{-1}]$ are the elements of the quantum monodromy matrix $\hat{T}(\lambda)$ [35] which form a 'quantum group' for particular λ and the idea of 'quantum groups' can be extended to arbitrary $\lambda \in \mathbb{C}$: in this extended sense the quantum group formed by the elements of $\hat{T}(\lambda)$ is also a Hopf algebra and it is a *non-commutative* Hopf algebra. (Hopf algebras have a 'comultiplication' as well as a multiplication and these quantum groups are both non-commutative and non-cocommutative [35,54,55,64]). The matrices $\hat{T}(\lambda)$ satisfy the matrix equations

$$R(\lambda,\mu)\hat{T}(\lambda) \otimes \hat{T}(\mu) = \hat{T}(\mu) \otimes \hat{T}(\lambda)R(\lambda,\mu); \tag{31}$$

$\otimes$ is direct ie. Kronecker product, and if $\hat{T}$ is 2x2, $R(\lambda,\mu)$ is a 4x4 matrix characteristic of the dynamical system. A quantum integrable system in one continuous space variable and one continuous time variable has an infinity of degrees of freedom. The matrix equation (31) is evidently a set of commutation relations for the elements of $\hat{T}(\lambda)$ which form the quantum group. These are commutation relations *induced* by a quantum group. The system is then quantum integrable because the matrix trace of (31) yields $[\hat{\Delta}(\lambda),\hat{\Delta}(\mu)] = 0$, and so a large number (a continuous infinity) of commuting operator constants $\hat{\Delta}(\lambda)$: $\hat{\Delta}(\lambda)\equiv\mathrm{Tr}\hat{T}(\lambda)$, and $\ell n\ \hat{\Delta}(\lambda)$ is a generator of quantum integrable Hamiltonians $\hat{H}$ describing integrable models, typically nonlinear integrable models, in one space and one time dimension [33-35,54,55,64].

A deformed Lie algebra $\tilde{g}_q$ is exemplified [33,35,54,55,64,68] by the deformation $su_q(2)$ of the Lie algebra su(2) : $su_q(2)$ is

$$[S^{\pm},S^{z}] = \mp S^{\pm},\ [S^{+},S^{-}] = [2S^{z}]. \tag{32}$$

The notation in the second expression is $[x] = (q^{x}-q^{-x})/(q-q^{-1})$, so with $q = e^{t}$

$$[2S^{z}] = (e^{2S^{z}t}-e^{-2S^{z}t})/(e^{t}-e^{-t}) = (\sinh 2S^{z}t)/(\sinh t). \tag{33}$$

When $q = 1$ (t=0) $su_q(2)$ is su(2). As noted the deformed loop algebra $\tilde{g}_q\otimes[\lambda,\lambda^{-1}]$ has the quantum group with the elements of $\hat{T}(\lambda)$ as its dual [35]. The $\hat{T}(\lambda)$ (the elements of $\hat{T}(\lambda)$) satisfy the commutation relations (31) and quantise an integrable quantum nonlinear (in general) field theory. However, this quantum group quantisation proves, for field theories, to be identical with the familiar canonical quantisation of these field theories. This seems to be true for all quantum integrable field theories in one space and one time dimension, and we can show it is true for the quantised Maxwell equations of normal q.e.d.

For discrete systems (integrable lattices [54,55,64,65]) and dynamical systems with a finite number of degrees of freedom this is not necessarily so. The Jaynes-Cummings model has two quantum degrees of freedom only. In [67] it is shown that a form of the Jaynes-Cummings model quantised in terms of q-bosons (q-quantised) and depending on q will display quantum revivals which depend on the value ($\neq 1$) of the q-parameter. The suggestion was implicitly made that observations of the revivals in the atomic inversion in the micromaser will indicate whether this system is q-quantised and there are q-bosons in the maser cavity or it is canonically quantised with ordinary bosons in the cavity. The analysis of [67] was marred by an assumption about the model that the coupling constant g (eqn.(29)) is field intensity dependent. Obviously it is important to see if there is any *physics* in the q-deformation theory. Therefore in [23] we worked out the q-boson theory of the q-deformed Jaynes-Cummings model. We now repeat this calculation here, but the new results are the Figs. 9 and 10 which show the effects of q-deformation on the revivals for $Q = \infty$.

The q-quantised Jaynes-Cummings model consists of one q-deformed 2-level atom and one q-deformed mode of the electromagnetic field, a q-boson [54,55,64,65,72-74]. Although the q-bosons [72-74] are new, the ideas were already available some 10 years ago [75-77]. The q-deformed 2-level atom has the $su_q(2)$ algebra (32) with the notation (33). But because it is *one* $S = \frac{1}{2}$ system $(2S^{z})^{2} = 1$ and $[2S^{z}]$, equation (33), is $2S^{z}$. The single q-deformed 2-level atom is therefore the ordinary 2-level atom. This is no longer true of $N_A>1$ 2-level atoms displaying collective action in black-body radiation as observed in [15] : for $N_A = 2$, $[S^{+},S^{-}] = 2(q+q^{-1})S^{z}$ for example, and so, in a comparison with §§3 and 4, we examine the case of giant oscillator formation, and its q-dependence, in black-body radiation in a low-Q cavity in more detail in this section (equations (42)-(47)).

It is not true that two or more q-deformed 2-level atoms in high Q cavities are the same as two or more ordinary 2-level atoms either. Thus if there were two or more atoms in the micromaser cavity at any one time this feature would need to be taken into account within the q-deformation theory.

It remains to deform the Heisenberg algebra $[a,a^\dagger] = 1$ of the cavity mode and obtain the q-bosons [54,55,64,65,72-74]. The algebra of q-bosons has the elements a^+, a^- and $\hat{n}$, and

$$[a^\pm, \hat{n}] = \mp a^\pm, \tag{34}$$

$$a^-a^+ - qa^+a^- = q^{-\hat{n}}. \tag{35}$$

The vacuum state is $|0\rangle$: $a^-|0\rangle = 0$. Then $\hat{n}$ has eigenvalues n and eigenstates $|n\rangle$, n = 0,1,... :

$$\hat{n}|n\rangle = n|n\rangle, \tag{36}$$

$$|n\rangle = ([n]!)^{-\frac{1}{2}} (a^+)^n|0\rangle \tag{37}$$

with $[n]! \equiv [1].[2]. \dots . [n]$ in which

$$[n] \equiv (q^n - q^{-n})/(q-q^{-1}). \tag{38}$$

On the Fock space spanned by the $|n\rangle$, when $q \to q^{-1}$ (35) is

$$a^-a^+ - q^{-1}a^+a^- = q^{\hat{n}}, \tag{39}$$

and from (39) and (35)

$$a^+a^- = [\hat{n}] \equiv (q^{\hat{n}} - q^{-\hat{n}})/(q-q^{-1}). \tag{40}$$

(There is a deeper argument for (39) and (40) which does not appeal to $q \to q^{-1}$ on the Fock space.) One also has $a^+|n\rangle = [n+1]^{\frac{1}{2}}|n+1\rangle$ and $a^-|n\rangle = [n]^{\frac{1}{2}}|n-1\rangle$, and $[n]$ replaces n in the usual (canonical) bose theory. There is now the Primakov-Holstein realisation of $su_q(2)$ which is [54,55,64], and compare (7) for ordinary bosons,

$$S^+ = a^+\sqrt{[2S-\hat{n}]}\ , \quad S^- = \sqrt{[2S-\hat{n}]}\, a^-\ , \quad S^z = \hat{n} - S \tag{41}$$

which satisfies (32) for any S.

In comparison with (21), we have derived the master equation

$$\frac{d\rho_a}{dt} = -i[\hat{H}_o,\rho_a] - \langle a^-a^+\rangle_{th}\, g^2\kappa^{-1}\{S^+S^-\rho_a - 2S^-\rho_aS^+ + \rho_aS^+S^-\}$$
$$- \langle a^+a^-\rangle_{th}\, g^2\kappa^{-1}\{S^-S^+\rho_a - 2S^+\rho_aS^- + \rho_aS^-S^+\} \tag{42}$$

for N_A q-deformed (q-quantised) 2-level atoms in a low-Q cavity : $S = \frac{1}{2}N_A$, and $\hat{H}_o = \hbar\omega_o S^z$. These N_A atoms are coupled to q-deformed black-body radiation through a single resonant cavity mode coupled to a heat bath at temperature $T = k_B^{-1}\beta^{-1}$. There are thus *two* q-parameters, $q = e^{\tilde{\gamma}}$(say) with $\tilde{\gamma} = \mu\hbar$ (real) for the atoms and $q = e^{\gamma}$, γ complex (in general) for the q-bosons of the black-body radiation. For $q=e^{i\gamma}$, $\gamma \in \mathbb{R}$, $0 < \gamma < \frac{1}{2}\pi$ we find $\langle a^+a^-\rangle_{th}$ plotted as a function of β has a maximum at $\beta = \beta_o = \omega^{-1}\ell n(1+\sqrt{2(1-\cos\gamma)}\ )$; $0 < \beta_o < \omega^{-1}\ell n(1+\sqrt{2})$. Also $\langle a^+a^-\rangle_{th}^{max} = [2(1-\cos\gamma)]^{-\frac{1}{2}}.\ [2+\sqrt{2(1-\cos\gamma)}]^{-1}$. For $q = e^{\gamma}$, $\gamma \in \mathbb{R}$, $\langle a^+a^-\rangle_{th} \longrightarrow \infty$ at $\beta_c = \gamma\omega^{-1}$ and is roughly the Planck function $(e^{\beta\hbar\omega}-1)^{-1}$ shifted to the right by amount β_c. However the mean number $\bar{n}_{th}$ of the Hamiltonian $\hat{H} = \hbar\omega\hat{n}$ for the q-bosons is

$$\bar{n}_{th} = Z^{-1}\sum_{n=0}^{\infty}\langle n|\hat{n}e^{-\beta\hbar\omega\hat{n}}\,|n\rangle = (e^{\beta\hbar\omega}-1)^{-1} \tag{43}$$

where Z is the partition function equal to $(1-e^{-\beta\omega})^{-1}$. Results (43) are the same as those for ordinary bosons once ω is chosen. On the other hand the exact results for $\langle a^+a^-\rangle_{th}$ and $\langle a^-a^+\rangle_{th}$ are

$$\langle a^+a^-\rangle_{th} = (e^{\beta\hbar\omega}-1)/(e^{\beta\hbar\omega}-q)(e^{\beta\hbar\omega}-q^{-1})\ , \tag{44}$$

$$\langle a^- a^+\rangle_{th} = \langle a^+ a^-\rangle_{th}\, e^{\beta\hbar\omega} . \tag{45}$$

The second relation is the usual Boltzmann relation, and $\bar{n}_{th}\, e^{\beta\hbar\omega} = \bar{n}_{th}+1$ from (16) the usual relation in this case. The results for $\langle a^+ a^-\rangle_{th}$ for $q = e^{\gamma}$ ($\gamma\in\mathbb{R}$) and $q = e^{i\gamma}$($\gamma\in\mathbb{R}$) derive from (44).

The Hamiltonian $\hat{H}_o$ in the master equation (42) is $\hbar\omega_o S^z = \hbar\omega_o(\hat{n} - \frac{1}{2} N_A)$ by (41), and ω_o is the atomic resonance frequency *as measured*. The energy levels are therefore equally spaced with spacing $\hbar\omega_o$. If these are on resonance with the q-dependent black-body radiation in a low-Q cavity the result (45) for the black-body radiation means that $\rho_a = Z^{-1} e^{-\beta\hat{H}_o}$ at equilibrium, where $Z = \mathrm{Tr}\, e^{-\beta\hat{H}_o}$. And this means that the mean number N_+ of atoms in their upper states at equilibrium is

$$N_+ = X\left[1-(N_A+1)\, X^{N_A} + N_A X^{N_A+1}\right]\left[1-X-X^{N_A+1}+X^{N_A+2}\right]^{-1} , \tag{46}$$

where $X = e^{-\beta\hbar\omega_o} = \bar{n}\,[1+\bar{n}]^{-1}$ and $\bar{n} = (e^{\beta\hbar\omega_o}-1)^{-1}$ is the (ordinary) Planck function deriving from (43) *for the atoms*. It is noteworthy that on resonance $\omega = \omega_o$, $X = \langle a^+ a^-\rangle_{th} / \langle a^- a^+\rangle_{th}$ for the q-deformed black-body radiation also - because of (45). Thus for $N_A = 1$, where $N_+ = X[1+X]^{-1}$, $N_+ = \bar{n}[1+2\bar{n}]^{-1} = \langle a^+ a^-\rangle_{th}\, [\langle a^- a^+\rangle_{th} + \langle a^+ a^-\rangle_{th}]^{-1}$, but when $N_A \to \infty$ where $N_+ = X[1-X]^{-1}$, $N_+ = \bar{n} = \langle a^+ a^-\rangle_{th}\, [\langle a^- a^+\rangle_{th} - \langle a^+ a^-\rangle_{th}]^{-1} = \langle[\hat{n}]\rangle_{th}\, [\langle a^- a^+\rangle_{th} - \langle a^+ a^-\rangle_{th}]^{-1}$.

The formula (46) is exactly that derived [33] in [6,9] and §4 for ordinary bosons. Thus the experiment of Ref.15, which observed the asymptote $\bar{n}$ does not distinguish between q-quantisation (the q-bosons) and canonical quantisation (the ordinary bosons). On the other hand, if the system *is* q-quantised the giant oscillator formed as $N_A \to \infty$ is a giant q-deformed oscillator (a giant q-boson). If $\bar{S}^+ \equiv [2S]^{-\frac{1}{2}}\, q^{\frac{1}{2}\hat{n}} S^-$ and $\bar{S}^- = [2S]^{-\frac{1}{2}} S^+ q^{\frac{1}{2}\hat{n}}$, the Primakov-Holstein transformation, equation (41) shows that, as $2S = N_A \to \infty$, $\bar{S}^+$ and $\bar{S}^-$ become a^- and a^+ respectively while

$$\bar{S}^+\, \bar{S}^- - q\, \bar{S}^-\, \bar{S}^+ = q^{-\hat{n}} , \tag{47}$$

the giant collective q-boson. For $S = \frac{1}{2}$ ($N_A=1$) there is the q-boson representation (41) identical with the ordinary boson representation of eqn. (8) and [33], and there is the q-fermion representation identical with the ordinary fermion representation (7). (The q-fermions have $aa^\dagger + qa^\dagger a = q^{\hat{n}}$, $aa^\dagger + q^{-1}a^\dagger a = q^{-\hat{n}}$ and $\hat{n}^2 = \hat{n}$ means $a^\dagger a = [\hat{n}] = \hat{n}$, $aa^\dagger = [1-\hat{n}] = 1-\hat{n}$.) Thus the experiments of [15] exemplify bose-fermi equivalence [33,35] or q-boson—q-fermion equivalence! One finds from (44) that

$$\langle\bar{S}^+\, \bar{S}^-\rangle = \langle[\hat{n}]\rangle = (e^{\beta\hbar\omega_o} - 1)/(e^{\beta\hbar\omega_o} - q)(e^{\beta\hbar\omega_o} - q^{-1}) \tag{48}$$

where q is the deformation parameter $q = e^{\tilde{\gamma}}$ in (47) which is for the atoms. It would be interesting to devise an experiment to check this result.

In contrast an experiment like that of [16] which observed super-radiant emission from a low-Q cavity will in principle show different results in the q-quantised and canonically quantised cases. From the master equation (42) in the Dicke basis $|S,m\rangle$, eigenstates of $\hat{H}_o$ in (42) with $-S \le$

$m \le +S$ and $S = \frac{1}{2}N_A$, the probability $P_m(t) = \langle S,m|\rho_a(t)|S,m\rangle$ of finding $N_+ = \frac{1}{2}N_A+m$ atoms excited at time t satisfies (compare eqn.(23))

$$\frac{d\, P_m(\bar{t})}{d\bar{t}} = -\{\nu_m\langle a^-a^+\rangle_{th} + \nu_{m+1}\langle a^+a^-\rangle_{th}\}\, P_m(t)$$

$$+\nu_m\langle a^+a^-\rangle_{th}\, P_{m-1}(\bar{t}) + \nu_{m+1}\, \langle a^-a^+\rangle_{th}\, P_{m+1}(\bar{t}) \tag{49a}$$

with $\bar{t} \equiv 2g^2\kappa^{-1}t$ which is also

$$\frac{dP_m(\bar{t})}{d\bar{t}} = \frac{(e^{\beta\hbar\omega}-1)}{(e^{\beta\hbar\omega}-q)(e^{\beta\hbar\omega}-q^{-1})}\left\{-P_m(\bar{t})[\nu_{m+1}+e^{\beta\hbar\omega}\nu_m]\right.$$

$$\left.+\nu_m P_{m-1}(\bar{t}) + e^{\beta\hbar\omega}\nu_{m+1}\, P_{m+1}(\bar{t})\right\}, \tag{49b}$$

and

$$\nu_m = [\tfrac{1}{2}N_A + m][\tfrac{1}{2}N_A - m + 1] \tag{49c}$$

and is q-deformed. Evidently the super-radiant emission now depends on $q = e^{\tilde{\gamma}}$ for the atoms. We hope to report numerical results for this emission, comparable with those computed for ordinary bosons in [9] and displayed in the Figs. 3 and 4, for example, elsewhere.

9. THE Q-DEFORMED Q-QUANTISED MICROMASER

In the case of very high Q ($Q=\infty$) the q-deformed Jaynes-Cummings model also leads to observationally different results from those for the ordinary Jaynes-Cummings model [32]. The q-deformed Jaynes-Cummings model studied here next will be that of [23] which has Hamiltonian $\hat{H}$ and number operator $\hat{N}$

$$\hat{H} = \omega_0\, S^z + \omega\hat{n} + g\,(S^+a^- + a^+S^-),\ \hat{N} = S^z + \hat{n} + \tfrac{1}{2}\,. \tag{50}$$

Since $[S^\pm,S^z] = \mp S^z$ and $[a^\pm,\hat{n}] = \mp a^\pm$ are unchanged from the ordinary Jaynes-Cummings model equation (10), $[\hat{H},\hat{N}] = 0$. Thus this q-deformed Jaynes-Cummings model is still a completely integrable quantum system. Note that there are obvious alternative choices: one can replace $\hat{n}$ by $\frac{1}{2}(a^+a^- + a^-a^+)$ in (50) either in $\hat{H}$ alone or in both of $\hat{H}$ and $\hat{N}$. In the second case $[\hat{H},\hat{N}] \neq 0$, while in the first the algebra becomes a little tedious compared with (50). Evidently (50) is the simplest q-deformed Jaynes-Cummings model we can consider.

From (50) for eigenvalue n of $\hat{N}$ and $n>0$, the eigenenergies of $\hat{H}$ are now the roots of

$$(\tfrac{1}{2}\omega_0+\omega(n-1)-\lambda)(-\tfrac{1}{2}\omega_0+\omega n-\lambda) - g^2[n] = 0. \tag{51}$$

These are the eigenenergies eqn. (25) of the undeformed Jaynes-Cummings model equation (29) [32], when $[n] \to n$ for $q \to 1$. When $n = 0$, $(-\frac{1}{2}\omega_0-\lambda) = 0$, so $\lambda = -\frac{1}{2}\omega_0$. This is the undeformed ground state energy of the Jaynes-Cummings model. When $n = 1$, $[n] = 1$ so

$$\lambda = \pm\, \tfrac{1}{2}\omega \pm \tfrac{1}{2}[(\omega_0-\omega)^2 + 4\, g^2]^{\frac{1}{2}}\,. \tag{52}$$

The vacuum field Rabi splitting is therefore unchanged in the q-deformed Jaynes-Cummings model : for zero detuning $\Delta \equiv (\omega_0-\omega) = 0$, $\lambda = \frac{1}{2}\omega_0 \pm g$ and the splitting is 2g as in the Jaynes-Cummings model. No check on the q-deformation theory is therefore possible in experiments like those of Haroche and colleagues [22] to measure the vacuum field Rabi splitting as described in §5. However, the four transition frequencies between the first two pairs of excited levels are, at resonance, $\omega_0 \pm g(q + q^{-1})^{\frac{1}{2}} \pm g$ and depend on q. The Fig.3 of [23] shows that two of these side bands, at $\omega_0 \pm (g(q+q^{-1})+g)^{\frac{1}{2}}$, might just about show up in a vacuum field Rabi splitting experiment with $Q = 10^8$, searching for any q-dependence. The experiments so far reported for $N_A \geq 3$ [22] still need considerable refinement to show up such side-bands.

The expectation value of the atomic inversion at resonance (and $Q = \infty$) is

$$\langle S^z(t)\rangle = \tfrac{1}{2} \sum_{n=0}^{\infty} p_n(1-2\sin^2(g[n+1]^{\frac{1}{2}}t)); \tag{53}$$

p_n is the probability of finding n photons in the mode at t = 0. The Fig. 9 shows the revivals in $\langle S^z(t)\rangle$ as a function of gt for an initial undeformed coherent state $|\alpha\rangle$ (the q-bosons have a representation in terms of ordinary bosons [54,64] so such a coherent state is a possible initial state for a q-boson system): $\Delta = \omega_0 - \omega = 0$ in Fig. 9 and $|\alpha|^2 = 2$. Since p_n is 'undeformed' the deformation in (53) is only in the resonant Rabi splittings $2g[n+1]^{\frac{1}{2}}$ (Note that, at t = 0, $\langle S^z(0)\rangle = +\frac{1}{2}$ and the appropriate eigenvalue

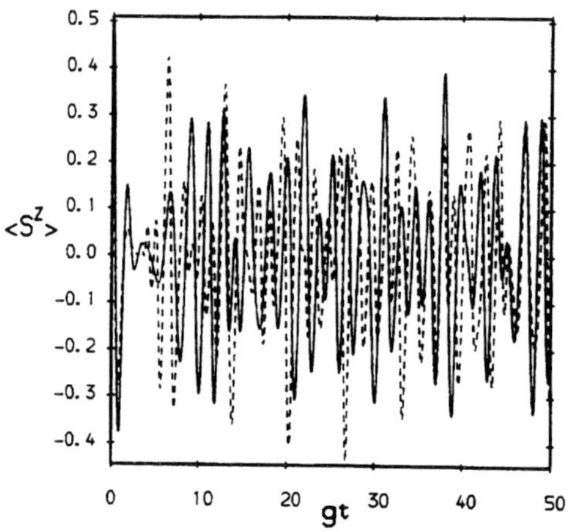

Fig.9 Revivals in the 'q-quantised' micromaser: $Q = \infty$ and detuning $\Delta = \omega_0 - \omega = 0$; initial state is an *ordinary*, undeformed, α-state with $|\alpha|^2 = 2$. Full curve, canonical quantisation (q=1); broken curve q = 1.5

of $\hat{N}$ is (n+1), not n, and the Rabi splitting is $2g[n+1]^{\frac{1}{2}}$). Thus for n = 1,2... in the sum in (53) the terms depend on q. The full curve in Fig. 9 is the canonically quantised case q = 1: the broken curve has the q-deformed Rabi splitting with q = 1.5. There is a considerable difference exhibited in the two cases, but we do not know if q = 1.5 makes any sense e.g. $q = e^{\gamma}$ and $\gamma = \lambda\hbar$ for some $\lambda \in \mathbb{R}$ (which we have no value for).

The natural normalised q-deformed α-state is the state

$$|\alpha\rangle = \sum_n \alpha^n([n]!)^{-\frac{1}{2}} \, |n\rangle \, \exp_q - \tfrac{1}{2}|\alpha|^2 \tag{54}$$

with the property $a^-|\alpha\rangle = \alpha|\alpha\rangle$. This has the q-Poisson distribution $p_n = (|\alpha|^{2n}/[n]!)\exp_q -|\alpha|^2$ in which $\exp_q X$ is the q-deformed exponential function of X in which $n! \to [n]!$. The Fig. 10 shows this case as initial state: $\Delta = 0$, $q = 1.5$, and $|\alpha|^2 = 2$. This is substantially changed from the dashed curve in Fig. 9 where $\Delta = 0$, $q = 1.5$, $|\alpha|^2 = 2$, and p_n is the usual Poisson distribution.

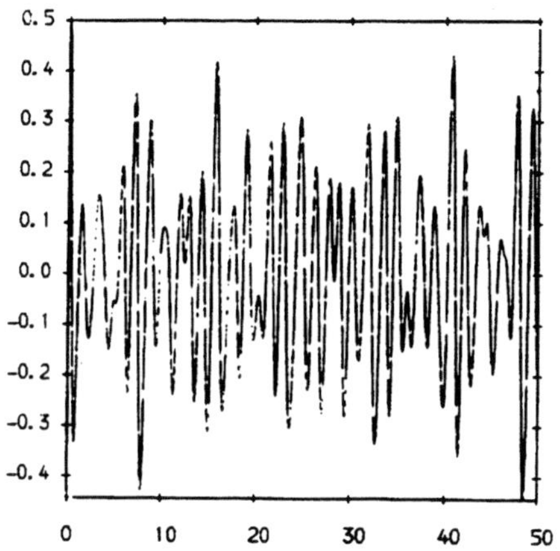

Fig. 10 Revivals in the 'q-quantised' micromaser - parameters as in Fig. 9; initial state a q-deformed α-state with $|\alpha|^2 = 2$, $q = 1.5$.

The possibility of detecting quantum group quantisation in the micromaser this way depends on the size of $\lambda\hbar$ for $q = e^{\lambda\hbar}$, and the effects of black-body radiation which as noted will also be q-deformed. Although we have computed revivals for the ordinary J-C model in an ordinary heat bath [24,25,78] we have not yet attempted such calculations in the deformed cases. The Fig. 5 of [25] shows revivals in $\langle\hat{n}\rangle = \langle a^\dagger a\rangle$ for the undeformed J-C model for $Q = 2 \times 10^{10}$ and an initial undeformed coherent state field $|\alpha\rangle$ with $|\alpha|^2 = 1$, $\bar{n}_{th} = 0.15$ and $\omega_0 = 2\pi \times 21.456$ GHz corresponding to the $63P_{3/2} \to 61D_{5/2}$ transitions for ^{85}Rb. That the few black-body photons move the positions of the revivals in a detectable fashion was already reported in [24]. The Fig. 6 of [25] shows 1-photon revivals in $C(t) \equiv [\langle a^\dagger(t)\ a^\dagger(t)\ a(t)\ a(t)\rangle / \langle a^\dagger(t)\ a(t)\rangle^2 - 1\}$ for the same parameters as the Fig. 5 ($C(t) = P_2 P_1^{-2}$ in the notation of §10 following). The results show that $C(t) < 0$ for much of the resonant motion and is < 0 off resonance during the revivals. Here the field is *antibunched* (sub-Poissonian): the Mandel Q-parameter $Q = \langle a^\dagger a^\dagger a\ a\rangle / \langle a^\dagger a\rangle - \langle a^\dagger a\rangle$ is < 0 .

We show how these calculations for the undeformed J-C model coupled to a heat-bath (which were also reported in the §5) were carried out in the next section.

10. SINGLE ATOM CANONICAL CAVITY Q.E.D. AT FINITE TEMPERATURES

The model is the undeformed J-C model in which the single mode is coupled to a heat bath at $T > 0$. The well known master equation is [14,18-20,24], and as given at eqn.(20),

$$\frac{d\rho}{dt} = - i\,[\hat{H},\rho] - (\bar{n}_{th}+1)\,\kappa[a^\dagger a\rho - 2a\rho a^\dagger + \rho a^\dagger a] - \bar{n}_{th}\,\kappa\,[aa^\dagger\rho - 2a^\dagger\rho a + \rho aa^\dagger], \qquad (55)$$

in which $\kappa = \frac{1}{2}\omega\, Q^{-1}$ and $\bar{n}_{th} = (e^{\beta\omega}-1)^{-1}$ ($\hbar = 1$) for *ordinary bosons:* $\hat{H}$ is the J-C Hamiltonian eqn.(29). The master equation (55) for the total density operator ρ and ordinary bosons compares with that in (42) for the reduced density operator ρ_a and given there for q-bosons. As noted master eqn. (55) was first introduced for ordinary bosons at eqn. (20) in §4.

The calculational problem for (55) (or (20)) rests in the infinite number of photon states $|n\rangle$ spanning the Hilbert space at finite temperatures: a complete basis is $|e\rangle|n-1\rangle$, $|g\rangle|n\rangle$ with $n = 1,2,\ldots$, and $|g\rangle|0\rangle$. Our procedure has been [24] to calculate the four one-body quantities $P_n \equiv \langle a^{\dagger n}a^n\rangle$, $Q_n \equiv \langle a^{\dagger n}a^n S^z\rangle$ (and $n \geq 0$), $S_n \equiv \langle S^+ a^{\dagger n-1}a^n + a^{\dagger n}a^{n-1}S^-\rangle$, $R_n \equiv i\,\langle S^+ a^{\dagger n-1}a^n - a^{\dagger n}a^{n-1}S^-\rangle$ (and $n > 0$), which satisfy a closed coupled system of equations infinite in number: $Q_0 \equiv \langle S^z\rangle$ and $P_1 = \langle a^\dagger a\rangle$, $P_2 = \langle a^\dagger a^\dagger aa\rangle$. The problem reduces [24] to the solution of the matrix equations

$$\frac{d\psi_n}{dt} = \mathcal{A}_n\,\psi_n + B_n\,\psi_{n+1} + C_n\,\psi_{n-1},\quad n \geq 0,\ C_0 = 0, \tag{56}$$

in which $\psi_n = [P_n, Q_n, R_n, S_n]^T$ and $\mathcal{A}_n$, B_n and C_n are 4×4 matrices. The Laplace transform on t yields a 3-term matrix recurrence relation solved by the matrix continued fraction techniques introduced by Risken [80]. We also calculate the 2-body correlation functions in a similar way [18-20]. These correlations at t and $t + \tau$ satisfy a closed system of the form (56) with larger matrices and τ replacing t. The fluorescence spectrum of the Fig. 3 of [23] at $Q = 10^8$ is calculated this way. Moreover, and as described in §5 we have calculated the fluorescence and cavity field spectra for $10^3 \lesssim Q \leq 10^8$ and watched their *qualitative* changes as Q increases: vacuum field Rabi splitting is first discernible for Q's $\sim 10^6$ as was noted. One can also expect to see the breakdown of the Born and Markov approximations with increasing Q [18].

We turn to the canonically quantised micromaser.

An instructive result found by the methods just described is shown in the Fig. 4 of [23]. This shows the evolution of the Fock state $|10\rangle$ in the presence of the single atom for $Q = 2 \times 10^{10}$ and $\bar{n}_{th} = 0.15$: $\kappa g^{-1} = 0.0000767$, and $63P_{3/2} \to 61D_{5/2}$ transitions of ^{85}Rb are assumed: detuning $\Delta = 0$. The scale of gt is that 9 units ≈ one atomic transit time for crossing the cavity (gt_{int}, §7). The Figure shows that after 8 such transit times the amplitude of the inversion has fallen some 10%. In the empty-of-atoms cavity time $gt_e = gT_p - gt_{int}$ of the micromaser (§§6,7 and Refs. [28,29]) this amplitude would be $\sim (0.9)^{36} = 0.02$. Thus if the Fock state $|10\rangle$ were created it would decay substantially in each period T_p. This suggests that a smaller T_p is needed for the creation and maintenance of Fock states in the micromaser. We investigate this suggestion in the next sections §§11 and 12. We look first of all at the fundamental canonically quantised quantum theory of the micromaser.

11. THE CANONICALLY QUANTISED MICROMASER AS A FUNDAMENTAL QUANTUM DEVICE

For long repitition rates $T_p \gg t_{int}$ there is, to a probability ≈ 1, at most one atom in the maser cavity. Moreover, during most of the time T_p the maser cavity is empty of atoms. Theories of the micromaser have been developed for $Q = \infty$ throughout all of the motion [32,71,81] and for the micromaser coupled to a heat bath and $Q < \infty$ when, but only when, there is no atom in the cavity [69,70]. When the atom is in the cavity during the short time t_{int}, the atom and field are supposed to evolve under the *unitary* transformation $\hat{U} = \exp[-it\hbar^{-1}\hat{H}]$ where $\hat{H}$ is the J-C Hamiltonian eqn. (29), i.e. the evolution during t_{int} is for $Q = \infty$ [69,70]. If an atom enters the maser cavity at $t = 0$ in its upper state $|e\rangle$ with the cavity field in the state $|n\rangle$ at $t = 0$, the unitary operator $\hat{U}$ transforms this state to $|e\rangle|n\rangle$ with probability [81]

$$a(n) = \cos^2 \sqrt{n+1}\, gt_{int}, \tag{57}$$

and to $|g\rangle\,|n+1\rangle$ with probability

$$b(n) = \sin^2 \sqrt{n+1}\, gt_{int}\ , \tag{58}$$

and $a(n) + b(n) = 1$. In the *sequence* (at $Q = \infty$) in which atoms enter the maser cavity successively the probability p_n of finding n photons in the cavity will depend on the number of atoms k which have passed through the cavity i.e. $p_n = p_n(k)$. Thus [71]

$$p_n(k+1) = a(n)\, p_n(k) + b(n-1)p_{n-1}(k). \tag{59}$$

Consequently, if $b(n-1) = 0$, $p_n(k+1) = a(n)\, p_n(k)$ (this is not correctly stated in Refs. [23,30,31]!), and if initially $p_n(0) = 0$ for all $n > n_o$, $p_n(k) = p_n(k+1) = 0$ for all $n > n_o$ and the p_n with $n > n_o$ never build up in the cavity field. In effect a barrier is created [71] at $n = n_o$. A crucial result demonstrated in [71] is then, that if $n = n_o$ is the zero of (58), and initially $p_n = p_n(0) = 0$ for all $n > n_o$, then $p_{n_o}(k) \longrightarrow p_{n_o}(k+1) \to 1$, i.e. a pure *Fock* state $|n_o\rangle$ is created in the cavity. In the present analysis we shall be primarily concerned to discover in detail under what conditions (if any) a Fock state could build up in a $Q = \infty$ cavity which initially contains black-body radiation and, more to the point for an experiment, under what conditions (if any) a Fock state could build up in a real maser cavity with $Q < \infty$ which is coupled all the time to a heat bath at a finite temperature T.

The zeros of b(n), eqn. (58), define 'trapping' states of the field [32,66,69–71]. It is convenient to consider these trapping states as the zeros of (58) labelled by n_o and r: $(n_o+1)^{\frac{1}{2}}\, gt_{int} = r\pi$ ([66] uses q for our r). Thus, as example, $gt_{int} = \frac{1}{2}\pi$ and $r = 1$, $n_o = 3$ defines a 'first' trapping state; then there are the second, third, etc trapping states with $n_o = 15, 35, 63, \ldots$. There are trapping states for *any* $gt_{int} = r\pi(n_o+1)^{-\frac{1}{2}}$, i.e. for gt_{int} the trapping states have the measure of the irrationals, the gt_{int} for trapping states are therefore mathematically 'dense', and there is a very large number of such states. Without attempting to make this argument rigorous one can therefore expect that the cavity field of the micromaser 'almost always', i.e. with probability 1, evolves to a Fock state. The Fock state is characterized by the variance σ, defined by eqn. (28), equal to zero. The experiments [28,29] show however that such evolution does not necessarily occur : *sub*-Poissonian fields with $0 < \sigma < 1$ were detected, and the observations conform to the very different predictions of the theoretical work [69,70] for $Q < \infty$ and finite temperatures.

At $Q = \infty$ one finds in any case [71] that evolution to a pure Fock state $|n_o\rangle$ will not occur if the cavity initially contains black-body radiation. If gt_{int} defines a first trapping state (n_o,r), then if (n_1,r), (n_2,r) etc. are the second, third, etc. trapping states, each of these will evolve as long as the initial state of the cavity field is $\sum_{n=0}^{N_{max}} \alpha_n |n\rangle$ ($\alpha_{N_{max}} \neq 0$) and the n_i of the trapping states have $n_i < N_{max}$. In the case of black-body radiation, $N_{max} = \infty$ so all of the sequence of trapping states (n_i, r) will evolve, *and no steady state will ever be attained!* However in [71] no time scale is given for this evolution. From a computer calculation with $Q = \infty$ and $gt_{int} = \frac{1}{2}\pi$, we find that after 91 inverted atoms have entered and excited successively, $p_3 = 0.9997105$, $p_{15} = 0.2659983 \times 10^{-3}$ and $p_{35} = 0.197007 \times 10^{-14}$. After 500 inverted atoms this becomes $p_3 = 0.9997105$, $p_{15} = 0.2894501 \times 10^{-3}$ and $p_{35} = 0.701391 \times 10^{-14}$. After 2000 inverted atoms only p_{35} changes (to $0.7019316 \times 10^{-14}$). All the figures then remain unchanged as 100,000 atoms have passed!

The second reason why Fock states do not occur at the values $Q = 3 \times 10^{10}$ and $T = 0.5^{\circ}$K of the experiments [28,29] is that the finite but small value $\bar{n}_{th} = 0.15$ of the black-body radiation induced by coupling to a heat bath at that T *qualitatively* (not just quantitatively) changes the action of the micromaser compared with its action at $Q = \infty$. The observations [28,29] which use T_p (~ 7,000 μ sec) >> t_{int} (~ 35μ sec), are consistent with the theoretical predictions [69,70]. We sketch the nature of these particular theories [69,70] which use unitary evolution by $\hat{U}$ in the short time t_{int} and couple in the heat bath only during the long time $t_e = T_p - t_{int}$.

The result (59) means that

$$p_n(k+1) - p_n(k) = a(n)\, p_n(k) + b(n-1)\, p_{n-1}(k) - p_n(k) \,. \tag{60}$$

If a dynamically steady state can be reached repeating each T_p then, with the average *rate* $R \equiv T_p^{-1}$, the change $\Delta\rho^f_{n,n}$ in the field density matrix during a time Δt in $t_{int} << \Delta t << T_p$ is

$$\Delta\rho^f_{n,n} = R\, \Delta t\, [p_n(k+1) - p_n(k)] \tag{61}$$

since the square bracket is the change for a single atom after k atoms have passed. But providing the situation *is* steady, so that we need not distinguish the number of atoms k which have passed through,

$$\begin{aligned}\Delta\rho^f_{n,n}\Big|_{gain} &= R\, \Delta t\, [(a(n)-1)\rho^f_{n,n} + b(n-1)\rho^f_{n-1,n-1}] \\ &\equiv \Delta t\, [-A(n+1)\, F(n+1)\rho^f_{n,n} + An\, F(n)\, \rho^f_{n-1,n-1}]\end{aligned} \tag{62a}$$

where $F(n) \equiv n^{-1}\sin^2\sqrt{n}gt_{int}$ and $A \equiv R$, and the $\rho^f_{n,n}$ do not distinguish the passage of atoms. Then by coupling in the heat bath during the long time t_e, one finds from the master equation (55) that

$$\begin{aligned}\Delta\rho^f_{n,n}\Big|_{loss} = \Delta t\, \{&(\bar{n}_{th}+1)\, C(n+1)\rho^f_{n+1,n+1} + \bar{n}_{th}\, Cn\rho^f_{n-1,n-1} \\ &- C\, [(n+1)\, \bar{n}_{th} + n(\bar{n}_{th}+1)]\, \rho^f_{n,n}\}\end{aligned} \tag{62b}$$

where $C = 2\kappa = T_{cav}^{-1}$. The total change $\Delta\rho^f_{n,n}$ is then given *additively* (i.e. 'by additive gain and loss')

$$\frac{\Delta\rho^f_{n,n}}{\Delta t} = \frac{\Delta\rho^f_{n,n}}{\Delta t}\Bigg|_{gain} + \frac{\Delta\rho^f_{n,n}}{\Delta t}\Bigg|_{loss}\,, \tag{63}$$

and the coarse grained expression $\Delta\rho^f_{n,n}/\Delta t$ is identified with the actual derivative $\partial\rho^f_{n,n}/\partial t$ for the field (N.B. $\Delta t >> t_{int}$). One then has the coarse grained master equation for the field density matrix $\rho^f_{n,n}$

$$\frac{\partial\rho^f_{n,n}}{\partial t} = - A(n+1)\, F(n+1)\, \rho^f_{n,n} + An\, F(n)\, \rho^f_{n-1,n-1}$$
$$+ (\bar{n}_{th}+1)\, C(n+1)\, \rho^f_{n+1,n+1} + \bar{n}_{th}\, Cn\, \rho^f_{n-1,n-1}$$
$$- C(n+1)\, \bar{n}_{th}\, \rho^f_{n,n} - Cn(\bar{n}_{th}+1)\, \rho^f_{n,n} \qquad (64)$$

and (compare [81] for the full density matrix ρ) *this is diagonal throughout the whole motion.* In the steady state $\partial\rho^f_{n,n}/\partial t = 0$, detailed balance means

$$Cn\,(\bar{n}_{th}+1)\, \rho^f_{n,n} = [An\, F(n) + Cn\, \bar{n}_{th}]\, \rho^f_{n-1,n-1} \qquad (65)$$

so

$$\rho^f_{n,n} = f(\bar{n}_{th}) \left[\frac{A}{C} \frac{F(n)}{\bar{n}_{th}} + 1 \right] \rho^f_{n-1,n-1} \qquad (66)$$

where $f(\bar{n}_{th}) \equiv \bar{n}_{th}/(1+\bar{n}_{th})$. Without detailed balance (64) depends on $\rho^f_{n+1,n+1}$, $\rho^f_{n,n}$, $\rho^f_{n-1,n-1}$ and in the steady state yields a 3-term recurrence relation (§10). Detailed balance yields (66) which is two term and easily solved to

$$\rho^f_{n,n} = C_0 \left(\frac{\bar{n}_{th}}{1+\bar{n}_{th}} \right)^n \prod_{k=1}^{n} \left[1 + \frac{A}{C} \frac{F(k)}{\bar{n}_{th}} \right] \qquad (67)$$

in which C_0 is a normalisation constant such that $\sum_{n=0}^{\infty} \rho^f_{n,n} = 1$: by definition $AC^{-1} = R\, T_{cav}$. The expression (67) for the micromaser was first obtained by Meystre [69] by a somewhat different route. The route *we* have used has been developed in such a way that A and C and F(n) can be interpreted in terms of ordinary single mode laser theory: if $A \equiv 2R\, \Omega^2 \gamma_{eg}^{-2}$, $C = T_{cav}^{-1}$ and $nF(n) = n[\, 1 + 4n\Omega^2\, \gamma_{eg}^{-2}]^{-1}$ in which Ω is the Rabi frequency and γ_{eg}^{-1} is the spontaneous emission time, we regain the essentials of the Scully-Lamb laser theory [82] as extended to include the contribution of black-body radiation at finite temperatures [83] : γ_{eg}^{-2} is replaced by a combination of γ_g, γ_e and γ_{eg} in [82] and in effect $f(\bar{n}_{th})$ replaces $\bar{n}_{th}$ inside the square bracket in [83] (however at sub-Kelvin temperatures $f(\bar{n}_{th}) \approx \bar{n}_{th}$). The threshold for laser operation is [82] A = C, and for A > C the photon distribution $p(n) = \rho^f_{n,n}$ has a peak at $n \neq 0$ (coherence). Moreover for large AC^{-1} (well above threshold) $\rho^f_{n,n} \approx \frac{|\alpha|^2}{n} \rho^f_{n-1,n-1}$ where (for large enough Ω^2) $F(n) = [4n\, \Omega^2\, \gamma_{eg}^{-2}]^{-1}$ and $|\alpha|^2 = A\, C^{-1}\, F(n)\, n = T_{cav} 2R\, \Omega^2\, \gamma_{eg}^{-2}/4\Omega^2\, \gamma_{eg}^{-2} = \frac{1}{2}\, T_{cav}\, R$.

Lugiato et al [70] in effect gave the route from (61) to reach (67) for the micromaser identically, with $A\, C^{-1} = R\, T_{cav}$. They make the same point that (64) for the micromaser is also conventional laser theory. The new feature in the micromaser theory is that instead of the spontaneous emission time γ_{eg}^{-1} controlling the theory, so that $\gamma_{eg}^{-1} \ll \Delta t \ll T_{cav}$, it is t_{int}, satisfying $t_{int} \ll \Delta t \ll T_{cav}$, which does so. This has the important consequence that, unlike laser theory where there is essentially only one parameter, AC^{-1} with $AC^{-1}nF(n) \sim RT_{cav}$, the micromaser theory yielding (67) has the *two* parameters $R\, T_{cav}$ and gt_{int}. It is this which enables the micromaser to perform as a strictly *quantum* device producing *sub*-Poissonian light in the cavity field (see below). Meystre [69] defined $R\,T_{cav} = N_{ex}$, the mean number of atoms traversing the cavity in a cavity decay time $T_{cav} = (2\kappa)^{-1}$, and gave a natural 'tipping angle' Θ for the atom in the cavity

$$\Theta \equiv (R\,T_{cav})^{1/2}\,gt_{int} = (N_{ex})^{1/2}\,gt_{int}\,. \tag{68}$$

The formula (67) with $A\,C^{-1} = N_{ex}$ will presumably be good for $T_p >> t_{int}$ as was assumed providing the temperature T is large enough. Evidently, as $T \to 0$, $\bar{n}_{th} \to 0$ and (67) yields

$$\rho^f_{n,n} = C_0\left(N_{ex}\right)^n \prod_{k=1}^{n} \frac{\sin^2 \sqrt{k}\,gt_{int}}{k}\,. \tag{69}$$

This is the exact solution of $C\,\rho^f_{n,n} = A\,F(n)\,\rho^f_{n-1,n-1}$ which derives from (64) and (65) in the steady state at $T = 0$: moreover $p_n = \rho^f_{n,n} = 0$ for $n \geq (n_0+1)$ where $\sin(n_0+1)^{1/2}\,gt_{int} = 0$. But now the theory is incomplete compared with that of (59), because no Fock state $p_{n_0} = 1$ is predicted; nor does the theory show that the higher trapping states will emerge in the presence of black-body radiation at $T > 0$ and that there is no steady state — (69) is still a steady state theory. At $Q = \infty$ (no damping) $C = 0$ in (64) and for a steady state

$$-\,A(n+1)\,F(n+1)\,\rho^f_{n,n} + An\,F(n)\,\rho^f_{n-1,n-1} = 0. \tag{70}$$

This is

$$-(1 - \cos^2\sqrt{n+1}\,gt_{int})\,\rho^f_{n,n} + (\sin^2\sqrt{n}\,gt_{int})\,\rho^f_{n-1,n-1} = 0, \tag{71}$$

namely (59) with no distinction made in the $\rho^f_{n,n}$ for the number of atoms k which have passed through the cavity ($\rho^f_{n,n}$ is independent of k). For the creation of Fock states through the 'trapping state' (n_0,r) the state of the field does depend significantly on the number k of atoms which have passed (as the Fig.11 below makes clear). *Apparently sequential evolution is fundamental to the creation of Fock states in the maser cavity* .

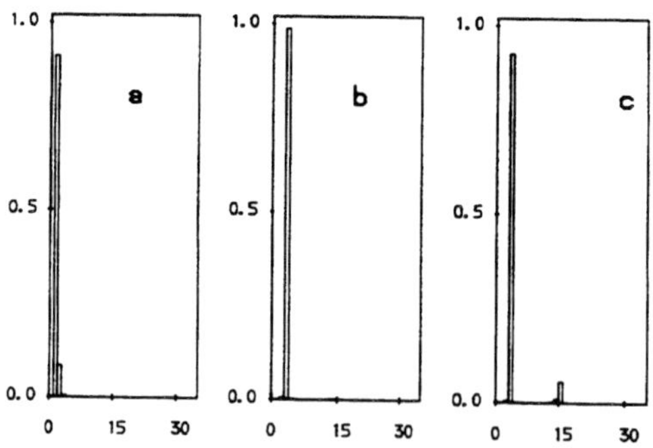

Fig. 11. Calculated histogram of $p_n = \rho_f(n)$ after 1(a), 40(b) and 500(c) atoms have passed through the cavity: $\omega = \omega_0$ and $t_{int} = \pi/2g$. See end this §11 for details.

The formula (67) for the micromaser, with $A\ C^{-1} = N_{ex} = R\ T_{cav}$, is sensitive to the value of $\bar{n}_{th}$. It is intuitive that for $T_p >> t_{int}$ it is a good approximation to permit unitary evolution under $\hat{U}$ in the short time t_{int} and to couple in the heat bath during the long time t_e. Predictions based on (67) [69,70] agree with the observations [28,29]. The conclusion is that this description of the effects of black-body radiation changes the theory *qualitatively* compared with the $Q = \infty$ theory based on (59).

To achieve Fock states in the cavity it seems [66] from (67) that temperatures $T \lesssim 0.2^{\circ}K$ and the current N_{ex}'s ~ 50 will be needed. However it seems we must give up (67) for $T = 0$ and use (59). Alternatively we can look again at the more exact equations for the total density matrix ρ for which (64) is both a projection to ρ^f and an approximation. The Fig. 4 of [23] based on the exact equations for ρ^f suggests that if we wish to create Fock states in the micromaser at finite $Q < \infty$ and $T > 0$ we should reduce T_p to $T_p \gtrsim t_{int}$. Evidently it is then necessary to couple in the heat bath throughout the motion. This coupling introduces an intrinsically *nonlinear* gain and loss mechanism: linear additive gain and loss eqn. (63) does not apply.

A simple minded approach to see this, which however is still concerned with t_e, the empty-of-atoms cavity time is as follows:- assume there *is* a steady state varying during T_p but repeating each T_p. Suppose $\langle \bar{n} \rangle \equiv \langle a^{\dagger} a \rangle$, the mean number of photons in the cavity, is n_1 as the inverted atom enters, is n_2 when it leaves after time t_{int}, and is n_3 before the next atom enters at T_p. For the steady state $n_3 = n_1$, and

$$\bar{n}_{th} + (n_2 - \bar{n}_{th})\, e^{-2\kappa(T_p - t_{int})} = n_3 = n_1 \qquad (72)$$

where the case n = 1 of the exact expression (74) below has been used. Since $n_2 \le n_1 + 1$ (only one photon can be dumped) then, for small enough $\bar{n}_{th}$ compared with n_1, n_2, n_3 (a numerically valid assumption in the present context though this removes $\bar{n}_{th}$ from the theory)

$$n_2 \le \left[1 - e^{-2\kappa(T_p - t_{int})}\right]^{-1} . \qquad (73)$$

It follows that for $t_{int} << T_p$ (the assumption of [69,70]) and $2\kappa T_p << 1$ (an assumption of [70]), $n_2 \le (2\kappa T_p)^{-1} = N_{ex}$. The equality is the equality of linearly additive gain and loss $R = T_p^{-1} = 2\kappa n_2$. More generally this simple minded analysis shows the gain-loss mechanism is nonlinear even in the empty-of-atoms time t_e. But it tells us nothing about the evolution during t_{int}, only that n_2 could become large for $T_p \approx t_{int}$.

We have used the methods sketched in the previous section to investigate this nonlinear gain and loss regime: we solve (55) for the total density operator ρ during t_{int} when the atom is in the cavity: the heat bath is thus coupled in. We then solve the same equation with the atom omitted during the time $t_e = T_p - t_{int}$ and the heat bath coupled in. This is done numerically or we can make use of the exact analytic formula derived from (55) when no atom is present: we define $P_n = (n!)^{-1} \langle a^{\dagger n} a^n \rangle$ and find (by solving equation (77) below) that

$$P_n(\tau) = (\bar{n})^n + \sum_{r=1}^{n} e^{-r\tau} M_{nr}(\bar{n})^{n-r} \sum_{j=1}^{r} (-1)^{r-j} M_{rj}(\bar{n})^{r-j}\{P_j(0) - (\bar{n})^j\} \qquad (74)$$

in which $M_{ij} = i!\ (j!(i-j)!)^{-1}$, $i \geq j$; $M_{ij} = 0$, $i<j$: κ enters through $\tau = 2\kappa t_e$ and (74) generalizes (72) with $\bar{n} \equiv \bar{n}_{th}$.

So far in these calculations both t_{int} and t_e repeat regularly with fixed T_p (a stochastic input of the atoms is not yet included). The empty of atoms cavity has initial condition $\rho_f(0) = Tr_a\rho(t_{int})$, where $\rho(t_{int})$ is ρ as the atom of current interest leaves the cavity and $\rho_f(0)$ evolves to $\rho_f(t_e)$ in the empty of atoms cavity time either by numerical solution for $\rho_f(t)$ from (55) with no atom present or through (74): $\rho_f(t_e)$ now serves as initial condition for the cavity field as the next atom enters so $\rho = |e\rangle \rho_f(T_p-t_{int}) \langle e|$ is the initial condition for ρ as the next atom enters (we here conveniently use ρ_f for the cavity field density operator not ρ^f).

The equations of motion during the interaction time t_{int} can be reduced to a 3-term recurrence relation solved once again by matrix continued fraction methods. Details will be published elsewhere [79,84]. A significant point is that although off-diagonal elements of ρ develop, $\rho_f(t_e)$ is diagonal if $\rho_f(-t_{int}) = Tr_a\rho(0)$ is initially diagonal. But this is the case since the first atom enters a cavity containing only black-body radiation. This, feature also arises in the unitary transformation of ρ during the time t_{int} [81], but the equations (64) for ρ_f are always diagonal.

Obviously the smallest empty-of-atoms time is zero, i.e. $T_p - t_{int} = 0$ and $T_p = t_{int}$. The Fig. 11 above shows the evolution of the probabilities p_n for finding n photons in the cavity as a function of the number of inverted atoms which have passed through the cavity in successive intervals $gt_{int} = \frac{1}{2}\pi$. This is the condition for evolution to a first trapping state $n_0 = 3$, $r = 1$ at $Q = \infty$ and the next trapping state has $n_0 = 15$. However, in these calculations $Q = 2 \times 10^{10} < \infty$ and $T = 0.5^\circ K$ much as in the experiments [28,29]. It will be seen that after 40 atoms have passed p_3 evolves to 0.988, while p_{15} now evolves and after 500 atoms have passed p_3 and p_{15} have become 0.924 and 0.056 respectively. The Fig. 2 of [78] also shows comparable results for $gt_{int} = \frac{1}{4}\pi$ and the first trapping state is at $n_0 = 15$: $p_{15} = 0.774$ after 400 atoms have passed and is still growing. Thus at this finite Q and T the system displays the action predicted for $Q = \infty$ in [71]. Presumably no steady state is reached as the higher trapping states at $n_0 = 35, 63,\ldots$ begin significantly to grow. Notice the dependences on the numbers of atoms which have passed: the theory is very different from the steady state theories based on eqns. (64).

Thus the major *conclusion must be that as $t_e = T_p - t_{int}$ becomes smaller the finite Q finite T system behaves more and more like a $Q = \infty$ system as described in* [32,71].

We support this conclusion by the further work described in the next section.

12. MICROMASER ACTION AS THE EMPTY OF ATOMS CAVITY TIME CHANGES

The numerical calculations described in §§10 and 11 are slow even on the Amdahl 5890 system available in Manchester. The matrix continued fraction methods used offer the very high accuracy and good stability necessary for the detailed investigation of black-body dynamics displayed in Fig. 11, but calculations are very long. We have developed faster, less accurate and less stable, methods for the PC. We described some of the results found so far through these in [23]. We amplify those results in [23] by the further results in the $T_p >> t_e$ regime described here. These particular calculations determine the moments P_n, Q_n, R_n and S_n mentioned below eqn. (55). On resonance we need P_n, Q_n and R_n only, and the equations of motion derived from (55), with the atom present in the cavity, are

$$\begin{aligned}
\dot{Q}_0 &= -R_1 \\
\dot{P}_n &= n R_n - 2\kappa g^{-1} n (P_n - \bar{n} P_{n-1}) \\
\dot{Q}_n &= -\tfrac{1}{2} n R_n - (n+1) R_{n+1} - 2n \kappa g^{-1} (Q_n - \bar{n} Q_{n-1}) \\
\dot{R}_n &= 4Q_n + P_{n-1} + 2Q_{n-1} - \kappa g^{-1} \{(2n-1)R_n - 2(n-1)R_{n-1} \bar{n}\}.
\end{aligned} \tag{75}$$

Conveniently we use $\bar{n}$ for $\bar{n}_{th}$ and have rescaled $P \longrightarrow (n!)^{-1}P$, etc. so with time t for gt

$$\begin{aligned}
P_n(t) &\equiv \mathrm{Tr}\,\{\rho(t)(a^\dagger)^n a^n\}/n! \\
Q_n(t) &\equiv \mathrm{Tr}\,\{\rho(t)(a^\dagger)^n a^n S^z\}/n! \\
R_n(t) &\equiv \mathrm{Tr}\,\{\rho(t)\ S^+(a^\dagger)^{n-1} a^n - \rho(t)(a^\dagger)^n a^{n-1} S^-\}/n!\ .
\end{aligned} \tag{76}$$

The empty of atoms cavity field evolves as

$$\dot{P}_n = -2n\,\kappa g^{-1} (P_n - \bar{n} P_{n-1}) \tag{77}$$

and the analytical solution of this is (74). When the atom is in, we close the system by $R_{m+1} \approx (m+1)^{-1} P_1 R_m$. We need $m >> \langle a^\dagger a\rangle \equiv P_1$ and there are 3m+1 equations.

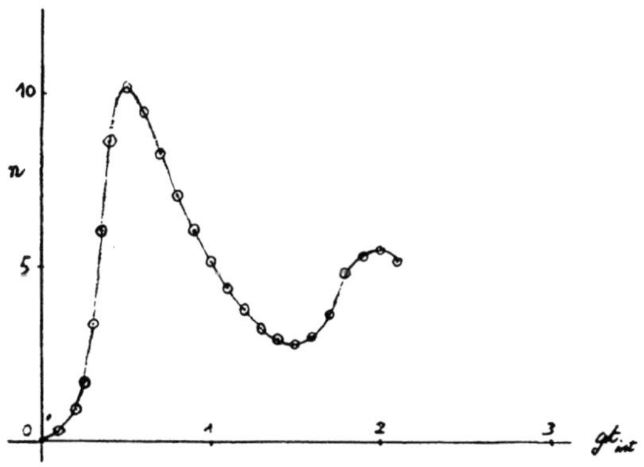

Fig.12 Plot from the PC of the 'long time' value $\langle\hat{n}\rangle = \langle a^\dagger a\rangle$ on atomic exit as a function of gt_{int} for $Q = 1.5\times10^8$ ($\kappa g^{-1} = 0.01$) and $\bar{n}_{th} = 0.15$: $gT_p = 5$ ($N_{ex} = 10$) and $\Theta = (3.162)\,gt_{int}$.

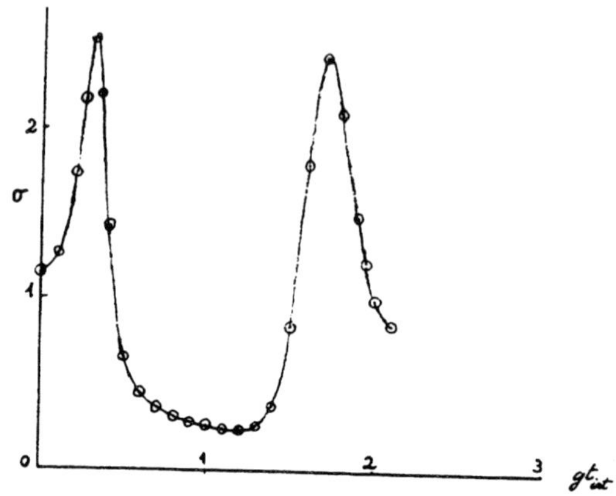

Fig.13 Plot of σ from the PC as a function of gt_{int} for the parameters of Fig.12 Note the super-Poissonian region where $gt_{int} \approx 1.8(\Theta \approx 5.69 \sim 2\pi)$.

The Figs. 12 and 13 show results achieved this way for $Q \doteq 1.5 \times 10^8$ $\bar{n}_{th} = 0.15$, $\omega = \omega_0$(exact resonance) and different gt_{int}. In the Fig. 13 there are pronounced sub-Poissonian regions where $\sigma < 1$ and there is a general qualitative agreement in this respect with the results predicted in [69,70] and confirmed in the experiments [28,29]. Qualitative agreement over the plotted range $gt_{int} \lesssim 2.5$ can be reached through $\Theta = (N_{ex})^{\frac{1}{2}} gt_{int} = (2\kappa T_p)^{-\frac{1}{2}} gt_{int} = \sqrt{10}\, gt_{int} \doteq 3.162\, gt_{int}$. Thus there is also the super-Poissonian region in Figs. 12,13 which occurs after the threshold and where $\Theta \approx 5.69$ ($\sim 2\pi$), and the $<$ is satisfied in (73) with $N_{ex} = 10$. The significance of these results compared with previous work [28,29,69,70] however is that $gt_{int} \lesssim 2.5$ and $gT_p = 5$, so the empty of atoms cavity time reaches up to as little as 50%. Agreement with [28,29,69,70] is thus reached by the rather low value (1.5×10^8) chosen for Q. Figs. 14 and 15 depict the steady state when the quality factor Q is improved so that $\kappa g^{-1} = 7 \times 10^{-4}$ and $\langle gt_e \rangle = g R^{-1} = 72$ (so $N_{ex} \approx 10$ as before). The error made in regarding t_e as a Poisson process rather than $(t_e + t_{int})$ is very small. In (74) it suffices to replace the exponentials $e^{-2\kappa r t_e}$ by $(1 + 2\kappa r R^{-1})^{-1}$ and all of Figs. 14 to 20 are calculated on this basis. The two sets of Figures, 12,13 and 14,15 are very similar for $0 < gt_{int} < 1.7$ although only Figs. 14, 15 approach the $T_p \gg t_{int}$ regime of the experiments [28,29]. In neither set is there any evidence of the trapping state $n_0 = 3$ at $gt_{int} = \frac{1}{2}\pi$ (for example) in agreement with the ideas of Ref.[66]. Instead the super-Poissonian region at $\Theta \sim 5.69$ occurs. Thus there are *three* important parameters: the empty-of-atoms cavity time t_e, $\bar{n}_{th}$, *and* Q.

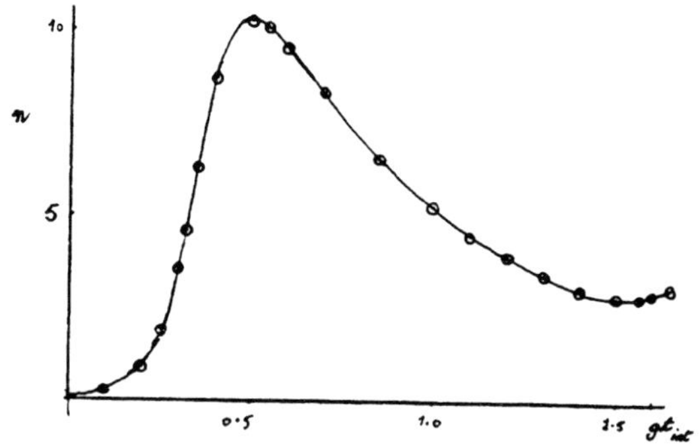

Fig. 14. Evolution of the cavity photon number: $Q \approx 2\times10^9$ ($\kappa g^{-1} = 7\times10^{-4}$) and $\langle gt_e\rangle_{Av} = g\,R^{-1} = 72$ (so $N_{ex} \approx 10$ as in Fig. 12): $\bar{n}_{th} = 0.15$.

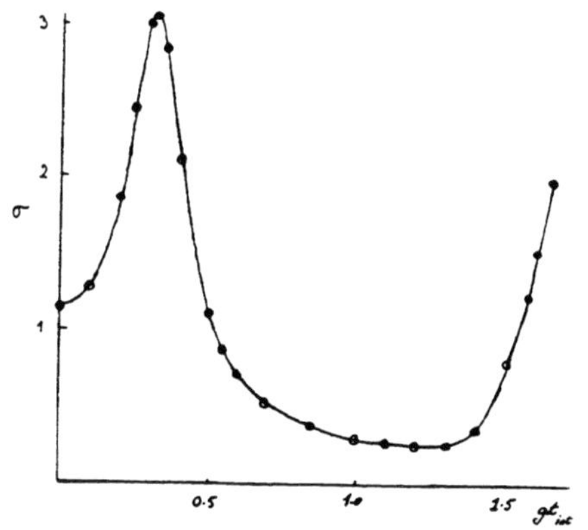

Fig 15. Plot of σ from the PC as a function of gt_{int} for the parameters of Fig 14. Note the super-Poissonian region again occurring where $gt_{int} \approx 1.8$

Notice that by setting $N_{ex} = 10$ for both Figs. 12, 13 and Figs. 14, 15 they show the very similar behaviours. The Figs. 8-11 in [23] show that, on the contrary, for large $Q \simeq 2 \times 10^{10}$ *and* gT_p small ($gT_p = gt_{int}$ and then gT_p increasing) plots of $\langle\hat{n}\rangle = \langle a^\dagger a\rangle$ as a function of time show that for $gt_{int} = \frac{1}{2}\pi$ the trend is that as $gt_e \rightarrow 72$, σ on final exit has increased from 0.053 (almost a Fock state) for $gt_e = 1$ ($gT_p = 1+gt_{int} = 1 + \frac{1}{2}\pi$) to $\sigma = 1.106$ which is the super-Poissonian ($\sigma > 1$) value much as in the observations [28,29] - though very many more atoms were used in the steady state for these observations.

Our conclusion from this is [23] that a relatively short empty of atoms cavity time t_e will help in the creation of Fock states at finite Q and T.

Evidently the nonlinear gain and loss mechanism enhances the cavity Q towards $Q = \infty$ (as of course does the reduction $T_p << T_{cav}$, but the Fig.11 shows there is no steady state with evolution on a time scale very short compared with that for the evolution $Q = \infty$).

The Figs. 16-20 support this conclusion. All of these figures are in the regime of large $gT_p >> gt_{int}$: $gt_{int} = 1.047$ and $gT_p = 216, 360, 504, 720$ and 1008. The σ at exit follow the trend $\sigma = 0.204, 0.216, 0.242, 0.281, 0.332$. These figures compare with $\sigma \geq 0.3$ as observed in the experiments [28,29] with $gT_p \sim 308$. Again very many more atoms were used in these experiments to achieve a steady state. At zero temperature, with $Q = \infty$, $gt_{int} = 1.047$ should give a steady oscillation between the two number states $|8\rangle$ and $|9\rangle$. Although, of the Figs. 16-20, Fig. 16 represents conditions where thermalisation effects are least (T_p is smallest), the expectation value $\langle\hat{n}\rangle = n$ on exit is much less than 8 and $\sigma \approx 0.2$ is well above zero. Nevertheless, we think Figs. 16-20 support our view that reducing T_p towards t_{int} will increase the trend towards pure Fock state evolution in the micromaser cavity field.

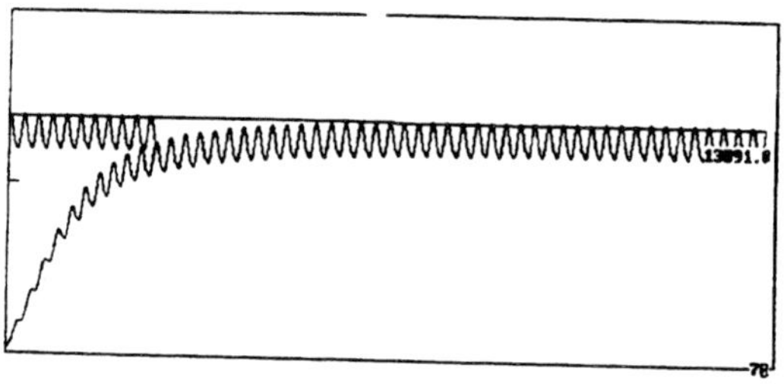

Fig. 16. Evolution of cavity photon number for $gt_{int} = 1.047$ chosen to minimise σ on exit : $g\,R^{-1} = 216$, $\kappa g^{-1} \sim 7 \times 10^{-5}$ ($Q = 2 \times 10^{10}$); $\bar{n}_{th} = 0.15$. For convenience gt (mod 72) is plotted rather than gt. After 65 atoms $\langle\hat{n}\rangle = 6.16$, $\sigma = 0.204$, $\langle S^z\rangle = 0.332$.

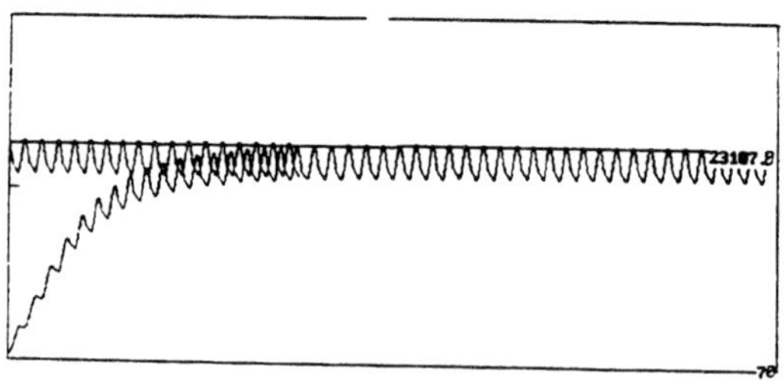

Fig. 17. Evolution of cavity photon number. Parameters as for Fig. 16 but $gR^{-1}=360$. After 65 atoms have passed $\langle\hat{n}\rangle = 5.68$, $\sigma = 0.216$, $\langle S^z\rangle = 0.221$.

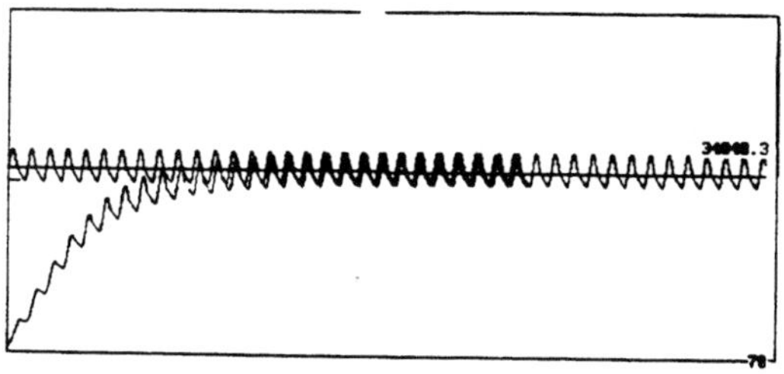

Fig. 18. Evolution of cavity photon number. Parameters as for Fig. 16 but $g R^{-1}$ = 504. After 70 atoms have passed $\langle\hat{n}\rangle$ = 5.31, σ = 0.242, $\langle S^z\rangle$ = 0.1720.

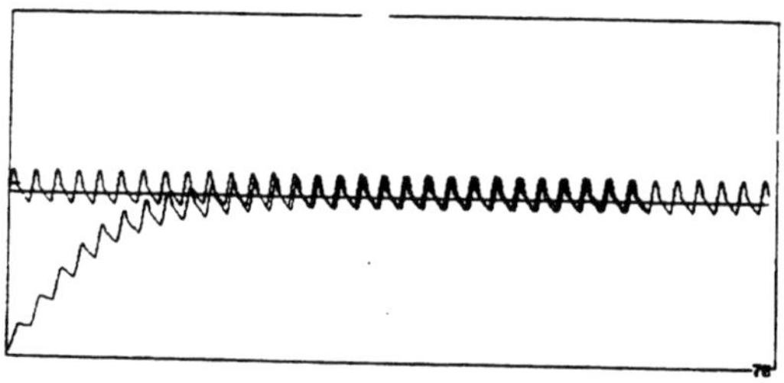

Fig. 19. Evolution of cavity photon number. Parameters as for Fig. 16 but $g R^{-1}$ = 720. After 65 atoms have passed $\langle\hat{n}\rangle$ = 4.92, σ = 0.281, $\langle S^z\rangle$ = 0.0510.

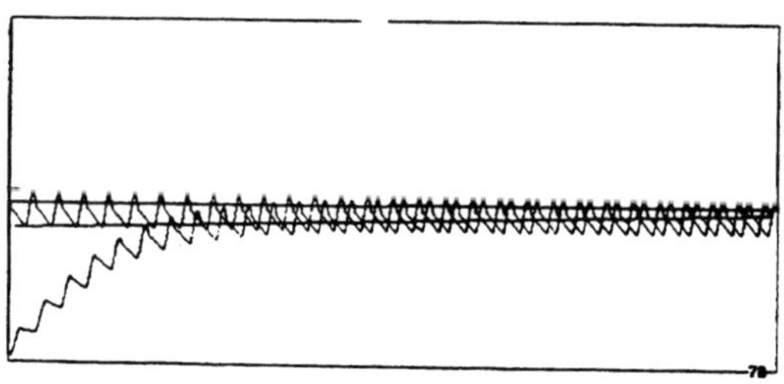

Fig. 20. Evolution of cavity photon number. Parameters as for Fig. 16 but $g R^{-1}$ = 1008. After 60 atoms have passed $\langle\hat{n}\rangle$ = 4.47, σ = 0.332, $\langle S^z\rangle$ = -0.0379.

We recognize [23] decreasing T_p increases the likelihood of more than one atom in the cavity at the same time and this considerable complication is being explored.

13. CONCLUSIONS

The conclusion for the canonically quantised micromaser is that reached previously [23]: the possibility of Fock state creation in this micromaser will be enhanced by reducing T_p towards t_{int} bearing in mind, however, that as $T_p \rightarrow t_{int}$ the increasing probability of more than one atom in the cavity at the same time could substantially change and complicate predictions.

The message for Fock state creation therefore seems to be:- Reduce T_p *to some extent* and reduce T. But it may then be necessary to follow a sequential evolution of the cavity field as the inverted atoms enter since no steady state is apparently reached in the presence of black-body radiation (of whatever small amount).

Further detailed studies of trapping state evolution in the canonically quantised micromaser at realistic T_p and T's < 0.2^{o}K are currently in hand.

As for the q-quantised micromaser - we have not carried out black-body photon studies of this maser but we expect that the contrast between $Q = \infty$ and $Q < \infty$ theory as established for the canonically quantised maser remains qualitatively unchanged. Therefore we now hope for *quantitative* comparisons between all aspects of the two micromaser theories and the experiments. These should include *quantitative* comparisons between the q-boson theory of revivals and the observations of such revivals. In this way (§9) bounds $\neq 1$ could be put on any possible values of the parameter q. Moreover (§6) quantitative studies of the dynamics of the micromaser can show how good a model the J-C model is and whether the J-C interaction $g(a^{+}S^{-}+S^{+}a)$ (eqn. (29)) is a sufficient description of the atom-field interaction. The q-boson interaction $g(S^{+}a^{-} + a^{+}S^{-})$ (eqn. (50)) adds additional non-linearities to the J-C interaction (§6) which is furthermore taken in rotating wave approximation. Moreover, detailed studies of the photon statistics of the cavity field (where possible) might show up q-Poisson or q-sub-Poisson statistics instead of Poisson and sub-Poisson statistics (§9). It is also possible to see q-boson effects in the side bands of the one-atom fluorescence spectra (§9) and in the super-radiance (§8). However it is easy to show that the n-photon pure Fock state is the same state for both q-bosons and ordinary bosons [65]. We therefore expect that some reduction of T_p towards t_{int} will facilitate Fock state evolution in either form of micromaser. Once that Fock state is observed, necessarily under the non-steady conditions described in §11, we can hope to look further into the question of q-bosons in the micromaser and the possible values of q!

14. FURTHER READING

Further helpful if rather different reading can be found in 'Cavity Quantum Electrodynamics' by Serge Haroche and Jean-Michel Raimond, Scientific American, April 1993 pp.34-66 and in the reading list of six articles given on the p.66 there.

One of us (RKB) is grateful to the Center for Nonlinear Studes, Los Alamos National Laboratory, Los Alamos NM 87545, USA for the opportunity finally to complete the write-up of these two Lectures, I and II, and would like to acknowledge this fact.

REFERENCES

1. S.S. Hassan, R.K. Bullough, R.R. Puri and S.V. Lawande, *Physica* Vol.103A (1980), p.213
2. S.S. Hassan and R.K. Bullough in 'Optical Bistability' (C.M. Bowden, M. Ciftan, H.R. Robl eds.) New York, Plenum Press (1981) p.367.
3. R.R. Puri, S.V. Lawande and S.S. Hassan, *Optics Communications* Vol.35 (1980), p.179.
4. S.V. Lawande, R.R. Puri and S.S. Hassan, *Journal of Physics B : Atomic and Molecular Physics* Vol.14 (1981) p.4171.
5. S.S. Hassan, G.P. Hildred, R.R. Puri and S.V. Lawande, *Journal of Physics B : Atomic and Molecular Physics* Vol.15 (1982) p.1029.
6. S.S. Hassan, G.P. Hildred, R.R. Puri and R.K. Bullough *Journal of Physics B : Atomic and Molecular Physics* Vol.15 (1982) p.2635.
7. S.S. Hassan, G.P. Hildred, R.R. Puri and R.K. Bullough, in 'Coherence and Quantum Optics 5' (L. Mandel and E. Wolf eds.) New York, Plenum Press (1984) p.491.
8. R.R. Puri, G.P. Hildred, S.S. Hassan and R.K. Bullough in 'Coherence and Quantum Optics 5' (L. Mandel and E. Wolf eds.) New York, Plenum Press (1984) p.527.
9. G.P. Hildred, R.R. Puri, S.S. Hassan and R.K. Bullough *Journal of Physics B : Atomic and Molecular Physics* Vol.17 (1984) L 535.
10. R.K. Bullough *Hyperfine Interactions* (J.C. Baltzer A.G. Scientific Publishing Company) Vol.37 (1987) p.71.
11. R.K. Bullough and 10 co-authors in 'Squeezed and Non-classical Light', (P. Tombesi and E.R. Pike eds.) New York, Plenum Press, NATO ASI Series B : Physics Vol.190 (1989) p.81.
12. R.H. Dicke, *Physical Review* Vol.93 (1954) p.99.
13. R.R. Puri and S.V. Lawande, *Physics Letters* Vol.72A (1979) p.200.
14. R.R. Puri and S.V. Lawande, *Physica* Vol.101A (1980) p.599.
15. J.M. Raimond, P. Goy, M. Gross, C. Fabre and S. Haroche, *Physical Review Letters* Vol.49 (1982) p.117.
16. J.M. Raimond, P. Goy, M. Gross, C. Fabre and S. Haroche *Physical Review Letters* Vol.49 (1982) p.1924.
17. P. Goy, J.M. Raimond, M. Gross and, S. Haroche *Physical Review Letters* Vol.50 (1983) p.1903.
18. G.S. Agarwal, R.K. Bullough and G.P. Hildred, *Optics Communications* Vol.59 (1986) p.23.
19. G.S. Agarwal, R.K. Bullough and N. Nayak, *Optics Communications* Vol.85 (1991) p.202.
20. G.S. Agarwal, R.K. Bullough, G.P. Hildred and N. Nayak, 'Cavity quantum electrodynamics at arbitrary Q and T'. To be published.
21. S. Haroche, Private Communication at the Davos, Switzerland European Science Foundation meeting, November 1990.
22. S. Haroche, Private Communication at the European Quantum Electronics Conference EQEC'91, Edinburgh, Scotland, August 1991.
23. R.K. Bullough, N.M. Bogoliubov, N. Nayak and B.V. Thompson in 'Quantum Measurements in Optics' (P. Tombesi and D.F. Walls eds.) New York, Plenum Press (1992) p.129.
24. N. Nayak, R.K. Bullough, B.V. Thompson and G.S. Agarwal, IEEE Journal of Quantum Electronics Vol.24, No.7 (1988) p.1331.
25. N. Nayak, R.K. Bullough and B.V. Thompson in 'Coherence and Quantum Optics' (J.H. Eberly, L. Mandel and E. Wolf eds.) New York, Plenum Press (1990) p.809.
26. D. Meschede, H. Walther and G. Müller, *Physical Review Letters* Vol.54 (1985) p.551.
27. G. Rempe, H. Walther and N. Klein, *Physical Review Letters* Vol.58 (1987) p.353.
28. G. Rempe, F. Schmidt-Kaler and H. Walther, *Physical Review Letters* Vol.44 (1990) p.2783.

29. Gerhard Rempe and Herbert Walther, *Physical Review* Vol.42A (1990) p.1650.
30. R.K. Bullough, N. Nayak and B.V. Thompson in 'Proceedings of the International Workshop on Quantum Optics' (G.S. Agarwal and R. Inguva eds.) New York, Plenum Press (1992).
31. R.K. Bullough, N.M. Bogoliubov, N. Nayak and B.V. Thompson in Proceedings of the Dubna meeting 'Problems in Quantum Optics (Sept-Oct 1991)' (A.S. Shumovsky ed.) *Quantum Nonlinear Phenomena* Vol.2. To appear 1993.
32. E.T. Jaynes and F.W. Cummings, Proceedings of the Institute of Electrical and Electronic Engineers Vol.51 (1963) p.89.
33. R.K. Bullough and J. Timonen in 'Differential Geometric Methods in Theoretical Physics' Springer Lecture Notes in Physics Vol.375 (C. Bartocci, U. Bruzzo and R. Cianci eds.) Heidelberg, Springer-Verlag (1991), p.71.
34. R.K. Bullough, N.M. Bogoliubov and J. Timonen *Physica* Vol.51D (1991) p.43.
35. R.K. Bullough, S. Olafsson, Yu-Zhong Chen and J. Timonen in 'Differential Geometric Methods in Theoretical Physics' (Ling-Lie Chau and Werner Nahm eds.) New York, Plenum Press (1990) p.47.
36. J.J. Sanchez-Mondragon, N.B. Navozhny and J.H. Eberly, *Physical Review Letters* Vol.51 (1983) and references in Ref.[18].
37. M. Brune, J.M. Raimond and S. Haroche, *Physical Review* Vol.35A (1987) p.154.
38. M. Brune, J.M. Raimond, P. Goy, L. Davidovich and S. Haroche, *Physical Review Letters* Vol.59 (1987) p.1899.
39. R.R. Puri and R.K. Bullough *Journal of the Optical Society of America* B Vol.5, No 10 (1988) p.2021.
40. R.R. Puri, R.K. Bullough and N. Nayak in 'Coherence and Quantum Optics' (J.H. Eberly, L. Mandel and E. Wolf eds.) New York, Plenum Press (1990), p.943.
41. R.R. Puri and R.K. Bullough. To be published.
42. B.M. Garraway, R.K. Bullough, S.S. Hassan and R.R. Puri in 'Coherence and Quantum Optics' (J.H. Eberly, L. Mandel and E. Wolf eds.) New York, Plenum Press (1990), p.383.
43. B.M. Garraway, R.K. Bullough, S.S. Hassan and R.R. Puri in 'ECOOSA '90 - Quantum Optics', Institute of Physics Conference Series Vol.115 : Section 1 (M. Bertolotti and E.R. Pike eds.) Bristol, Institute of Physics (1991) p.69.
44. M.R.B. Wahiddin, B.M. Garraway, and R.K. Bullough, Journal of Modern Optics Vol.34, Nos. 6/7 (1987), p.1007.
45. M.R.B. Wahiddin, S.S. Hassan, J. Timonen and R.K. Bullough in 'Coherence and Quantum Optics' (J.H. Eberly, L. Mandel and E. Wolf eds.) New York, Plenum Press (1990) p.1195.
46. M.R.B. Wahiddin, S.S. Hassan, R.R. Puri and J. Timonen in 'ECOOSA '90 - Quantum Optics', Institute of Physics Conference Series Vol.115 : Section 1 (M. Bertolotti and E.R. Pike eds.) Bristol, Institute of Physics (1991) p.85.
47. S.S. Hassan, R.K. Bullough and H.A. Batarfi 'Optical dispersion relations and complex refractive index for atoms in squeezed vacua'. Paper in these proceedings.
48. R.K. Bullough, D.J. Pilling and J. Timonen in 'Solitons' (M. Lakshmanan ed.) Springer Series in Nonlinear Dynamics, Heidelberg. Springer-Verlag (1988) p.356.
49. R.K. Bullough, Yu-zhong Chen and J. Timonen, in 'Nonlinear World Vol.2' (V.G. Bar'yakhtar, V.M. Chernousenko, N.S. Erokhin, A.G. Sitenko and V.E. Zakharov eds.) Singapore, World Scientific (1990), p.1377.
50. S. Coleman *Physical Review* Vol.11D (1975) p.2088.
51. The canonical transformation is due to E.A. Power and S. Zienau - see R.K. Bullough and F. Hynne 'Quantum theory of dielectrics'. To be published 1993.

52. M. Tavis and F.W. Cummings, *Physical Review* Vol 170 (1968) p.179.
53. R.K. Bullough 'Optical Solitons, Chaos and All That : Quantum Optics and Nonlinear Phenomena' in Association of Asian Pacific Physical Societies Bulletin, Vol.2, No.2 (1992) p.23.
54. R.K. Bullough and N.M. Bogoliubov in 'Proceedings of the XXth International Conference on Differential Geometric Methods in Theoretical Physics' (S. Catto and A. Rocha eds.) Singapore, World Scientific (1992) p.485.
55. R.K. Bullough and N.M. Bogoliubov in 'Non-linear Processes in Physics' (A.S. Fokas, D.J. Kaup, A.C. Newell and V.E. Zakharov) Springer Series in Nonlinear Dynamics, Berlin, Springer-Verlag (1993) p.232.
56. R.K. Bullough, R.R. Puri and S.S. Hassan in 'Molecular and Biological Physics of Living Systems' (R.K. Mishra ed.) Amsterdam, Kluwer Academic Publishers (1990) p.1.
57. E.M. Purcell *Physical Review* Vol.69 (1946) p.681.
58. R. Saunders, S.S. Hassan and R.K. Bullough *Journal of Physics A : Mathematical and General* Vol.9 (1976), p.1725.
59. R.K. Bullough and R. Saunders in 'Co-operative effects in Matter and Radiation' (C.M. Bowden, D.W. Howgate and H.R. Robl eds.) New York, Plenum Press (1977), p.209 and other articles in this volume.
60. R.K. Bullough, R. Saunders and F. Ahmad, *Journal of Physics A*, Vol.8 (1975) p.759.
61. S.S. Hassan, M.R. Wahiddin, R.K. Bullough and R. Saunders, 'Atoms in a broad band squeezed vacuum: I, Master equation for the Dicke model'. To be published.
62. M.R.B. Wahiddin, Ph.D. Thesis, University of Manchester (August 1989).
63. J.I. Cirac, H. Ritsch and P. Zoller, *Physical Review* Vol.44A, No.7 (1991), p.4541.
64. N.M. Bogoliubov and R.K. Bullough, *Journal of Physics A : Mathematical and General* Vol.25 (1992) p.4057.
65. N.M. Bogoliubov, R.K. Bullough and G.D. Pang, *Physical Review* Vol.47B. To appear May 1993.
66. P. Meystre, G. Rempe and H. Walther, *Optics Letters* Vol.13 (1988) p.1078.
67. M. Chaichan, D. Ellinas and P. Kulish, *Physical Review Letters* Vol.65 (1990), p.980.
68. L.A. Takhtajan *Advanced Studies in Pure Mathematics* Vol.19 (1989), p.435.
69. P. Filipowicz, J. Javainen and P. Meystre, *Physical Review A* Vol.34 (1986) p.3077.
70. L.A. Lugiato, M.O. Scully and H. Walther, *Physical Review A* Vol.36 (1987) p.740.
71. F.W. Cummings and A.K. Rajagopal *Physical Review A* Vol.39 (1989) p.3414.
72. A.J. Macfarlane, *Journal of Physics A : Mathematical and General*, Vol.22 (1989), p.4581.
73. L.C. Biedenharn, *Journal of Physics A : Mathematical and General*, Vol.22 (1989), L873.
74. P.P. Kulish and E.V. Damaskinsky, *Journal of Physics A : Mathematical and General* Vol.33 (1990), L415.
75. A. Jannusis and 5 co-authors, *Hadronic Journal* Vol.6 (1983), p.1563 and references.
76. A. Jannusis, 'Quantum group and the Lie-Admissible Q-Algebra' in 'Proceedings of the 5th International Conference on Hadronic Mechanics and Nonpotential Interactions' and references. To appear 1991.
77. A Jannusis, G. Brodimas, S. Baskontas and D. Vavougios 'Quantum Group for Linear Combination of Bose and Fermi Operators' and references. Preprint 1990. Department of Physics, University of Patras, Greece.
78. N. Nayak, B.V. Thompson and R.K. Bullough in 'ECOOSA '90-Quantum

Optics' Institute of Physics Conference Series Vol.115 : Section 1 (M. Bertolotti and E.R. Pike eds.) Bristol, Institute of Physics (1991) p.81.

79. N. Nayak, B.V. Thompson and R.K. Bullough 'Cavity quantum electrodynamics - finite temperature theories of the ^{55}Rb atom maser'. To be published.

80. H. Risken, 'Fokker-Planck Equations', Berlin, Springer-Verlag (1984).

81. J. Kruse, M.O. Scully and H. Walther, *Physical Review A* Vol.36 (1987), p.4547.

82. M.O. Scully and Willis E. Lamb, Jr., *Physical Review* Vol.159 (1967), p.159.

83. M.O. Scully, Dae M. Kim and Willis E.Lamb, Jr., *Physical Review A* Vol.2 (1970) p.2529.

84. N. Nayak, B.V. Thompson and R.K. Bullough. To be published.

85. L. Allen and J.H. Eberly 'Optical Resonance and Two-level Atoms' New York, Dover (1987).

86. R.K. Bullough, 'Quantum theory of linear and non-linear refractive index'. Three lectures at Prescott, Arizona, 1970 in 'Physics of Quantum Electronics 1970' Optical Sciences Technical Report No.66, Tucson, University of Arizona (1971) pp.247-384.

87. S.S. Hassan and R.K. Bullough, *Physica A* Vol. 151 (1988) p.397.

88. S.S. Hassan and R.K. Bullough, *Physica A* Vol. 151 (1988) p.425.

89. S.S. Hassan, R.K. Bullough and R.R. Puri, *Physica A* Vol. 163 (1990) p.625.

Note added in proof:-

Of course it should have been pointed out in the §3 of the Lecture I, following eqns. (2) - (8), that if $aa^\dagger + a^\dagger a = 1$, and $(a)^2 = (a^\dagger)^2 = 0$ for fermions, then $a^\dagger, a$ and $\hat{n} \equiv a^\dagger a - \frac{1}{2}$ form a representation of su(2) for $S = \frac{1}{2}$ since $[a^\dagger, \hat{n}] = -a^\dagger$, $[a, \hat{n}] = a$, $[a^\dagger, a] = 2\hat{n}$. Moreover if $a^\dagger \equiv \sum_{j=1}^{N_A} a_j^\dagger$, $a = \sum_{j=1}^{N_A} a_j$, $\hat{n} = \sum_{j=1}^{N_A} (a^\dagger a - \frac{1}{2}) = (\sum_{j=1}^{N_A} a^\dagger a) - S$, then if the fermion-like operators $a_j^\dagger, a_j$ *commute*, rather than anticommute, for $i \neq j$, i.e. $[a_j^\dagger, a_j] = 0$, etc. rather than $a_j^\dagger a_j + a_j a_j^\dagger = 0$, etc., $a^\dagger$, a and $\hat{n}$ form an N_A+1 representation of su(2). Then, following eqn. (16), if $\bar{a} = (2S)^{-\frac{1}{2}} a$, $\bar{a}^\dagger = (2S)^{-\frac{1}{2}} a^\dagger$, $[\bar{a}, \bar{a}^\dagger] = 1$ for $S = \frac{1}{2} N_A \to \infty$; and this is the H-W algebra. However, because, for $i \neq j$, $a_j^\dagger, a_j$ etc. commute rather than anti-commute, this is not an N_A-fermions representation of su(2) for $N_A < \infty$, and it is not a fermion representation of the giant quantum oscillator for $S = \frac{1}{2} N_A \to \infty$.

Erratum: factors $\hbar$ are mistakenly omitted in eqn. (26).

Studies in Classical
and Quantum Nonlinear Optics

ISBN 1-56072-168-5

Transverse Nonlinear Optics

W. J. Firth
Department of Physics and Applied Physics
University of Strathclyde
Glasgow G4 0NG
Scotland

1. INTRODUCTION

The formation and evolution of spatial patterns is an interesting feature of many branches of science, ranging across biology, chemistry, hydrodynamics, and optics. Spontaneous formation of spatial patterns implies a spontaneous breaking of the symmetry of the system, which in turn implies the coexistence of several - or many - stable output states for a given input. Thus complex patterns are indicative of potential for image recognition and data processing capacity, an idea that is discussed later in this article.

Studies of optical systems are undergoing an expansion of nonlinear dynamics and chaos from the time domain into the space-time domain, a development which is essential to describe real, rather than idealised, nonlinear systems. The significance is much wider, however, because the space-time dynamics of lasers and associated nonlinear optical systems is also important for applications. Technological demand for high-power laser beams of high spatial coherence and quality is raising questions and demanding answers as to how lasers can be controlled more subtly than by strangling their output with apertures. Miniaturised and integrated nonlinear optical systems involving all-solid-state lasers, especially semiconductor lasers, are making great progress. Because of low cost and great robustness these systems promise to transform and greatly expand laser applications. But this increasing complexity of nonlinear optoelectronic devices and systems raises serious questions of stability and functionality. Because monolithic devices have few of the manual adjustments on which table-size lasers depend, these questions must be addressed at the design stage, and understanding of the operational characteristics and of potential problems assumes key importance.

Many research groups now focus their attention on systems where the spatial profile of the electromagnetic field assumes importance. A pair of review articles (with extensive bibliography of such studies) has recently been published as part of the Proceedings of the 1991 Solvay Conference [1,2]. These extend and update a review published in JOSA B [3] as part of a special issue on transverse effects in nonlinear optics.

These and related topics will be reviewed below without a great deal of detail. A very simple system, namely a single slice of Kerr medium coupled to a single feedback mirror, which we have analysed in detail and which is generating a substantial body of experimental observations, is discussed at rather more length. Linear stability analysis of this system [4] reveals both static and dynamic instabilities. Nonlinear analysis and numerical simulations [5,6] in two transverse dimensions are presented and interpreted, including in particular the spontaneous appearance of hexagonal patterns.

2. COEXISTENCE OF SPATIAL PATTERNS

A common first contact with laser-related spatial patterns is a set of transverse mode patterns [7] which have been widely reproduced in laser textbooks. These patterns, which quite closely reproduce the theoretical Hankel-Gaussian TEM_{mn} patterns, display many features of relevance for the present work.

First, these patterns were not truly spontaneous, in that deliberate misalignment was used to break the spatial symmetry of the cavity and select particular patterns. On the other hand, these adjustments were slight, so that the whole suite of patterns were in a sense "close" in the phase space of the laser. More recent work [8] has demonstrated patterns of comparable and indeed even greater complexity without misalignment, rather by adjusting the focus of a telescope internal to the cavity. Similarly, Weiss [2] shows that fine tuning of the cavity length can dramatically affect experimental laser patterns. In these works the breaking of the cylindrical symmetry seems to be truly spontaneous, which implies the coexistence of several output patterns consistent with the same laser conditions.

Coexistence, rather than mere existence, of spatial patterns in optics is the feature of most fundamental and practical interest in this topic. It generates the complexity of the patterns themselves, and through sensitivity to initial conditions allows chaos, variously spatial, transient, or truly dynamic. It also promises information capacity. Just as a bistable element can store one bit of information, so a multistable system can store many bits [9], provided the problems of coding, reading, and writing can be overcome.

Another very important feature of the Kogelnik-Rigrod [7] and later work is the relative simplicity of the patterns generated. This clearly represents a limit on the information capacity of these systems to a byte or two. The reason is basically one of available drive power in comparison with the nonlinearity of the optical system. Of course laser technology has made great strides since 1962, but it is still the case that the energy required to produce large-scale patterns using only "all-optical" techniques is often rather high. "All-optical" here means involving only light and atomic or similar quantum states directly coupled to the light, as in a gas laser. Much more complex patterns can be achieved when less directly optical systems are used, such as liquid crystal light valves. These have a complex coupling to light which induces changes in optical properties. Such changes with appropriate feedback can cause spatial and other instabilities [10], displaying richly complex patterns at modest input power levels.

Diode laser arrays typically also possess multiple coexisting states, frequently unstable. Here the applications objective is to limit the system to just one dominant spatial pattern. We can hope and expect that the analytical and simulation techniques which emerge from the studies of spatial patterns can find a useful role in attempts to improve the performance of laser arrays.

3. LINEAR AND NONLINEAR ANALYSIS

Most problems involving susbstantial nonlinearity defy analytic solution, even without spatial degrees of freedom. Approximate analytic methods are therefore essential, both as a preliminary and as a complement to numerical work. Modulational instability analysis, in which one examines the stabiltiy of a given exact solution with respect to perturbations with a certain wave vector K and/or a frequency Ω, is a key technique in this regard. The hope is that this wave vector, calculated by linearisation about the exact solution, will dominate the fully developed nonlinear solution and thus give the fundamental scale of the pattern, while Ω controls the basic frequency spectrum of a dynamically-evolving pattern .

The technique can only be applied where an essentially exact basic solution is known. In most cases, this restricts one to analysis of the stability of constant, homogeneous (plane-wave) solutions to the system equations. Even so, one frequently finds that the transverse wave vectors emerging from plane wave modulational instability analysis dominate the formation of patterns on more realistic finite beams, such as the gaussian beams characterisitic of lasers and laser-driven systems.

In optics, the modulational instability technique is well and usefully developed for unidirectional propagation in a Kerr medium. This is a medium for which the refractive index changes linearly with field intensity, increasing for what is called a self-focusing medium: and decreasing for a self-defocusing medium. The Kerr effect in a self-focusing medium tends to narrow the beam with consequent increase in intensity - a process which can run away, leading to catastrophic damage.

The theory is best understood for the case of only a single transverse dimension, which corresponds to propagation in a planar waveguide. This problem is formally identical to dispersive nonlinear waves in a monomode optical fibre, both being describable to a good approximation by the Nonlinear Schrodinger Equation (NLS), one of the best studied and understood of all nonlinear partial differential equations:

$$i\frac{\partial E}{\partial z}+\frac{1}{2}\frac{\partial^2 E}{\partial x^2}+|E|^2 E=0 \qquad (1)$$

Here the first and second order derivatives respectively describe propagation and diffraction (or dispersion), while the term in $|E|^2$ describes the Kerr effect, assumed positive here. All physical parameters have been scaled away in this dimensionless form, which is thus able to describe nonlinear waves in many areas of science [11].

The NLS has an exact homogeneous solution of constant amplitude, the nonlinearity only leading to a change of phase with propagation – self phase modulation. One readily finds that perturbations with transverse wave vector K grow on propagation over a range of K^2 extending from zero to $4I$, where I is the intensity of the homogeneous solution. This implies that arbitrarily weak plane waves are unstable in self-focusing Kerr media, but because the gain per unit length scales as I one needs a high intensity to see the effect in media of finite length. It is also easy to show that a negative or defocusing Kerr medium is modulationally stable.

Unusually for nonlinear partial differential equations, the NLS is exactly integrable, in the sense that given any initial field E(x,0) there is an exact algorithm (known as the inverse scattering method) by which the subsequent evolution E(x,z) of the field can be determined. In particular, there are a set of "soliton" solutions, which are asymptotically filaments or pulses of hyperbolic secant shape. These can propagate and even pass through each other, without permanent disturbance [11]. It is precisely these properties which make optical solitons so attractive a vehicle for optical comunications, but it does limit their interest from the point of view of pattern formation and complex dynamics. Nevertheless, modulational instability analysis provides useful information on the evolution, giving the scale of the periodic solutions (pulse trains or bundles of filaments), and some insight into the soliton solutions also.

Extension of the NLS to two dimensions fails for the self-focusing case, due to formation of a singularity as the beam self-focuses down to a point. Some dissipation or saturation of the Kerr effect, or both, is required to forestall such an unphysical occurrence. For a saturable medium enclosed in a cavity simulations suggest that that the "solitons" become rings, as might be expected, but on a somewhat longer timescale these ring structures are themselves modulationaly unstable, and the cylindrical symmetry spontaneously breaks with the formation of bright "spots" [12]. Even in a Kerr medium these bright spots will not collapse as they would in an extended medium, because the high spatial frequencies linked with collapse are continuously dissipated in a cavity of finite finesse, the input pump energy being supplied only at low spatial frequencies (in a broad gaussian beam).

More recently, studies of a simplified "mean field limit" [13] of a nonlinear cavity system with transverse effects showed qualitatively similar behaviour, in that the homogeneous solution may be modulationally unstable against pattern formation. In this sytem, however, the simulations indicate that *hexagonal* patterns are preferred, a result which is confirmed by analysis [14].

The number of bright spots in these systems seems approximately proportional to the area, so we might wish to consider each spot as a bit or pixel in an information store or processor. There is the very considerable practical problem of finding a suitable material system in which to generate these structures - most strongly nonlinear materials have the wrong sign of nonlinear coefficient, or other disadvantages such as poor spatial resolution. There are two further basic problems which have to be solved, first that the spots may move around chaotically, which is obviously unacceptable in an information system, and second that they are not themselves bistable: following switch-up of a broad beam, they fill the whole beam with spots.

What is needed is a method of pinning the spots in position, and making them individually bistable and addressable as binary pixels. It turns out that both these problems can be solved together, and rather simply, by applying a small constant spatial modulation to the pump beam, with a spatial scale comparable to a typical spot separation, and in any desired array pattern [9]. It was shown, in one transverse dimension, that 20 "soliton slots" could be individually addressed and switched. This implies that the system possesses over one million coexistent stable states for a single constant, spatially periodic input field.

4. COUNTERPROPAGATING BEAMS

Many, even most, nonlinear optical systems incorporate counterpropagating beams, especially those based on Fabry-Perot etalons, as are most lasers and many bistable devices. The extra degrees of freedom which bidirectional beams confer greatly increase the scope for complexity and pattern formation. No cavity structure is required, as Grynberg and co-workers have shown [15], with in particular observations of hexagonal patterns in the far field of laser beams counterpropagating in a sodium vapour cell with no mirror or other external feedback. The far field is the Fourier transform of the pattern within the medium itself, and we can thus infer spontaneous symmetry breaking within the medium leading to hexagonal patterns, albeit necessarily constrained by the power, and thus diameter, limitations of the input beams.

Grynberg has shown how cross-coupling of the phases of counterpropagating beams can lead to spatial instabilities [15]. Our group [17,18] has extended and developed such models to include detailed analysis and computer simulations, finding (in contrast to unidirectional propagation) instabilities even for defocusing media. Modulational stability analysis captures rather well the thresholds and dominant K-vectors of these patterns. The process was identified as a generalisation of the phenomenon of phase-conjugate self-oscillation [19] as had been shown much earlier in a limited regime by Soviet authors [20]. Numerical studies reveal much more, including hexagons [21]. Thresholds are relatively modest, and consistent with those observed. Hexagonal patterns have been observed directly, i.e. in the near field, in more recent experiments by the Grynberg group using rubidium vapour [22]. This experiment used *cw* excitation, and was thus able to categorise the pattern formation as a function of optical tuning around the resonance lines of rubidium.

Why should hexagons appear in such experiments and simulations? The basic reason is the same as for other physical systems such as fluid convection [23], which readily gives rise to hexagonal convection cells. Such cells in lava resulted in the Giant's Causeway in Ireland and other columnar basalt formations. Modulational instability usually determines only the magnitude, and not the direction, of the unstable wave vector – as for the NLS considered above. If by accident or design one particular direction is favoured, a pattern of stripes is formed, usually termed *rolls* from the convection case. Coupling of rolls through nonlinearity can give rise to more complex patterns. For example, three rolls forming an equilateral triangle may couple so as to mutually reinforce, thus forming the basic unit of a hexagonal pattern.

A little analysis shows that such coupling arises very naturally in the presence of a quadratic nonlinearity – the product of two roll amplitudes generating a source term for the third. Now the Kerr efect is a third-order (cubic) nonlinearity, as the NLS (1) shows, so how can it generate hexagons? The answer is that when such equations are rephrased in terms of the *deviations* from the homogeneous solution, quadratic terms normally arise. The main exception is where a symmetry of the system annuls such terms. This occurs for counterpropagating beams in a defocusing Kerr medium, and simulations indicate that squares, not hexagons, are the preferred pattern for that system [24].

In earlier experiments on sodium vapour [25] a counterpropagating beam was provided by an external lens/mirror combination. Patterns exhibiting complex dynamics on the time scale of the round trip to the mirror, but also on longer time scales, were observed, though limited in complexity by the input power available. A much simpler feedback mirror configuration, with Kerr medium and plane mirrors is described in the next section.

5. KERR SLICE WITH FEEDBACK MIRROR

This section reports very complex space-time behaviour in a rather simple optical system, as described in two recent papers [5,6] following a modulational stability analysis by the author [4]. Hexagon formation is prevalent close to threshold (Fig. 1). Several recent experiments on this kind of system have revealed pattern formation, often with hexagonal structures In addition, the system is functionally closely related to a feedback system in which a liquid crystal light valve is used to synthesise a Kerr nonlinearity [10].

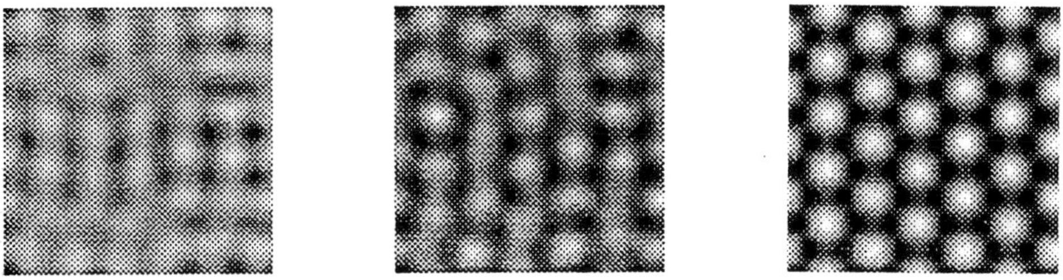

Figure 1. Spontaneous formation of a regular hexagon pattern in a defocusing Kerr slice with plane feedback mirror. Time increases from left to right, and the intensity of the feedback field is plotted: dark is dark and white is bright. (After [6]).

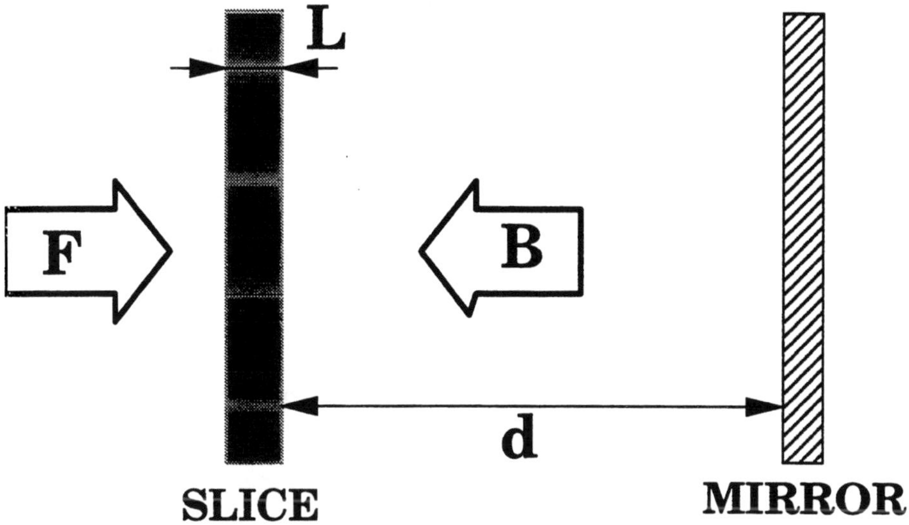

Figure 2. Schematic set-up of Kerr slice with feedback mirror. The forward amplitude F is smooth on input, but in pattern formation acquires a phase modulation in the Kerr slice through interaction with the feedback field B. B is formed from F by diffractive propagation to the mirror and back. (After [6]. See also Equations (2, 3).)

The basic model, as stated above, is very simple, namely a thin slice of Kerr medium irradiated from one side by a spatially smooth beam - indeed a plane wave in the analysis and these numerical studies - with a feedback mirror behind to generate a counterpropagating beam in the Kerr slice (Fig. 2). In this respect it somewhat resembles the Tucson feedback mirror experiment [25], and indeed this experiment was a motivation for the model. A further motivation lay in the counterpropagation instabilities described above, *viz.* the somewhat negative motivation that describing diffraction and nonlinearity in the same medium leads to computationally demanding models, so that it is of interest to examine models in which these effects are separated the better to understand their respective roles.

We describe the system as comprising a thin slice of Kerr-like medium in which the equation obeyed by the refractive index change Δn is:

$$(-\mathbf{l}_D^2\nabla^2 + \mathbf{t}\partial_t + \mathbf{1})\ \mathbf{Dn(r,t)} = |\mathbf{F}|^2 + |\mathbf{B}|^2 \quad (2)$$

Here F and B are respectively the amplitudes of the forward and backward travelling fields in the slice, whose intensities act as a source for the index change Δn. We neglect diffraction within the slice, so that these fields are simply phase modulated in traversing the slice, by an amount proportional to Δn. The Kerr excitation itself is assumed to relax with time constant τ and to diffuse transversely with diffusion length l_D. Here and throughout the Laplacian is purely transverse. In free space, the amplitude E of a field propagating in the +z direction obeys:

$$\frac{\partial \mathbf{E}}{\partial \mathbf{z}} = \frac{\mathbf{i}}{\mathbf{2k}}\nabla^2\mathbf{E} \quad (3)$$

Integrating this equation from slice to mirror and back determines B given F on exit from the slice, which is the input amplitude F_{in} modulated in phase in proportion to Δn. Equations (2,3) thus define the system of Fig. 2, and are the basis of both analysis and simulations [4, 5, 6]. Obviously the steady-state plane wave solution is trivially found. Modulational stability analysis shows that this solution becomes unstable for nonlinear phase shifts of around half a radian, for a feedback mirror of reasonably high relectivity.[4]. In particular, integrating this system for a defocusing medium gives rise to the hexagon formation shown in Fig. 1.

Spontaneous spatial structures do not usually arise in defocusing media, and this is confirmed by modulational stability analysis of plane wave solutions in such systems. Yet in some experiments [10, 25,26] spatial structures do appear in defocusing media. The answer lies not in the nonlinear medium itself but in the free-space propagation, and is thus well illustrated by the feedback mirror model.

What happens is that a smooth beam traversing a slice of Kerr medium gets no spatial structure of any kind from self-action, but can be *phase* modulated by a counterpropagating beam if that beam is *amplitude* modulated. Indeed only if that beam is amplitude modulated, because the Kerr effect is insensitive to the phases of the fields involved. A closed positive feedback loop is thus requires that the phase modulated input field returns from the mirror as an amplitude modulated beam. This interconversion of phase and amplitude modulation is which is precisely what diffractive propagation does as has been known for well over a century, since Talbot showed that a spatially modulated beam will self-image due to diffraction after a distance now called the *Talbot length.* This length T is equal to $2\Lambda^2/\lambda$ where Λ is the pattern period and λ the optical wavelength.

More relevant here is that a beam amplitude modulated at Λ becomes phase modulated after a distance $T/4$ and vice versa, i.e after a relative phase shift of $\pi/2$ as compared to the 2π of the full Talbot length. Accordingly in the feedback mirror system one could expect the wave vector of the spontaneous pattern to be that which makes the mirror distance just $T/8$. Indeed this is the case, both in the theory and the recent experiments by Tamburrini et al [27] employing a liquid crystal as Kerr-like medium. These authors pointed out the relation of the pattern formation to the Talbot effect, and also noted that the experimental scaling is more accurately Talbot-like than in the present author's theory, in which transverse diffusion of the nonlinearity causes an increase in the scale of the pattern.

To be precise the roundtrip distance is $T/4$ only where the Kerr effect is focusing. For a defocusing medium, the phase modulation is reversed, and thus one might expect that three-quarters of the Talbot length would be the optimum round-trip. Indeed this is the case in both theory and experiment. In fact the Rome group [27] were able to check both signs of nonlinearity by putting the mirror at a *negative* distance. This was achieved by interposing a lens between slice and mirror, which could be translated so as to project the image on either side of the slice. For a collimated input beam it is only the product of Kerr phase shift and diffraction distance which matters, and so a positive nonlinearity at a negative distance behaves just like a negative nonlinearity at positive distance.

The Talbot effect depends upon the beam being much broader than the pattern imposed upon it, since any significant whole-beam diffraction would diminish, and eventually destroy, the effect. Thus the pattern formation in the feedback mirror system has a certain global character, depending on constructive interference on feedback of wavelets originating in different parts of the slice. This global nature is especially significant for a defocusing medium, where both diffraction and nonlinearity are tending to disperse, not concentrate, the light. The spots forming the hexagonal pattern on the feedback beam act as negative lenslets for the input beam, pushing the input energy away from their centres. This is strikingly evident in Figure 3, which shows the evolution of the field amplitude as the beam propagates to the mirror and back in such a case [6]. The negative lensing causes a honeycomb pattern to form as the energy repelled from adjacent spots reinforces halfway between them, before continued diffraction causes at each spot constructive interference of light originating from neighbouring spots.

Using perturbation techniques from applied mathematics [11], nonlinear analysis of the Kerr slice with feedback mirror yields predictions, including hexagon formation, in reasonable ageement with simulations in the near-threshold regime [6]. These techniques cannot be relied upon for strong excitation, however, and both quantitative and qualitative deviations from the theory develop. In particular the hexagons break up and what appears to be optical turbulence ensues [6].

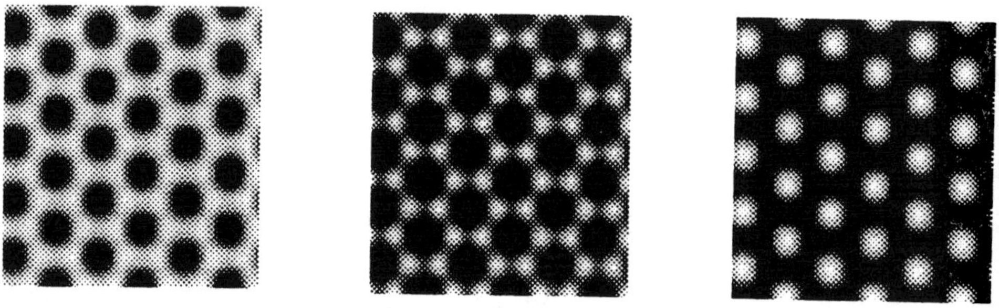

Figure 3. Evolution of the field amplitude E propagating under (2) from the slice to the mirror and back, for the case of a stable hexagonal pattern in a defocusing medium – the final frame of Fig. 1. The first frame is just after the slice, and shows the intial stages of the conversion from phase to amplitude modulation. The second frame shows the field close to the mirror, with a "honeycomb" structure. The rightmost frame is the backward field at the slice (as in Fig. 1), showing that the bright spots of the hexagonal structure cause the dark regions of the other two frames, because they act as negative lenses. (After [6]).

As well as the experiments already mentioned, Macdonald and Eichler demonstrated spot patterns using a liquid crystal cell with feedback mirror [28]. These patterns showed the Talbot scaling described above, but no unambiguous evidence for hexagonal structures. Honda reported hexagonal patterns in the far field of a photorefractive crystal with reflective feedback [29]. Grynberg (private communication) has observed hexagonal and other structures in a rubidium vapour cell with feedback mirror: one extra complcation was that the vector nature of the field seems to play a strong role, perhaps not surprisingly in view of the multilevel nature of alkali vapours.

A further complication, shared by most experiments, is that the input field had a gaussian-beam, rather than a plane-wave, distribution. This makes comparisons with theory and simulations less direct than one would like, and so it is desirable to try to analyse gaussian beam problems. From a numerical point of view, gaussian beams might seem easier to simulate than plane waves, but in reality plane waves are economically and effectively simulated by imposing periodic boundary conditions on a large but finite domain. While gaussian beams can be handled similarly, the need to avoid spurious effects due to the boundary makes for a larger domain of integration, much of which is "dark".

Finite beam effects are much more serious for the analyst, because there is no longer an exact solution whose modulational stability can be tested. It is for just this reason that most analysis concentrates on plane wave excitation, trusting that real beams will behave similarly, if broad enough.

One way forward is to concentrate on features of the beam which are qualitatively distinct from the plane wave, such as its *symmetry* [30]. In particular, while the extensive suite of translational and rotational symmetries possessed by the field equations are preserved under plane wave excitation, the *translational* symmetry is lost for gaussian beam input. Applying techniques of symmetry analysis, Papoff *et al* were able to categorise the possible bifurcations through which the surviving rotational symmetry could be broken. Not only were polygonal patterns predicted, including some – such as pentagons – which are forbidden in infinitely extended systems, but statements could be made about the *secondary bifurcations* through which such polygons lose stability [30]. These include the onset of bodily rotation of polygons (observed in the experiments of Giusfredi *et al* [25]), as well as chaotic itinerancy, in which there is an erratic succession of imperfect polygons. These analytic results were in good accord with the simulations reported in the same work [30], of which Figure 4 is a sample.

Further experiments involving two-dimensional optical feedback have been reported by Akhmanov and co-workers [10,31]. They used a system containing a liquid crystal light valve (LCLV) together with ring feedback which provides long-range transverse coupling. Recalling for a moment the Kerr slice system of Figure 2, it will be observed that the only effect of the feedback beam is to modulate the phase of the input beam – there is no energy exchange between the input and feedback beams. It follows that one could instead *synthesise* a Kerr effect through any kind of transducer which detects the intensity of the feedback beam and encodes the phase of the input beam in proportion. Spatial phase modulators in general, and liquid crystal light valves in particular, perform just this function. There is thus a close functional similarity between LCLV and Kerr slice systems. In particular one should expect hexagonal patterns for the case where the long range coupling in the feedback loop is entirely due to diffraction. A recent experiment of this kind by the Darmstadt group [32] shows a very large area pattern with a very strong hexagonal coordination over a substantial central domain.

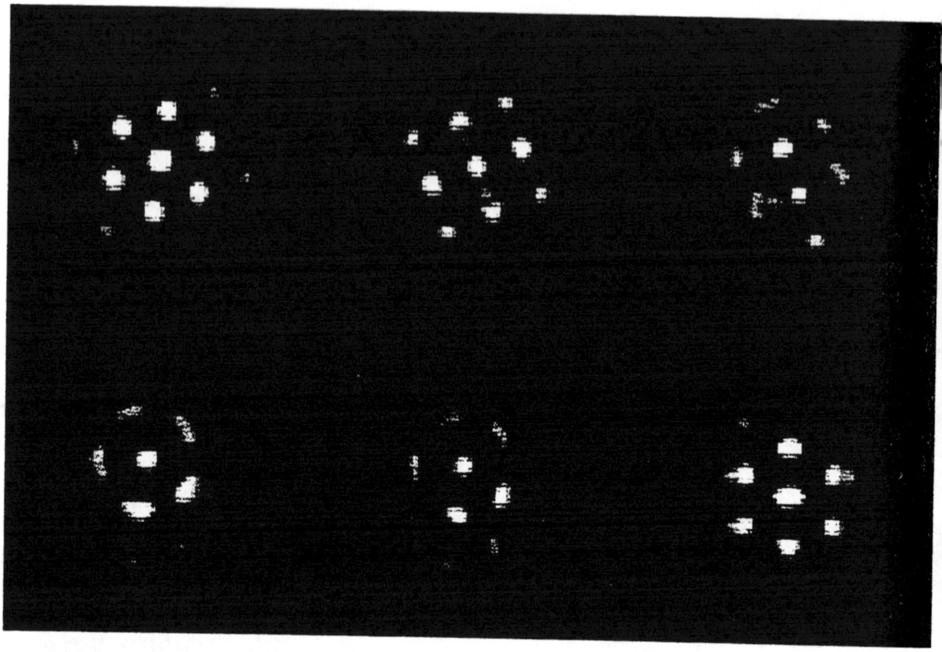

Figure 4. A set of frames showing chaotic itinerancy in the Kerr slice with feedback mirror for the case of gaussian beam input. The six frames (read from left to right) show an approximate sixfold symmetry evolving through stages of near fourfold, two-fold, one-fold and threefold symmetries before returning close to, but not quite, its original state. *(Courtesy F. Papoff).*

The LCLV system also points the way to a rich generalisation of pattern formation, since diffraction is by no means the only means of inducing long-range transverse coupling in the feedback loop. Aperturing and spatial filtering are others, but the most dramtic effects are achieved through bodily rotation of the feedback through a fixed angle. The Moscow group have observed strikingly complex and beautiful patterns, with domain structures, spiral waves, and even apparently turbulent character [10,31]. From a qualitative point of view, these patterns are amazingly similar to those seen in the video feedback experiments[33], where a TV camera is pointed at its own monitor, possibly rotated around the viewing axis. In a sense both are video feedback experiments, the latter being indirect via electronic processing in the TV system.

These last remarks, linking the "all-optical" Kerr slice systems to LCLV and video feedback experiments is important not only in stressing the universality of many pattern-forming processes, but also suggestive of routes to application. The Moscow group have already demonstrated optical logic and adaptive optics in the LCLV system, and there are clear links also to neural networks [34]. Some of these ideas may be developable in more conventional, and faster, optical materials such as semiconductors through exploitation of these functionality analogies.

6. TRANSVERSE PATTERNS IN LASERS

In line with the author's interests and experience, the dominant theme of this article has been *passive* nonlinear optical systems, systems which are externally driven by a coherent optical field (usually from a laser), and which exploit passive nonlinearities such as the Kerr effect. Many of the methods and techniques can, however, be carried over to *active* media such as lasers, and the review by Weiss [2] can be consulted for that purpose. In the Introduction, however, I stressed the technical importance of understanding and controlling the spatial instabilities leading to pattern formation in lasers. It is therefore appropriate to close with a few remarks on the similarities and differences between active and passive systems.

Perhaps the key difference is that hexagon formation *via* the quadratic coupling mechanism described above is generally absent in lasers. This can be traced to the fact that a free-running laser has an arbitrary phase. This symmetry disallows quadratic coupling, and means that laser patterns are dominated by the cubic couplings.

The arbirariness in the phase of the laser field implies the possible existence of *topological defects* in the field pattern. Regions of the laser medium remote from each other may start up with different phases, and there is no reason that these differences need anneal out as the coherence develops. Instead, the phase may have a singularity, increasing or decreasing by a multiple of 2π around a point defect, at which point the amplitude is necessarily zero. Such points are termed *optical vortices* by Coullet *et al* [35], and are topological in the sense that adding a constant to the field does not destroy the zero, but merely moves it:. Vortices are only created and destroyed in pairs, like electrical charges. Defects are zeroes of the field, and thus dark spots of the intensity pattern, but the converse is not necessarily true, especially when the term "defect" is deemed to imply a certain dynamical significance, e.g. as the mechanism for generating turbulence [35]. There is some debate over whether the phase singularities of the standard Gauss-Laguerre or Gauss-Hermite patterns [7] are defects in this strict sense [2].

Optical vortices occur only on what is conventionally described as the negative detuning side of the laser gain resonance line. On the positive detuning side, instead, curved-mirror resonators possess well-known transverse-mode resonances. For plane mirrors and infinitely extended gain profile, however, no properly confined transverse modes exist. It has been shown that such resonators exhibit, on the positive detining side, *exact* solutions in the form of transverse travelling waves [36]. These can be viewed as an accommodation to the peak gain wavelength being too long to fit the cavity (which is the case of positive detuning) by tilting the wave to fit. Indeed the lowest threshold mode occurs exactly at the wavelength of peak gain.

The pure travelling wave solutions are exact, as noted above, but of uniform amplitude. They thus give rise to patterns in the intensity only where they become unstable, or where they have defects, such as occurs at a boundary. Since all real lasers have finite transverse extent, the effect of boundaries on these modes, and by extension the manner in which they connect with the conventioanl transverse modes of curved-mirror resonators, is an interesting and important question. The development of broad area and surface-emitting semiconductor diode lasers in search of high power makes these questions technologically important, both because such lasers have plane endfaces and because of the need to understand and control their spatial properties.

7. CONCLUSION

In this paper I have tried to give evidence for my thesis that spatial patterns are interesting, complex and widespread in optics. Investigation of such patterns does present certain problems. On the experimental side there is the need for adequate power, preferably continuous-wave, over relatively large areas. As I have argued, this indicates a use of strong, and therefore usually slow, nonlinearities involving significant and maybe complex excitation dynamics within the material medium, while still retaining significant optical coupling to and with the medium. On the theoretical side, the problem is one of a lack of analytical tools, leading inevitably to numerical simulations, which in this field are extremely resource-hungry.

To control computer demand, one may consider one of a number of reduction strategies, to try to capture essentials within the simplest adequate model. Such strategies include the conception and investigation of the simplest possible physical systems, of which I believe the Kerr slice with feedback mirror described above to be a good example. They also include reduction of the number of equations necessary to describe the system through techniques such as adiabatic elimination, which is especially important in problems of the present kind, where very different timescales for different dynamical variables is the norm. A recent paper shows some of the possibilities and subtleties of adiabatic elimination in optical systems [37].

A further reduction strategy is the identification of organising structures which capture the essentials of ostensibly complex pattern formation and evolution. A fine example is soliton perturbation theory [38], which reduces problems of interacting solitary waves to the much simpler problem of Newtonian particle dynamics. Similarly, it is possible to consider some complex laser outputs to be describable as the interaction of optical vortices [35,39].

Lastly, these problems also represent opportunities. At the fundamental level experiments and theory on spatial patterns in optics may lead to new insights into the difficult and topical problems of morphogenesis, spatio-temporal chaos and turbulence. At the device level, the trend to increasingly complex optical systems such as diode laser arrays leads to problems of instability and spatial complexity which demand, and will respond to, the same kind of techniques as those fundamental investigations. Lastly, spatial complexity is a necessary feature of information processing systems, and the search for optical devices of this kind, especially in configurations such as neural networks which require a certain plasticity of design or operation, seems ripe for a symbiosis with studies of spontaneous pattern formation in optics.

ACKNOWLEDGMENT

Assistance from G. D'Alessandro, G. Harkness, A.J. Scroggie, G-L. Oppo and F. Papoff is gratefully acknowledged. Collaboration with J.V. Moloney and co-workers, E.M. Wright and C. Paré have contributed much to the material presented here. The author, however, accepts full responsibility for any errors of fact or attribution in this work. This work is supported in part by SERC grants GR/F 49811, GR/F 76502 and GR/G 12665, through NATO, and in part through ESPRIT BRA 7118 (TONICS).

REFERENCES

1. L.A. Lugiato, *Physics Reports* **219**, 1992 pp. 293-310.
2. C.O. Weiss, *Physics Reports* **219**, 1992 pp. 311-338.
3. N.B. Abraham and W.J. Firth, *J. Opt. Soc. Am.* **B7**, 1990 pp. 951
4. W.J. Firth, *J. Mod. Opt.* **37** 1990 pp. 151-154
5. G. D'Alessandro and W J Firth, *Phys. Rev. Lett.* **66**, 1991 pp. 1597-1600.
6. G. D'Alessandro and W J Firth, *Phys. Rev.* **A46**, 1992 pp. 537-548.
7. H. Kogelnik and W.W. Rigrod, *Proc. IRE* **50**, 1962 p. 220.
8. C. Green, G.B. Mindlin, E.J. D'Angelo, H.G. Solari and J.R. Tredicce, *Phys. Rev. Lett.* **65**, 1990 pp. 3125-3128.
9. G.S. McDonald and W.J. Firth, *J. Opt. Soc. Am.* **B7**, 1990 pp. 1328-1335.
10. S.A. Akhmanov, M.A. Vorontsov and V.Y. Ivanov, *JETP Lett.* **47**, 1988 pp. 707-.
11. A.C. Newell, *Solitons in Mathematics and Physics*, Philadelphia: SIAM, 1987.
12. J.V. Moloney, H. Adachihara, D.W. McLaughlin and A.C. Newell, *Chaos, Noise and Fractals*, E R Pike and L A Lugiato (Ed.), London: Hilger 1987 pp. 137-186.
13. L.A. Lugiato and R. Lefever, *Phys. Rev. Lett.* **59**, 1987 pp. 2208-2211.
14. W.J. Firth, A.J. Scroggie, G.S. McDonald and L.A. Lugiato, *Phys. Rev.* **A46**, 1992 pp. R3609-R3612.

15. G. Grynberg, E. LeBihan, P. Verkerk, P. Simoneau, J.R.R. Leite, D. Bloch, S. LeBoiteux and M. Ducloy, *Opt. Commun.* **67**, 1988 pp. 363-368.
16. G. Grynberg, *Opt. Commun.* **66**, 1988 pp. 321-324.
17. W.J. Firth and C. Paré, *Opt. Lett.* **13**, 1988 pp. 1096-1098.
18. W.J. Firth, A. Fitzgerald and C. Paré, *J. Opt. Soc. Am.* **B7**, 1990 pp. 1087
19. D.M. Pepper and A. Yariv, *Optical Phase Conjugation*, R.A. Fisher (Ed.), New York: Academic 1983 *and op cit.*
20. S.N. Vlasov and V.I. Talanov, in *Optical Phase Conjugation in Nonlinear Media*, V.I. Bespalov (Ed.), Gorki: Inst. of Appl. Phys., 1979.
21. R. Chang, W.J. Firth, R. Indik, J.V. Moloney and E.M. Wright, *Opt. Commun.* **88**, 1992 pp. 167-172.
22. A. Petrossian, M. Pinard, A. Maitre, J.Y. Courtois and G. Grynberg, *Europhys. Lett.*, **18**, 1992 pp. 689-694.
23. S. Ciliberto, P. Coullet, J. Lega, E. Pampaloni and C. Perez-Garcia, *Phys. Rev. Lett.* **65**, 1990 pp. 2370-2373.
24. J.B. Geddes, R.A. Indik, J.V. Moloney and W.J. Firth, submitted to *Phys. Rev. A* 1993.
25. G. Giusfredi, J.F. Valley, R. Pon, G. Khitrova and H.M. Gibbs, *J. Opt. Soc. Am.* **B5**, 1988 pp. 1181-.
26. R. Neubecker, M. Kreuzer, H. Gottschling and T. Tschudi, in *ECOOSA 90–Quantum Optics*, M. Bertolotti and E.R. Pike (Ed.), *IOP Conf. Series 115*, IOPP: Bristol 1991, pp. 161-164.
27. M. Tamburrini, M. Bonavita, S. Wabnitz and E. Santamato, *Opt. Lett.* **18**, 1993 pp. 855-857. E. Ciamramella, M. Tamburrini amd E. Santamato, submitted to *Appl. Phys. Lett.* 1993.
28. R. Macdonald and H.J. Eichler, *Opt. Commun.* **89**, 1992 pp. 289-294.
29. T. Honda, *Opt. Lett.* **18**, 1993 pp. 598-600.
30. F. Papoff, G. D'Alessandro, G.L. Oppo and W.J. Firth, *Phys. Rev.* **A48**, 1993 pp. 634-641.
31. S.A. Akhmanov, M.A. Vorontsov, V.Y. Ivanov, A.V. Larichev and N.I. Zheleznykh, *J. Opt. Soc. Am.* **B9**, 1992 pp. 78-90.
32. B. Thüring, R. Neubecker and T. Tschudi, *Opt. Commun.* 1993 *(in press).*
33. J. Crutchfield, *IEEE Trans. Circuits and Systems* **35**, 1988 pp. 776-.
34. D. Psaltis, D. Brady and K. Wagner, *Appl. Opt.* **27**, 1988 pp. 1752-.
35. P. Coullet, L. Gil and F. Rocca, *Opt. Commun.* **73**, 1989 pp. 403-408.
36. P.K. Jakobsen, J.V. Moloney, A.C. Newell and R. Indik, *Phys. Rev.* **A45**, 1992 pp. 8129-8137.
37. G.L. Oppo, G. D'Alessandro and W.J. Firth, *Phys. Rev.* **A44**, 1991 pp. 4712-4720.

38. A.B. Aceves, P. Varatharajah, A.C. Newell, E.M. Wright, G.I. Stegeman, D.R. Heatley, J.V. Moloney and H. Adachihara, *J. Opt. Soc. Am.* **B7**, 1990 pp. 963-974.
39. F.T. Arecchi, G. Giacomelli, P.L. Ramazza and S. Residori, *Phys. Rev. Lett.* **65**, 1990 pp. 2531-4.

Studies in Classical
and Quantum Nonlinear Optics

ISBN 1-56072-168-5

MOLECULAR SPECTROSCOPY AND DYNAMICS MODIFIED BY A METALLIC SURFACE

Thomas F. George
Departments of Chemistry and Physics,
Washington State University, Pullman Washington 99164-1046, USA

Henk F. Arnoldus
Department of Physics, Mendel Hall, Villanova University,
Villanova, Pennsylvania 19085, USA

ABSTRACT

Spectroscopic properties of atoms and molecules are altered by the presence of a dielectric or metallic surface. Also, the dynamical behavior changes when an atom or molecule is close to such a medium, even when it is physically separated from the material. Starting from Heisenberg's equation of motion for the density operator of an atom in vacuum, but near a medium, we derive a general expression for the atomic relaxation operator that accounts for spontaneous decay. The coefficients in this operator are the relaxation constants which equal the inverse lifetimes of excited states. These constants depend on the details of the environment, and they contain information about the medium and the geometry. In a photoabsorption experiment, these parameters are equal to the linewidths of spectral lines, and this provides a method of measuring the relaxation constants. On the other hand, the relaxation operator appears in the equation of motion for the atomic density operator, and therefore it affects the time evolution of the atomic state. This dynamical dependence on the presence of a medium influences, for instance, the spectral distribution of emitted radiation (fluorescence), which provides an alternative way of observing the surface-modified dynamics of atoms or molecules.

1. INTRODUCTION

In a series of experiments in the mid-70's by Drexhage [1], it was demonstrated that the decay rate of excited dye molecules is different near a metal substrate, as compared to decay in empty space. In these experiments, the molecules were held at a fixed distance from a metal surface by depositing first a thin layer of neutral material (fatty acid) on the metal. In this way, the distance between the molecule and the metal could be controlled. The decay rate of the molecules was then measured as a function of the separation distance, and it was found that the lifetime of an excited state could be either enhanced or reduced. In more recent experiments, lifetimes of excited atoms near a mirror, in between two mirrors, and in a microwave cavity were measured, and both inhibition and enhancement of spontaneous decay was observed [2-5]. These results were in good qualitative agreement with theory [6-10].

Relaxation constants, or inverse lifetimes, have been calculated in many ways: semiclassical radiation theory, linear-response theory, Green's function techniques, Fermi's Golden Rule, reservoir theory, etc. We shall adopt the reservoir approach because it not only gives the relaxation constants, but also the equation of motion for the atomic density operator. From this equation, atomic correlation functions which determine the probe absorption profile can be obtained. The spectral distribution of the emitted fluorescence during the decay can also be evaluated from the equation of motion for the atomic density operator. Also determined by this equation are the temporal photon correlations and the photon statistics of emitted radiation, either in spontaneous decay or in resonance fluorescence.

2. LIOUVILLE FORMALISM

The quantum state of an atom or molecule near a medium can be represented by a density operator $\rho(t)$. Spontaneous decay of an atom is brought about by the interaction between the atom and the radiation field of the vacuum, and therefore the density operator must also include the state of this radiation field. On one hand, we could consider the emission of fluorescence during spontaneous decay, regardless of what happens to the atom. In this paper, however, we will be interested in the state and behavior of the atom, regardless of the state of the radiation field. This is accounted for by the partial density operator of the atom, which is defined as

$$\rho_a(t) = Tr_r\, \rho(t) \quad . \tag{1}$$

Here the trace runs over a complete set of states of the radiation field only, and therefore $\rho_a(t)$ is an operator in the Hilbert space of the atom. The purpose of relaxation theory is to derive an equation of motion for $\rho_a(t)$, which takes into consideration that there is a coupling to the electromagnetic field, leading to spontaneous decay, but such that the photons do not appear explicitly anymore in this equation. The starting point is the equation of motion of the total system

$$i\hbar\frac{d\rho}{dt} = [H, \rho] \quad , \qquad H = H_a + H_r + H_{ar} \quad , \tag{2}$$

where H_a, H_r and H_{ar} are the Hamiltonians of the atom, the radiation, and the interaction, respectively. Equation (2) is subject to the constraints

$$\rho^{\dagger} = \rho \quad , \quad Tr\,\rho = 1 \quad , \tag{3}$$

and when we take the trace over the states of the radiation field in both equations, this implies

$$\rho_a^{\dagger} = \rho_a \quad , \quad Tr_a\,\rho_a = 1 \tag{4}$$

for the partial atomic density operator.

It will turn out that spontaneous decay can not be accounted for by an effective Hamiltonian, leading to an equation of type (2) for the atomic density operator. It is possible, however, to represent the relaxation by a linear operation on the reduced density operator. It is therefore convenient to introduce the concept of Liouville space. In this linear vector space, sometimes called superspace, a vector is identified with an operator in Hilbert space. That is, there is by definition a one-to-one correspondence between the operators in Hilbert space and the vectors in this Liouville space. For instance, the three Hamiltonians in (2) are vectors, and so is the density operator and the reduced density operator. Operators in Liouville space (Liouvillians) act on these vectors, and the result of such an operation is again a vector. Operators can be added, multiplied, exponentiated, etc. in the usual way, and these operations follow the general rules of a linear vector space. It should be noted that when the dimension of the Hilbert space is N, then the dimension of the corresponding Liouville space is N^2.

As a first example we define the Liouvillian L by its action on an arbitrary Liouville vector Π,

$$L\Pi = \frac{1}{\hbar}[H, \Pi] \quad , \tag{5}$$

where H is the total Hamiltonian of the system. Similarly, the operators L_a, L_r, and L_{ar} are the commutators with the corresponding Hamiltonians. Then the equation of motion (2) can be written as

$$i\frac{d\rho}{dt} = L\rho \quad , \tag{6}$$

or equivalently

$$i\frac{d\rho}{dt} = (L_a + L_r + L_{ar})\rho \quad . \tag{7}$$

The solution of (2) is

$$\rho(t) = e^{-\frac{i}{\hbar}Ht}\rho(0)e^{\frac{i}{\hbar}Ht} \quad , \tag{8}$$

and the solution of (6) is

$$\rho(t) = e^{-iLt}\rho(0) \quad . \tag{9}$$

Comparison of (8) and (9) then leads to the important operator identity

$$e^{-iLt}\Pi = e^{-\frac{i}{\hbar}Ht}\Pi e^{\frac{i}{\hbar}Ht} \tag{10}$$

for Π arbitrary (as it will be from now on). Relation (10) holds of course for any Liouvillian which is the commutator with a Hilbert space operator, divided by $\hbar$.

3. RELAXATION THEORY

In this section we formulate the general theory of relaxation in the Liouville formalism. The electromagnetic field is considered as a large reservoir, or heat bath, and the atom as a small system which interacts with the reservoir. An integral of (7) is

$$\rho(t) = e^{-i(L_a+L_r)t}\rho(t_o) - i\int_0^t dt' \, e^{-i(L_a+L_r)(t-t')}L_{ar}\rho(t') \quad , \tag{11}$$

as can be verified by differentation. Then we substitute this back into (7), which yields

$$i\frac{d}{dt}\rho(t) = (L_a + L_r)\rho(t) + L_{ar}e^{-i(L_a+L_r)t}\rho(t_o)$$

$$-iL_{ar}\int_0^t dt' \, e^{-i(L_a+L_r)(t-t')}L_{ar}\rho(t') \quad . \tag{12}$$

Then we take the trace over the states of the radiation field, and use

$$Tr_r \, L_r\Pi = 0 \quad , \tag{13}$$

which follows from the fact that L_r is a commutator. This gives

$$i\frac{d}{dt}\rho_a(t) = L_a\rho_a(t) + Tr_r \, L_{ar}e^{-i(L_a+L_r)t}\rho(t_o)$$

$$-iTr_r \, L_{ar}\int_0^t dt' \, e^{-i(L_a+L_r)(t-t')}L_{ar}\rho(t') \tag{14}$$

as an exact integral of the equation of motion.

Equation (14), however, is not a closed equation for the atomic density operator because the last two terms on the right-hand side still contain the full density operator of the system. The first term on the right-hand side is the free evolution of the atom, without coupling to the radiation field. The radiation field is a large reservoir, and we shall assume it to be in its thermal-equilibrium state $\overline{\rho}_r$. This operator has the properties

$$\overline{\rho}_r^{\dagger} = \overline{\rho}_r \quad , \quad Tr_r\, \overline{\rho}_r = 1 \quad , \quad [H_r , \overline{\rho}_r] = 0 \quad . \tag{15}$$

It can be shown that in the density operator $\rho(t_o)$ in (14) can be factorized according to

$$\rho(t_o) = \rho_a(t_o)\overline{\rho}_r \quad , \tag{16}$$

and similarly $\rho(t')$ in the last term can be factorized. This involves a slight approximation (the Markov approximation), but for the present problem this is absolutely negligible. Then we assume that the interaction is such that

$$Tr_r\, L_{ar}(\rho_a(t_o)\overline{\rho}_r) = 0 \quad , \tag{17}$$

which expresses that the average interaction energy is equal to zero at t_0. For spontaneous decay, this is exact for all times t. From (15) - (17) it then follows that the second term on the right-hand side of (14) vanishes. It can be shown that in the last term the major contribution to the integral comes from the region t' = t in the integrand, and we can safely replace the lower limit by $-\infty$. Then we make the substitution $t - t' = \tau$ which gives

$$i\frac{d}{dt}\rho_a(t) = L_a\rho_a(t) - iTr_r\, L_{ar}\int_0^{\infty} d\tau\, e^{-i(L_a+L_r)\tau} L_{ar}(\rho_a(t-\tau)\overline{\rho}_r) \quad . \tag{18}$$

Next, under the integral sign we replace the atomic density operator by its free evolution value during the time τ, e.g.

$$\rho_a(t-\tau) = e^{iL_a\tau}\rho_a(t) \quad , \tag{19}$$

and this yields

$$i\frac{d}{dt}\rho_a(t) = L_a\rho_a(t) - iTr_r\, L_{ar}\int_0^{\infty} d\tau\, e^{-i(L_a+L_r)\tau} L_{ar} e^{iL_a\tau}(\rho_a(t)\overline{\rho}_r) \quad . \tag{20}$$

Then we define a Liouville operator Γ in atomic Liouville space by its action on an arbitrary atomic operator or Liouville vector Π_a:

$$\Gamma\Pi_a = Tr_r\, L_{ar}\int_0^\infty d\tau\, e^{-i(L_a+L_r)\tau}L_{ar}e^{iL_a\tau}(\Pi_a\overline{\rho}_r) \quad . \tag{21}$$

This time-independent operator Γ accounts for the relaxation of the atomic density operator, and the equation of motion for $\rho_a(t)$ becomes

$$i\frac{d\rho_a}{dt} = (L_a - i\Gamma)\rho_a \quad . \tag{22}$$

Given an initial state $\rho_a(0)$, the solution of this equation is

$$\rho_a(t) = e^{(-iL_a-\Gamma)t}\rho_a(0) \quad . \tag{23}$$

4. SPONTANEOUS DECAY

In order to evaluate the relaxation operator Γ explicitly, we have to specify the three parts of the Hamiltonian and the density operator of the thermal reservoir. We shall assume that the atom is located in empty space, somewhere near a medium. Then $\overline{\rho}_r$ is the density operator of the electromagnetic field. This operator is therefore the density operator of black-body radiation, and for the visible region of the optical spectrum this is simply the vacuum state. Consequently, we have

$$\overline{\rho}_r = |0\rangle\langle 0| \quad , \tag{24}$$

with $|0\rangle$ the multimode zero-photon Fock state. For visible radiation the wavelength is much larger than the size of an atom, and therefore the interaction between the atom and the electromagnetic field is governed by the electric dipole coupling. Let the atom have a dipole operator $\vec{\mu}$ and be located at position $\vec{r} = \vec{h}$ in space. The electric field operator is $\vec{E}(\vec{r})$, and the dipole interaction Hamiltonian is

$$H_{ar} = -\vec{\mu}\bullet\vec{E}(\vec{h}) = -\sum_{i=x,y,z}\mu_i E(\vec{h})_i \quad . \tag{25}$$

In the product $\mu_i E(\vec{h})_i$ the first factor is pure atomic and the Cartesian component of the electric field acts only in the Hilbert space of the radiation. Therefore, these two operators commute, and this can be used to simplify the expression for Γ considerably. Working out the commutators that define the Liouvillians and applying identity (10) a few times then yields

$$\Gamma\Pi_a = \sum_i [\mu_i\, , Q_i\Pi_a - \Pi_a Q_i^\dagger] \quad . \tag{26}$$

The Hilbert space operators Q_i are defined by

$$Q_i = \int_0^\infty d\tau\, e^{-iL_r\tau} \sum_j f_{ij}(\tau)\mu_j \tag{27}$$

in terms of the functions

$$f_{ij}(\tau) = \frac{1}{\hbar^2} < 0|\{e^{iL_r\tau} E(\vec{h})_i\} E(\vec{h})_j|0 > \quad . \tag{28}$$

We have also used that μ_i and $E(\vec{h})_i$ are Hermitian operators. In expression (26) for Γ, all operators act in atomic space only, and the same holds for Q_i in (27). The radiation field only appears through the functions $f_{ij}(\tau)$, which can be considered as the vacuum correlation functions. Hence all properties of the surrounding medium, its geometry, and the way it modifies the vacuum are incorporated in these nine (complex-valued) functions. On the other hand, the functions $f_{ij}(\tau)$ are determined completely by the medium, and are independent of the presence of the atom. In this way, the problem of spontaneous decay of any atom (or molecule) in the vicinity of media is reduced to the calculation of these vacuum correlation functions.

A great simplification arises when the geometry has a cylindrical symmetry about, say, the z-axis and when the atom is located on this axis. Then it can be shown that

$$f_{ij}(\tau) = 0 \quad , \quad i \neq j \quad , \tag{29}$$

and that

$$f_{xx}(\tau) = f_{yy}(\tau) \quad . \tag{30}$$

Consequently, there are only three nonzero correlation functions, and only two different ones. We shall write $f_{\parallel}$ for f_{xx} and $f_{\perp}$ for f_{zz}. This situation arises, for instance, for any flat substrate. Then the notation $\parallel$ indicates that these functions couple to the parallel component of the dipole moment with respect to the surface, which is the direction is perpendicular to the symmetry axis, and vice versa.

5. TWO - LEVEL SYSTEM

In most practical situations, only one particular level of an atom will be excited, either by collisions or by resonant excitation with a laser beam. Then the atom can decay from this state to the ground state under emission of a photon, and the relaxation of its density operator is described by (22). With P_e and P_g as the projectors onto the excited state and ground state, respectively, the atomic Hamiltonian can be represented as

$$H_a = \hbar\omega_e P_e + \hbar\omega_g P_g \quad , \tag{31}$$

so that $\hbar\omega_e$ and $\hbar\omega_g$ are the respective energies of the levels. Both levels can possibly be degenerate. When only two levels take part in the dynamics, then the projectors obey the closure relation:

$$P_e + P_g = 1 \quad . \tag{32}$$

With the use of (32) the Liouvillian for the free atomic evolution becomes

$$L_a \Pi = \omega_o [P_e , \Pi] \quad , \tag{33}$$

where $\omega_o = \omega_e - \omega_g$ is the level separation. From now on we drop the subscript *a* on the atomic density operator and on the arbitrary operator Π, as in (33).

Assuming cylinder symmetry, the operator Q_i reduces to

$$Q_i = \int_0^\infty d\tau \, f_{ii}(\tau) e^{-iL_a\tau} \mu_i \quad . \tag{34}$$

The exponential in (34) can be expanded in projectors as

$$e^{-iL_a\tau}\Pi = \sum_{\alpha\beta} e^{-i(\omega_\alpha - \omega_\beta)\tau} P_\alpha \Pi P_\beta \quad , \tag{35}$$

were the summation runs over the four combinations of *e* and *g*. For an atom, the dipole operator $\vec{\mu}$ has only transition matrix elements between the excited state and the ground state, so that

$$P_e \mu_i P_e = P_g \mu_i P_g = 0 \quad . \tag{36}$$

A molecule can have a permanent dipole moment in both states, but it can be shown that this would not contribute to the relaxation operator. From (35) and (36) we then find

$$e^{-iL_a\tau}\mu_i = e^{-i\omega_o\tau} P_e \mu_i P_g + e^{i\omega_o\tau} P_g \mu_i P_e \quad . \tag{37}$$

Substituting (37) into (34) and carrying out the integration then yields

$$Q_i = \bar{f}_{ii}(-\omega_o) P_e \mu_i P_g + \bar{f}_{ii}(\omega_o) P_g \mu_i P_e \quad , \tag{38}$$

where $\bar{f}(s)$ is the Fourier-Laplace transform of the vacuum correlation function:

$$\bar{f}(s) = \int_0^\infty d\tau \, e^{is\tau} f(\tau) \quad . \tag{39}$$

As we shall see later on, the function $\bar{f}(s)$ vanishes for $s < 0$. This gives for Q_i

$$Q_i = \bar{f}_{ii}(\omega_o)P_g\mu_i P_e \quad , \tag{40}$$

and we notice that only the Fourier-Laplace transform of the correlation function, evaluated at the atomic transition frequency, appears in the expression for the relaxation operator.

From (26) and (40) we then obtain for Γ

$$\Gamma\Pi = \sum_{i=x,y,z}\{\bar{f}_{ii}(\omega_o)\mu_i^2 P_e\Pi + \bar{f}_{ii}(\omega_o)^*\Pi P_e\mu_i^2 - 2\mu_i P_e\Pi P_e\mu_i Re\,\bar{f}_{ii}(\omega_o)\} \quad . \tag{41}$$

Here we have dropped the so-called nonsecular terms, since they contribute negligibly to the relaxation in the visible domain of the spectrum [11]. It will appear that the function $\bar{f}_{ii}(\omega_o)$ is real, and therefore we can simplify (41) slightly to

$$\Gamma\Pi = \sum_{i=x,y,z}\bar{f}_{ii}(\omega_o)\{\mu_i^2 P_e\Pi + \Pi P_e\mu_i^2 - 2\mu_i P_e\Pi P_e\mu_i\} \quad . \tag{42}$$

From (32) and (36) we can obtain identities like $P_e\mu_i P_g = P_e\mu_i$, which have been used frequently in the derivation of (41).

6. TWO - STATE SYSTEM

When the excited state has an angular-momentum quantum number j_e then this state is $(2j_e+1)$-fold degenerate, and similarly the ground state will be $(2j_g+1)$-fold degenerate. For such a situation, (42) can be worked out in terms of Clebsch-Gordan coefficients and the reduced dipole matrix element as it appears in the Wigner-Eckart theorem for vector operators [12]. In many experimental situations an atom is effectively prepared as a two-state system by selective optical pumping [13]. Then the single excited state can be denoted by $|e\rangle$, the ground state is $|g\rangle$, and the projectors are

$$P_e = |e\rangle\langle e| \quad , \quad P_g = |g\rangle\langle g| \quad . \tag{43}$$

The dipole operator obeys the equality

$$\vec{\mu} = P_e\vec{\mu}P_g + P_g\vec{\mu}P_e \tag{44}$$

because of (36), and for a two-state system this can be written as

$$\vec{\mu} = \vec{\mu}_{eg}d + \vec{\mu}_{eg}^{*}d^{\dagger} \quad , \tag{45}$$

in terms of the atomic raising and lowering operators

$$d = |e\rangle\langle g| \quad , \quad d^{\dagger} = |g\rangle\langle e| \quad . \tag{46}$$

In (45), $\vec{\mu}_{eg} = \langle e|\vec{\mu}|g\rangle$ is the transition dipole matrix element. We shall assume that the two states are chosen such that this matrix element is either parallel or perpendicular, in the same sense as for the function $f(\tau)$.

With these notations, (42) can be simplified to

$$\Gamma\Pi = \tfrac{1}{2}A\{P_e\Pi + \Pi P_e - 2d^{\dagger}\Pi d\} \quad , \tag{47}$$

with A as the Einstein coefficient for spontaneous emission (or decay), defined as

$$A = 2\bar{f}(\omega_o)\vec{\mu}_{eg} \bullet \vec{\mu}_{eg}^{*} \quad . \tag{48}$$

There are two possible values for A, depending on the two possibilities for the function $\bar{f}(\omega_o)$, and we shall indicate them by $A_{\parallel}$ and $A_{\perp}$, in obvious notation. In the more general case that the dipole matrix element has both a parallel and perpendicular component, *A* in (47) has to be written as $A_{\parallel}$ and a line identical to the right-hand side has to be added, but with $A_{\parallel}$ replaced by $A_{\perp}$. For an atom in empty space, the Einstein coefficient is

$$A = \frac{\omega_o^{\;3}}{3\pi\varepsilon_o\hbar c^3}\vec{\mu}_{eg} \bullet \vec{\mu}_{eg}^{*} \quad , \tag{49}$$

and this should be the limiting value for the case where the atom is far away from any medium.

7. EQUATION OF MOTION

For a two-state atom in the vicinity of any medium, but without an external field or collisions with other particles, the equation of motion is (22)

$$i\frac{d\rho}{dt} = (L_a - i\Gamma)\rho \quad , \tag{50}$$

with L_a from (33) and the operator Γ given by (47). Let us indicate the populations of the two levels by n_e and n_g. They are equal to the matrix elements

$$n_e = \langle e|\rho|e\rangle \quad , \quad n_g = \langle g|\rho|g\rangle \tag{51}$$

of the atomic density operator. They are subject to the constraint

$$n_e + n_g = 1 \quad , \tag{52}$$

as follows from $Tr\,\rho = 1$. When we take matrix element <e|...|e> of (50), we find

$$\frac{dn_e}{dt} = -An_e \quad , \tag{53}$$

which is an equation for the population of the excited state only. The solution is

$$n_e(t) = n_e(0)e^{-At} \quad , \tag{54}$$

which is simple exponential decay from the excited state to the ground state. The time constant for the decay is $1/A$, which equals the lifetime of the excited state. The population of the ground state then follows from (52). Taking the matrix element <e|...|g> of (50) gives

$$\frac{d\rho_{eg}}{dt} = (-i\omega_o - \tfrac{1}{2}A)\rho_{eg} \quad , \tag{55}$$

which is an equation for the coherence $\rho_{eg} = \langle e|\rho|g \rangle$ between the two levels. The solution is

$$\rho_{eg}(t) = \rho_{eg}(0)e^{-i\omega_o t}e^{-\frac{1}{2}At} \quad , \tag{56}$$

showing that the coherence decays at a rate A/2, which is twice as slow as the population n_e. The coherence also oscillates with the atomic transition frequency.

Observable fluorescence during the decay has a photon emission rate proportional to the population of |e>, and the absorption profile for weak external radiation is determined by the regression of the atomic coherence. Consequently, all measurable influence by a medium on the dynamics and the spectroscopic properties of an atom must be contained in the Einstein coefficient A, and therefore in the Fourier-Laplace transform of the vacuum correlation function, $\bar{f}(\omega_o)$.

8. DIELECTRIC LAYER

In order to illustrate the dramatic effect that the presence of a medium can have on the Einstein *A* coefficient, we work out the situation where the medium is a slab of dielectric material with a dielectric constant equal to ε. The boundaries of the medium will be the planes $z = 0$ and $z = -\Delta$, and the atom will be located on the positive z-axis at $\bar{r} = h\bar{e}_z$, $h > 0$. This geometry is illustrated in Fig. 1. We need an expression for the electric field operator in the region $z > 0$ and the Hamiltonian of the radiation field in order to evaluate $f_{ii}(\tau)$ from (28).

In empty space, the electric field in a quantization box with volume V is given by

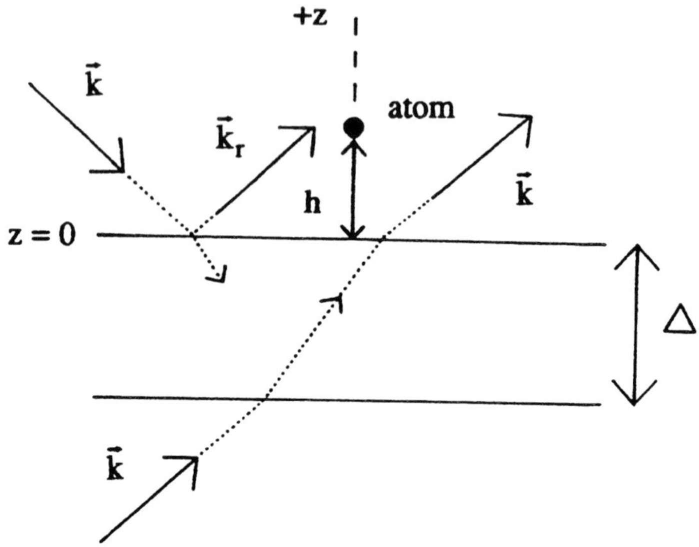

Fig. 1. An incoming vacuum wave from the positive z-direction reflects at the surface, and partially refracts. A wave from the negative z-direction is partially transmitted through the layer, and contributes to the field in z > 0.

$$\vec{E}(\vec{r},t)_{empty} = \sum_{\vec{k}\sigma} E_{\vec{k}\sigma} \vec{e}_{\vec{k}\sigma} e^{i(\vec{k}\bullet\vec{r}-\omega t)} + \text{H.c.} \quad , \tag{57}$$

where the t-dependence signifies the Heisenberg picture and operator $E_{\vec{k}\sigma}$ is defined as

$$E_{\vec{k}\sigma} = \sqrt{\frac{\hbar\omega}{2\varepsilon_o V}} a_{\vec{k}\sigma} \quad , \tag{58}$$

with $a_{\vec{k}\sigma}$ as the annihilation operator for a photon in a mode with wave vector $\vec{k}$ and polarization σ, and $\omega = ck$. We shall take the two possible transverse polarization directions as surface (*s*) polarized and plane (*p*) polarized.

The vacuum field for empty space is a superposition of plane-wave modes, and these waves are incident upon the dielectric layer. Since Maxwell's equations also hold in quantum mechanics (in the Heisenberg picture), these plane waves reflect at the dielectric in the same way as in classical optics. The atomic dipole is located in z > 0, and therefore we only consider the field in that region of space. As illustrated in Fig. 1, an incident vacuum wave from the region z > 0 reflects partially as a specular wave and is partially refracted into the medium. Waves that enter the medium from the region $z < \Delta$ are partially transmitted through the layer, and they contribute to the field in $z > 0$. From the boundary condition at $z = 0$ it follows that the specular wave must have a wave vector with the same parallel component with respect to the surface as the incident wave. Therefore, if we write for the wave vector of the incident wave

$$\vec{k} = \vec{k}_{\parallel} + \vec{k}_{\perp} \quad , \tag{59}$$

then the specular wave has the wave vector

$$\vec{k}_r = \vec{k}_{\parallel} - \vec{k}_{\perp} \quad . \tag{60}$$

The wave vector of the transmitted wave, which originated from $z < \Delta$, is the same as of the incident wave from $z < \Delta$. If an incident wave is σ (*s* or *p*) polarized, then the specular wave or the transmitted wave is also σ-polarized. Given the parallel component of the wave vector of an incident wave, for the *s*-polarized unit polarization vectors we take

$$\vec{e}_{\vec{k}s} = \frac{1}{k_{\parallel}} \vec{k}_{\parallel} \times \vec{e}_z \quad . \tag{61}$$

For *p*-polarization we take the unit vectors of the incident and transmitted waves as

$$\vec{e}_{\vec{k}p} = \frac{1}{k} \vec{k} \times \vec{e}_{\vec{k}s} \quad , \tag{62}$$

in terms of the corresponding *s*-polarization unit vectors, and for the specular wave we take

$$\vec{e}_{\vec{k}_r p} = \frac{1}{k} \vec{k}_r \times \vec{e}_{\vec{k}s} \quad . \tag{63}$$

The amplitude of a specular wave or a transmitted wave is simply a Fresnel coefficient $R_{\vec{k}\sigma}$ or $T_{\vec{k}\sigma}$, respectively, times the amplitude of the corresponding incident wave. Combining everything then gives for the electric field operator in $z > 0$

$$\vec{E}(\vec{r},t) = \sum_{\vec{k}\sigma} E_{\vec{k}\sigma} \{ \vec{e}_{\vec{k}\sigma} e^{i(\vec{k}\bullet\vec{r}-\omega t)} + R_{\vec{k}\sigma} \vec{e}_{\vec{k}_r\sigma} e^{i(\vec{k}_r\bullet\vec{r}-\omega t)} \}$$
$$+ \sum_{\vec{k}\sigma} E_{\vec{k}\sigma} T_{\vec{k}\sigma} \vec{e}_{\vec{k}\sigma} e^{i(\vec{k}\bullet\vec{r}-\omega t)} \quad + \text{H.c.} \quad . \tag{64}$$

Here, the first summation runs over wave vectors with $k_z < 0$, and the second summation has only wave vectors with $k_z > 0$.

9. VACUUM CORRELATION FUNCTIONS

Equation (64) gives the electric field operator of the vacuum near the dielectric in the Heisenberg picture. For $t = 0$ and $\vec{r} = \vec{h} = h\vec{e}_z$ this becomes the Schrödinger-picture electric field operator at the position of the atom, as it appears in the interaction Hamiltonian (25) and the vacuum correlation function (28). The quantity $\{\exp(iL_r\tau)\}E(\vec{h})_i$ in (28) is effectively the i-th component of the Heisenberg-picture electric field operator at the position of the atom, and evaluated at $t = \tau$. When we insert the two expressions for the field in (28) and take the quantum expectation value in the vacuum using

$$<0|a_{\vec{k}\sigma}a_{\vec{k}'\sigma'}|0> = <0|a_{\vec{k}\sigma}^{\dagger}a_{\vec{k}'\sigma'}^{\dagger}|0> = <0|a_{\vec{k}\sigma}^{\dagger}a_{\vec{k}'\sigma'}|0> = 0 \quad , \tag{65}$$

$$<0|a_{\vec{k}\sigma}a_{\vec{k}'\sigma'}^{\dagger}|0> = \delta_{\vec{k}\vec{k}'}\delta_{\sigma\sigma'} \quad , \tag{66}$$

we obtain for the correlation function with $j = i$

$$f_{ii}(\tau) = \sum_{\vec{k}\sigma, k_z<0} \frac{\omega}{2\varepsilon_o \hbar V} |\{\vec{e}_{\vec{k}\sigma} e^{i\vec{k}\bullet\vec{h}} + R_{\vec{k}\sigma}\vec{e}_{\vec{k}_r\sigma} e^{i\vec{k}_r\bullet\vec{h}}\}_i|^2 e^{-i\omega\tau}$$
$$+ \sum_{\vec{k}\sigma, k_z>0} \frac{\omega}{2\varepsilon_o \hbar V} |\{T_{\vec{k}\sigma}\vec{e}_{\vec{k}\sigma} e^{i\vec{k}\bullet\vec{h}}\}_i|^2 e^{-i\omega\tau} \quad . \tag{67}$$

In the limit $V \to \infty$ the first summation over the wave vectors can be replaced by an integration according to

$$\frac{1}{V}\sum_{\vec{k}\sigma, k_z<0}(\ldots) = \frac{1}{8\pi^3}\int_{k_z<0} d^3\vec{k}\,(\ldots) \quad , \tag{68}$$

and in spherical coordinates the integral can be evaluated as

$$\int_{k_z<0} d^3\vec{k}\,(\ldots) = \int_0^\infty dk \int_{\pi/2}^{\pi} d\theta \int_0^{2\pi} d\phi\, k^2 \sin\theta\,(\ldots) \quad . \tag{69}$$

The second summation in (67) can be transformed in a similar way. In spherical coordinates we have $\vec{k} = k\vec{e}_r$, and with (60) and the fact that $\vec{h}$ has only a perpendicular component, we conclude that $\vec{k}_r \bullet \vec{h} = -\vec{k} \bullet \vec{h}$. With our phase conventions (61) - (63) the polarization vectors become

$$\vec{e}_{\vec{k}s} = \vec{e}_{\vec{k}_r s} = -\vec{e}_\phi \quad , \tag{70}$$

$$\vec{e}_{\vec{k}p} = \vec{e}_\theta \quad , \quad \vec{e}_{\vec{k}_r p} = -\vec{e}_\theta - 2\vec{e}_z \sin\theta \quad . \tag{71}$$

The Fresnel reflection and transmission coefficients depend on the polar angle θ, but not on the angle ϕ. For $k_z > 0$ the polar angle is equal to the angle of incidence θ_i, but for $k_z < 0$ the two angles are related as $\theta_i = \pi - \theta$. Furthermore, the Fresnel coefficients depend on θ_i through $\cos\theta_i$, and we shall write

$$P_{\vec{k}\sigma} = P_\sigma(\omega, \cos\theta_i) \quad , \quad T_{\vec{k}\sigma} = T_\sigma(\omega, \cos\theta_i) \quad , \tag{72}$$

For the integrations over θ and k we make the substitutions $u = \cos\theta_i$ and $\omega = ck$, respectively. We then obtain for the two correlation functions

$$f_\perp(\tau) = \frac{1}{8\pi^2 \varepsilon_o \hbar c^3} \int_0^\infty d\omega\, \omega^3\, e^{-i\omega\tau} \int_0^1 du\, (1 - u^2)$$

$$\times \left\{ \left| 1 + R_p(\omega, u) e^{2ih\omega u/c} \right|^2 + \left| T_p(\omega, u) \right|^2 \right\} \quad , \tag{73}$$

$$f_\parallel(\tau) = \frac{1}{16\pi^2 \varepsilon_o \hbar c^3} \int_0^\infty d\omega\, \omega^3\, e^{-i\omega\tau} \int_0^1 du$$

$$\times \left\{ \left| 1 + R_s(\omega, u) e^{2ih\omega u/c} \right|^2 + \left| T_s(\omega, u) \right|^2 \right\}$$

$$+ \frac{1}{16\pi^2 \varepsilon_o \hbar c^3} \int_0^\infty d\omega\, \omega^3\, e^{-i\omega\tau} \int_0^1 du\, u^2$$

$$\times \left\{ \left| 1 - R_p(\omega, u) e^{2ih\omega u/c} \right|^2 + \left| T_p(\omega, u) \right|^2 \right\} \quad . \tag{74}$$

The perpendicular correlation function is determined by the Fresnel coefficients for p-waves only, but the parallel function depends on both s-waves and p-waves. Also notice that the geometrical factor in the u-integrands is different in all three integrals. Furthermore, if we would have taken $i \neq j$ in (67), then the integral over ϕ would give zero, and thus this correlation function equals zero.

As shown in section 5, we need the Fourier-Laplace transform (39) of the correlation functions, and evaluated at $s = \omega_o$. The τ dependence in (73) and (74) is only through the exponentials in the integrands, and therefore the Fourier-Laplace transformation leads to the following integral

$$\int_0^\infty d\tau\, e^{i(\omega_o - \omega)\tau} = \pi\delta(\omega_o - \omega) + P\frac{i}{\omega_o - \omega} \quad , \tag{75}$$

where P stands for principal value. This principal value part gives an imaginary contribution to the relaxation constants, which leads to an effective level shift (the Lamb shift). In the present nonrelativistic theory this imaginary part diverges in the upper integration limit over ω. For these high frequencies the dipole approximation (which is the long-wavelength limit) breaks down, and also relativistic effects have to be taken into account. The final result is a very small value for this imaginary part, and we shall therefore omit it. With (75) the Fourier-Laplace transforms of (73) and (74) become

$$\tilde{f}_{\perp}(\omega_o)=\frac{\omega_o^3}{8\pi\varepsilon_o\hbar c^3}\int_0^1 du\,(1-u^2)\left\{\left|1+R_p(\omega_o,u)e^{2ih\omega_o u/c}\right|^2+\left|T_p(\omega_o,u)\right|^2\right\} \tag{76}$$

$$\tilde{f}_{\parallel}(\omega_o)=\frac{\omega_o^3}{16\pi\varepsilon_o\hbar c^3}\int_0^1 du\,\left\{\left|1+R_s(\omega_o,u)e^{2ih\omega_o u/c}\right|^2+\left|T_s(\omega_o,u)\right|^2\right\}$$
$$+\frac{\omega_o^3}{16\pi\varepsilon_o\hbar c^3}\int_0^1 du\,u^2\left\{\left|1-R_p(\omega_o,u)e^{2ih\omega_o u/c}\right|^2+\left|T_p(\omega_o,u)\right|^2\right\}. \tag{77}$$

The Fresnel coefficients now only appear with their value at the atomic resonance frequency ω_o, and the integration is over the (cosine of the) angle of incidence.

The Fresnel coefficients appearing in (76) and (77) can be calculated from Maxwell's equations in the usual way. They are found to be

$$R_s=(e^{2i\delta v}-1)\frac{v^2-u^2}{D_s}\quad,\qquad R_p=(e^{2i\delta v}-1)\frac{v^2-(u\varepsilon)^2}{D_p}\quad, \tag{78}$$

$$T_s=\frac{4uv}{D_s}e^{i(v-u)\delta}\quad,\qquad T_p=\frac{4uv\varepsilon}{D_p}e^{i(v-u)\delta}\quad, \tag{79}$$

with

$$u=\cos\theta_i\quad,\qquad v=\sqrt{\varepsilon-1+u^2}\quad, \tag{80}$$

and

$$D_s=(v+u)^2-(v-u)^2e^{2i\delta v}\quad, \tag{81}$$

$$D_p=(v+u\varepsilon)^2-(v-u\varepsilon)^2e^{2i\delta v}\quad, \tag{82}$$

and the dimensionless layer thickness

$$\delta = k\Delta = \frac{\omega_o \Delta}{c} \quad . \tag{83}$$

The u-dependence appears to be too complicated to perform the integrations in (76) and (77) explicitly. For the dielectric constant ε we have to take its value at frequency ω_o. It can be verified easily from (78) and (79) that for $\varepsilon \geq 1$ the reflection and transmission coefficients are related as

$$|R_\sigma|^2 + |T_\sigma|^2 = 1 \quad , \quad \sigma = s, p \quad , \tag{84}$$

where both Fresnel coefficients are evaluated at the same angle of incidence.

10. EINSTEIN COEFFICIENTS

The perpendicular and parallel Einstein coefficients for spontaneous decay can now be found from (48). We shall write

$$A_i = A b_i \quad , \quad i = \perp, \| \quad , \tag{85}$$

where *A* is the value (49) for decay in empty space. The correction factors due to the presence of the surface are then

$$b_\perp = \frac{3}{4}\int_0^1 du\,(1-u^2)\left\{\left|1 + R_p e^{i\beta u}\right|^2 + \left|T_p\right|^2\right\} \quad , \tag{86}$$

$$b_\| = \frac{3}{8}\int_0^1 du\,\left\{\left|1 + R_s e^{i\beta u}\right|^2 + \left|T_s\right|^2 + u^2\left[\left|1 - R_p e^{i\beta u}\right|^2 + \left|T_p\right|^2\right]\right\} \quad , \tag{87}$$

with

$$\beta = \frac{2\omega h}{c} \quad . \tag{88}$$

In the absense of the medium we have effectively $R_s = R_p = 0$ and $T_s = T_p = 1$, as follows from (78) and (79) with $\varepsilon = 1$ or $\delta = 0$. Then (86) and (87) give $b_\perp = b_\| = 1$, as it should.

For a dielectric we have $\varepsilon \geq 1$, and we can use (84) to simplify the expressions for $b_\perp$ and $b_\|$. Eliminating the transmission coefficients gives

$$b_\perp = \frac{3}{2}\int_0^1 du\,(1-u^2)\,\mathit{Re}\left(1 + R_p e^{i\beta u}\right) \quad , \tag{89}$$

$$b_\| = \frac{3}{4}\int_0^1 du\,\mathit{Re}\left(1 + R_s e^{i\beta u} + u^2\left[1 - R_p e^{i\beta u}\right]\right) \quad , \tag{90}$$

which can be simplified to

$$b_\perp = 1 + \frac{3}{2}\int_0^1 du\,(1-u^2)\,\mathit{Re}R_p e^{i\beta u} \quad , \tag{91}$$

$$b_\parallel = 1 + \frac{3}{4}\int_0^1 du\,\mathit{Re}(R_s - u^2 R_p)e^{i\beta u} \quad . \tag{92}$$

The dependence on the atom-surface normal distance h is incorporated in the parameter β. When the atom is far away from the surface ($\beta \to \infty$), then the exponentials oscillate very fast as a function of u, and they integrate out to zero. This gives $b_\perp = b_\parallel = 1$, as expected.

11. METAL SURFACE

A special case of practical importance is when the layer is a metal. This is characterized by $\varepsilon < 0$, and this makes the parameter v from (80) positive imaginary. In addition we shall assume that the layer thickness is much larger than an optical wavelength ($\delta >> 1$). Then the Fresnel coefficients simplify to

$$T_s = T_p = 0 \quad , \quad R_s = \frac{u-v}{u+v} \quad , \quad R_p = \frac{u\varepsilon - v}{u\varepsilon + v} \quad , \tag{93}$$

and since v is imaginary and u, is real this gives

$$|R_\sigma| = 1 \quad , \quad \sigma = s,p \quad . \tag{94}$$

Therefore, the reflectivity at a metal surface (mirror) equals unity, and there can only be a phase shift upon reflection. If we define the angles

$$\phi_s = \arctan\left(\frac{\sqrt{1-\varepsilon-u^2}}{u}\right) \quad , \tag{95}$$

$$\phi_p = \arctan\left(\frac{\sqrt{1-\varepsilon-u^2}}{u\varepsilon}\right) \quad , \tag{96}$$

then the Fresnel coefficients can be represented as

$$R_\sigma = e^{-2i\phi_\sigma} \quad . \tag{97}$$

For a metal, the relation (84) also holds, and therefore the Einstein coefficients relative to empty space become with (91) and (92):

$$b_\perp = 1 + \frac{3}{2}\int_0^1 du\,(1-u^2)\,Re\,e^{i(\beta u - 2\phi_p)} \quad , \tag{98}$$

$$b_\parallel = 1 + \frac{3}{4}\int_0^1 du\,Re\,(e^{-2i\phi_s} - u^2 e^{-2i\phi_p})e^{i\beta u} \quad . \tag{99}$$

Metals like silver are nearly perfect conductors, and they can be modelled effectively by $\varepsilon \to -\infty$. Then the phase angles from (95) and (96) become

$$\phi_s = \pi/2 \quad , \quad \phi_p = 0 \quad , \tag{100}$$

and the Fresnel reflection coefficients are

$$R_s = -1 \quad , \quad R_p = 1 \quad . \tag{101}$$

With these values, the integrals in (98) and (99) can be performed analytically, and the result is

$$b_\perp = 1 - 3\left[\frac{\cos\beta}{\beta^2} - \frac{\sin\beta}{\beta^3}\right] \quad , \tag{102}$$

$$b_\parallel = 1 - \frac{3}{2}\left[\frac{\sin\beta}{\beta} + \frac{\cos\beta}{\beta^2} - \frac{\sin\beta}{\beta^3}\right] \quad . \tag{103}$$

For $\beta \to \infty$ we find again $b_\perp = b_\parallel = 1$, as it should. When the atom is very close to the surface (within a fraction of a wavelength), we find

$$b_\perp = 2 \quad , \quad b_\parallel = 0 \quad . \tag{104}$$

In this limit, an atom with its transition dipole moment perpendicular to the surface has twice the Einstein coefficient for decay, compared to an atom in empty space, and therefore the excited state has half the lifetime. On the other hand, when the dipole is parallel to the surface, the Einstein coefficient approaches zero, and this indicates that the atom in its excited state would not decay spontaneously at all. In Fig. 2 the behavior of the Einstein coefficients as a function of the distance between the atom and the surface is shown. The oscillatory behavior of these functions is due to interference between radiation which is emitted directly during the decay and radiation which is reflected by the metal. In (98) and (99), the factor $\exp(i\beta u)$ accounts for the retardation in this process.

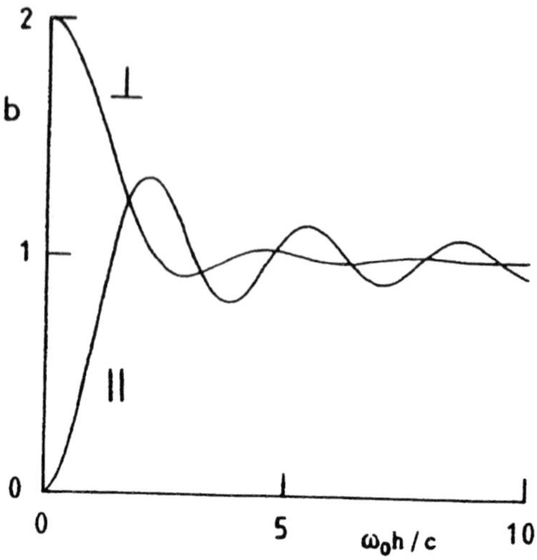

Fig. 2. Relative Einstein coefficients as a function of the atom-surface distance.

ACKNOWLEDGMENTS

This work was supported in part by the National Science Foundation under Grant CHE - 9196214. We also acknowledge the National Science Foundation for a Research Opportunity Award.

REFERENCES

1. K. H. Drexhage, in *Progress in Optics XII* (E. Wolf, Ed.), Amsterdam: North-Holland, 1974.
2. P. Goy, J. M. Raimond, M. Gross, and S. Haroche, *Phys. Rev. Lett.* 50 (1983) 1903.
3. R. G. Hulet, E. S. Hilfer, and D. Kleppner, *Phys. Rev. Lett.* 55 (1985) 2137.
4. W. Jhe, A. Anderson, E. A. Hinds, D. Meschede, L. Moi, and S. Haroche, *Phys. Rev. Lett.* 58 (1987) 666.
5. D. J. Heinzen, J. J. Childs, J. E. Thomas, and M. S. Feld, *Phys. Rev. Lett.* 58 (1987) 1320.
6. H. Morawitz, *Phys. Rev.* 187 (1969) 1792.
7. G. S. Agarwal, *Phys. Rev. Lett.* 32 (1974) 703.
8. M. R. Philpott, *J. Chem. Phys.* 62 (1975) 1812.
9. B. N. J. Persson, *Solid State Commun.* 27 (1978) 417.
10. R. R. Chance, A. Prock, and R. Silbey, *Adv. Chem. Phys.* 39 (1978) 1.
11. C. Cohen-Tannoudji, in *Frontiers in Laser Spectroscopy*, Proceedings of the 27th Les Houches Summer School (R. Balian, S. Haroche, and S. Liberman, Eds.), Amsterdam: North Holland, 1977.
12. H. F. Arnoldus and T. F. George, *Phys. Rev. A* 37 (1988) 761.
13. J. A. Abate, *Opt. Commun.* 10 (1974) 269.

Studies in Classical
and Quantum Nonlinear Optics

ISBN 1-56072-168-5

NONLINEAR OPTICS IN DROPLETS

Steven C. Hill
Army Research Laboratory, White Sands Missile Range, NM 88002
New Mexico State University
Department of Electrical and Computer Engineering
Las Cruces, New Mexico 88003

Richard K. Chang
Yale University
Department of Applied Physics and Center for Laser Diagnostics
New Haven, Connecticut 06520

1 INTRODUCTION

Nonlinear optical processes in micrometer-sized droplets were reviewed in a 1990 article [1] dedicated to Professor Nicholaas Bloembergen's 70th birthday. Morphology-dependent resonances (MDRs) which support nonlinear processes in droplets were reviewed in 1988 [2]. The present article emphasizes the nonlinear optical phenomena observed and theory developed since the two previous reviews.

The interpretations of experimental results prior to these reviews [1], [2] relied mainly on the following: 1) nonlinear optical effects in liquids and gases in an optical cell or in a hollow core fiber [3]–[6]; 2) models of elastic light scattering by droplets [7]–[10], including MDRs [11]–[17]; and 3) models of radiation from sources inside spheres [18].

Now, improved expressions for MDR positions [14]–[17], widths and strengths [17] have been developed. Many of the more recent experiments have been designed with a better understanding of: 1) radiation from sources inside spheres [18]–[31], including modifications of transition rates (quantum electrodynamic or QED effects) [20]–[26], and transfer of energy within droplets [30], [31]; 2) internal fields of droplets [32]–[44], including fields generated with pulsed laser beams on or near resonance [43], [44] or with Gaussian beams [38]–[42], [44]; and 3) resonances of perturbed [45]–[54] or inhomogeneous [55]–[62] spheres. Recent work has also been furthered by the development of nonlinear optical models specific to micrometer-sized droplets. The present intertwining of experimental and theoretical work has motivated RKC (an experimentalist) and SCH (a modeler) to write a review intended for scientists interested in the unique features of nonlinear optical scattering resulting from the droplet optical cavity.

Fluorescence [63]–[68] and spontaneous Raman emission [69]–[75] from droplets have been measured. A variety of nonlinear optical emission processes have also been studied: lasing [76]–[91], stimulated Raman scattering (SRS) [92]–[117], stimulated Brillouin scattering (SBS) [118], [119], coherent Raman mixing [120], coherent anti-Stokes Raman scattering (CARS) [120], second-harmonic generation (SHG) [121], third-order sum-frequency generation (TSFG) [121]–[123], stimulated anti-Stokes Raman scattering (SARS) [124], and stimulated librational scattering (SLS) [125], [126]. The incident intensities required for these processes are lower than those needed in bulk media. The volumes required in droplets are smaller than in bulk. Other nonlinear optical effects such as laser-induced shape distortions and "nonlinear Mie scattering" [127]–[132], as well as optical bistability [133], [134], have also been observed in droplets. Energy transfer within spheres [135], [136] and other internal absorption and scattering effects [137]–[146], including two-photon absorption effects [137] have been reported in droplets. In solid particles [147]–[153], lasing [147], [148], [150]–[152], bistability [149], and second harmonic generation (SHG) [153], have been demonstrated. Nonlinear optical effects and other resonance effects have been investigated in single micrometer-sized droplets [76]–[146], [154]–[174] that are flowing in a linear stream or are levitated, and in numerous droplets forming a dilute or dense spray [175], [176].

Many of the experimental studies of nonlinear optics in droplets have been motivated by the importance of characterizing droplets and sprays. Techniques for measuring droplet evaporation rate, temperature, surface tension and viscosity [154]–[164] are now beginning to be applied to combustion-related problems [162], [164]. Fluorescence and Raman emission from droplets [63]–[75] continue to be applied to droplet characterization. SRS spectra have been used to measure shape perturbations of droplets falling through air [165]–[167]. There is increasing interest in using optical scattering and emission to determine chemical species concentrations within droplets. Unique assignments of MDRs is important for the most accurate size and refractive index determinations [90], [169], [171], [173], [174]. SRS, fluorescence emission, and lasing have been used to characterize sprays [175], [176] for combustion applications.

To understand these experimental observations, several groups of researchers have developed models for nonlinear optical processes in droplets. These models combine the fundamentals of nonlinear optics [3]–[6] (studied extensively in bulk media and in fibers) with an analysis of electromagnetic waves in a spherical cavity. Models have been developed for: SBS [177]–[181], SRS [182]–[183], CARS [184], [185], SHG [186], TSFG [187], bistability [133], [134], [188], and two-photon absorption [55]. SHG has been modeled in solid anisotropic particles [186]. One problem in relating these models to experiments is that nonlinear effects in droplets are usually very dependent upon the particular MDRs which support most of the nonlinear processes. However, there are so many possible MDRs, or combinations of input and output MDRs, that detailed interpretations of experimental observations are not possible. Advances in determining the specific input and output MDRs supporting nonlinear processes will facilitate more rigorous comparisons between theory and experiment.

Nonlinear optical effects are particularly strong in droplets. First, the pump intensities at some positions inside a droplet can be thousands of times larger than the intensity of the incident wave when the input frequency is near a MDR [32], [33], and can be hundreds of times larger when the incident wave is not at a resonance frequency. The droplet morphology greatly modifies the spatial distribution of the input wave. With plane wave illumination the internal intensity is most concentrated along the axis of the droplet and is largest in the outer part of the droplet on the side of the shadow-face. Second, QED effects increase the nonlinear gain (for lasing, SRS, or SBS), because the density-of-final photon states is proportional to the internal-intensity enhancement factor [26]. The transition rates of molecules can be greatly enhanced at positions inside the droplets where the mode intensities are large [20], [21], [23]. Third, the droplet-air interface provides efficient feedback at the MDR frequencies (this feedback is necessary, but not sufficient for a large QED enhancement). Fourth, the generating and output waves in a droplet are confined to a very small volume, and so the spatial overlaps between the waves can be large.

The concentration of the pump radiation, the enhancement of the gain, the provision of optical feedback, and the confinement of the input and output waves all contribute to strengths of nonlinear optical processes in single micrometer-sized droplets. Many of the post-1990 experiments were planned to verify and extend our present understanding of the nonlinear optical processes unique to the droplets' morphology as well as to apply nonlinear spectroscopy to the diagnostics of droplets in either the atmosphere, a combustor, or a flow field.

Since MDRs are so important for understanding nonlinear optics in droplets, we begin by reviewing the properties of MDRs of spheres and perturbed spheres. We then describe observations of nonlinear optics in droplets, emphasizing the work of one of us (RKC). Finally we review some of the models which have been developed for nonlinear optics in droplets, emphasizing the work of one of us (SCH).

2 MODES AND MORPHOLOGY-DEPENDENT RESONANCES OF SPHERES AND PERTURBED SPHERES

2.1 Modal Expansion of Fields in Spheres

The electric fields scattered or radiated from a sphere can be expressed as [8]

$$\mathbf{E}(kr) = \sum_{n,m}^{\infty} c_M(n,m) h_n^{(1)}(kr) \mathbf{X}_{n,m} + c_E(n,m) \nabla \times (h_2^{(1)}(kr) \mathbf{X}_{n,m}), \tag{1}$$

where $h_n^{(1)}(kr)$ is the spherical Hankel function, $k = \omega/c$ is the wavenumber outside the sphere, and $\mathbf{X}_{n,m}(\theta,\phi)$ is the vector spherical harmonic (VSH) (equivalent to the $\mathbf{Y}_{n,n,m}$ of Chew et al. [18]), i.e.,

$$\mathbf{X}_{nm}(\theta,\phi) = -\sqrt{\frac{(2n+1)(n-m)!}{4\pi n(n+1)(n+m)!}} \left(i\hat{\phi} \frac{\partial P_n^m}{\partial \theta} + \hat{\theta} m \frac{P_n^m}{\sin\theta} \right) e^{im\phi}, \tag{2}$$

where $P_n^m(\cos\theta)$ is the associated Legendre function. The incident fields can be expressed similarly but with the $h_n^{(1)}(kr)$ replaced by $j_n(kr)$, and with the c_M and c_E replaced by a_M and a_E. For the fields induced within the sphere, the $h_n^{(1)}(kr)$ is replaced by the spherical Bessel function $j_n(k_t r)$, where $k_t = \mathrm{m}(\omega)\omega/c$ is the wavenumber inside the sphere, $\mathrm{m}(\omega)$ is the index of refraction inside the sphere at ω, and the c_M and c_E are replaced by b_M and b_E. Note that m in *Italics* indicates azimuthal mode number and $\mathrm{m}(\omega)$ in Roman script indicates index of refraction.

The coefficients of the scattered fields and the fields induced inside the sphere are proportional to,

$$f_n = \frac{i/k_t a}{\mathrm{m}(\omega)^2 j_n(k_t a) \left[ka h_n^{(1)}(ka)\right]' - h_n^{(1)}(ka) \left[k_t a j_n(k_t a)\right]'} \tag{3}$$

for the transverse magnetic modes (the b_E and c_E) and to,

$$g_n = \frac{i/k_t a}{j_n(k_t a) \left[ka h_n^{(1)}(ka)\right]' - h_n^{(1)}(ka) \left[k_t a j_n(k_t a)\right]'}, \tag{4}$$

for the transverse electric modes (the b_M and c_M), where a is the droplet radius, $ka = x = \omega a/c$ is the size parameter, the primes ($'$) on j_n and $h_n^{(1)}$ indicate differentiation with respect to the argument.

In a dielectric sphere, the angular patterns of the fields of the modes are determined by the $\mathbf{X}_{nm}(\theta,\phi)$ and $\nabla \times \mathbf{X}_{nm}(\theta,\phi)$. The modes are labeled by the mode number n and azimuthal mode number m. [The n used here is typically labeled

l in quantum mechanics, where it is termed the ***azimuthal*** or ***orbital-angular-momentum*** quantum number, and the m used here is labeled the ***magnetic*** quantum number; the degeneracy of the m modes can be lifted by applying a magnetic field. The total angular momentum is $\sqrt{n(n+1)}\hbar$, and the z–component of the angular momentum is $m\hbar$.]

The vector spherical harmonics, $\mathbf{M}^1_{omn}$, $\mathbf{M}^1_{emn}$, $\mathbf{N}^1_{omn}$, $\mathbf{N}^1_{emn}$, which are often used in analyses of light scattering by particles [9], [10], are related to the $\mathbf{X}_{nm}(\theta,\phi)$ by the following expressions,

$$j_n(kr)\frac{\mathbf{X}_{nm}(\theta,\phi)+\mathbf{X}^*_{nm}(\theta,\phi)}{2} = -\sqrt{\frac{(2n+1)(n-m)!}{4\pi n(n+1)(n+m)!}}\mathbf{M}^1_{omn} \tag{5}$$

$$j_n(kr)\frac{\mathbf{X}_{nm}(\theta,\phi)-\mathbf{X}^*_{nm}(\theta,\phi)}{2} = i\sqrt{\frac{(2n+1)(n-m)!}{4\pi n(n+1)(n+m)!}}\mathbf{M}^1_{emn} \tag{6}$$

and the $\mathbf{N}_{mn} = \nabla \times \mathbf{M}_{mn}/k$.

2.2 Modes and Morphology-Dependent Resonances–Theory

MDRs have been observed in optical radiation pressure and elastic scattering [154], as well as in the fluorescence, Raman, lasing, and SRS spectra mentioned in Section 1. The MDR frequencies depend on the morphology (size, shape and refractive index) of the particles. These resonances have also been termed whispering gallery modes [147], [149], and give rise to the ripple structure in scattering and absorption spectra [9], [11].

The resonance frequencies are the poles of the coefficients of the field, i.e., the poles of the f_n and g_n coefficients (Eqs. 3 and 4). Since the fields are expanded in the virtual modes, the field coefficients have the same denominators whether the incident field is an evanescent wave, a Gaussian beam [40], [41], or the field from a dipole inside the sphere [18]. The denominators of the scattering, emission, or absorption coefficients are independent of the incident fields, as long as the droplet morphology is not modified by the incident fields. The terms "resonance frequencies" and "resonance size parameters" are used interchangeably since the size parameter is a normalized frequency.

The quality factor, or Q, of a resonance is defined as

$$Q = \frac{2\pi(\text{Stored energy})}{\text{Energy lost per cycle}}.$$

The average lifetime of photons in the resonant mode is

$$\tau = \frac{Q}{\omega_0},$$

where ω_0 is the resonance frequency. Electromagnetic energy couples out of a high-Q mode relatively slowly, but by reciprocity it also couples into the mode slowly. The calculated Q's of idealized small spheres can be enormous (greater than 10^{100}) [12], [17], [147]. Although no real spheres are perfect, the measured lifetime ($\tau = Q/\omega$) of optical energy stored in a droplet can be longer than 10 ns [95].

Some properties of MDRs are: 1) the feedback provided by the MDRs allows the internal energies to build to large values; 2) the large density of final states for an emission process near an MDR can enhance the transition rates of molecules at positions inside the droplets where the energy per unit volume is largest [19]–[26], [86], [87]; 3) the waves in resonance with MDRs can have specific phase velocities [123]; and 4) the density of MDRs per spectral interval is large, e.g., more than 4 per 10 cm^{-1} interval for CCl_4 droplets with $x \approx 100$ [13].

In a closed lossless cavity the modes and the resonant modes are identical. A dielectric sphere is not a closed system, but is "open," or "leaky" [20], [21], [24]. The MDRs are the resonances of the modes. It is only at the resonant frequencies that these modes are "quasi-normal modes," MDRs, or "resonant" modes. The droplet is in some aspects analogous to a linear-laser cavity with spherical mirrors. The transverse modes of the linear cavity are described by Gaussian-Hermite functions labeled with subscripts n and m [5]. The resonance frequencies are determined by the longitudinal variations of the fields of these modes. For any n, m combination there are infinitely many possible longitudinal field patterns, and these field patterns depend on the spatial distribution of the polarization which is the source of the energy in the modes.

In a sphere the resonances are determined by the radial variations of the fields, which are analogous to the longitudinal variations in the linear cavity. Each mode (n, m) can have infinitely many resonances. The MDRs are specified by the n and m of the mode, along with a mode order, l, which is the number of radial peaks in the angle-averaged internal intensity distribution. The incident fields at a particular frequency, generated by sources either internal or external to the sphere, can be expanded in the modes (n, m). Because the fields must be matched at the boundaries, the induced fields of the resonant modes (at the particular frequency of the source) are large. At any frequency, on-resonance or off-resonance, the mode is specified by the n, m, but the radial variations of the fields are not uniquely specified. These radial variations depend upon the particular sources for these modes. However, in all cases the radial variations are determined by combinations of the spherical Bessel and Hankel functions.

A simple physical interpretation of MDRs is that of rays or beams propagating around the sphere, confined by total internal reflection. After circumnavigating the sphere the beam returns to its starting position in phase. A beam approaches the internal surface of the sphere beyond the critical angle and so is almost totally reflected.

Near MDRs the lineshapes can be approximated well by complex Lorentzian functions as,

$$g_n(x, x_i, \Delta x_i) \approx \frac{G_i \Delta x_i^{1/2}}{(x - x_i) + i x_i / 2Q_i} \tag{7}$$

where x_i is the real part of the complex resonance frequency of the ith MDR. The subscript i is a contracted index which ranges over the possible combinations of n, l, and (when necessary in a perturbed sphere) m. The full-width at half maximum (FWHM) is $\Delta x_i = x_i / Q_i$ where Q_i is the quality factor of the mode. The expression for f_n is the same as for g_n but with G_i replaced by F_i. The squared magnitudes are $|g_n|^2 \approx G_i^2 L_i(x, x_i, \Delta x_i)$ and $|f_n|^2 \approx F_i^2 L_i(x, x_i, \Delta x_i)$, where the normalized Lorentzian lineshapes are,

$$L_i(x, x_i, \Delta x_i) = \frac{\Delta x_i}{(x - x_i)^2 + (\Delta x_i/2)^2}. \tag{8}$$

Resonance locations can be found by searching for the zeros of the denominators of the field coefficients [12], [90], or the zeros of the imaginary parts of the field coefficients [171]. For large spheres, asymptotic expressions have been developed for spectral locations of MDRs [15], [17].

An expression for the size parameter spacing between resonances of the same order and polarization is [14],

$$x_{n+1} - x_n = \frac{x \arctan[(\mathrm{m}x/n)^2 - 1]^{1/2}}{n[(\mathrm{m}x/n)^2 - 1]^{1/2}} \tag{9}$$

This type of expression is often used in interpreting measured resonance spectra. An explicit expression for mode spacings which does not depend upon x, but depends only upon the mode number, mode order, and polarization may be particularly useful [17].

The widths Δx (FWHM) of MDRs are well approximated by [17],

$$\Delta x = \frac{2}{[N x^2 y_n(x)^2]} \tag{10}$$

where y_n is the spherical Neumann function, and

$$N = \begin{cases} \mathrm{m}^2 - 1 & \text{for TE modes} \\ (\mathrm{m}^2 - 1)\left[\left(\dfrac{n + \frac{1}{2}}{x}\right)^2 \left(1 + \dfrac{1}{\mathrm{m}^2}\right) - 1\right] & \text{for TM modes.} \end{cases} \tag{11}$$

2.3 Radiation from Molecules Inside a Sphere–Theory

2.3.1 Radiation from a Source Inside a Sphere–Theory

The radiation from a time-harmonic polarization, $\mathbf{p}(\mathbf{r}', \omega)$, inside a sphere has been treated in detail [18], [27], [28]. The $\mathbf{p}(\mathbf{r}', \omega)$ may have an arbitrary spatial dependence, where the source position is $\mathbf{r}'$. The fields directly generated by the $\mathbf{p}(\mathbf{r}', \omega)$ induce additional fields which are required to satisfy the boundary conditions at the surface of the sphere. If the source of the fields at ω is the internal polarization $\mathbf{p}(\mathbf{r}', \omega)$, then the coefficients of the total field far from the sphere are [18], [27]:

$$\int a_E(n, m) d^3r' = 4\pi k^2 \int \mathbf{p}(\mathbf{r}', \omega) \cdot \{\nabla' \times [j_n(k_t r')\mathbf{X}^*_{nm}(\theta', \phi')]\} \, d^3r' \tag{12}$$

and

$$\int a_M(n, m) d^3r' = 4\pi i k^3 \int \mathbf{p}(\mathbf{r}', \omega) \cdot j_n(k_t r')\mathbf{X}^*_{nm}(\theta', \phi') d^3r' \tag{13}$$

if there is no external incident field at frequency ω.

2.3.2 Modifications of Transition Rates–Theory

The above expressions for calculating the radiation from a source inside a sphere can be used to determine the effects of position $\mathbf{r}$ and frequency ω on the transition rates of molecules inside a droplet [19], [23]. Transition rates can be enhanced when the frequency is near a resonance at positions where the internal fields of the MDR are relatively large. The transition rates can be suppressed when the frequency is either far from resonance or at the null locations of the internal field. Conceptually, if the fields emitted from the dipole are reflected by the sphere surface, and return to the source in such a way that they help drive the oscillating dipole, the transition rates can be enhanced. In this classical analysis, the problem is similar to that of an antenna radiating in a metal box. It is difficult to drive an antenna which is placed at a node of the cavity, but easy to drive an antenna at an anti-node.

Similar modifications of transition rates have been determined by analyzing how the cavity modifies the density of photon states [21], [24], [26]. An advantage of the quantum-mechanically based analysis is that it is applicable to problems for which a classical analysis is not suitable [24]. "Spontaneous" emission is actually stimulated by vacuum fluctuations of the fields. The droplet modifies these vacuum fluctuations. Near a resonance, the vacuum fluctuations are large at positions where the fields of the resonant mode are large. The droplet cavity modifies the density-of-final photon states into which a molecule can radiate [20], [26]. Since the possible final-states are redistributed into a linewidth δx, the enhancement is proportional to the $Q = x/\delta x$ of the MDR. If the radiation stays inside the droplet for such a long time that it is reabsorbed before it leaves the cavity, the total emission is not proportional to the Q [24].

2.4 Internal Intensities–Theory

2.4.1 Plane Wave Illumination–Theory

The internal intensities of spheres have been studied using wave theory (separation of variables) and geometrical optics [32]–[37]. Geometrical optics can help in understanding the internal intensity distributions. Analyses in terms of a Debye-series expansion and the comparison of these results with geometrical optics results are particularly useful for clarifying the origins of the internal distributions [36]. The internal intensities are maximal at internal caustics of the droplets. The Debye-series expansion allows the wave solutions, corresponding to the rays that have reflected a given number of times, to be separated from the total solution [36]. The correspondence [36], [37] between the wave and ray pictures is gratifying.

2.4.2 Non-plane Wave Illumination–Theory

Although droplets are often illuminated with plane waves, some nonlinear effects are best observed with non-plane wave illumination. The internal fields of spheres illuminated with Gaussian beams can be very different from those generated by plane waves [38]–[41]. When the droplet is illuminated with a tightly-focused Gaussian beam [41] the internal intensity $(\mathbf{E} \cdot \mathbf{E}^*)$ can be predominantly Gaussian, as illustrated in Fig. 1, for an incident beam focused at $y_0 = 0.75\ a$. When the droplet

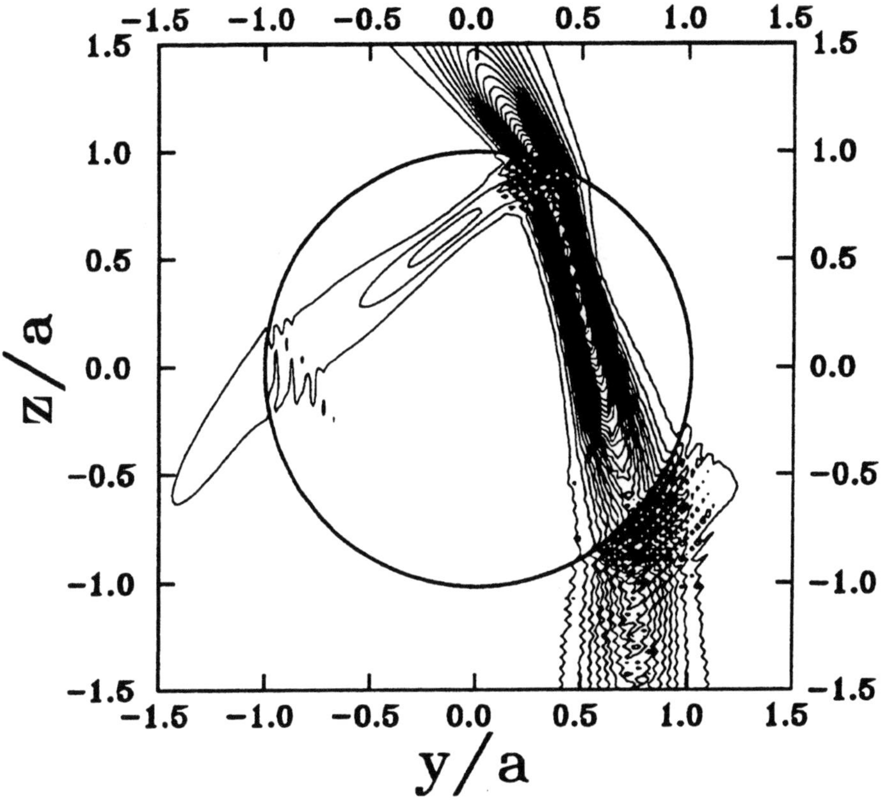

Figure 1: Contour plot of the intensity in the y–z plane inside and near a sphere illuminated with an off-axis Gaussian beam. The beam propagates in the $+z$– direction. The parameters are: $x = 79.627952$ (the resonance frequency of the $TE_{94,2}$ mode), m= 1.36, the spot size at the beam waist is $\approx 0.149a$). The peak intensity of the incident beam occurs at $x_0 = 0$, $y_0 = 0.75a$ and $z_0 = 0$. The maximum value of the internal intensity, $|E|^2_{max}$, is 1.36. The value of $|E|^2_{max}$ depends on the number and positions of the data points considered for the plots. (From Ref. [41].)

is illuminated with a tightly-focused Gaussian beam focused at the edge or outside the droplet, the internal intensity tends to be more like that of a resonant mode. Figure 2 shows the internal intensity for a beam focused at the rim of the droplet, i.e., at $y_0 = a$.

2.4.3 Time-Dependent Internal Fields–Theory

Transient intensities inside a large dielectric sphere resulting from excitation with plane-wave pulses [43] and pulsed Gaussian beams [44] have been computed. The pulses have a Gaussian time-dependence. The center frequency of the pulse is either on or near a MDR.

The time-dependent electric field at a location $\mathbf{r}$ inside the sphere is computed by Fourier transforming the incident field amplitude, multiplying by the Fourier transform of the impulse response function (the transfer function for the linear system) at the particular $\mathbf{r}$ for each of the three field components, and then inverse Fourier transforming the individual field components to obtain the total time-dependent field, i.e.,

$$\mathbf{E}^{\mathrm{int}}(\mathbf{r},t) = \mathcal{F}^{-1}\{E^{\mathrm{inc}}(\omega)\mathbf{E}^{\delta}(\mathbf{r},\omega)\}, \tag{14}$$

where $\mathbf{E}^{\delta}(\mathbf{r},\omega)$ is the Fourier transform of the impulse response function at $\mathbf{r}$ and $E^{\mathrm{inc}}(\omega)$ is the Fourier transform of the incident pulse, given by,

$$\begin{aligned} E^{i}(\omega) &= \mathcal{F}\{E^{\mathrm{inc}}(t)\} \\ &= \mathcal{F}\left\{\hat{E}^{\mathrm{inc}}(t)e^{-i\omega_0 t}\right\}. \end{aligned} \tag{15}$$

The Fourier transform of the impulse response function, $\mathbf{E}^{\delta}(\mathbf{r},\omega)$, at a location $\mathbf{r} = (r,\theta,\phi)$ is obtained by evaluating the internal field amplitude at $\mathbf{r}$ for a unit-amplitude plane wave or Gaussian beam incident on the sphere at different frequencies.

In a sphere, both the internal spectrum and the associated time-dependence vary with spatial location, particularly when the incident frequency is near a MDR. The time dependence of the intensity at an internal location near the surface shows an exponential tail with a time constant of $1/\Delta\omega_r$, where $\Delta\omega_r$ is the resonant linewidth of the MDR, as long as the incident spectrum overlaps significantly with the MDR. The primary factors determining the time dependence of the internal intensity are the pulse width of the incident Gaussian intensity, the lifetime of the nearest high-Q MDR, the detuning, and the position inside the droplet.

Figure 3 shows the three locations inside the sphere where the time dependence was calculated. These points are superimposed on the contour plots of the off-resonant and resonant internal energy distributions for the $TE_{94,2}$ MDR along with surface plots [10] of the intensities for the two cases. At location (a) the intensity is small unless the incident field is on a MDR of the sphere. Point (b) is on the front hot-line where the intensity is relatively large. The contribution of the resonant field at point (b) is negligible. In a geometrical optics picture the intensity at point (b) is caused by the intersection of rays internally reflected once from the sphere surface [36], [37], [110]. Point (c) is at a location where the intensity is relatively small for either the on-resonance or off-resonance case.

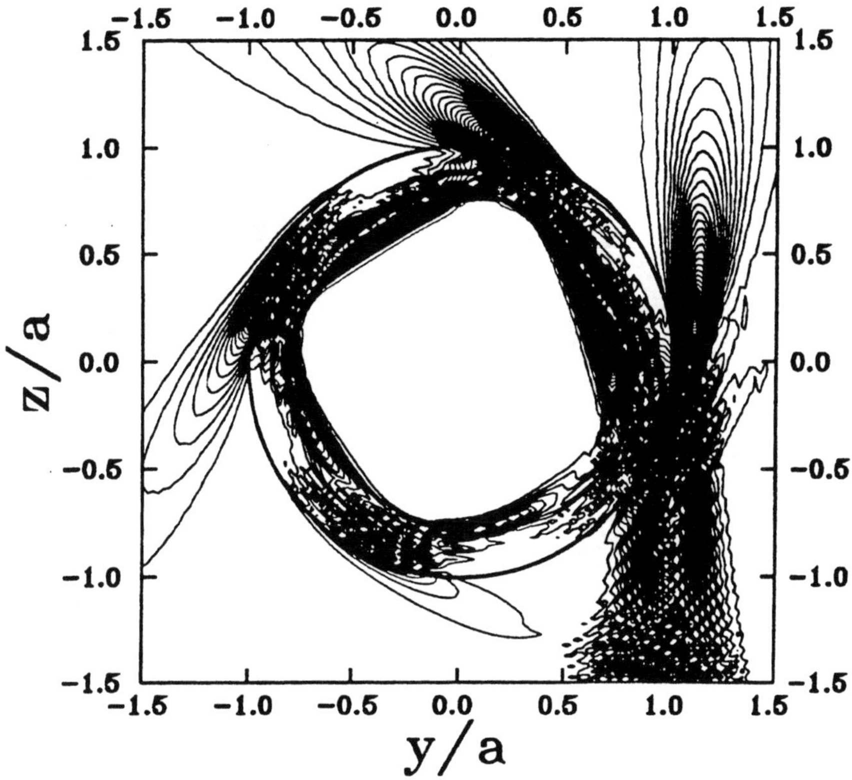

Figure 2: Surface and contour plots for the same case as in Fig. 1 except $y_0 = a$. The maximum intensity inside the sphere is 402 times the incident intensity. (From Ref. [41].)

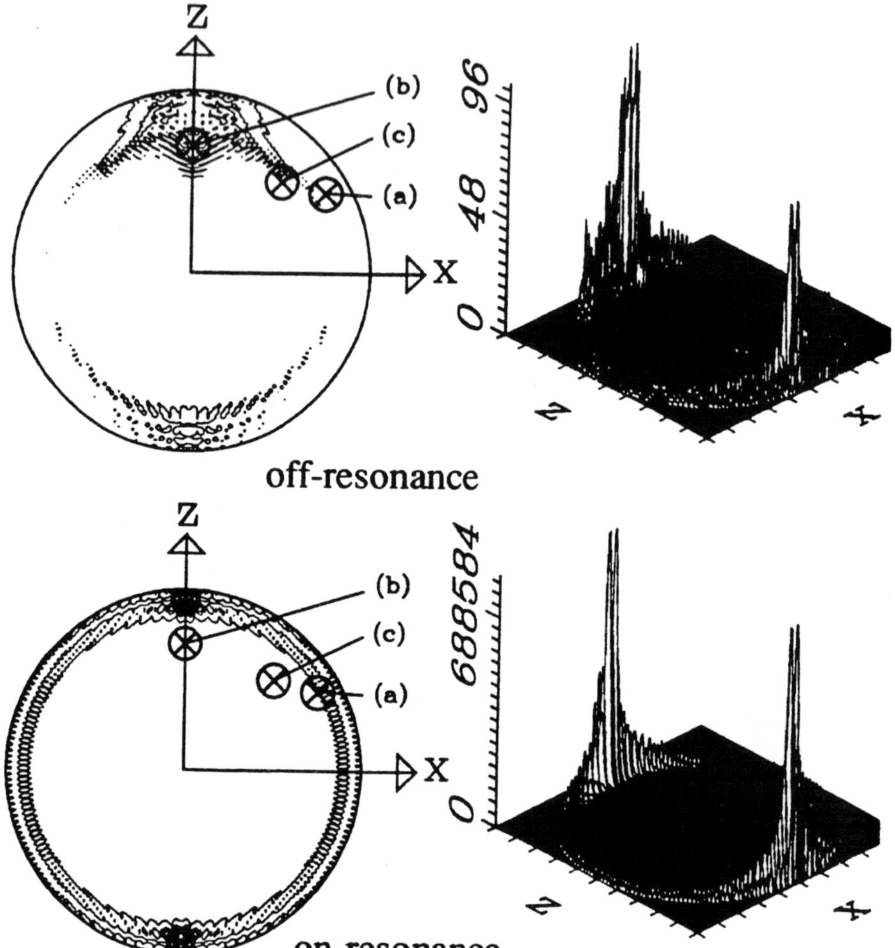

Figure 3: Points in the x–z plane where the internal intensities are computed. The coordinates of the points are (a) $r = 0.87a$, $\theta = 60°$, $\phi = 0°$; (b) $r = 0.7a$, $\theta = 0°$, $\phi = 0°$; (c) $r = 0.707a$, $\theta = 45°$, $\phi = 0°$. Points (a), (b), and (c) are superimposed upon contour plots of the internal intensities generated by on-resonance and off-resonance y-polarized incident plane waves. Surface plots of the internal intensities for on-resonance and off-resonance incidence are also shown. The MDR is $TE_{94,2}$, which has a $Q = 8.8 \times 10^6$ and a resonant size parameter $x_0 = 79.62795$. For an incident wavelength of 0.532 μm the resonant lifetime τ_r is 2.5 ns and the radius of the sphere is 6.74 μm. The refractive index of the sphere is 1.36. (From Ref. [43].)

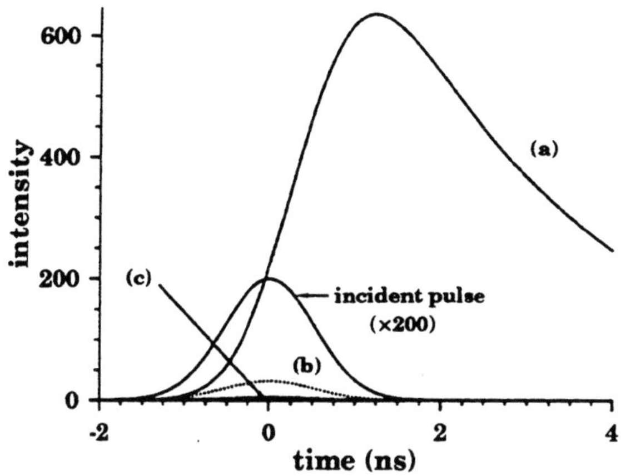

Figure 4: Transient internal intensities at the three locations shown in Fig. 3. The incident y-polarized plane wave has a Gaussian time dependence and propagates along the z axis. The scaled (×200) incident pulse is also shown. Curve (a) is at point (a), (b) is at point (b), and (c) is at point (c). The incident Gaussian pulse is on-resonance with a pulse duration of 1.25 ns. The incident pulse is on the $TE_{94,2}$ resonance and the pulse duration τ_0 is one half the resonant lifetime $\tau_r = 2.5$ ns. (From Ref. [43].)

Figure 4 shows the transient intensity at the three different locations shown in Fig. 3 for a y–polarized incident Gaussian pulse of duration $\tau_0 = 1.25$ ns. Curve (a) shows the transient intensity at point (a) of Fig. 3. The intensity at this location is significant only near resonance. So, curve (a) shows the transient build-up of the resonant field inside the sphere. Curve (b) shows the transient intensity at point (b) of Fig. 3, which is on the front hot-line. For a sphere of radius 7 μm and refractive index of 1.36, the rays cross at the forward hot-line about 0.03 ps after the incident rays would have passed the center of the sphere, a time that is much smaller than the resonant lifetime for the sphere. So curve (b) follows the input pulse in the time scale shown on the plot. Curve (c) shows the transient intensity at point (c) of Fig. 3. The intensity at this location also follows the incident pulse.

2.5 Perturbed and Inhomogeneous Spheres–Theory

Perfectly homogeneous and spherical droplets exist only in a world of mathematical abstraction. Falling droplets are perturbed into spheroids or other shapes [45]. Liquid droplets may contain solid inclusions [143] or gas bubbles [138], [144]. Evaporative cooling may cause temperature and refractive index gradients in droplets [58], [90]. Differential evaporation rates of multicomponent droplets may cause refractive index gradients [58]. Droplets evaporating in high pressure combustion environments may have large radial gradients.

When nonlinear optical processes occur in droplets, the illumination intensities are so high that the internal fields are intensity-dependent (see Section 2.7).

Consequently, even if droplets were initially perfectly spherical and homogeneous, the incident optical fields (which induce the nonlinear effects) would generate either inhomogeneities or shape perturbations [46] in the droplets. The inhomogeneities may be caused by an intensity-dependent real part of the refractive index (e.g., by molecular reorientation or thermal effects and by other effects) or by an intensity-dependent gain or loss (e.g., 2-photon absorption). A variety of methods have been developed for computing the fields of inhomogeneous or otherwise-perturbed spheres.

2.5.1 T-Matrix and Related Methods

The T-matrix method [10], [47] is useful for computing the scattering by homogeneous or layered particles. Although it is applicable to arbitrarily-shaped objects [47], it has been applied primarily to axisymmetric objects. In the T-matrix method, the incident, internal, and scattered fields are expanded in the vector spherical harmonics (VSH). The scattered coefficients are related to the incident coefficients by the T-matrix, which contains all the information about the droplet or particle. The internal coefficients are related to the incident coefficients by the A-matrix, which again, depends only upon the particle. The T-matrix and the A-matrix are computed using a set of electric and magnetic surface currents. These currents are the sources of fields which cancel the incident fields throughout a circumscribed sphere inside the droplet, and which radiate the fields scattered by the particle. Evaluations of the T-matrix and the A-matrix require evaluations of integrals of products of the VSH over the surface of the droplet. The method gives excellent results for spheroids having axial ratios of up to approximately three, when the optical size of the spheroid is not too large. Using the T-matrix method it was found that resonance frequencies (which are degenerate in a sphere) split into multiple resonance frequencies when the droplet is perturbed into a spheroid [49].

A related method [50] can also be used to find the relations between the incident and scattered coefficients. The method appears to be applicable to near-spherical droplets where the deviation from sphericity is less than a few tenths of a radius [50]. In this method the relations are obtained by matching the fields on the surface of the perturbed sphere. It is assumed that the VSH are approximately correct both inside and outside the non-spherical droplet even at the surface of the droplet. The results obtained depend upon the validity of this assumption.

2.5.2 Time-Independent Perturbation Method

A time-independent perturbation method (TIPM) has proved useful for analyzing the resonance locations [51], Q's [52] and photon lifetimes [53] of shape-perturbed and inhomogeneous [54] droplets. Some key results of the TIPM [51], [54] for an axisymmetric perturbation are reviewed here for the case where the unperturbed system is a homogeneous lossless sphere of radius a. The relative permittivity of the unperturbed sphere is $\epsilon_o = \mathrm{m}_r^2$. The resonant size parameter is $x_o = 2\pi a/\lambda$. Although the TIPM is applicable to shape perturbations and to perturbations of the real or imaginary part of the refractive index, for simplicity, only the case of a change in the imaginary part of the index is considered here.

The perturbed refractive index is

$$\mathrm{m}(\mathbf{r}) = \mathrm{m}_r + i\mathrm{m}_i(\mathbf{r}), \tag{16}$$

where the associated change in m_r (specified by the Kramers-Kronig relations) is omitted since m_i is small.

The perturbed relative permittivity is

$$\epsilon(\mathbf{r}) = \epsilon_o + \epsilon_1(\mathbf{r}) = \mathrm{m}^2(\mathbf{r}) = \mathrm{m}_r^2 - \mathrm{m}_i^2(\mathbf{r}) + i2\mathrm{m}_r\mathrm{m}_i(\mathbf{r}). \tag{17}$$

In the first-order TIPM the shift (ω_1) in the complex resonance location (for a mode with unperturbed complex resonance location ω_o), is

$$\frac{\omega_1}{\omega_0} = -\frac{V}{2G}, \tag{18}$$

where the overlap integral, V, and the normalization integral, G, are defined for TE modes as [54]

$$V = \int_R \epsilon_1(\mathbf{r})\{\phi(r)\}^2 |\mathbf{X}_{n,m}|^2 d^3r, \tag{19}$$

$$G = (\mathrm{m}_r^2 - 1)\frac{a^3}{2}[h_n^{(1)}(x_0)]^2. \tag{20}$$

where $\phi(r)$ is proportional to $j_n(\mathrm{m}kr)$ inside the droplet and to $h_n(kr)$ outside the droplet.

The complex pole for the resonance changes from the unperturbed value $\omega_o = \omega_{or} - i\omega_{oi}$ to the perturbed value $\omega_p = (\omega_{or} + \omega_{1r}) - i(\omega_{oi} - \omega_{1i})$. The total perturbed Q is then

$$Q = \frac{\omega_{or} + \omega_{1r}}{2(\omega_{oi} - \omega_{1i})}. \tag{21}$$

For small perturbations in the imaginary part of the refractive index, ω_{1r} is usually much less than a fraction of a linewidth and can be neglected. For example, for the $\mathrm{TE}_{37,1}$ mode of a homogeneous sphere with $\mathrm{m}_i(\mathbf{r}) = 10^{-3}$, ω_{1r} is about 10^{-3} times the linewidth. The sign of ω_{1i} is negative for absorption and positive for gain.

The resonance locations computed using the TIPM compare very well with those computed using the T-matrix method for either shape-perturbed homogeneous droplets [51], or for droplets which contain an off-centered internal sphere [54]. The changes in the Q computed with the two methods are typically within a factor of 2 or 3 of each other when the change from the unperturbed Q is less than 50 percent [54].

It is found that when a sphere is distorted into a spheroid, the shift in the resonance frequency of a MDR labeled by n, m is [51],

$$\delta x = \frac{xe}{6}\left(\frac{3m^2}{n(n+1)} - 1\right) \tag{22}$$

where $e = (r_p - r_e)/a$, r_p and r_e are the polar and equatorial radii, and a is the radius of the equivalent sphere.

For a 45 μm radius droplet, which is typical of droplets used in many experiments, e can be approximately 0.001 when the distortion results from inertial effects on droplets flowing through air [165]. For m nearly equal to n, δx is approximately $3.3 \times 10^{-4}x$. A mode having a Q of 10^6 has a linewidth $\Delta x = 10^{-6}x$. Thus, relative to the MDR linewidth, the frequency spacing between adjacent azimuthal modes is 330 for MDRs with $Q = 10^6$.

An effect of the splitting is that each azimuthal mode in Eq. (1) has its own Lorentzian function $L(x)$, and so the f_n and g_n in Eqs. (3) and (4) should be labeled as $f_{n,m}$ and $g_{n,m}$. That is, each MDR specified by n, m has a different resonance frequency.

2.5.3 Effective-Average Gain and Loss Coefficients

Methods have been developed [55] for calculating the Q's and the effective-average gain and loss coefficients for the modes of a dielectric sphere which has a spatial perturbation of the imaginary part of the refractive index [see Eq. (16)]. The real part of the refractive index is assumed constant. The method is similar to that used for computing the gain or loss coefficients of linear lasers [5]. The Q's of absorbing spheres are determined from the effective-average loss coefficients, and can be compared with results from the TIPM [51]. The basic difference between the effective-average method and the TIPM is in the treatment of the fields outside the sphere. In the TIPM, the field squared $(\mathbf{E}(\mathbf{r}) \cdot \mathbf{E}(\mathbf{r}))$is used, not the electric energy density.

Two kinds of effective-average values of the imaginary part of the refractive index are important. One effective-average value of m_i can be used to compute the internal or scattered field coefficients for the modes of an absorbing sphere; the other can be used to compute the Q and the effective-average gain or loss coefficients for the modes. Since each mode has a different spatial profile, each mode may have a different effective-average gain or absorption coefficient.

The effective-average imaginary part of the refractive index which can be used to compute the internal or scattered field coefficients for the modes of an absorbing sphere is labeled $\mathrm{m}_{i_s}^{\mu}$, where s suggests the volume of the sphere. This $\mathrm{m}_{i_s}^{\mu}$ is the average of $\mathrm{m}_i(\mathbf{r})$, weighted by the electric energy density of the mode μ *within* the sphere volume,

$$\mathrm{m}_{i_s}^{\mu} = \frac{\displaystyle\int_{V_s} \tfrac{1}{2}\mathrm{m}_i(\mathbf{r})\epsilon_r(\mathbf{r})|\mathbf{E}^{\mu}(\mathbf{r})|^2 dv}{\displaystyle\int_{V_s} \tfrac{1}{2}\epsilon_r(\mathbf{r})|\mathbf{E}^{\mu}(\mathbf{r})|^2 dv}, \tag{23}$$

where $\mathbf{E}^{\mu}(\mathbf{r})$ is the unperturbed electric field of mode μ, $\epsilon_r(\mathbf{r}) = \mathrm{Re}\{\epsilon(\mathbf{r})\} = \mathrm{Re}\{\mathrm{m}^2(\mathbf{r})\}$, and the volume V_s is the volume of the sphere. The integrand of the numerator is proportional to the rate of growth or decay of energy per unit volume inside the sphere, while the denominator is proportional to the total energy of the mode inside the sphere. A sphere with homogeneous loss has $\mathrm{m}_i(\mathbf{r}) = \mathrm{m}_i = \mathrm{m}_{i_s}^{\mu}$. The internal and scattered field coefficients corresponding to the mode μ were computed using $\mathrm{m}_{i_s}^{\mu}$ for the case of two-photon absorption [55].

The other type of effective-average imaginary part of the refractive index, $\mathrm{m}_{i_r}^{\mu}$, can be used to compute the Q's and the effective-average gain or loss coefficients.

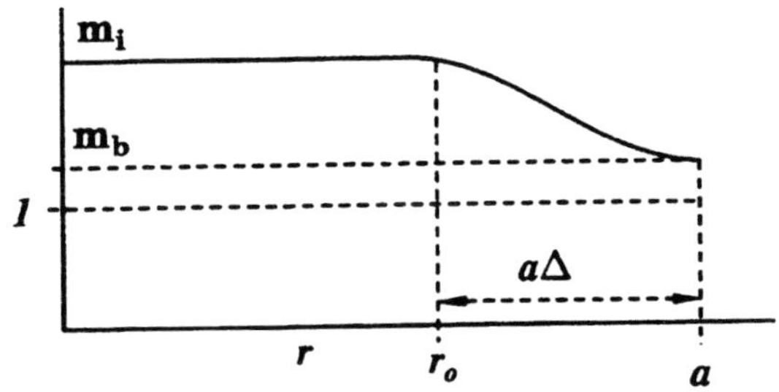

Figure 5: Radial dependence of the refractive index simulating a smooth roll-off. The quantity $a\Delta$ is equal to $a - r_0$ since $\Delta = (1 - r_0/a)$. (From Ref. [57].)

The sub-subscript r indicates that $m^{\mu}_{i_r}$ is most directly related to the rate of loss (gain) of energy and hence to the Q. This $m^{\mu}_{i_r}$ is the average of $m_i(\mathbf{r})$, weighted by the electric energy density of the mode [5] (including the energy trapped outside the sphere),

$$m^{\mu}_{i_r} = \frac{\int_{V_s} \frac{1}{2} m_i(\mathbf{r}) \epsilon_r(\mathbf{r}) |\mathbf{E}^{\mu}(\mathbf{r})|^2 dv}{\int_{V_m} \frac{1}{2} \epsilon_r(\mathbf{r}) |\mathbf{E}^{\mu}(\mathbf{r})|^2 dv}, \tag{24}$$

where V_m is the mode volume, which includes V_s and the evanescent waves of that mode trapped outside the sphere. Both $m^{\mu}_{i_r}$ and $m^{\mu}_{i_s}$ were used in modeling the scattering by a sphere which contains a two-photon absorber [55].

2.5.4 Radially Inhomogeneous Spheres

Resonant frequencies and Q's of radially-inhomogeneous spherical particles were computed [57], [58] using the separation-of-variables method [7], [56]. Only the radial equations differ from the equations for the homogeneous sphere. These radial equations were solved using a Runge-Kutta method [57], [58].

Figure 5 shows the geometry and the terminology used to define the distribution of refractive index considered here. The refractive index is constant and equal to m_i ($i \equiv$ internal) for $r < r_0$, where r is the radial distance from the center of the sphere, and is

$$m(r) = m_b + (m_i - m_b)\cos^2\left[\frac{\pi(r - r_0)}{2(a - r_0)}\right],$$

for $r_0 \le r \le a$, where m_b ($b \equiv$ boundary) is the refractive index at the surface.

For a fixed value of the refractive index at the boundary of the sphere, the first-order resonance location shifts towards higher values of size parameter as Δ increases. This shift is expected because the average refractive index seen by the MDR is reduced as Δ increases; the resonance locations are known to shift to a higher size parameter (shorter wavelength) as the refractive index decreases [101]. Also, for a fixed Δ, the average refractive index decreases as m_b decreases because the radius a is fixed, and so the resonant size parameters increase.

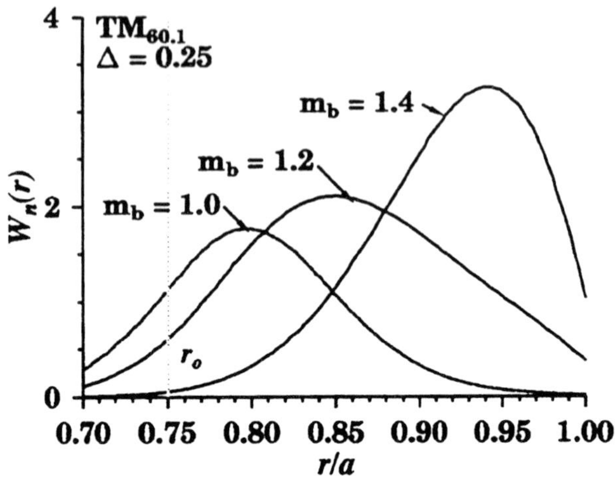

Figure 6: Radial functions $W_n(r)$ for $\Delta = 0.25$ vs. the normalized radial distance r/a, where a is the sphere radius. The core refractive index is 1.5. (From Ref. [57]).

Numerical results for modes with Q's in the range of 500 to 10^{16} show that when the change in refractive index occurs in only the outer few percent of the droplet radius, the change in the Q is less than 20 percent.

The radial dependence of $W_n(r)$ is shown in Fig. 6 for $\Delta = 0.25$. The peak in the radial distribution moves inward with decreasing m_b. Also, as the value of m_b decreases the slope of the radial distribution at the sphere boundary decreases.

2.5.5 Spheres with Inclusions or with Particles on the Surface

The T-matrix method can be used to compute the fields of multilayered particles [48], [54]. This method works well when the inclusions are situated near the center of the droplet, and when each inclusion is inside the next larger one. Methods have been developed for computing the fields in spherical droplets which contain arbitrarily-situated, spherical inclusions [61], [62]. In these methods, the fields in each spherical region are expanded in the VSH centered about the origin of each sphere. Then an addition theorem for the VSH is used to express all the fields in terms of a single coordinate system, and the fields are matched at each of the surfaces. Calculated results have been obtained for an included sphere which moved from one side of the larger sphere to the other [61].

Scattered fields and resonance frequencies have also been obtained for a large sphere which has a second sphere either nearby or adhering to the surface [60]. The large sphere was centered at the origin and illuminated with a plane wave propagating along z. The effects on the MDR (frequency, height, and shape) were largest when the adhering sphere at the surface was situated on or near the z-axis (i.e., endfire illumination). When the adhering sphere was on the x-axis (broadside incidence) the effects on the MDR were far smaller or even negligible.

3 EXPERIMENTAL OBSERVATIONS OF INELASTIC EMISSION AND NONLINEAR OPTICS IN DROPLETS

3.1 Fluorescence Emission

The fluorescence spectrum of a droplet containing a laser dye consists of numerous sharp but shallow peaks superimposed on a broadband fluorescence. This broadband emission is well known for an optical cell containing laser dyes. The sharp but shallow peaks (commonly referred to as output MDRs) can be assigned to specific MDRs [90], [170]. These peaks are a manifestation of an enhanced Einstein A coefficient [26] for dye molecules located within an internal region where the energy of the output mode is largest. The strongest fluorescence comes from regions which have a large spatial overlap between the intensity distribution of the excitation radiation and the output MDR. Because the spatial distribution of MDRs does not extend to radius $r < a/\mathrm{m}(\omega)$, the dye molecules within the inner core (with $r < a/\mathrm{m}(\omega)$) emit a continuum fluorescence spectrum. The sharp peak features in the fluorescence spectrum are associated with the QED enhanced Einstein A coefficient for molecules located in the region $a/\mathrm{m}(\omega) < r < a$ where the intensity distribution of the MDRs is nonzero. The sharpness of a fluorescence peak, which is usually limited by the spectrograph resolution (e.g., 1 cm^{-1}), is related to the Q value of the MDR. The relative fluorescent height is proportional to the intensity enhancement factor of the MDR. Consequently, should all the dye molecules be localized in the region with $r \geq a/\mathrm{m}(\omega)$, then the intensity ratio of the sharp peaks and the continuum would increase relative to that for the uniform distribution of dye molecules.

The role of QED enhancement of the Einstein A coefficient is confirmed by observing significant lifetime modification (both shortening and lengthening) of the spontaneously emitted fluorescence from chelated Europium ions in 10 μm diameter dimethylformamide-alcohol droplets [86], [87] and from Rodamine 6G in glycerol droplets [68]. Several cavity QED phenomena, such the spectral and lifetime modifications, that have been observed in microdroplets have been reviewed [87], [88].

The relative height of an individual fluorescence peak is proportional to the volume fraction of the dye molecules within the region of spatial overlap between the excitation radiation intensity and the specific MDR intensity. The spatial overlap is highly sensitive to the illumination geometry and to the wavelength of the excitation radiation, both of which determine the coupling of the input radiation with a MDR. For example, regardless of the wavelength of the input beam, MDRs are not excited by a tightly focused beam aimed through the principal diameter of the droplet [38], [40]. Efficient coupling to tightly focused beams is achieved when the beam is aimed at the droplet rim or even outside the droplet rim, provided that the input wavelength or the droplet radius is tuned near or on the size parameter corresponding to a MDR [38], [40]. When the input radiation couples to a MDR, then an "input resonance" has been excited and the relative height of each fluorescence peak is proportional to the overlap integral of two MDRs. Such "double resonance" conditions can greatly reduce the lasing threshold and the thresholds for other nonlinear effects.

The identification of input MDRs can be determined by monitoring the wavelength-integrated fluorescence intensity while varying the excitation laser wavelength (for a fixed radius droplet) or by varying droplet radius [174] (for a fixed input wavelength). Both the fluorescence-excitation spectrum and the fluorescence emission spectrum are sensitive to the MDR internal-field coefficients that become large for high-Q MDRs [13]. The integrated area under the square of the internal field coefficients is approximately constant as the linewidth decreases. By contrast, the integrated area under the resonances of the elastic scattering spectra tends toward zero as the linewidth decreases. Consequently, low Q, but not extremely high-Q MDRs are observed in elastic scattering spectra [2], [13].

Persistent hole burning in the fluorescent spectrum in a random collection of microparticles has been demonstrated [67]. These results are the first demonstration of room-temperature, persistent-spectral hole-burning, and they suggest the possibility of an all optical memory.

3.2 Spontaneous Raman Emission

The Raman gain coefficient is enhanced by as much as the Einstein A and B coefficients. Analogously, the enhancement factor of the Raman scattering cross section is dependent on the MDR Q values and on the spatial overlap between the internal distributions of the pump laser beam and the MDR that is providing the feedback. Evidence for the enhanced Raman scattering cross section and hence, the Raman gain for SRS, was confirmed by the sharp but shallow peaks in the spontaneous Raman scattering spectrum of water droplets [69]–[71] and, more recently, in droplets of other liquids [72], [73], [75]. In the spontaneous Raman scattering spectrum of levitated and slowly evaporating droplets, the presence of input and output resonances is considered a "nuisance". When an input resonance is eventually found, (commonly referred to as the "double resonance condition") the spatial overlap between the input and output resonances greatly affects the height of the Raman peaks on MDRs. In interpreting the vibrational modes, the input and output resonance peaks associated with MDRs can be misunderstood as vibrational peaks of the liquid. Consequently, whenever possible, the spontaneous Raman scattering spectrum is integrated over a long time period in order to average out the input and output resonances as the droplet size continues to decrease due to evaporation.

From the lineshape of the spontaneous Raman scattering spectrum in the O–H stretching mode region of H_2O droplets, temperatures (with an accuracy of $\pm 1°$ C) have been determined in the range from 10° to 65° C [161]. Raman spectroscopy of single droplets has been comprehensively reviewed [74].

3.3 Lasing

Lasing in droplets occurs when the round-trip gain exceeds the round-trip loss as the amplified fluorescent waves circumnavigate the droplet rim. The Einstein B coefficient is enhanced in the region that has a good spatial overlap between the intensity distribution of the pump and the distribution of the MDR that is providing the optical feedback [21], [26]. For a microsphere that is highly transparent at the fluorescent wavelength, the main loss is associated with the radiation leakage of the particular MDRs within the fluorescence gain profile. The combination of the enhanced Einstein B coefficient and the low round-trip loss contribute to a low lasing threshold. The high Q MDRs of droplets have led to the observation of thresholdless laser emission of Rhodamine 6G in methanol droplets. The emission increased linearly as the cw pumping intensity increased [84]. When two fluorescent transitions have different Einstein B coefficients, lasing at wavelengths corresponding to the weaker gain also can be achieved [150], [151].

With cw Argon-ion laser pumping (at $\approx$ 35 W/cm^2), the lasing threshold can be exceeded in micrometer-sized ethanol droplets ($a \approx 30$ μm) containing 10^{-4} to 10^{-5} M Rhodamine 6G [76]. Significantly different lasing spectra can be observed when the pump beam (from a single-longitudinal mode of a Q-switched Nd:YAG laser) is made less than the droplet diameter, suggesting that the spatial overlap between the pump beam and the output MDR is important for the lasing case [81].

The lasing threshold in droplets is so low that even with two-photon absorption (TPA) of pulsed laser radiation, laser emission has been reported for ethanol droplets containing dye molecules (10^{-2} M Coumarin and 2.5×10^{-4} –10^{-3} M Rhodamine 6G) [89]. Because the TPA of the higher S_2 band of Rhodamine 6G is larger than that of the lower S_1 band, lasing emission in the 605 nm region was achieved at 1 GW/cm^2 when pumped with 697 nm radiation. By contrast, when pumped with 1064 nm radiation, $\approx$10 GW/cm^2 was needed to observe lasing in the 605 nm region. However, at pump intensities above 1 GW/cm^2, the plasma produced by laser-induced breakdown introduced severe losses and subsequently quenched lasing within the droplet.

Time resolved measurements of lasing and elastic scattering have been made in Rhodamine 6G droplets. The laser emission occurred almost simultaneously with the elastic scattering and had a shorter growth time than SRS. The radiation lifetime of the lasing emission from droplets was shorter than the corresponding bulk-liquid fluorescence lifetime [79].

The lasing spectrum of Rhodamine 6G in water solution droplets consists of MDRs belonging to several mode orders. This multiplicity of lasing at several MDRs of different mode orders is consistent with spatial hole burning within the outer rim of the droplets [83].

3.4 Stimulated Raman Scattering (SRS)

Besides lasing in droplets, SRS from droplets [92]–[117] and, more recently, from a single liquid jet in the form of a cylinder [116], is the most studied of all the nonlinear optical effects. The SRS spectra from both droplets and cylinders exhibit MDR peaks. In the optical-cell case, the single-pass amplification starting from the spontaneous Raman "noise" near the entrance window requires the triple product of the gain, I_{in}, and the length to reach e^{30}. In the droplet case, an SRS signal at a MDR appears when the round-trip Raman gain exceeds the round-trip loss.

The Raman-gain coefficient is enhanced by as much as the Einstein A and B coefficients. The QED enhanced Raman gain is so high, and the double resonance condition is so efficient in storing the input and SRS radiation, that cw stimulated Raman scattering has been reported [114]. The pump intensity can be as low as 8 kW/cm^2 for benzene droplets and 24 kW/cm^2 for toluene droplets [107]. Even for the non-resonant input MDR case, the QED-enhanced Raman gain and the concentration of the input-pump radiation at a localized region give rise to a large imaginary part of the third-order susceptibility, and, consequently, an intensity-dependent $\Delta m(\omega)$ [105]. Recent experiments have shown that, as the droplet radius is continuously tuned (by as much as 0.3 percent), spectral locking of a sharp SRS peak is observed [105]. The mode pulling is proposed to be associated with the large Raman gain. This large gain not only affects the imaginary part of the third-order susceptibility, but through the Kramers-Kronig relations, also affects the real part of the third-order susceptibility or $\Delta \mathrm{m}(\omega, I_{\mathrm{in}})$. It should be noted that the Raman-gain-related $\Delta \mathrm{m}(\omega, I_{\mathrm{in}})$ (as large as 0.002) is distinct from the standard intensity-dependent index of refraction change associated with molecular reorientation, where $\Delta \mathrm{m}(\omega, I_{\mathrm{in}}) = \mathrm{m}_2 I_{\mathrm{in}}(t)$.

Double resonance conditions have been deliberately investigated in SRS in order to identify the mode numbers and mode orders of the MDRs [107]. Double resonance has also been used to demonstrate that the SRS threshold can be lowered when the input resonance condition is reached [99]. The MDRs that provide the feedback for the SRS radiation have been uniquely identified [107] by determining precisely the droplet radius and refractive index. Elastic scattering at 632.8 nm by accurately-controlled droplets has been used to determine a and $\mathrm{m}(\omega)$. For small-diameter ethanol droplets ($a = 11.5\mu$m), mode orders as low as 2 and 3 are involved in the SRS process. The double resonance condition, when both the input-pump wave and the resultant SRS waves are on MDRs is actually easier to treat theoretically since the spatial overlap of a pair of MDRs is involved. In the off-input resonance condition, the spatial overlap involves the overlap of the intensity of an MDR with the total internal intensity (which includes a sum over all modes).

By finely tuning the droplet radius to achieve the double resonance condition, the Q of the MDR has been indirectly deduced from the linewidth of the MDR [99], [106]. A Q of the MDR that is supporting the SRS wave has been directly determined, by measuring the decay time τ of the SRS intensity after the 100 ps input-pump laser beam is off, since $Q = \omega\tau$ [95].

Another manifestation of the large Raman gain is the cascade SRS process. Up to 14th-Stokes order [93] and recently up to the 20th-Stokes order [123] (see Fig. 7) have been observed in CCl_4 droplets. In attempting to investigate what is the highest order of Stokes SRS, CCl_4 is usually chosen because of its low vibrational frequency (459 cm^{-1}). The internal SRS intensities of various Stokes orders can build so high that even third-order sum-frequency generation (TSFG) involving the laser field and the various order Stokes SRS fields has been observed [121]–[123]. Combinations of higher-order Stokes radiation from two different vibrational modes, one with a larger and another with a weaker spontaneous Raman cross section, can be observed from droplets with a single component [93] or binary liquid [108].

The SRS intensity on a MDR of mode number $n = m$ has a $\cos^2 m\phi$ angular dependence when a measurement is made in a plane containing the droplet equator, i.e., in the x–y plane, where the laser propagation direction is along the x–axis and the droplet flow is along the z–axis. Because n is a large number for large droplets, the angular structure is commonly referred as the "fine structure" in the angular dependence. The experimental arrangement for the measurements is shown in Fig. 8, and the experimental results are given in Fig. 9.

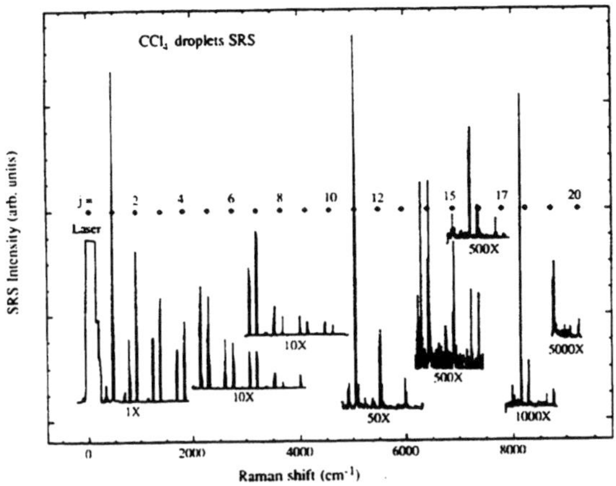

Figure 7: Multiorder SRS spectra from CCl_4 droplets (with $a \approx 26\mu m$) detected with a CCD camera and plotted as a function of the Raman shift. The integer j (above the points) corresponds to the Raman shifted frequency $\omega_{js} = \omega_L - j\omega_{\rm vib}$, where $\omega_{\rm vib} = 459$ cm^{-1} is the frequency of the ν_1 vibrational mode of CCl_4 and ω_L is the input-laser frequency corresponding to an incident laser wavelength of 0.532 μm. Cascade multiorder Stokes SRS up to $j = 20$ is detected. Combinations and overtones of the ν_2 and ν_4 vibrational modes are also detected. The CCD camera is relatively insensitive to SRS photons having wavelengths greater than 1 μm. (From Ref. [123].)

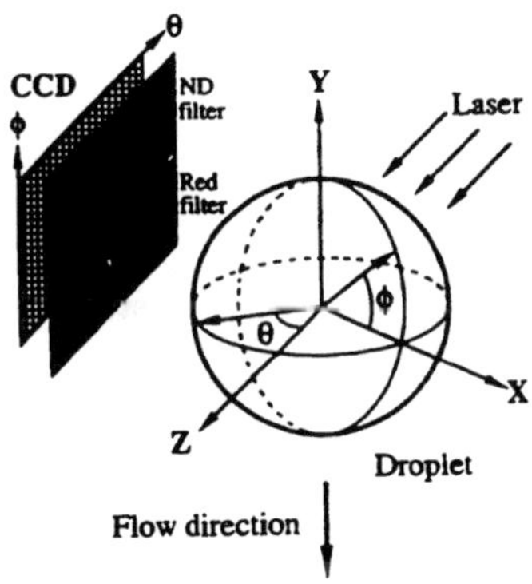

Figure 8: Experimental arrangement used to measure simultaneously the angular distribution of the elastic scattering (through a neutral density filter) and the red SRS (through a red filter) when the droplets are illuminated with a green laser beam. No lenses are used in the collection optics. (From Ref. [109].)

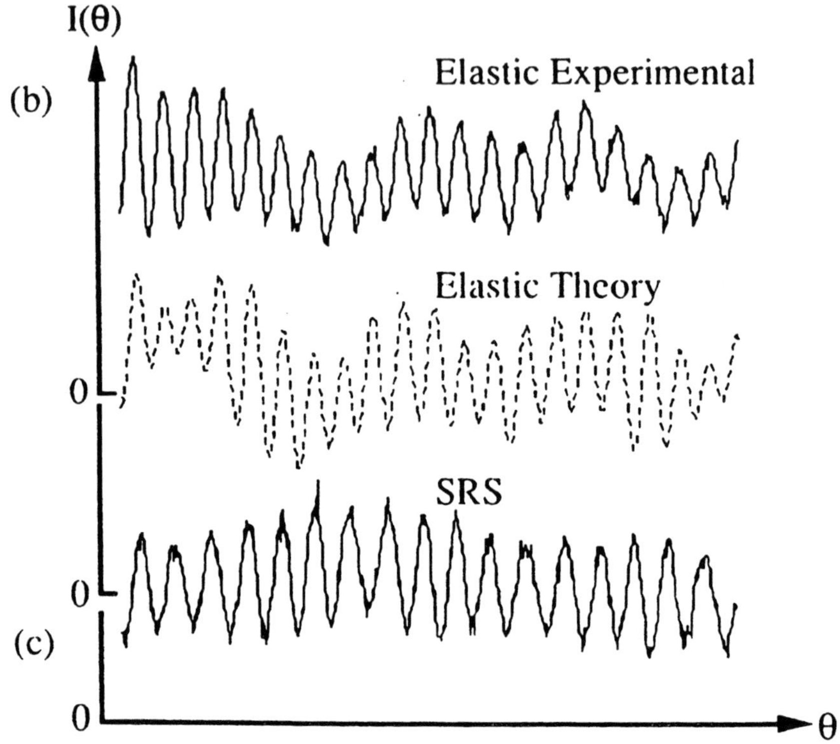

Figure 9: Single-shot angular distributions of both the elastic scattering and the SRS from ethanol droplets. The experimental distributions (both elastic and SRS) are recorded with a CCD camera. The theoretical elastic scattering distribution is calculated using Mie theory. The droplet radii are $a \approx 31\ \mu$m. In all cases, the input-laser beam intensity is set just above the SRS threshold so that a single MDR peak appears in the SRS spectrum. The illuminating laser beam is aimed at the droplet center. The number of SRS peaks and valleys in 180° is equal to the MDR mode number n. The existence of peaks and valleys in the angular distribution implies that the two counterpropagating SRS waves are phased locked during the duration of the SRS. In other data, not shown, where the focused laser beam is aimed at the droplet rim, more angular peaks are observed. (From Ref. [109].)

The existence of an angular structure implies that the SRS forms a standing wave which is composed of two coherent counterpropagating waves having approximately equal amplitudes [109]. The waves are phase locked throughout the entire temporal development of the SRS pulse. The SRS occurs as a standing wave because the Raman gain is the same for waves traveling in either the forward or the backward direction. The Brillouin gain, by contrast, is largest in the backward direction and is nearly zero in the forward direction.

The MDRs which support SRS are a function of the position of the incident laser beam. The SRS threshold intensity depends upon the spatial overlap between the MDR-intensity distribution and the internal-intensity distribution of the incident beam. Consequently, the particular MDR which supports the SRS (among the several MDRs within the Raman gain profile) depends upon the illumination geometry of the focused input-laser beam. When the illuminating beam is focused at the edge or outside the droplet, the SRS tends to be supported by low-order MDRs (e.g., mode order of 3) [109]. Calculations show that when the beam is focused far enough outside the droplet, the laser energy tends to couple primarily into a few low-order modes, even though the laser frequency may be hundreds of linewidths from the nearest resonance frequencies [40]. By contrast, when the incident beam is focused at the droplet center, the measured angular dependence indicates that a MDR with mode number 325 (and hence a mode order of 15) is selected [109].

At higher input intensity, multiorder-Stokes SRS and more than one MDR-related peak are observed within the first-order Stokes SRS spectrum. At these intensities, even if the multiorder-Stokes SRS is blocked with a color glass filter, the combination of the rapidly varying angular-dependent fields of two or more MDRs (with different n's) "washes out" the fine structure. Indeed, no SRS fine-structure angular pattern was observed from water droplets, because the Raman gain of water is broad and more than one MDR can provide the feedback for the SRS [100].

3.5 Stimulated Brillouin Scattering (SBS)

Direct experimental evidence of SBS is provided by a linear photodiode array placed behind a Fabry-Perot interferometer [118] as shown in Fig. 10. When a single-mode Q-switched laser of 10 ns duration is used to irradiate a transparent droplet, the spectrally resolved peak (on the Stokes side of the input laser frequency) is observed to be dependent on the liquid velocity of sound. When the incident beam is focused at the droplet's rim, the Brillouin gain favored backscattered SBS and the Brillouin shift is at its maximum, corresponding to a scattering angle of 180°. When the incident beam is focused along the droplet's principal diameter, corresponding to a scattering angle of 90°, the Brillouin shift decreases. Conspicuously absent is the SBS from CS_2 droplets, particularly when CS_2 in an optical cell is known to have the narrowest spontaneous Brillouin linewidth and, hence, the largest Brillouin gain. The absence of SBS of CS_2 droplets is attributed to the MDR shifts and phase-modulation of the SBS fields, both of which are caused by the large intensity-dependent index of refraction $\Delta m(\omega, I_{\text{in}}) = \mathrm{m}_2 I_{\text{in}}(t)$ of CS_2 [118]. Note in Fig. 11 that the SRS and SBS pulses in all three curves are temporally correlated; i.e., the decay of the first SBS pulse is followed by the growth of the first SRS pulse and the regrowth of the $(n+1)$-th SBS pulse occurs after the decay of the n-th SRS

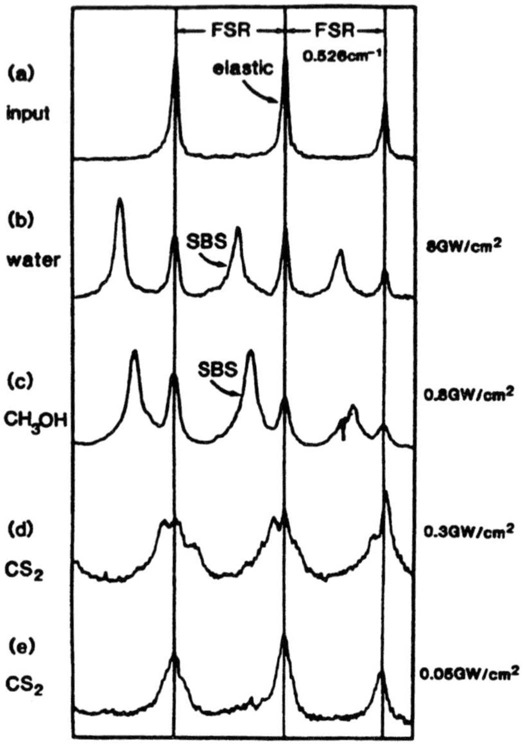

Figure 10: Single-laser-shot spectra of the SBS and the elastic scattering detected by an intensified linear photodiode array placed behind a Fabry-Perot interferometer having a free spectral range (FSR) of 0.526 cm^{-1}. The spectra shown are for three interferometric fringes. The single-mode input-laser beam is focused on the droplet rim and the input intensity for each spectrum is indicated on the right. The elastic scattering spectrum (a) indicates the spectral resolution of the interferometer. The SBS spectra from water droplets (b), and from methanol droplets (c) indicate that only the Stokes component is detected. (From Ref. [118].)

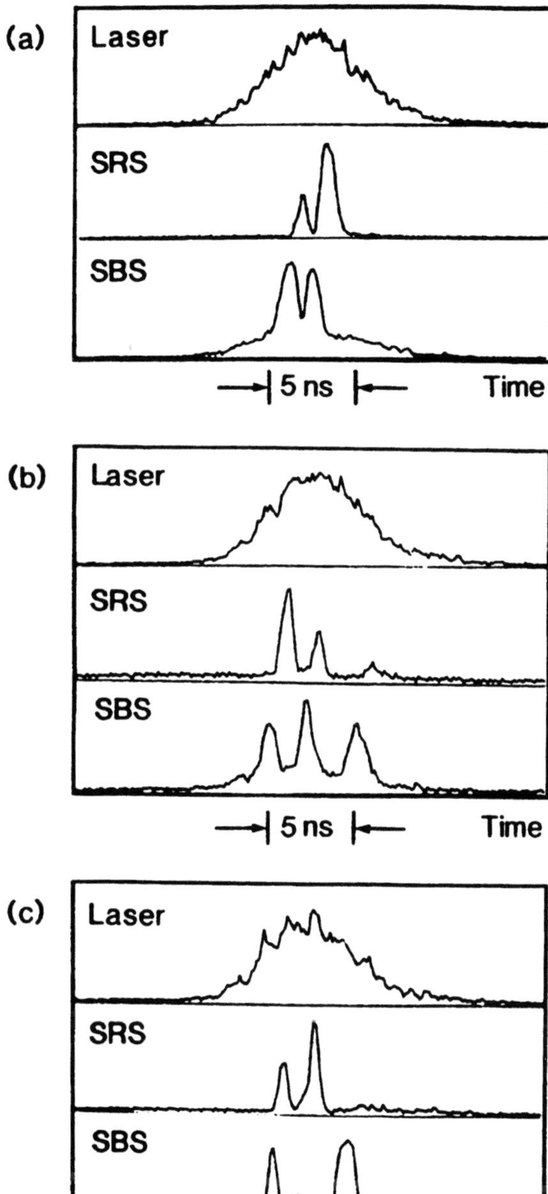

Figure 11: Single-shot time profiles of the input-laser beam, the SRS, and the SBS. All three signals are simultaneously measured from ethanol droplets (with $a \approx 45\mu m$) that are illuminated with 1 GW/cm^2. (From Ref. [103].)

pulse. Based on this observation, the conclusion is drawn that the single-mode laser pumps the SBS and subsequently the SBS pumps the SRS.

Indirect evidence for the existence of SBS is provided by the following observations when a green laser beam is incident on a microdroplet: 1) the scattered light intensity is nonlinearly dependent on the input intensity, i.e., there exists a definite intensity threshold; 2) the temporal response of the scattered light follows the temporal behavior of the input radiation at low-input intensity and has a long-decay behavior at high-input intensity [95]; 3) the peaks and valleys in the temporal behavior of the green-scattered light and the SRS are anticorrelated, suggesting that the SBS is directly pumping the SRS [103]; 4) when no SBS is generated with a multimode laser, the SRS threshold is 3× higher, another confirmation that the SBS, which has a 3× larger gain than SRS, is pumping the SRS [103]; 5) the green and red SRS emerge from the same edge regions of the droplet and have the same arc lengths, confirming that different m-value MDRs of the SBS pump similar m-value MDRs of the SRS; 6) when the input laser beam (with 10 ns pulse duration) is focused on the droplet rim, the intensity-dependent green scattered light circumnavigates the droplet rim in the backward direction, implying a backward-traveling SBS; 7) when the input laser beam (with a series of mode-locked 100 ps pulses) is focused on the droplet rim, the intensity-dependent green scattered light circumnavigates the droplet rim in the near-forward direction, implying a nearly forward-traveling SBS and long-wavevector acoustic phonons [119]; and 8) the retention of the long wavevector acoustic phonons is confirmed by delaying two groups of mode-locked pulses by as long as 171.6 ns [119] (see Fig. 12). The occurrence of SBS is often unavoidable and can complicate other nonlinear optical interactions which may be of primary interest to the investigator.

3.6 Coherent Anti-Stokes Raman Scattering (CARS) and Coherent Raman Mixing (CRM)

The phase-matching conditions for lasing, SRS, and SBS processes are automatically satisfied. Consequently, for these processes the droplet morphology provides all the advantages (high internal intensities, large Q-factor feedback, and QED-enhanced gain coefficients) without any restrictions imposed by the phase-matching condition. In a droplet, the phase-matching condition for plane waves propagating along one dimension in an optical cell is replaced by the spatial overlap condition of several standing waves or MDRs. Observations of nonlinear optical interactions which require phase matching preceded the present conceptual understanding and numerical calculations of spatial overlaps of MDRs.

Coherent anti-Stokes Raman scattering (CARS) in droplets was investigated with a monochromatic pump beam (second harmonic of a Q-switched Nd:YAG laser) and with a Stokes beam from a broadband dye laser [94], [120]. No MDR-related peaks were detected in the broadband CARS. The phase-matched angle for the pump and Stokes beams incident on the droplet were essentially equal to that for the same liquid in an optical cell. The broadened and asymmetric phase-matched angle that was reported can be explained by the spreading of the wave

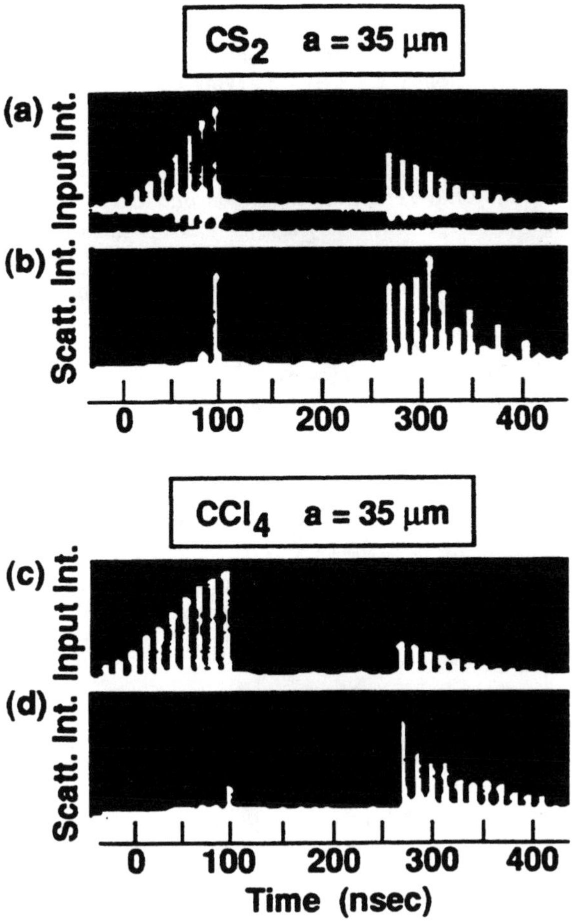

Figure 12: Time profiles of the input-laser radiation are shown in (a) and (c). The Pockels cell pulse selector separates the mode-locked input-laser pulse train into two groups of pulses which are separated by by 171.6 ns. The time profiles of the near-forward SBS are shown from CS_2 droplets in (b) and from CCl_4 droplets in (d). The near-forward SBS intensity associated with the second group of input-laser pulses is always larger than with the first group of input-laser pulses. The long-wavelength acoustic phonons generated by first group of pulses are retained during the 171.6 ns gap. (From Ref. [119].)

vectors inside the droplet and by spatial overlaps of the non-input-resonant internal intensity distributions of the pump and Stokes radiation just within the droplet shadow face.

In view of our present understanding of spatial overlaps of MDRs, the lack of MDR-related peaks in the CARS spectra and the observed phase-matched angles for the pump and Stokes beams may be attributed to having observed the CARS signal at the "wrong" detection angle. By observing the CARS signal at the phase-matched angle for a cell, it is possible that the experiment was biased toward detecting the CARS that does not require output MDRs. Had the detection angle emphasized the radiation that exits tangentially to the droplet rim, we suspect that the CARS signal may have emphasized the output MDRs. The lack of MDR-related peaks in the CARS spectrum may also be attributed to the use of a broadband Stokes wave (centered at the spontaneous Raman band) that encompassed several to many MDRs. Possibly with a narrowband Stokes beam (centered within the Raman band), MDRs would be observed in the CARS output as the Stokes beam frequency or droplet size was tuned.

Coherent Raman mixing (CRM), which is observed through an increased gain in the SRS signal, has been demonstrated in droplets [94], [120]. The droplets were illuminated with two different pump beams, one in the green and another in the near infrared. Again, the phased-matched angle for the two pump beams emphasized internal intensity distributions which would not tend to excite output MDRs. In view of our present understanding of phase-matching, and its equivalence to spatial overlaps of MDRs, if MDRs are to be excited, the two pump beams should be focused at, or outside, the droplet rim. If both the green and the near-infrared beams had been focused at the outer edge of the droplet, the spatial overlaps between the waves and the SRS should have been greater, and a larger coherent Raman gain should have been observed.

An experimental difference between the CARS measurements and the recent TSFG measurements is that the CARS spectra were measured with plane wave illumination, while in the TSFG measurements the incident beam was focused at the edge or outside the droplet. The latter illumination geometry deposits more input-laser energy into the near-resonant modes.

3.7 Sum-Frequency Generation

Third-harmonic generation (THG), at 3× the incident laser frequency, and third-order sum-frequency generation (TSFG), involving the triple product of the various order Stokes SRS fields and/or of the incident laser field, were readily observed from droplets of various liquids [121]–[123]. In droplets, MDRs can provide optical feedback for both the generating SRS waves and for the resultant TSFG fields. The THG signals associated with the multiorder-Stokes SRS fields (which are always on MDRs) were generally larger than the THG signal associated with the input-laser field (which is not likely to be on an MDR, particularly when the droplet radius is not intentionally tuned). Because the vibrational frequency of CCl_4 molecules is low (459 cm^{-1}) and many multiorder-Stokes SRS frequencies can be generated, the TSFG spectrum of CCl_4 droplets could, in principle, consist of many peaks. In the experiment where the droplet size was held constant, only a few of the many

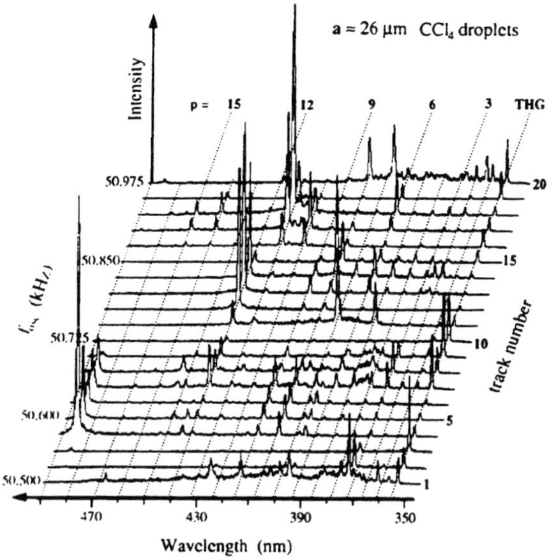

Figure 13: Detected TSFG spectra from CCl_4 droplets at 20 consecutive vibrating orifice frequencies f_{osc}. The horizontal abscissa indicates the wavelength increasing from $3\omega_L$. The series of dotted lines correspond to $\omega_{TSFG} = 3\omega_L - p\omega_{vib}$, where p is an integer. The detector is a position-sensitive resistive-anode device. (From Ref. [123].)

possible TSFG peaks were detected. The TSFG intensity at a particular wavelength was found to be highly sensitive to the droplet radius (see Fig. 13). As the droplet size was changed incrementally, the TSFG intensity peaked at another wavelength. As droplet radius was monotonically increased or decreased, there was no obvious trend regarding which of the many possible wavelengths of TSFG appeared [123].

In order to simplify the TSFG spectra, D_2O droplets were chosen. Only the first-order Stokes SRS can be generated in D_2O; consequently, there can only be two TSFG peaks (involving two laser fields and one SRS field or involving the laser field and twice the SRS fields) and two THG peaks (involving the triple product of the laser field or the triple product of the SRS field) [122], [123]. All THG and TSFG investigations suggest that a necessary but not sufficient condition for generating an observable TSFG signal is that the particular TSFG be near a MDR. The sufficient condition requires the particular TSFG or THG be near a MDR that has a good spatial overlap with the triple product of MDRs with which the three driving fields are in resonance [187].

In droplets, TSFG is much weaker than SRS, lasing, or SBS. When TSFG is detectable, it is four to six orders of magnitude weaker than typical multiorder SRS signals from droplets. The intensity of TSFG per unit volume of liquid is orders of magnitude greater in droplets than it is in an optical cell.

Attempts have been made to observe second harmonic generation (SHG) from droplets with molecules that have and do not have inversion symmetry. The liquid as a whole has no inversion symmetry [121]. An extremely weak SHG signal (at 2× the incident laser frequency) was occasionally observed from benzene droplets when the droplet radius was discretely tuned. Because the benzene molecules and liquid both have an inversion symmetry, the presence of SHG suggests that the signal may be from the surface layer of benzene [3].

3.8 Stimulated Anti-Stokes Raman Scattering (SARS)

Stimulated anti-Stokes Raman scattering (SARS) has been observed in CCl_4, ethanol, and water droplets [124]. The large difference between the first- and second-order SARS intensities and the small difference between the higher-order SARS intensities of CCl_4 droplets indicate that the first-order SARS process involves a resonant $\chi^{(3)}$ and the higher-order SARS process involves a nonresonant $\chi^{(3)}$. For ethanol droplets, the first-order SARS intensity is 10^4 times lower than the first-order SRS intensity. A change in the droplet size affects the first-order SARS intensity by over 2 orders of magnitude for a constant first-order SRS intensity. The dependence of the SARS intensity on the droplet size, and the occasional lack of correlation between the SRS and SARS spectra, indicate that the SARS intensities may be enhanced by the SARS frequencies being near MDRs [124] (see Fig. 14).

3.9 Stimulated Librational Scattering (SLS)

Phase modulation broadening of the CS_2 SRS peak has been attributed to the picosecond time derivative of the phase change associated with a multimode mode laser [94], [125]. By using a single-mode laser beam to irradiate the CS_2 droplets, a series of MDR peaks extending to some 500 cm^{-1} on the Stokes side of the SRS peaks has been observed [126] (see Fig. 15). The role of phase modulation in providing the needed gain so far from main SRS peak has recently been questioned.

Stimulated librational scattering (SLS) is now proposed as the cause of the MDR-related peaks on the Stokes side of the elastic scattering, as well as on the Stokes side of the various multiorder Stokes SRS. Librational scattering is associated with the orientational alignment of anisotropic molecules in the presence of an optical field. However, in liquids, inter-molecular interactions tend to oppose the aligning torque and create a restoring force. The growth of the SLS field is analogous to the growth of the SRS field. Instead of the vibrational modes, the SLS involves the parametric interaction of the input-laser field with the librational mode which in turn is driven by the beating of the input-laser field and the SLS field [126].

3.10 Laser- and Electric-Field-Induced Distortions and Nonlinear Mie Scattering

Electric-field-induced quadrupolar shape distortions (one part in 10^5) have been measured by using elastic scattering near a MDR [129]. The laser-induced electrostrictive process can cause shape distortions in a highly transparent droplet, long after the laser pulse has ended [46], [128]. Electrostriction causes a pulling force

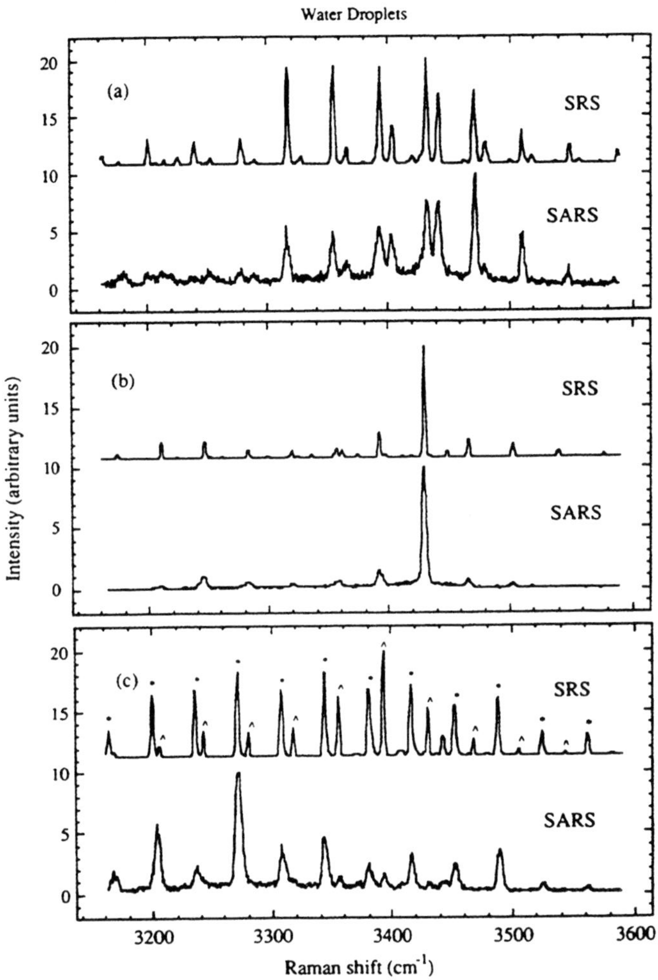

Figure 14: Stimulated anti-Stokes Raman scattering (SARS) and SRS spectra detected simultaneously with two different spectrographs and multi-array detectors. A train of mode-locked pulses (with a maximum intensity of ≈ 0.5 GW/cm^2) illuminates the rim of the water droplets (having $a \approx 35$ μm). Many of the SARS spectra mirror the SRS spectra [e.g., (a) and (b)]. However, some of the SARS spectra miss a set of MDR-related peaks that exist in the SRS spectra [e.g., (c)]. Because phase matching is required for the SARS process, a good spatial overlap among the modes of the SARS, SRS, and SBS is essential. (From Ref. [124].)

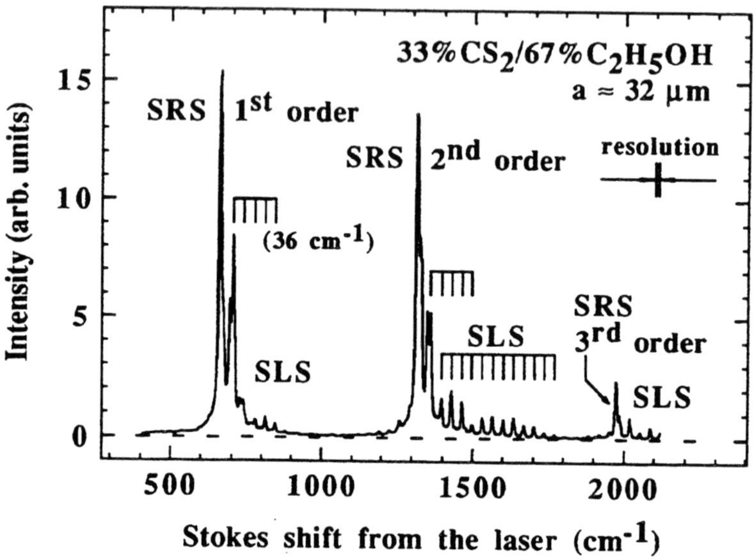

Figure 15: SLS (which is pumped by SRS) and three orders of SRS detected simultaneously from a single CS_2 droplet ($a \approx 30\ \mu m$) which is illuminated by a single-mode, frequency-doubled, Q-switched Nd:YAG laser (pulse duration of 7 ns). The strong peaks at 656 cm^{-1}, 1312 cm^{-1}, and 1968 cm^{-1} are respectively, the first, second and third order SRS lines. In a cell, the spontaneous linewidth of the Raman line of CS_2 is 0.5 cm^{-1}. However, in the droplet, extensive asymmetric broadening (> 500 cm^{-1}) towards lower frequencies of the multi-order CS_2 SRS lines is seen. The proposed cause of the SRS line broadening is SLS pumped by the SRS. The periodic structure seen on the broadened portion of the SRS line is associated with the MDRs of the droplet. Thus, the MDR-related peaks suggest an exponential gain for the SLS process. The long time duration of the incident light (7 ns) precludes the possibility of self-phase modulation as the mechanism for broadening. In the figure, the broadening of the second-order SRS line is more extensive than either the first- or third-order lines, suggesting that the SLS broadening of each of the multi-order SRS lines is independent of the SLS from the other SRS lines. (From Ref. [126].)

on the droplet surface and, thereby, excites surface waves and induces a surface motion at later times. The observed temporal development of the shape distortion [128] is remarkably well reproduced by a calculation [46] that includes both the electrodynamic and hydrodynamic equations pertinent to the large spheres used in the experiment. The laser-induced electrostrictive process can induce density gradients within the droplet volume. This process can also cause nonlinear Mie scattering of the incident radiation that is on an input MDR [130]. There are experimental indications that, after the passage of the intense pump beam, the laser-induced electrostrictive pressure causes bubble formation. Static pressure is applied to the vibrating orifice generator. Because the liquid is under higher gas pressure before the droplets are generated, bubbles can form more readily in the droplets [138], [144]. The gas bubbles inside the droplet can cause additional elastic scattering of the radiation which is circumnavigating around the droplet on a MDR [144].

There are other possible mechanisms for extra light leakage from droplets, i.e., for leakage that is not accounted for by homogeneous absorption or by $e^{-(\omega/Q)t}$ losses, where the Q is calculated for a spherical droplet. Ray optics and Lorenz-Mie calculations of the internal- and surface-intensity distributions have shown that some of the input rays of a parallel plane wave beam, are refracted by the spherical illuminated face, and converge to a ring on the droplet shadow face. This ring of highest surface intensity is referred to as the Descartes ring [34]. The rays reflected at the shadow face converge to a small region near the droplet shadow face [34], [36], [37]. This localized interior region is commonly termed the "internal-focal line" or "internal-focal spot."

Laser-induced losses can occur at the internal-focal spot because of the following mechanisms [131], [132]: 1) elastic scattering associated with a temperature-dependent index of refraction change $\Delta m(\omega, \Delta T) = (\partial m/\partial T)\Delta T$, if there is a non-zero absorption coefficient; 2) elastic scattering associated with an intensity-dependent index of refraction change $\Delta m(\omega, I_{in}) = m_2 I_{in}(t)$; 3) elastic scattering associated with a local density change resulting from the electrostrictive pressure generated by the nonlinear beating of the laser field with the SBS field; and 4) elastic scattering associated with an index of refraction change caused by the plasma that is generated after laser-induced breakdown. Laser-induced extra losses can also occur on the droplet surface at the Descartes ring or elsewhere where the surface intensity is a maximum. Light tripping can occur at the surface bulges that result from the following laser-induced perturbations: 5) laser-induced electrostriction of a transparent droplet; and 6) localized laser heating and subsequent lowering of the surface tension.

The possible surface and volume perturbation mechanisms, 1) through 6), mentioned above, can also be grouped according to their temporal response, i.e., to whether or not the perturbations are quasi-instantaneous or cumulative [131], [132]. Quasi-instantaneous perturbations, such as mechanisms 2) and 3), are essentially dependent only on the instantaneous irradiance of the internal laser radiation. They are not dependent on the history of the previous pulses. Cumulative perturbations, such as mechanisms 1), 4), 5), and 6), are dependent on the time integral of the input-laser pulse or the pulse fluence; they can continue to develop even after the laser pulse has ended.

To distinguish between the various possible mechanisms of extra light leakage, two types of experiments were conducted on lasing droplets [131], [132]. One set of experiments used a variable number of 100 ps pulses, each with low fluence but high irradiance. Cumulative perturbations became evident as the number of 100 ps pulses increased. Another set of experiments used one green laser beam and one near-IR laser beam, both with 7 ns duration and high fluence. The lasing droplets were absorbing in the green and transparent in the near-IR. Hence, the extra leakage from surface perturbations caused by the laser-induced temperature rise could, in principle, be isolated from other forms of laser-induced perturbations. In spite of various experimental studies, complete understanding of the various perturbation mechanisms has not been reached.

Even though the extra leakage at a specific location on the droplet surface (e.g., at the Descartes ring) or in the droplet interior (e.g., at the internal focal point or line) can appear brighter than the normal MDR leakage tangential to the droplet rim, geometric considerations must be considered. The degradation of the overall Q value can be small. The MDR leakage is distributed over the droplet surface while extra laser-induced leakage is confined to a small region on the surface or is caused by a small perturbation in the droplet interior. Selective laser-induced perturbations at specific locations are important for future attempts to "cavity dump" the internally-circulating radiation, i.e., to force the internally-trapped radiation to emerge at a specific location on the droplet surface or from inside the droplet. Study of laser-induced perturbations in the droplet interior and/or on the droplet surface will continue, because both the MDR feedback and the QED gain enhancement factor are dependent on the overall Q of the droplet cavity.

3.11 Bistability

A photothermal optical bistablity has been observed in a single aqueous droplet containing an absorbing dye (e.g., 10^{-5} M crystal violet) [133], [134]. At an intensity of 50 W/cm^2, the elastic scattering spectrum of a levitated droplet recorded with increasing input wavelength is decidedly different from that recorded with decreasing input wavelength. The mechanism for this effect is associated with the photothermal evaporation which accompanies the increased absorption at wavelengths corresponding to MDRs.

3.12 Energy Transfer, Absorption and Internal Scattering Effects

Lasing of dyed submicrometer-sized hydrosols (considered as "acceptors") in a liquid droplet containing Fluorescein 548 dye molecules (considered as "donors") has been observed [142], and confirms the original experimental findings of a greatly enhanced energy-transfer coefficient between donors and acceptors inside a droplet [135]. Evidence for enhanced energy transfer between Fluorescein 548 dye in ethanol and fluorescent hydrosols is provided by the observation of lasing of the submicron polystyrene latex hydrosols. The input green laser pumped the Fluorescein beyond its lasing threshold and then the light emitted by the Fluorescein pumped the latex hydrosols beyond their lasing threshold [142]. Because several MDRs span the donor fluorescent profile, the enhancement of the energy-transfer coefficient between donors and acceptors is proportional to the Q values of the MDRs. Equivalently, the enhancement is proportional to the donor-photon cavity lifetime, which is much longer than the donor fluorescent lifetime in an optical cell [30], [31].

The addition of an absorbing dye (having an imaginary part of m(ω) less than 10^{-5}) into the droplet does not destroy the MDR peaks in the elastic scattering spectrum [140]. When the droplets contain dye molecules which absorb in the lasing-wavelength region, the Q values of the fluorescent MDRs are lowered. By purposely adding an absorptive dye at the fluorescent wavelengths, the inverse of the overall Q value is found to be the sum of the inverse Q's for absorption and leakage ($1/Q_{\text{overall}} = 1/Q_{\text{absorptive}} + 1/Q_{\text{leakage}}$, where $Q_{\text{absorptive}} = 2\pi \text{m}(\omega)/\lambda\alpha$ and α is the absorption coefficient) [139].

Elastic scattering losses must also be included in models of droplets which contain hydrosols. The inverse Q for elastic scattering ($1/Q_{\text{scattering}}$) can be added along with the inverse Q's associated with other independent loss mechanisms. The addition of latex hydrosols to the droplets suppresses the Rhodamine 6G lasing when a certain total-threshold latex hydrosol surface area is reached in the droplet. The threshold is independent of the hydrosol size [143]. These results suggest that the dye molecules are adsorbed onto the surfaces of the latex hydrosols, where the decreased intermolecular separation of dye molecules facilitates nonradiative decay of the excited singlet state [143].

The addition of latex hydrosols can lower the overall Q factor within the droplet to such an extent that other forms of coherent emission, such as SRS, can also be suppressed [146]. Thermally-induced surface ripples that have amplitudes of a few Angstrom do not affect Q values which are lower than 10^8 [53]. However, inhomogeneously distributed losses on the droplet surface and in the interior must be accounted for (in addition to the uniformly-distributed losses discussed above for the normal MDR leakage, dye absorption, and hydrosol scattering). Additional laser-induced losses at specific locations of high input-laser intensity on the droplet surface [110], [131], [132] and within the droplet interior [132] have been investigated recently.

Quenching of the SRS by two-photon absorption, involving one pump photon and one SRS photon, has also been reported [113], [137].

3.13 Nonlinear Optics in Solid Particles

Lasing was observed from the rim of a flashlamp-pumped mm-sized solid sphere of $CaF_2:Sm^{++}$ in liquid He [147]. The emission exhibited a sharp threshold. Continuous wave ascillation in a large Nd-YAG sphere has been observed [148]. Lasing at 855–860 nm was observed in a Nd:glass sphere ($a = 25$ μm) pumped with a 1 W Argon-ion laser [150], [151]. The high-Q MDRs in the 855–860 nm range significantly increased the optical feedback so that lasing threshold could be exceeded even for the lower-gain transitions of the Nd^{3+} ions doped in a glass host [150], [151]. Optically pumped laser emission in "thumb tack" or disk-shaped semiconductors has been reported [152]. The lasing threshold in these lasers is greatly reduced because of the efficient optical trapping of the internally-generated stimulated emission. The Q's of the MDRs associated with microdisks are several hundreds.

Second harmonic generation (SHG) has been observed from individual microcrystals of $LiNbO_3$ (in an optical cell) illuminated with 10–70 mW of the 1.064 μm radiation from a cw Nd:YAG laser [153]. No attempt was made to investigate the SHG intensity dependence on the particle size or the incident wavelength. The relevance of MDRs to this SHG observation is not clear. This experiment motivated the theory of SHG in spheres that are large enough to exhibit MDR-related effects [186].

A photothermal optical bistability has been observed in a fused-quartz sphere, known to have a low thermal conductivity and an intensity-dependent index of refraction [149]. A quartz sphere (20 to 200 μm in radius) was placed close to a prism and illuminated with a single-mode He-Ne laser. The prism efficiently coupled energy into a MDR. The Q factor ($\approx 1.5 \times 10^8$) of the MDR was determined by tuning the laser frequency within the He-Ne gain curve [149]. The elastic scattering spectra of the sphere (with radius of 100 μm) showed marked differences as the He-Ne laser was tuned through its 50 MHz gain curve. The hysterisis in the spectral response resulted from a thermally-induced bistability.

4 CHARACTERIZATION OF DROPLETS AND SPRAYS

4.1 Measurements of Evaporation Rate, Temperature, Surface Tension and Viscosity

The evaporation and condensation rates of droplets in a continuous stream of closely spaced droplets have been determined by noting the spectral shifts of MDRs in the lasing spectra [155]. The relative evaporation rates of closely spaced droplets in a segmented stream have been determined by measuring the spectral shifts in the lasing spectra of each of the flowing droplets [162]. Measured shifts in the MDR-related peaks in the elastic scattering spectra were used to determine the evaporation rates of optically-levitated or electrodynamically-trapped droplets [156], [157], [160]. Resonances of layered spheres [59] have been used to monitor the evaporation of a glycerol coat over a glass core [158].

Fluorescence and lasing peaks have been assigned to specific MDR mode numbers and mode orders for very reproducible microdroplets containing laser dyes [85], [90]. The high precision assignments of line positions revealed the need to include optical dispersion of $m(\omega)$ and $\Delta m(\omega, \Delta T) = (\partial m/\partial T)\Delta T$ changes. The latter allow estimation of the droplet cooling due to evaporation and of the small temperature gradient near the droplet surface [90].

The temperatures and emissivities of carbon and aluminum oxide particles have been determined from the radiative decay curves measured after the particles were illuminated with a short pulse from a CO_2 laser [163]. When the particles were in a low-pressure environment, the emissivities were in good agreement with values calculated using Mie theory.

The surface tension and dynamic viscosity of droplets have been determined by observing the oscillation frequencies and the damping times of the dye lasing peaks [127]. The increased evaporation rate of a droplet immediately after the passage of a shock wave has also been determined by measuring the spectral shifts of MDRs in the dye lasing spectra [164]. Since a shock wave can induce shape deformations which can also cause spectral shifts, there is some ambiguity in the interpretation of the shifts. In the near future, it is expected that more groups will apply the sensitivity of MDR spectral locations to determine physical and thermodynamical properties of droplets and sprays relevant to the combustion environment. SRS has been used to measure the evaporation of fuel sprays [175].

4.2 Measurements of Droplet Shape Deformation

MDRs which are degenerate in a sphere have been predicted theoretically to be split into multiple MDRs in droplets having slight shape deformations [51]. These predictions stimulated several experimental investigations. In one measurement the SRS from distorted droplets was spectrally analyzed. Since the flowing droplets (at velocity of 5 m/sec) were only slightly deformed by the inertial force [45], and since the resultant frequency splittings were small [51], a Fabry-Perot interferometer was used to analyze the SRS spectra [165], [166] (see Fig. 16). The Fabry-Perot interferometer has a free-spectral-range of 0.5 cm^{-1}.

If MDRs having different mode numbers n were excited, then the interpretation of the spectral peaks in the interferometer could be ambiguous. To ensure that the SRS was supported by an MDR having only one mode number n, a spectrograph was used simultaneously with the interferometer. When I_{in} is low (0.01 GW/cm^2), the SRS threshold can be reached for MDRs having different $m = n - k$ (where k is a small integer). These different $m = n - k$ MDRs have different spatial overlaps with the input-laser internal-intensity distribution. With plane-wave illumination, the $m = n$ MDR tends to have the largest overlap with the input-laser internal-intensity distribution, and the overlaps decrease as k increases. That few $m \approx n$ MDRs are involved is consistent with the visual observation that the SRS emerges from two short vertical arcs at the droplet edge and is centered at the equatorial plane [103].

The theory that predicts frequency splitting [51] can also predict the precession frequency of a group of MDRs having the same mode number n and a narrow range of azimuthal number m [166], [167]. Consequently, the degree of shape deformation can be determined either from a frequency- or time-domain measurement. When the input-laser beam is focused below the droplet equatorial plane (at $\theta + 90°$ measured from the vertical z-axis), the SRS involves MDRs with $m \approx n \cos\theta$ (see Fig. 17). Two diametrically opposite SRS arcs, inclined at angles $\theta + 90°$ and θ are expected on the droplet rim. However, for droplets deformed by inertial forces, four and not two SRS arcs are observed, located at $\pm(\theta + 90°)$ and $\pm\theta$. Four SRS arcs, instead of two, are observed because the MDRs precess with normals inclined at angle θ (see Fig. 18).

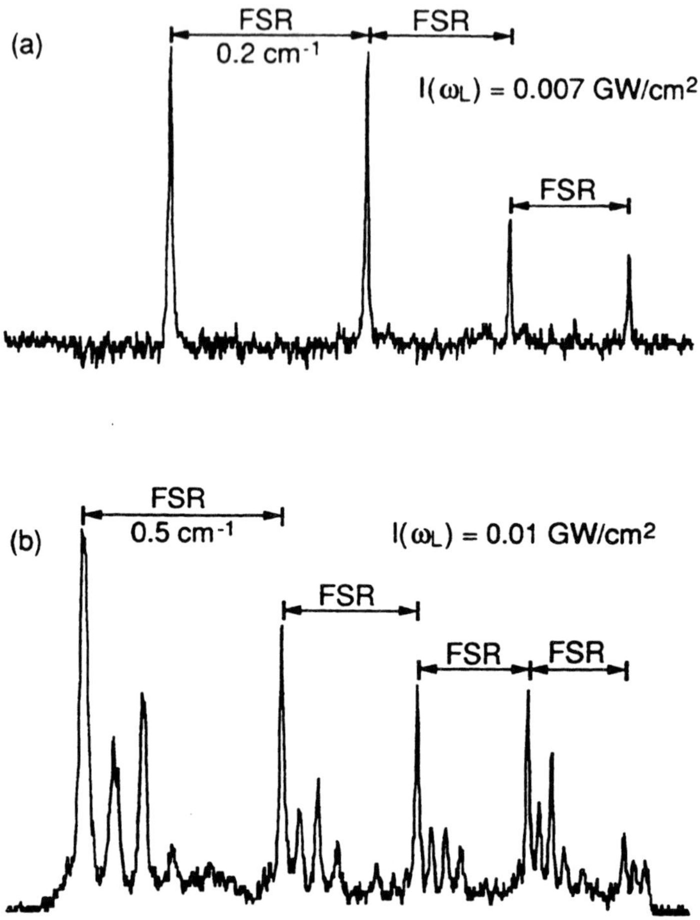

Figure 16: (a) The SRS spectrum measured with a Fabry-Perot interferometer having a free spectral range of 0.2 cm^{-1}. The input-laser intensity is 0.007 GW/cm^2 which is just above the SRS threshold. A single discrete SRS peak corresponds to the $m = n$ MDR that has the best overlap with the pump intensity distribution that is confined to the droplet equatorial plane. (b) The SRS spectrum measured with an interferometer having a free spectral range of 0.5 cm^{-1}. The input laser intensity is 0.01 GW/cm^2. Four MDRs with $m = n$, $(n - 2)$, $(n - 4)$, and $(n - 6)$ have enough overlap with the pump intensity distribution, and hence four discrete SRS peaks are observed. The perturbation theory for a slightly deformed spheriod relates the frequency separation of these observed peaks to the difference between the semi-major and semi-minor axes. (From Ref. [165].)

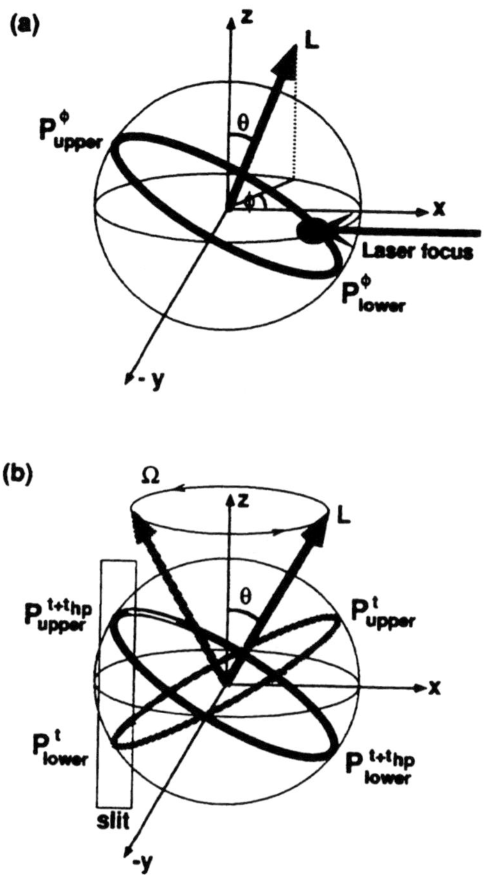

Figure 17: (a) Spherical droplet illuminated with a focused laser beam propagating in the $-x$ direction. The focus beam intersects the sphere at a location below the equator. The SRS circulates around the droplet in a great circle that has a normal inclined at θ. (b) Oblate spheroidal droplet illuminated with a focused laser beam. The normal of the great ellipse precesses around the axisymmetric z-axis at frequency Ω. The entrance slit of the streak camera is also shown. The camera simultaneously detects the SRS leaking from two spots, $P^{t+t_{hp}}_{\text{upper}}$ and P^{t}_{lower} on one side of the droplet. The precession period is $t_{hp} = \pi/\Omega$. (From Ref. [167].)

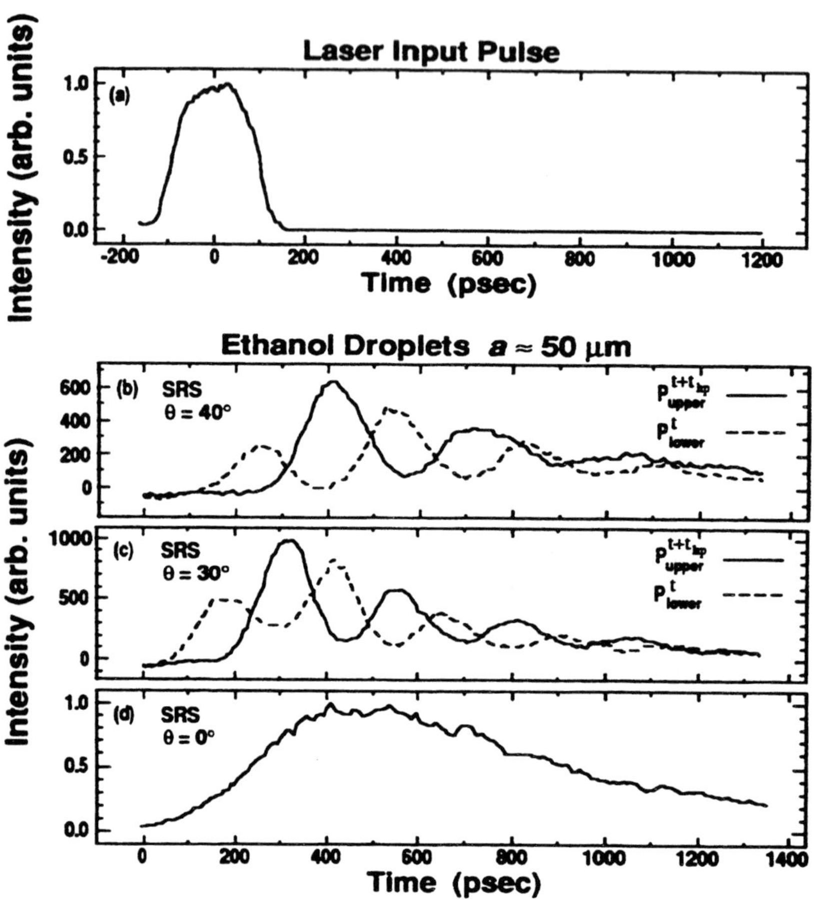

Figure 18: The time profiles of (a) the incident-laser beam and (b)–(d) the SRS leakage from one side of the droplet. The incident-laser beam is centered (b) 40° below, (c) 30° below, and (d) on the equatorial plane. The time profiles from these two spots are 180° out of phase. The droplet distortion can be deduced from the precession frequency. (From Ref. [167].)

4.3 Determination of Chemical Species in Droplets

In addition to the determination of the physical and thermal properties of a droplet, experiments have been directed toward the determination of chemical species within droplets. The fact that the vibrational levels of molecules determine the spectral "fingerprints" of the molecules has spurred research in the infrared absorption, fluorescence, and Raman emission from droplets.

The infrared absorption spectrum of a single droplet can be deduced from photothermally induced changes of the MDR-related elastic scattering spectrum [168]. The droplet is heated by a modulated IR beam and the heating causes size changes in the droplet.

4.3.1 Spontaneous Raman and Fluorescence Emission

Spontaneous Raman scattering spectra in the visible wavelength region can provide direct evidence for specific chemical species within the droplet [69]–[75]. However, the extremely-small spontaneous Raman scattering cross section (on the order of $10^{-30}/\text{cm}^2$) requires an integration time of at least several seconds. By contrast, fluorescence spectra are much less specific for particular molecules; but the fluorescence cross section is at least 12 orders of magnitude larger than the Raman cross section. Single molecules of Rhodamine 6G contained in a single glycerol-water droplet (radius of $\simeq 6.5\mu\text{m}$) have been detected with 514.5 nm excitation from an argon ion laser [66]. The wavelength shifts and splittings of MDRs within the fluorescent and Raman scattering linewidths are then used to provide droplet size and shape deformation information.

4.3.2 Stimulated Raman Scattering and Fluorescence Seeding of Stimulated Raman Scattering

Stimulated Raman scattering (SRS) appears to be a sensitive technique for species-specific determination of molecules within a droplet. The chemical species identification and the concentration of various species of a multicomponent fuel droplet can be determined from the frequency shift of the SRS signal from the laser frequency and from the relative SRS intensity [101], [102]. However, the species concentration must be large and the input-pump intensity must be high (above 0.1 GW/cm^2) before the Raman gain can exceed the leakage and other losses within the droplet. Without fluorescence seeding or resonance-enhanced Raman effects, SRS cannot be used to detect trace quantities of species within a droplet. Another main drawback of the SRS technique is that the SRS intensity is a nonlinear function of the species concentration. Quantitative determination of the species, once the species identification has been made through the Raman shifts, is at best difficult and inaccurate.

Because the Raman gain is linearly proportional to the species concentration [101], [102], [175], the minimum concentration needed for the round-trip Raman gain to overcome the round-trip losses is about 0.1 M. However, the spontaneous Raman scattering cross section can be resonantly enhanced when the incident-pump laser wavelength is within the electronic band. A delicate balance must be achieved between increasing the resonant Raman scattering gain and increasing the absorption as the incident laser wavelength is further tuned into the absorption band.

The vibrational Raman modes of 10^{-3} M Rhodamine 6G in ethanol have recently been measured in microdroplets irradiated with a pump-laser beam at a wavelength of 550 nm [117]. The particular choice of Rhodamine 6G as the trace species makes it difficult to separate the three following possible mechanisms which may increase the SRS intensity: 1) fluorescence seeding, because Rhodamine 6G has a high fluorescence efficiency [111]; 2) extra gain at the Raman wavelength provided by the population inversion of the Rhodamine 6G molecules [91]; and 3) resonant enhanced Raman gain, because the pump-laser wavelength is within the Rhodamine 6G absorption band [117]. Only a limited class of dye molecules can achieve population inversion. It remains to be determined whether or not fluorescence seeding and/or resonant Raman enhancement, without the extra lasing gain provided by the population inversion, are sufficient for the generation of SRS from trace species components. By working in the UV, where many trace molecules will fluoresce, it is expected that the inevitable fluorescence and the possible resonant enhancement of the Raman scattering cross section will decrease the detectivity level of minority species in a multicomponent droplet by SRS spectroscopy.

In the absence of external seeding by injecting radiation at the Stokes wavelength, SRS builds up from the spontaneous Raman signal, which is commonly referred to as "noise". The SRS process can be considered as the amplification of spontaneous Raman noise. From shot to shot, the SRS spectra from D_2O and H_2O droplets are noted to be highly random and lead to the conjecture that such randomness is a manifestation of chaos [115]. In an optical cell, external seeding is achieved by a second laser beam at the Stokes wavelength, which is determined by the wavelength of the strong pump beam. Recently, internal seeding of SRS within a droplet was achieved with the fluorescent photons emitted by a dissolved dye in the same droplet [111]. Selective enhancement of a particular Raman peak (e.g., with weaker Raman gain) can be achieved if the input-laser wavelength and the dye molecules are adjusted so that there is a good spectral overlap between the fluorescence and the particular Raman peak which has the weak gain. Since the fluorescence intensity of dye molecules is usually much larger than the spontaneous Raman scattering intensity of the host liquid, the SRS builds up from the seeded fluorescence signal. However, the Raman gain is still determined by the spontaneous Raman cross section of the liquid and the input-laser intensity [111]. When the Raman peak with the strongest gain coincides with the fluorescence profile, the extra Raman gain of the host can actually deplete the inverted population [91]. The extra Raman gain in a narrow wavelength region of a homogeneously-broadened fluorescence gain profile causes the lasing at all wavelengths to be suppressed and the SRS peak to be greatly enhanced.

4.3.3 Coherent Anti-Stokes Raman Scattering

The CARS technique may be the best nonlinear diagnostic technique for chemical species concentration. Its advantage results from the fact that the CARS intensity at a particular frequency is proportional to the square of the concentration of that species. In addition, the CARS intensity can be increased by increasing the intensities of the pump and Stokes beams. The intensity of CARS depends upon the spatial overlap of the pump, Stokes, and output waves. The strong dependence of the CARS intensity upon the overlap of these waves suggests that it should be possible to determine location-specific, species-specific concentrations inside of droplets. If either or both of the pump wave and the Stokes waves are tightly focused Gaussian beams, and the focal position(s) of the beam(s) relative to the droplet can be determined, then it should be possible to obtain position-dependent species information which could be useful for inhomogeneous droplets. CARS has been used to characterize combustion in gases [3], [4], [6] and is readily detectable in droplets [120]. However, CARS has not yet been applied to the characterization of droplets and sprays. Part of the reason for its lack of use in droplets is that to clearly determine chemical species, a detailed knowledge of the internal fields at the pump and Stokes frequencies is required.

In a typical CARS measurement, the sample is illuminated with a pump beam (at frequency ω_1), and with a Stokes beam (at the frequency (ω_2) of the first-order Stokes Raman scattered radiation). The CARS signal is emitted at frequency $\omega_{AS} = \omega_1 - \omega_2$. In gases, CARS has been used to determine temperatures and species concentrations. In liquids, however, CARS cannot be used to determine temperatures; a droplet reaches its boiling point before reaching the temperatures where vibrational CARS is applicable.

Since the CARS signal from gases has a narrow ($\sim$1 cm^{-1}) linewidth, while the CARS linewidths from liquids are broader ($\sim$10 cm^{-1}), by analyzing the CARS spectra with a high spectral resolution it should be possible to distinguish between the CARS from the surrounding vapor, and the CARS from the liquid droplet. A second way to distinguish CARS from droplets and the CARS from vapor may be to examine the output intensity as a function of the tuning angle between the pump wave and the Stokes wave. CARS from a gas is a more sensitive function of the tuning angle than is CARS from a droplet [120].

4.4 Characterization of Sprays

Two-dimensional orange-fluorescence and red-lasing images of a spray containing Rhodamine-6G dye have been used to distinguish the vapor phase from the liquid phase [176]. The liquid is in the form of sheets, ligaments, small droplets, and large droplets. The red-lasing image of large droplets that have high Q's can be distinguished from the orange fluorescence of liquid sheets, ligaments, and small droplets that have lower Q's. As with lasing, SRS is only obtainable in large droplets. The red SRS of the pure ethanol droplets and red lasing radiation from the Rhodamine-6G dye-doped water droplets are used to detect selectively the presence of large droplets in different regions of the sprays.

Lasing images require much less input-laser intensity than the SRS images and are therefore preferred when the inclusion of fluorescent dye dopants is permissible. Because the temporal resolution is defined by the input-laser pulse duration (15 ns), the incident-laser pulse freezes the droplet motion. Thus the fluorescence, lasing, and SRS imaging techniques reported are well suited to the study of high-velocity injection events. By using multiple incident-laser pulses, the inelastic images may allow one to monitor the temporal evolution of: 1) droplet formation from the liquid sheets and ligaments; 2) the breakup of larger droplets into smaller droplets; 3) the coalescence of smaller droplets into larger droplets; and 4) the evaporation of smaller and larger droplets. The decrease in droplet size caused by evaporation can also be monitored by detecting the lasing (or SRS) spectra and by measuring the wavelength spacing of MDR-related peaks. By employing a proper combination of inelastic imaging techniques, it is possible to determine droplet sizes and number densities in a two-dimensional cross section of fuel spray [176].

5 MODELS OF NONLINEAR OPTICS IN DROPLETS

5.1 Stimulated Brillouin Scattering–Theory

The theory of SBS and the resultant acoustic normal modes has been developed in detail [177]–[181]. In SBS, the spatial overlap of the intensities (not the phase relations between the input and output waves) is important. The present theories of SBS emphasize the doubly-resonant case in which the frequencies of both the laser field and the SBS field fall within a MDR. This leads to enhancements of both the input and SBS fields. The standard wavevector selection rule of the bulk medium is replaced by angular momentum conservation involving the MDR and the acoustic waves. Because the SBS field must be on a MDR, its wavevector is always nearly tangent to the droplet.

5.2 Stimulated Raman Scattering–Theory

A time-dependent, one-dimensional heuristic model for the growth and coupling of SRS and multi-order Stokes SRS in droplets has been reported [183]. The standard one-dimensional nonlinear-wave equations are modified to accommodate the Raman waves that are on MDRs and thus circulate around the outer part of the droplet. Radiation losses are included for: 1) the nonzero absorption; 2) leakage commensurate with MDRs; and 3) intensity depletion in generating the next higher-order Stokes SRS. When the input laser is multimodal, the laser radiation directly pumps the first-order Stokes SRS. The temporal growth and decay of the first- through fourth-order Stokes SRS are calculated using the Raman gain coefficient and the phase-matching length as two adjustable parameters. In principle, the exact enhancement of the Raman gain coefficient ought to be predictable for specific MDRs within the spontaneous Raman scattering profile. Furthermore, the growth of the SRS need not be considered as a traveling wave that grows as it circumnavigates the droplet rim. The problem could also be treated by determining the nonlinear source polarization ($\mathbf{P}^{NLS}$) which directly pumps the various MDRs within the Raman gain profile.

In the heuristic model [183], the Raman gain coefficient determines the delay time of the first-order Stokes SRS while the phase-matching length affects the temporal correlation of the multiorder Stokes SRS and the peak-to-valley ratios of the intensity versus time profiles. When the input laser is in a single mode, the laser radiation directly pumps the SBS, because the gain coefficient for the SBS process is larger than that for the SRS process. The SBS radiation, being on a MDR, then pumps the first-order Stokes SRS, which in turns pumps the second-order Stokes

SRS and the cascade first-order Raman process repeats until the fifth-order Stokes considered in the heuristic model. For both the multimode and single mode laser pumping, the calculated delay time and the correlated growth and decay of the various order Stokes SRS and of the SBS compare favorably with those of experimental observations [96], [97], [103], [104].

5.3 Nonlinear Mixing Processes–Theory

We review second-order and third-order polarizations, discuss some general features of nonlinear mixing processes, and then discuss the two nonlinear mixing processes that have been modeled in most detail: CARS and TSFG.

5.3.1 Nonlinear Polarization and Phase Matching

The third-order nonlinear source polarization at frequency ω and at position $\mathbf{r}$ is

$$\mathbf{P}_i^{NLS}(\mathbf{r},\omega) = D\sum_{jkl} \chi^{(3)}_{ijkl}(\omega = \omega_a \pm \omega_b \pm \omega_c)E_j(\mathbf{r},\omega_a)E_k(\mathbf{r},\omega_b)E_l(\mathbf{r},\omega_c) \qquad (25)$$

where $\chi^{(3)}_{ijkl}$ $(\omega = \omega_a \pm \omega_b \pm \omega_c)$ is the third-order susceptibility; $\mathbf{E}_a(\mathbf{r},\omega_a)$, $\mathbf{E}_b(\mathbf{r},\omega_b)$, $\mathbf{E}_c(\mathbf{r},\omega_c)$ are the amplitudes of the electric fields at frequencies ω_a, ω_b, ω_c, respectively; i, j, k, l range over the three orthogonal coordinate directions; and the degeneracy D is the number of distinct permutations of $\omega_a, \omega_b, \omega_c$ [6]. In general, all the possible $\omega = \omega_a \pm \omega_b \pm \omega_c$ contribute to the $\mathbf{P}^{NLS}$.

The expression for the second-order nonlinear source polarization is similar, but has two electric fields, and uses the second-order susceptibility $\chi^{(2)}_{ijk}$ $(\omega = \omega_a \pm \omega_b)$. In the electric-dipole approximation, the second-order susceptibility $\chi^{(2)}$, is zero in materials which have a center of inversion symmetry. Since the molecules in liquids are usually randomly oriented, $\chi^{(2)}$ for liquids is zero in this approximation. However, at the surface of a droplet there is an abrupt discontinuity and so the inversion symmetry is broken. Second harmonic generation (SHG) can be generated within $\lambda/2\pi$ of a surface [3], where λ is the wavelength. However, SHG and second-order sum-frequency generation (SSFG) are extremely weak processes in droplets [121].

The third-order susceptibility $\chi^{(3)}$ is nonzero in any material. In an isotropic liquid, $\chi^{(3)}$ has only three independent elements [6]. Often in CARS measurements the objective is to measure the Raman susceptibility $\chi^{(3)}_{ijkl}(\omega_{AS} = \omega_a + \omega_b - \omega_R)$, which can be large near Raman-allowed vibrational bands at frequencies ω_R. In the case of TSFG, the relevant $\chi^{(3)}_{ijkl}$ are those with $\chi^{(3)}_{ijkl}(\omega = \omega_a + \omega_b + \omega_c)$ when ω_a, ω_b, and ω_c are positive.

The $\mathbf{P}^{NLS}$ for nonlinear wave-mixing processes may have either a broad or narrow frequency spectrum. For example, the $\mathbf{P}^{NLS}$ which is generated by SRS waves has a very narrow frequency spectrum because the SRS, which is supported by high-Q MDRs, has a spectrum with a very narrow linewidth. Consequently, the center frequency ($\omega = 2\omega_a$ or $\omega = 3\omega_a$) of the $\mathbf{P}^{NLS}$ is rarely within a linewidth of a high-Q output MDR. Other nonlinear mixing processes which are generated directly by the fields of the laser, (e.g., CARS) may have a more broad frequency spectrum. In droplets, SRS and lasing are typically *at* the frequencies of MDRs. CARS and TSFG will only rarely be within a linewidth of a high-Q output MDR.

Nonlinear processes which are dependent upon the phase relations between the input and output fields can be divided into two categories. In the first category, the fields which generate the nonlinear polarization are at the incident laser frequency(ies) and can be computed using linear electromagnetic theory (i.e., Mie theory in the case of spherical droplets). Examples of this category are the simplest cases of SSFG, TSFG, and CARS when the intensities are low enough that SRS is not generated. In the second category, at least some of the fields are generated by nonlinear processes, and so a complete analysis requires an analysis of the nonlinear processes which generate these fields within the droplet. For example, CARS is coupled [6] particularly strongly when it is phase matched to internally generated SRS. Consequently, when the pump and Stokes intensities are large enough to generate SRS, a model of CARS should consider this coupling.

In bulk liquids the amount of nonlinear output is limited by the coherence length between the generating waves and the resultant wave. The coherence length is related to the spatial overlap of the waves along the propagation direction (one dimensional) in the bulk medium. Because the generating and resultant waves travel with different phase velocities, these waves become out of phase after traveling one coherence length.

At a given r and θ, the time-dependence of a mode having an azimuthal mode number m is given by $e^{-i(\omega t - m\phi)}$. A point of constant phase travels at the rate ω/m. So, if the generating and output waves have the same ratios of ω/m, these waves rotate around the droplet axis with the same phase velocity and, hence, keep a constant phase relation. Given this same ratio of ω/m, however, the generating polarization and the output fields may be in phase or out of phase at other positions r,θ. For good phase matching, the generating nonlinear polarization and the fields of the output mode should be generally in phase throughout the volume of the sphere. Whenever the spatial overlap integrals are relatively large, the phases of the generating and output fields are also relatively matched. The "phase matching" problem can be conceptualized as a spatial overlap problem (in which the phases are included), or as a conservation of angular momentum problem. The results are the same.

5.3.2 Coherent Anti-Stokes Raman Scattering

A model of CARS has been developed [184], [185] in which a third-order polarization is generated by the interaction of the pump wave and the Stokes wave, and then this third-order polarization radiates inside the sphere in a manner similar to that described previously [18], [27]. In this model both the pump and Stokes waves are plane waves which are computed exactly using Mie theory. It is found that the CARS signal can be relatively large in the backward direction. The problem is computationally intensive (particularly at the time the work was reported, 1982), and so the spheres studied were relatively small (up to 1.3 μm radius). Resonances are not prominent in such small spheres.

5.3.3 Third-Order Sum-Frequency Generation

A model has been developed for TSFG in droplets where the third-order polarization is generated by SRS, or by SRS in combination with fields at the incident laser frequency [187]. The model was developed to help explain experimental observations of TSFG [121]–[123]. Since the SRS waves are known to be supported by MDRs, the frequency of the third-order polarization will only rarely be within a few linewidths of a high-Q output MDR. For the illumination geometry used in the experimental measurements, where the laser is focused at the droplet edge or outside the droplet [123], the waves at the laser frequency are largest in the low-order modes.

The observed energy is the radiated power integrated over frequency, time interval, and solid angle. The integral for total power emitted can be written,

$$P^T_{n_3,m_3} = 2\pi ck^4 A^2_{M_a} A^2_{M_b} A^2_{M_c} |K_j|^2 |\chi^{(3)} K_Y|^2 K_L, \tag{26}$$

where the nonlinear polarization is assumed to be generated by three standing-wave TE fields which are labeled n_s, m_s, where $s = a$, b, or c; the frequencies of the generating waves, ω_a, ω_b, and ω_c, are centered around the resonance frequencies of the MDRs; and where $|K_j|^2$, K_L, and K_Y^2 are as follows: 1) The radial component of the spatial overlap is

$$|K_j|^2 = \left| \int_0^a j_{n_a}(k_{at}r') j_{n_b}(k_{bt}r') j_{n_c}(k_{ct}r') j_{n_3}(k_t r') r'^2 dr' \right|^2, \tag{27}$$

which varies slowly enough with frequency that it makes no difference whether $j_{n_3}(k_k r')$ is evaluated at w_3, or at $\omega_{1a} + \omega_{1b} + \omega_{1c}$. 2) The angular part of the spatial overlap is

$$|\chi^{(3)} K_Y|^2 = \left| \int \left[\chi^{(3)} \left(\frac{\mathbf{X}_{n_a m_a} + \mathbf{X}^*_{n_a m_a}}{2} \right) \left(\frac{\mathbf{X}_{n_b m_b} + \mathbf{X}^*_{n_b m_b}}{2} \right) \left(\frac{\mathbf{X}_{n_c m_c} + \mathbf{X}^*_{n_c m_c}}{2} \right) \right] \cdot \mathbf{X}^*_{n_3 m_3} d\Omega \right|^2. \tag{28}$$

The azimuthal-angle portion of the overlap integral provides a strict selection rule for four-wave mixing processes, i.e., $m_3 = \pm m_a \pm m_b \pm m_c$. 3) The spectral overlap integral, K_L, is,

$$\int_0^\infty \left| \int_{\omega_b=0}^{\omega} \int_{\omega_a=0}^{\omega-\omega_b} g_{n_a,m_a}(\omega_a) g_{n_b,m_b}(\omega_b) g_{n_c,m_c}(\omega_c = \omega - \omega_a - \omega_b) d\omega_a d\omega_b \right|^2 |g_{n_3,m_3}(\omega)|^2 \, d\omega. \tag{29}$$

The integrands of K_Y and K_j can vary from positive to negative (since the phases of the waves are included), and so $|K_Y|^2$ and $|K_j|^2$ can be very small. For certain combinations of input and output modes, $|K_Y|^2$ can be zero. K_L, on the other hand, is a product of two squared functions and, therefore, is always positive. The spatial integrals of the products of four different electric field distributions are typically orders of magnitude lower than the integrals of products of wave intensities which would be used in modeling lasing or SRS. Consequently, because of the positive and negative values of each term in the integrand, TSFG is expected to be much weaker than SRS or lasing. TSFG has been observed to be very weak, i.e., 10^{-4} to 10^{-6} times the SRS intensity [123].

The θ-components of the output TE fields and the polarization are much larger than the ϕ-components, since the intensities are largest near $\theta = 90°$. Since the field that is generated by the polarization is largest near $\theta \approx \sin^{-1}(m_3/n_3)$, the largest overlap tends to occur when $m_1/n_1 \approx m_3/n_3$. This fact, combined with the selection rule $m_3 = \pm 3m_1$, indicates that the angular overlap integral $|K_Y|^2$ is largest when n_3 is near $3n_1$. As seen in the radial overlap results, the requirement that $n_3 \approx 3n_1$ can be satisfied in droplets, since $x_3 = 3x_1$ for the THG process, and MDRs with Q's > 500 have $x_1 < n_1 < m(\omega_1)x_1$ and have $x_3 < n_3 < m(\omega_3)x_3$.

Figure 19 shows a term proportional to the integrand of the dominant contribution to K_Y, plotted vs. θ for three combinations of n_1 and n_3. The curves help in visualizing THG and in understanding why some angular overlap integrals are small.

TSFG in droplets is much weaker than SRS or lasing because: 1) The spatial phases of the fields are important in the overlap integrals, K_Y and K_j. Consequently, these integrals can be very small or zero. In the case of SRS or lasing on the other hand, the overlap integrals involve products of the intensities, which are always positive since they are the squares of the magnitudes of the fields. Consequently, the overlap integrals required for TSFG are typically orders of magnitude lower than the integrals of products of intensities required for SRS or lasing. 2) In cases with detectable TSFG, the linewidth of the third-order polarization is orders of magnitude smaller than the linewidth of a typically-broad spontaneous Raman or fluorescence spectrum. With such a narrow linewidth in the TSFG case, there is only a small probability of finding an output MDR with both a good spatial overlap and with a resonance frequency within even hundreds of linewidths of the third-order polarization. When the frequency spacing is large, the frequency overlap integral K_L tends to be relatively small. For SRS or lasing, on the other hand, the gain profiles are broad enough that there is always a MDR within the gain profile, not just within hundreds of linewidths. 3) The real part of the third-order susceptibility for TSFG is usually very small.

5.4 Bistability–Theory

A catastrophe-based graphic method was used to explain the observed photothermal optical bistability of a dye-doped absorbing liquid droplet levitated in an electrodynamic trap [133], [134]. In the experiments modeled, the droplet contained a salt solution and was in equilibrium in a particular vapor pressure. Then the laser frequency was swept in one direction. When the input fields were near an MDR, more energy was absorbed. Consequently the droplet evaporated slightly, and decreased slightly in size. If the laser wavelength is decreased as the droplet shrinks, then the droplet will be on-resonance for a longer period of time. If the laser wavelength is increased as the droplet shrinks, then the droplet quickly falls off-resonance. When the laser is tuned toward larger wavelengths, the measured spectrum near resonances is decidedly different from when the laser is tuned toward smaller wavelengths.

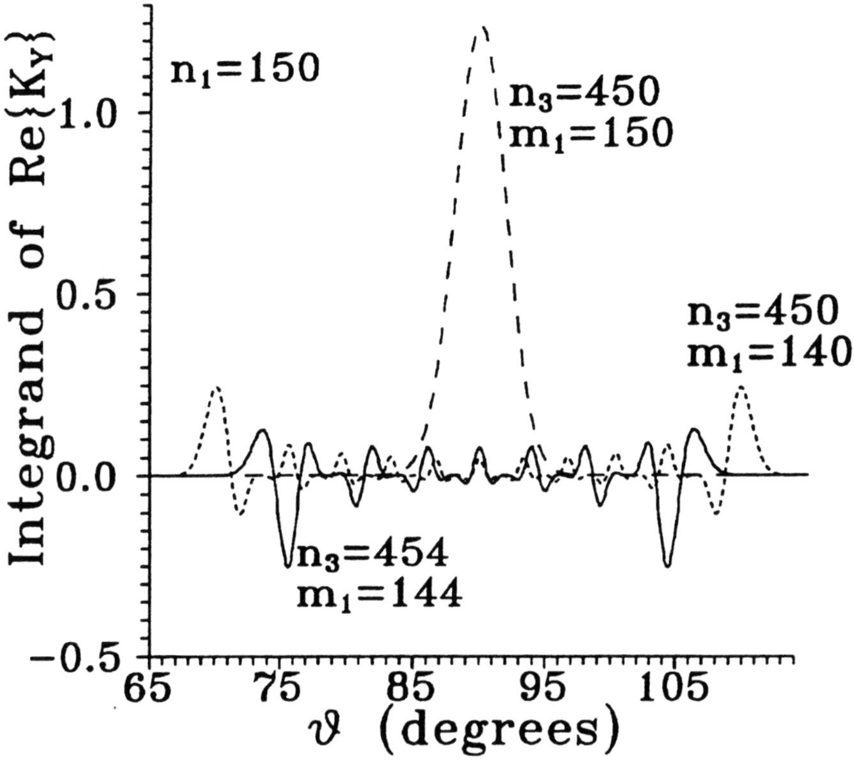

Figure 19: A term proportional to the integrand of the dominant contribution to K_Y plotted vs. θ for three combinations of n_1 and n_3, with $m_3 = 3m_1$. All the curves have $n_1 = 150$. The curve which would integrate to the largest value of K_Y is for the case n_3= 450 and $m_1 = m_3/3 = 150$ (see long-dashed curve). In this curve both the nonlinear polarization and the output fields are largest in the equatorial plane and the integrand is always positive. When $n_3 = 454$ and $m_1 = 144$ (the solid line) the integrand alternates from positive to negative and consequently integrates to a very small value. When $n_3 = 450$ and $m_1 = m_3/3$ =140 (see short-dashed curve), the integrand also oscillates from positive to negative, but is enough more positive that it integrates to a positive value. (From Ref. [187].)

Bistability has been analyzed [149], [188] in spheres which have an intensity-dependent refractive index of the form,

$$m(\mathbf{r}) = m_o + m_2 I(\mathbf{r}) \tag{30}$$

where m_o is the homogeneous refractive index (assumed real), and $I(\mathbf{r})$ is the internal intensity at position $\mathbf{r}$.

In Ref. [188] all the modes of the sphere were used to calculate $I(\mathbf{r})$, and develop an expression for bistability. The TIPM [51] was used to calculate the shifts in resonance locations. Since all the modes were included, this general bistability expression is applicable even when the incident frequency is more than 50 linewidths away from a high-Q ($Q > 10^4$) MDR.

In the analysis [188], the small perturbation of the refractive index is axisymmetric, as it would be if the sphere were illuminated with a circularly-polarized plane wave. Two features facilitating the analysis of bistability in large axisymmetric spheres are: 1) for small intensity-dependent changes in the refractive index the internal field coefficients of all modes, except the near-resonant mode, are essentially unaltered; 2) the resonance frequency can shift by many linewidths before there is a significant change in the quality factor Q.

The general expression for bistability was then simplified by assuming that the contribution of the nonresonant modes is negligible when the incident frequency is close enough to an MDR. The resulting expressions [188] are similar to those obtained assuming an evanescent excitation wave with a variable coupling efficiency [149]. The expressions are similar to those obtained for bistability in ring and Fabry-Perot cavities.

Bistability is obtained by varying the intensity or the frequency of the driving fields. When m_2 is positive, bistability is observed only when the incident frequency is less than, but near, a MDR frequency of the sphere. The total internal energy density is shown in Fig. 20 as a function of the incident plane-wave intensity. The internal-energy density exhibits bistable behavior. Results are shown for two different modes and for two different ratios of the resonant-mode energy. The all-mode and the single-mode results agree well when most of the energy is in the resonant mode. The agreement decreases as the contributions of the non-resonant modes increase. In the single-mode calculations, the contribution of the non-resonant modes has been neglected and consequently the threshold intensity is higher than that for the all-mode case. Thus the single-mode computations give an upper estimate of the threshold intensity required for bistability.

Also, a quasi-steady-state approximation was used to calculate the effects of bistability on the backscattering from a CS_2 sphere illuminated with a plane wave which has a Gaussian time-dependence [188].

5.5 Two-photon absorption–Theory

A model of two-photon absorption has been developed [55]. An effective-average gain or loss coefficient, similar to those used in modeling linear lasers, is defined in spherical dielectric cavities which have a spatially varying gain or loss (see Section 2.7.3). The evanescent waves outside the sphere are included by integrating

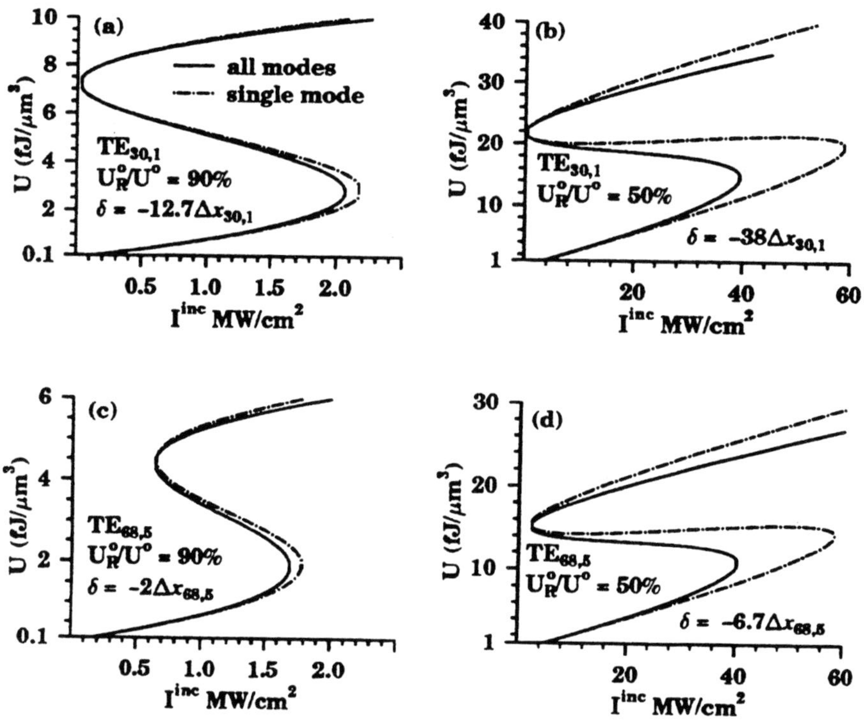

Figure 20: The total internal energy density as a function of the incident plane-wave intensity for two different modes and for two different ratios of the resonant-mode energy. Energy densities calculated using all modes and using a single near-resonant mode are shown. a) and b) show the computed energy densities for the $TE_{30,1}$ mode at detunings $-12.7\Delta x_{30,1}$ and $-38\Delta x_{30,1}$. c) and d) show the energy densities for the $TE_{68,5}$ mode for detunings $-2\Delta x_{68,5}$ and $-6.7\Delta x_{68,5}$. (From Ref. [188].)

the energy density up to the first zero of the Neumann function. The method has been applied to the modeling of an effective-average two-photon absorption loss for an MDR in the case of quasi-steady-state time-dependent scattering.

The sphere is illuminated with a circularly polarized plane wave. With this illumination the internal intensity is axisymmetric and the overlap integrals in Eq. (24) only need be evaluated in r, θ. Only modes with $m = 1$ are required. The analysis and results are for a single TE MDR. In general, all the modes contribute to the two-photon absorption loss. However, near a high-Q MDR the majority of the energy in the sphere can be in a single mode. For example, the peak energy density $(\mathbf{E} \cdot \mathbf{E}^*)$ of the $TE_{37,1}$ MDR on-resonance is 1000. Off-resonance the peak energy density inside the sphere is only 50. The on-resonance energy density is a much more sensitive function of m_i than is the off-resonance energy density. Consequently, as long as the incident field is near a MDR having a Q of a few thousand or greater, the on-axis intensity is much smaller than the intensity in the resonant mode, even with a small amount of absorption. Also, the energy density of a low-order mode is more localized in the outer part of the droplet than is the peak of the non-resonant intensity (which is closer to the origin and localized to a small volume). Hence, the spatial overlap between the non-resonant intensity and the field of the resonant mode tends to be small. Only a very small error is introduced by ignoring the contribution to two-photon loss resulting from the overlap of the off-resonance intensity with the resonant fields.

The intensity I of a plane wave travelling in the z direction through a two-photon absorbing medium is governed by [3]

$$\frac{dI}{dz} = -\beta I^2, \tag{31}$$

where β is the two-photon absorption coefficient.

In a droplet, an expression for $m_{i_s}^{\mu}$ is

$$\frac{m_{i_s}^{\mu}}{|c_{o1n}(m_{i_s}^{\mu})|^2} = \frac{\lambda \beta m_r I_{inc}}{4\pi {D_{n1}}^2} \frac{\int_{V_s} |\mathbf{X}_{n1}|^4 |j_n(mkr)|^4 dv}{\int_{V_s} |\mathbf{X}_{n1}|^2 |j_n(mkr)|^2 dv}. \tag{32}$$

A quasi-steady-state approximation was used to compute the ratio of the backscattered intensity to the incident intensity, $R = I_s(t)/I_i(t)$, for the case of a circularly polarized incident plane wave which has a Gaussian time dependence. The sphere contains a two-photon absorbing dye. The slowly-varying envelope of the incident pulse is $I_i(t) = I_p e^{-(t/T)^2}$, where I_p is the peak intensity of the pulse and T is the half-width at the I_p/e point. The values are: $T = 50$ ns; the center frequency of the pulse is on the $TE_{37,1}$ MDR; and $I_s(t) \simeq I_s^{\mu}(t)$, where $I_s^{\mu}(t)$ is the backscattered intensity of the TE $_{37,1}$ mode. Since the MDR lifetime $\tau = 35.8$ ps $\ll T = 50$ ns, the quasi-steady-state approximation is valid.

Figure 21 shows the ratio $I_s(t)/I_i(t)$ for three different values of I_p: 100, 500, and 1000 MW/cm^2. It is difficult to extract the effect of the two-photon absorption simply by examining the temporal profile of the scattered light, because the change in the shape of the scattering-intensity curves is small. The ratio $R = I_s(t)/I_i(t)$

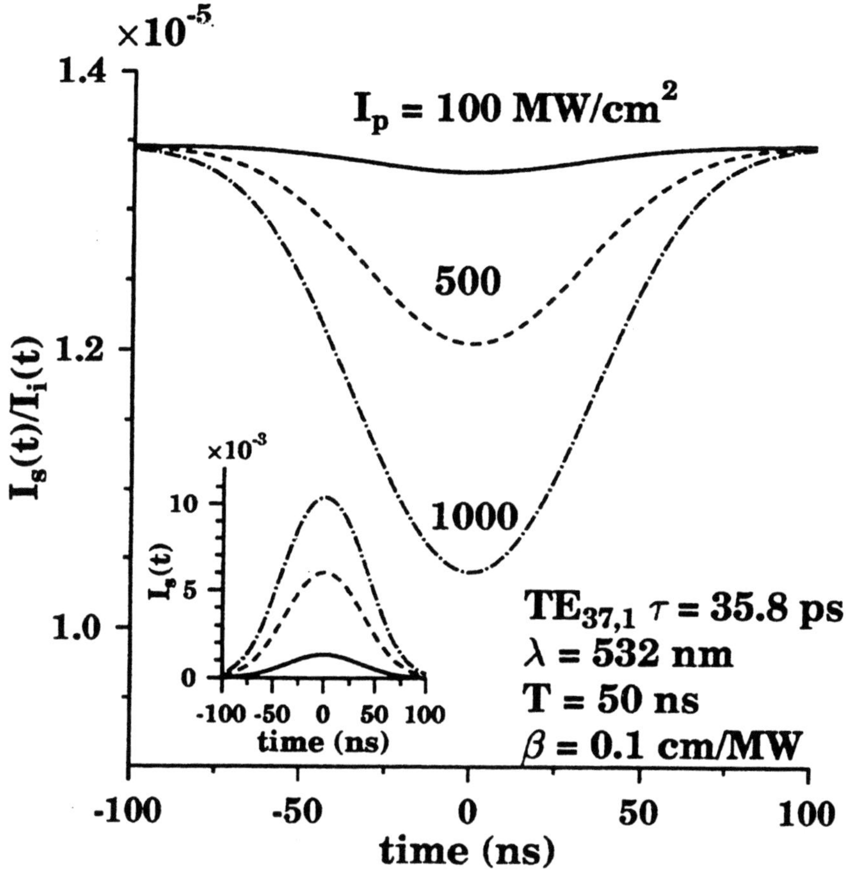

Figure 21: The backscattered intensity relative to the incident intensity ($I_s(t)/I_i(t)$) for three different values of the peak intensity I_p: 100, 500, and 1000 MW/cm^2, for a sphere which contains a two-photon absorbing dye. The insert shows the temporal profile of the backscattered intensity. (From Ref. [55].)

should be a useful quantity for experimental verification of two-photon absorption effects. R is a constant in the absence of intensity-dependent absorption even if there are temporal fluctuations in the incident intensity. If $\partial R/\partial t \neq 0$, and $|\partial R/\partial t|$ increases with intensity, then the scattering will suggest a nonlinear effect, which is induced because of two-photon absorption. For the $TE_{37,1}$ mode when $I_p =$ 1000 MW/cm^2 the peak I_s is 42.3 smaller than when $\beta = 0$. To obtain the same percentage change in the transmitted intensity in a bulk media with an incident plane wave intensity of 1000 MW/cm^2, the length of the medium must be 73 μm which is 30 times the sphere radius.

6 SUMMARY

It is expected that many more nonlinear optical effects will be observed in single micrometer-sized droplets. The pace will hasten now that there are many researchers interested in various aspects of droplet cavity resonances. The combined efforts of researchers will further the basic understanding of the electromagnetic properties of MDRs of perturbed droplets, the QED enhancement and suppression of molecules within the droplets, and the coupling among various nonlinear optical processes that require phase matching. A main impetus of the research will be to use linear and nonlinear spectroscopy of droplets to unravel the chemical, thermal, and physical properties of stationary or flowing droplets. There is a compelling need to extend single-droplet diagnostic techniques to sprays, particularly to dense sprays.

7 ACKNOWLEDGEMENTS

We gratefully acknowledge the partial support of this research by the U.S. Army Research Office (contract numbers DAAL03-92-G-0308 to New Mexico State University, and DAAL03-91-G-0042 to Yale University) and by the Air Force Office of Scientific Research (Grant No. 91-0150 to Yale University). Part of this work was done while SCH held a National Research Council–Army Research Laboratory Research Associateship. We acknowledge collaborations and helpful discussions with Peter Barber, Kenneth Young, P. T. Leung, H.-M. Lai, and helpful discussions with Ronald G. Pinnick and Stephen Arnold. We thank our former/present students who have worked with us on this topic through the years since 1984. Students of RKC: William P. Acker, Gang Chen, Janice L. Cheung, Wen-Feng Hsieh, Kimberly A. Juvan, Alfred S. Kwok, David H. Leach, Ali Serpenguzel, J. Christian Swindal, Huey-Ming Tzeng, and Jian-Zhi Zhang. Students of SCH: Dipakbin Q. Chowdhury, E. Esam M. Khaled, Md. Mohiuddin Mazumder and Hasan I. Saleheen. We also recognize the valuable help of post-doctoral researchers who have explored this area with us: Robert E. Benner, Dipakbin Q. Chowdhury, Johannes H. Eickmans, James F. Owen, Shi-Xiong Qian, Judith Snow, Frederick H. Wirth, Carol F. Wood, and Jia-Biao Zhang. We thank M. Elizabeth Hernandez for editing and latexing the manuscript.

References

Previous Reviews

[1] R. K. Chang, "Some of Bloembergen's nonlinear optical effects revisited in single micrometer-sized droplets," in *Resonances*, M. D. Levenson, E. Mazur, P. S. Pershan, and Y. R. Shen, eds. (World Scientific, Singapore, 1990), p. 200.

[2] S. C. Hill and R. E. Benner, "Morphology-dependent resonances," in *Optical Effects Associated with Small Particles*, P. W. Barber and R. K. Chang, eds. (World Scientific, Singapore, 1988), p. 3.

Books on Nonlinear Optics in Bulk

[3] Y. R. Shen, *The Principles of Nonlinear Optics* (John Wiley, New York, 1984).

[4] A. C. Eckbreth, *Laser Diagnostics for Combustion Temperature and Species* (Abacus Press, Cambridge, 1988).

[5] A. Yariv, *Quantum Electronics*, 3rd ed., (John Wiley, New York, 1989).

[6] R. W. Boyd, *Nonlinear Optics* (Academic, San Diego, 1992).

2 MODES AND MDRS OF SPHERES AND PERTURBED SPHERES

2.1 Modal Expansion of Fields in Spheres

[7] M. Kerker, *The Scattering of Light and Other Electromagnetic Radiation* (Academic, New York, 1969).

[8] J. D. Jackson, *Classical Electrodynamics*, 2nd ed., (Wiley, New York, 1975), Ch. 16.

[9] C. F. Bohren and D. R. Huffman, *Absorption and Scattering of Light by Small Particles* (Wiley, New York, 1983).

[10] P. W. Barber and S. C. Hill, *Light Scattering by Particles: Computational Methods* (World Scientific, Singapore, 1990).

2.2 Modes and Morphology-Dependent Resonances–Theory

[11] P. Chylek, "Partial-wave resonances and the ripple structure in the Mie normalized extinction cross section," J. Opt. Soc. Am. **66**, 285 (1976).

[12] P. R. Conwell, P. W. Barber and C. K. Rushforth, "Resonant spectra of dielectric spheres," J. Opt. Soc. Am. A **1**, 62 (1984).

[13] S. C. Hill and R. E. Benner, "Morphology-dependent resonances associated with stimulated processes in microspheres," J. Opt. Soc. Am. B **3**, 1509 (1986).

[14] P. Chylek, "Resonance structure of Mie scattering: distance between resonances," J. Opt. Soc. Am. A **7**, 1609 (1990).

[15] S. Schiller and R. L. Byer, "High-resolution spectroscopy of whispering gallery modes in large dielectric spheres," Opt. Lett. **16**, 1138 (1991).

[16] V. V. Datsyuk, "Some characteristics of resonant electromagnetic modes in a dielectric sphere," Appl. Phys. B **54**, 184 (1992).

[17] C. C. Lam, P. T. Leung, and K. Young, "Explicit asymptotic formulas for the positions, widths, and strengths of resonances in Mie Scattering," J. Opt. Soc. Am B **9**, 1585 (1992).

2.3 Radiation from Molecules Inside a Sphere–Theory

2.3.1 Radiation from a Source Inside a Sphere–Theory

[18] H. Chew, P. J. McNulty, and M. Kerker, "Model for Raman and fluorescent scattering by molecules embedded in small particles," Phys. Rev. A **19**, 396 (1976).

2.3.2 Modification of Transition Rates–Theory

[19] H. Chew, "Transition rates of atoms near spherical surfaces," J. Chem. Phys. **87**, 1355 (1987).

[20] S. C. Ching, H. M. Lai, and K. Young, "Dielectric microspheres as optical cavities: thermal spectrum and density of states," J. Opt. Soc. Am. B **4**, 1995 (1987).

[21] S. C. Ching, H. M. Lai, and K. Young, "Dielectric microspheres as optical cavities: Einstein A and B coefficients and level shifts," J. Opt. Soc. Am. B **4**, 2004 (1987).

[22] H. Chew, "Total fluorescent scattering cross sections," Phys Rev. A **37**, 4107 (1988).

[23] H. Chew, "Radiation lifetimes of atoms inside dielectric particles," Phys. Rev. A **38**, 3410 (1988).

[24] H. M. Lai, P. T. Leung, and K. Young, "Electromagnetic decay into a narrow resonance in an optical cavity," Phys. Rev. A **37**, 1597 (1988).

[25] P. T. Leung and K. Young, "Time-independent perturbation theory for quasinormal-mode solutions in quantum mechanics," Phys. Rev. A **44**, 3152 (1991).

[26] H. M. Lai, P. T. Leung, S. Y. Liu, and K. Young, "Cavity QED in microdroplets," Proceedings of Shanghai International Symposium on Quantum Optics, SPIE **1726**, 27 (1992).

2.3.3 Fluorescence, and Spontaneous Raman and Brillouin Emission–Theory

[27] H. Chew, M. Sculley, M. Kerker, P. J. McNulty, and D. D. Cooke, "Raman and fluorescent scattering by molecules embedded in small particles: results for coherent optical processes," J. Opt Soc. Am. **68**, 1686 (1978).

[28] M. Kerker, P. J. McNulty, M. Sculley, H. Chew, and D. D. Cooke, "Raman and fluorescent scattering by molecules embedded in small particles: results for incoherent optical processes," J. Opt Soc. Am. **68**, 1676 (1978).

[29] S. C. Ching, P. T. Leung, and K. Young, "Spontaneous Brillouin scattering in a microdroplet," Phys. Rev. A **41**, 5026 (1990).

2.3.4 Energy Transfer Within Spheres–Theory

[30] S. D. Druger, S. Arnold and L. M. Folan, "Theory of enhanced energy transfer between molecules embedded in spherical dielectric particles," J. Chem. Phys. **87**, 2649 (1987).

[31] P. T. Leung and K. Young, "Theory of enhanced energy transfer in an aerosol particle," J. Chem. Phys. **89**, 2894 (1988).

2.4 Internal Intensities–Theory

2.4.1 Plane Wave Illumination–Theory

[32] P. Chylek, J. D. Pendleton and R. G. Pinnick, "Internal and near-surface scattered field of a spherical particle at resonance conditions," Appl. Opt. **24**, 3490 (1985).

[33] D. S. Benincasa, P. W. Barber, J.-Z. Zhang, W.-F. Hsieh, and R. K. Chang, "Spatial distribution of the internal and near-field intensities of large cylindrical and spherical scatterers," Appl. Opt. **26**, 1348 (1987).

[34] M. A. Jarzembski and V. Srivastava, "Electromagnetic field enhancement in small liquid droplets using geometric optics," Appl. Opt. **28**, 4962 (1989).

[35] H. M. Lai, P. T. Leung, K. L. Poon, and K. Young, "Characterization of the internal energy density in Mie scattering," J. Opt. Soc. Am. A **8**, 1553 (1991)

[36] J. A. Lock and E. A. Hovenac, "Internal caustic structure of illuminated liquid droplets," J. Opt. Soc. Am. A **8**, 1541 (1991).

[37] D. Q. Chowdhury, P. W. Barber, and S. C. Hill, "Energy-density distribution inside large nonabsorbing spheres via Mie theory and geometrical optics," Appl. Opt. **31**, 3518 (1992).

2.4.2 Nonplane Wave Illumination–Theory

[38] J. P. Barton, D. R. Alexander, and S. A. Schaub, "Internal and near-surface electromagnetic fields for a spherical particle irradiated by a focused laser beam," J. Opt. Soc. Am. A **64**, 1632 (1988).

[39] J. P. Barton, D. R. Alexander, and S. A. Schaub, "Internal fields of a spherical particle illuminated by a tightly focused laser beam: Focal point positioning effects at resonance," J. Appl. Phys. **65**, 2900 (1989).

[40] E. E. M. Khaled, S. C. Hill, P. W. Barber, and D. Q. Chowdhury, "Near-resonance excitation of dielectric spheres with plane waves and off-axis Gaussian beams," Appl. Opt. **31**, 1166 (1992).

[41] E. E. M. Khaled, S. C. Hill, and P. W. Barber, "Scattered and internal intensity of a sphere illuminated with a Gaussian beam," IEEE Trans. Antennas and Propagat. **AP-41**, scheduled for March 1993.

[42] J. A. Lock, "Excitation of morphology-dependent resonances by an off-axis Gaussian beam," J. Opt. Soc. Am. A (submitted).

2.4.3 Time-Dependent Internal Fields–Theory

[43] D. Q. Chowdhury, S. C. Hill, and P. W. Barber, "Time-dependence of internal intensity of a dielectric sphere on and near resonance," J. Opt. Soc. Am. A **9**, 1364 (1992).

[44] E. E. M. Khaled, D. Q. Chowdhury, S. C. Hill, and P. W. Barber, "Internal and scattered time-dependent intensity of a dielectric sphere illuminated with a pulsed Gaussian beam," J. Appl. Phys. (submitted).

2.5 Perturbed and Inhomogeneous Spheres–Theory

[45] T. D. Taylor and A. Acrivos, "On the deformation and drag of a falling viscous drop as low Reynolds number," J. Fluid Mech. **18**, 466 (1964).

[46] H. M. Lai, P. T. Leung, K. L. Poon, and K. Young, "Electrostrictive distortion of a micrometer-sized droplet by a laser pulse," J. Opt. Soc. Am. B **6**, 2420 (1989).

2.5.1 T-Matrix and Related Methods

[47] P. W. Barber and C. Yeh, "Scattering of electromagnetic waves by arbitrarily shaped dielectric bodies," Appl. Opt. **14**, 2864 (1975).

[48] D. S. Wang and P. W. Barber, "Scattering by inhomogeneous dielectric bodies,' Appl. Opt., **18**, 1190 (1979).

[49] P. W. Barber and S. C. Hill, "Effects of particle nonsphericity on light scattering," in *Optical Particle Sizing: Theory and Practice*, G. Gouesbet and G. Grehan, eds. (Plenum Press, New York, 1988), p. 43.

[50] J. P. Barton and D. R. Alexander, "Electromagnetic fields for an irregularly shaped, near-spherical particle illuminated by a focused laser beam," J. Appl. Phys., **69**, 7973 (1991).

2.5.2 Time-Independent Perturbation Method

[51] H. M. Lai, P. T. Leung, K. Young, S. C. Hill and P. W. Barber, "Time-independent perturbation for leaking electromagnetic modes in open systems with application to resonances in microdroplets," Phys. Rev. A **41**, 5187 (1990).

[52] H. M. Lai, C. C. Lam, P. T. Leung, and K. Young, "The effect of perturbations on the widths of narrow morphology-dependent resonances in Mie scattering," J. Opt. Soc. Am. B **8**, 1962 (1991).

[53] H. M. Lai, P. T. Leung, and K. Young, "Limitations on the photon storage lifetime in electromagnetic resonances of highly transparent microdroplets," Phys. Rev. A **41**, 5199 (1990).

[54] M. M. Mazumder, S. C. Hill, and P. W. Barber, "Morphology-dependent resonances in inhomogeneous spheres: comparison of layered T-matrix and time-independent perturbation method," J. Opt. Soc. Am. A **9**, 1844 (1992).

2.5.3 Effective-Average Gain and Loss Coefficients

[55] D. Q. Chowdhury, S. C. Hill, and M. M. Mazumder, "Quality factors and effective-average modal gain or loss in inhomogeneous spherical resonators: application to two-photon absorption," IEEE J. Quantum Electron. (to be published).

2.5.4 Radially Inhomogeneous Spheres

[56] P. J. Wyatt, "Scattering of electromagnetic plane waves from inhomogeneous spherically symmetric objects," Phys. Rev., **127**, 1837 (1962), "errata: Scattering of electromagnetic plane waves from inhomogeneous spherically symmetric objects," Phys. Rev. **134**, AB1 (1964).

[57] D. Q. Chowdhury, S. C. Hill, and P. W. Barber, "Morphology-dependent resonances in radially inhomogeneous spheres," J. Opt. Soc. Am. A **8**, 1702 (1991).

[58] M. Schneider, E. D. Hirleman, H. Saleheen, D. Q. Chowdhury, and S. C. Hill, "Light scattering by radially inhomogeneous fuel droplets evaporating in a high-temperature environment," Proceedings of Conference on Laser Applications in Combustion amd Combustion Diagnostics, SPIE **1862**, (1993).

2.5.5 Spheres with Inclusions or with Particles on the Surface

[59] R. L. Hightower and C. B. Richardson, "Resonant Mie scattering from a layered sphere," Appl. Opt. **27**, 4850 (1988).

[60] K. A. Fuller, "Optical resonances and two-sphere systems," Appl. Opt. **30**, 4716 (1991).

[61] F. Borghese, P. Denti, R. Saija, and O. I. Sindoni, "Optical properties of spheres containing a spherical eccentric inclusion," J. Opt. Soc. Am. A **9**, 1327 (1992).

[62] K. A. Fuller, "Scattering and absorption by inhomogeneous spheres and sphere aggregates," Proceedings of Conference on Laser Applications in Combustion amd Combustion Diagnostics, SPIE **1862**, (1993).

3 EXPERIMENTAL OBSERVATIONS OF INELASTIC EMISSION AND NONLINEAR OPTICS IN DROPLETS

3.1 Fluorescence Emission

[63] R. E. Benner, P. W. Barber, J. F. Owen, and R. K. Chang, "Observation of structure resonances in the fluorescence spectra from microspheres," Phys. Rev. Lett. **44**, 475 (1980).

[64] J. F. Owen, R. K. Chang, and P. W. Barber, "Morphology-dependent resonances in Raman scattering, fluorescence emission, and elastic scattering from microparticles," Aerosol Sci. Technol. **1**, 293 (1982).

[65] S. C. Hill, R. E. Benner, C. K. Rushforth, and P. R. Conwell, "Structural resonances observed in the fluorescence emission from small spheres on substrates," Appl. Opt. **23**, 1680 (1984).

[66] W. B. Whitten, J. M. Ramsey, S. Arnold, and B. V. Bronk, "Single-molecule detection limits in levitated microdroplets," Anal. Chem **63**, 1027 (1991).

[67] S. Arnold, J. Comunale, W. B. Whitten, J. M. Ramsey, and K. A. Fuller, "Room-temperature microparticle-based persistent hole-burning spectoscopy," J. Opt. Soc. Am. B **9**, 819 (1992).

[68] M. D. Barnes, W. B. Whitten, S. Arnold, and J. M. Ramsey, "Homogeneous linewidths of Rhodamine 6G at room temperature from cavity-enhanced spontaneous emission rates," J. Chem Phys. (to be published).

3.2 Spontaneous Raman Emission

[69] R. Thurn and W. Kiefer, "Raman-microsampling technique applying optical levitation by radiation pressure," Appl. Spectrosc. **38**, 78 (1984).

[70] R. Thurn and W. Kiefer, "Structural resonances observed in the Raman spectra of optically levitated liquid droplets," Appl. Opt. **24**, 1515 (1985).

[71] T. R. Lettieri and R. L. Preston, "Observation of sharp resonances in the spontaneous Raman spectrum of a single optically levitated microdroplet," Opt. Comm. **54**, 349 (1985).

[72] G. Schweiger, "Observation of input and output structural resonances in the Raman spectrum of a single spheroidal dielectric microparticle," Opt. Lett. **15**, 156 (1990).

[73] G. Schweiger, "Observation of morphology-dependent resonances caused by the input field in the Raman spectrum of microdroplets," J. Raman Spectr. **21**, 165 (1990).

[74] G. Schweiger, "Raman scattering on single aerosol particles and on flowing aerosols: a review," J. Aerosol Sci. **21**, 483 (1990).

[75] M. F. Buehler, T. M. Allen, and E. J. Davis, "Microparticle Raman spectroscopy of multicomponent aerosols," J. Colloid and Interface Sc. **146**, 79 (1991).

3.3 Lasing

[76] H. M. Tzeng, K. F. Wall, M. B. Long, and R. K. Chang, "Laser emission from individual droplets at wavelengths corresponding to morphology-dependent resonances," Opt. Lett. **9**, 499 (1984).

[77] S.-X. Qian, J. B. Snow, H.-M. Tzeng, and R. K. Chang, "Lasing droplets: highlighting the liquid-air interface by laser emission," Science **231**, 486 (1986).

[78] H.-B. Lin, A. L. Huston, B. L. Justus, and A. J. Campillo, "Some characteristics of a droplet whispering-gallery-mode laser," Opt. Lett. **11**, 614 (1986).

[79] A. Biswas, H. Latifi, R. L. Armstrong, and R. G. Pinnick, "Time-resolved spectroscopy of laser emission from dye-doped droplets," Opt. Lett. **14**, 214 (1989).

[80] H. Latifi, A. Biswas, R. L. Armstrong, and R. G. Pinnick, "Lasing and stimulated Raman scattering in spherical liquid droplets: time, irradiance, and wavelength dependence," Appl. Opt. **29**, 5387 (1990).

[81] J. C. Knight, H. S. T. Driver, and G. N. Robertson, "Effect of the pumping geometry on the half-widths of the lasing peaks observed from Rhodamine 6G ethanol droplets," Opt. Lett. **15**, 980 (1990).

[82] A. J. Campillo, J. D. Eversole, and H.-B. Lin, "Cavity quantum electrodynamics enhancement of stimulated emission in microdroplets," Phys. Rev. Lett. **67**, 437 (1991).

[83] H.-B. Lin, J. D. Eversole, and A. J. Campillo, "Spectral properties of lasing microdroplets," J. Opt. Soc. Am. B **9**, 43 (1992).

[84] K.-H. Lin and W.-F. Hsieh, "Transient response of a thresholdless microdroplet dye laser," Opt. Lett. **16**, 1608 (1991).

[85] J. D. Eversole, H.-B. Lin, and A. J. Campillo, "Cavity-mode identification of fluorescence and lasing in dye-doped microdroplets," Appl. Opt. **31**, 1982 (1992).

[86] H.-B. Lin, J. D. Eversole and A. J. Campillo, "Cavity-modified spontaneous-emission rates in liquid microdroplets," Phys. Rev. A **45**, 6756 (1992).

[87] A. J. Campillo, J. D. Eversole and H.-B. Lin, "Cavity quantum electrodynamic enhancement of spontaneous and stimulated emission in microdroplets," Modern Phys. Lett. **B6**, 447 (1992).

[88] A. J. Campillo, J. D. Eversole and H.-B. Lin, "Cavity QED enhanced spontaneous and stimulated emission in liquid droplets," Proceedings of Shanghai International Symposium on Quantum Optics, SPIE **1726**, 59 (1992).

[89] A. S. Kwok, A. Serpenguzel, W.-F. Hsieh, R. K. Chang, and J. B. Gillespie, "Two-photon pump lasing in microdroplets," Opt. Lett. **17**, 1435 (1992).

[90] J. D. Eversole, H.-B. Lin, A. L. Huston, A. J. Campillo, P. T. Leung, S. Y. Liu, and K. Young, "High precision identification of morphology-dependent resonances in optical processes in microdroplets," J. Opt. Soc. Am. B (to be published).

[91] A. S. Kwok and R. K. Chang, "Suppression of lasing by stimulated Raman scattering in microdroplets," extended abstract of Joint Meeting of QELS/CLEO '93, Baltimore, MD (to be published).

3.4 Stimulated Raman Scattering (SRS)

[92] J. B. Snow, S.-X. Qian, and R. K. Chang, "Stimulated Raman scattering from individual water and ethanol droplets at morphology-dependent resonances," Opt. Lett. **10**, 37 (1985).

[93] S.-X. Qian and R. K. Chang, "Multi-order Stokes emission from micrometer-sized droplets," Phys. Rev. Lett. **56**, 926 (1986).

[94] R. K. Chang, S.-X. Qian, and J. Eickmans, "Stimulated Raman scattering, phase modulation, and coherent anti-Stokes Raman scattering from single micrometer-size liquid droplets," in *Methods of Laser Spectroscopy*, Y. Prior, A. Ben-Reuven, and M. Rosenbluh, eds. (Plenum Press, New York, 1986), p. 249.

[95] J.-Z. Zhang, D. H. Leach, and R. K. Chang, "Photon lifetime within a droplet: temporal determination of elastic and stimulated Raman scattering," Opt. Lett. **13**, 270 (1988).

[96] R. G. Pinnick, A. Biswas, P. Chylek, R. L. Armstrong, H. Latifi, E. Creegan, V. Srivastava, M. Jarzembski, and G. Fernandez, "Stimulated Raman scattering in micrometer-sized droplets: time resolved measurements," Opt. Lett. **13**, 494 (1988).

[97] W.-F. Hsieh, J.-B. Zhang, and R. K. Chang, "Time Dependence of Multi-order Stimulated Raman Scattering from Single Droplets," Opt. Lett. **13**, 497 (1988).

[98] M. Golombok and D. B. Pye, "Multimode stimulated Raman scattering in fuel droplets," Chem. Phys. Lett. **151**, 161 (1988).

[99] A. Biswas, H. Latifi, R. L. Armstrong, R. G. Pinnick, "Double-resonance stimulated Raman scattering from optically levitated glycerol droplets," Phys. Rev. A **40**, 7413 (1989).

[100] R. G. Pinnick, A. Biswas, R. L. Armstrong, H. Latifi, E. Creegan, V. Srivastava, and G. Fernandez, "Stimulated Raman scattering in micrometer-sized droplets: measurements of angular scattering characteristics," Opt. Lett. **13**, 1099 (1989).

[101] W. P. Acker, A. Serpenguzel, R. K. Chang, and S. C. Hill, "Stimulated Raman scattering of fuel droplets: chemical concentration and size determination," Appl. Phys. B **51**, 9 (1990).

[102] A. Serpenguzel, G. Chen, R. K. Chang, "Stimulated Raman scattering of aqueous droplets containing ions: concentration and size determination," Particulate Sci. and Tech. **8**, 1 (1990).

[103] J.-Z. Zhang, G. Chen, and R. K. Chang, "Pumping of stimulated Raman scattering by stimulated Brillouin scattering within a single droplet: input laser linewidth effects," J. Opt. Soc. Am. B **7**, 108 (1990).

[104] S.-X. Qian, S. Yuan, Y. Li, H. Wang, X. Huang, and Z. Yu. "Comparison between the temporal characteristics of picosecond SRS from the cell and SRO from the droplet," Opt. Commun. **74** , 414 (1990).

[105] H.-B. Lin, J. D. Eversole, and A. J. Campillo, "Frequency pulling of stimulated Raman scattering in microdroplets," Opt. Lett. **15**, 387 (1990).

[106] H.-B. Lin, J. D. Eversole, and A. J. Campillo, "Double resonance stimulated Raman scattering in micrometer-sized droplets," J. Opt. Soc. Am. B **10**, 2079 (1990).

[107] H.-B. Lin, J. D. Eversole, and A. J. Campillo, "Identification of morphology-dependent resonances in stimulated Raman scattering from microdroplets," Opt. Comm. **77**, 407 (1990).

[108] A. Biswas, R. L. Armstrong, and R. G. Pinnick, "Stimulated Raman scattering threshold behavior of binary mixture micrometer-sized droplets," Opt. Lett. **15**, 1191 (1990).

[109] G. Chen, W. P. Acker, R. K. Chang, and S. C. Hill, "Angular fine structure in the stimulated Raman scattering from single droplets," Opt. Lett. **16**, 117 (1991).

[110] V. Srivastava and M. A. Jarzembski, "Laser-induced stimulated Raman scattering in the forward direction of a droplet: comparison of Mie theory with geometric optics," Opt. Lett. **16**, 126 (1991).

[111] A. S. Kwok and R. K. Chang, "Fluorescence seeding of weaker-gain Raman modes in microdroplets: enhancement of stimulated Raman scattering," Opt. Lett. **17**, 1262 (1992).

[112] J. C. Knight, H. S. T. Driver, and G. N. Robertson, "Nonlinear light scattering from liquid droplets," South African J. Sc. **88**, 162 (1992).

[113] R. K. Chang and A. Serpenguzel, "Characteristics and Applications of Stimulated Raman Scattering in Microdroplets," in *XIIIth International Conference on Raman Spectroscopy*, 1992, Wurzburg, Germany, W. Kiefer, M. Cardona, G. Schaack, F. W. Schneider, and H. W. Schrotter, eds. (John Wiley, New York, 1992) p. 33.

[114] H.-B. Lin, J. D. Eversole and A. J. Campillo, "Continuous-wave stimulated Raman scattering in microdroplets," Opt. Lett. **17**, 828 (1992).

[115] R. L. Armstrong, J.-G. Xie, T. E. Ruekgauer, J. Gu, and R. G. Pinnick, "Observation of random stimulated Raman scattering emission from liquid water microdroplets," submitted to Opt. Lett.

[116] R. G. Pinnick, G. L. Fernandez, J.-G. Xie, T. Ruekgauer, J. Gu, and R. L. Armstrong, "Stimulated Raman scattering and lasing in micrometer-sized cylindrical liquid jets: time and spectral dependence," J. Opt. Soc. Am. B **6**, 865 (1992).

[117] A. S. Kwok and R. K. Chang, "Stimulated resonance Raman scattering of strongly fluorescent molecules," extended abstract of Joint Meeting of QELS/CLEO, Baltimore, MD May 2–7, 1993.

3.5 Stimulated Brillouin Scattering (SBS)

[118] J.-Z. Zhang and R. K. Chang, "Generation and suppression of stimulated Brillouin scattering in single liquid droplets," J. Opt. Soc. Am. B **6**, 151 (1989).

[119] F. H. Wirth, K. A. Juvan, D. H. Leach, J. C. Swindal, R. K. Chang, and P. T. Leung, "Phonon-retention effects on stimulated Brillouin scattering from micrometer droplets illuminated with multiple-short laser pulses," Opt. Lett. **17**, 1334 (1992).

See also Ref. [103]

3.6 Coherent Anti-Stokes Raman Scattering (CARS) and Coherent Raman Mixing(CRM)

[120] S.-X. Qian, J. B. Snow, and R. K. Chang, "Coherent Raman mixing and coherent anti-Stokes Raman scattering from micrometer-size droplets," Opt. Lett. **10**, 499 (1985).

3.7 Sum-Frequency Generation

[121] W. P. Acker, D. H. Leach, and R. K. Chang, "Third-order optical sum-frequency generation in micrometer-sized liquid droplets," Opt. Lett. **14**, 402 (1989).

[122] D. H. Leach, W. P. Acker, and R. K. Chang, "The effect of the phase velocity and spatial overlap of spherical resonances on sum-frequency generation in droplets," Opt. Lett. **15**, 894 (1990).

[123] D. H. Leach, R. K Chang, W. P. Acker, and S. C. Hill, "Third-order sum-frequency generation in droplets: experimental results," J. Opt. Soc. Am. B **10**, 34 (1993).

3.8 Stimulated Anti-Stokes Raman Scattering (SARS)

[124] D. H. Leach, R. K. Chang, and W. P. Acker, "Stimulated anti-Stokes Raman scattering in microdroplets," Opt. Lett. **17**, 387 (1992).

3.9 Stimulated Librational Scattering (SLS)

[125] S.-X. Qian and R. K. Chang, "Phase-modulation-broadened line shapes from micrometer-size CS_2 droplets," Opt. Lett. **11**, 371 (1986).

[126] J. L. Cheung, A. S. Kwok, and R. K. Chang, "Stimulated librational scattering in microdroplets," Extended Abstracts of the Joint Meeting of QELS/CLEO, Baltimore, MD May 2-7, 1993.

3.10 Laser- and Electric-Field-Induced Distortions and Nonlinear Mie Scattering

[127] H.-M. Tzeng, M. B. Long, R. K. Chang, and P. W. Barber, "Laser-induced shape distortions of flowing droplets deduced from morphology-dependent resonances in fluorescence spectra," Opt. Lett. **10**, 209 (1985).

[128] J.-Z. Zhang and R. K. Chang, "Shape distortion of a single water droplet by laser-induced electrostriction," Opt. Lett. **13**, 916 (1988).

[129] S. Arnold, D. E. Spock, and L. M. Folan, "Electric-field-modulated light scattering near a morphological resonance of a trapped aerosol particle," Opt. Lett. **15**, 1111 (1990).

[130] A. L. Huston, H.-B. Lin, J. D. Eversole, and A. J. Campillo, "Nonlinear Mie scattering: electrostrictive coupling of light to droplet acoustic modes," Opt. Lett. **15**, 1176 (1990).

[131] J.-G. Xie, T. E. Ruekgauer, J. Gu, R. L. Armstrong, R. G. Pinnick, and J. D. Pendleton, "Physical basis for Descartes ring scattering in laser-irradiated microdroplets," Opt. Lett. **16**, 1817 (1992).

[132] G. Chen, D. Q. Chowdhury, R. K. Chang, and W.-F. Hsieh, "Laser-induced radiation leakage from microdroplets," J. Opt. Soc. Am. B **10**, 620 (1993).

3.11 Bistability

[133] S. Arnold, K. M. Leung, and A. Pluchino, "Optical bistability of an aerosol particle," Opt. Lett. **11**, 800 (1986).

[134] S. Arnold, T. R. O'Keeffe, K. M. Leung, L. M. Folan, T. Scalese, and A. Pluchino, "Optical bistability of an aqueous aerosol particle detected through light scattering: theory and experiment," Appl. Opt. **29**, 3473 (1990).

3.12 Energy Transfer, Absorption, and Internal Scattering Effects

[135] L. M. Folan, S. Arnold, and S. D. Druger, "Enhanced energy transfer within a microparticle," Chem. Phys. Lett. **118**, 322 (1987).

[136] S. Arnold and L. M. Folan, "Energy transfer and the photon lifetime within an aerosol particle," Opt. Lett. **14**, 387 (1989).

[137] M. Golombok, F. J. Bergin, and D. B. Pye, "Quenching of stimulated Raman scattering by two-photon absorption in fuels," J. Raman Spectr. **20**, 805 (1989).

[138] N. F. Bunkin and V. B. Karpov, "Optical cavitation of transparent liquid exposed to broadband laser light," JETP Lett. **52**, 18 (1990).

[139] P. Chylek, H.-B. Lin, J. D. Eversole, and A. J. Campillo, "Absorption effects on microdroplet resonant emission structure," Opt. Lett. **16**, 1723 (1991).

[140] P. Chylek, D. Ngo, and R. G. Pinnick, "Resonance structure of composite and slightly absorbing spheres," J. Opt. Soc. Am. A **9**, 775 (1992).

[141] H.-B. Lin, A. L. Huston, J. D. Eversole, A. J. Campillo, and P. Chylek, "Internal scattering effects on microdroplet resonant emission structure," Opt. Lett. **17**, 970 (1992).

[142] R. L. Armstrong, J. G. Xie, T. E. Ruekgauer, and R. G. Pinnick, "Energy-transfer-assisted lasing from microdroplets seeded with fluorescent sol," Opt. Lett. **17**, 943 (1992).

[143] R. L. Armstrong, J.-G. Xie, T. E. Ruekgauer, J. Gu, and R. G. Pinnick, "Effects of submicrometer-sized particles on microdroplet lasing," Opt. Lett. **18**, 119 (1993).

[144] A. L. Huston, H.-B. Lin, J. D. Eversole, and A. J. Campillo, "Effect of nanometer bubble formation on microdroplet cavity quality factor," Opt. Lett. (to be published).

[145] R. L. Armstrong, J.-G. Xie, T. E. Ruekgauer, and R. G. Pinnick, "Double resonance lasing and fluorescence emission from microdroplets seeded with fluorescent sol," Opt. Lett. (to be published).

[146] J.-G. Xie, T. E. Ruekgauer, R. L. Armstrong, and R. G. Pinnick, "Suppression of stimulated Raman scattering from microdroplets by seeding with nm-sized latex particles," Opt. Lett. (to be published).

3.13 Nonlinear Optics in Solid Particles

[147] C. G. B. Garrett, W. Kaiser, and W. L. Long, "Stimulated emission into optical whispering modes of spheres," Phys. Rev. **124**, 1807 (1961).

[148] T. Baer, "Continuous wave oscillation in a Nd-YAG sphere," Opt. Lett. **12**, 392 (1987).

[149] V. B. Braginsky, M. L. Gorodetsky, and V. S. Ilchenko, "Quality-factor and nonlinear properties of optical whispering-gallery modes," Phy. Lett. A **137**, 393 (1989).

[150] Y. Z. Wang, Y. Q. Li, Y. S. Liu and B. L. Lu, "A Nd-glass micro-spherical cavity laser induced by cavity QED effects," Proceedings of Conference on Advanced Laser Concepts and Application, SPIE **1501**, 40 (1991).

[151] Y. Z. Wang, Y. Q. Li, Y. S. Liu and B. L. Lu, "Observation of cavity QED effects in a Nd-glass microspherical cavity," Laser Spectroscopy X, by M. Ducloy, E. Giacobino and G. Camy, eds. **205** (World Scientific, 1991).

[152] S. L. McCall, A. F. J. Levi, R. E. Slusher, S. J. Pearton, and R. A. Logan, "Whispering-gallery mode microdisk lasers," Appl. Phys. Lett. **60**, 289 (1992).

[153] S. Sato and H. Inaba, "Observation of second-harmonic generation from optically trapped microscopic $LiNO_3$ particles using Nd:YAG laser," Electronics Lett. **28**, 286 (1992).

4 CHARACTERIZATION OF DROPLETS AND SPRAYS

4.1 Measurement of Evaporation Rate, Temperature, Surface Tension, and Viscosity

[154] A. Ashkin and J. M. Dziedzic, "Observations of optical resonances of dielectric spheres by light scattering," Appl. Opt. **20**, 1803 (1981).

[155] H. M. Tzeng, K. F. Wall, M. B. Long, and R. K. Chang, "Evaporation and condensation rates of liquid droplets deduced from structure resonances in the fluorescence spectra," Opt. Lett. **9**, 273 (1984).

[156] C. B. Richardson, H.-B. Lin, R. McGraw, and I. N. Tang, "Growth rate measurement for single suspended droplets using the optical resonance method," Aerosol Sc. and Tech. **5**, 10 (1986).

[157] C. B. Richardson, R. L. Hightower, and A. L. Pigg, "Optical measurement of the evaporation of sulfuric acid droplets," Appl. Opt. **25**, 1226 (1986).

[158] R. L. Hightower, C. B. Richardson, H. B. Lin, J. D. Eversole, and A. J. Campillo, "Measurement of scattering of light from layered microspheres," Opt. Lett. **13**, 946 (1988).

[159] M. Golombok and D. B. Pye, "Droplet evaporation measured by nonlinear Raman method," J. Phy. D: Appl. Phys. **23**, 1103 (1990).

[160] A. K. Ray, A. Souyri, E. J. Davis, and T. M. Allen, "Precision of light scattering techniques for measuring optical properties of microspheres," Appl. Opt. **30**, 3974 (1991).

[161] R. Vehring and G. Schweiger, "Optical determination of the temperature of transparent microparticles," Appl. Spectr. **46**, 25 (1992).

[162] G. Chen, A. Serpenguzel, R. K. Chang, and W. P. Acker, "Relative evaporation rates of droplets in a segmented stream determined by droplet cavity fluorescence peak shifts," Proceedings of the Conference on Laser Applications in Combustion amd Combustion Diagnostics, SPIE, **1862** (1993).

[163] D. W. Pack, A. Pluchino, and D. E. Masturzo, "Emissivity measurements of single levitated micro-particles," Proceedings of the Conference on Laser Applications in Combustion amd Combustion Diagnostics, SPIE, **1862** (1993).

[164] T. J. Anderson and M. Winter, "Measurement of droplet vaporization rate enhancement caused by acoustic disturbances," J. Comb. Sc. and Technology (to be published).

4.2 Measurement of Droplet Shape Deformation

[165] G. Chen, R. K. Chang, S. C. Hill, and P. W. Barber, "Frequency splitting of degenerate spherical cavity modes: stimulated Raman spectrum of deformed droplets," Opt. Lett. **16**, 1269 (1991).

[166] G. Chen, J. C. Swindal, and R. K. Chang, "Frequency splitting and precession of cavity modes of a droplet deformed by inertial forces," Proceedings of Shanghai International Symposium on Quantum Optics, SPIE **1726**, 291 (1992).

[167] J. C. Swindal, D. H. Leach, R. K. Chang, and K. Young, "Precession of morphology-dependent resonances in non-spherical droplets," Opt. Lett. **18**, 191 (1993).

4.3 Determination of Chemical Species in Droplets

[168] S. Arnold, E. K. Murphy, and G. Sageev, "Aerosol particle molecular spectroscopy," Appl. Opt. **24**, 1048 (1985).

See also Refs. [66] (fluorescence); [69], [74], [75] (spontaneous Raman scattering); [92], [98], [101], [102], [108], [111], [112] [125] (SRS); and [120] (CARS).

4.4 Assignment of Resonances

[169] P. Chylek, V. Ramaswamy, A. Ashkin, and J. M. Dziedzic, "Simultaneous determination of refractive index and size of spherical dielectric particles from light scattering data," Appl. Opt. **22**, 2302 (1983).

[170] P. R. Conwell, C. K. Rushforth, R. E. Benner, and S. C. Hill, "An efficient automated algorithm for the sizing of dielectric microspheres using the resonance spectrum," J. Opt. Soc. Am. A **1**, 1181 (1984).

[171] S. C. Hill, C. K. Rushforth, R. E. Benner, and P. R. Conwell, "Sizing dielectric spheres and cylinders by aligning measured and computed resonance locations: algorithm for multiple orders," Appl. Opt. **24**, 2380 (1985).

[172] M. Golombok and D. B. Pye, "Droplet sizing in fuel injections by stimulated Raman scattering," Opt. Lett. **15**, 872 (1990).

[173] J. D. Eversole, H.-B. Lin, A. L. Huston, and A. J. Campillo, "Spherical-cavity-mode assignments of optical resonances in microdroplets using elastic scattering," J. Opt Soc. Am. A **7**, 2159 (1990).

[174] H.-B. Lin, J. D. Eversole, and A. J. Campillo, "Vibrating orifice droplet generator for precision optical studies," Rev. Sci. Instrum. **61**, 1018 (1990).

See also Ref. [90]

4.5 Characterization of Sprays

[175] M. Golombok and D. B. Pye, "Stimulated Raman scattering in diesel injected fuel sprays," J. Phys. D: Appl. Phys. **22**, 851 (1989).

[176] A. Serpenguzel, J. C. Swindal, R. K. Chang, and W. P. Acker, "Two-dimensional imaging of sprays with fluorescence, lasing, and stimulated Raman scattering," Appl. Opt. **31**, 3543 (1992).

5 MODELS OF NONLINEAR OPTICS IN DROPLETS

5.1 Stimulated Brillouin Scattering–Theory

[177] S. M. Chitanvis and C. D. Cantrell, "Simple approach to stimulated Brillouin scattering in glass aerosols," J. Opt. Soc. Am. B **6**, 1326 (1989).

[178] C. D. Cantrell, "Theory of nonlinear optics in dielectric spheres. II. Coupled-particle-wave theory of resonant, resonantly pumped stimulated Brillouin scattering," J. Opt. Soc. Am. B **8**, 2158 (1991).

[179] C. D. Cantrell, "Theory of nonlinear optics in dielectric spheres. III. Partial wave-index dependence of the gain for stimulated Brillouin scattering," J. Opt. Soc. Am B **8**, 2181 (1991).

[180] C. D. Cantrell, "An overview of stimulated Brillouin scattering in microdroplets," Proceedings of Conference on Nonlinear Optics and Materials SPIE, **1497**, 28 (1991).

[181] P. T. Leung and K. Young, "Doubly resonant stimulated Brillouin scattering in a microdroplet," Phys. Rev. A **44**, 593 (1991).

5.2 Stimulated Raman Scattering–Theory

[182] M. Golombok and D. B. Pye, "Simulation of stimulated Raman scattering in droplets," J. Phys. D: Appl. Phys. **23**, 1109 (1990).

[183] A. Serpenguzel, G. Chen, R. K. Chang, and W.-F. Hsieh, "Heuristic model for the growth and coupling of nonlinear processes in droplets," J. Opt. Soc. Am. B **9**, 871 (1992).

5.3 Nonlinear Mixing Processes–Theory

[184] J. Cooney and A. Gross, "Coherent anti-Stokes Raman scattering by droplets in the Mie size range," Opt. Lett. **7**, 218 (1982).

[185] A. Gross, "Coherent anti-Stokes Raman scattering by droplets in the Mie size range," Ph. D. Dissertation, Drexel University, Philadelphia, PA (1982).

[186] K. Hayata and M. Koshiba, "Theory of surface-emitting second-harmonic generation from optically trapped microspheres," Phys. Rev. A **46**, 6104 (1992).

[187] S. C. Hill, D. H. Leach and R. K. Chang, "Third-order sum-frequency generation in droplets: model with numerical results for third-harmonic generation," J. Opt. Soc. Am. B **10**, 16 (1993).

5.4 Bistability–Theory

[188] M. M. Mazumder, S. C. Hill, and D. Q. Chowdhury, "Dispersive optical bistability in a dielectric sphere," Phys. Rev. A (submitted).

See also Refs. [133], [134], [149].

Studies in Classical
and Quantum Nonlinear Optics

ISBN 1-56072-168-5

NONLINEAR QUANTUM BEAT SPECTROSCOPY

Jorn M. Hvam
Fysisk Institute, Odense Universitet
DF 5230 Odense M, Denmark

1. INTRODUCTION

When light-matter interactions are investigated on a time scale that is short compared to the dephasing time of the real optical transitions, the strong resonance enhancement of nonlinear optical processes may be fully exploited.[1] Coherent interactions between the light-field and the excitations of the medium [2] as well as coherent interactions between different excited states of the medium [3] may take place. A number of nonlinear wave-mixing experiments can be performed to investigate the dynamics of the excitations (excitons), including quantum interference phenomena and the dephasing processes themselves.[4] Close-lying electronic resonances, that are coherently excited by a short laser pulse, may exhibit quantum interferences or quantum beats, that can be observed in linear optical spectroscopy, such as resonance fluorescence and transmission, or in nonlinear wave mixing experiments.[3]

I shall here concentrate on third order nonlinearities, mixing three input fields either from the same or from different laser beams, to yield a third order nonlinear polarization that will either modify the original beams in a nonlinear fashion or create new collimated beams propagating in different directions.[4] If the input beams are of ultrashort pulses, and the system is composed of independent two-level atoms, the outgoing nonlinear signal may either be in the form of a free induction decay or a photon echo, depending on whether the system is homogeneously or inhomogeneously broadened, respectively.[5] In solids, however, excitation interactions, or polarization interactions, result in nonlinear signals whose time dependences may be distinctly different from the above cases.[6][7]

Quantum beat spectroscopy was first introduced in atomic and molecular physics as a high-resolution linear spectroscopy working in the time domain.[3] With modest time resolution, only extremely small splittings, in the megahertz to gigahertz range, can be observed with this technique. As faster and faster lasers have been developed, splittings in the terahertz range can be measured even in media where dephasing times are short, e.g. in liquids [8] and solids. Thus, recently quantum beats have been observed in a number of solids, particularly involving excitons in semiconductors.[9]

In section 2, I shall briefly discuss the concept of quantum beat spectroscopy exemplified by resonance fluorescence spectroscopy in single or many atom systems. Then, section 3 shall introduce nonlinear spectroscopy in the form of degenerate four-wave mixing, photon echo or nonlinear transmission, particularly for excitons in semiconductors. Section 4 shall combine the two, and discuss quantum beats appearing in wave mixing and nonlinear transmission experiments. Examples will be given from free and bound excitons in a bulk semiconductor (AgBr, CdS, CdSe) as well as from excitons in low-dimentional GaAs quantum wells. Propagation effects including quantum interferences between propagating polariton modes will be discussed in Section 5, before conclusions are made in Section 6.

2. QUANTUM BEAT SPECTROSCOPY

Linear quantum beat spectroscopy (QBS) was developed already in the sixties as a high resolution spectroscopy operating in the time domain.[3] When two or more close-lying atomic levels are resonantly excited by the same coherent light pulse, the atom is prepared in a mixed state that is a coherent superposition of pure states. This will give rise to an interference term in the wave function that will manifest itself, e.g. by an oscillation of the fluorescence decay observed after the short-pulse excitation.

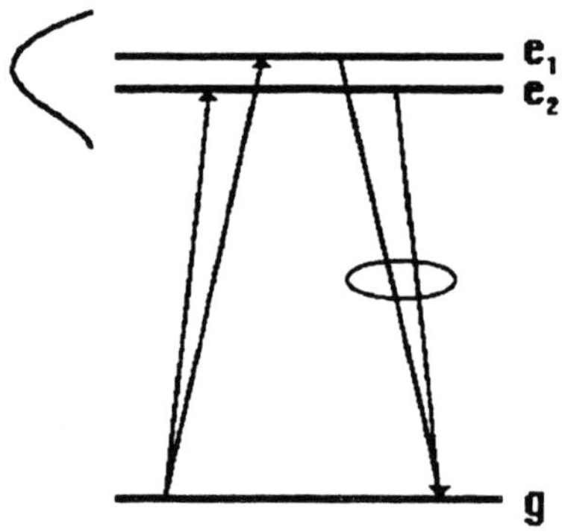

Figure 1 Resonance fluorescence in nearly degenerate two-level system

The situation is illustrated by the nearly degenerate two-level system in Fig. 1, where a simultaneous excitation from the ground state $|g\rangle$ to the close-lying excited levels $|e_1\rangle$ and $|e_2\rangle$ is provided by the same coherent laser pulse. The total Hamiltonian for the system is [2]

$$H = \hbar\omega_g |g\rangle\langle g| + \sum_l \hbar\omega_l |e_l\rangle\langle e_l| + \sum_k \hbar\Omega_k a_k^+ a_k$$
$$+ \hbar\sum_{k,l} [g_{lk} a_k^+ |g\rangle\langle e_l| + adj.] \tag{1}$$

where $\hbar$ is Planck's constant, $\hbar\omega_j$ is the energy of the atomic state $|j\rangle$, $\hbar\Omega_k$ is the energy of the electromagnetic mode k, and a_k^+ and a_k are the corresponding creation and annihilation operators, respectively. The light-atom dipole interaction is specified by the coupling coefficients

$$g_{lk} = -\frac{\mu_l \mathcal{E}_k}{2\hbar} U_k \tag{2}$$

where μ_i is the transition dipole moment of the atomic level $|i\rangle$ and $\mathcal{E}_k$ is the electric field 'per photon' of the surrounding electromagnetic mode k, characterized by the mode function U_k. The electric field operator for a single mode is thus

$$E_k(r, t) = E_k^+(r, t) + E_k^-(r, t) = \mathcal{E}_k [a(t) + a^+(t)] U_k(r) \tag{3}$$

where E^+ and E^- refer to the positive and negative frequency components of the field operator, respectively.

After excitation with a short pulse laser, the system is left in a coherent superposition of excited states $|e_i\{0\}\rangle$ with dipole allowed transitions from the ground state and no photons in any of the electromagnetic modes, i.e. with an initial wavefunction $|\psi(t)\rangle$

$$|\psi(0)\rangle = \sum_l c_{lo}(0) |e_l\{0\}\rangle \tag{4}$$

where the coefficients c_{io} contain the dipole moments as well as the integrated field strength of the laser pulse. At a later time t, one of the excited atomic states may have returned to the ground state and a photon emitted in mode k, as described by the wavefunction

$$|\psi(t)\rangle = \sum_l c_{lo}(t)\, e^{-i\omega_l t} |e_l\{0\}\rangle + \sum_k c_{gk}(t)\, e^{-i\omega_k t} |g\,\{1_k\}\rangle \tag{5}$$

where $\omega_k = \omega_g + \Omega_k$.

The rate equations for the time dependent coefficients in Eq. (5) are obtained from the Schrödinger equation, and following the standard Wigner-Weisskopf approximations for the theory of spontaneous emision, one finds the solutions [2]

$$\frac{dc_{ik}(t)}{dt} = -\frac{1}{2}\gamma c_{ik}(t) \tag{6}$$

and

$$c_{gk}(t) = \sum_{i=1}^{2} c_{i0}(0) \frac{i g_{ik}}{\frac{1}{2}\gamma_i + i\delta_{ik}} \left[e^{-(\frac{1}{2}\gamma_i + i\delta_{ik})t} - 1 \right] \tag{7}$$

where $\delta_{ik} = \omega_i - \omega_g - \Omega_k \equiv \omega_{ig} - \Omega_k$, and each level is charaterized by a single decay rate γ specifying the dephasing time $T_2 = 2/\gamma$ as well as the lifetime $T_1 = 1/\gamma$.

The intensity of spontaneous emission, on the detector at position r is expressed by $I(t) = \langle\psi(t)| E^-(r)E^+(r) |\psi(t)\rangle$ which, with the state vector of Eq.(5) transforms into

$$I(t) = \left|\sum_k c_{gk}^*(t)\, e^{ik\cdot r}\right|^2 = \frac{\Theta(t')}{r^2c^2}\left[\sum_{i=1}^{2} |g_{ik}\omega_{ig}c_{i0}(0)|^2 e^{\gamma_i t'} + \left(g_{1k}g_{2k}^*\omega_{1g}\omega_{2g}c_{10}^*(0)c_{20}(0)e^{-\frac{1}{2}(\gamma_1+\gamma_2)t'} e^{-i\omega_{12}t'} + c.c.\right)\right] \tag{8}$$

where in the last expression only two excited states i=1,2 are considered and the retarded time $t' = t - r/c$ at the detector position has been introduced. $\Theta(t')$ is the Heaviside step function. The first two terms in the square brackets of Eq. (8) represent the exponential decays of the two excited states with decay constants γ_1 and γ_2 and the two last terms reveal the interference between the two excited states, the so-called quantum beat. With coupling and expansion coefficients that are real, the fluorescence decay in Eq. (8) is expressed by

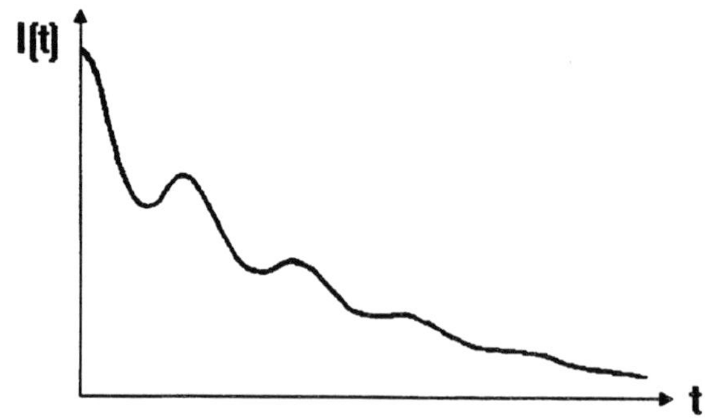

Figure 2 Resonance fluorescence decay with quantum beats as in Eq. (9)

$$I(t) = \Theta(t)\, I_0\, [\mu_1^4 e^{-\gamma_1 t} + \mu_2^4 e^{-\gamma_2 t} + 2\mu_1^2\mu_2^2 e^{-\frac{1}{2}(\gamma_1 + \gamma_2)t} \cos\omega_{12} t\,] \qquad (9)$$

where I_0 is proportional to the integrated intensity of the exciting laser pulse arriving at t = 0. This decay is illustrated in Fig. 2.

From Eq. (9), it is seen that the overall resonance fluorescence decay is dominated by the level with the strongest oscillator strength and the smallest decay rate or damping. The interference term, however, is strongest when both the oscillator strengths are large, and the damping of the modulations is governed by the fastest dephasing rate. The modulation frequency directly reveals the energy splitting between the excited states

$$\nu_{12} = \frac{\omega_{12}}{2\pi} = \frac{E_1 - E_2}{h} \qquad (10)$$

Thus, in systems where the dephasing time is long enough, so that several oscillation periods can be observed in the coherent fluorescence decay, this is a very precise way to measure energy splittings of nearly degenerate systems in the time domain. At the same time, the coherence and dephasing time of the system can be measured in a simple linear fluorescence experiment.[3]

The quantum beat modulation, as described here, is a very general quantum mechanical interference phenomenon, that is related to the fact, that in the present experiment there is no detailed knowledge of the path of the excitation-decay cycle. The system starts in the ground state and ends there again via two or more possible excited states, but we do not know which way it went. If the laser is only exciting one state resonantly, or if the emission is beeing spectrally resolved to better than the energy splitting between the excited states, we know which way, and the quantum beats dissapear. This is just like in any other interference phenomenon, as e.g. the double slit experiment.

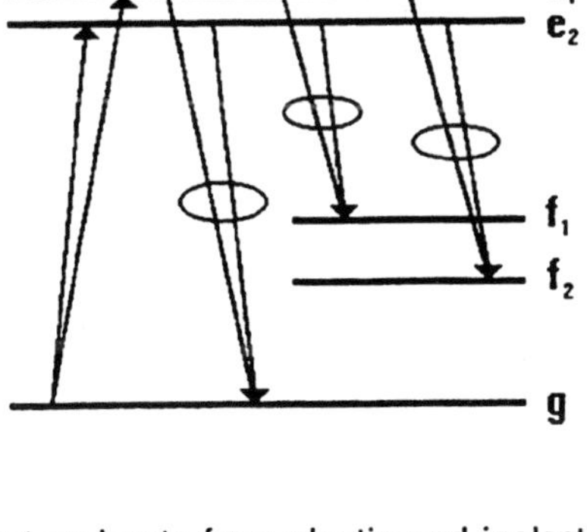

Figure 3 Quantum beats from elastic and inelastic scattering. No beats from split final state levels.

It is interesting to note in detail the conditions for observing single atom quantum beats as described above. The final state for the fluorescence decay may be an excited state different from the initial ground state without destroying the quantum beats, i.e. both elastic and inelastic scattering (fluorescence) is possible. However, quantum beats between different final states cannot be observed, because is this case it can be checked in which state the system is left after the decay. Again, only indistinguisable pathways give rise to interference, as illustrated in Fig. 3.

In the above, the quantum interference is occurring within the individual independent atoms. In many-atom systems collective quantum beats may be observed if participation of different atoms cannot be distinguished.[3] A collective contribution is only possible for elastic scattering, but a many-atom ground state splitting may give rise to quantum beats as illustrated in Fig.4 for two atoms initially in state $|g_1g_2\rangle$. After coherent excitation, a mixed state may be formed $|\psi(0)\rangle = \alpha_1 |g_1e\rangle + \alpha_2 |eg_2\rangle$. If the final state after the decay is another excited state f, there will be no quantum beat since $|g_1f\rangle$ and $|fg_2\rangle$ are distinguisable. If the final state is again the ground state $|g_1g_2\rangle$, a collective quantum beat may appear (Fig. 4).

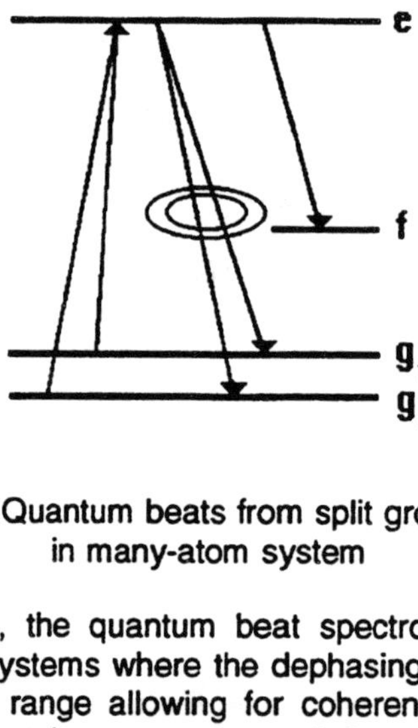

Figure 4 Quantum beats from split ground state in many-atom system

As already mentioned, the quantum beat spectroscopy was first used on independent single atom systems where the dephasing times are in the microsecond or even millisecond range allowing for coherent transient measurements without the need of ultrafast laser sources and detectors. Correspondingly, the energy splittings that can be investigated are in the megahertz to gigahertz range, i.e. ultrahigh spectral resolution from an optical point of view. Quantum beat spectroscopy is typically used to measure hyperfine splittings and Zeeman effects in atomic excitations in various environments.[3] Since such nearly degenerate atomic levels can be rather complex with a number of optically active levels, it may be necessary to Fourier transform the fluorescence decay before an identification of the various beat levels can take place. Figure 5 shows an example of such a complex quantum beat signal and the corresponding Fourier transform for the hyperfine structure of the $7P_{3/2}$ level of Cs vapor excited from the F = 3 hyperfine ground state by a pulsed dye laser.[10] Figure 5a shows the fluorescence decays, $I_\parallel$ and $I_\perp$, with polarization vector parallel and perpendicular to the polarization direction of the incident laser, respectively. The two signals contain the same beat structure, however, with opposite phase. The Fourier transform of the difference signal $I_\parallel - I_\perp$ is shown in Fig. 5b and reveals the quantum beats between the F = 2, 3 and 4 hyperfine excited states, ν_{32}, ν_{43} and $\nu_{42} = \nu_{43} + \nu_{32}$.[10]

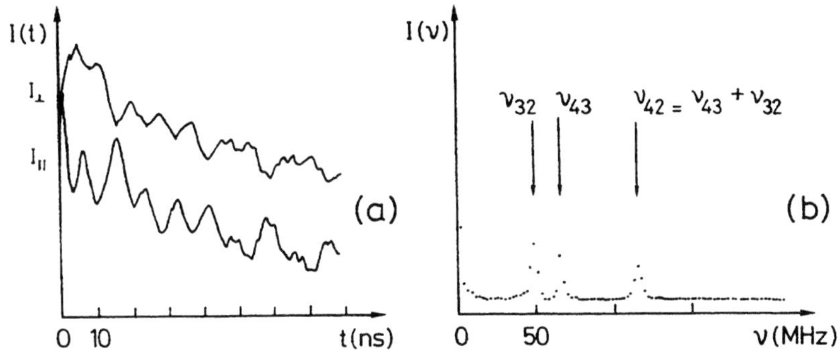

Figure 5 Quantum beat signals (a) and Fourier spectrum (b) in fluorescence from the $7P_{3/2}$ level of Cs, excited from the F=3 hyperfine ground state (Ref. 10).

3. NONLINEAR OPTICAL SPECTROSCOPY

In a resonance fluorescence experiment as described above, the intensity of the incident laser pulse may become sufficiently high, so that a strong nonlinear component to the polarization appears due to resonance enhancements of the nonlinear optical susceptibility. In a nearly degenerate to-level system, the quantum interferences will show up also in the nonlinear signals [11][12] and lay the ground for a nonlinear quantum beat spectroscopy (NQBS).

In a system with inversion symmetry, the lowest order nonlinearity is expressed by the third order nonlinear polarization[1][2]

$$P^{(3)}(\boldsymbol{k},\omega) = \chi^{(3)}(\omega;\omega_1,\omega_2,\omega_3)\ E_1(\boldsymbol{k}_1,\omega_1)E_2(\boldsymbol{k}_2,\omega_2)E_3(\boldsymbol{k}_3,\omega_3) \tag{11}$$

where $\chi^{(3)}$ is the third order nonlinear susceptibility ($\chi^{(3)} \propto \mu^3$), and E_i (i = 1, 2, 3) are the electric fields of up to three mutually coherent laser beams. Equation (11) expresses in general four-wave-mixing processes, where (up to) three input waves create a nonlinear polarization, that in turn acts as an antenna for a fourth outgoing wave, as specified by energy and momentum conservation

$$\omega = \sum_l \pm\omega_l \wedge \boldsymbol{k} = \sum_l \pm\boldsymbol{k}_l \qquad (12)$$

where +/- are entered for absorbed/emitted waves. Due to the fixed dispersion relation $\omega = \omega(\mathbf{k})$ in any medium, the two equations in (12) are very restrictive. They are only met under special conditions, called phase matching, and then ususally only for one particular polarization wave at a time. Phase-matching insures that the interacting waves and the nonlinear polarization propagate with the same speed through the medium, so that an appreciable energy transfer between waves can take place over a certain length.

In degenerate four-wave-mixing (DFWM), all the optical frequencies, including that of the outgoing wave, are identical $\omega = \omega_1 = \omega_2 = \omega_3$ ($= -\omega_1 + \omega_2 + \omega_3$, e.g.), and the phase matching condition $\mathbf{k} = -\mathbf{k}_1 + \mathbf{k}_2 + \mathbf{k}_3$ will depend strongly on the number of independent beams involved in the mixing process. With one beam only, $\mathbf{k} = \mathbf{k}_1 = \mathbf{k}_2 = \mathbf{k}_3$, there is always perfect phase match. The nonlinear polarization takes the form

$$P^{(3)}(\boldsymbol{k},\omega) = \chi^{(3)}(\omega;-\omega,\omega,\omega)|E(\boldsymbol{k},\omega)|^2 E(\boldsymbol{k},\omega) \qquad (13)$$

and is usually recognized through intensity dependent optical constants. The intensity dependent refractive index $n(I) = n_0 + n_2 I$, where $n_2 \propto \mathrm{Re}\chi^{(3)}$, is responsible for e.g. self focussing effects [1]. Similarly, the intensity dependent absorption coefficient $\alpha(I) = \alpha_0 + \alpha_2 I$, where $\alpha_2 \propto \mathrm{Im}\chi^{(3)}$, describes saturable absorption as well as induced absorption, depending on the sign of α_2 [1]. With two laser beams at an angle $\theta = \arccos(\mathbf{k}_1 \cdot \mathbf{k}_2 / k^2)$, there is still perfect phase match in the directions $\mathbf{k} = \mathbf{k}_1$ and $\mathbf{k}' = \mathbf{k}_2$ (Bragg diffraction) and near phase-match in the directions $\mathbf{k} = 2\mathbf{k}_2 - \mathbf{k}_1 - \Delta\mathbf{k}_\perp$ and $\mathbf{k}' = 2\mathbf{k}_1 - \mathbf{k}_2 - \Delta\mathbf{k}_\perp$, where $\Delta\mathbf{k}_\perp$ is the unavoidable phase mismatch that in a slab geometry will be perpendicular to the slab, as shown in Fig. 6.[1]

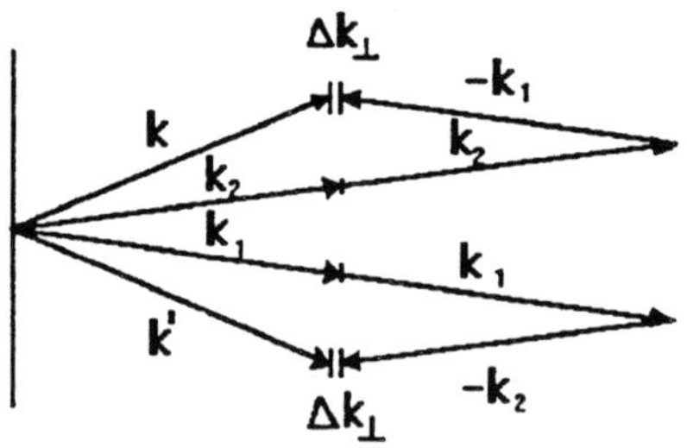

Figure 6 Wave vector conservation for two-beam DFWM.

The latter directions are particularly interesting from a nonlinear-optics-spectroscopy point of view, since here the nonlinear signal takes the form of a collimated beam, propagating in a direction with no linear signal. The nonlinear signal is therefore easily isolated with simple spatial filtering. The intensity I_s of the signal in such a direction is given by [1]

$$I_s \propto \frac{\sin^2(\frac{\Delta k_\perp d}{2})}{(\frac{\Delta k_\perp d}{2})^2} d^2 \qquad (14)$$

where d is the sample thickness, or the nonlinear interaction length. For thin sample conditions $\Delta k_\perp d \ll 1$ there is near phase-match and the signal $I_s \propto d^2$. For very thin samples ($kd \ll 1$) near phase-match can even be obtained in the backscattering direction.[1]

The two-beam configuration above is often called first order self-diffraction because it can be considered as an induced grating, set up by the interference of the two beams with a grating vector $\pm(\mathbf{k}_2 - \mathbf{k}_1)$ that will scatter the input beams themselves in the near phase-match directions discussed above. Alternatively, a third input beam can be involved in the DFWM experiment giving rise to third order nonlinear diffraction in the directions $\pm \mathbf{k}_1 \pm \mathbf{k}_2 \pm \mathbf{k}_3$, where all three signs cannot be identical. For the coherent nonlinear spectroscopy discussed here, however, little is obtained by adding the third beam, so I shall discuss only the two-beam configuration in the following.

In a pulsed nonlinear optical experiment with two beams, the dynamical information is obtained with the pump-probe technique. The first pulse excites the sample and the response (transmission or diffraction) from the second pulse is monitored, usually time integrated, as a function of the delay τ between the first and the second pulse (see Fig. 7). In a pulsed DFWM experiment on a system of two-level atoms characterized by the longitudinal and transverse relaxation times T_1 and T_2 ($T_2 \leq 2T_1$), the dephasing time T_2 is easily measured if the laser pulse length T_ℓ is short compared to the dephasing time. [5]

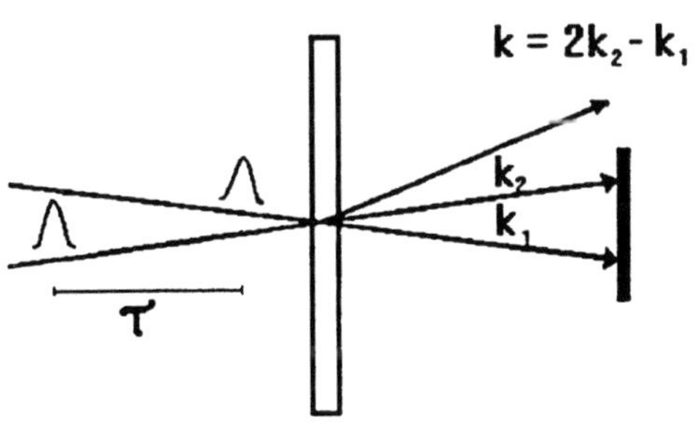

Figure 7 Time resolved degenerate four-wave mixing

If the two-level system is only homogeneously broadened, a free polarization decay $P \propto \exp\{-t/T_2\}$ will be observed in the diretion $2\mathbf{k}_2 - \mathbf{k}_1$ as a funtion of time t after the arrival of the second pulse (#2). The integrated intensity of this signal I_{hom} will depend on the coherent overlap of the polarization from pulse #1 and the field from pulse #2, i.e. on the delay τ. For $\tau > T_t$, one obtains [1][5]

$$I_{hom}(\tau) \propto \Theta(\tau)\, I_0^3\, \mu^8\, e^{-\frac{2\tau}{T_2}} \tag{15}$$

where again $\Theta(\tau)$ is the Heaviside function, I_0 is the integrated intensity of the identical input pulses and τ is positive when pulse #1 arrives before pulse #2.

For an inhomogeneously broadened system of two-level atoms, the nonlinear signal appears as a photon echo at the time $t = \tau$ after the arrival of pulse #2 and propagating in the direction $2\mathbf{k}_2 - \mathbf{k}_1$. Hence, the integrated intensity of the nonlinear signal, or photon echo, will decay faster with the delay between the input pulses, i.e. according to [1][5]

$$I_{inhom}(\tau) \propto \Theta(\tau)\, I_0^3\, \mu^8\, e^{-\frac{4\tau}{T_2}} \tag{16}$$

In systems that are not well described by independent two-level atoms, like for example semiconductors, the situation is much more complicated. Coulomb and exchange interactions between photoexcited elctron-hole pairs, or excitons, has to be taken into account, and the result is that the DFWM signal behaves neither as a free induction decay, nor as a photon echo. The temporal behaviour of the nonlinear signal is somewhat intermediate between the two cases above, [7] and as a result the integrated intensities of the nonlinear signal in Eqs. (15) and (16) also have contributions for negative delays, i.e. for pulse #2 arriving before pulse #1.[6]

4. NONLINEAR QUANTUM BEATS

The DFWM decays, or correlation traces, of Eqs. (15) and (16) assumed the sytems to be two-level like. If, on the other hand, there is a multiplicity of nearly degenerate levels, the decay of the nonlinear signal will be affected much the same way as in resonance fluorescence, described above in Sec. 2.[3] However, the quantum interferences, or beats, will in this case appear in the correlation traces, [11][12] as indicated in Fig. 8. The first pulse (E_1) arrives at the sample at t = 0 and sets up a (linear) polarization that will decay according to the dephasing time and show interference modulations in the case of nearly degenerate coherently excited states (Fig. 8a). This free induction decay is responsible for the fluorescence decay as well as the nonlinear interaction with the second pulse (E_2) arriving at $t = \tau$. It is intuitively obvious that the integrated intensity of the nonlinear signal will carry over the beat modulation into the correlation trace of the nonlinear signal as shown in Fig. 8b.

In Fig. 8 is assumed a homogeneously broadened system with a free induction decay, but a similar modulation of the correlation trace is obtained with inhomogeneous broadening. It was first shown by Lampert, Compaan and Abella, [11] that the

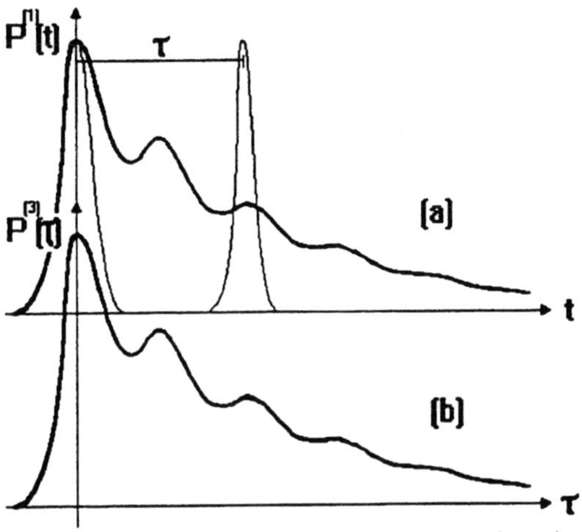

Figure 8 Quantum beats in the linear polarization decay (a) and in the correlation trace of the nonlinear signal (b).

photon echo correlation trace $I_{pe}(\tau)$ for a nearly degenerate system combines the features of Eqs. (9) and (16) into

$$I_{pe}(\tau) = \Theta(\tau)\, I_0^3\, [\mu_1^4 e^{-\gamma_1 \tau} + \mu_2^4 e^{-\gamma_2 \tau} + 2\mu_1^2 \mu_2^2 e^{-\frac{1}{2}(\gamma_1 + \gamma_2)\tau} \cos\omega_{12}\tau\,]^2 \tag{17}$$

where again the beat frequency is given by Eq. (10) and the other symbols have their usual meaning. Since in DFWM experiments, scattering is always elastic, a collective contribution may reveal also splitting in the ground state of the system.[3] In more complicated interacting systems, the nonlinear correlation traces may not take the nice analytical form of Eq. (17), but numerical calculations show the same features of the nonlinear quantum beats concerning the intensity dependence, the beat periods, the damping rates, and the dependence of the modulation depth on the oscillator strengths.[13][14] If polarization interaction or two-photon resonances are present, again the Heaviside function is relaxed and signals as well as quantum beats may be observed even for negative delay, $\tau < 0$.[6][15]

In the early days, linear as well as nonlinear quantum beat spectroscopy was applied to atomic systems in various environments, e.g. to determine the hypefine splitting of different atomic states.[3][16] Figure 9 shows an example of the photon-echo modulation pattern of the $^3P_0$-$^3H_4$ transition of a crystal of LaF_3:Pr^{3+} (0.03 at.%).[17] The upper trace is the experiment and the lower trace shows how well a theoretical calculation reproduces the complicated beat pattern. The actual level splittings and dampings are revealed by Fourier transforming the correlation trace, as shown in Fig. 10. The observed beat frequencies correspond to the level splitting of the $^3H_4$ ground state in the Pr^{3+} ions. The deduced linewidths of the nuclear transitions are of the order of 200 kHz.[17]

The reason why NQBS was restricted to noninteracting atomic systems lies in the fundamental conditions for observing quantum beats

$$\gamma < \nu_{12} < \Delta f \tag{18}$$

where $\Delta f = 1/T_\ell$ is the band width of the exciting laser. In isolated atoms, dampings are in the megaherz range, so that only moderately short laser pulses are required. In liquids [8] or solids, e.g. semiconductors,[9] electronic damping is in the terahertz range, so that picosecond or subpicosecond laser pulses are required to study possible quantum beats.[13][14][15]

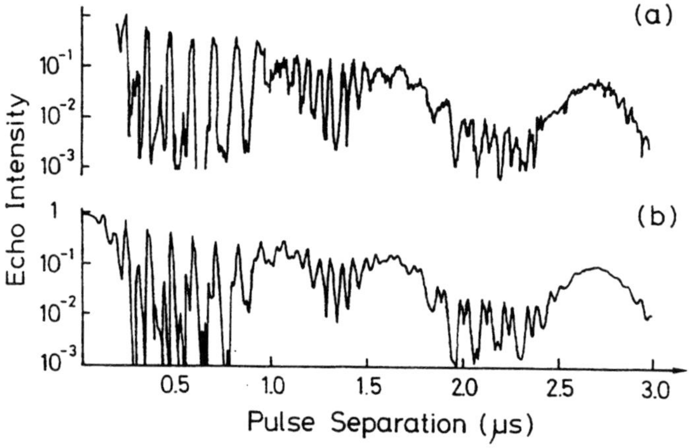

Figure 9 Experimental (a) and theoretical (b) photon echo modulation pattern in LaF_3:Pr^{3+} after Ref. 17.

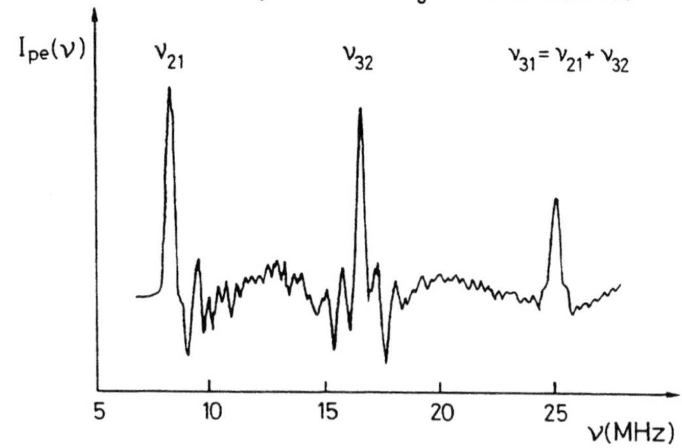

Figure 10 Fourier spectrum of the echo modulation in Fig.9 (from Ref. 17)

Another aspect of strong interatomic interactions in solids is the formation of electronic energy bands that will impede the observation of interferences between nearly degenerate discrete energy levels. In semiconductors, however, Coulomb interaction between photoexcited electron-hole pairs form hydrogen-like bound states, excitons, that can be either freely moving in the crystal [18] or bound to local impurities [19] (free or bound excitons). These excitonic states are the lowest elementary electronic excitations of semiconductors and introduce some narrow resonant optical transitions just below the fundamental band gap, that are very important for the optical properties, and in particular the nonlinear optical properties, of semiconductors in this spectral range.[1] Thus, recently a number of observations of quantum beats have been made on excitons in semiconductors,[9][14][18][19] particularly in GaAs quantum well structures.[13][15][20][21][22][23][24][25][26] In the following section, I shall discuss some examples of NQBS on excitons in semiconductors.

4.1 Bulk Semiconductors

The first observations of quantum beats from free excitons in semiconductors were made by Langer, Stolz, and von der Osten [18] performing linear resonance fluorescence on the indirect free excitons in AgBr with an applied magnetic field. Quantum beats were observed between the Zeeman split exciton levels, corroborating previously determined exchange interactions and electron and hole g-factors. The dephasing times were unusually long (T_2 = 300 - 400 ps), comparable to the energy relaxation time and slightly dependent on the magnetic field strength. The same group later applied their QBS to bound excitons in CdS [19] and in this case they found surprisingly short dephasing times ($T_2 \approx$ 20 ps) of excitons bound to neutral acceptors and donors, possibly due to a relatively high doping level.

We have performed NQBS by means of DFWM in the spectral range of bound excitons in CdSe without an applied magnetic field.[27][28][29] By applying a synch-

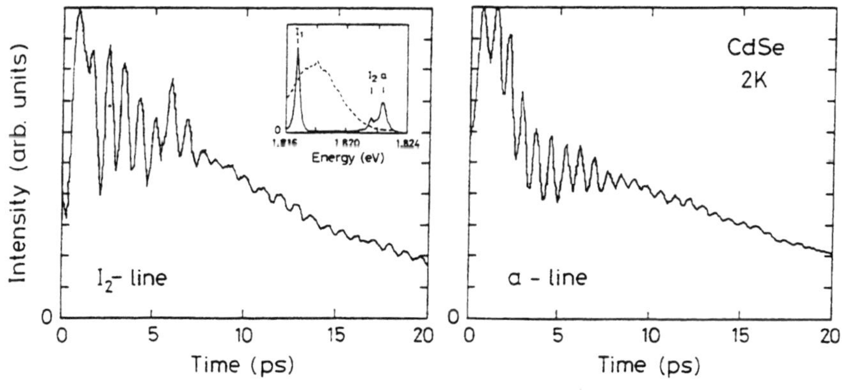

Figure 11 Modulation of the nonlinear signal from bound excitons in CdSe.

ronously pumped picosecond dye laser with a coherence time of less than 500 fs, and a corresponding spectral width of about 4 meV, we can excite simultaneously, and coherently, several bound (and free) exciton states and observe quantum beats in the resonantly enhanced DFWM signal. Figure 11 shows two examples of strongly modulated correlation traces when the spectrally broad laser is centered near the I_1-line (exciton bound to neutral acceptor). The DFWM signal is then strongly enhanced at the positions of the I_1-line, the I_2-line (exciton bound to neutral donor) and an unidentified a-line, as shown in the inset. The correlation trace to the left in Fig. 11 is detected at the spectral position of the I_2-line and the trace to the right is detected in the a-line. It is easily seen that they are distinctly different. Hence, it turns out that spectral resolution of the nonlinear signal is an important key for the identification of the levels involved in the beatings.

Figure 12 shows the corresponding Fourier transform of the correlation traces in Fig. 11. Assuming that spectral resolution of the nonlinear signal determines one of the levels involved in the observed beats, the level scheme shown in the insets of Fig. 11 can be established. The vertical lines indicate the beat frequencies that are actually observed. It can be seen that the NQBS reveals a fine structure in the levels involved in the nonlinear transitions, e.g. the I_2'-I_2'' splitting, that is not spectrally resolved, probably due to inhomogenous broadening. Dephasing times of the transitions involved in the beat signal can be determined from the linewidths in the beat spectra in Fig. 12. More details about these result are reported separately in this volume.[29]

One should stress the importance of spectral resolution in the NQBS, since it is different from linear QBS. In linear resonance fluorescence the quantum beats dissappear if the signal is spectrally resolved.[3] This is not the case in NQBS because here the time integrated nonlinear signal is detected, and the time resolution is obtained from the input pulses. The maximum information allowed by the uncertainty relations is obtained with transform limited input pulses and transform limited spectral resolution of the time integrated output signal.[9]

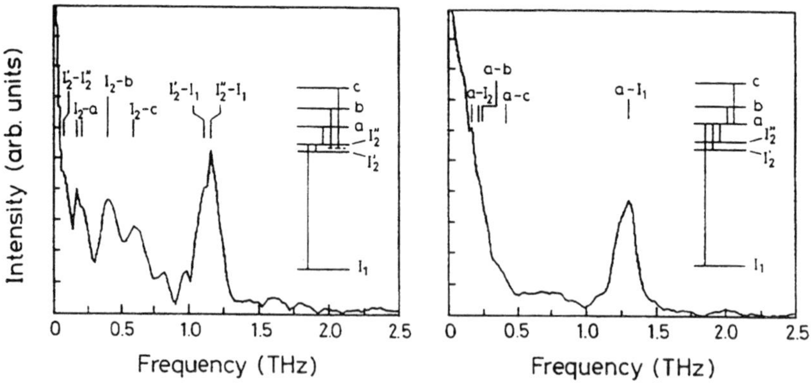

Figure 12 Fourier transforms of the beat traces in Fig. 11

4.2 Quantum Well Structures

GaAs/AlGaAs multiple quantum well (MQW) structures have been studied intensively for the last decade or more. One reason is that they have very interesting linear and nonlinear optical properties associated with excitonic features.[30] In Bulk GaAs, the exciton binding energy and Bohr radius are $E_{xb} = 4$ meV and $a_x = 300$ Å, respectively. This means that excitonic features are only significant at the lowest temperatures. Accordingly, the biexciton, composed of two excitons, is calculated to be very weakly bound by the energy $E_{xxb} = 0.1$ meV and is never observed experimentally. In a quantum well with a width of about 100 Å, the electron-hole pair is squeezed into a quasi two dimensional (2D) exciton with a binding energy of 10 meV, and the binding energy of the biexciton is calculated to be 1-2 meV.[31]

In a perfect quantum well, the 2-D excitons are free to move in the plane of the well. If there is some interface roughness over a length scale that is larger than the exciton diameter there will be a tendency of localization of the excitons in islands with largest well widths and thereby lowest confinement energy of the exciton.[32][33] This leads to inhomogeneous broadening of the exciton lines and a Stokes shift of the exciton luminescence relative to the absorption line (exciton relaxation). Interface roughness on a length scale smaller than the exciton diameter gives rise to individual scattering of electrons and holes and thereby contributes to the dephasing of the excitons. Hence, nonlinear optical spectroscopy of excitons in MQW's is useful, among other things to characterize the interface quality of the quantum structures.[32][33]

Figure 13 shows linear absorption and luminescence spectra at low temperature of a GaAs/AlGaAs MQW sample with 20 layers of 100 Å wells with 150 Å barriers. The absorption spectrum reveals the two lowest 2-D exciton levels associated with the heavy-hole valence band (hhx) and the light-hole valence band (lhx), respectively. The luminescence spectrum has a small Stokes shift mainly due to reabsorption in the sample (transmission geometry) and an additional intensity and

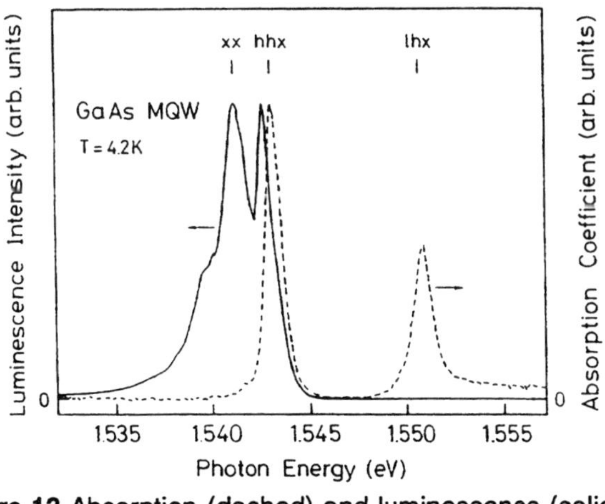

Figure 13 Absorption (dashed) and luminescence (solid) spectra of GaAs/AlGaAs multiple quantum wells.

temperature dependent line on the low energy side of the hhx-line, which we ascribe to the recombination of biexcitons. The small Stokes shift and the modest exciton line widths indicate good interface quality with negligible exciton localization.

Experiments of DFWM in a two-beam geometry show very strong enhancements of the nonlinear signal at the hhx and the lhx resonances, as can be seen in Fig. 14.

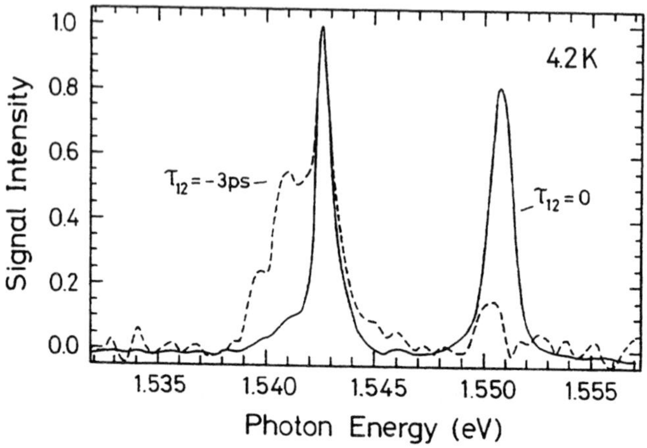

Figure 14 Excitation spectrum of nonlinear signal for $\tau = 0$ (solid) and for $\tau < 0$ (dashed).

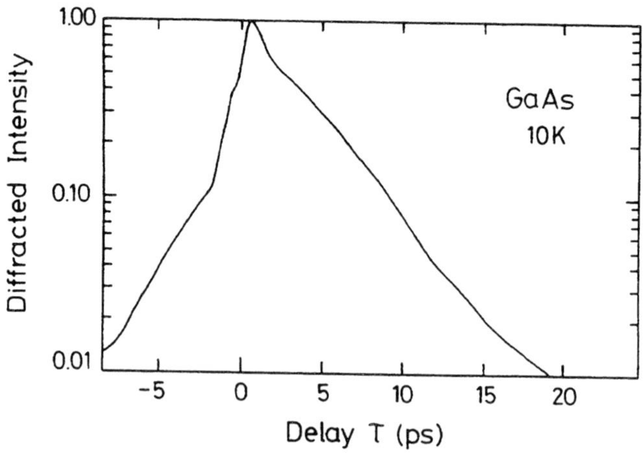

Figure 15 Correlation trace of DFWM signal in hhx resonance.

A correlation trace of the DFWM signal in the hhx resonance is shown in Fig. 15. From the near-exponential decay for positive delay, a dephasing time of $T_2 = 16$ ps is determined, assuming a slight inhomogeneous broadening of the hhx resonance. [34][35] This dephasing time is comparable to the bulk value for GaAs,[36] indicating again that the interface quality is good in the investigated sample. The contribution for negative delay indicates exciton interaction to be of importance.[6] In Fig. 14, it is seen also that the DFWM signal for negative delay has contributions from the spectral range below the hhx resonance where biexciton transitions are resonant.[37] With a dephasing time of several picoseconds it is possible to observe quantum beats between the different excitonic levels in GaAs MQW's, and a number of such observations have in fact been made.[20][21][22][23][24][25][26]

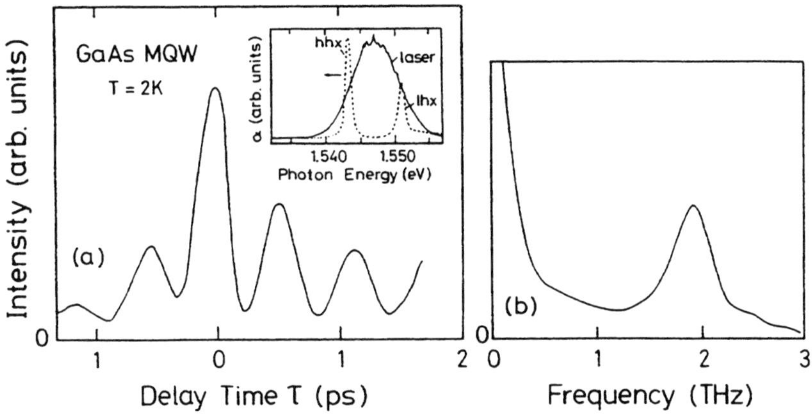

Figure 16 Correlation trace (a) and Fourier spectrum (b) from GaAs MQW's showing hhx-lhx quantum beats.

Figure 16a shows strong modulations of the correlation trace of the DFWM or photon echo signal when the spectrally broad laser is tuned between the hhx and lhx resonances (see inset) and exciting simultaneously and coherently the two exciton levels. The Fourier transform of the correlation trace, shown in Fig. 16b, clearly reveals a single beat frequency of $\nu = 1.93$ THz, corresponding to an energy splitting of $\Delta E = h\nu = 8.0$ meV, which is just the energy separation of the hh and lh excitons. From the width of the beat spectrum $\Delta\nu = 450$ GHz, one can determine a dephasing time of 4.4 ps again assuming a photon echo, i.e. an inhomogeneous broadening of the resonances. This dephasing time is smaller than determined above for the hhx and is probably to be ascribed to the lhx resonance.

If the laser is tuned to the spectral range of the biexciton transitions, i.e. just below the hhx resonance, quantum beats are again observed (Fig. 17a), but now with a beat frequency $\nu = 425$ GHz (Fig. 17b) corresponding to an energy separation of 1.76 meV. This separation is equal to the energy separation between the hhx line and the luminescence line (xx), due to biexciton recombination above (Fig. 13), and also in accordance with the resonance enhancement of the DFWM signal for negative delay (Fig. 14).

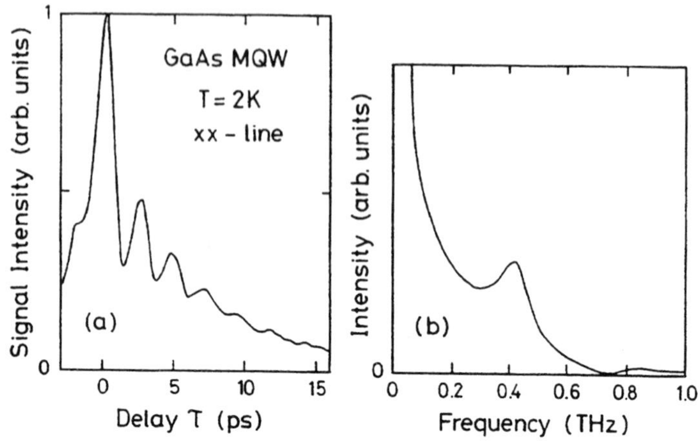

Figure 17 Quantum beats (a) and Fourier spectrum (b) below the hhx resonance in GaAs MQW's.

The observed beat is interpreted as a quantum beat between the ground state to exciton transitions and the exciton to biexciton transitions and the beat frequency therefore reveals the binding energy of the biexciton in the GaAs quantum wells. A necessary condition for such a beat to occur is that a coherent mixture of excitons and biexcitons is formed. This, however, is done efficiently via two-photon transitions

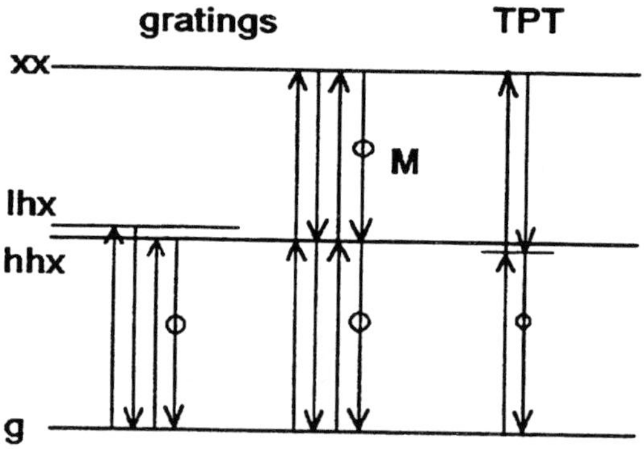

Figure 18 Level diagram for excitons and biexcitons in GaAs quantum wells with different four-wave mixing processes.

either directly to the biexciton states or via the hhx states. A level diagram is shown in Fig. 18, indicating also the different FWM processes, some of which are indistinguishable in a given experiment, giving rise to the observed quantum beats.

Similar experiments were performed by Lovering et al.[26] obtaining equivalent quantum beats from GaAs MQW's. However, they deduce a biexciton binding energy of about 1.1 meV and ignore modulations of about half the period. The Fourier transform of our beat signal cannot justify an interpretation in terrm of two different beats showing a single well-defined beat frequency with a width $\Delta\nu$ = 120 GHz. The width of the beat spectrum is consistent with the dephasing time of 16 ps found above for the hhx resonance indicating that dephasing of excitons and biexcitons occur via the same scattering processes, i.e. interface roughness scattering at low temperatures and excitation densities, phonon scattering at higher temperatures and exciton-exciton scattering at higher densities. There are other reports of nonlinear quantum beats in the same spectral range of GaAs MQW's that have ascribed to beats between free and bound excitons[23] and between excitons in different well widths either within the same well or in adjacent wells.[20][24]

4.3 Quantum Beats Versus Polarization Beats

In the presentation of the experimental results above on nonlinear quantum beats, there has been no clear distinction whether the observed modulations were due to real quantum mechanical interferences in a three(or more)-level system or due to interferences in the macroscopic polarization from two (or more) independent two-level systems.[3] In a DFWM experiment both phenomena can occur and may be very difficult to distinguish from one another. For ordinary quantum beat spectroscopy this is not too important, since both phenomena contain the same information about level splittings, oscillator strengths and dephasing rates.[28] If, on

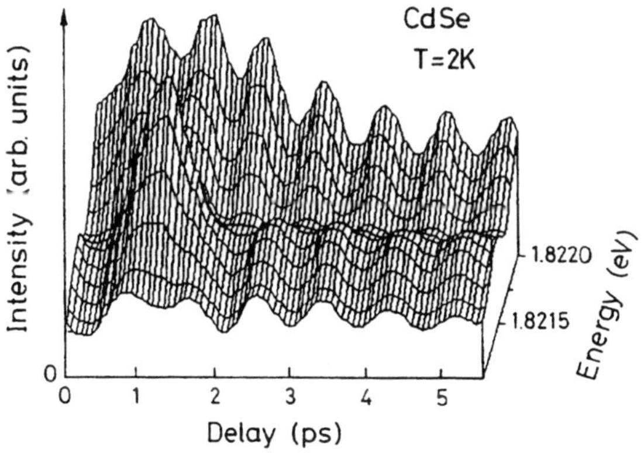

Figure 19 Spectral dependence of nonlinear beat signal around the I_2 bound exciton in CdSe

the other hand, the issue is to determine whether different electronic states are interacting or noninteracting, it is very important to be able to distinguish between quantum beats and polarization beats.

It turns out that spectral resolution of a time integrated DFWM signal (correlation trace)[38] or temporal resolution of the spectrally integrated DFWM signal[39][40] may provide the necessary distinction between quantum beats and polarization beats. Tokizaki and coworkers have simulated the two cases in a numerical calculation with spectral resolution.[38] For quantum beats, the amplitude and phase of the modulations are spectrally independent, when going through the resonances, whereas for polarization beats the modulation undergoes a phase shift in the resonances. Figure 19 shows a case of spectrally resolved correlation traces with beat modulations in the bound exciton region of CdSe. It it seen that there is a clear phase shift of the modulations of about π at the spectral position of I_2-resonance. This suggests that the bound exciton transitions beating with the I_2-transition are not interacting strongly with it. Thus, beats between bound exciton transitions, at low impurity concentrations, are due to polarization interferences, whereas beats between free and bound excitons are real quantum beats. Such spectral dependences have not been observed in the interferences observed in the GaAs quantum well system, which have also been interpreted in a multi-level scheme (Fig. 18). Recently, an experimental distinction has been made between quantum beats and polarization beats in specially designed multiple quantum well structures, by direct time resolution of the DFWM beat signal.[40]

5. PROPAGATION EFFECTS

The nonlinear quantum beat spectroscopy of excitons in semiconductors, treated in Sec. 4, has assumed the excitons to be immobile on the time scale of the optical transit times. This is clearly the case for bound excitons and for excitons that are

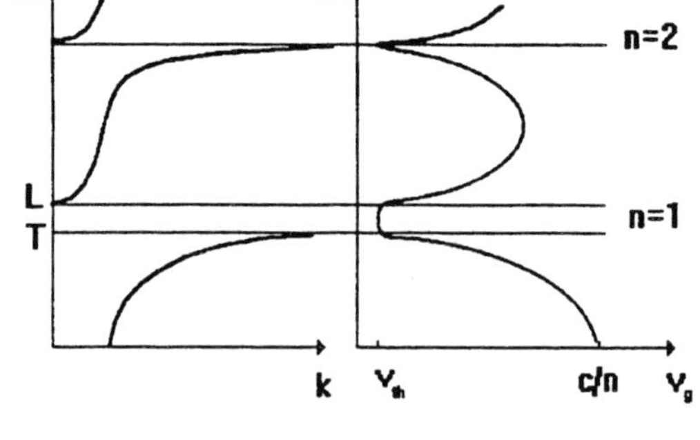

Figure 20 Polariton dispersion and group velocity dispersion near n = 1, 2 exciton resonances in semiconductor.

quantum confined in the direction of light propagation. For free excitons in bulk semiconductors, however, this is not trivially true. The strong coupling between light and excitons result in polariton modes [41] propagating with velocities that change continuously from the speed of light to thermal exciton velocities, when approaching the free exciton resonance (see Fig. 20). This may lead to special interference effects and quantum beats between propagating polaritons in the spectral range around, or between, free exciton resonances where the medium shows strong dispersion. These effects have been observed in linear optical transmission [42][43] as well as in DFWM experiments [14][27][44].

One problem of observing the polariton interferences is that in the region of strong polariton dispersion, the media are usually also strongly absorptive, so that extremely thin samples and/or relatively weak coupling to light are required. A general feature to be expected for the beat modulations is a time (and frequency) dependent modulation period.[42][43][44] Figure 21 shows, from Ref. 42, the temporal evolution of a 30-ps laser pulse, tuned to the 1S exciton resonance in Cu_2O, as transmitted through a crystal of thickness d = 0.91 mm. The points are experimental and the curve is the result of a numerical calculation with the oscillator strength and a phenomenological damping constant as fitting parameters. The resonance is chosen because it is only quadrupole allowed so that near-resonant light (polaritons) can be transmitted through a relatively thick sample. The modulations are intuitively understood as beats between polaritons on each side of the resonance arriving simultaneously at the exit of the sample. With increasing time, the interference will be between slower and slower polaritons, i.e closer and closer to the resonance and to each other. Hence, the modulation period will increase with time, as observed. Similar result have been reported from FWM quantum beats of propagating polaritons in InGaAs/InAlAs multiple quantum wells.[44]

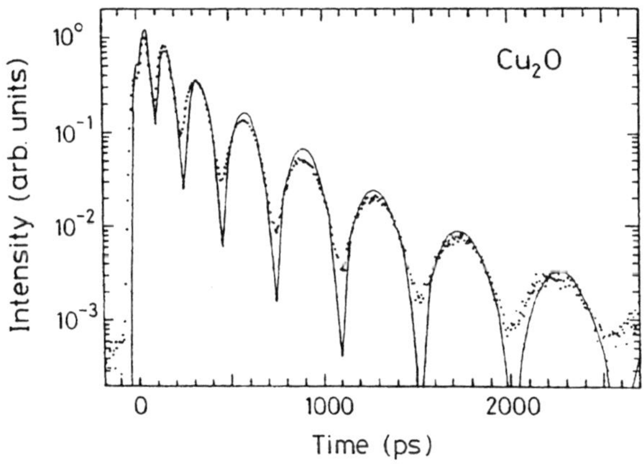

Figure 21 Transmission quantum beats of polaritons in 1S exciton resonance of Cu_2O, from Ref. 42.

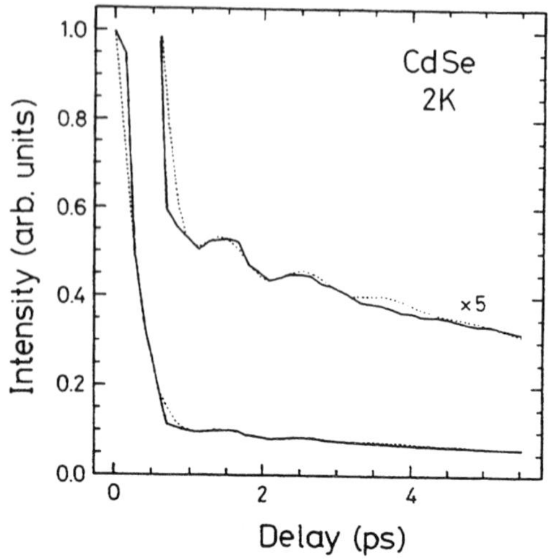

Figure 22 Quantum beats in the correlation trace of propagating polaritons in CdSe.

We have investigated quantum beats of propagating polaritons in the transmission window between the n = 1 and the n = 2 excitons resonances in thin crystals of CdSe. When performing a two-beam DFWM experiment in this spectral range, we observe nonequidistant oscillations in the correlation trace of the nonlinear signal, as shown in Fig. 22. In this case, the oscillation period decreases with time because the interfering (slower) polaritons are further apart as time goes on.

The solid curve in Fig. 22 is the experiment and the dashed curve is the result of a calculation on a modified two-level model including the polariton effects.[24] From the fit are determined the damping coefficients of the n = 1 and the n = 2 excitons, assumed homogeneously damped. In the calculations, the dispersion of the input fields as well as of the nonlinear signal is taken into account. With a more elaborate model one could hope to get more detailed information about the energy dependent damping of polaritons in semiconductors.[45]

6. SUMMARY AND CONCLUSIONS

Coherent optical spectroscopy in the form of linear and nonlinear quantum beat spectroscopy has for several decades played a significant role as a ultrahigh-resolution spectroscopy of noninteracting atoms and molecules. Although working in the time-domain, this spectroscopy imposed modest requirements on the time resolution, because of the fine and hyperfine splittings, that were investigated, and because of the long coherence and dephasing times typical for noninteracting atomic and molecular systems.

The recent developments in ultrafast laser systems and spectroscopies, however, have made it possible to work with femtosecond time resolution so that coherent optical spectroscopies, including quantum beat spectroscopy, can be performed on systems, e.g. liquids and solids, with strong interactions and dephasing times in the picosecond range. Correspondingly, the energy splittings to be studied by quantum beat spectroscopy are in the THz, or meV, range.

In the last few years, there has been a revival of quantum beat spectroscopy, particularly nonlinear quantum beat spectroscopy, to investigate oscillator strengths, energy splittings and interaction mechanisms of free, bound and confined excitons in various semiconductor materials and structures. In most of these works, the quantum beat oscillations have been observed as modulations of the correlation traces in degenerate four-wave mixing experiemnts in the two-beam configuration.

In the present article is reviewed nonlinear quantum beat spectroscopy on free and bound excitons in bulk semiconductors such as AgBr, CdS and CdSe and on quasi two-dimensional excitons in GaAs/AlGaAs multiple quantum wells. The distinction between nonlinear quantum beats from a three-level system and the polarization beats from two independent two-level systems is being discussed. It turns out that the distinction experimentally can be done either by spectral resolution or by real-time resolution (nonlinear up-conversion). It is found that beats between noninteracting bound excitons, at low impurity concentrations, can be considered as polarization beats between independent two-level system, whereas beats between free exciton states or between free and bound exciton states behave as quantum beats in a multi-level system.

Particularly in the GaAs/AlGaAs quantum well system, quantum beats are observed between a variety of free and bound exciton states. The formation of quasi two-dimensional biexcitons in GaAs/AlGaAs quantum wells has been finally confirmed and a binding energy of the biexcitons is determined to 1.8 meV in wells with a width of 100 Å. As a monitor of the dephasing time, nonlinear quantum beat spectroscopy is also used to measure the quality of the interfaces, since at low temperature and low excitation intensity interface roughness scattering is the most important contribution to dephasing.

Finally, propagation effects have been considered by investigation of quantum beats between propagating polaritons in bulk semiconductors and quantum well systems. It turns out that in strongly dispersive media, the usual thin sample phase-matching conditions break down and quantum interferences take place between different polariton modes propagting with the same group velocity. Nonlinear quantum beat spectroscopy of propagating polaritons may in the future give more detailed information about damping of exciton-light polariton modes in semiconducting materials and structures.

In general, it is suggested that nonlinear quantum beat spectroscopy will play a signicant role in the near future as a useful coherent time-domain spectroscopy, that with the advent of ultrafast femtosecond laser systems, also can be applied to strongly interacting systems such as liquids and semiconductors.

ACKNOWLEDGMENTS

The author is indebted to K.-H. Pantke, V.G. Lyssenko, B.S. Razbirin, P. Schillak, D. Oberhauser, J. Erland and I. Balslev for their many contributions to the present work, which was supported by the Danish Natural Science Research Council.

REFERENCES

1. J.M. Hvam, C. Dörnfeld, and C.R. Paton: "Coherent nonlinear optical resonances in II-VI semiconductors," *SPIE Vol. 1216 Nonlinear Optical Materials and Devices for Photonic Switching*, 1990, pp. 176-189.
2. P. Meystre and M. Sargent III, *Elements of Quantum Optics*, Berlin, Heidelberg: Springer Verlag, 1990, 484 pp.
3. S. Haroche, "Quantum beats and time-resolved fluorescence spectroscopy" in: *High Resolution Laser Spectroscopy*, K. Shimoda (Ed.), Berlin, Heidelberg: Springer Verlag, 1976, pp. 253-313.
4. Y.R. Shen, *The Principles of Nonlinear Optics*, New York: John Wiley & Sons, 1984, 555 pp.
5. T. Yajima and Y. Taira, "Spatial optical parametric coupling of picosecond light pulses and transverse relaxation effects in resonant media," *J. Phys. Soc. Japan*, Vol. 47, No. 5, 1979, pp. 1620-1626.
6. K. Leo, M. Wegener, J. Shah, D.S. Chemla, E.O. Göbel, T.C. Damen, S. Schmitt-Rink, and W. Schäfer, "Effects of coherent polarization interactions on time-resolved degenerate four-wave mixing," *Phys. Rev. Lett.* Vol. 65, No. 11, 1990, pp. 1340-1343.
7. D.-S. Kim, J. Shah, T.C. Damen, W. Schäfer, F. Jahnke, S. Schmitt-Rink, and K. Köhler, "Unusually slow temporal evolution of femtosecond four-wave mixing signals in intrinsic GaAs quantum wells: Direct evidence for dominance of interaction effects," *Phys. Rev. Lett.* Vol. 69, No. 18, 1992, pp. 2725-2728.
8. P.C. Becker, H.L. Fragnito, J.Y. Bigot, C.H. Brito Cruz, R.L. Fork, and C.V. Shank, "Femtosecond photon echoes from molecules in solution," *Phys. Rev. Lett.* Vol. 63, No. 5, 1989, pp. 505-507.
9. H. Stolz, "Picosecond time-resolved light scattering from coherent exciton states," in *Festkörperprobleme / Advances in Solid State Physics* Vol. 31, U. Rössler (Ed.), Braunsweig: Vieweg, 1991, pp. 219-241.
10. S. Haroche, J.A. Paisner, and A.L. Schawlow, "Hyperfine quantum beats observed in Cs vapor under pulsed dye laser excitation," *Phys. Rev. Lett.* Vol. 30, No. 20, 1973, pp. 948-951.
11. L.Q. Lampert, A. Compaan, and I.D. Abella, "Effects of nearly degenerate states on photon-echo behaviour," *Phys. Rev. A*, Vol. 4, No. 5, 1971, pp. 2022-2024.
12. R.L. Shoemaker and F.A. Hopf, "Observation of quantum beats in photon echoes and optical nutation," *Phys. Rev. Lett.* vol. 33, No. 26, 1974, pp. 1527-1530.
13. B .F. Feuerbacher, J. Kuhl, R. Eccleston, and K. Ploog, "Quantum beat between the light and the heavy hole excitons in a quantum well," *Solid State Commun.* Vol. 74, No. 12, 1990, pp. 1279-1283.
14. K.-H. Pantke, P. Schillak, B.S. Razbirin, V.G. Lyssenko, and J.M. Hvam, "Nonlinear quantum beats of propagating polaritons," *Phys. Rev. Lett.* Vol. 70, No. 3, 1993, pp. 327-330.
15. K.-H. Pantke, D. Oberhauser, V.G. Lyssenko, J.M. Hvam, and G. Weimann, "Coherent generation and interference of excitons and biexcitons in GaAs/ $Al_xGa_{1-x}As$ quantum wells," *Phys. Rev. B.* Vol.47, No. 2, 1993, 4pp.

16. L.Q. Lampert, "Effect of superhyperfine interactions on photon-echo behavior in dilute ruby," *Phys. Rev. B* Vol. 7, No. 5, 1973, pp. 1834-1846.
17. Y.C. Chen, K. Chiang, and S.R. Hartmann, "Spectroscopic and relaxation character of the $^3P_0$-$^3H_4$ transition in LaF_3:Pr^{3+} measured by photon echoes," *Phys. Rev. B* Vol. 21, No. 1, 1980, pp. 40-47.
18. V. Langer, H. Stolz, and W. von der Osten, "Observation of quantum beats in the resonance fluorescence of free excitons," *Phys. Rev. Lett.* Vol. 64, No. 8, 1990, pp. 854-857.
19. H. Stolz, V. Langer, E. Schreiber, S. Permogorov, and W. von der Osten, "Picosecond quantum-beat spectroscopy of bound excitons in CdS," *Phys. Rev. Lett.* Vol. 67, No. 6, 1991, pp. 679-682.
20. E.O. Göbel, K. Leo, T.C. Damen, J. Shah, S. Schmitt-Rink, W. Schäfer, J.F. Müller, and K. Köhler, "Quantum beats of excitons in quantum wells," *Phys. Rev. Lett.* Vol. 64, No. 15, 1990, pp. 1801-1804.
21. W.A.J.A. van der Poel, A.L.G.J. Severens, and C.T. Foxon, "Quantum beats in the exciton emission of type II GaAs/AlGaAs quantum wells," *Optics Commun.* Vol. 76, No. 2, 1990, pp. 116-120.
22. K. Leo, T.C. Damen, J. Shah, E.O. Göbel, and K. Köhler, "Quantum beats of light hole and heavy hole excitons in quantum wells," *Appl. Phys. Lett.* Vol. 57 No. 1, 1990, pp. 19-21.
23. K. Leo, T.C. Damen, J. Shah, and K. Köhler, "Quantum beats of free and bound excitons in GaAs/$Al_xGa_{1-x}As$ quantum wells," *Phys. Rev. B* Vol. 42, No. 17, 1990, pp. 11359-11361.
24. K. Leo, J. Shah, E.O. Göbel, T.C. Damen, S. Schmitt-Rink, W. Schäfer, and K. Köhler, "Coherent oscillations of a wave packet in a semiconductor double-quantum-well structure," *Phys. Rev. Lett.* Vol.66, No. 2, 1991, pp. 201-204.
25. S. Bar-Ad and I. Bar-Joseph, "Absorption quantum beats of magnetoexcitons in GaAs heterostructures," *Phys. Rev. Lett.* Vol. 66, No. 19, 1991, pp. 2491-2494.
26. D.J. Lovering, R.T. Phillips, G.J. Denton, and G.W. Smith, "Resonant generation of biexcitons in a GaAs quantum well," *Phys. Rev. Lett.* Vol. 68, No. 12, 1992, pp. 1880-1883.
27. K.-H. Pantke, P. Schillak, J. Erland, V.G. Lyssenko, B.S. Razbirin, and J.M. Hvam, "Nonlinear quantum beats of excitons in CdSe," *phys. stat. sol. (b)* Vol. 173, No.1, 1992, pp. 91-98.
28. K.-H. Pantke, V.G. Lyssenko, B.S. Razbirin, J. Erland, and J.M. Hvam, "Nonlinear quantum beat spectroscopy in CdSe," in *Proc. 21st Int. Conf. Phys. Semicond.*, Beijing 1992, K. Huang and L.L. Chang (Eds.), Singapore: World Scientific Publishing Co., 1993, 8 pp.
29. J. Erland, K.-H. Pantke, B.S. Razbirin, V.G. Lyssenko, and J.M. Hvam, "Quantum interferences of excitons in CdSe," in this volume.
30. L. Banyai, I. Galbraith, and H.Haug, "Biexciton nonlinearity in GaAs/$Ga_xAl_{1-x}As$ quantum wells and quantum well wires," *Phys. Rev. B* Vol 38, No. 6, 1988, pp. 3931-3936.
31. D.A. Kleinman, "Binding energy of biexcitons and bound excitons in quantum wells," *Phys. Rev. B* Vol. 28, No. 2, 1983, pp. 871-879.
32. D. Oberhauser, K.-H. Pantke, W. Langbein, V.G. Lyssenko, H. Kalt, J.M. Hvam, G. Weimann, and C. Klingshirn, "Coherent and incoherent exciton dynamics in $Al_{1-y}Ga_yAs$/GaAs multiple quantum wells," *phys. stat. sol. (b)* Vol. 173, No. 1, 1992. pp. 53-68.

33. D. Oberhauser, K.-H. Pantke, J.M. Hvam, G. Weimann, and C. Klingshirn, "Exciton scattering in quantum wells at low temperature," *Phys. Rev. B,* in press.
34. L. Schultheis, A. Hononld, J. Kuhl, K. Köhler, and C.W. Tu, "Optical dephasing of homogeneously broadened two-dimensional exciton transitions in GaAs quantum wells," *Phys. Rev. B* Vol. 34, No. 12, 1986, pp. 9027-9030.
35. M.D. Webb, S.T. Cundiff, and D.G. Steel, "Observation of time-resolved picosecond stimulated photon echoes and free polarization decay in GaAs/Al-GaAs multiple quantum wells," *Phys. Rev. Lett.* Vol. 66, No. 7, 1991, pp. 934-937.
36. L. Schultheis, J. Kuhl, A. Hononld, and C.W. Tu, "Picosecond phase coherence and orientational relaxation of excitons in GaAs," *Phys. Rev. Lett.* Vol.57, No. 14, 1986, pp. 1797-1800.
37. B.F. Feuerbacher, J. Kuhl, and K. Ploog, "Biexcitonic contribution to the degenerate-four-wave-mixing signal from a $GaAs/Al_xGa_{1-x}As$ quantum well," *Phys. Rev. B* Vol. 43, No. 3, 1991, pp. 2439-2441.
38. T. Tokizaki, A. Nakamura, Y. Ishida, T. Yajima, I. Akai, and T. Karasawa, "Femtosecond quantum beats in excitons of a layered semiconductor BiI_3," in *Springer Series in Chemical Physics, Vol.53: Ultrafast Phenomena VII,* C.B. Harris, E.P. Ippen, G.A. Mourou, and A.H. Zewail (Eds.), Berlin, Heidelberg: Springer Verlag, 1990, pp. 253-255.
39. E.O. Göbel, M. Koch, J. Feldmann, G. von Plessen, T. Meier, A. Schulze, P. Thomas, S. Schmitt-Rink, K. Köhler, and K. Ploog, "Time resolved four wave mixing in GaAs/AlAs quantum well structures," *phys. stat. sol (b),* Vol. 173, No.1, 1992, pp. 21-30.
40. M. Koch, J. Feldmann, G. von Plessen, E.O. Göbel, P. Thomas, and K. Köhler, "Quantum beats versus polarization interference: An experimental distinction," to be published.
41. J.J. Hopfield, "Theory of the contribution of escitons to the complex dielectric constant of crystals," *Phys. Rev.* Vol. 112, No. 5, 1958, pp. 1555-1567.
42. D. Fröhlich, A. Kulik, B. Uebbing, A. Mysyrowicz, V. Langer, H. Stolz, and W. von der Osten, "Coherent propagation and quantum beats of quadrupole polaritons in Cu_2O," *Phys. Rev. Lett.* Vol. 67. No. 17, 1991, pp. 2343-2346.
43. D. Fröhlich, A. Kulik, B. Uebbing, V. Langer, H. Stolz, and W. von der Osten, "propagation beats of quadrupole polaritons in Cu_2O," *phys. stat. sol. (b),* Vol. 173, No. 1, 1992, pp. 31-40.
44. T. Rappen, G. Mohs, and M. Wegener, "Coherent polariton dynamics in quantum wells," *phys. stat. sol (b),* Vol. 173, No. 1, 1992, pp. 77-82.
45. M. Dagenais and W.F. Sharfin, "Measurement of the damping dispersion of exciton polaritons in CdS," *Phys. Rev. Lett.*, Vol. 58, No. 17, 1987, pp. 1776-1779.

Studies in Classical
and Quantum Nonlinear Optics

ISBN 1-56072-168-5

LOCAL-FIELD STUDIES IN THE NONLINEAR OPTICS OF MESOSCOPIC SYSTEMS

Ole Keller
Institute of Physics, University of Aalborg
Pontoppindanstraede 103, DK-9220 Aalborg Øst, Denmark

1 INTRODUCTION

The physics of mesoscopic systems has attracted substantial attention in recent years [1]. From the point of view of fundamental physics the attractiveness is associated with the fact that the discreteness of the energy spectrum and the interference effect of the electron waves lead to a number of fascinating and exotic phenomena not observed in bulk samples. The understanding of the basic physical properties of the mesoscopic systems has developed hand in hand with the progress in the so-called nanostructure technology, and the latest developments seem to indicate that a particular fruitful and comprehensive interplay between basic physics, applied physics and technology is to appear in the future.

A subdomain in the field of mesoscopic physics is that of mesoscopic optics. Stimulated by the pioneering work of Kubo [2] on the electronic properties of small metallic particles, a work which maybe is the very earliest one in mesoscopic physics, the quantum theory of Kawabata and Kubo [3] gave rise to a widespread interest in studying the possibilities of observing quantum size effects related to the optical properties of small particles. A general review of quantum size effects in metal particles recently has been presented by Halperin [4], and a survey of the particle-size dependence of the optical and far infrared absorption has been given by Kreibig and Genzel [5]. From a theoretical point of view the central concept in the mesoscopic physics of small particles is the linear polarizability of the particles [6]-[13]. Also, the linear optical response of small microcrystallites has been studied in recent years [14]-[21]. When all the three dimensions of the mesoscopic system are smaller than the Fermi wavelength of the electrons, so that the electrons essentially are confined in all space coordinates, the mesoscopic system is called a quantum dot. Quantum dots containing only a few electrons have been fabricated recently [22], and the far-infrared properties, e.g. resonances, studied [23]-[25]. When the electron motion of the mesoscopic system is confined in two of the spatial coordinates, we speak of a quantum wire, and if the motion is confined in one coordinate only the mesoscopic system is named a quantum well in optics. Confined motion of the electrons in one direction can occur for instance if an ultrathin, i.e. a (few) monolayer thick, metal or semiconductor film is placed on top of a dielectric or semiconducting substrate. The linear optical reflection properties of these quantum well systems also have been studied experimentally [26, 27] and theoretically [28, 29]. If, instead of a single layer, a periodic array of quantum wells are produced one has a so-called multi-quantum well structure, and if in addition the wave functions of the various wells are allowed to overlap the rigorous confinement is more or less invalidated. A recent review of the linear optical properties of these mesoscopic media, called quantum lattices, and of the multi-quantum well structures has been given by Raj and Tilley [30]. Theoretical studies aiming at an understanding of the electrodynamics of superlattices on the microscopic level are presented in Refs. [31]-[33].

In the present work I shall focuse my attention on a theoretical study of the local field inside a mesoscopic medium, emphasizing the approach used by my group in recent years to calculate various linear and nonlinear optical properties of mesoscopic systems. I shall adopt the point of view that in order to understand on the microscopic level many of the electrodynamic properties of mesoscopic systems it is necessary to have a knowledge of the prevailing, local electromagnetic field inside the mesoscopic medium under investigation. The main effort will be devoted to studies of the nonlinear electrodynamics of the mesoscopic systems focusing, for simplicity, the analysis on second-order phenomena, viz. optical second-harmonic generation and the photon drag.

In section 2, the scattering theory formalism for the local field is described. Since a mesoscopic system usually in one way or another is in contact with a macrosystem I shall make use of so-called pseudo-vacuum propagators, i.e. electromagnetic Green's functions which in their propagation characteristics take into account the electrodynamic properties of the macrosystem. As examples of mesoscopic systems in contact with macrosystems one could mention for instance (i) small particles, possibly atoms and molecules, near surfaces [34]-[42], a subject of interest in its own right now a days not least in relation to studies in near-field optics [43]-[48], and (ii) microcrystallites and arrays of quantum dots [49]-[52] in a solid matrix. Although the explicit construction of the pseudo-vacuum propagator can be done for simple geometrical configurations only [53, 54] [which fortunately also are the most common ones used in the studies of the mesoscopic systems], the propagator formalism leads to a conceptually clear and in practice often adequate integral equation for the local field inside the mesoscopic system. In the integral-equation formalism the local field is driven by the background field i.e. in the linear case the electromagnetic field prevailing in space if the mesoscopic system is absent and in the nonlinear case the forced part of the field stemming from the radiation accompanying the forced part of the nonlinear current density in the mesoscopic medium. The kernel of the integral equation in a compact notation gives an account of the two fundamental types of nonlocality, i.e. the electromagnetic (through the propagation of the field in space) and the electronic (through the spatial extension of the material wave function). To exemplify in a simple manner the basic principles the integral equation formalism is applied to a study of systems exhibiting infinitesimal translational invariance in time and space.

In section 3, the spatial structure of the microscopic linear conductivity response tensor is analyzed. Since I have limited myself to investigations of the nonlinear response of mesoscopic systems within the framework of a parametric description, the linear response function plays a crucial role for a selfconsistent calculation of the second-harmonic as well as the *dc* local field. by means of a conventional density-matrix operator approach the tensor-product structure of the paramagnetic and diamagnetic parts of the conductivity tensor is analyzed. It is demonstrated and stressed that even in a many-body calculation the paramagnetic conductivity tensor can be separated into a sum of tensor products where the two vectors appearing in the individual products are functions of the observation and source coordinates, respectively. This fact together with the fact that the summation, which runs over all possible single-particle transitions, for a mesoscopic system can be limited to a few terms because of the smallness of the mesoscopic medium often make an explicit calculation of the local field feasible.

In cases where the paramagnetic interaction dominates a separation techniques known also from analysis of the electrodynamics of quasi-two-dimensional electron systems [55], of the eigenmodes of quantum lattices [31, 33], and of local-field phenomena attached to the electrodynamics of an atom interacting with a normal or phaseconjugating mirror [56] can be adopted. Section 4 is terminated by a discussion of the eigenmode condition for a mesoscopic system and a treatment of the moment expansion of the so-called field-field response tensor. The moment expansion leads to the well known multipole (electric dipole, magnetic dipole, electric quadropole, etc.) picture of the radiation from the sources inside the mesoscopic medium [56, 57]. It is stressed that the multipole expansion, often used incorrectly when studying for instance the radiation associated with the electronic transition between highly delocalized states (Bloch-like, for instance), has a rigorous basis in all cases if the local field is eliminated in favour of the external field before making the moment expansion, the reason being that the external field in the optical wavelength range is slowly varying across the mesoscopic domain in contrast to the local field. In some cases, e.g. for thin films excited by s-polarized light [28], also the local field can be almost constant across the mesoscopic system.

In section 5 and the subsequent sections particular attention is paid to studies of local field phenomena associated with second-order nonlinear effects in mesoscopic systems. For small metal particles and for droplets the second [58] and third [59]-[61] order optical nonlinearities have been investigated in recent years. The photon-drag effect has been studied in two-dimensional electron gas systems [62, 63], in semiconductor heterostructures with emphasis on the intersubband contribution [64], and in single-level metallic quantum well with special reference to the role of the local-field effects [65]. The nonlinear optics of semiconductor quantum wells has been reviewed by Abram [66], and with main emphasis on the paramagnetic contribution the optical second-harmonic generation from niobium quantum wells is analysed in this volume [67].

Using the density-matrix-operator approach the nonlinear response tensors responsible for the optical rectification process and the second-harmonic generation are calculated and discussed in section 5. The even-order nonlinear field-matter interactions can cause a net transfer of momentum from the light to the conduction electrons of a metal or semiconductor and thus generate a *dc* current in the particle system. To investigate the role of local-field effects for this so-called photon-drag effect in mesoscopic systems, in section 6 we investigate the drag phenomenon in a one-level metallic quantum well. I have limited myself to systems with a single bound state to stress the conceptual aspects of the problem. Subsequent to calculations of the photon-drag response tensor and the forced part of the nonlinear *dc* current density, the prevailing local *dc* fields and current densities are obtained in a selfconsistent manner. Since a selfconsistent calculation of the local field at the fundamental frequency is so important for the prediction of the magnitude of the photon-drag effect, a detailed study of this field also is undertaken in section 6, paying special attention to the local-field resonances. A detailed discussion of the solution of the basic integral equation for the local field at the fundamental frequency is given for two simple models of conceptual and practical interest. In the so-called slave model [29] the electron dynamics perpendicular to the quantum well is slaved by the dynamics along the well, and in the self-field model, only the (local) self-field part of the electromagnetic propagator is kept. In section 7, the optical second-harmonic generation in mesoscopic systems is treated. Taking as a starting point the microscopic expression for the forced current density the moment expansion of the nonlinear background response tensor is presented, and the nonlinear background field calculated. The present article is terminated by a heuristic discussion of the second-harmonic generation from a paramagnetic two-level metallic quantum well. Adopting the random-phase-approximation model and the electromagnetic self-field approximation the local field at the second-harmonic frequency is obtained and the local-field resonances located. Finally, on the basis of an explicit calculation of the field-field response tensor, the driven part of the nonlinear current density is obtained in the local limit and the associated background field determined.

2 SCATTERING THEORY FORMALISM FOR THE LOCAL FIELD

In the present section we shall establish an integral equation formalism for the local electric field of central importance for our studies of nonlocal phenomena associated with the nonlinear electrodynamics of mesoscopic systems. As we shall realize, this formalism provides us with an intuitive and appealing picture of the basic features of the underlying physics. As a starting point for our approach we take the microscopic Maxwell-Lorentz equations for the electromagnetic field, i.e.,

$$\vec{\nabla} \times \vec{E}(\vec{r},t) = -\frac{\partial \vec{B}(\vec{r},t)}{\partial t} , \tag{1}$$

$$\vec{\nabla} \times \vec{B}(\vec{r},t) = \mu_0 \vec{J}(\vec{r},t) + \frac{1}{c_0^2}\frac{\partial \vec{E}(\vec{r},t)}{\partial t} , \tag{2}$$

$$\vec{\nabla} \cdot \vec{E}(\vec{r},t) = \frac{1}{\epsilon_0}\rho(\vec{r},t) , \tag{3}$$

$$\vec{\nabla} \cdot \vec{B}(\vec{r},t) = 0 \tag{4}$$

in standard notation. By combining Eqs. (1) and (2) one obtains in dyadic notation the usual wave equation for the electric field, $\vec{E}(\vec{r},t)$, namely

$$\left[\overleftrightarrow{U}(\nabla^2 - \frac{1}{c_0^2}\frac{\partial^2}{\partial t^2}) - \vec{\nabla}\vec{\nabla}\right] \cdot \vec{E}(\vec{r},t) = \mu_0 \frac{\partial \vec{J}(\vec{r},t)}{\partial t} , \tag{5}$$

where $\overleftrightarrow{U}$ is the unit dyade of dimension 3 × 3.

2.1 Source and System Dynamics

Let us now divide the total current density, $\vec{J}(\vec{r},t)$, into two parts, viz.,

$$\vec{J}(\vec{r},t) = \vec{J}_{ext}(\vec{r},t) + \vec{J}_{syst}(\vec{r},t) \ . \tag{6}$$

The so-called external part, $\vec{J}_{ext}(\vec{r},t)$, stems from a charge density distribution, $\rho_{ext}(\vec{r},t)$, which we shall identify as the source distribution driving the coupled system of particles and fields. The part $\vec{J}_{syst}(\vec{r},t)$, represents the current density distribution induced in our system by the source field. As a part of the basic framework of our theory we shall assume that the source dynamics develops independently of the induced system dynamics (see Fig. 1). In other words we shall assume that the external electric, $\vec{J}_{ext}(\vec{r},t)$, and magnetic, $\vec{B}_{ext}(\vec{r},t)$, fields satisfy the Maxwell-Lorentz equations in Eqs. (1)-(4) if $\vec{J}(\vec{r},t)$ and $\rho(\vec{r},t) = \rho_{ext}(\vec{r},t) + \rho_{syst}(\vec{r},t)$ are replaced by $\vec{J}_{ext}(\vec{r},t)$ and $\rho_{ext}(\vec{r},t)$. In the context of nonlinear optics one normally identifies the external current density as the forced part of the nonlinear current density, i.e. the part driven by the fundamental field. Essentially, this means that our approach is limited to the parametric regime of dynamics. It is possible, and often necessary, to lift this restriction. However, to emphasize simple conceptual aspects associated with local-field effects we shall sustain from doing so in this work.

2.2 Electromagnetic Pseudo-Vacuum Propagator

The scattering theory formalism for the local field is based on the introduction of the electromagnetic propagator concept. Conventionally, one makes use of the well known dyadic vacuum propagator (Green's function) $\overleftrightarrow{G}_0 (\vec{r}-\vec{r},t-t')$ which is the solution to the "delta-source" wave equation

$$\left[\overleftrightarrow{U} (\nabla^2 - \frac{1}{c_0^2}\frac{\partial^2}{\partial t^2}) - \vec{\nabla}\vec{\nabla}\right] \cdot \overleftrightarrow{G}_0 (\vec{r}-\vec{r}',t-t') = \overleftrightarrow{U} \delta(\vec{r}-\vec{r})\delta(t-t') \tag{7}$$

satisfying the Silver-Müller radiative boundary conditions at infinity. However, in a number of studies, e.g. those on quantum wells, it is convenient to introduce so-called pseudo-vacuum propagators. By a pseudo-vacuum propagator we mean an electromagnetic propagator which in its propagation characteristics incorporate a part of the electromagnetic response of the system under study. Different kinds of pseudo-vacuum propagators can be introduced depending on what is appropriate in the actual case. As a general guideline for the introduction of a pseudo-vacuum propagator one looks for the possibilities of dividing the system current density into two parts, i.e. a part which is associated either with a particular "geographical" subpart of the system or with the response one would obtain in a simple model calculation, and a part which contains the interesting (and difficult) problem to be investigated. The use of pseudo-vacuum propagators instead of vacuum propagators often also leads to the advantage that the resulting integral equation for the effective local field will be easier to solve. As an example, the use of pseudo-vacuum propagators can narrow the spatial domain over which a selfconsistent numerical calculation has to be carried out, cf. Refs. [53,54],[68]-[70].

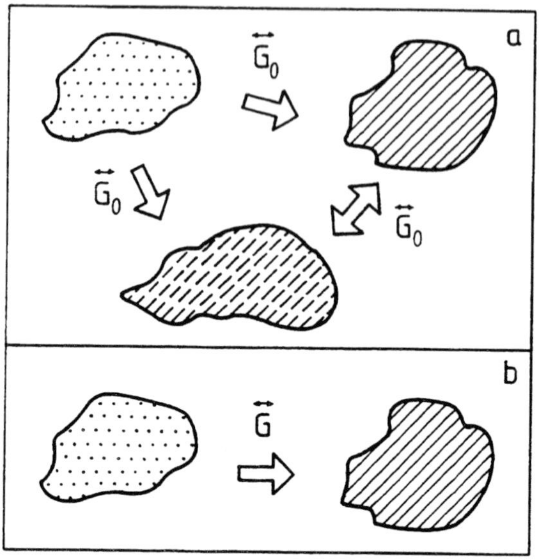

Figure 1. *Schematic diagrams displaying the division of the total current density into parts belonging to the source region (dotted domain), the mesoscopic system (hatched domain), and the macrosystem in contact with the mesoscopic medium under consideration (dashed domain). As indicated by the arrows it is assumed that the source (external) dynamics develops independently of that of the meso and macrosystems. The different parts are assumed to be coupled only electromagnetically, and the interaction is described either in terms of the vacuum propagator, $\overleftrightarrow{G}_0$ (Fig. a) or, if the pseudo-vacuum (macro system) dynamics is induced in a screened Green's function, by means of the pseudo-vacuum propagator $\overleftrightarrow{G}$ (Fig. b).*

In the view of the considerations above the current density induced in our system is divided as follows:

$$\vec{J}_{syst}(\vec{r},t) = \vec{J}_{pseudo}(\vec{r},t) + \vec{J}_{ind}(\vec{r},t) \ , \tag{8}$$

where $\vec{J}_{pseudo}(\vec{r},t)$ represents the response to be included in the pseudo-vacuum propagator characteristics and $\vec{J}_{ind}(\vec{r},t)$ is the remaining part of the induced system response. Since we are going to develop a response formalism based on the use of pseudo-vaccuum propagators we have stressed by the subscript *ind* that $\vec{J}_{ind}(\vec{r},t)$ represents the response induced by the external field distribution prevailing in a pseudo-vacuum in which the field propagation is described by the pseudo-vacuum propagator (Figs. 1 and 2). By assuming that the pseudo-vacuum current density is linearly, but possibly nonlocally in time and space, related to the prevailing electric field, $\vec{E}(\vec{r},t)$, one has

$$\vec{J}_{pseudo}(\vec{r},t) = \int \overleftrightarrow{\sigma}_{pseudo}\ (\vec{r},\vec{r}\,',t,t') \cdot \vec{E}(\vec{r}\,',t')d^3r'dt' \ , \tag{9}$$

where $\overleftrightarrow{\sigma}_{pseudo}(\vec{r},\vec{r}\,',t,t')$ is the pseudo-vacuum conductivity tensor. By combining Eqs. (5), (6), (8), and (9) the wave equation for the prevailing electric field takes the form

$$\left[\overleftrightarrow{U}(\nabla^2 - \frac{1}{c_0^2}\frac{\partial^2}{\partial t^2}) - \vec{\nabla}\vec{\nabla}\right]\cdot\vec{E}(\vec{r},t) - \mu_0\frac{\partial}{\partial t}\int\overleftrightarrow{\sigma}_{pseudo}(\vec{r},\vec{r}\,',t,t')\cdot\vec{E}(\vec{r}\,',t')d^3r'dt'$$

$$= \mu_0\frac{\partial}{\partial t}(\vec{J}_{ext}(\vec{r},t) + \vec{J}_{ind}(\vec{r},t)) \,. \tag{10}$$

The pseudo-vacuum dyadic electromagnetic propagator, denoted by $\overleftrightarrow{G}(\vec{r},\vec{r}\,',t,t')$, is introduced now as the solution to the integro-differential equation

$$\left[\overleftrightarrow{U}(\nabla^2 - \frac{1}{c_0^2}\frac{\partial^2}{\partial t^2}) - \vec{\nabla}\vec{\nabla}\right]\cdot\overleftrightarrow{G}(\vec{r},\vec{r}\,',t,t')$$

$$-\mu_0\frac{\partial}{\partial t}\int\overleftrightarrow{\sigma}_{pseudo}(\vec{r},\vec{r}\,'',t,t'')\cdot\overleftrightarrow{G}(\vec{r}\,'',\vec{r}\,',t'',t')d^3r''dt'' = \overleftrightarrow{U}\delta(\vec{r}-\vec{r}\,')\delta(t-t')$$

$$\tag{11}$$

satisfying boundary conditions appropriate for the problem at hand. In terms of the pseudo-vacuum propagator the solution of Eq. (10) for the prevailing electric field, including the appropriate boundary conditions, is given by

$$\vec{E}(\vec{r},t) = \mu_0\int\overleftrightarrow{G}(\vec{r},\vec{r}\,',t,t')\cdot\frac{\partial}{\partial t'}(\vec{J}_{ext}(\vec{r}\,',t') + \vec{J}_{ind}(\vec{r}\,',t'))d^3r'dt' \,. \tag{12}$$

2.3 Basic Integral Equation for the Local Field

Let us suppose that we have succeeded in constructing the pseudo-vacuum propagator, $\overleftrightarrow{G}(\vec{r},\vec{r}\,',t,t')$. Since also the external current density distribution, $\vec{J}_{ext}(\vec{r}\,',t')$, is assumed to be known, it follows that the so-called background (B) field

$$\vec{E}_B(\vec{r},t) = \mu_0\int\overleftrightarrow{G}(\vec{r},\vec{r}\,',t,t')\cdot\frac{\partial\vec{J}_{ext}(\vec{r}\,',t')}{\partial t'}d^3r'dt' \,, \tag{13}$$

can be considered as a prescribed quantity. From a physical point of view the background field is the field that would prevail everywhere in space if the induced current density $\vec{J}(\vec{r},t)$ was absent. The form of $\vec{E}_B(\vec{r},t)$ as appearing from Eq. (13) has an obvious physical meaning. Thus, from every infinitesimal time-space volume element $d^3r'dt'$ centered at $(\vec{r}\,',t')$ inside the source distribution a field having an associated source strenght $\left[\partial\vec{J}_{ext}(\vec{r}\,',t')/\partial t'\right]d^3r'dt'$ is radiated out in the pseudo-vacuum. The field propagation in the pseudo-vacuum is described in terms of the pseudo-vacuum propagator $\overleftrightarrow{G}(\vec{r},\vec{r}\,',t,t')$ relating, essentially, the field at $(\vec{r},t)$ to the current density fluctuations $\partial\vec{J}_{ext}(\vec{r}\,',t')/\partial t'$ at $(\vec{r}\,',t')$.

In the following it is assumed that also the induced current density and the local electric field are related via a linear and nonlocal constitutive equation, viz.,

$$\vec{J}_{ind}(\vec{r},t) = \int\overleftrightarrow{\sigma}(\vec{r},\vec{r}\,',t,t')\cdot\vec{E}(\vec{r}\,',t')d^3r'dt' \,, \tag{14}$$

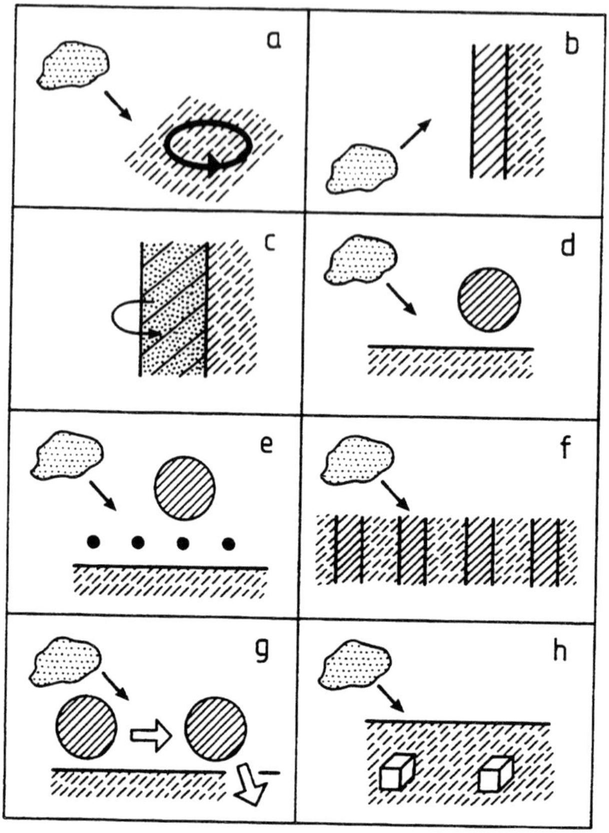

Figure 2. *Various examples of mesoscopic systems of interest for local-field studies in linear and nonlinear optics. (a). A micron-sized metallic ring embedded in a solid. It is predicted that the second-harmonic generation from the ring will be accompanied by an induced dc current inside the ring. Provided the phase coherence length of the electron exceeds the circumference of the ring the current can flow virtually without attenuation, like in a superconductor. (b) and (c). An ultrathin metallic film (quantum well) placed on top of a substrate. In linear electrodynamics (Fig. b) the source region is located outside the quantum well whereas it lies inside the thin film in the nonlinear case (Fig. c). (d). A small spherical metal particle placed near a surface. Local-field formalisms can be used to study the surface-dressed linear and nonlinear polarizabilities of this mesoscopic object. (e). A mesoscopic system consisting of a small particle plus a number of point dipoles placed on top of a flat surface can be used to improve our understanding of optical near-field interactions (optical near-field microscopy). (f). Local-field effects are of importance also in multiple quantum well structures and quantum lattices. (g). Local-field calculations are important for understanding the electrodynamics of mesoscopic particles moving parallel to or penetrating*

a flat structure. An understanding of the electrodynamics of this kind of system is of value for instance for the description of the radiation accompanying electron tunnelling between a small tip and a surface. (h). Local-field formalisms can be used to describe the optical properties of a single quantum dot and the interaction between two (or more) dots in a solid matrix. In all figures the hatching of the various domains follows the convention of Fig.1.

where $\overleftrightarrow{\sigma}(\vec{r},\vec{r}\,',t,t')$ is the so-called conductivity response tensor of our system imbedded in the pseudo-vacuum.

By inserting Eqs. (13) and (14) into Eq. (12) it is realized that the local electric field is determined by the integral equation

$$\vec{E}(\vec{r},t) = \vec{E}_B(\vec{r},t) + \int \overleftrightarrow{K}(\vec{r},\vec{r}\,',t,t') \cdot \vec{E}(\vec{r}\,',t')d^3r'dt' \ , \tag{15}$$

the kernel $\overleftrightarrow{K}(\vec{r},\vec{r}',t,t')$ being given by

$$\overleftrightarrow{K}(\vec{r},\vec{r}',t,t') = \mu_0 \int \overleftrightarrow{G}(\vec{r},\vec{r}'',t,t'') \cdot \frac{\partial}{\partial t''} \overleftrightarrow{\sigma}(\vec{r}'',\vec{r}',t'',t')d^3r''dt'' \ . \tag{16}$$

The physical interpretation of the content of the integral equation in (15) with the kernel given by Eq. (16) is as follows: The local field at a given time-space point $(\vec{r},t)$ is the sum of the background field at this point and the field radiated from the various points of the moving charge distribution (current density $\vec{J}_{ind}(\vec{r}\,'',t''))$ of our system. The strength of the radiation from the induced charge motion is determined in a selfconsistent manner by the prevailing field, hence the integral equation for the local field.

In the following sections we shall limit ourselves to local-field calculations in systems exhibiting translational invariance in time. For such systems one has $\overleftrightarrow{G}(\vec{r},\vec{r}'',t,t'') = \overleftrightarrow{G}(\vec{r},\vec{r}'',t-t'')$ and $\overleftrightarrow{\sigma}(\vec{r}'',\vec{r}',t'',t') = \overleftrightarrow{\sigma}(\vec{r}'',\vec{r}',t''-t')$ so that a Fourier transformation in time can be used to simplify Eqs. (15) and (16). Thus, in Fourier-($\Omega-$) space the basic integral equation for the local field takes the form

$$\vec{E}(\vec{r};\Omega) = \vec{E}_B(\vec{r};\Omega) + \int \overleftrightarrow{K}(\vec{r},\vec{r}';\Omega) \cdot \vec{E}(\vec{r}';\Omega)d^3r' \tag{17}$$

with a kernel given by

$$\overleftrightarrow{K}(\vec{r},\vec{r}';\Omega) = -i\mu_0\Omega \int \overleftrightarrow{G}(\vec{r},\vec{r}'';\Omega) \cdot \overleftrightarrow{\sigma}(\vec{r}'',\vec{r}';\Omega)d^3r'' \ . \tag{18}$$

2.4 Electronic and Electromagnetic Nonlocality

Formally, the integral equation in (15) has the solution

$$\vec{E}(\vec{r},t) = \int \overleftrightarrow{\Gamma}(\vec{r},\vec{r}',t,t') \cdot \vec{E}_B(\vec{r}',t')d^3r'dt' \ , \tag{19}$$

where the nonlocal field-field response function $\overleftrightarrow{\Gamma}(\vec{r},\vec{r}',t,t')$ has to be obtained from the integral equation

$$\overleftrightarrow{\Gamma}(\vec{r},\vec{r}',t,t') = \overleftrightarrow{U}\,\delta(\vec{r}-\vec{r}')\delta(t-t')$$

$$+\int \overleftrightarrow{K}(\vec{r},\vec{r}'',t,t'')\cdot \overleftrightarrow{\Gamma}(\vec{r}'',\vec{r}',t'',t')d^3r''dt'' \ . \tag{20}$$

For systems with underlying translational invariance in time, Eqs. (19) and (20) are simplified to

$$\vec{E}(\vec{r};\Omega) = \int \overleftrightarrow{\Gamma}(\vec{r},\vec{r}';\Omega)\cdot \vec{E}_B(\vec{r}';\Omega)d^3r' \ , \tag{21}$$

and

$$\overleftrightarrow{\Gamma}(\vec{r},\vec{r}';\Omega) = \overleftrightarrow{U}\delta(\vec{r}-\vec{r}') + \int \overleftrightarrow{K}(\vec{r},\vec{r}'';\Omega)\cdot \overleftrightarrow{\Gamma}(\vec{r}'',\vec{r}';\Omega)d^3r'' \ . \tag{22}$$

If one inserts Eq. (19) into Eq. (14) one obtains the following nonlocal constitutive equation between the induced current density $\vec{J}_{ind}(\vec{r},t)$ and the prescribed background field $\vec{E}_B(\vec{r}',t')$:

$$\vec{J}_{ind}(\vec{r},t) = \int \overleftrightarrow{\sigma}_B(\vec{r},\vec{r}',t,t')\cdot \vec{E}_B(\vec{r}',t')d^3r'dt'' \ , \tag{23}$$

where the so-called background conductivity response tensor is given by

$$\overleftrightarrow{\sigma}_B(\vec{r},\vec{r}',t,t') = \int \overleftrightarrow{\sigma}(\vec{r},\vec{r}'',t,t'')\cdot \overleftrightarrow{\Gamma}(\vec{r}'',\vec{r}',t'',t')d^3r''dt'' \ . \tag{24}$$

The background conductivity response tensor is of conceptual interest because in a compact form it allows us to determine the current density induced by the background field itself. Thus, both propagation and screening effects are included in $\overleftrightarrow{\sigma}_B(\vec{r},\vec{r}',t,t')$. In practice it is normally a difficult (if not impossible) task to establish an explicit expression for $\overleftrightarrow{\sigma}_B(\vec{r},\vec{r}',t,t')$. As one would expect a nonlocal (in time and space) relation exists between $\vec{J}_{ind}$ and $\vec{E}_B$. What one should stress, however, is the fact that this nonlocality is composed of both "electronic" and "electromagnetic" nonlocality (Fig. 3). The electronic nonlocality, basically associated with the

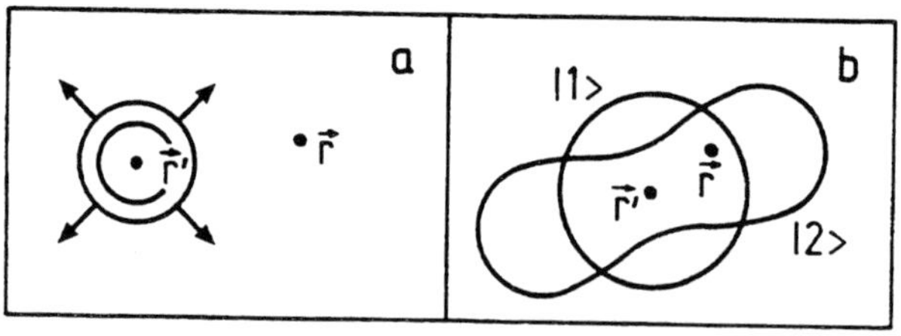

Figure 3. *Schematic illustration showing the nonlocal coupling between a source ($\vec{r}'$) and observation ($\vec{r}$) point. In (a), an outgoing electromagnetic wave is responsible for the interaction (electromagnetic nonlocality), and in (b), optical excitation (de-excitation) of the particle system between the orbitals $\mid 1>$ and $\mid 2>$ effects the coupling (electronic nonlocality).*

nonlocal character of the wave functions, appears in $\overleftrightarrow{\sigma}(\vec{r},\vec{r}',t,t')$, and the electromagnetic nonlocality, stemming from the electromagnetic radiation process, appears in $\overleftrightarrow{G}(\vec{r},\vec{r}',t,t')$ and thus is hidden in $\overleftrightarrow{\Gamma}(\vec{r},\vec{r}',t,t')$. As long as $\overleftrightarrow{G}(\vec{r},\vec{r}',t,t')$ is a real pseudo-vacuum propagator (and not just $\overleftrightarrow{G_0}$) it should be emphasized that electronic nonlocality also appears in $\overleftrightarrow{G}(\vec{r},\vec{r}',t,t')$ via the electronic screening effects incorporated in the propagation of the field in the pseudo-vacuum.

2.5 Local-Field Resonances

The background field gives rise to excitations in our system placed in the pseudo-vacuum. These excitations are expected to be pronounced when they occur under local-field resonance conditions. The condition for having a local-field resonance in the system is that a local field can exist in the system in the absence of the background field (Fig. 4). Hence, the resonance field, $\vec{E}_{res}(\vec{r},t)$, has to satisfy the homogeneous integral equation

$$\int \left[\overleftrightarrow{U}\delta(\vec{r}-\vec{r}')\delta(t-t') - \overleftrightarrow{K}(\vec{r},\vec{r}',t,t')\right] \cdot \vec{E}_{res}(\vec{r}',t')d^3r'dt' = 0 , \tag{25}$$

cf. Eq. (15). For time-translationally invariant systems Eq. (25) is equivalent to

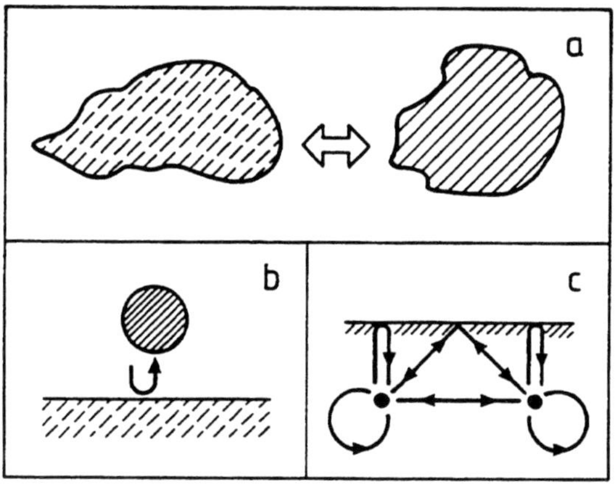

Figure 4. *The local-field resonances of a mesoscopic system dressed by a macrosystem are equivalent to the selfsustaining oscillations (eigenmodes) of the coupled meso-macro system (compare (a) and Fig.1). (b). The determination of the local-field resonances of a surface-dressed mesoscopic particle is of importance for our understanding of the electrodynamics of particles near (or on) surfaces. (c). As indicated by the arrows many channels of electromagnetic propagation contribute if one aims at calculating the local-field resonances of two coupled quantum dots placed near a flat surface inside a solid.*

$$\int \left[\overleftrightarrow{U}\,\delta(\vec{r}-\vec{r}\,')-\overleftrightarrow{K}(\vec{r},\vec{r}\,';\Omega)\right]\cdot \vec{E}_{res}(\vec{r}\,';\Omega)d^3r'=0\ . \tag{26}$$

After having studied in Section 3 the structure of the kernel $\overleftrightarrow{K}(\vec{r},\vec{r}\,';\Omega)$ we shall in Section 4 demonstrate how the resonance condition in Eq. (26) can be expressed solely in terms of the properties of the (known) kernel $\overleftrightarrow{K}(\vec{r},\vec{r}\,';\Omega)$.

2.6 Time and Space Translationally Invariant Systems

The concepts discussed in the subsections 2.2-2.5 take a particular simple form in systems exhibiting translational invariance in both time and space. Although the mesoscopic systems, towards which the discussion of the local-field effects in this article is aimed, do not in general exhibit translational invariance against displacements in all three Cartesian directions in space it is instructive from a heuristic point of view to consider systems possessing four-dimensional translational invariance.

Denoting the spatial Fourier vector (wave vector) by $\vec{Q}$, in a four-dimensional translationally invariant pseudo-vacuum the Fourier transformed electromagnetic propagator $\overleftrightarrow{G}(\vec{Q},\Omega)$ has to satisfy the algebraic dyadic equation

$$\left\{\overleftrightarrow{U}\left[(\frac{\Omega}{c_0})^2-Q^2\right]+\vec{Q}\vec{Q}+i\mu_0\Omega\,\overleftrightarrow{\sigma}_{pseudo}(\vec{Q},\Omega)\right\}\cdot\overleftrightarrow{G}(\vec{Q},\Omega)=\overleftrightarrow{U}\ . \tag{27}$$

By introducing the transverse ($\sigma^T_{pseudo}(Q,\Omega)$) and longitudinal ($\sigma^L_{pseudo}(Q,\Omega)$) pseudo-vacuum conductivity response functions, related to $\overleftrightarrow{\sigma}_{pseudo}(\vec{Q},\Omega)$ via

$$\overleftrightarrow{\sigma}_{pseudo}(\vec{Q},\Omega)=\left(\overleftrightarrow{U}-\frac{\vec{Q}\vec{Q}}{Q^2}\right)\sigma^T_{pseudo}(Q,\Omega)+\frac{\vec{Q}\vec{Q}}{Q^2}\sigma^L_{pseudo}(Q,\Omega)\ , \tag{28}$$

the pseudo-vacuum propagator takes the explicit form

$$\overleftrightarrow{G}(\vec{Q},\Omega)=\frac{\overleftrightarrow{U}-\frac{\vec{Q}\vec{Q}}{Q^2}}{N^T_{pseudo}(Q,\Omega)}+\frac{\frac{\vec{Q}\vec{Q}}{Q^2}}{N^L_{pseudo}(Q,\Omega)}\ , \tag{29}$$

where

$$N^T_{pseudo}(Q,\Omega)=\left(\frac{\Omega}{c_0}\right)^2\left[1+\frac{i\sigma^T_{pseudo}(Q,\Omega)}{\epsilon_0\Omega}\right]-Q^2\ , \tag{30}$$

and

$$N^L_{pseudo}(Q,\Omega)=(\frac{\Omega}{c_0})^2\left[1+\frac{i\sigma^L_{pseudo}(Q,\Omega)}{\epsilon_0\Omega}\right]\ . \tag{31}$$

The transverse and longitudinal dispersion relations of the pseudo-vacuum are determined by the conditions $N^T_{pseudo}(Q,\Omega)=0$ and $N^L_{pseudo}(Q,\Omega)=0$, respectively.

In the four-dimensional Fourier space the selfconsistent equation for the local field takes the algebraic form

$$\vec{E}(\vec{Q},\Omega)=\vec{E}_B(\vec{Q},\Omega)+\overleftrightarrow{K}(\vec{Q},\Omega)\cdot\vec{E}(\vec{Q},\Omega) \tag{32}$$

with a kernel given by

$$\overleftrightarrow{K}(\vec{Q},\Omega) = -i\mu_0\Omega \overleftrightarrow{G}(\vec{Q},\Omega)\cdot \overleftrightarrow{\sigma}(\vec{Q},\Omega)\ , \tag{33}$$

as one readily realizes from Eqs. (17) and (18). By combining Eqs. (32) and (33), the relation between the local field and background field becomes

$$\vec{E}(\vec{Q},\Omega) = \overleftrightarrow{\Gamma}(\vec{Q},\Omega)\cdot \vec{E}_B(\vec{Q},\Omega)\ , \tag{34}$$

where

$$\overleftrightarrow{\Gamma}(\vec{Q},\Omega) = \left[\overleftrightarrow{U} + i\mu_0\Omega \overleftrightarrow{G}(\vec{Q},\Omega)\cdot \overleftrightarrow{\sigma}(\vec{Q},\Omega)\right]^{-1}\ . \tag{35}$$

By introducing the transverse (T) and longitudinal (L) conductivity response functions $\sigma^T(Q,\Omega)$ and $\sigma^L(Q,\Omega)$ in the same manner as done for the pseudo-vacuum conductivity in Eq. (28), defining $N^T(Q,\Omega)$ and $N^L(Q,\Omega)$ in relation to $\sigma^T(Q,\Omega)$ and $\sigma^L(Q,\Omega)$ equivalently to the pseudo-vacuum relations in Eqs. (30) and (31), and by making use of the decomposition for $\overleftrightarrow{G}(\vec{Q},\Omega)$ given in Eq. (29), it can be shown by straightforward calculations that the relation in Eq. (34) can be written as follows:

$$\vec{E}(\vec{Q},\Omega) =$$

$$\left[\frac{\overleftrightarrow{U} - \frac{\vec{Q}\vec{Q}}{Q^2}}{1 + \frac{i\mu_0\Omega}{N^T_{pseudo}(Q,\Omega)N^T(Q,\Omega)}} + \frac{\frac{\vec{Q}\vec{Q}}{Q^2}}{1 + \frac{i\mu_0\Omega}{N^L_{pseudo}(Q,\Omega)N^L(Q,\Omega)}}\right]\cdot \vec{E}_B(\vec{Q},\Omega)\ . \tag{36}$$

As one would expect beforehand, it appears immediately from Eq. (36) that the relation between the local field and background field can be split into two independent ones among the transverse fields, $\vec{E}^T = (\overleftrightarrow{U} - \vec{Q}\vec{Q}/Q^2)\cdot\vec{E}$ and $\vec{E}^T_B = (\overleftrightarrow{U} - \vec{Q}\vec{Q}/Q^2)\cdot \vec{E}_B$, and the longitudinal fields, $\vec{E}^L = \vec{Q}\vec{Q}\cdot\vec{E}/Q^2$ and $\vec{E}^L_B = \vec{Q}\vec{Q}\cdot\vec{E}_B/Q^2$. The associated scalar functions $\Gamma^T = E^T/E^T_B$ and $\Gamma^L = E^L/E^L_B$ are given by

$$\Gamma^T(Q,\Omega) = 1 + \frac{i\mu_0\Omega}{N^T_{pseudo}(Q,\Omega)N^T(Q,\Omega)}\ , \tag{37}$$

and

$$\Gamma^L(Q,\Omega) = 1 + \frac{i\mu_0\Omega}{N^L_{pseudo}(Q,\Omega)N^L(Q,\Omega)}\ . \tag{38}$$

The background conductivity response given by the tensor

$$\overleftrightarrow{\sigma}_B(\vec{Q},\Omega) = \overleftrightarrow{\sigma}(\vec{Q},\Omega)\cdot \overleftrightarrow{\Gamma}(\vec{Q},\Omega) \tag{39}$$

can, using the decomposition form in Eq. (28), be characterized by the following transverse and longitidinal scalar quantities:

$$\sigma^T_B(Q,\Omega) = \sigma^T(Q,\Omega)\left[1 + \frac{i\mu_0\Omega}{N^T_{pseudo}(Q,\Omega)N^T(Q,\Omega)}\right]^{-1}\ , \tag{40}$$

and

$$\sigma_B^L(Q,\Omega) = \sigma^L(Q,\Omega)\left[1 + \frac{i\mu_0\Omega}{N_{pseudo}^L(Q,\Omega)N^L(Q,\Omega)}\right]^{-1} , \tag{41}$$

respectively.

For systems exhibiting four-dimensional translational invariance the resonant local field $\vec{E}_{res}(\vec{Q},\Omega)$ fulfils the dyadic algebraic equation

$$\left[\overleftrightarrow{U} - \overleftrightarrow{K}(\vec{Q},\Omega)\right] \cdot \vec{E}_{res}(\vec{Q},\Omega) = 0 . \tag{42}$$

It follows from this equation that the resonance condition for the Fourier transformed local field is

$$Det\left\{\overleftrightarrow{\Gamma}^{-1}(\vec{Q},\Omega)\right\} = 0 , \tag{43}$$

where $Det\{\cdots\}$ means the determinant of $\cdots$. Equation (43) is equivalent to

$$\left(\Gamma^T(Q,\Omega)\right)^{-2}\left(\Gamma^L(Q,\Omega)\right)^{-1} = 0 , \tag{44}$$

i.e. an equation which splits into the two separate conditions $N_{pseudo}^T(Q,\Omega)N^T(Q,\Omega) = 0$ (double degenerated), and $N_{pseudo}^L(Q,\Omega)N^L(Q,\Omega) = 0$. As one would expect the four dispersion relations $N_{pseudo}^T(Q,\Omega) = N_{pseudo}^L(Q,\Omega) = N^T(Q,\Omega) = N^L(Q,\Omega)$are thus precisely equivalent to the local-field resonance condition. It is not surprising that the resonances for the pseudo-vacuum and the remaining part of our system are uncoupled. If not, we would not in the first instance be able to make the decomposition of the system current density given in Eq. (8).

3 SPATIAL STRUCTURE OF THE LINEAR CONDUCTIVITY RESPONSE TENSOR

To make progress in the analysis of the basic integral equation for the local field (Eq. (17)) one has to investigate the spatial properties of the background field, $\vec{E}_B(\vec{r};\Omega)$, and the kernel, $\overleftrightarrow{K}(\vec{r},\vec{r}\,';\Omega)$. As far as the kernel is concerned this analysis involves an investigation of the spatial structure of the linear conductivity response tensor $\overleftrightarrow{\sigma}(\vec{r},\vec{r}\,';\Omega)$ and the electromagnetic pseudo-vacuum propagator $\overleftrightarrow{G}(\vec{r},\vec{r}\,';\Omega)$. The analysis of $\overleftrightarrow{G}(\vec{r},\vec{r}\,';\Omega)$, among other things, requires a study of the pseudo-vacuum conductivity response tensor, $\overleftrightarrow{\sigma}_{pseudo}(\vec{r},\vec{r}\,';\Omega)$. Since we have assumed the particles of the pseudo-vacuum and of the remaining part of the system to be electronically uncoupled, the analysis of the spatial structure of $\overleftrightarrow{\sigma}(\vec{r},\vec{r}\,';\Omega)$ and $\overleftrightarrow{\sigma}_{pseudo}(\vec{r},\vec{r}\,';\Omega)$ can be carried out along the same lines. To be specific, and to emphasize that mainly the spatial structure of $\overleftrightarrow{\sigma}(\vec{r},\vec{r}\,';\Omega)$ is of importance for the present conceptual discussion of local-field calculations in mesoscopic systems, the treatment of this sesction is devoted to an analysis of $\overleftrightarrow{\sigma}(\vec{r},\vec{r}\,';\Omega)$.

3.1 Density-Matrix Operator Approach

Our calculation of the linear conductivity response tensor proceeds along a well kvown path, but as we go along we focuse our attention especially on the spatial $\vec{r}$ and $\vec{r}\,'$ dependences of $\overleftrightarrow{\sigma}(\vec{r},\vec{r}\,';\Omega)$.

The conductivity tensor is obtained from a calculation of the induced current density, $\vec{J}_{ind}(\vec{r};\Omega)$, as a function of the selfconsistent field, $\vec{E}(\vec{r}\,';\Omega)$, cf. the Fourier analysed version of Eq. (14) for time invariant systems. The induced current density $\vec{J}_{ind}(\vec{r},t)$ is given in general by [71]-[73]

$$\vec{J}_{ind}(\vec{r},t) = Tr\left\{\rho\vec{\mathcal{J}}\right\} , \tag{45}$$

where ρ is the many-body density-matrix operator of our system, and $\vec{\mathcal{J}}$ is the intrinsic one-body electron current density operator. The operation $Tr\{\cdots\}$ stands for trace of$\cdots$. The density-matrix operator obeys the Liouville equation of motion, i.e. [71]-[73],

$$i\hbar\frac{\partial\rho}{\partial t} = [\mathcal{H},\rho] , \tag{46}$$

where

$$\mathcal{H} = \mathcal{H}_F + \mathcal{H}_I \tag{47}$$

is the Hamilton operator of the system, composed of a free (F) part $\mathcal{H}_F$, and a part $\mathcal{H}_I$ responsible for the interaction (I) with the selfconsistent electromagnetic field.

In the presence of a monochromatic selfconsistent field, described in a gauge where the time-dependent part of the scalar potential is zero by the vector potential

$$\vec{A}(\vec{r},t) = \frac{1}{2}\left[\vec{A}(\vec{r};\Omega)e^{-i\Omega t} + \vec{A}^*(\vec{r};\Omega)^{i\Omega t}\right] , \tag{48}$$

it is natural to make a series expansion

$$\rho = \rho_F + \rho_0 + \frac{1}{2}\sum_{n=1}^{\infty}\left[\rho_n e^{-in\Omega t} + \rho_n^\dagger e^{in\Omega t}\right] \tag{49}$$

of the many-body density-matrix operator. In Eq. (49), ρ_F and ρ_0 stand for the free (field unperturbed) and nonlinear *dc* parts of the density-matrix operator, respectively. The Fourier amplitudes of ρ and their Hermitian conjugate(†) ones are denoted by ρ_n and $\rho_n^\dagger$. Since, as we shall see in the following subsection, the interaction Hamiltonian is given by

$$\mathcal{H}_I = \mathcal{H}_0 + \frac{1}{2}\sum_{m=1}^{2}\left[\mathcal{H}_m e^{-im\Omega t} + \mathcal{H}_m^\dagger e^{im\Omega t}\right] , \tag{50}$$

insertion of the expansions in Eqs.(49) and (50) into Eq.(46), and linearization, lead to the following well known equation of motion for ρ_1:

$$\hbar\Omega\rho_1 = [\mathcal{H}_F,\rho_1] + [\mathcal{H}_1,\rho_F] . \tag{51}$$

To proceed our calculation we use the fact (see subsection (3.2)) that the one-body current density operator has the general form

$$\vec{\mathcal{J}}(\vec{r},t) = \vec{\mathcal{J}}_F(\vec{r}) + \frac{1}{2}\left[\vec{\mathcal{J}}_1(\vec{r};\Omega)e^{-i\Omega t} + \vec{\mathcal{J}}_1^\dagger(\vec{r};\Omega)e^{i\Omega t}\right] , \tag{52}$$

where $\vec{\mathcal{J}}_F(\vec{r})$ is its free part, and $\vec{\mathcal{J}}_1(\vec{r};\Omega)$ is the field-perturbed part. A field-perturbed contribution to $\vec{\mathcal{J}}$ is necessary to keep the description of the electrodynamic field-matter interaction gauge invariant. By making use of the series expansions for ρ (Eq.49) and $\vec{\mathcal{J}}$ (Eq.52), and linearization, the first-harmonic part of the induced current density is seen to be given by

$$\vec{J}_{ind}(\vec{r};\Omega) = Tr\left\{\rho_F \vec{\mathcal{J}}_1(\vec{r};\Omega)\right\} + Tr\left\{\rho_1 \vec{\mathcal{J}}_F(\vec{r})\right\} \tag{53}$$

The first term, $Tr\left\{\rho_F \vec{\mathcal{J}}_1(\vec{r};\Omega)\right\}$, on the right hand side of Eq.(53) is the so-called diamagnetic term. The diamagnetic contribution to $\vec{J}_{ind}$ will always be present in order to obtain a gauge invariant theory for the electrodynamic particle-field coupling [74, 75]. In a sense the diamagnetic term is structure independent or better phrased structureless because, as we shall realize, it is a function only of the particle density in the field undisturbed state. The paramagnetic term, on the other hand, in general is strongly structure dependent. As an example, when the electron system in a *BCS* superconductor makes a phase transition from the normal to the superconducting state the paramagnetic contribution to $\vec{J}_{ind}(\vec{r};\Omega)$ is strongly reduced (at the absolute zero in temperature it is even zero) so that only the diamagnetic term, responsible for the linear Meissner effect, survives [76, 77].

3.2 Many-Body Calculation; the Tensor-Product Structure of the Paramagnetic Conductivity

In the present subsection I shall summarize the many-body calculation of the linear conductivity response tensor, and, in a view of our special interest in the spatial structure of the conductivity, pay special attention to the tensor-product form of the paramagnetic term. Also, I shall stress the conceptual possibility of a tensor-product decomposition of the diamagnetic contribution to the conductivity, although such a decomposition normally seems of limited value for the local-field calculations on microscopic and mesoscopic few level systems. Even though in my opinion nonlinear spin interaction phenomena can be of significant importance for local-field effects in mesoscopic few-level systems, we shall limit ourselves here to a treatment of spin-independent interactions.

Thus, using second quantized notation as a starting point one takes the following minimal-coupling interaction Hamiltonian [78]:

$$\mathcal{H}_I(t) = \sum_{\vec{k},\vec{k}'}\left\{ < \vec{k}' \mid \frac{e}{2m}(\vec{p}\cdot\vec{A} + \vec{A}\cdot\vec{p}) + \frac{e^2}{2m}\vec{A}\cdot\vec{A} \mid \vec{k} > \sum_{\sigma} a^{\dagger}_{\vec{k}',\sigma} a_{\vec{k},\sigma} \right\}, \tag{54}$$

where $\mid k >$ and $\mid \vec{k}' >$ are one-electron stationary state vectors, in principle chosen in an arbitrary manner but in practice carefully selected to fit the problem at hand, cf. subsection 3.3. The $\vec{k}$ and $\vec{k}'$ summations each runs over the complete set of one-body eigenstates characterized by the quantum indices collectively denoted by $\vec{k}$ (and $\vec{k}'$). The spin index is denoted by σ, and $a_{\vec{k}\sigma}$ and $a^{\dagger}_{\vec{k}',\sigma}$ are the one-electron annihilation and creation operators belonging to the states $(\vec{k},\sigma)$ and $(\vec{k}',\sigma)$, respectively. Finally, $-e$, m, and $\vec{p}$ are the electron charge, mass, and momentum operator.

It is important for our local-field studies to calculate the interaction Hamiltonian in the $\vec{r}$-representation. Hence, by means of the usual Schrödinger wave functions

$$\psi_{\vec{k}}(\vec{r}) = <\vec{r} \mid \vec{k}> , \tag{55}$$

the three characteristic factors of the interaction Hamiltonian in Eq. (50) take the explicit forms

$$\mathcal{H}_1 = \sum_{\vec{k},\vec{k}'} \left\{ \left[-\int \vec{j}_{\vec{k}\to\vec{k}'}(\vec{r}) \cdot \vec{A}(\vec{r};\Omega) d^3r \right] \sum_\sigma a^\dagger_{\vec{k}',\sigma} a_{\vec{k},\sigma} \right\} , \tag{56}$$

$$\mathcal{H}_2 = \sum_{\vec{k},\vec{k}'} \left\{ \left[\frac{e^2}{4m} \int \psi^*_{\vec{k}'}(\vec{r})\psi_{\vec{k}}(\vec{r})\vec{A}(\vec{r};\Omega) \cdot \vec{A}(\vec{r};\Omega) d^3r \right] \sum_\sigma a^\dagger_{\vec{k}',\sigma} a_{\vec{k},\sigma} \right\} , \tag{57}$$

and

$$\mathcal{H}_0 = \sum_{\vec{k},\vec{k}'} \left\{ \left[\frac{e^2}{4m} \int \psi^*_{\vec{k}'}(\vec{r})\psi_{\vec{k}}(\vec{r})\vec{A}(\vec{r};\Omega) \cdot \vec{A}^*(\vec{r};\Omega) d^3r \right] \sum_\sigma a^\dagger_{\vec{k}',\sigma} a_{\vec{k},\sigma} \right\} , \tag{58}$$

where

$$\vec{j}_{\vec{k}\to\vec{k}'}(\vec{r}) = \frac{e\hbar}{2mi} \left[\psi_{\vec{k}}(\vec{r})\vec{\nabla}\psi^*_{\vec{k}'}(\vec{r}) - \psi^*_{\vec{k}'}(\vec{r})\vec{\nabla}\psi_{\vec{k}}(\vec{r}) \right] \tag{59}$$

is the one-electron transition current density from state $\mid \vec{k} >$ to state $\mid \vec{k}' >$. To calculate ρ_1 only the expression for $\mathcal{H}_1$ is needed. In order to determine the linear contribution to the induced current density one has to have explicit expressions for $\vec{\mathcal{J}}_F(\vec{r})$ and $\vec{\mathcal{J}}_1(\vec{r};\Omega)$. To this end, one begins with the second quantized expression for the current density operator, i.e.,

$$\vec{\mathcal{J}}(\vec{r},t) = \sum_{\vec{k},\vec{k}'} \left\{ <\vec{k}' \mid -\frac{e}{2m} [\vec{p}\delta(\vec{r}-\vec{r}_e) + \delta(\vec{r}-\vec{r}_e)\vec{p}] - \frac{e^2}{m}\vec{A}\delta(\vec{r}-\vec{r}_e) \mid \vec{k}> \right.$$

$$\left. \times \sum_\sigma a^\dagger_{\vec{k}',\sigma} a_{\vec{k},\sigma} \right\} , \tag{60}$$

where $\vec{r}_e$ is the electron position vector. The term containing $\vec{A}$ is responsible for the diamagnetic effect. In the $\vec{r}$-representation a calculation based on Eq. (60) leads to

$$\vec{\mathcal{J}}_F(\vec{r}) = \sum_{\vec{k},\vec{k}'} \left\{ \vec{j}_{\vec{k}\to\vec{k}'}(\vec{r}) \sum_\sigma a^\dagger_{\vec{k}',\sigma} a_{\vec{k},\sigma} \right\} , \tag{61}$$

and

$$\vec{\mathcal{J}}_1(\vec{r};\Omega) = -\frac{e^2}{m} \sum_{\vec{k},\vec{k}'} \left\{ \psi^*_{\vec{k}'}(\vec{r})\psi_{\vec{k}}(\vec{r})\vec{A}(\vec{r};\Omega) \sum_\sigma a^\dagger_{\vec{k}',\sigma} a_{\vec{k},\sigma} \right\} . \tag{62}$$

Before embarking on the final part of the many-body calculation of the linear conductivity response tensor, let us introduce the relevant quantities of the field-unperturbed many-body system. Thus, the eigenvalue problem for the unperturbed Hamilton operator reads

$$\vec{\mathcal{H}}_F \mid I >= E_I \mid I > , \tag{63}$$

where $\mid I >$ is the Ith many-body energy eigenstate and E_I the associated energy. By assuming the unperturbed system to be in thermal equilibrium furthermore one has

$$\rho_F \mid I >= P_I \mid I > , \tag{64}$$

where P_I is the normalized probability for finding the system in the many-body eigenstate $\mid I >$. In thermal equilibrium only the diagonal matrix elements of ρ_F are different from zero.

3.2.1 Diamagnetic Conductivity Response Tensor

The diamagnetic contribution to the induced current density according to Eq. (53) is given by

$$\vec{J}^{dia}_{ind}(\vec{r};\Omega) = \sum_{I,J} < I \mid \rho_F \mid J >< J \mid \vec{\mathcal{J}}_1(\vec{r};\Omega) \mid I > , \tag{65}$$

$\mid J >$ denoting the Jth many-body state of the field-unperturbed system. Utilizing Eq. (64) and the orthonormality of the many-body states, i.e. $< I \mid J >= \delta_{I,J}$ where $\delta_{I,J}$ is the Krönicker delta, Eq. (65) is reduced to

$$\vec{J}^{dia}_{ind}(\vec{r};\Omega) = \sum_{I} P_I < I \mid \vec{\mathcal{J}}_1(\vec{r};\Omega) \mid I > . \tag{66}$$

By inserting in this equation, the expression for $\vec{\mathcal{J}}_1(\vec{r};\Omega)$ (Eq. (62)), making use of the relation $\vec{E}(\vec{r};\Omega) = i\Omega\vec{A}(\vec{r};\Omega)$ (in the chosen gauge), and by comparing the resulting relation between $\vec{J}^{dia}_{ind}(\vec{r};\Omega)$ and $\vec{E}(\vec{r};\Omega)$ to the general form

$$\vec{J}^{dia}_{ind}(\vec{r};\Omega) = \int \overleftrightarrow{\sigma}^{dia}(\vec{r},\vec{r}\,';\Omega)\cdot\vec{E}(\vec{r}\,';\Omega)d^3r' , \tag{67}$$

it appears that the diamagnetic contribution $\overleftrightarrow{\sigma}^{dia}(\vec{r},\vec{r}\,';\Omega)$, to the conductivity response tensor is given by

$$\overleftrightarrow{\sigma}^{dia}(\vec{r},\vec{r}\,';\Omega)$$

$$= \frac{ie^2\overleftrightarrow{U}}{m\Omega}\sum_{\vec{k},\vec{k}'}\sum_{I}\left\{P_I\psi^*_{\vec{k}'}(\vec{r})\psi_{\vec{k}}(\vec{r}) < I \mid \sum_{\sigma} a^\dagger_{\vec{k}',\sigma}a_{\vec{k},\sigma} \mid I >\right\}\delta(\vec{r}-\vec{r}\,') . \tag{68}$$

The expression in Eq. (68) can be simplified by noting that $< I \mid a^\dagger_{\vec{k}',\sigma}a_{\vec{k},\sigma} \mid I >= 0$ for $\vec{k}' \neq \vec{k}$. If $\vec{k}' = \vec{k}$, $\mathcal{N}_{\vec{k}\sigma} \equiv a^\dagger_{\vec{k},\sigma}a_{\vec{k},\sigma}$ is just the number operator for particles in state $(\vec{k},\sigma)$ satisfying the eigenvalue equation

$$\mathcal{N}_{\vec{k},\sigma} \mid I >= N^I_{\vec{k},\sigma} \mid I > \tag{69}$$

with $N^I_{\vec{k},\sigma} = 1$ or 0 depending on whether the single-particle state $(\vec{k},\sigma)$ in $| I >$ is occupied or empty, respectively. By means of the considerations above the expression for $\overset{\leftrightarrow}{\sigma}^{dia}(\vec{r},\vec{r}\,';\Omega)$ is reduced to

$$\overset{\leftrightarrow}{\sigma}^{dia}(\vec{r},\vec{r}\,';\Omega) = \frac{ie^2}{m\Omega}\delta(\vec{r}-\vec{r}\,')\overset{\leftrightarrow}{U}\sum_{\vec{k},\sigma}\left\{| \psi_{\vec{k}}(\vec{r}) |^2 \sum_I N^I_{\vec{k},\sigma} P_I\right\} . \tag{70}$$

Finally, since

$$N_{\vec{k},\sigma} = \sum_I N^I_{\vec{k},\sigma} P_I \tag{71}$$

is the number of particles found in the single-particle state $(\vec{k},\sigma)$ when our system is in the mixed many-body state characterized via the P_I's, and since in turn

$$N(\vec{r}) = \sum_{k,\sigma} N_{\vec{k},\sigma} | \psi_{\vec{k}}(\vec{r}) |^2 \tag{72}$$

is the electron density at $\vec{r}$, the diamagnetic linear conductivity tensor takes the simple but general form

$$\overset{\leftrightarrow}{\sigma}^{dia}(\vec{r},\vec{r}\,';\Omega) = \frac{ie^2}{m\Omega}N(\vec{r})\delta(\vec{r}-\vec{r}\,')\overset{\leftrightarrow}{U} . \tag{73}$$

As mentioned earlier the diamagnetic conductivity response tensor is structureless in the sense that it depends only on the electron density distribution. Furthermore, $\overset{\leftrightarrow}{\sigma}^{dia}$ is isotropic. Finally, one notices that the diamagnetic response essentially is of *local* character meaning that the induced current density, $\vec{J}^{dia}_{ind}(\vec{r};\Omega)$, at $\vec{r}$ does not depend on the selfconsistent electric field $\vec{E}(\vec{r}\,';\Omega)$ in the neighbourhood of $\vec{r}$ but only on the field precisely at $\vec{r}$. In contrast to this we shall realize that the paramagnetic response is inherently *nonlocal*. In some cases, however, the paramagnetic response can appear as if it were local. This happens when the selfconsistent field (or in the case of $\overset{\leftrightarrow}{\sigma}_B$ the background field) essentially is constant across the range in space in which the response tensor is effectively different from zero. In these cases, one is faced with what I have called *hidden nonlocality*.

The universal character of the diamagnetic response stemming from the aforementioned gauge invariance also transpires from the fact that only an electron density different from zero at $\vec{r}$ is required to have a nonzero effect at this point. Also, the isotropy of the response points to its universal character.

3.2.2 Paramagnetic Conductivity Response Tensor

To obtain the paramagnetic contribution

$$\vec{J}^{par}_{ind}(\vec{r};\Omega) = \sum_{I,J} < I | \rho_1 | J >< J | \vec{\mathcal{J}}_F(\vec{r}) | I > \tag{74}$$

to the induced current density one starts by expressing the matrix element $< I | \rho_1 | J >$ in terms of $< I | \mathcal{H}_1 | J >$ using Eq. (51). Hence, one finds in view of the relations in Eqs. (63) and (64)

$$\vec{j}^{par}_{ind}(\vec{r};\Omega) = \sum_{I,J} \frac{P_J - P_I}{\hbar\Omega + E_J - E_I} < I \mid \mathcal{H}_1 \mid J >< J \mid \vec{\mathcal{J}}_F(\vec{r}) \mid I > \ . \tag{75}$$

By inserting in this equation the expressions for $\mathcal{H}_1$ and $\vec{\mathcal{J}}_F(\vec{r})$ taken from Eqs. (56) and (61), and $\vec{A}(\vec{r};\Omega) = \vec{E}(\vec{r};\Omega)/(i\Omega)$, one gets

$$\vec{j}^{par}_{ind}(\vec{r};\Omega) = \sum_{\substack{\vec{k},\vec{k}' \\ \vec{k}'',\vec{k}'''}} A^{\vec{k}'''\to\vec{k}''}_{\vec{k}'\to\vec{k}}(\Omega)\vec{j}_{\vec{k}'\to\vec{k}}(\vec{r}) \int \vec{j}_{\vec{k}'''\to\vec{k}''}(\vec{r}') \cdot \vec{E}(\vec{r}';\Omega)d^3r' \tag{76}$$

with the abbreviation

$$A^{\vec{k}'''\to\vec{k}''}_{\vec{k}'\to\vec{k}}(\Omega) = \frac{i}{\Omega}\sum_{I,J} \frac{P_J - P_I}{\hbar\Omega + E_J - E_I} < I \mid \sum_{\sigma'} a^{\dagger}_{\vec{k}'',\sigma'} a_{\vec{k}''',\sigma'} \mid J >$$

$$\times < J \mid \sum_{\sigma} a^{\dagger}_{\vec{k},\sigma} a_{\vec{k}',\sigma} \mid I > \ . \tag{77}$$

By comparing Eq. (76) and the general formula for the paramagnetic response, i.e.,

$$\vec{j}^{par}_{ind}(\vec{r};\Omega) = \int \overleftrightarrow{\sigma}^{par} (\vec{r},\vec{r}';\Omega) \cdot \vec{E}(\vec{r}';\Omega)d^3r' \ , \tag{78}$$

it is realized that the paramagnetic conductivity tensor is given by

$$\overleftrightarrow{\sigma}^{par} (\vec{r},\vec{r}';\Omega) = \sum_{\vec{k},\vec{k}'} \vec{j}_{\vec{k}'\to\vec{k}}(\vec{r})\vec{S}_{\vec{k}\to\vec{k}'}(\vec{r}';\Omega) \ , \tag{79}$$

where

$$\vec{S}_{\vec{k}\to\vec{k}'}(\vec{r}';\Omega) = \sum_{\vec{k}'',\vec{k}'''} A^{\vec{k}'''\to\vec{k}''}_{\vec{k}'\to\vec{k}}(\Omega)\vec{j}_{\vec{k}'''\to\vec{k}''}(\vec{r}') \ . \tag{80}$$

The result in Eq. (79) is the basic one for our purpose. It demonstrates that the paramagnetic conductivity tensor, which is a function of the two variables $\vec{r}$ and $\vec{r}'$ can be decomposed into a sum of tensor products. The vectors, $\vec{j}_{\vec{k}'\to\vec{k}}(\vec{r})$ and $\vec{S}_{\vec{k}\to\vec{k}'}(\vec{r}';\Omega)$, appearing in the individual tensor products are functions of $\vec{r}$ and $\vec{r}'$ alone, and the summation runs over all possible single-particle transitions.

The final expression for the conductivity response tensor which is obtained by adding the results of Eqs. (73) and (79) thus becomes, cf. Fig. 5

$$\overleftrightarrow{\sigma} (\vec{r},\vec{r}';\Omega) = \frac{ie^2}{m\Omega}N(\vec{r})\delta(\vec{r}-\vec{r}') \overleftrightarrow{U} + \sum_{\vec{k},\vec{k}'} \vec{j}_{\vec{k}'\to\vec{k}}(\vec{r})\vec{S}_{\vec{k}\to\vec{k}'}(\vec{r}';\Omega) \ . \tag{81}$$

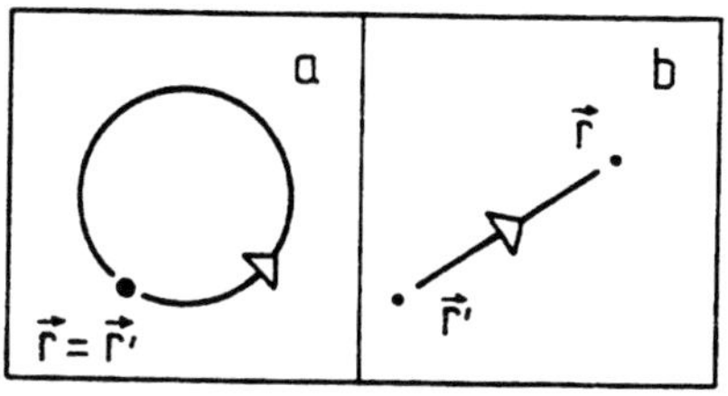

Figure 5. *Schematic diagrams showing the two basic contributions to the microscopic linear conductivity response tensor relating the current density at space point $\vec{r}$ to the electric field at $\vec{r}\,'$. (a). The local diamagnetic term. (b). The essentially nonlocal paramagnetic term.*

3.3 Random-Phase Approximation

For many purposes it is sufficient to calculate the conductivity under the assumption that the eigenstates of the many-electron Hamiltonian $\mathcal{H}_F$ consist of direct products of single-particle-like states. Within this so-called random-phase approximation (RPA) [78] the ground state $\mid G>$ is given by

$$\mid G>= \prod_{k<k_F,\sigma} a^\dagger_{\vec{k},\sigma} \mid 0> , \tag{82}$$

where $\mid 0>$ denotes the particle-vacuum state. The notation $k < k_F$ is meant to indicate that only the lowest lying energy eigenstates are filled (in a condensed-matter system k_F is the Fermi wave number). In the RPA-approach the excited eigenstates are of the particle-hole pair type, i.e.

$$\mid \vec{k}\,', \sigma'; \vec{k}, \sigma >= a^\dagger_{\vec{k}\,',\sigma'} a_{\vec{k},\sigma} \mid G> . \tag{83}$$

In turn, this means that $\vec{k}\,''' = \vec{k}$, $\vec{k}\,'' = \vec{k}\,'$, and $\sigma' = \sigma$ in Eq. (77). The energy difference $E_J - E_I$ must be equal to $E_J - E_I = \epsilon_{\vec{k}} - \epsilon_{\vec{k}\,'}$ where $\epsilon_{\vec{k}}$ and $\epsilon_{\vec{k}\,'}$ are the energies of the single-particle-like states $(\vec{k}, \sigma)$ and $(\vec{k}\,', \sigma)$ satisfying a single-particle Schrödinger equation

$$\left(-\frac{\hbar^2}{2m}\nabla^2 + V_{eff}\right) \psi_{\vec{k}}(\vec{r}) = \epsilon_{\vec{k}} \psi_{\vec{k}}(\vec{r}) , \tag{84}$$

the particles being subjected to an effective potential V_{eff}. Since $\mid J > \propto a^\dagger_{\vec{k},\sigma} a_{\vec{k}\,',\sigma} \mid I >$, one has $P_J - P_I = P_I^{\vec{k},\sigma}(1 - P_I^{\vec{k}\,',\sigma}) - P_I^{\vec{k}\,',\sigma}(1 - P_I^{\vec{k},\sigma}) = P_I^{\vec{k},\sigma} - P_I^{\vec{k}\,',\sigma}$ where $P_I^{\vec{\kappa},\sigma}$ is the probability that there *is* an electron in the one-electron level $(\vec{\kappa}, \sigma)$ of the many-body state $\mid I >$. Utilizing next that $\sum_I \left(P_I^{\vec{k},\sigma} - P_I^{\vec{k}\,',\sigma}\right) = f_{\vec{k}} - f_{\vec{k}\,'}$, where $f_{\vec{\kappa}}$ is the Fermi-Dirac distribution function, $f_{\vec{\kappa}} = \{exp[(\epsilon_{\vec{\kappa}} - \mu)/(kT)] + 1\}^{-1}$, one obtains in the RPA-calculation (the factor of 2 originating from the spin summation)

$$\mathcal{A}_{\vec{k}'\to\vec{k}}^{\vec{k}'''\to\vec{k}''}(\Omega) = \frac{2i}{\Omega}\frac{f_{\vec{k}} - f_{\vec{k}'}}{\hbar\Omega + \epsilon_{\vec{k}} - \epsilon_{\vec{k}'}}\delta_{\vec{k}''',\vec{k}}\delta_{\vec{k}'',\vec{k}'} , \tag{85}$$

and thus the standard result

$$\overleftrightarrow{\sigma}_{RPA}^{par}(\vec{r},\vec{r}';\Omega) = \frac{2i}{\Omega}\sum_{\vec{k},\vec{k}'}\frac{f_{\vec{k}} - f_{\vec{k}'}}{\hbar\Omega + \epsilon_{\vec{k}} - \epsilon_{\vec{k}'}}\vec{j}_{\vec{k}'\to\vec{k}}(\vec{r})\vec{j}_{\vec{k}\to\vec{k}'}(\vec{r}') . \tag{86}$$

The diamagnetic conductivity tensor in RPA still has the form given in Eq. (73), but with an electron density calculated from the formula

$$N_{RPA}(\vec{r}) = 2\sum_{\vec{k}} f_{\vec{k}} \mid \psi_{\vec{k}}(\vec{r}) \mid^2 . \tag{87}$$

Let me close this section by stressing once again that as long as the electrons of the pseudo-vacuum and of the remaining part of the system are electronically decoupled expressions can be set up for $\overleftrightarrow{\sigma}_{pseudo}$ which are identical in form to those obtained for $\overleftrightarrow{\sigma}$ in this section. If the two parts of the system are coupled, expressions of the form given here holds for the system as a whole.

4 LOCAL-FIELD CALCULATIONS IN MESOSCOPIC SYSTEMS

In section 2, the basic integral equation for the local electric field (Eq. 17) was established, and in section 3, the paramagnetic (Eq. 79) and diamagnetic (Eq. 73) contributions to the conductivity response tensor which, together with the electromagnetic pseudo-vacuum propagator constitute the structure of the dyadic kernel (Eq. 18) of the integral equation, were investigated. In the present section we start to discuss with main emphasis on the paramagnetic interaction, how to calculate the local field itself in mesoscopic systems where the optical response often can be obtained taking into account a few quantum levels only. We proceed with analyses of the field-field and background conductivity response tensors, and with the moment expansion of these response functions useful for slowly varying background fields. The section is terminated via a discussion of the resonance conditions for the local field.

4.1 Dominating Paramagnetic Interaction

Let us assume that the diamagnetic contribution to the local-field effects is neglible. In this case the basic integral equation for the local field takes the form

$$\vec{E}(\vec{r};\Omega) = \vec{E}_B(\vec{r};\Omega) + \sum_{\vec{k}\to\vec{k}'}\vec{F}_{\vec{k}'\to\vec{k}}(\vec{r};\Omega)\int\vec{S}_{\vec{k}\to\vec{k}'}(\vec{r}';\Omega)\cdot\vec{E}(\vec{r}';\Omega)d^3r' , \tag{88}$$

where

$$\vec{F}_{\vec{k}'\to\vec{k}}(\vec{r};\Omega) = -i\mu_0\Omega\int\overleftrightarrow{G}(\vec{r},\vec{r}'';\Omega)\cdot\vec{j}_{\vec{k}'\to\vec{k}}(\vec{r}'')d^3r'' . \tag{89}$$

Since it is assumed that the pseudo-vacuum propagator and the single-particle transition current density are known quantities, $\vec{F}_{\vec{k}'\to\vec{k}}(\vec{r};\Omega)$ is a known function of $\vec{r}$. In turn, this implies that the local field *inside the system* is equal to the background field plus a sum of contributions stemming from the individual single-particle transitions, $\vec{k}' \to \vec{k}$. The spatial form of these individual contributions is given by $\vec{F}_{\vec{k}'\to\vec{k}}(\vec{r};\Omega)$ and the strength by the yet unknown *number* $\int \vec{S}_{\vec{k}\to\vec{k}'}(\vec{r}';\Omega) \cdot \vec{E}(\vec{r}';\Omega)d^3r'$. The physical picture of this result is obvious. Thus, when our system is deexcited from the state $|\vec{k}'>$ to the state $|\vec{k}>$ a current density distribution $\vec{j}_{\vec{k}'\to\vec{k}}(\vec{r}'')d^3r''$ is generated. This current density distribution gives rise to a radiated field which at a given point $\vec{r}$ inside the system can be calculated by means of the pseudo-vacuum propagator $\overleftrightarrow{G}(\vec{r},\vec{r}'';\Omega)$. Apart from a proportionality factor, the total field radiated in relation to the electronic transition $\vec{k}' \to \vec{k}$ is obtained by integration over $\vec{r}''$, cf. Eq. (89). To obtain the strength of this radiation one needs to know the excitation amplitude for the reverse ($\vec{k} \to \vec{k}'$) transition. This amplitude depends on the prevailing local field and is given by $\int \vec{S}_{\vec{k}\to\vec{k}'}(\vec{r}';\Omega) \cdot \vec{E}(\vec{r}';\Omega)d^3r'$. For a given many-body system the excitation process is a complicated one involving in general a sum of many single-particle transitions $\vec{k}''' \to \vec{k}''$ (see Eq. (80)) each of an amplitude strength $\mathcal{A}^{\vec{k}'''\to\vec{k}''}_{\vec{k}'\to\vec{k}}(\Omega)$ given by the comprehensive expression in Eq. (77).

For brevity, we shall in the following leave out the superfluous reference to a given Ω from our notation, make the replacements $\vec{k} \Rightarrow m$ and $\vec{k}' \Rightarrow n$, and just write $\vec{k} \to \vec{k}'$ as mn. Each of the indices m and n of course stands for the complete set of quantum numbers needed (apart from the spin index) and used to characterize the given state. Introducing also

$$\beta_{mn} = \int \vec{S}_{mn}(\vec{r}) \cdot \vec{E}(\vec{r})d^3r \ , \tag{90}$$

Eq. (88) is written as follows:

$$\vec{E}(\vec{r}) = \vec{E}_B(\vec{r}) + \sum_{m,n} \beta_{mn}\vec{F}_{nm}(\vec{r}) \ . \tag{91}$$

Once the unknown constants β_{mn} have been determined, Eq. (91) constitutes the solution for the local field. To obtain the β_{mn}'s, the ansatz in Eq. (91) is inserted back into Eq. (90). This gives

$$\beta_{mn} - \sum_{o,p} N^{mn}_{po}\beta_{op} = H_{mn} \ , \tag{92}$$

where

$$N^{mn}_{po} = -i\mu_0\Omega \int \vec{S}_{mn}(\vec{r}) \cdot \overleftrightarrow{G}(\vec{r},\vec{r}') \cdot \vec{j}_{po}(\vec{r}')d^3r'd^3r \ , \tag{93}$$

and

$$H_{mn} = \int \vec{S}_{mn}(\vec{r}) \cdot \vec{E}_B(\vec{r})d^3r \ , \tag{94}$$

in principle are known numbers. Letting the indices m and n run through the different possibilities, Eq. (92) gives a set of inhomogeneous linear algebraic equations among the unknown β_{mn}'s. Conceptually, the set consists of infinitely many equations (since the complete set of single-particle eigenstates contains an infinite number of states), but for a mesoscopic system it is often sufficient to take into account only a few single-particle-like levels especially if the set is chosen conveniently. Thus the procedure used here to obtain the local field is a promising one for few-level systems. Since $\beta_{mm} = 0$ for all m, the set of equations in (92) is among $k(k-1)$ in general complex unknown β_{mn}'s if the number of relevant levels is k.

Once the local field inside the system has been obtained, the field outside is obtained directly by carrying out the integration in Eq. (89) letting $\vec{r}$ in the pseudo-vacuum propagator be the relevant point of observation outside our system. In passing we note that the integrations to be performed in Eqs. (88) and (89) to obtain $\vec{E}(\vec{r})$ outside the system is over the spatial domain of our system , only.

4.2 Diamagnetic Limit

Let us address now the question of how to calculate the local field when the diamagnetic effect dominates the electrodynamic interaction. This situation, for instance, can occur in cases where the excitation energy $\hbar\Omega$ is much smaller than the energy differences between all pairs of relevant one-particle levels.

For the purely diamagnetic case, the integral equation for the local field is given by

$$\vec{E}(\vec{r}) = \vec{E}_B(\vec{r}) + \frac{\mu_0 e^2}{m} \int N(\vec{r}\,') \overleftrightarrow{G}(\vec{r},\vec{r}\,') \cdot \vec{E}(\vec{r}\,') d^3 r' \,, \tag{95}$$

as one readily realizes by combining Eqs. (17), (18) and (73). Due to the inherently local form of the diamagnetic interaction the separation technique used in the previous subsection *a priori* seems not adequate here, although in principle it can be used as we shall demonstrate at the end of this subsection. It is thoughtful that the universal diamagnetic effect with its fundamentally simple and structureless local form often causes a more difficult problem for local-field calculations than the paramagnetic effect with its complicated and structure-rich nonlocal form.

In some cases one can take advantage of the fact that the pseudo-vacuum propagator is a slowly varying function of $\vec{r}\,'$ across the mesoscopic system. In these cases it is adequate to make a Taylor series expansion of $\overleftrightarrow{G}(\vec{r},\vec{r}\,')$ in $\vec{r}\,'$ around $\vec{r}\,' = 0$ (the system being placed in the vicinity of $\vec{r}\,' = 0$). Thus, by inserting

$$\overleftrightarrow{G}(\vec{r},\vec{r}\,') = \sum_{\alpha=0}^{\infty} \frac{1}{\alpha!} (\vec{r}\,' \cdot \vec{\nabla}\,')^{\alpha} \overleftrightarrow{G}(\vec{r},\vec{0}) \tag{96}$$

into Eq. (95) and retaining only a few terms one is faced with a problem which is accessible for a calculational procedure similar to that used in the paramagnetic case. To exemplify this let us consider just the simplest case, i.e. the one for which one can take $\overleftrightarrow{G}(\vec{r},\vec{r}\,') \simeq \overleftrightarrow{G}(\vec{r},\vec{0})$. With the abbrevation

$$\overleftrightarrow{F}_0(\vec{r}) = \frac{\mu_0 e^2}{m} \overleftrightarrow{G}(\vec{r},\vec{0}) \tag{97}$$

the local field has to fulfil the equation

$$\vec{E}(\vec{r}) = \vec{E}_B(\vec{r}) + \overleftrightarrow{F}_0(\vec{r}) \cdot \vec{\beta}_0 \,, \tag{98}$$

where the unknown vector $\vec{\beta}_0$ is given by

$$\vec{\beta}_0 = \int N(\vec{r})\vec{E}(\vec{r})d^3r \tag{99}$$

By inserting the ansatz in Eq. (98) into Eq. (99) one finds that $\vec{\beta}_0$ is given by

$$\vec{\beta}_0 = \left[\overleftrightarrow{U} - \int N(\vec{r})\overleftrightarrow{F}_0(\vec{r})d^3r\right]^{-1} \cdot \int N(\vec{r})\vec{E}_B(\vec{r})d^3r \ . \tag{100}$$

In certain cases it is advantageous to split off the so-called self-field (SF) contribution $\overleftrightarrow{g}^{SF}(\vec{r},\vec{r}\,') = \overleftrightarrow{g}^{SF}(\vec{r})\delta(\vec{r}-\vec{r}\,')$ to the pseudo-vacuum propagator. Thus, by inserting

$$\overleftrightarrow{G}(\vec{r},\vec{r}\,') = \overleftrightarrow{\mathcal{G}}(\vec{r},\vec{r}\,') + \overleftrightarrow{g}^{SF}(\vec{r})\delta(\vec{r}-\vec{r}\,') \ , \tag{101}$$

where $\overleftrightarrow{\mathcal{G}}$ is the nonlocal part of $\overleftrightarrow{G}$, into Eq. (95) one obtains

$$\left[\overleftrightarrow{U} - \frac{\mu_0 e^2}{m} N(\vec{r})\overleftrightarrow{g}^{SF}(\vec{r})\right] \cdot \vec{E}(\vec{r}) = \vec{E}_B(\vec{r})$$

$$+ \frac{\mu_0 e^2}{m} \int N(\vec{r}\,')\overleftrightarrow{\mathcal{G}}(\vec{r},\vec{r}\,') \cdot \vec{E}(\vec{r}\,')d^3r' \ . \tag{102}$$

To solve this integral equation beyond the often dominating self-field approximation i.e. the one in which one sets $\overleftrightarrow{\mathcal{G}}(\vec{r},\vec{r}\,') = \overleftrightarrow{0}$, it is in some problems possible to make use of a physically appealing tensor-product decomposition of $\overleftrightarrow{\mathcal{G}}(\vec{r},\vec{r}\,')$, namely [54]

$$\overleftrightarrow{\mathcal{G}}(\vec{r},\vec{r}\,') = \sum\!\!\!\!\!\!\int_n^{R^{(n)}(\vec{r},\vec{r}\,')} \vec{\Gamma}_O^{(n)}(\vec{r})\vec{\Gamma}_S^{(n)}(\vec{r}\,') \ , \tag{103}$$

where $\vec{\Gamma}_O^{(n)}(\vec{r})$ and $\vec{\Gamma}_S^{(n)}(\vec{r}\,')$ are functions, respectively, of the observation (O) and source (S) coordinates $\vec{r}$ and $\vec{r}\,'$, only. The generalized summation symbol $\sum\!\!\!\!\!\int_n$ is meant to indicate that the tensor-product superposition can contain both a summation $\sum_n$ over discrete n values and in integration $\int dn$ over a continuous n spectrum. The superscript $R^{(n)}(\vec{r},\vec{r}\,')$ added to $\sum\!\!\!\!\!\int$ indicates that $\vec{r}$ and $\vec{r}\,'$ in the individual tensor products normally are restricted (R) to specified regions of our system. The presence of this restriction in general prevents us from solving the Eq. (102) (with (103)) completely by the kernel separation technique. However, leaving out the non-separable part of the $\overleftrightarrow{\mathcal{G}}$-kernel, one can use the separation technique to solve the remaining part of the integral equation. After having done this, it might be possible to obtain a good solution to the complete problem using the Born-series iteration technique, taking as the zeroth-order solution the "exact" solution for a kernel consisting essentially of the self-field propagator plus the separable $\overleftrightarrow{\mathcal{G}}$-parts of the propagator. As an example of the use of this technique the reader is referred to Ref. [70].

Let me return now to the basic integral equation for the local field, i.e. Eq. (17) with Eq. (18) and $\overleftrightarrow{\sigma} = \overleftrightarrow{\sigma}^{dia}$. If we use the completeness theorem for the single-particle wave functions $\psi_{\vec{k}}(\vec{r})$, i.e.,

$$\delta(\vec{r}-\vec{r}\,') = \sum_m \psi_m^*(\vec{r})\psi_m(\vec{r}\,') \tag{104}$$

to decompose $\overleftrightarrow{\sigma}^{dia}(\vec{r},\vec{r}\,')$ of Eq. (73) into a sum over $\vec{k}$ of scalar products containing two factors of which one is a function of $\vec{r}$ the other a function of $\vec{r}\,'$, the basic integral equation for the pure diamagnetic case takes the form

$$\vec{E}(\vec{r}) = \vec{E}_B(\vec{r}) + \sum_m \overleftrightarrow{F}_m(\vec{r}) \cdot \vec{\beta}_m , \tag{105}$$

where

$$\overleftrightarrow{F}_m(\vec{r}) = \frac{\mu_0 e^2}{m} \int \psi_m^*(\vec{r}\,')N(\vec{r}\,')\overleftrightarrow{G}(\vec{r},\vec{r}\,')d^3r' , \tag{106}$$

and

$$\vec{\beta}_m = \int \psi_m(\vec{r})\vec{E}(\vec{r})d^3r . \tag{107}$$

The unknown $\vec{\beta}_m$-vectors are obtained by solving the super-matrix equation

$$\sum_n \left[\overleftrightarrow{U}\delta_{m,n} - \int \psi_m(\vec{r})\overleftrightarrow{F}_n(\vec{r})d^3r\right] \cdot \vec{\beta}_n = \int \psi_m(\vec{r})\vec{E}_B(\vec{r})d^3r . \tag{108}$$

By a super-matrix equation we mean a matrix equation where each vector component in itself is a vector and each matrix element in itself a matrix. In so far that it is a good approximation to apply the completeness theorem to a truncated few-level system the local field is given by Eq. (105) with the $\vec{\beta}_m$'s determined by Eq. (108) in the pure diamagnetic case.

4.3 Competing Para and Diamagnetic Interactions

It is outside the scope of this article to undertake a comprehensive analysis of the cases where it is necessary to take into account simultaneously the para and diamagnetic couplings. However, let me stress that provided it is justified in the mesoscopic system under study to use the completeness relation in Eq. (104) the general solution for the local electric field has the form, cf. the results in Eq. (91) and (105),

$$\vec{E}(\vec{r}) = \vec{E}_B(\vec{r}) + \sum_{m,n} \beta_{mn}\vec{F}_{nm}(\vec{r}) + \sum_m \overleftrightarrow{F}_m(\vec{r}) \cdot \vec{\beta}_m . \tag{109}$$

To determine the unknown numbers, β_{mn}, and vectors, $\vec{\beta}_m$, for a mesoscopic system containing k levels the number of linear and inhomogeneous algebraic equations among unknown numbers is $k(k-1)+3k = k(k+2)$, $k(k-1)$ for the "paramagnetic constants", and $3k$ for the "diamagnetic constants". Thus, for a two and three-level system one has, respectively, 8 and 15 unknowns in general. For a one-level system, in which only diamagnetic effects can be present, the solution based on use of the completeness relation will always be too bad in practice.

4.4 Moment Expansion of the Field-Field Response Tensor

In the preceeding three subsections we have studied various schemes for calculating the local field $\vec{E}(\vec{r};\Omega)$ as a function of the background field, $\vec{E}_B(\vec{r};\Omega)$. It appears from these investigations that the relation between $\vec{E}$ and $\vec{E}_B$ precisely is of the integral type given in Eq. (21). In certain cases the background field varies slowly in space across the domain of nonlocality, i.e. the region in $\vec{r}'$-space around a given point $\vec{r}$ in which the field-field response tensor $\overleftrightarrow{\Gamma}(\vec{r},\vec{r}')$ effectively is different from zero. It is important to mention that in linear optics the background field almost always is slowly varying across the domain of nonlocality, whereas in nonlinear optics this is usually not the case. In fact, one of the main reasons that local-field effects *a priori* are more interesting in nonlinear than in linear optics stems from the rapid variation of $\vec{E}_B$ across the range of nonlocality. This statement will be underlined by our subsequent investigations of the local-field effects associated with the photon drag and second-harmonic generation in a metallic quantum well.

If the background field is slowly varying in space it is adequate to make a Taylor series expansion of $\vec{E}_B(\vec{r}')$ around $\vec{r}$ in Eq. (21). Doing this one obtains

$$\vec{E}(\vec{r}) = \overleftrightarrow{\Gamma}_l(\vec{r}) \cdot \vec{E}_B(\vec{r}) + \overleftrightarrow{\Gamma}_{nl}(\vec{r}) : \vec{\nabla}\vec{E}_B(\vec{r}) + \cdots , \tag{110}$$

where the second-rank tensor

$$\overleftrightarrow{\Gamma}_l(\vec{r}) = \int \overleftrightarrow{\Gamma}(\vec{r},\vec{r}')d^3r' , \tag{111}$$

is the so-called local (l) field-field response function, and the third-rank tensor

$$\overleftrightarrow{\Gamma}_{nl}(\vec{r}) = \int \overleftrightarrow{\Gamma}(\vec{r},\vec{r}')(\vec{r}-\vec{r}')d^3r' , \tag{112}$$

is named the near-local (nl) field-field response function. Both these response functions (and the higher-rank ones appearing in Eq. (110)) are functions of $\vec{r}$ only. Although these thus appear to be local response functions they possess *hidden* nonlocality, due to the underlying integrations over $\vec{r}'$. The series of response tensors appearing in Eq. (110) are obtained as tensor moments of the microscopic field-field response tensor $\overleftrightarrow{\Gamma}(\vec{r},\vec{r}')$ around $\vec{r}$. The price we have paid to prevent working with an essentially nonlocal (but compact) response tensor is that we have to use often several (essentially local) tensors of increasing rank. When the range of nonlocality is small the expansion in Eq. (110) can be quite useful.

To be specific, let us illustrate the considerations above by discussing the local and near-local response tensors, $\overleftrightarrow{\Gamma}_l^{par}(\vec{r})$ and $\overleftrightarrow{\Gamma}_{nl}^{par}(\vec{r})$, belonging to the case where the paramagnetic interaction dominates. It appears from Eq. (94) that the background field enters the field-field response tensor through H_{mn}. The relevant expansion thus is

$$H_{mn} = \left[\int \vec{S}_{mn}(\vec{r}')d^3r'\right] \cdot \vec{E}_B(\vec{r})$$

$$+ \left[\int \vec{S}_{mn}(\vec{r}')(\vec{r}-\vec{r}')d^3r'\right] : \vec{\nabla}\vec{E}_B(\vec{r}) + \cdots . \tag{113}$$

From a physical point of view, the expansion in Eq. (113) has an obvious meaning. Hence, from the expression for $\vec{S}_{mn}$ given in Eq. (80) it is seen that a superposition of integrals of the form $\int \vec{J}_{mn}(\vec{r}')d^3r'$ enters the local term. It can be shown [57] that these integrals are proportional to the electric-dipole (ED) transition matrix element between the single-particle states $| m >$ and $| n >$. In the term proportional to $\nabla \vec{E}_B$ in addition to ED-integrals now integrals of the type $\int \vec{j}_{mn}(\vec{r}')\vec{r}'d^3r'$ enter. These integrals are proportional to a weighted sum of the electric-quadropole (EQ) and magnetic-dipole (MD) transition matrix elements between the $| m >$ and $| n >$ states, cf. [57]. In terms of the expansion in Eq. (113), the local and near-local field-field response tensors take the forms

$$\overleftrightarrow{\Gamma}_l^{par}(\vec{r}) = \overleftrightarrow{U} + \sum_{m,n} \left\{ \beta^0_{mn} \vec{F}_{nm}(\vec{r}) \int \vec{S}_{mn}(\vec{r}')d^3r' \right\} , \tag{114}$$

and

$$\overleftrightarrow{\Gamma}_{nl}^{par}(\vec{r}) = \sum_{m,n} \left\{ \beta^0_{mn} \vec{F}_{nm}(\vec{r}) \int \vec{S}_{mn}(\vec{r}')(\vec{r} - \vec{r}')d^3r' \right\} , \tag{115}$$

respectively, the common factors β^0_{mn} being determined from the solution of Eq. (92). In a given few-level mesoscopic system it can happen that it is sufficient to calculate the local field on the basis of the ED-approximation in Eq. (114), or in cases where all relevant electric dipole transitions are forbidden on the basis of Eq. (115), alone. We terminate this subsection by citing the result obtained for the pure diamagnetic case in the approximation where $\overleftrightarrow{G}(\vec{r}, \vec{r}') \simeq \overleftrightarrow{G}(\vec{r}, \vec{0})$, i.e.

$$\overleftrightarrow{\Gamma}_l^{dia}\left(\vec{r}; [\overleftrightarrow{G}(\vec{r}, \vec{0})]\right) = \overleftrightarrow{U} + N_0 \overleftrightarrow{F}_0(\vec{r}) \cdot \left[\overleftrightarrow{U} - \int N(\vec{r}) \overleftrightarrow{F}_0(\vec{r})d^3r \right]^{-1} , \tag{116}$$

where $N_0 = \int N(\vec{r})d^3r$ is the total number of electrons in our system, and

$$\overleftrightarrow{\Gamma}_{nl}^{dia}\left(\vec{r}; [\overleftrightarrow{G}(\vec{r}, \vec{0})]\right) = \overleftrightarrow{F}_0(\vec{r}) \cdot \left[\overleftrightarrow{U} - \int N(\vec{r}) \overleftrightarrow{F}_0(\vec{r})d^3r \right]^{-1} \vec{N}_1(\vec{r}) , \tag{117}$$

$\vec{N}_1(\vec{r}) = \int N(\vec{r}')(\vec{r} - \vec{r}')d^3r'$ being the first-order moment of the electron density around $\vec{r}$.

4.5 Resonance Conditions

In subsection 4.5, I touched upon the subject of local-field resonances and in subsection 2.6 the resonance condition was set up in the case where the system and pseudo-vacuum each exhibit four-dimensional translational invariance. Let us now consider the resonance conditions for spatially inhomogeneous systems, and let us to be specific study the paramagnetic case. It readily appears that the resonance condition in this case is given by the condition that the determinant of the system of equations in (92) is zero, i.e.

$$Det \left\{ \delta_{mn,op} - N^{mn}_{po}(\Omega) \right\} = 0 . \tag{118}$$

Although the resonance condition in Eq. (118) is conceptually simple to set up the detailed analysis of it often will be involved even in mesoscopic few-level systems because it necessitates a calculation of complicated integrals of the type given in Eq. (93). It is clear that the resonance condition of our system in general will depend on the radiative properties and the self-field response of the pseudo-vacuum. This dependence is described in a compact manner via the pseudo-vacuum propagator, occuring in N_{po}^{mn}.

In the pure diamagnetic case the local-field resonance condition can be set up in different ways depending on the approximate method used to calculate the local field, cf. the discussion in subsection 4.2. In the simple case where $\overleftrightarrow{G}(\vec{r},\vec{r}') \simeq \overleftrightarrow{G}(\vec{r},\vec{0})$, Eq. (100) (with Eq. (97)) immediately shows that the resonance condition is given by

$$Det\left\{\overleftrightarrow{U} - \frac{\mu_0 e^2}{m}\int N(\vec{r})\,\overleftrightarrow{G}(\vec{r},\vec{0})d^3r\right\} = 0\ , \tag{119}$$

which in the case where $\overleftrightarrow{G}(\vec{r},\vec{0})$ is essentially constant across the mesoscopic system $\left(\overleftrightarrow{G}(\vec{r},\vec{0}) \simeq \overleftrightarrow{G}(\vec{0},\vec{0})\right)$ simply becomes $Det\left\{\overleftrightarrow{U} - (\mu_0 e^2 N_0/m)\,\overleftrightarrow{G}(\vec{0},\vec{0})\right\} = 0$.

5 TWO SECOND-ORDER BACKGROUND RESPONSE FUNCTIONS

Up to this point our analysis has been essentially linear. The investigation of the local field in the linear limit is of importance for parametric interaction processes in nonlinear optics for two reasons. First of all, local-field effects enter at the mixing frequency. Secondly, to each of the fundamental beam frequencies a local-field calculation is attached. In this section we enter the nonlinear local-field calculation through an investigation of the background field stemming in the parametric limit from the driven part of the nonlinear current density. Even though the use of the parametric approximation in a huge number of interesting nonlinear optical studies is inadequate, I have limited myself here to the parametric regime because it constitutes the conceptually simplest framework for local-field studies in the nonlinear optics of mesoscopic systems, and because even this "simple" situation leads to exotic local-field phenomena not at all found in the linear optics of mesoscopic media. Roughly speaking, the reason that local-field effects can enter the nonlinear analyses of parametric systems in a dramatic manner stems from the fact that the background field usually is rapidly varying across the spatial domain of the mesoscopic system. In the linear case this field usually is almost constant, so that the moment expansion (see subsection 4.4) can be applied.

To make our framework even simpler, but in a manner where we still are able to capture basic concepts of the nonlinear local-field analysis, I shall restrict myself to a treatment of second-order nonlinear effects, and assume the existence of only one fundamental beam. Hence, the investigation is limited to analyses of optical second-harmonic generation and rectification. Even within these restricted conditions a number of exotic and esoteric local-field phenomena occur in mesoscopic systems. In the succeeding two sections emphasis is devoted to studies of the photon-drag effect (section 6) and optical second-harmonic generation (section 7) in a one- and two-level metallic quantum well, respectively.

5.1 Optical Rectification

It follows from the density-matrix operator approach presented in subsection 3.1 that the dc-part, $\vec{J}_{ind}(\vec{r};\Omega = 0) \equiv \vec{J}_0(\vec{r})$, of the forced nonlinear current density can be found from the expression

$$\vec{J}_0(\vec{r}) = Tr\left\{\rho_0 \vec{\mathcal{J}}_F(\vec{r})\right\} + \frac{1}{4} Tr\left\{\rho_1 \vec{\mathcal{J}}_1^\dagger(\vec{r})\right\} + \frac{1}{4} Tr\left\{\rho_1^\dagger \vec{\mathcal{J}}_1(\vec{r})\right\} , \tag{120}$$

where ρ_1 already has been determined from Eq. (51), and the nonlinear dc-part, ρ_0, of the density-matrix operator can be obtained from the Liouville equation

$$[\mathcal{H}_F, \rho_0] + [\mathcal{H}_0, \rho_F] + \frac{1}{4}[\mathcal{H}_1, \rho_1^\dagger] + \frac{1}{4}[\mathcal{H}_1^\dagger, \rho_1] = 0 . \tag{121}$$

A straightforward calculation gives the following expressions for the individual terms of Eq. (120):

$$Tr\left\{\rho_0 \vec{\mathcal{J}}_F\right\} = \sum_{I,J} \frac{P_J - P_I}{E_J - E_I} < I \mid \mathcal{H}_0 \mid J >< J \mid \vec{\mathcal{J}}_F \mid I >$$

$$+\frac{1}{4} \sum_{I,J,M} \frac{1}{E_J - E_I} \left\{ \left[\frac{P_I - P_M}{E_M - E_I + \hbar\omega} + \frac{P_J - P_M}{E_J - E_M - \hbar\omega} \right] \right.$$

$$\times < I \mid \mathcal{H}_1 \mid M >< M \mid \mathcal{H}_1^\dagger \mid J >$$

$$\left. + \left[\frac{P_I - P_M}{E_M - E_I - \hbar\omega} + \frac{P_J - P_M}{E_J - E_M + \hbar\omega} \right] < I \mid \mathcal{H}_1^\dagger \mid M >< M \mid \mathcal{H}_1 \mid J > \right\}$$

$$\times < J \mid \vec{\mathcal{J}}_F \mid I > , \tag{122}$$

and

$$Tr\left\{\rho_1 \vec{\mathcal{J}}_1^\dagger\right\} = \left[Tr\left\{\rho_1^\dagger \vec{\mathcal{J}}_1\right\}\right]^*$$

$$= \sum_{I,J} \frac{P_J - P_I}{E_J - E_I + \hbar\omega} < I \mid \mathcal{H}_1 \mid J >< J \mid \vec{\mathcal{J}}_1^\dagger \mid I > . \tag{123}$$

The summations in Eqs. (122) and (123) run over all the exact many-body states ($\mid I >, \mid J >, \mid M >$) and the fundamental frequency has been denoted by ω.

Let us denote the contribution to the nonlinear dc current density from the second term on the right hand side of Eq. (120) by $\frac{1}{2}\vec{J}_0^C(\vec{r})$. The spatial structure of this contribution is readily obtained by inserting the expressions for $\mathcal{H}_1$ (Eq. (56)) and $\vec{\mathcal{J}}_1^\dagger$ (the Hermitian conjugate of Eq. (62)) into Eq. (123)). Thus, one finds

$$\vec{J}_0^C(\vec{r}) = \sum_{\substack{\vec{k},\vec{k}' \\ \vec{k}'',\vec{k}'''}} \mathcal{C}_{\vec{k}'\to\vec{k}}^{\vec{k}'''\to\vec{k}''}(\omega)\psi_{\vec{k}}^*(\vec{r})\psi_{\vec{k}'}(\vec{r})\vec{E}^*(\vec{r};\omega)\int \vec{j}_{\vec{k}'''\to\vec{k}''}(\vec{r}')\cdot\vec{E}(\vec{r};\omega)d^3r' , \tag{124}$$

where

$$\mathcal{C}_{\vec{k}'\to\vec{k}}^{\vec{k}'''\to\vec{k}''}(\omega) = \frac{e^2}{2im\omega}\mathcal{A}_{\vec{k}'\to\vec{k}}^{\vec{k}'''\to\vec{k}''}(\omega) . \tag{125}$$

It appears from Eq. (124) that the relation between $\vec{J}_0^C$ and the dyade $\vec{E}\vec{E}^*$ in triadic notation is given by the semi-local third-rank response tensor

$$\overleftrightarrow{\Sigma}_0^C(\vec{r},\vec{r}',\vec{r}'') = \delta(\vec{r}'-\vec{r})\,\overleftrightarrow{U}\sum_{\vec{k},\vec{k}'}\psi_{\vec{k}}^*(\vec{r})\psi_{\vec{k}'}(\vec{r})\vec{\mathcal{T}}_{\vec{k}\to\vec{k}'}^C(\vec{r}'';\omega) \tag{126}$$

with

$$\vec{\mathcal{T}}_{\vec{k}\to\vec{k}'}^C(\vec{r}'';\omega) = \sum_{\vec{k}'',\vec{k}'''}\mathcal{C}_{\vec{k}'\to\vec{k}}^{\vec{k}'''\to\vec{k}''}(\omega)\vec{j}_{\vec{k}'''\to\vec{k}''}(\vec{r}'') . \tag{127}$$

Due to the inherent semi-local structure, the nonlinear response tensor $\overleftrightarrow{\Sigma}_0^C(\vec{r},\vec{r}',\vec{r}'')$ is no more difficult to handle than the paramagnetic term in the linear conductivity response tensor.

The contribution to $\vec{J}_0(\vec{r})$ from $Tr\left\{\rho_0\vec{\mathcal{J}}_F(\vec{r})\right\}$ we divide into two, one, named $\vec{J}_0^A(\vec{r})$, which is associated with the terms containing $\mathcal{H}_0$ in Eq. (122) and one, $\vec{J}_0^B(\vec{r})$, stemming from the remaining terms in this equation. A straightforward calculation performed along the same lines as the previous one shows that the relation between $\vec{J}_0^A$ and $\vec{E}\vec{E}^*$ is given by the so-called simultaneous two-photon response tensor

$$\overleftrightarrow{\Sigma}_0^A(\vec{r},\vec{r}',\vec{r}'') = \left[\sum_{\vec{k},\vec{k}'}\vec{j}_{\vec{k}'\to\vec{k}}(\vec{r})\mathcal{T}_{\vec{k}\to\vec{k}'}^A(\vec{r}';\omega)\right]\overleftrightarrow{U}\,\delta(\vec{r}''-\vec{r}') , \tag{128}$$

where

$$\mathcal{T}_{\vec{k}\to\vec{k}'}^A(\vec{r}';\omega) = \sum_{\vec{k}'',\vec{k}'''}\mathcal{B}_{\vec{k}'\to\vec{k}}^{\vec{k}'''\to\vec{k}''}(\omega)\psi_{\vec{k}''}^*(\vec{r}')\psi_{\vec{k}'''}(\vec{r}') \tag{129}$$

with

$$\mathcal{B}_{\vec{k}'\to\vec{k}}^{\vec{k}'''\to\vec{k}''}(\omega) = \frac{e^2}{4im\omega^2}\lim_{\omega\to 0}\left[\omega\mathcal{A}_{\vec{k}'\to\vec{k}}^{\vec{k}'''\to\vec{k}''}(\omega)\right] . \tag{130}$$

Also $\overleftrightarrow{\Sigma}_0^A$ is a semi-local response function of the same structural complexity as the linear, paramagnetic conductivity tensor.

The response function $\overleftrightarrow{\Sigma}_0^B (\vec{r},\vec{r}',\vec{r}'')$ which relates $\vec{J}_0^B$ to $\vec{E}\vec{E}^*$ (and $\vec{E}^*\vec{E}$) has a more complex structure as one would expect immediately from the expression in Eq. (122). It is a so-called double-nonlocal response tensor relating the nonlinear *dc* current density at the point $\vec{r}$ to the product of the electric field and its complex conjugate taking in general at two different neighbouring points $\vec{r}'$ and $\vec{r}''$. The $\vec{r}$-dependence of $\overleftrightarrow{\Sigma}_0^B$ is given by a weighted sum of the transition current density $\vec{j}_{\vec{k}'\to\vec{k}}(\vec{r})$ over the various combinations of $\vec{k}'$ and $\vec{k}$. The individual weight factors themselves are linear combinations of dyadic products of the form $\vec{j}_{\vec{k}''\to\vec{k}'''}(\vec{r}'')\vec{j}_{\vec{k}^{IV}\to\vec{k}^{V}}(\vec{r}')$ over the possible combinations of pairs $(\vec{k}''\to\vec{k}''',\vec{k}^{IV}\to\vec{k}^{V})$ of transitions. Although the $\overleftrightarrow{\Sigma}_0^B (\vec{r},\vec{r}',\vec{r}'')$ response tensor in general is of substantial interest for the analysis of optical rectification processes in mesoscopic systems we shall sustain from writing its explicit expression down here, because it is a long one and because it is not needed for our study of the photon-drag effect in a one-level metallic quantum well in the subsequent section.

For the optical rectification process, the nonlinear part of the local-field problem essentially is a magnetostatic one. This implies that for the background problem the central quantity is the forced current density and not the associated electric field. In fact, in the case of a mesoscopic superconducting system (or a practically collisionfree metal in the normal state) the *dc* electric field is zero.

If the system under study exhibits translational invariance and provided that the fundamental field from the source consists of a single plane wave it can be shown [79] that the response tensors $\overleftrightarrow{\Sigma}_0^A$ and $\overleftrightarrow{\Sigma}_0^B$ are both zero. For a translationally invariant system the contribution from the semi-local (sl) processes to the driven current density is given by

$$\vec{J}_0^{sl}(\vec{r}) = \frac{1}{2}\int \overleftrightarrow{\Sigma}_0^C (\vec{r}-\vec{r}',\vec{r}-\vec{r}'') : \vec{E}(\vec{r}'';\omega)\vec{E}^*(\vec{r}';\omega)d^3r''d^3r'$$

$$+c.c. \, , \tag{131}$$

or, if one emphasizes the semi-local structure by writing $\overleftrightarrow{\Sigma}_0^C$ in the form

$$\overleftrightarrow{\Sigma}_0^C (\vec{r}-\vec{r}',\vec{r}-\vec{r}'') = \delta(\vec{r}-\vec{r}')\, \overleftrightarrow{U}\overleftrightarrow{R}_0 (\vec{r}-\vec{r}'') \, , \tag{132}$$

by

$$\vec{J}_0^{sl}(\vec{r}) = \frac{1}{2}\vec{E}^*(\vec{r};\omega)\int \overleftrightarrow{R}_0 (\vec{r}-\vec{r}'')\cdot\vec{E}(\vec{r}'';\omega)d^3r'' + c.c. \, . \tag{133}$$

In Fourier space, Eq. (133) is replaced by

$$\vec{J}_0^{sl}(\vec{Q}) = \frac{1}{2}\frac{1}{(2\pi)^3}\int \vec{E}^*(\vec{\kappa}-\vec{Q},\omega)\overleftrightarrow{R}_0(\vec{\kappa})\cdot\vec{E}(\vec{\kappa},\omega)d^3\kappa + c.c. \, . \tag{134}$$

If the background field at the fundamental frequency consists of only one Fourier component, say $\vec{q}$, one has $\vec{E}(\vec{\kappa},\omega) = \vec{E}(\vec{q},\omega)\delta(\vec{\kappa}-\vec{q})$ and thus $\vec{E}^*(\vec{\kappa}-\vec{Q},\omega) =$

$\vec{E}^*(\vec{q},\omega)\delta(\vec{Q})$. By inserting these expressions into Eq. (134), one obtains the following spatial spectrum for the current denstiy:

$$\vec{J}_0^{sl}(\vec{Q}) = \frac{1}{2}\frac{1}{(2\pi)^3}\vec{E}^*(\vec{q},\omega)\vec{R}_0(\vec{q}) \cdot \vec{E}(\vec{q},\omega)\delta(\vec{Q}) + c.c. \,. \tag{135}$$

For a single plane-wave background field at the fundamental frequency the associated forced nonlinear *dc* current density thus is space-independent, as one would have anticipated beforehand.

If in addition to the translational symmetry our system exhibits inversion symmetry one has [79, 80]

$$\vec{R}_0(-\vec{q}) = -\vec{R}_0(\vec{q}) \,. \tag{136}$$

The implication of this is relation $\vec{R}_0(\vec{0}) = -\vec{R}_0(\vec{0})$ which means that the forced nonlinear *dc* current density vanishes in the long-wavelength limit. The photon-drag effect, to be discussed for a metallic quantum well in section 6, is an example of an effect which does not exist in the local ($\vec{q} \rightarrow 0$) limit.

5.2 Optical Second-Harmonic Generation

The second-harmonic part of the forced nonlinear current density, $\vec{J}_{ind}(\vec{r};\Omega = 2\omega) \equiv \vec{J}_2(\vec{r})$, can be obtained from the formula

$$\vec{J}_2(\vec{r}) = Tr\left\{\rho_2\vec{\mathcal{J}}_F(\vec{r})\right\} + \frac{1}{2}Tr\left\{\rho_1\vec{\mathcal{J}}_1(\vec{r})\right\} \,, \tag{137}$$

cf. the density-matrix operator approach outlined in section 3.1. To obtain its explicit form, the Liouville equation of motion for the second-harmonic part ρ_2 of ρ, i.e.

$$2\hbar\omega\rho_2 = [\mathcal{H}_F, \rho_2] + [\mathcal{H}_2, \rho_F] + \frac{1}{2}[\mathcal{H}_1, \rho_1] \,, \tag{138}$$

is used to express the matrix elements of ρ_2 between the exact many-body states as functions of the matrix elements of $\mathcal{H}_1$ and $\mathcal{H}_2$. During the course of calculation the expression for ρ_1 given in Eq. (51) has to be used also. As a result of this calculation we obtain

$$\vec{J}_2(\vec{r}) = \vec{J}_2^A(\vec{r}) + \vec{J}_2^B(\vec{r}) + \vec{J}_2^C(\vec{r}) \,, \tag{139}$$

where

$$\vec{J}_2^A(\vec{r}) = \sum_{I,J} \frac{P_J - P_I}{2\hbar\omega + E_J - E_I} < I \mid \mathcal{H}_2 \mid J >< J \mid \vec{\mathcal{J}}_F(\vec{r}) \mid I > \tag{140}$$

originates in the terms in $Tr\left\{\rho_2\vec{\mathcal{J}}_F\right\}$ containing $\mathcal{H}_2$,

$$\vec{J}_2^B(\vec{r}) = \frac{1}{2}\sum_{I,J} \frac{1}{2\hbar\omega + E_J - E_I}$$

$$\times \left\{ \sum_M \left[\frac{P_J - P_M}{\hbar\omega + E_J - E_M} + \frac{P_M - P_I}{\hbar\omega + E_M - E_I} \right] < I \mid \mathcal{H}_1 \mid M >< M \mid \mathcal{H}_1 \mid J > \right\}$$

$$\times < J \mid \vec{\mathcal{J}}_F(\vec{r}) \mid I > \tag{141}$$

stems from the remaining terms in $Tr\left\{\rho_2 \vec{\mathcal{J}}_F\right\}$, and

$$\vec{J}_2^C(\vec{r}) = \frac{1}{2}\sum_{I,J} \frac{P_J - P_I}{\hbar\omega + E_J - E_I} < I \mid \mathcal{H}_1 \mid J >< J \mid \vec{\mathcal{J}}_1(\vec{r}) \mid I > \tag{142}$$

comes from $\frac{1}{2}Tr\left\{\rho_1 \vec{\mathcal{J}}_1\right\}$. We have given the superscripts A,B, and C to the various contributions to indicate that their spatial structure resemble those having corresponding superscripts in the forced nonlinear *dc* current density (Fig. 6). The conceptual discussion thus parallels that for $\vec{J}_0^A$, $\vec{J}_0^B$, and $\vec{J}_0^C$. For a two-level mesoscopic system, $\vec{J}_2^B$ will vanish. In the paper by A. Liu and myself in this volume, the second-harmonic generation process in a two-dimensional metallic quantum well having two bound levels only and exhibiting inversion symmetry parallel to the barriers is investigated. Also, the treatment is restricted in the sense that the background current density is taken in the local (electric dipole) limit. Since, with these limitations $\vec{J}_2^B$ will vanish we shall not proceed to a general analysis of the associated response function here. Also $\vec{J}_2^A$ will be zero in the local limit due to the fact that when $\vec{A}(\vec{r};\Omega)$ in Eq. (57) is constant across the domain of integration, the orthonormality condition $\int \psi_{\vec{k}'}^* \psi_{\vec{k}} d^3r = \delta_{\vec{k}',\vec{k}}$ implies that $\mathcal{H}_2$ is proportional to the number operator, so that in turn the eigenstates $\mid I >$ and $\mid J >$ in Eq. (140) must be identical and hence

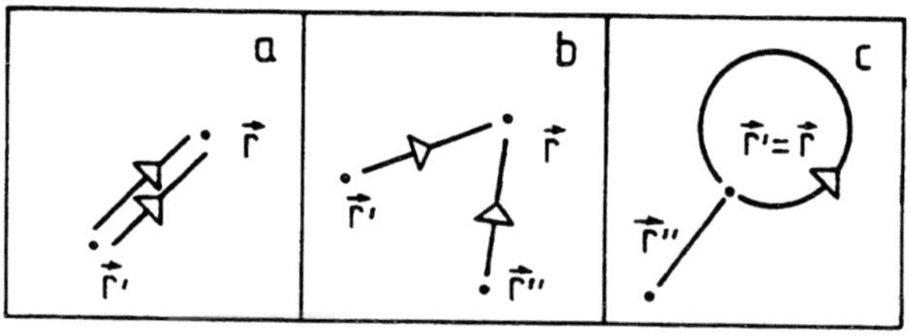

Figure 6. *Schematic diagrams displaying the three basic contributins to the second-order nonlinear conductivity response tensor relating the forced part of the nonlinear current density at space point $\vec{r}$ to the local electric fields at $\vec{r}'$ and $\vec{r}''$. (a). The so-called simultaneous two-photon channel. (b). The double-nonlocal channel. (c). The semi-local channel.*

$\vec{J}_2^A = \vec{0}$. Due to the fact that thus both $\vec{J}_2^A$ and $\vec{J}_2^B$ will vanish for the system to be analysed by Liu and me in this book, and not least for brevity, we focuse the subsequent discussion on $\vec{J}_2^C$ and the electric background field generated by this current density distribution.

Let us write the current density operator in Eq. (62) in the form ($\Omega = 2\omega$, here)

$$\vec{\mathcal{J}}_1(\vec{r};2\omega) = O(\vec{r})\vec{A}(\vec{r};2\omega) \ . \tag{143}$$

Since the operator O is Hermitian, i.e. $O^\dagger = O$, it follows that

$$\vec{\mathcal{J}}_1^{\,\dagger}(\vec{r};2\omega) = O(\vec{r})\vec{A}^*(\vec{r};2\omega) \ . \tag{144}$$

By comparing Eqs. (123) and (142), and making use of Eqs. (143) and (144), it immediately follows that the nonlinear conductivity response tensor $\overleftrightarrow{\Sigma}_2^C(\vec{r},\vec{r}\,',\vec{r}\,'')$, introduced via the constitutive equation

$$\vec{J}_2^C(\vec{r}) = \int \overleftrightarrow{\Sigma}_2^C(\vec{r},\vec{r}\,',\vec{r}\,'') : \vec{E}(\vec{r}\,'';\omega)\vec{E}(\vec{r}\,';\omega)d^3r''d^3r' \ , \tag{145}$$

is related to the corresponding *dc* response tensor simply by

$$\overleftrightarrow{\Sigma}_2^C(\vec{r},\vec{r}\,'\vec{r}\,'') = -\overleftrightarrow{\Sigma}_0^C(\vec{r},\vec{r}\,',\vec{r}\,'') \ , \tag{146}$$

the minus sign stemming from the replacements of the vector potentials by the electric fields ($\vec{E} = i\omega\vec{A}$ and $\vec{E}^* = -i\omega\vec{A}^*$).

The background field, $\vec{E}_B^C(\vec{r};2\omega)$, originating in the current density distribution $\vec{J}_2^C(\vec{r})$ is obtained from the equation

$$\vec{E}_B^C(\vec{r};2\omega) = -2i\mu_0\omega \int \overleftrightarrow{G}(\vec{r},\vec{r}\,';2\omega)\cdot\vec{J}_2^C(\vec{r}\,')d^3r' \ , \tag{147}$$

cf. Eq. (13). By inserting in this equation the expression for $\vec{J}_2^C(\vec{r})$ taken from Eq. (145), and by utilizing Eqs. (146) and (126), one gets

$$\vec{E}_B^C(\vec{r};2\omega) = 2i\mu_0\omega \sum_{\vec{k},\vec{k}\,'} \left\{ \left[\int \psi_{\vec{k}}^*(\vec{r}\,')\psi_{\vec{k}\,'}(\vec{r}\,')\overleftrightarrow{G}(\vec{r},\vec{r}\,';2\omega)\cdot\vec{E}(\vec{r}\,';\omega)d^3r' \right] \right.$$

$$\left. \times \left[\int \vec{\mathcal{T}}_{\vec{k}\to\vec{k}\,'}^C(\vec{r};\omega)\cdot\vec{E}(\vec{r};\omega)d^3r \right] \right\} \ . \tag{148}$$

I have mentioned that the local field effects may be especially pronounced in the nonlinear optics of a mesoscopic system because of a possible rapid variation of the background field across the system. A rapid variation occurs in cases where the self-field part of the propagator (see Eq. (101)) is present. For a given transition ($\vec{k}\,' \Rightarrow \vec{k}$), the contribution from the self-field to the background field varies in space in the manner $\psi_{\vec{k}}^*(\vec{r})\psi_{\vec{k}\,'}(\vec{r})\overleftrightarrow{g}^{SF}(\vec{r};2\omega)\cdot\vec{E}(\vec{r};\omega)$. In the random-phase approximation the vector $\vec{\mathcal{T}}_{\vec{k}\to\vec{k}\,'}^C(\vec{r};\omega)$ takes the simple form

$$\vec{\mathcal{T}}_{\vec{k}\to\vec{k}\,'}^C(\vec{r};\omega) = \frac{e^2}{m\omega^2}\frac{f_{\vec{k}} - f_{\vec{k}\,'}}{\hbar\omega + \epsilon_{\vec{k}} - \epsilon_{\vec{k}\,'}}\vec{\jmath}_{\vec{k}\to\vec{k}\,'}(\vec{r}) \ . \tag{149}$$

The considerations concerning the forced current density and the background field put forward in the section terminate the general analysis of the self-consistent field calculations of second-order nonlinear phenomena in the optics of mesoscopic media. In the subsequent two sections we shall apply these considerations in essentially one-dimensional form to studies of a metallic quantum well.

6 PHOTON DRAG IN A ONE-LEVEL METALLIC QUANTUM WELL

In the preceding sections the basis for studying local-field phenomena attached to second-order nonlinearities in mesoscopic systems was established. In the following, we shall see how the fundamental theory can be applied to investigate the photon-drag effect in a metallic quantum well. From a practical point of view such a mesoscopic system is especially adequate to treat, because, if the quantum well is excited by a monochromatic plane wave and the metallic properties along the well are treated within the framework of the jellium approximation, the local field becomes a function of only one space coordinate. In contrast to this essentially one-dimensional system, studies of local fields in for instance small (spherical) particles and quantum dots in general necessitate three-dimensional calculations. To emphasize the conceptual aspects of the local-field effects in the context of the photon-drag phenomenon in a metallic quantum well, furthermore, I shall assume that there exists only one bound state in the quantum well and that the energy of this state lies below the Fermi energy.

6.1 Photon-Drag Response Tensor and Forced Current Density

Let us consider now a quantum well in the form of a (few) monolayer thin plane-parallel metal film of thickness $\sim d$ placed on top of a semiinfinite nonconducting, isotropic and homogeneous substrate occupying the halfspace $z > 0$ in a Cartesian xyz-coordinate system (see Fig.7), and let us assume that the electrons in their motion parallel to the plane of the film exhibit free-electron-like behaviour with an effective mass m. Limiting ourselves to a one-electron description, the particles in the quantum well have stationary-state wave functions of the form

$$\psi_{\vec{k}}(\vec{r}) = (2\pi)^{-1}\psi_n(z)e^{i\vec{k}_{\parallel}\cdot\vec{r}}, \tag{150}$$

where $\psi_n(z)$ denotes the one-dimensional wave function of the nth energy eigenstate of the quantum well, and $(2\pi)^{-1}exp(i\vec{k}_{\parallel}\cdot\vec{r})$ describes the free-electron behaviour of a particle having a wave vector $\vec{k}_{\parallel}$ along the film plane. The state $\vec{k}$ thus is characterized by the quantum indices n and $\vec{k}_{\parallel}$. The wave function $\psi_n(z)$ satisfies the time-independent Schrödinger equation

$$\left[-\frac{\hbar^2}{2m}\frac{d^2}{dz^2} + V_{eff}(z)\right]\psi_n(z) = \epsilon_n\psi_n(z) \ , \tag{151}$$

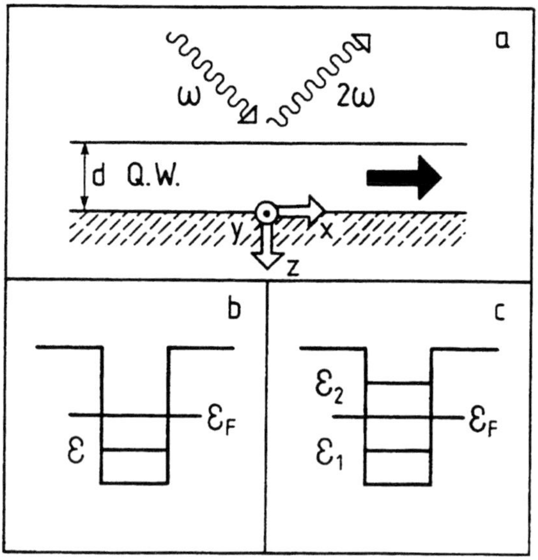

Figure 7. *Schematic diagrams showing in (a) the optical second-harmonic generation and the photon-drag current (black arrow) from a quantum well of thickness d placed on top of a semiinfinite substrate, and in (b) and (c), quantum wells with one or two bound states, respectively. the one-level system with a bound-state energy (ϵ) below the Fermi energy (ϵ_F) is used in this article to study the pure diamagnetic photon-drag phenomenon, and the two-level system, having energies $\epsilon_1 < \epsilon_F$ and $\epsilon_2 > \epsilon_F$, is employed in an investigation of the paramagnetic second-harmonic generation.*

where $V_{eff}(z)$ is the effective one-electron potential of the quantum well and ϵ_n is the energy of eigenstate number n. The total energy of the state $\vec{k}$ is given by $\epsilon_{\vec{k}} = \epsilon_n + \frac{\hbar^2}{2m}k_{\|}^2$. It follows from Eqs. (126),(127),(125), and (85) that the semi-local response tensor is given by

$$\overleftrightarrow{\Sigma}_0^C(\vec{r},\vec{r}',\vec{r}'') = \frac{e^2\delta(\vec{r}-\vec{r}')}{m\omega^2}\overleftrightarrow{U}\sum_{\vec{k},\vec{k}'}\frac{f_{\vec{k}} - f_{\vec{k}'}}{\hbar\omega + \epsilon_{\vec{k}} - \epsilon_{\vec{k}'}}\psi_{\vec{k}}^*(\vec{r})\psi_{\vec{k}'}(\vec{r})\vec{j}_{\vec{k}\to\vec{k}'}(\vec{r}'') \tag{152}$$

in the RPA approach. By taking advantage of the fact that our system exhibits infinitesimal translational invariance against displacements parallel to the xy-plane, the prevailing local field stemming from a plane polarized incident electromagnetic field takes the form

$$\vec{E}(\vec{r};\omega) = \vec{E}(z;\vec{q}_{\|},\omega)e^{i\vec{q}_{\|}\cdot\vec{r}} , \tag{153}$$

where $\vec{q}_{\|}$ is the component of the wave vector of the incident field parallel to the surface. It can be shown that the infinitesimal translation invariance in the xy-plane alone makes the response tensors $\overleftrightarrow{\Sigma}_0^A$ and $\overleftrightarrow{\Sigma}_0^B$ zero, cf. the remarks in subsection 5.1. This means that only the semi-local response tensor contributes to the induced *dc* current density in the quantum well. The translational invariance plus the plane-wave ansatz in Eq. (153) imply that the semi-local *dc* current density, $\vec{J}_0^{sl}(\vec{r})$, becomes independent of the x and y-coordinates, i.e. $\vec{J}_0^{sl}(\vec{r}) = \vec{J}_0^{sl}(z)$. In the following, for brevity, we shall omit the superfluous reference to the indices $\vec{q}_\parallel$ and ω from the notation and thus let $\vec{E}(z,\vec{q}_\parallel,\omega) \Rightarrow \vec{E}(z)$, $\overleftrightarrow{\Sigma}_0^C(z,z',z'';\vec{q}_\parallel,\omega) \Rightarrow \overleftrightarrow{\Sigma}_0^C(z,z',z'')$, etc. Also, we shall write $\vec{J}_0^{sl}(z)$ just as $\vec{J}_0(z)$. Making use of Eqs. (150),(152), and (153) a straightforward but albeit lenghty calculation shows that the induced *dc* current density is given by

$$\vec{J}_0(z) = \frac{1}{2}\int_{-\infty}^{\infty} \overleftrightarrow{\Sigma}_0^C(z,z',z'') : \vec{E}(z'')\vec{E}(z')dz''dz' \quad + \text{c.c.}\,, \tag{154}$$

where

$$\overleftrightarrow{\Sigma}_0^C(z,z',z'') = -\frac{\hbar e^3\delta(z-z')\,\overleftrightarrow{U}}{4m^2\omega^2}$$

$$\times\sum_{n,n'}\int_{-\infty}^{\infty}\frac{f(\epsilon_n+\frac{\hbar^2}{2m}k_\parallel^2)-f(\epsilon_{n'}+\frac{\hbar^2}{2m}\mid\vec{k}_\parallel+\vec{q}_\parallel\mid^2)}{\hbar\omega+\epsilon_n-\epsilon_{n'}+\frac{\hbar^2}{2m}k_\parallel^2-\frac{\hbar^2}{2m}\mid\vec{k}_\parallel+\vec{q}_\parallel\mid^2}\psi_n^*(z)\psi_{n'}(z)$$

$$\times\left\{(2\vec{k}_\parallel+\vec{q}_\parallel)\psi_n(z'')\psi_{n'}^*(z'')+\vec{e}_z\left[\psi_{n'}^*(z'')\frac{d\psi_n(z'')}{dz''}-\psi_n(z'')\frac{d\psi_{n'}^*(z'')}{dz''}\right]\right\}\frac{d^2k_\parallel}{(2\pi)^2}, \tag{155}$$

$\vec{e}_z$ being a unit vector in the z-direction. In Eq. (155), the Fermi-Dirac distribution function $f_{\vec{k}}$ has been denoted by $f(\epsilon_n+\frac{\hbar^2}{2m}k_\parallel^2)$, and an analogous notation has been used for the state $(n',\vec{k}_\parallel+\vec{q}_\parallel)$, i.e. $f(\epsilon_n'+\frac{\hbar^2}{2m}\mid\vec{k}_\parallel+\vec{q}_\parallel\mid^2)$. Limiting myself now to the case where there exists only one bound state of energy ϵ and with a wave function $\psi(z)$ in the quantum well, by assuming that the frequency of the incident light is so low that photoemission processes can be neglected, and by choosing $\psi(z)$ to be real, the expression for the nonlinear response tensor in Eq. (155) is reduced to

$$\overleftrightarrow{\Sigma}_0^C(z,z',z'') = -\frac{\hbar e^3\,\overleftrightarrow{U}}{4m^2\omega^2}\psi^2(z)\psi^2(z'')\delta(z-z')$$

$$\times\int_{-\infty}^{\infty}\frac{f(\epsilon+\frac{\hbar^2}{2m}\mid\vec{k}_\parallel-\vec{q}_\parallel\mid^2)-f(\epsilon+\frac{\hbar^2}{2m}k_\parallel^2)}{\hbar\omega+\frac{\hbar^2}{2m}\mid\vec{k}_\parallel-\vec{q}_\parallel\mid^2-\frac{\hbar^2}{2m}k_\parallel^2}(2\vec{k}_\parallel-\vec{q}_\parallel)\frac{d^2k_\parallel}{(2\pi)^2}\,. \tag{156}$$

When writing down Eq. (156) a substitution $\vec{k}_\parallel \Rightarrow \vec{k}_\parallel - \vec{q}_\parallel$ has been invoked, also.

For a quantum well in the normal, i.e. nonsuperconducting, metallic state usually it is an excellent approximation to assume that the distribution function is a step function. i.e. $f(E) = 1$ for $E < \epsilon_F$, ϵ_F being the Fermi energy of the electrons, and $f(E) = 0$, else. Using this $T = 0K$ approximation it is possible to carry out the double integral in Eq. (156) by analytical calculations. Doing this, one obtains under the assumption that the bound energy level in the well lies below the Fermi level ($\epsilon < \epsilon_F$) after a tedious calculation the following result:

$$\overleftrightarrow{\Sigma}_0^C (z, z', z'') = \frac{\hbar e^3}{4m^2\omega^2}\psi^2(z)\psi^2(z'')\delta(z - z')(R_+ - R_-)\,\overleftrightarrow{U}\,\vec{e}_x \ , \tag{157}$$

where

$$R_\pm = \frac{1}{2\pi\beta}\left\{\frac{1}{\beta}\left(\frac{2\alpha_\pm}{\beta} \mp q_\parallel\right)\left[\alpha_\pm - \left(\alpha_\pm^2 - \beta^2\kappa_\parallel^2\right)^{1/2}\right] - \kappa_\parallel^2\right\} \tag{158}$$

with

$$\alpha_\pm = \hbar\omega \pm \frac{\hbar^2}{2m}q_\parallel^2 \ , \tag{159}$$

$$\beta = \frac{\hbar^2}{m}q_\parallel \ , \tag{160}$$

and

$$\kappa_\parallel = \left[\frac{2m}{\hbar^2}(\epsilon_F - \epsilon)\right]^{1/2} \ . \tag{161}$$

In deriving Eq. (157) we have chosen $\vec{q}_\parallel$ to ly along the x-axis, i.e. $\vec{q}_\parallel = q_\parallel \vec{e}_x$, where $\vec{e}_x$ is a unit vector in the positive x-direction. In passing one should notice that the last term, $\kappa_\parallel^2$, in the curly bracket of Eq. (158) does not contribute to the difference $R_+ - R_-$ entering Eq. (157). It appears from Eq. (157), and it holds even if the low-temperature approximation is not made that the semilocal third-rank response tensor has only three nonvanishing (and identical) tensor elements, i.e. $\vec{e}_x\vec{e}_x\vec{e}_x$, $\vec{e}_y\vec{e}_y\vec{e}_x$, and $\vec{e}_z\vec{e}_z\vec{e}_x$ in triadic notation. The symmetry scheme and the spatial structure of the nonlinear *dc*-response tensor of the quantum well are shown schematically in Fig.8. By inserting Eq. (157) into (154), the *dc* current density induced by the fundamental electromagnetic field takes the form

$$\vec{J}_0(z) = \frac{1}{2}\left\{K(q_\parallel, \omega)\psi^2(z)\vec{E}^*(z)\int_{QW}\psi^2(z')E_x(z')dz' \ + c.c.\right\} \tag{162}$$

with

$$K(q_\parallel, \omega) = \frac{\hbar e^3}{4m^2\omega^2}(R_+ - R_-) \ . \tag{163}$$

By the subscript QW at the foot of the integral sign we have indicated that the integration extends over the quantum well (QW). The dynamics of the two-dimensional free-electron gas essentially is described in terms of $K(q_\parallel, \omega)$. It is readily demonstrated that $R_+ - R_- \to 0$ for $q_\parallel \to 0$, so that

$$K(q_\parallel \to 0, \omega) = 0 \ . \tag{164}$$

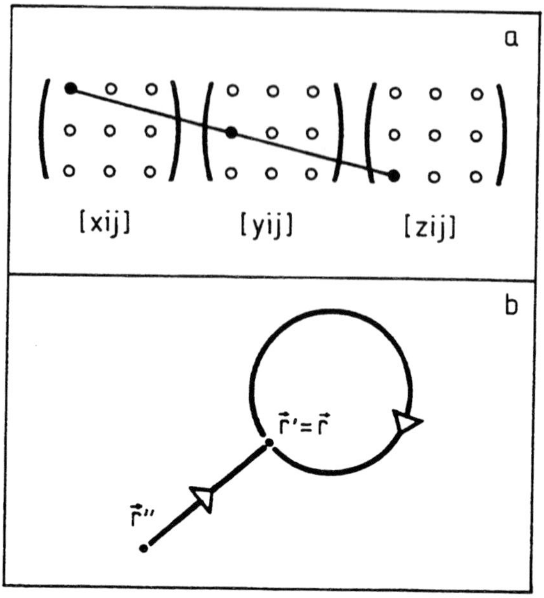

Figure 8. *Schematic diagrams showing in (a) the symmetry scheme of the third-rank photon-drag response tensor, and in (b) the semi-local structure of the photon-drag response.*

This means that the induced current density vanishes in the longwavelength (local) limit, an expected result in view of the fact that there cannot be any net momentum exchange between the electrons and the light in this limit and because the quantum well has only one bound state. Since a net exchange of momentum between the light and particle fields is required in order to induce a *dc* current density in the quantum well, this nonlinear phenomenon usually is named the photon-drag effect. It appears from Eq.(162) that the photon-drag effect is absent if the prevailing fundamental electromagnetic field is polarized perpendicular to the plane of incidence (*s*-polarized), cf. the analysis in subsection 6.3. A *p*-polarized field will give rise to a photon-drag effect as long as it has an x-component, as will a field consisting of superposition of *s* and *p*-components.

6.2 Prevailing Local *dc* Fields and Current Densities

In the previous subsection an expression (Eq. (162) with (163)) was derived for the *forced* part of the photon-drag current density. In order to obtain a selfconsistent solution to the combined set of the Schrödinger and Maxwell equations including the appropriate boundary conditions of the problem one has to add an as yet unknown *free* (F) part, $\vec{J}_F$, to the forced part of the *dc* current density. The infinitesimal translational invariance of the quantum well system parallel to the xy-plane implies that the prevailing *dc* magnetic field, B_0, and the *free* part of the *dc* currrent density are functions of z only. In turn, it follows from the Maxwell equation

$$\vec{\nabla} \times \vec{B}_0(z) = \mu_0(\vec{J}_F(z) + \vec{J}_0(z)) \tag{165}$$

that the total current density in the direction perpendicular to the quantum well vanishes, i.e.

$$\vec{e}_z \cdot (\vec{J}_F(z) + \vec{J}_0(z)) = 0 \ . \tag{166}$$

If the electrons are allowed to flow freely (i e. without externally impressed *dc* fields) along the quantum well the associated (selfconsistent) *dc* current density will be $(\overleftrightarrow{U} - \vec{e}_z\vec{e}_z) \cdot \vec{J}_0(z)$.

In order for the component of the total *dc* current density in the z-direction to vanish, a z-dependent *dc* electric field, $\vec{E}_0(z) = E_{0,z}(z)\vec{e}_z$, necessarily must be created across the quantum well. To determine this field one makes use of a result from the subsequent subsection, namely that the free current density in the z-direction is related to the unknown field via a constitutive equation of the form

$$J_{F,z}(z) = \int_{QW} \sigma_{zz}(z, z'; \omega = 0) E_{0,z}(z') dz' \ , \tag{167}$$

where $\sigma_{zz}(z, z'; \omega = 0)$ is the relevant (zz-component) of the linear *dc* conductivity response function. For the single-level quantum well under study this response function in the relaxation time approximation (see subsection 6.3) is given by

$$\sigma_{zz}(z, z'; \omega = 0) = \frac{e^2 \tau}{m} n(z) \delta(z - z') \ , \tag{168}$$

where

$$n(z) = \frac{m}{\pi \hbar^2} (\epsilon_F - \epsilon) \psi^2(z) \tag{169}$$

is the relevant electron density in the metallic quantum well, and τ is the momentum relaxation time of the electrons. By combining Eqs. (162), and (166)-(169) one finds that the *dc* electric field induced by the photon-drag effect across the quantum well is

$$E_{o,z}(z) = \frac{\pi \hbar^2}{2e^2 \tau(\epsilon - \epsilon_F)} \left\{ K(q_\parallel, \omega) E_z^*(z) \int_{QW} \psi^2(z') E_x(z') dz' \ + \ c.c. \right\} \ . \tag{170}$$

In order to obtain a finite value for $E_{0,z}$ it is necessary to incorporate irreversible loss mechanisms in the electron dynamics at $\omega = 0$. This was done as far as the linear conductivity response tensor $\sigma_{zz}(z, z'; \omega = 0)$ is concerned by means of the naive relaxation time approach, cf. Eq. (168). To be consistent one should also take into account relaxation effects when deriving the expressions for the nonlinear response tensor $\overleftrightarrow{\Sigma}_0^C$, and the fundamental field (see section 6.3). Within the framework of the relaxation time approach, the expression for $\overleftrightarrow{\Sigma}_0^C$ in Eq. (156) still holds provided one makes the replacement $\omega \Rightarrow \omega + i/\tau$ in the denominator inside the integral sign. In consequence, with the replacement $\omega \Rightarrow \omega + i/\tau$ in the expression for $\alpha_\pm$ in Eq. (159), the formula in Eq. (157) for $\overleftrightarrow{\Sigma}_0^C$ is the correct one in the relaxation time approximation.

6.3 Local Field at the Fundamental Frequency

6.3.1 Conductivity Response Tensor

In order to achieve an explicit determination of the *dc* field generated across the quantum well in the photon-drag phenomenon one needs to calculate the local field at the fundamental frequency. The determination of this field in itself is a formidable task requiring an extended use of the formalism established in sections 2,3, and 4. In the following I shall indicate some of the main steps to be taken and the conclusions involved in this task. For a detailed account the reader is referred to the litterature cited at the proper places below. As a first step towards the calculation of $\vec{E}(z)$ one has to obtain the expression for the linear conductivity response tensor $\overleftrightarrow{\sigma}$ (Eq. (81)) of the metallic quantum well. Within the framework of the random-phase approximation in the relaxation time approximation, one can, taking as the starting point Eq. (86) for the paramagnetic part and Eq. (73) (with Eq. (87)) for the diamagnetic part of the conductivity tensor, and using the procedure outlined in subsection 6.1, derive the expression for the relevant conductivity response tensor $\overleftrightarrow{\sigma}(z,z';\vec{q}_{\parallel},\omega)$ [For brevity we leave out the subscript RPA on $\overleftrightarrow{\sigma}$]. The conductivity tensor $\overleftrightarrow{\sigma}(z,z';\vec{q}_{\parallel},\omega)$, obtained from $\overleftrightarrow{\sigma}(\vec{r},\vec{r}';\omega)$ by Fourier transformations in the x and y-coordinates obeys the constitutive equation

$$\vec{J}_{ind}(z;\vec{q}_{\parallel},\omega) = \int \overleftrightarrow{\sigma}(z,z';\vec{q}_{\parallel},\omega)\cdot\vec{E}(z';\vec{q}_{\parallel},\omega)dz' \tag{171}$$

between the induced current density $\vec{J}_{ind}(z;\vec{q}_{\parallel},\omega)[\vec{J}_{ind}(\vec{r};\omega) = \vec{J}_{ind}(z;\vec{q}_{\parallel},\omega) \times exp(i\vec{q}_{\parallel}\cdot\vec{r})]$ and the fundamental field $\vec{E}(z;\vec{q}_{\parallel},\omega)$. For the single-level metallic quantum well under study, one obtains in the low-temperature limit ($T \to 0$) the following result for the linear response tensor $\overleftrightarrow{\sigma}(z,z';\vec{q}_{\parallel},\omega)$, leaving out the reference to $\vec{q}_{\parallel}$ and ω from the notation [65]:

$$\overleftrightarrow{\sigma}(z,z') = \begin{pmatrix} \sigma_{xx}(z,z') & 0 & 0 \\ 0 & \sigma_{yy}(z,z') & 0 \\ 0 & 0 & \sigma_{zz}(z,z') \end{pmatrix}, \tag{172}$$

where

$$\sigma_{xx}(z,z') = \frac{e^2\tau}{m(1-i\omega\tau)}n(z)\delta(z-z') + \frac{2ie^2\hbar^2}{m^2\omega}\psi^2(z)\psi^2(z')(S_- - S_+), \tag{173}$$

$$\sigma_{yy}(z,z') = \frac{e^2\tau}{m(1-i\omega\tau)}n(z)\delta(z-z') + \frac{2ie^2\hbar^2}{m^2\omega}\psi^2(z)\psi^2(z')(T_- - T_+), \tag{174}$$

and

$$\sigma_{zz}(z,z') = \frac{e^2\tau}{m(1-i\omega\tau)}n(z)\delta(z-z') \tag{175}$$

with $n(z)$ being given by Eq. (169). In Eqs. (173)-(175) the terms containing $n(z)$ originate in the diamagnetic part of the response tensor. The wave-vecor ($\vec{q}_{\parallel}$) dependence of $\overleftrightarrow{\sigma}(z, z'; \vec{q}_{\|}, \omega)$ is contained in the quantities

$$S_{\pm} = \frac{1}{2\pi\beta^2}\left(\frac{\alpha_{\pm}}{\beta} \mp \frac{q_{\|}}{2}\right)^2 \left[\alpha_{\pm} - \left(\alpha_{\pm}^2 - \beta^2\kappa_{\|}^2\right)^{1/2}\right] - \frac{\kappa_{\|}^2}{4\pi\beta}\left(\frac{\alpha_{\pm}}{\beta} \mp q_{\|}\right) , \quad (176)$$

and

$$T_{\pm} = \frac{\alpha_{\pm}}{2\pi\beta^2}\left\{\frac{1}{3\beta^2\alpha_{\pm}}\left[\left(\alpha_{\pm}^2 - \beta^2\kappa_{\|}^2\right)^{3/2} - \alpha_{\pm}^3\right] + \frac{\kappa_{\|}^2}{2}\right\} \quad (177)$$

appearing in the paramagnetic contributions to the response tensor. One should notice that the linear conductivity tensor is diagonal. Let me stress that this property stems from the fact that only one level is assumed to exist in the quantum well. In general, i.e. in cases where more than one level are present and intersubband transitions thus allowed, the elements σ_{xz} and σ_{zx} will be different from zero. The elements σ_{xy}, σ_{yx}, σ_{yz}, and σ_{zy}, however, always vanish since the s and p-polarized excitations are uncoupled. The reason that σ_{xz} and σ_{zx} are zero originates in the observation that when only one level is present the z-component of the paramagnetic transition current density vanishes (the diamagnetic effect always gives a diagonal response). Since the z-component of the transition current density is zero, $\sigma_{zz}(z, z'; \omega)$ only consists of the local, $\vec{q}_{\|}$-independent, diamagnetic contribution. In the static limit ($\omega\tau << 1$), σ_{zz} of Eq. (175) is reduced to the expression cited in Eq. (169). The paramagnetic terms entering the expressions for σ_{xx} (Eq. (173)) and σ_{yy} (Eq.(174)) are different ($S_{\pm} \neq T_{\pm}$) since the induced electron motions along the x and y-directions are not equivalent when $q_{\|} \neq 0$. It is easy to demonstrate that the paramagnetic contributions to the conductivity response tensor vanish in the local limit, i.e. $S_{\pm}(q_{\|} = 0, \omega) = T_{\pm}(q_{\|} = 0, \omega) = 0$.

6.3.2 Slave Approximation

By inserting Eq.(18) into Eq. (17) and by performing a Fourier analysis in the coordinates parallel to the surface it is realized that the local electric field at the fundamental frequency can be obtained by solving the basic integral equation

$$\vec{E}(z) = \vec{E}^B(z) - i\mu_0\omega \int_{QW} \overleftrightarrow{G}(z, z') \cdot \overleftrightarrow{\sigma}(z', z'') \cdot \vec{E}(z'')dz''dz' . \quad (178)$$

In Eq. (178) the superfluous reference to $\vec{q}_{\|}$ and ω has been omitted from all quantities. The background field $\vec{E}^B(z)$ consists of the sum of the incident field and the field reflected from the substrate, and $\overleftrightarrow{G}(z, z')$ is the electromagnetic propagator of the vacuum/substrate pseudo-vacuum. Explicit expressions are known for the various tensor elements of this propagator [54, 69]. The tensorial forms of the Green's function and the conductivity tensor (Eq. (172)) allow us to decouple the vectorial integral equation into one scalar equation for the s-polarized part of the field and two mutually coupled scalar equations for the p-component. Thus,

$$E_y(z) = E_y^B(z) - i\mu_0\omega \int_{QW} G_{yy}(z, z')\sigma_{yy}(z', z'')E_y(z'')dz''dz' , \quad (179)$$

and

$$E_x(z) = E_x^B(z) - i\mu_0\omega \int_{QW} [G_{xx}(z,z')\sigma_{xx}(z',z'')E_x(z'')$$

$$+G_{xz}(z,z')\sigma_{zz}(z',z'')\, E_z(z'')]\, dz''dz' \; , \qquad (180)$$

$$E_z(z) = E_z^B(z) - i\mu_0\omega \int_{QW} [G_{zx}(z,z')\sigma_{xx}(z',z'')E_x(z'')$$

$$+G_{zz}(z,z')\sigma_{zz}(z',z'')E_z(z'')]\, dz''dz' \; , \qquad (181)$$

respectively.

Let us consider the solution of the integral equation for the s-polarized part of the field. Since for optical frequencies the background field and the propagator are essentially constant across the quantum well, i.e. $E_y^B(z) \simeq E_y^B(0) \equiv E_y^B$, and $G_{yy}(z,z') \simeq G_{yy}(0,0)$, the local field as well must be approximately constant across the well, i.e. $E_y(z) \simeq E_y(0) \equiv E_y$. In turn, this implies that the solution of Eq. (179) is given by, cf. Refs. [28] and [65]

$$E_y = \frac{E_y^B}{1 - K_{yy}} \; , \qquad (182)$$

where

$$K_{yy} = e^2 G_{yy}(0,0) \left[\frac{\mu_0\omega(\epsilon_F - \epsilon)}{\pi\hbar^2(\omega + i/\tau)} + \frac{2\mu_0\hbar^2(T_- - T_+)}{m^2} \right] \; . \qquad (183)$$

For practical purposes one has to insert the expressions $E_y^B = E_y^0(1 + r_s)$ and $G_{yy}(0,0) = (1+r_s)/(2iq_\perp^0)$ into the equations above E_y^0, r_s, and $q_\perp^0 = \left[(\omega/c_0)^2 - q_\parallel^2\right]^{1/2}$ being the amplitude of the s-polarized incident field, the s-polarixed amplitude reflection coefficient of the substrate, and the component of the wave vector of the incident field perpendicular to the surface, respectively. Let me mention here that a generalization of the result obtained above to cases with more than one bound state in the quantum well recently has been applied succcessfully to interpret linear s-polarized infrared reflection data for a niobium quantum well deposited on a crystalline quartz substrate [28, 81] as illustrated in Fig.9.

The coupled integral equations for the p-polarized part of the local field in general have to be solved by numerical methods even if the slow variation of (i) the p-polarized background field and (ii) a number of elements (terms) in the electromagnetic pseudo-vacuum propagator is taken into account. Instead of embarking on the general analysis let us consider the solution of the set of integral equations in Eq. (180) and (181) in a limiting case of both conceptual and practical interest. The absence of the paramagnetic transition current density perpendicular to the quantum well in a one-level system implies that the x and z-components of the local field are coupled only through the off-diagonal elements of the electromagnetic propagator, cf. Eqs.(180) and (181).

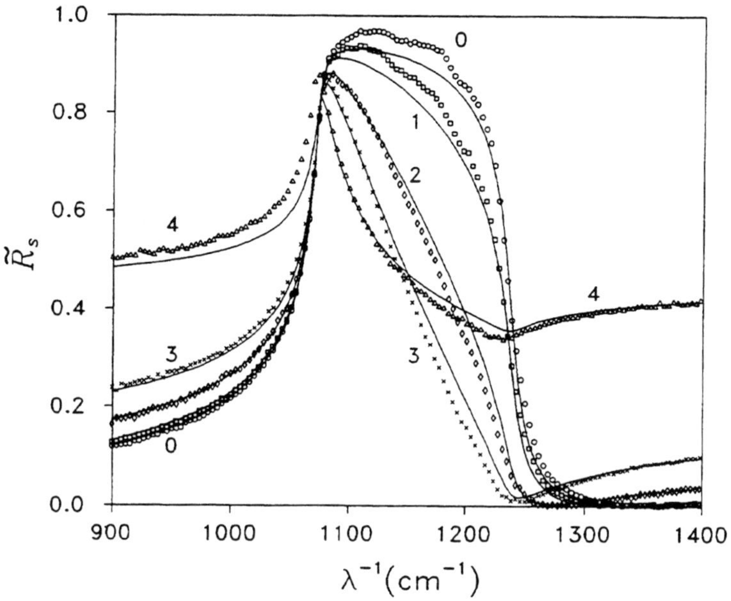

Figure 9. *The linear energy reflection coefficient for s-polarized light, $\tilde{R}_s$, as a function of the reciprocal vacuum wavelength, λ^{-1}, for a bare quartz substrate (0), and for quartz covered with niobium films (quantum wells) of the following thicknesses (in Å) : (1) 3, (2) 12, (3) 25, and (4) 50. The points (with various symbols) represent the experimental data of Alieva et al. [26]. The solid lines are the result of a theoretical local-field calculation [28, 81].*

In an approximation which I have called the slave approximation [29, 81] and which at least in the $q_{\parallel} = 0$ limit has turned out to give results for the local field in extremely close agreement with those obtained by exact numerical calculations [29, 81], the term containing G_{xz} in Eq. (180) is neglected. By neglecting this term the integral equation for the local field along the quantum well becomes independent of $E_z(z)$, i.e.

$$E_x(z) = E_x^B(z) + K_{xx} \int_{QW} \psi^2(z')E_x(z')dz' , \tag{184}$$

where

$$K_{xx} = e^2 G_{xx}(0,0) \left[\frac{\mu_0 \omega(\epsilon_F - \epsilon)}{\pi \hbar^2 (\omega + i/\tau)} + \frac{2\mu_0 \hbar^2 (S_- - S_+)}{m^2} \right] \tag{185}$$

with $G_{xx}(z, z') \approx G_{xx}(0,0) = (c_0/\omega)^2(1 - r_p)q_\perp^0/(2i)$, r_p being the p-polarized amplitude reflection coefficient of the substrate. Since, at optical frequencies $E_x^B(z) \simeq E_x^B(0) = E_x^0(1 - r_p) \equiv E_x^B$ [E_x^0 being the x-component of the amplitude of the incident field] $E_x(z)$ essentially is independent of z, $E_x(z) \simeq E_x(0) \equiv E_x$, so that

$$E_x = \frac{E_x^B}{1 - K_{xx}} \ . \tag{186}$$

As one could have anticipated the solutions for E_x Eq. ((186)) and E_y (Eq. (182)) take the same form in the slave approximation.

Turning our attention towards the determination of the z-component of the local field, let us extract the self-field part, $(c_0/\omega)^2\delta(z - z')$, of the background (pseudo-vacuum) propagator from G_{zz}, i.e.

$$G_{zz}(z, z') = (\frac{c_0}{\omega})^2\delta(z - z') + P_{zz}(z, z'), \tag{187}$$

where $P_{zz}(z, z') \simeq P_{zz}(0,0) = (c_0/\omega)^2(1 + r_p)q_\|^2/(2iq_\perp^0)$ is the "propagating" part of the Green's function. As indicated this part is almost constant across the quantum well. Neglecting slowly varying terms in $G_{zx}(z, z')$ this is given by [54]

$$G_{zx}(z, z') \simeq \frac{q_\|}{2i}(\frac{c_0}{\omega})^2\mathcal{F}_-(z, z') \ , \tag{188}$$

where

$$\mathcal{F}_-(z, z') = \Theta(z' - z) - \Theta(z - z') - r_p \ , \tag{189}$$

Θ being the heaviside unit step function. By inserting Eqs. (173),(175)[both with (169)], and (186)-(188)[with (189)] into Eq. (181) one obtains the following integral equation for $E_z(z)$:

$$(1 - V\psi^2(z))E_z(z) = E_z^{B,eff}(z) + W\int_{QW}\psi^2(z')E_z(z')dz' \ , \tag{190}$$

where

$$E_z^{B,eff}(z) = E_z^B + \frac{(A + U)E_x^B}{1 - K_{xx}}\int_{QW}\mathcal{F}_-(z, z')\psi^2(z')dz' \tag{191}$$

is an effective (eff) background field consisting of the impressed background field, E_z^B, plus the field created by the selfconsistently induced motion of the electrons along the quantum well (in an approximation where the effect of the field induced across the well is negligible). The slave approximation bears its name from the principle that the dynamics perpendicular to the quantum well is slaved by the dynamics along the well, cf. Eq. (191). In the two equations above we have introduced the abbreviations

$$V = \frac{e^2\tau(\epsilon_F - \epsilon)}{\pi i\epsilon_0\omega\hbar^2(1 - i\omega\tau)} \ , \tag{192}$$

$$W = e^2P_{zz}(0,0)\frac{\mu_0\omega(\epsilon_F - \epsilon)}{\pi\hbar^2(\omega + i/\tau)} \ , \tag{193}$$

$$A = \frac{q_\| e^2\tau(\epsilon - \epsilon_F)}{2\pi\epsilon_0\omega\hbar^2(1 - i\omega\tau)} \ , \tag{194}$$

and

$$U = \frac{e^2\hbar^2 q_{\|}(S_- - S_+)}{i\epsilon_0 m^2\omega^2} \ . \tag{195}$$

It is apparent that the solution of Eq. (190) must have the form

$$E_z(z) = \frac{E_z^{B,eff}(z) + WC}{1 - V\psi^2(z)} \ , \tag{196}$$

where

$$C = \int_{QW} \psi^2(z')E_z(z')dz' \tag{197}$$

is a yet unknown constant. This constant is readily obtained by inserting Eq. (196) into (197). Doing this, one gets

$$C = \left[1 - W\int_{QW} \frac{\psi^2(z)dz}{1 - V\psi^2(z)}\right]^{-1} \int_{QW} \frac{\psi^2(z)E_z^{B,eff}(z)dz}{1 - V\psi^2(z)} \ . \tag{198}$$

Having thus obtained C, the z-component of the local field across the quantum well is given by Eq. (196).

6.3.3 Self-Field Approximation

If one neglects all parts of the electromagnetic background propagator with the exception of the self-field part, $(c_0/\omega)^2\delta(z - z')\vec{e}_z\vec{e}_z$, in the basic integral equations in (179)-(181), the local field *inside* the quantum well is given by

$$\vec{E}_{\|} = \vec{E}_{\|}^B \ , \tag{199}$$

and

$$E_z(z) = \frac{E_z^B}{1 - V\psi^2(z)} \ , \tag{200}$$

where $\vec{E}_{\|}$ and $\vec{E}_{\|}^B$ denote the projections of the local and background fields on the xy-plane, respectively. Although the self-field approximation certainly is too crude to be able to account for the linear p and s-polarized reflectivities which almost exclusively depend on the local-field corrections along the quantum well, this approximation describes the z-component of the local field quite well at least in thin quantum wells [81].

In a sense, the self-field approximation allows us to make contact to macroscopic electrodynamics as far as the electromagnetic field analysis is concerned. Thus, in the self-field approximation the integral equation (181) for $E_z(z)$ can be written

$$\int_{QW} \left[\delta(z - z') + \frac{i}{\epsilon_0\omega}\sigma_{zz}(z, z')\right] E_z(z')dz' = E_z^B(z) \ . \tag{201}$$

By introducing the zz-component, $\epsilon_{zz}(z, z')$, of the relative dielectric tensor

$$\epsilon_{zz}(z, z') = \delta(z - z') + \frac{i}{\epsilon_0\omega}\sigma_{zz}(z, z') \tag{202}$$

it readily appears that the left hand side of Eq. (201) is just the z-component of the electric displacement field, $\vec{D}(z)$, i.e.

$$D_z(z) = \epsilon_0 \int_{QW} \epsilon_{zz}(z,z')E_z(z')dz' , \tag{203}$$

divided by ϵ_0. This means that Eq. (201) is equivalent to the relation

$$D_z(z) = \epsilon_0 E_z^B(z) . \tag{204}$$

Since the background field is almost constant across the quantum well, it appears from Eqs. (199) and (204) that the tangential component of the prevailing electric field, $\vec{E}_{\|}$, and the normal component of the electric displacement field, D_z, are constant across the well. If one considers an interface between two semi-infinite media (here the vacuum and the substrate) one notices that in macroscopic electrodynamics $\vec{E}_{\|}$ and, if there is no induced charge at the interface, also D_z are constant across the boundary.

The relation in Eq. (204) was derived under the assumption that the conductivity tensor (Eq. (172) (and hence also the dielectric response tensor) was diagonal. It holds, however, in general within the framework of the self-field approximation. To prove this, we take the basic integral equation in (178) in the self-field approximation, viz.

$$E_z(z) = E_z^B(z) - \frac{i}{\epsilon_0\omega}\vec{e}_z \cdot \int_{QW} \overleftrightarrow{\sigma}(z,z') \cdot \vec{E}(z')dz' , \tag{205}$$

and rewrite it as follows

$$\vec{e}_z \cdot \int_{QW} \left[\overleftrightarrow{U}\,\delta(z-z') + \frac{i}{\epsilon_0\omega}\overleftrightarrow{\sigma}(z,z')\right] \cdot \vec{E}(z')dz' = E_z^B(z) . \tag{206}$$

By comparing this equation with the general relation

$$\vec{D}(z) = \epsilon_0 \int_{QW} \overleftrightarrow{\epsilon}(z,z') \cdot \vec{E}(z')dz' , \tag{207}$$

where

$$\overleftrightarrow{\epsilon}(z,z') = \overleftrightarrow{U}\,\delta(z-z') + \frac{i}{\epsilon_0\omega}\overleftrightarrow{\sigma}(z,z') \tag{208}$$

is the relative dielectric tensor, an equation analogous to that in (204) appears immediately.

It is of interest to compare the result obtained for the z-component of the electric displacement field in the self-field approximation (Eq. (204)) with the one found in the slave approximation. Upon a simple contemplation, it appears that

$$D_z(z) = \epsilon_0(1 - V\psi^2(z))E_z(z) , \tag{209}$$

cf. Eqs. (175) [with (169)], (202),(203), and (192). A comparison to Eq. (196) in turn shows that

$$D_z = \epsilon_0(E_z^{B,eff}(z) + WC) . \tag{210}$$

The equation above is the one which replaces Eq. (204) in the slave approximation. Often the term WC is negligible so that $D_z = \epsilon_0 E_z^{B,eff}$. Whereas the electric displacement is constant across the quantum well in the self-field approximation, in general, it varies appreciable with the position in the slave approximation, cf. Eqs. (189) and (191).

6.4 Photon Drag in the Slave and Self-Field Approximations

Having analysed the local field at the fundamental frequency let us return to a brief discussion of the photon-drag effect within the framework of the slave and self-field approximations.

6.4.1 Drag Current and Voltage

From an experimental point of view the integrated photon-drag current density, $\vec{I}_0$, given by

$$\vec{I}_0 = (\overleftrightarrow{U} - \vec{e}_z \vec{e}_z) \cdot \int_{QW} \vec{J}_0(z) dz \tag{211}$$

with $\vec{J}_0(z)$ taken from Eq. (162), is of particular importance. The integrated drag current density is independent of the z-component of the local field at the fundamental frequency, and given within the slave approximation in explicit form as follows:

$$\vec{I}_0 = \frac{1}{2} \left\{ K(q_\parallel, \omega) \frac{E_x^B}{1 - K_{xx}} \left[\frac{E_x^B \vec{e}_x}{1 - K_{xx}} + \frac{E_y^B \vec{e}_y}{1 - K_{yy}} \right]^* \quad + \quad c.c. \right\} , \tag{212}$$

as one readily realizes by using the normalization of the wave function and the expressions for the local-field components displayed in Eqs. (182) and (186). The *dc* voltage, $V_0 = -\int E_{0,z}(z) dz$, induced across the quantum well is readily obtained in the slave approximation by inserting the appropriate formula (Eqs. (186) and (196)) for the local-field components into Eq. (170) and performing the integration over the well domain. Thus,

$$V_0 = -\frac{\pi \hbar^2}{2e^2 \tau(\epsilon - \epsilon_F)}$$

$$\times \left\{ K(q_\parallel, \omega) \frac{E_x^B}{1 - K_{xx}} \int_{QW} \frac{(E_z^{B,eff}(z))^* + W^* C^*}{1 - V^* \psi^2(z)} dz \quad + c, .c. \right\} \tag{213}$$

Crude approximate expressions for $\vec{I}_0$ and V_0 readily can be obtained on the basis of the self-field model. Hence, setting $K_{xx} = K_{yy} = 1$ one obtains

$$\vec{I}_0 = \frac{1}{2} \left[K(q_\parallel, \omega) E_x^B (\vec{E}_\parallel^B)^* + \text{c.c.} \right] , \tag{214}$$

or in particular if the incident field is *p*-polarized

$$\vec{I}_0^p = \vec{e}_x \mid E_x^B \mid^2 Re\, K(q_\parallel, \omega) . \tag{215}$$

Furthermore,

$$V_0 = -\frac{\pi \hbar^2}{2e^2 \tau(\epsilon - \epsilon_F)} \left\{ K(q_\parallel, \omega) E_x^B (E_z^B)^* \int_{QW} \frac{dz}{1 - V^* \psi^2(z)} \quad + \text{c.c.} \right\} . \tag{216}$$

6.4.2 Local-Field Resonances

In subsection 2.5, I discussed in a general mode of expression the condition for obtaining selfsustaining oscillations in the local field of a system, and in subsection 4.5, resonance conditions in mesoscopic media were described. In the context of the photon-drag effect, the local-field resonances are those associated with the fundamental field. Thus, for s-polarized excitation of the quantum well the local-field resonance occurs for

$$K_{yy} = 1 , \tag{217}$$

cf. Eq. (182). In the p-polarized case, the condition for obtaining a local-field resonance is a common one for the x and z-components of the field when these are coupled nonparametrically. Within the framework of the slave approximation, in which the x-component of the local field is independent of the z-component, the resonance condition splits into two, viz. one for field along the quantum well and one for the field component perpendicular to the well. It readily appears from Eq. (186) that the resonance condition for $E_x(z)$ is given by

$$K_{xx} = 1 , \tag{218}$$

an expected result in view of the equivalence of the integral equations for E_x and E_y in the slave approximation. The resonance condition for the z-component of the local field can be obtained from Eq. (190) setting $E_z^{B,eff} = 0$, because when the background field is absent $E_z^{B,eff} = 0$, as can be seen from Eq. (191). For the homogeneous part of Eq. (190) to have a nonzero solution it is realized that one must require

$$\int_{QW} \frac{\psi^2(z)dz}{1 - V\psi^2(z)} = W^{-1} . \tag{219}$$

The resonance condition in Eq. (219), of course, could have been obtained immediately from Eq. (198), setting the denominator of C equal to zero. Let me emphasize at this place that when the resonance condition in Eq. (219) is fulfilled the z-component of the local field is resonantly enhanced *in every point in space*. Another type of resonance can occur at specific depths, $z = z_0$, inside quantum well if the condition

$$V\psi_0^2(z) = 1 \tag{220}$$

is met. By means of Eqs. (169) and (192), Eq. (220) can be written in the form

$$1 - \frac{\omega_p^2(z_0)}{\omega(\omega + i/\tau)} = 0 , \tag{221}$$

where $\omega_p(z_0) = \left[n(z_0)e^2/(m\epsilon_0)\right]^{1/2}$. The relation in Eq. (221) can be recognized as the condition for excitation of a localized plasmon having a cyclic frequency $\omega_p(z_0)$. The localized plasmon normally does not cause a pronounced resonance in the photon-drag voltage because it enters via some weighted integrals over the quantum well, cf. Eqs. (198) and (213). The localized plasmon resonance is absent in the integrated photon-drag current density, $\vec{I}_0$. Once again, let me stress that the localized plasmon resonance and the local-field resonance are two different concepts, a fact often misunderstood in the litterature. It is interesting to notice that within the framework

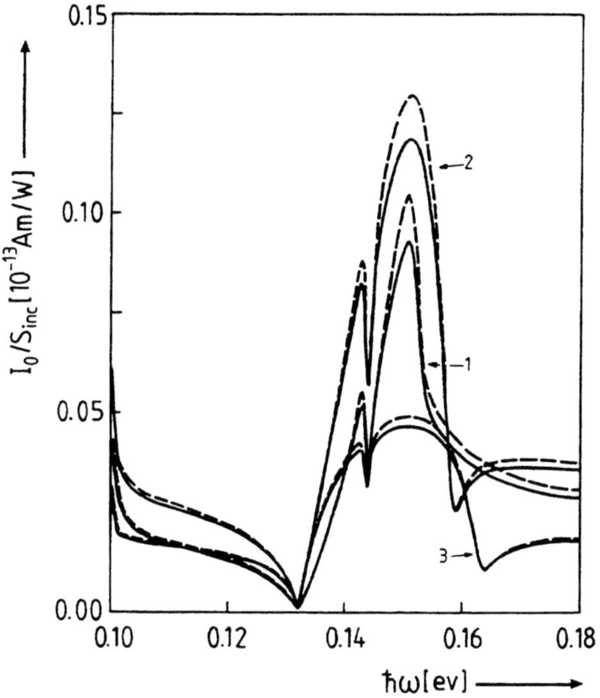

Figure 10. *Magnitude of the ratio between the integrated photon-drag current density* (I_0) *and the time-averaged electromagnetic Poynting vector* (S_{inc}) *of the incident field as a function of the photon energy* ($\hbar\omega$) *for three different angles of incidence* (θ), *viz.* $\theta = 20°(1)$, $50°(2)$, *and* $70°(3)$. *The incident field is p-polarized, and the mesoscopic system consists of a niobium quantum well of thickness* $d = 3Å$ *deposited on a crystalline quartz substrate. Fully drawn curves: Slave model; dashed curves: Self-field model. The figure is taken from Ref. [65] in which also a detailed account of the results is given.*

of the self-field approximation only the localized plasmon resonance is present (see Eq. (200)). From a different point of view the localized plasmon resonance equivalently can be characterized as a self-field resonance. The relations in Eqs. (217)-(219), constitute the eigenmode conditions of the quantum well system. Especially, the condition in Eq. (218) in implicit form gives the various branches of the surface-wave dispersion relation of the quantum well system. For a niobium quantum well at a vacuum/quartz interface the surface-wave dispersion relation has been investigated recently [81]. The eigenmode conditions in Eq. (217) and (219) has not yet been explored.

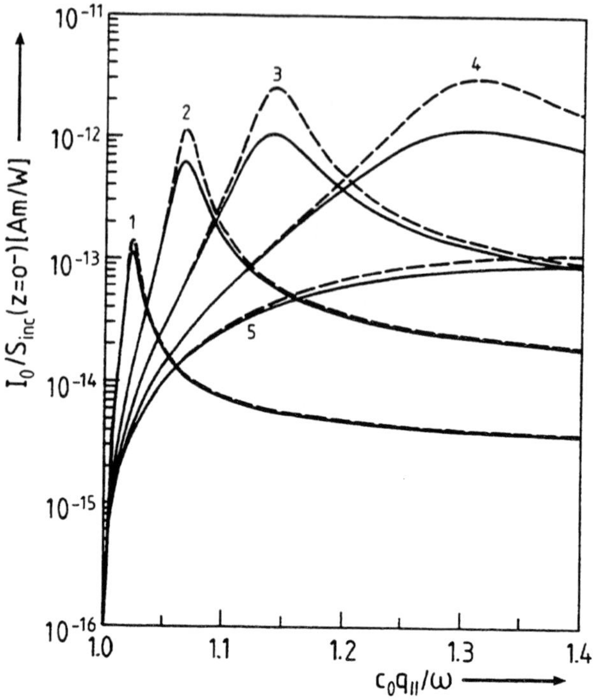

Figure 11. *Magnitude of the ratio between the integrated photon-drag current density* (I_0) *and the time-averaged electromagnetic Poynting vector* (S_{inc}) *of the incident field as a function of the normalized (real) wave vector* ($c_0q_{\|}/\omega$) *parallel to the plane of the quantum well (Nb/quartz system with d= 3Å). The incident field is that belonging to a surface electromagnetic wave (SEW) excited in the Otto configuration and* S_{inc} *is the value of the electromagnetic energy flow just outside the substrate. Curves are plotted in the vicinity of the SEW-resonance for the full system for five different photon energies, viz.* $\hbar\omega$ = *0.134 eV (1), 0.136 eV (2), 0.139 eV (3), 0.141 eV (4), and 0.144 eV (5). Fully drawn curves: Slave model; dashed curves: Self-field model. By comparison to Fig. 10 it appears that the loccal-field resonance (related to the condition* $K_{xx} = 1$ *in this case) enhances the photon-drag current density by two orders of magnitude. The reason that SEW-excitation is used stems from the fact that to fulfil the resonance condition* $K_{xx} = 1$, *the real part of* $q_{\|}$ *must exceed* ω/c_0. *For a detailed account of the theory behind Fig. 11 see Ref. [65].*

In the context of the photon-drag effect, it is realized (see Eqs. (212)) that the integrated drag current density can be resonantly enhanced if $K_{xx} = 1$ or $K_{yy} = 1$. To enhance the drag voltage one would seek at fulfilling either the condition $K_{xx} = 1$ or the relation in Eq. (219). As far as the *dc* part, $E_{0,z}(z)$, of the local field is concerned a spatially localized resonance can occur if the resonance condition for the local plasmon excitation in Eq. (220) is fulfilled, cf. Eqs. (170) and (196). In Ref. [81] the spatially localized field resonance is examined numerically for the vacuum/niobium/quartz quantum well system.

Numerical results for the photon-drag current density in a 3Å thick Nb quantum well are presented in Figs. 10 and 11.

7 OPTICAL SECOND-HARMONIC GENERATION IN A MESOSCOPIC MEDIUM

In the present section I shall treat the specific local-field problems which are associated with the optical second-harmonic generation from mesoscopic systems, including metallic quantum wells. As in the previous section, in the quantum well case it will be assumed that the jellium exhibits free-electron-like behaviour along the quantum well, and in the wake of a general discussion of the nonlinear background field, we shall limit ourselves to a heuristic study of two-level quantum wells. For two-level mesoscopic media, the paramagnetic interaction is of particular importance, and in the article by A. Liu and myself in this book the significance of the intersubband transition for the optical second-harmonic generation from niobium quantum wells is analysed. In most studies, a phenomenological multipole expansion is used when calculating the forced part of the nonlinear current denstiy. The expansion usually is terminated when electric dipole and electric quadropole plus magnetic dipole terms have been included. Since the expansion in previous works always has been made in the local field the multipole approach appears to be an unjustified procedure for metallic systems dominated by strongly delocalized (Bloch-like) wave functions. With emphasis on mesoscopic systems, I shall indicate below that the multipole expansion scheme can be meaningful even in metallic systems provided that the expansion is made not in the prevailing field but in the external field. In this manner, new mircroscopic expressions can be derived for the phenomenological constants of the local and near-local (hydrodynamic) models.

7.1 Nonlinear Background Dynamics

7.1.1 Forced Current Density

It is well known [57] that the forced part of the second-harmonic current density $\vec{J}_2(\vec{r}) = \vec{J}(\vec{r};2\omega)$, discussed in subsection 5.2 always can be written in the form

$$\vec{J}(\vec{r};2\omega) = \int \overleftrightarrow{\Sigma}_2 \,(\vec{r},\vec{r}_1,\vec{r}_2;\omega \rightarrow 2\omega) : \vec{E}(\vec{r}_2;\omega)\vec{E}(\vec{r}_1;\omega)d^3r_2d^3r_1 \ , \tag{222}$$

where the nonlinear response tensor

$$\overleftrightarrow{\Sigma}_2 = \overleftrightarrow{\Sigma}_2^A + \overleftrightarrow{\Sigma}_2^B + \overleftrightarrow{\Sigma}_2^C \ , \tag{223}$$

composed of three contributions denoted by superscripts A,B, and C, can be derived from the respective equations (140)-(142). By means of the field-field response tensor $\overleftrightarrow{\Gamma}\,(\vec{r},\vec{r}\,';\omega)$, the local field in Eq. (222) can be eliminated in favour of the background field at the fundamental frequency, cf. Eq. (21). Thus leaving out ω from the notation one obtains

$$\vec{J}_2(\vec{r}) = \int \overleftrightarrow{\Lambda}_2 (\vec{r}, \vec{r}_1, \vec{r}_2) : \vec{E}^B(\vec{r}_2)\vec{E}^B(\vec{r}_1) d^3r_2 d^3r_1 \ , \tag{224}$$

where the so-called nonlinear background response function $\overleftrightarrow{\Lambda}_2$, a tensor of third rank, is given by

$$\overleftrightarrow{\Lambda}_2 (\vec{r}, \vec{r}_1, \vec{r}_2) = \int \overleftrightarrow{\Sigma}_2 (\vec{r}, \vec{r}_3, \vec{r}_4) \overset{\circ}{\circ} \overleftrightarrow{\Gamma} (\vec{r}_4, \vec{r}_2) \ \overleftrightarrow{\Gamma} (\vec{r}_3, \vec{r}_1) d^3r_4 d^3r_3 \ . \tag{225}$$

The special symbol $\overset{\circ}{\circ}$ in Eq.(225) is meant to indicate that to obtain the tensor element $\Lambda_{2,ijk}$ one has to perform the operation $\sum_{2,ilm} \Gamma_{mk}\Gamma_{lj}$ with summation over repeated indices to be understood.

For the "one-dimensional" quantum well system, the nonlinear constitutive equation in (224) takes the form

$$\vec{J}_2(z) = \int \overleftrightarrow{\Lambda}_2 (z, z_1, z_2) : \vec{E}^B(z_2)\vec{E}^B(z_1) dz_2 dz_1 \tag{226}$$

with

$$\overleftrightarrow{\Lambda}_2 (z, z_1, z_2) = \int \overleftrightarrow{\Sigma}_2 (z, z_3, z_4) \overset{\circ}{\circ} \overleftrightarrow{\Gamma} (z_4, z_2) \ \overleftrightarrow{\Gamma} (z_3, z_1) dz_4 dz_3 \ . \tag{227}$$

In the equations above the quantities appearing are those obtained from Eqs. (224) and (225) upon Fourier analyses in the x and y-coordinates. As usual, the reference to $\vec{q}_{\parallel}$ and ω has been left implicit in the equations. In the mixed Fourier space, the standard nonlinear constitutive equation is given by

$$\vec{J}_2(z) = \int \overleftrightarrow{\Sigma}_2 (z, z_1, z_2) : \vec{E}(z_2)\vec{E}(z_1) dz_2 dz_1 \ . \tag{228}$$

7.1.2 Moment Expansion of the Nonlinear Background Response Tensor

Due to the slow variation of the background field at the fundamental frequency across the domain of a mesoscopic system it is adequate to make a Taylor series expansion of $\vec{E}^B(\vec{r}_\alpha)[\alpha = 1 \text{ or } 2]$ around $\vec{r}$ in Eq. (224). This leads to the series

$$\vec{J}_2(\vec{r}) = \overleftrightarrow{\Lambda}_{2,l} (\vec{r}) : \vec{E}^B(\vec{r})\vec{E}^B(\vec{r})$$

$$+ \overleftrightarrow{\Lambda}_{2,nl} (\vec{r}) \vdots \vec{\nabla}\vec{E}^B(\vec{r}) \mid \vec{E}^B(\vec{r}) \quad + \cdots \ , \tag{229}$$

where the vertical bar ($|$) means that the nabla operator is to act only to the left of $|$. The local nonlinear (third-rank) response tensor is given by

$$\overleftrightarrow{\Lambda}_{2,l} (\vec{r}) = \int \overleftrightarrow{\Lambda}_2 (\vec{r}, \vec{r}_1, \vec{r}_2) d^3r_2 d^3r_1 \ , \tag{230}$$

the integrations extending as usual over all space. The near-local nonlinear response function, which is a fourth-rank tensor, is obtained from $\overleftrightarrow{\Lambda}_2 (\vec{r}, \vec{r}_1, \vec{r}_2)$ as follows:

$$\overleftrightarrow{\Lambda}_{2,nl}(\vec{r}) = \int \Big[\overleftrightarrow{\Lambda}_2(\vec{r},\vec{r}_1,\vec{r}_2)(\vec{r}_2-\vec{r})$$

$$+ \tilde{\overleftrightarrow{\Lambda}}_2(\vec{r},\vec{r}_1,\vec{r}_2)(\vec{r}_1-\vec{r})\Big] d^3r_2 d^3r_1 \ , \tag{231}$$

where $\sim$ indicates that $\overleftrightarrow{\Lambda}_2$ is to be transported in the last two indices, i.e. $\tilde{\Lambda}_{2,ijk} = \Lambda_{2,ikj}$.

By combining Eqs. (111), (225), and (230) it is realized that the local response tensor, $\overleftrightarrow{\Lambda}_{2,l}(\vec{r})$ can be written in the alternative form

$$\overleftrightarrow{\Lambda}_{2,l}(\vec{r}) = \int \overleftrightarrow{\Sigma}_2(\vec{r},\vec{r}_3,\vec{r}_4) \overset{\circ}{\circ} \overleftrightarrow{\Gamma}_l(\vec{r}_4)\,\overleftrightarrow{\Gamma}_l(\vec{r}_3) d^3r_4 d^3r_3 \ . \tag{232}$$

For the one-dimensional system a moment expansion gives

$$\vec{J}_2(z) = \overleftrightarrow{\Lambda}_{2,l}(z) : \vec{E}^B(z)\vec{E}^B(z) + \overleftrightarrow{\Lambda}_{2,nl}(z) : \frac{d\vec{E}^B(z)}{dz}\vec{E}^B(z) \tag{233}$$

with

$$\overleftrightarrow{\Lambda}_{2,l}(z) = \int \overleftrightarrow{\Lambda}_2(z,z_1,z_2) dz_2 dz_1 \ , \tag{234}$$

and

$$\overleftrightarrow{\Lambda}_{2,nl}(z) = \int \left[\overleftrightarrow{\Lambda}_2(z,z_1,z_2)(z_2-z) + \tilde{\overleftrightarrow{\Lambda}}_2(z,z_1,z_2)(z_1-z)\right] dz_2 dz_1 . \tag{235}$$

Although it could have been done there is no need to make a Taylor series expansion in the coordinates along the surface since all quantities have the same and simple $\exp(i\vec{q}_{\parallel}\cdot\vec{r})$-dependence. A detailed account of the moment expansion scheme based, albeit, on a Taylor series expansion of the local field is presented in [57].

7.1.3 Nonlinear Background Field

The background field driving the second-harmonic response of a mesoscopic system stems from the forced part of the second-harmonic current density and is given by

$$\vec{E}^B(\vec{r};2\omega) = -2i\mu_0\omega \int \overleftrightarrow{G}(\vec{r},\vec{r}_1;2\omega)\cdot\vec{J}(\vec{r}_1;2\omega) d^3r_1 \ . \tag{236}$$

The nonlinear background field usually varies rapidly across the domain of the mesoscopic system due to the presence of a self-field part, $\overleftrightarrow{g}^{SF}$, of the propagator $\overleftrightarrow{G}(\vec{r},\vec{r}_1;2\omega)$. The self-field propagator is composed of so-called transverse and longitudinal parts, viz. [56]

$$\overleftrightarrow{g}^{SF}(\vec{r},\vec{r}_1;2\omega) = \frac{1}{3}\left(\frac{c_0}{2\omega}\right)^2 \overleftrightarrow{\delta}^T(\vec{r}-\vec{r}_1) + \left(\frac{c_0}{2\omega}\right)^2 \overleftrightarrow{\delta}^L(\vec{r}-\vec{r}_1) \ , \tag{237}$$

where $\overleftrightarrow{\delta}^T$ and $\overleftrightarrow{\delta}^L$ are the transverse and longitudinal delta functions, respectively. If we split off the self-field part of the Green's function in Eq. (236) one obtains

$$\vec{E}^B(\vec{r};2\omega) = -2i\mu_0\omega \int \overleftrightarrow{\mathcal{G}}(\vec{r},\vec{r}_1;2\omega)\cdot\vec{J}(\vec{r}_1;2\omega)d^3r_1$$

$$+\frac{1}{3i\epsilon_0(2\omega)}\vec{J}^T(\vec{r};2\omega)+\frac{1}{i\epsilon_0(2\omega)}\vec{J}^L(\vec{r};2\omega)\,, \tag{238}$$

where $\vec{J}^T$ and $\vec{J}^L$ denote the transverse and longitudinal parts of the forced current density in Eq. (224) [or equivalently in Eq. (222)] and $\overleftrightarrow{\mathcal{G}}$ stands for the "propagating" part of $\overleftrightarrow{G}$. The background field thus contains terms which vary in space as the T and L-parts of the forced nonlinear current density.

For the quantum well case the relevant background field takes the form

$$\vec{E}^B(z;2\vec{q}_\parallel,2\omega) = -2i\mu_0\omega \int \overleftrightarrow{G}(z,z_1;2\vec{q}_\parallel,2\omega)\cdot\vec{J}(z_1;2\vec{q}_\parallel,2\omega)dz_1 \tag{239}$$

with $\vec{J}(z_1;2\vec{q}_\parallel,2\omega)\equiv\vec{J}(z_1)$ being given by Eq. (226)[or (228)]. Dividing, once more, $\overleftrightarrow{G}$ into its self-field and propagating parts, $\vec{E}^B(z;2\vec{q}_\parallel,2\omega)$ is given by

$$\vec{E}^B(z;2\vec{q}_\parallel,2\omega) = -2i\mu_0\omega \int \overleftrightarrow{\mathcal{G}}(z,z_1;2\vec{q}_\parallel,2\omega)\cdot\vec{J}(z_1;2\vec{q}_\parallel,2\omega)dz_1$$

$$+\frac{1}{i\epsilon_0(2\omega)}\vec{e}_z\vec{e}_z\cdot\vec{J}(z;2\vec{q}_\parallel,2\omega)\,. \tag{240}$$

In many cases it will be sufficient to calculate the background field in the local limit, i.e. in an approach where a local relation (with hidden nonlocality) exists between the second and first harmonic background fields. Thus by means of Eqs. (229) and (236) one obtains

$$\vec{E}_l^B(\vec{r};2\omega) = -2i\mu_0\omega\left[\int \overleftrightarrow{G}(\vec{r},\vec{r}_1;2\omega)\cdot\overleftrightarrow{\Lambda}_l(\vec{r};2\omega)d^3r_1\right]:\vec{E}^B(\vec{r};\omega)\vec{E}^B(\vec{r};\omega)\,, \tag{241}$$

with $\overleftrightarrow{\Lambda}_{2,l}(\vec{r}_1)\equiv\overleftrightarrow{\Lambda}_l(\vec{r}_1;2\omega)$ in the general three-dimensional case. In the quantum well system, the corresponding relation becomes

$$\vec{E}_l^B(z;2\vec{q}_\parallel,2\omega)$$

$$= -2i\mu_0\omega\left[\int \overleftrightarrow{G}(z,z_1;2\vec{q}_\parallel,2\omega)\cdot\overleftrightarrow{\Lambda}_l(z_1;2\vec{q}_\parallel,2\omega)dz_1\right]:\vec{E}^B(z;\vec{q}_\parallel,\omega)\vec{E}^B(z;\vec{q}_\parallel,\omega)\,, \tag{242}$$

allowing yet $\vec{q}_\parallel$ to be different from zero.

7.2 Paramagnetic Two-Level Metallic Quantum Well. A Heuristic Approach

To illustrate the course of action attached to solving local-field problems in the optical second-harmonic generation from quantum well systems in this final section I shall consider a simple two-level metal quantum well, with the one level above and the other below the Fermi level, and assume that the paramagnetic coupling dominates the particle-field interaction in the frequency range of interest. Aiming at a heuristic description I shall confine myself to the RPA approach, assume that the electrons of the quantum well exhibit free-particle-like behaviour along the well, study only the long-wavelength ($\vec{q}_{\|} \rightarrow 0$) limit, and treat the electromagnetic field propagation within the framework of the self-field approximation.

7.2.1 Local Field at the Second-Harmonic Frequency

To determine the prevailing field at the second-harmonic frequency one takes at a starting point the basic integral equation for the local field $\vec{E}_z(z)$, viz.

$$\vec{E}_2(z) = \vec{E}_2^B(z) - 2i\mu_0\omega \int_{QW} \overleftrightarrow{G}_2(z,z') \cdot \overleftrightarrow{\sigma}_2(z',z'') \cdot \vec{E}_2(z'')dz''dz' . \tag{243}$$

subscripts 2 are added to the various quantities to stress that these belong to the frequency $\Omega = 2\omega$, and as usual the reference to $\vec{q}_{\|}$ and ω has been omitted from the notation. In the self-field approximation, where $\overleftrightarrow{G}_2(z,z')$ is replaced by the term $[c_0/(2\omega)]^2\delta(z-z')\vec{e}_z\vec{e}_z$, Eq. (243) is reduced to the form

$$\vec{E}_2(z) = \vec{E}_2^B + \frac{1}{2i\epsilon_0\omega}\vec{e}_z\vec{e}_z \cdot \int_{QW} \overleftrightarrow{\sigma}_2(z,z') \cdot \vec{E}_2(z')dz' . \tag{244}$$

To proceed from here we make use of the fact that the linear paramagnetic conductivity tensor $\overleftrightarrow{\sigma}_2^{par}(z,z')$ only has nonvanishing diagonal elements in the long-wavelength ($\vec{q}_{\|} \rightarrow 0$) limit. Within the framework of the RPA description, the nonzero elements of the long-wavelength paramagnetic conductivity tensor of a two-level system are given by

$$\sigma_{xx}^{par}(z,z';\vec{Q}_{\|} \rightarrow 0,\Omega) = \sigma_{yy}^{par}(z,z';\vec{Q}_{\|} \rightarrow 0,\Omega) = a(\Omega)\phi(z)\phi(z') , \tag{245}$$

and

$$\sigma_{zz}^{par}(z,z';\vec{Q}_{\|} \rightarrow 0,\Omega) = b(\Omega)\Phi(z)\Phi(z') , \tag{246}$$

at $T = 0K$ in the relaxation-time approximation, where

$$a(\Omega) = \frac{ie^2}{\pi\hbar^2\Omega}\,\frac{(\epsilon_2-\epsilon_1)(\epsilon_F-\epsilon_1)^2}{[\hbar(\Omega+i/\tau)]^2-(\epsilon_2-\epsilon_1)^2} , \tag{247}$$

$$b(\Omega) = \frac{ie^2}{2\pi m\Omega}\,\frac{(\epsilon_2-\epsilon_1)(\epsilon_F-\epsilon_1)}{[\hbar(\Omega+i/\tau)]^2-(\epsilon_2-\epsilon_1)^2} , \tag{248}$$

and

$$\phi(u) = \psi_1(u)\psi_2(u), \qquad u = z \text{ or } z' \tag{249}$$

$$\Phi(u) = \psi_1(u)\frac{d\psi_2(u)}{du} - \psi_2(u)\frac{d\psi_1(u)}{du}. \qquad u = z \text{ or } z' \tag{250}$$

In the equations above $\psi_n(u)$ $(n = 1, 2)$ denote the (real) wave functions of the lower $(n = 1)$ and upper $(n = 2)$ energy eigenstates, the associated energies being $\epsilon_1(< \epsilon_F)$ and $\epsilon_2(> \epsilon_F)$, respectively. Furthermore, $\vec{Q}_\parallel = \vec{q}_\parallel$ or $2\vec{q}_\parallel$, and $\Omega = \omega$ or 2ω, as appropriate. In the following only $\sigma_{zz}^{par}(z.z'; \vec{Q}_\parallel \to 0, \Omega)$ is needed. Thus, setting $\overleftrightarrow{\sigma}_2(z, z') = \overleftrightarrow{\sigma}^{par}(z, z,'; 2\vec{q}_\parallel \to 0, 2\omega)$ in Eq. (244), one obtains

$$\vec{E}_{2,\parallel}(z) = \vec{E}^B_{2,\parallel}(z) , \tag{251}$$

and

$$E_{2,z}(z) = E^B_{2,z}(z) + \frac{b(2\omega)}{2i\epsilon_0\omega} C\Phi(z) , \tag{252}$$

where

$$C = \int_{QW} \Phi(z')E_{2,z}(z')dz' . \tag{253}$$

In the present approximation the local field parallel (||) to the plane of the quantum well thus equals the nonlinear background field (Eq. (251)). To determine the unknown coefficient C, Eq. (252) is inserted in Eq. (253). This gives

$$C = \left[1 - \frac{b(2\omega)}{2i\epsilon_0\omega}\int_{QW} \Phi^2(z')dz'\right]^{-1} \int_{QW} \Phi(z')E^B_{2,z}(z')dz' , \tag{254}$$

and hence the following result for the z-component of the local field appears:

$$E_{2,z}(z) = E^B_{2,z}(z) + \frac{\int_{QW} \Phi(z')E^B_{2,z}(z')dz'}{\frac{2i\epsilon_0\omega}{b(2\omega)} - \int_{QW} \Phi^2(z')dz'}\Phi(z) . \tag{255}$$

It appears from Eq. (255) that the local-field correction to the z-component of the nonlinear background field has a spatial dependence proportional to $\Phi(z)$, which in turn is proportional to the z-component of the transition current density between the two states. The strength of the correction is given by the factor in front of $\Phi(z)$, and the local-field resonance condition for $E_{2,z}$ is

$$\frac{b(2\omega)}{2i\epsilon_0\omega}\int_{QW} \Phi^2(z)dz = 1 . \tag{256}$$

In passing, let me emphasize that in the paramagnetic case even the self-field approximation can lead to an essential local-field resonance, i.e. a resonantly enhanced field in every point in space. In comparison it was shown in section 6.4.2 that only a localized plasmon resonance was present in the diamagnetic case if the self-field approximation is adopted.

7.2.2 Background Field at the Second-Harmonic Frequency

The next step in our heuristic analysis aims at a calculation of the nonlinear background field, $\vec{E}^B_2(z)$. In the self-field approximation one immediately obtains from Eq. (240)

$$\vec{E}^B_2(z) = \frac{1}{2i\epsilon_0\omega}\vec{e}_z\vec{e}_z \cdot \vec{J}_2(z) . \tag{257}$$

This means that the nonlinear background field along the surface vanishes. Thus, in the present model one finds

$$\vec{E}_{2,\parallel}(z) = 0 \ , \tag{258}$$

as one might have anticipated. To obtain the z-component of the background field one needs to determine the z-component of the forced current density in the long-wavelength limit, i.e. $\vec{e}_z \cdot \vec{J}_2(z) = J_{2,z}(z; 2\vec{q}_\parallel \to 0, 2\omega)$.

7.2.3 Driven Current Density at the Second-Harmonic Frequency

To calculate the driven (forced) current density I shall assume that the linear (optical) background field, $\vec{E}^B$, varies slowly across the quantum well and hence make use of the local expression

$$\vec{J}_2(z) = \overleftrightarrow{\Lambda}_{2,l}(z) : \vec{E}^B \vec{E}^B \tag{259}$$

for the current density, $\vec{E}^B(z) \simeq \vec{E}^B(0) \equiv \vec{E}^B$. According to Eq. (232) the local nonlinear background response function is given by

$$\overleftrightarrow{\Lambda}_{2,l}(z) = \int_{QW} \overleftrightarrow{\Sigma}_2(z, z', z'') \mathring{\circ} \overleftrightarrow{\Gamma}_l(z'') \overleftrightarrow{\Gamma}_l(z') dz'' dz' \ , \tag{260}$$

where $\overleftrightarrow{\Gamma}_l$ is the local field-field response tensor at ω. The fully nonlocal field-field response tensor, $\overleftrightarrow{\Gamma}(z, z'; \vec{q}_\parallel, \omega)$, obeys the equation

$$\vec{E}(z; \vec{q}_\parallel, \omega) = \int_{QW} \overleftrightarrow{\Gamma}(z, z'; \vec{q}_\parallel, \omega) \cdot \vec{E}^B(z'; \vec{q}_\parallel, \omega) dz' \ . \tag{261}$$

Since equations of the same form as those in (251) and (255) hold for the fundamental field, it is obvious that the field-field response tensor in Eq. (261) is given by

$$\overleftrightarrow{\Gamma}(z, z'; \vec{q}_\parallel \to 0, \omega) = \overleftrightarrow{U} \delta(z - z')$$

$$+ \left[\frac{i\epsilon_0 \omega}{b(\omega)} - \int_{QW} \Phi^2(z) dz \right]^{-1} \Phi(z)\Phi(z') \vec{e}_z \vec{e}_z \tag{262}$$

in the self-field approximation and long-wavelength limit. In turn, the corresponding local field-field response tensor

$$\overleftrightarrow{\Gamma}_l(z, z'; \vec{q}_\parallel \to 0, \omega) = \int_{QW} \overleftrightarrow{\Gamma}(z, z'; \vec{q}_\parallel \to 0, \omega) dz' \ , \tag{263}$$

becomes

$$\overleftrightarrow{\Gamma}_l(z) = \overleftrightarrow{U} + M(\omega)\Phi(z)\vec{e}_z\vec{e}_z \ , \tag{264}$$

where

$$M(\omega) = \left[\frac{i\epsilon_0 \omega}{b(\omega)} - \int_{QW} \Phi^2(z) dz \right]^{-1} \frac{2m}{\hbar^2} (\epsilon_2 - \epsilon_1) P_{2\to 1} \ . \tag{265}$$

In deriving Eq. (264) [with (265)] use have been made of the result

$$\int_{QW} \Phi(z)dz = \frac{2m}{\hbar^2}(\epsilon_2 - \epsilon_1)P_{2\to 1} , \tag{266}$$

where

$$P_{2\to 1}(= P_{1\to 2}) = \int_{QW} \psi_2(z) z \psi_1(z) dz \tag{267}$$

is the electric-dipole transition matrix element between the states 1 and 2.

After having calculated $\overleftrightarrow{\Gamma}_l(z)$, let us focuse our attention on the tensorial form of $\overleftrightarrow{\Sigma}_2$ in the $\vec{q}_\parallel \to 0$ limit. As mentioned previously $\overleftrightarrow{\Sigma}_2^B$ will vanish in a two-level system. Furthermore, straightforward symmetry considerations of the type considered in Ref. [82] for instance show that the remaining parts of the nonlinear conductivity tensor in the long-wavelength limit have the tensorial forms

$$\overleftrightarrow{\Sigma}_2^A (z, z', z''; \vec{q}_\parallel \to 0, \omega) = \Sigma_2^A(z, z', z''; \omega)\vec{e}_z \overleftrightarrow{U} , \tag{268}$$

and [cf. also the analysis in subsections 5.1 and 5.2]

$$\overleftrightarrow{\Sigma}_2^C (z, z', z''; \vec{q}_\parallel \to 0, \omega) = \Sigma_2^C(z, z', z''; \omega) \overleftrightarrow{U} \vec{e}_z . \tag{269}$$

By means of Eqs. (264),(268), and (269) it follows after a little algebra that

$$\vec{e}_z \cdot \left(\overleftrightarrow{\Sigma}_2^A + \overleftrightarrow{\Sigma}_2^C \right) \overset{\circ}{\circ} \overleftrightarrow{\Gamma}_l(z'') \overleftrightarrow{\Gamma}_l(z')$$

$$= \Sigma_2^A(z, z', z''; \omega) \left\{ \overleftrightarrow{U} + \vec{e}_z\vec{e}_z \left[(1 + M(\omega)\Phi(z''))(1 + M(\omega)\Phi(z')) - 1 \right] \right\}$$

$$+ \Sigma_2^C(z, z', z''; \omega)\vec{e}_z\vec{e}_z(1 + M(\omega)\Phi(z''))(1 + M(\omega)\Phi(z')) . \tag{270}$$

Since it is known [82] that

$$\int_{QW} \Sigma_2^A(z, z', z''; \omega) dz'' dz' = 0 , \tag{271}$$

it is realized via Eqs. (259),(260), and (270) that the z-component of the forced current density is given by

$$J_{2,z}(z) = \mathcal{W}(z)(E_z^B)^2 , \tag{272}$$

where

$$\mathcal{W}(z) = \int_{QW} \left[\Sigma_2^A(z, z', z''; \omega) \right.$$

$$\left. + \Sigma_2^C(z, z', z''; \omega) \right] \left[1 + M(\omega)\Phi(z'') \right] \left[1 + M(\omega)\Phi(z') \right] dz'' dz' . \tag{273}$$

Putting together all the pieces, one obtains the following final result for the second-harmonic local field inside the quantum well:

$$\vec{E}_2(z)$$

$$= \frac{\vec{e}_z}{2i\epsilon_0\omega}\left[\mathcal{W}(z) + \frac{\int_{QW}\Phi(z')\mathcal{W}(z')dz'}{\frac{2i\epsilon_0\omega}{\delta(2\omega)} - \int_{QW}\Phi^2(z')dz'}\Phi(z)\right](E_z^B)^2 \ . \tag{274}$$

It was mentioned in subsection 7.2.1 that a local-field resonance occurs when the condition in Eq. (256) is met, this resonance stemming from the selfconsistent solution at 2ω. A similar resonance with 2ω replaced by ω appears from the selfconsistent field solution at ω. This last resonance enters $\vec{E}_2(z)$ via $M(\omega)$, cf. Eq. (265).

A detailed study of the optical second-harmonic generation from a paramagnetic quantum well which goes beyond the self-field approximation is given in the article by A. Liu and myself in this book, albeit with the prescription that the momentum expansion of the nonlinear current density is made in the local field and not in the background field at the fundamental frequency. In the above-mentioned article, a numerical calculation of the second-harmonic generation from a niobium quantum well showing the two local-field resonances is performed also.

The expression for $\mathcal{W}(z)$ can be put into a more explicit form inserting the low-temperature expressions for Σ_2^A and Σ_2^C, i.e.

$$\Sigma_2^A(z,z',z'';\omega)$$

$$= \frac{ie^3(\epsilon_F-\epsilon_1)}{4\pi m\omega^2}\frac{2\omega+i/\tau}{[\hbar(2\omega+i/\tau)]^2-(\epsilon_2-\epsilon_1)^2}\delta(z''-z')\phi(z')\Phi(z) \ , \tag{275}$$

and

$$\Sigma_2^C(z,z',z'';\omega)$$

$$= \frac{ie^3(\epsilon_F-\epsilon_1)}{2\pi m\omega^2}\frac{\omega+i\tau}{[\hbar(\omega+i/\tau)]^2-(\epsilon_2-\epsilon_1)^2}\delta(z'-z)\phi(z)\Phi(z'') \ . \tag{276}$$

By insertion of these results into Eq. (273) one obtains

$$\mathcal{W}(z) = \mathcal{W}^A(z) + \mathcal{W}^C(z) \ , \tag{277}$$

where

$$\mathcal{W}^A(z) = \frac{ie^3(\epsilon_F-\epsilon_1)}{4\pi m\omega^2}\frac{2\omega+i/\tau}{[\hbar(2\omega+i/\tau)]^2-(\epsilon_2-\epsilon_1)^2}M(\omega)\Phi(z)$$

$$\times\int_{QW}\phi(z')\Phi(z')\left[2+M(\omega)\Phi(z')\right]dz' \ , \tag{278}$$

and

$$\mathcal{W}^C(z) = \frac{ie^3(\epsilon_F-\epsilon_1)}{2\pi m\omega^2}\frac{\omega+i/\tau}{[\hbar(\omega+i/\tau)]^2-(\epsilon_2-\epsilon_1)^2}\phi(z)\left[1+M(\omega)\Phi(z)\right]$$

$$\times\int_{QW}\Phi(z')\left[1+M\Phi(z')\right]dz' \ . \tag{279}$$

Once the two wave functions and the associated energies have been determined the local field at the second-harmonic frequency can be calculated numerically.

References

[1] H. Fukuyama and T. Ando (Eds); *Transport phenomena in mesoscopic systems*, Berlin: Springer-Verlag, 1992, 283 pp.

[2] R. Kubo, "Electronic properties of fine metallic particles. I.", *J.Phys.Soc. Japan*, Vol. 17 No. 6, 1962, pp. 975-986.

[3] A. Kawabata and R. Kubo, "Electronic properties of fine metallic particles. II. Plasma resonance absorption", *J.Phys.Soc. Japan*, Vol.21, 1966, pp. 1765-1772.

[4] W.P. Halperin, "Quantum size effects in metal particles", *Rev.Mod.Phys.*, Vol. 58 No. 3, 1986, pp. 536-606.

[5] U. Kreibig and L. Genzel, "Optical absorption of small metallic particles", *Surf.Sci.*, Vol. 156, 1985, pp. 678-700.

[6] M.J. Rice, W.R. Schneider, and S. Strässler, "Electronic polarizabilities of very small metallic particles and thin films", *Phys.Rev. B.*, Vol. 8 No. 2, 1973, pp. 474-482.

[7] U. Kreibig, "Electronic properties of small silver particles: the optical constants and their temperature dependence", *J.Phys.F: Metal Phys.*, Vol. 4, 1974, pp. 999-1014.

[8] D.M. Wood and N.W. Ashcroft, "Quantum size effects in the optical properties of small metallic particles", *Phys.Rev. B*, Vol. 25 No. 10, 1982, pp. 6255-6274.

[9] W. Ekardt, "Dynamical polarizability of small metal particles: self-consistent sperical jellium background model", *Phys.Rev.Lett.*, Vol. 52 No. 21, 1984, pp. 1925-1928.

[10] D.E. Beck, "Self-consistent calculation of the polarizability of small jellium spheres", *Phys.Rev. B*, Vol. 30, no. 12, 1984, pp. 6935-6942.

[11] M.J. Puska, R.M. Nieminen, and M. Manninen, "Electric polarizability of small metal spheres", *Phys.Rev. B*, Vol. 31 No. 6, 1985, pp. 3486-3495.

[12] W. Ekardt, "Size-dependent photoabsorption and photoemission of small metal particles", *Phys.Rev. B*, Vol. 31 No. 10, 1985, pp. 6360-6370.

[13] E. Zaremba, and B.N.J. Persson, "Dynamic polarizability of small jellium spheres", *Phys.Rev. B*, Vol. 35 No. 2, 1987, pp. 596-606.

[14] K.P. Charlé, F. Frank, and W. Schulze, "The optical properties of silver microcrystallites in dependence on size and the influence of the matrix environment", *Ber.Bunsenges.Phys.Chem.*, Vol. 88, 1984, pp. 350-354.

[15] J. Warnock and D.D. Awschalom, "Quantum size effects in simple colored glass", *Phys.Rev. B*, Vol. 32 No. 8, 1985, pp. 5529-5531.

[16] B.G. Potter and J.H. Simmons, "Quantum size effects in optical properties of CdS-glass composites", *Phys.Rev. B*, Vol. 37 No. 18, 1988, pp. 10838-10845.

[17] P.E. Lippens and M. Lannoo, "Calculation of the band gab for small CdS and ZnS crystallites", *Phys.Rev. B*, Vol. 39 No. 15, 1989, pp. 10935-10942.

[18] P.D. Persans, A. Tu, Y.- J. Wu, and M. Lewis, "Size-distribution-dependent optical properties of semiconductor microparticle composites", *J.Opt.Soc.Amer. B*, Vol. 6 No. 4, 1989, pp. 818-823.

[19] P.E. Lippens and M. Lannoo, "Comparison between calculated and experimental values of the lowest excited electronic state of small CdSe crystallites", *Phys.Rev. B*, Vol. 41 No. 9, 1990, pp. 6079-6085.

[20] T. Inokuma and T. Arai, "Size effects on the temporal dynamics of edge emission in CdSe microcrystals embedded in a germanate glass matrix", *Phys.Rev. B*, Vol. 42 No. 17, 1990, pp. 11093-11098.

[21] G. Mei, S. Carpenter, L.E. Felton, and P.D. Persans, "Size effects on optical transition energies in CdS_xSe_{1-x} semiconductor nanocrystal glass composities", *J.Opt.Soc.Amer. B*, Vol. 9 No. 8, 1992, pp. 1394-1400.

[22] Ch. Sikorski and U. Merkt, "Spectroscopy of electronic states in InSb quantum dots", *Phys.Rev.Lett.*, Vol. 62 No. 18, 1989, pp. 2164-2167.

[23] L. Brey, N.F. Johnson, and B.I. Halperin, "Optical and magneto-optical absorption in parabolic quantum wells", *Phys.Rev. B*, Vol. 40 No. 15, 1989, pp. 10647-10649.

[24] A. Kumar, S.E. Laux, and F. Stern, "Electron states in GaAs quantum dot in a magnetic field", *Phys.Rev. B*, Vol. 42 No. 8, 1990, pp. 5166-5175.

[25] P. Maksym and T. Chakraborty, "Quantum dots in a magnetic field: Role of electron-electron interactions", *Phys.Rev.Lett.*, Vol. 65 No. 1, 1990, pp. 108-111.

[26] E.V. Alieva, E.I. Firsov, L.A. Kuzik, V.A. Yakovlev, and F.A. Pudonin; "Optical properties of superthin niobium films deposited on crystalline quartz", *Phys.Lett. A*, Vol. 152 No. 1,2, 1991, pp. 89-95.

[27] M. Yamamoto and T. Namioka, "In situ ellipsometric study of optical properties of ultrathin films", *Appl.Opt.*, Vol. 31 No. 10, 1992, pp. 1612-1621.

[28] O. Keller and A. Liu, "Optical diamagnetic response of a niobium quantum well", *Phys.Lett. A*, Vol. 167 No. 3, 1992, pp. 301-305.

[29] O. Keller and A. Liu, "Optical diamagnetic response of a niobium quantum well: *p*-polarized reflection coefficient", *Phys.Lett. A*, Vol. 172, No. 4, 1993, pp. 299-303.

[30] N. Raj and D.R. Tilley, "The electrodynamics of superlattices", In: *The dielectric function of condensed systems*, L.V. Keldysh, D.A. Kirzhnitz, and A.A. Maradudin (Eds.), Amsterdam: North-Holland, 1989, pp. 459-541.

[31] A.C. Tselis and J.J. Quinn, "Theory of collective excitations in semiconductor superlattice structures", *Phys.Rev. B*, Vol. 29 No. 6, 1984, pp. 3318-3335.

[32] R.D. King-Smith and J.C. Inkson, "Real-space inversion of the dielectric-response function of a superlattice", *Phys.Rev. B*, Vol. 33 No. 8, 1986, pp. 5489-5493.

[33] R.D. King-Smith and J.C. Inkson, "Electrodynamics of superlattices: Formalism", *Phys.Rev. B*, Vol. 36 No. 9, 1987, pp. 4796-4806.

[34] K.H. Drexhage, "Interaction of light with monomolecular dye lasers", In: *Progress in Optics XII*, E. Wolf (Ed.), Amsterdam: North-Holland, 1984, pp. 163-232.

[35] H. Morawitz and M.R. Philpott, "Coupling of an excited molecule to surface plasmons", *Phys.Rev. B*, Vol. 10 No. 12, 1974, pp. 4863-4868.

[36] E.H. Hellen and D. Axelrod, "Fluorescence emission at dielectric and metal-film interfaces", *J.Opt.Soc.Amer. B*, Vol. 4 No. 3, 1987, pp. 337-350.

[37] R. Fuchs and R.G. Barrera, "Dynamical response of a dipole near the surface of a nonlocal metal", *Phys.Rev. B*, Vol. 24 No. 6, 1981, pp. 2940-2950.

[38] G. Korzeniewski, T. Maniv, and H. Metiu, "The interaction between an oscillating dipole and a metal surface described by a jellium model and the random phase approximation", *Chem.Phys.Lett.*, Vol. 73, 1980, pp. 212-217.

[39] O. Keller and P. Sønderkær, "Elastic scattering of light from a few atomic dipoles on a flat metal surface", *SPIE*, Vol. 954, 1988, pp. 344-363.

[40] R.W. Rendel, D.J. Scalapino, and B. Muhlschlegel, "Role of local plasmon modes in light emission from small-particle tunnel junctions", *Phys.Rev.Lett.*, Vol. 41 No. 25, 1978, pp. 1746-1750. .

[41] R. Ruppin, "Optical absorption by a small sphere above a substrate with inclusion of nonlocal effects", *Phys.Rev. B*, Vol. 45 No. 19, 1992, pp. 11209-11215.

[42] S. Efrima and H. Metiu, "Classical theory of light scattering by an adsorbed molecule. I. Theory", *J.Chem.Phys.*, Vol. 70 No. 4, 1979, pp. 1602-1613.

[43] B. Labani, C. Girard, D. Courjon, and D. van Labeke, "Optical interaction between a dielectric tip and a nanometric lattice: Implications for near-field microscopy", *J.Opt.Soc.Amer. B*, Vol. 7 No. 6, 1990, pp. 936-943.

[44] C. Girard and D. Courjon, "Model for scanning tunnelling optical microscopy: A microscopic self-consistent approach", *Phys.Rev. B*, Vol. 42 No. 15, 1990, pp. 9340-9349.

[45] C. Girard and X. Bouju, "Self-consistent study of dynamical and polarization effects in near-field optical microscopy", *J.Opt.Soc.Amer. B*, Vol. 9 No. 2, 1992, pp. 298-305.

[46] O. Keller, M. Xiao, and S. Bozhevolnyi; "Configurational resonances in optical near-field microscopy: A rigorous point-dipole approach", *Surf.Sci.*, Vol. 280, 1993, pp. 217-230. .

[47] M. Specht, J.D. Pedarnig, N.M. Heckl, and T.W. Hänsch, "Scanning plasmon near-field microscope", *Phys.Rev.Lett.*, Vol. 68, No. 4, 1992, pp. 476-479.

[48] O. Keller, "Optical near-field interaction: On the local field inside a quantum tip", In: *Near-field optics*, D.W. Pohl (Ed.)[to be published by Kluwer Academic Publishers].

[49] W. Que and G. Kirczenow, "Theory of collective excitations in a two-dimensional array of quantum dots", *Phys.Rev. B*, Vol. 38 No. 5, 1988, pp. 3614-3617.

[50] J. Alsmeier, E. Batke, and J.P. Kotthaus, "Voltage-tunable quantum dots on silicon", *Phys.Rev. B*, Vol. 41 No. 3, 1990, pp. 1699-1702.

[51] A. Lorke, J.P. Kotthaus, and K. Ploog, "Coupling of quantum dots on GaAs", *Phys.Rev.Lett.*, Vol. 64 No. 21, 1990, pp. 2559-2562.

[52] W. Que, G. Kirczenow, and E. Castaño, "Nonlocal theory of collective excitations in quantum-dot arrays", *Phys.Rev. B*, Vol. 43 No. 17, 1991, pp. 14079-14090.

[53] O. Keller, "Screened electromagnetic propagators in nonlocal metal optics", *Phys.Rev. B*, Vol. 34 No. 5, 1986, pp. 3883-3899.

[54] O. Keller, "Tensor-product structure of a new electromagnetic propagator for nonlocal surface optics of metals", *Phys.Rev. B*, Vol. 37 No. 18, 1988, pp. 10588-10607.

[55] D.A. Dahl and L.J. Sham, "Electrodynamics of quasi-two-dimensional electrons", *Phys.Rev. B*, Vol. 16 No. 2, 1977, pp. 651-661.

[56] O. Keller, "Nonlocal electrodynamics of an atom in front of a normal or phasec-onjugating mirror", *J.Quantum Nonlinear Phenomena*, 1993 (in press).

[57] O. Keller, "Moment expansion of the optical second-harmonic response tensor of condensed media", *phys.stat.sol.(b)*, Vol. 157, 1990, pp. 459-471.

[58] O.A. Aktsipetrov, I.M. Baranova, E.M. Dubinina, S.S. Elovikov, P.V. Elyutin, D.A. Esikov, A.A. Nikulin, and N.N. Fominykh, "The new mechanism of surface-enhanced second harmonic generation in small metallic particles", *Phys.Lett. A*, Vol. 117 No. 5, 1986, pp. 239-242.

[59] F. Hache, D. Ricard, and C. Flytzanis, "Optical nonlinearities of small metal particles: surface-mediated resonance and quantum size effects", *J.Opt.Soc.Amer. B*, Vol. 3 No. 12, 1986, pp. 1647-1655.

[60] S.C. Hill, D.H. Leach, and R.K. Chang, "Third-order sum-frequency generation in droplets: model with numerical results for third-harmonic generation", *J.Opt.Soc.Amer. B*, Vol. 10 No. 1, 1993, pp 16-33.

[61] D.H. Leach, R.K. Chang, W.P. Acker, and S.C. Hill, "Third-order sum-frequency generation in droplets: experimental results", *J.Opt.Soc.Amer. B*, Vol. 10 No. 1, 1993, pp. 34-45.

[62] F.T. Vas'ko, "Photon-drag of two-dimensional electrons", *Sov.Phys.Semicond.*, Vol. 19 No. 7, 1985, pp. 808-809.

[63] S. Luryi, "Photon-drag effect in intersubband absorption by a two-dimensional electron gas", *Phys.Rev.Lett.*, Vol. 58 No. 21, 1987, pp. 2263-2266.

[64] M.I. Stockman, L.N. Pandey, and T.F. George, "Light-induced drift of quantum-confined electrons in semiconductor heterostructures", *Phys.Rev.Lett.*, Vol. 65 No. 27, 1990, pp. 3433-3436.

[65] O. Keller, "Photon drag in a single-level metallic quantum well", to be published.

[66] I.I. Abram, "The nonlinear optics of semiconductor quantum wells: physics and devices", In: *Nonlinear optics in solids*, O. Keller (Ed.), Berlin: Springer-Verlag, 1990, pp. 190-212.

[67] A. Liu and O. Keller, "Optical second-harmonic generation from niobium quantum wells", In: *this volume.*

[68] A. Bagchi, R.G. Barrera, and A.K. Rajagopal, "Perturbative approach to the calculation of the electric field near a metal surface", *Phys.Rev. B*, Vol. 20 No. 12, 1979, pp. 4824-4838.

[69] J.E. Sipe, "New Green-function formalism for surface optics", *J.Opt.Soc.Amer. B*, Vol. 4 No. 4, 1987, pp. 481-489.

[70] O. Keller, "Propagator study of the selvedge field in linear and nonlinear nonlocal jellium optics", *Phys.Rev. B*, Vol. 38 No. 12, 1988, pp. 8041-8060.

[71] W.H. Louisell, *Quantum statistical properties of radiation*, New York: Wiley, 1973, 527 pp.

[72] C. Cohen-Tannoudji, B. Diu and F. Laloë, *Quantum mechanics*, Vol. 1-2, New York: Wiley, 1977, 1524 pp.

[73] M. Weissbluth, *Quantum mechanics*, Tokyo: McGraw-Hill, 1989, 544 pp.

[74] L.I. Schieff, *Quantum mechanics*, Tokyo: McGraw-Hill, 1968, 544 pp.

[75] E.A. Power, *Introductory quantum electrodynamics*, London: Longmans, 1964, 147 pp.

[76] J.R. Schrieffer, *Theory of superconductivity*, Reading: Benjamin/Cummings, 1983, 332 pp.

[77] G. Rickayzen, *Theory of superconductivity*, New York: Interscience, 1964, pp. 483.

[78] A.L. Fetter and J.D. Walecka, *Quantum theory of many-particle systems*, New York: McGraw-Hill, 1971, pp. 601.

[79] O. Keller, "Theory of nonlinear electromagnetic rectification in a BCS superconductor", *Phys.Rev. B*, Vol. 42 No. 10, 1990, pp. 6049-6061.

[80] O. Keller, "Nonlinear electromagnetic rectification of BCS-paired electrons at a superconductor surface", *Phys.Rev. B*, Vol. 43 No. 13-A, 1991, pp. 10293-10309.

[81] O. Keller and A. Liu, "Local-field calculation of the optical diamagnetic response of a metallic quantum well", to be published.

[82] O. Keller, "Random-phase-approximation study of the response function describng optical second-harmonic generation from a metal selvedge", *Phys.Rev. B*, Vol. 33 No. 2, 1986, pp. 990-1009.

Studies in Classical
and Quantum Nonlinear Optics

ISBN 1-56072-168-5

NON-LINEAR WAVE MIXING WITH SPATIAL STRUCTURE AND DIFFRACTION EFFECTS

G. Lera and M. Nieto-Vesperinas
Instituto de Optica C.S.I.C.
Serrano 121
28006 Madrid Spain

ABSTRACT

Second harmonic and difference frequency generation are studied with pumping beams of arbitrary shape. The influence of the coherence of the input beams is analyzed, for different values of the coherence radious and both with and without taking into account depletion of the pumping beams.

INTRODUCTION

With the rapid development of Non-linear Optics and coherence theory, a natural question that arises is how do they interact with each other.

The influence of coherence properties in statistical beam propagation has been noticed in recent works in second harmonic generation (SHG)[1-2], and they predict a significant increase of the energy transferred to the harmonic when there is no complete coherence in the input wave. We are interested in the relationship between the statistical properties of the wave generated in the non-linear process and those of the input beams. Calculations of this kind have been made by Zubairy et al.[1-2], in SHG, by assuming two different models for the partially coherent input wave; beams generated by a "Gaussian Schell source", or by passing light through a "phase screen". Disregarding the details, the conclusion in both cases is that the conversion efficiency may be enhanced by up to a factor of two with respect to that of a coherent deterministic input beam.

These works are based on the assumption of some approximations, as the Parametric (PA) and Paraxial (PXA) approximations. However, there are situations in which these kind of approaches fail. We have developed a method that does not assumes neither the PA nor the PXA[3]. First, we will show how it works with perfectly coherent beams with Gaussian profiles, under conditions that invalid the other methods. Then we shall apply it to partially coherent beams.

Two beams with frequencies ω_2 and ω_3 are introduced into an unbounded uniaxial non linear crystal, propagating in the z-axis direction. To simplify the calculations, we assume one-dimensional sections. The properties of the medium and direction of the optical axis (c) are selected in order to obtain a third wave with frequency ω_1 which is the difference of the other two (Difference frequency generation).

METHOD OF SOLUTION

By means of the Angular Spectrum representation of the fields, it is possible to transform the Maxwell equations into the following system of ordinary differential equations:

$$\frac{\partial E_1(k_x, z)}{\partial z} = \frac{2\pi i \omega_1^2}{c^2 k_{1z}} \hat{o}_1(k_x)\vec{P}_1(k_x, z)e^{-ik_{1z}z} \tag{1a}$$

$$\frac{\partial E_2(k_x, z)}{\partial z} = \frac{2\pi i \omega_2^2}{c^2 (k_{2z})_{eff}} \hat{e}_2(k_x)\vec{P}_2(k_x, z)e^{-ik_{2z}z} \tag{1b}$$

$$\frac{\partial E_3(k_x, z)}{\partial z} = \frac{2\pi i \omega_3^2}{c^2 k_{3z}} \hat{o}_3(k_x)\vec{P}_3(k_x, z)e^{-ik_{3z}z} \tag{1c}$$

In Eqs. (1) $E_i(k_x,z)$ is the complex amplitude of the k_x plane wave for the ith-wave. $\hat{o}(k_x)$, $\hat{e}(k_x)$ are the ordinary and extraordinary polarization vectors, respectively; and $\vec{P}_i(k_x,z)$ is the polarization.

These equations are like those of three plane wave mixing, except that, here, we have one system for each component of the Spectrum. Furthermore the polarization that appears here is the 1-D Fourier transform of the spatial one, which is proportional to the product of the corresponding fields. For SHG we have only (1a) and (1b) with $\omega_2=2\omega_1$.

To derive these equations, the backward as well as the ordinary and extraordinary components of the spectra which are not nearly matched were neglected. A particular form for the profiles of the input waves is needed to solve the equations . We consider them to be Gaussian.

As a test of the efficiency of the method, we used the Manley-Rowe relation which provide us with several conserved magnitudes. In all our calculations these magnitudes remain constant within a high degree of accuracy.

NUMERICAL RESULTS

In Fig. 1 we show the photon conversion efficiency versus normalized length (L=Z/Zq) for an axial phase mismatch (apm) of -1.9 cms^{-1} and a walkoff angle (ρ) of -2 degrees.

In order to make it possible to compare with other curves for different values of the parameters, the depletion length Zq takes into account the energy of the input beams, the susceptibility and refractive indices of the medium, and the frequencies of the waves.

$$Z_q = \frac{D}{2\pi\,\chi_{eff}\,(\beta_1\,\beta_2\,\beta_3\,[\Sigma\,N_i^{(0)}])^{1/2}} \qquad (2)$$

$$\beta_i = \frac{\omega_i^2}{c^2 k_{iz}}$$

In Eq. (2) D is a constant that defines the scale; χ_{eff} is the effective nonlinear susceptibility, and $N_i(0)$ is the initial number of photons for the beam at frequency ω_i.

It can be seen here how the solid line, which belongs to our calculations, departs from the other two approximative curves for value L of one half of Zq. The deviations grow with the distance L.

These differences are due to the appearance of an out-of-axis phase matching peak, because of the non zero value of the axial phase mismatch. This peak shows large fraction of photons taken from the pumping beams, and thus, it makes the PA to fail. Also, as it is not placed in the near axis zone, the PXA underestimates the conversion efficiency.

Because of these differences, it is important to know how both, depletion and diffraction affect the behaviour of the system in the case of partially coherent light.

STATISTICAL BEAMS

Now we simulate a partially coherent beam by solving the propagation equations for a number of samples with different profiles. The statistics of these samples is controlled

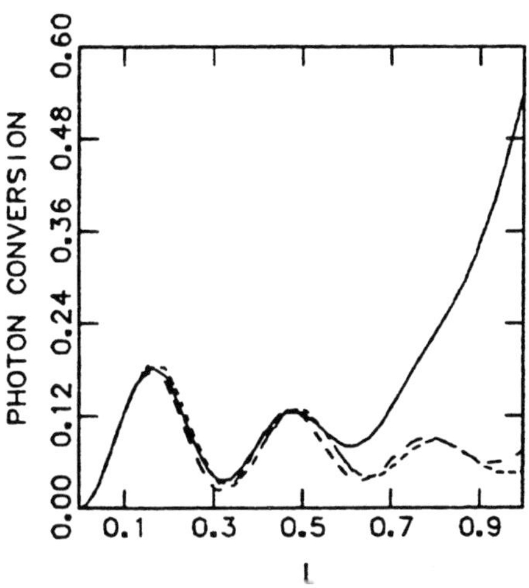

Fig. 1 Photon conversion efficiency for apm =-1.9, ρ=2. Solid line belongs to our results, while dashed and dotted line to the PXA and PA respectively.

We have generated them in such a way that, for the input beam at the z=0 plane, the mutual coherence function is given by:

$$\Gamma(x_1, x_2) = I_0\, e^{-(x_1^2+x_2^2)/W^2}\; e^{-(x_1-x_2)^2/(2\sigma_g^2)} \qquad (3)$$

Where W is the spot size and σ_g is the correlation length.

This is achieved by generating at each point of the profile real and imaginary parts of the field as independent Gaussian random variables, but each of them correlated for different points x. This leads to an exponential distribution for the intensity at each point[4]:

$$P(I) = \frac{1}{\langle I \rangle} e^{-I/\langle I \rangle} \qquad (5)$$

We have generated samples for two different values of the correlation length, namely ($\beta = s_g^2/W^2$) $\beta = 0.1$ and $\beta = 25$. and we have studied, for these samples, SHG using both the PA and our method. In order to simplify the problem we have taken zero axial phase mismatch and no walk-off.

Fig. 2 (in the case of lower coherence radious, and for the output of using our method) shows the transversal distribution of output intensity for one of the samples (dashed line) and the mean intensity over the ensemble of samples (solid line).

Fig. 3 shows the transversal distribution of output intensity for a sample (dashed line) and the mean intensity over the ensemble (solid line) for $\beta = 25$ obtained from the PA. Both figures are in arbitrary units.

In figure 2 we note that the mean intensity is not completely symmetric, this lack of symmetry is due to the fact that we have taken only 250 samples. As the coherence radious is much greater in the figure 3, we are not able to see these irregularities in the mean intensity, which is completely symmetric.

One important difference

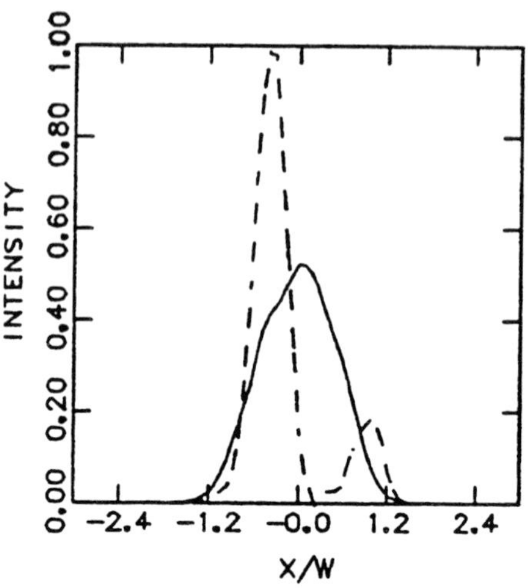

Fig. 2 Transversal distribution of output intensity for $\beta = 0.1$ in the NPC. Solid and dashed line represent mean and one sample intensity, respectively.

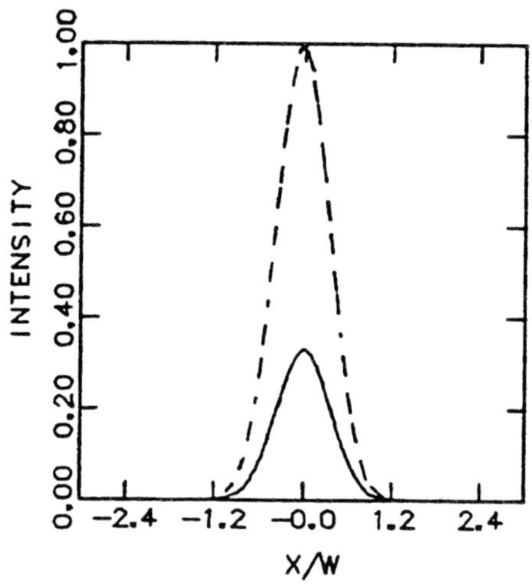

Fig. 3 Transversal distribution of output intensity for $\beta = 25$ in the PC. Solid and dashed line represent mean and one sample intensity, respectively.

between the two cases is that, due to the limit in the energy transfer in the case that considers depletion (non parametric case, NPC), there is a saturation effect in the intensity of the maximum, that remains constant or even decreases while the intensity in other parts of the transversal profile still grows. As there is not such a limitation in the case that does not take depletion into account (parametric case, PC), the intensity always grow, in all the parts of the profile.

This effect makes the width of the output in the first case greater than that of the second case.

In order to see the differences with the usual case with random profiles, we show in fig. 4 the photon conversion efficiency for $\beta=25$ and for the case of coherent beams in both the PC and NPC. As usual, the PC curves show a parabolic shape.

It is easier to see the difference in fig. 5 where we plot the ratio between the conversion efficiency in the non coherent case and in the coherent one versus crystal length. It can be seen that for low values of Z/Zq this ratio is approximately 2, due to the particular statistics we have chosen for our input beam. For the PC and $\beta=25$. this enhancement in the conversion efficiency remains constant along the crystal, but, for the NPC the enhancement for Z=Zq is lower than 20%, limiting the usefulness of degrading the coherence to enhance the conversion.

One important magnitude that can be calculated with this approach is the probability

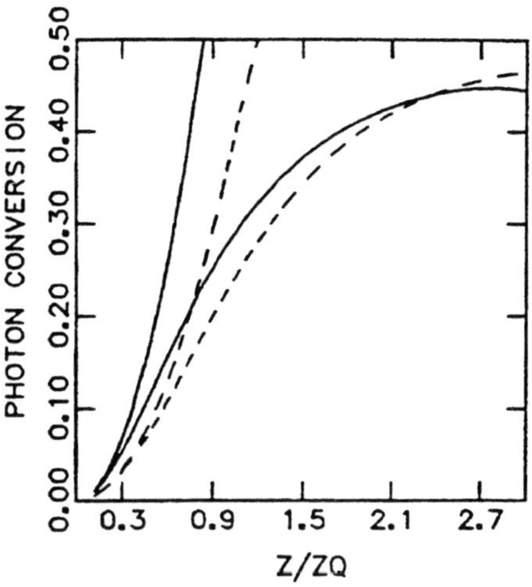

Fig. 4 Photon conversion efficiency. Solid lines represent the partially coherent case ($\beta=25$, PC and NPC) and dashed lines the coherent case. The PC curves have parabolic shape.

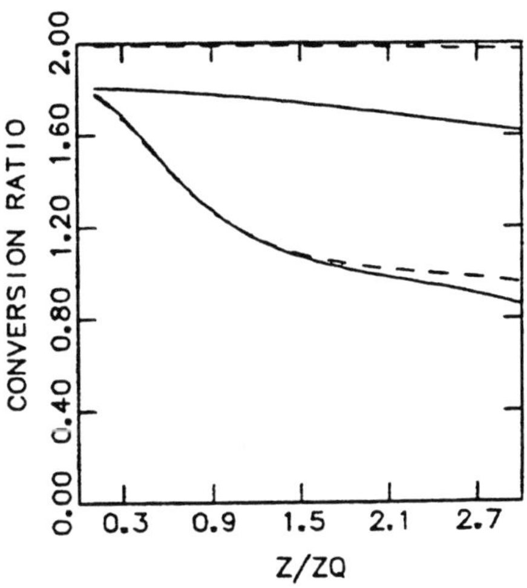

Fig. 5 Photon conversion ratio between the partially coherent and the coherent case. Solid and dashed lines belongs to $\beta=0.1$ and $\beta=25$, respectively. Upper curves belong to PC.

density of intensity for each point of the beam profile. This magnitude give us all the information about the statistical properties of the intensity.

For the input beam at the plane z=0 this distribution follows a negative exponential (4) at all points.

We can see for two different points (x=0,x=W) the probability histogram in the NPC for β=0.1 (fig. 6). Due to the above saturation effect the statistics for the central part of the profile (x=0) is completely different that for other points.(fig. 7 x=W).

However, for β=25 in the PC we can see in Figs. 8 and 9 how the histogram is practically equal for all points of the profile.

The differences between the NPC and PC (for x=0) are due to the fact that in the former there is no limit for the energy that the second harmonic can take, so when the initial energy is great, as the process is non-linear, it is much more amplified than when it is low, so the output intensity for these samples and the mean intensity ⟨I⟩ are very large. As there are still many samples with low initial intensity we find that the distribution is concentrated at values of I< ⟨I⟩.

Due to the saturation of the energy transfer in the NPC, the intensity for the second harmonic does not grow so much when the intensity of the input beam is initially great, hence the intensity mean is not so

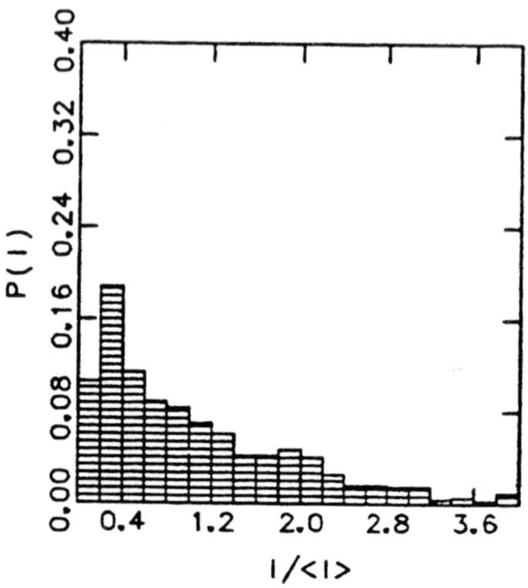

Fig. 6 Probability histogram in the NPC for β=0.1 and x=0.

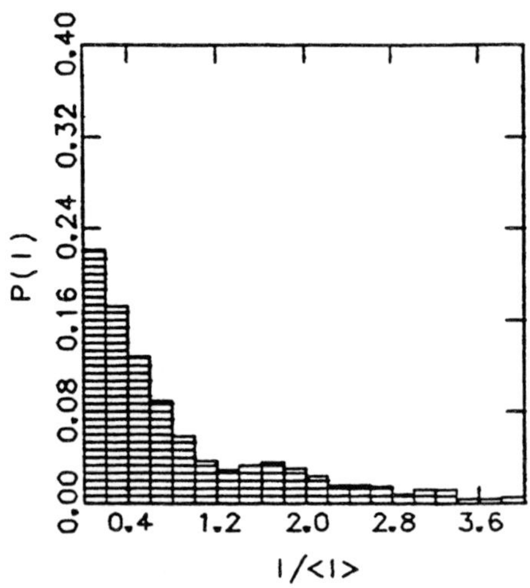

Fig. 7 Probability histogram in the NPC for β=0.1 and x=W.

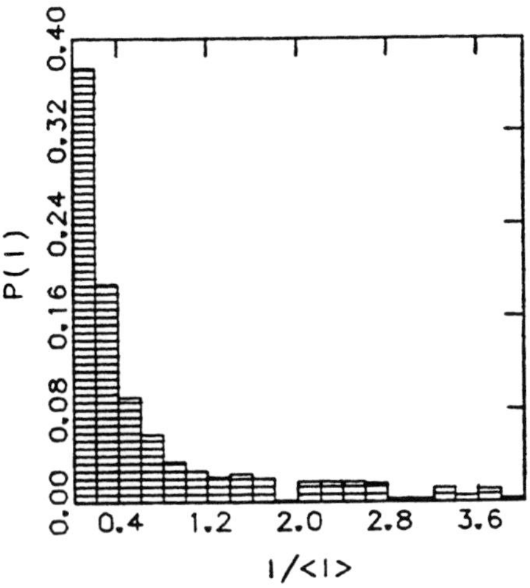

Fig. 8 Probability histogram in the PC for β=25 and x=0.

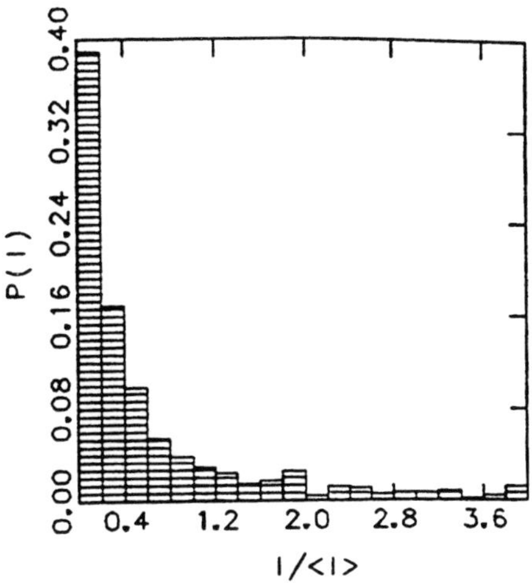

Fig. 9 Probability histogram in the PC for β=25 and x=W.

large in this case, and the distribution is not so concentrated in the $I < \langle I \rangle$ region.

For points out of the axis, the initial values in the intensity of the input field are lower, therefore the energy transferred limit is less important, and thus the distribution in this case reensambles better that of the PC.

REFERENCES

1 - M. S. Zubairy and J. K. McIver, Phys. Rev. A 3 (1985) 856.
2 - N. A. Ansary and M. S. Zubairy, Optics. Comm. 59 (1986) 390.
3 - M. Nieto-Vesperinas and G. Lera, Optics. Comm. (accepted for publication).
4 - V. A. Aleshkevich, S. S. Lebedev, and A. N. Matveev, Sov. Phys. JETP 56(4) (1982) 715.

Studies in Classical
and Quantum Nonlinear Optics

ISBN 1-56072-168-5

KERR-LIKE OPTICAL NONLINEARITIES IN LIQUID CRYSTALS

R. Macdonald and H.J. Eichler
Optisches Institut Technische Universität Berlin

CONTENTS

1. INTRODUCTION

The development of high intensity lasers and the discovery of materials with large optical nonlinearities in the last decades have extended the scope of nonlinear processes dramatically, resulting in considerable scientific and also practical interest [1]. Applications like phase-conjugation and photonic switching of light are important aspects in modern laser physics and optical information technologies. Organic materials like liquid crystals and polymers presently receive strong attention as promising candidates for nonlinear optical components because of large nonlinearities and the possibilty to manufacture devices with large apertures in thin films with high optical quality. The optical and other physical properties can be varied widely in these materials by molecular engineering.

Liquid crystals are anisotropic fluids which exhibit one or more ordered thermodynamic phases between the solid and the isotropic liquid state [2, 3, 4]. They usually consist of steric organic molecules which are more or less uniformly aligned but show a fluid behaviour at the same time. As a consequence these materials are *liquids* in which many physical properties show macroscopic anisotropy like in *crystals*. At higher temperatures liquid crystals become ordinary isotropic liquids. At lower temperatures solidification occurs. The intermediate phases are called *liquid crystalline* or *mesophases*. Since the orientational order in liquid crystals is relatively weak, it can be easily manipulated and changed with the help of modest fields. This leads e.g. to large electro- and magneto-optic effects [5] which are well-known and useful in many applications of today's display technology. In addition, orientational order can be also easily affected by optical fields, leading to strong optical nonlinearities.

The optical nonlinearities discussed in the following are describable by light-induced changes of the birefringence or the refractive index for a given polarization, respectively, and are therefore called [1] Kerr-like optical nonlinearities. Processes like harmonic generation which have been also observed and investigated in non-centrosymmetric liquid crystals will not be treated here.

In the present paper the main mechanisms leading to Kerr-like optical nonlinear behaviour are reviewed. For this purpose we will start with some liquid crystal fundamentals and discuss the main (linear) optical properties of the most important liquid crystalline phases in the following section 2. The effects of optical field induced reorientation using low power cw-lasers and thermally induced birefringe changes are then treated in section 3.

In section 4, we will discuss the dynamical response of nonlinear optical effects to short laser pulses obtained in dynamic grating experiments using picosecond laser pulses. It is shown that relatively high intensities and strong optical fields lead to new phenomena which are of interest not only for basic liquid crystal research but also for speeding up optical switching times. Section 5 finishes the article with concluding remarks.

It should be noted at this point, that the present selection of topics does not claim completeness. It is more like a personal view and reflects the authors' own involvement in the field. Other aspects and some different details of optical nonlinearities in liquid crystals are reviewed e.g. by N.V. Tabiryan et al. [6], I.C. Khoo [7], I. Janossy [8] and P. Palffy-Muhoray [9].

2. LIQUID CRYSTALLINE PHASES AND OPTICAL PROPERTIES

Liquid crystals usually consist of large "rod-" or "disc-like" organic molecules which tend to align themselves with a long range order due to anisotropic intermolecular forces. If we restrict ourselves to rod-like molecules in the following, three classes of liquid crystalline phases [10], shown in Fig.1, may be distinguished. In the *nematic* phases the molecules are more or less aligned in one direction but without any positional order.

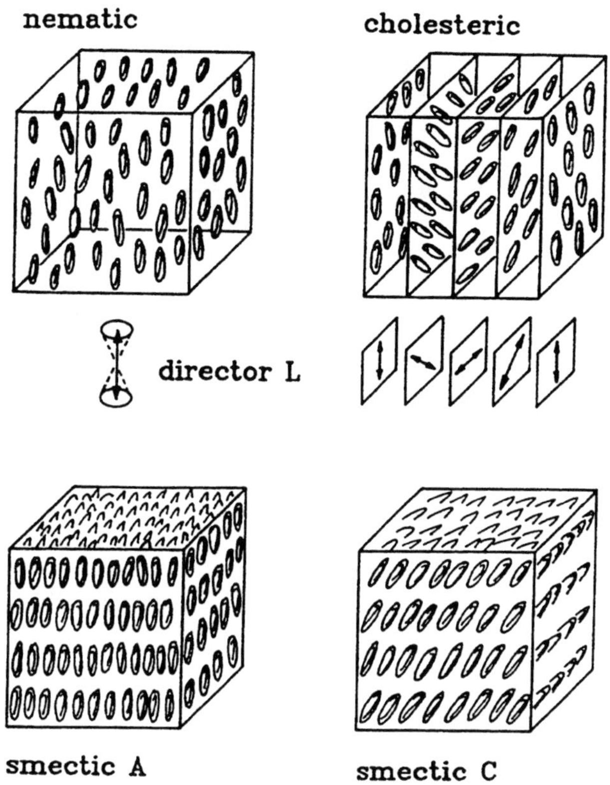

Fig. 1: Schematic representation of different liquid crystalline phases

The direction of preferred molecular alignment can be described by a unit-vector, the so-called *director* L. The orientation of the individual molecule, however, usually differs significantly from that direction and the director must be more correctly defined as the symmetry axis of the orientational distribution of molecules. Nematics are generally *uniaxial*, which means that the orientational distribution is rotationally symmetric around the director. The fluctuations in the alignment can be described by an orientational scalar *order parameter* S in this case [2]

$$S = \langle P_2 (\cos \vartheta) \rangle \qquad (1)$$

where $P_2(\cos\vartheta) = 3(\cos^2\vartheta - 1/3)/2$ is the second Legendre polynomial and ϑ the angle between the individual moleular axis and the director **L**. The brackets $\langle ... \rangle$ indicate averaging over ϑ. The order parameter is $S = 1$ for perfectly aligned liquid crystals and $S = 0$ in the isotropic phase (no alignment).

Cholesteric liquid crystals are very similiar to the nematic phases but consist of chiral molecules, i.e. the molecules differ from their mirror image. In cholesterics the molecules tend to form a helical structure with the helical axis perpendicular to the director. They can be considered as a stack of thin nematic layers which are twisted with respect to each other to form a helix.

The *smectic* phases are characterized by additional degrees of positional order. Generally, the molecules are concentrated in layers which can be considered as a one-dimensional density wave in these liquid crystals. In the smectic A (Sm A) and the smectic C (Sm C) phases there is no positional order within the layers. There are further smectic phases in which the molecules form two-dimensional lattices inside each layer, which is, however, beyond the scope of this paper. The Sm A phase is uniaxial and the director aligned perpendicular to the layers. In Sm C phases the director is tilted with respect to the layer normal axis, leading to a weak biaxiality. The Sm C phases may form a helical structure, which is denoted as Sm C*, when the molecules are chiral. The smectic layers are twisted with respect to each other in this case and the helical axis is parallel to the layer normal. It should be noted at this point, that in contrast to all uniaxial phases the reduced symmetry of biaxial phases of chiral molecules (like Sm C*) allows the appearence of spontaneous macroscopic polarization. As a consequence these particular phases are *ferroelectric* [11] which currently results in strong interest for both basic research and technical applications.

Most physical properties of liquid crystals are anisotropic and must be described by a symmetric tensor rather than a scalar. The anisotropic behaviour may be observed e.g. in the magnetic susceptiblity, in heat diffusion or optical birefringence. Additionally, there are new physical qualities accompanied with the orientational order which do not appear in simple isotropic liquids. Examples are the elastic forces and the rotational viscosity in connection with director deformations as will be shown in section 3.1. A detailed description of the physical properties of liquid crystals may be found in the monographs of P.G. de Gennes [2], S. Chandrasekhar [3] and W.H. de Jeu [4]. We will emphasize here the optical anisotropy which may be helpful in understanding the optical nonlinear effects.

The optical properties of most liquid crystals are similiar to those of birefringent crystals and can be described for uniaxial phases by a dielectric tensor

$$\varepsilon_{\mu\nu} = \varepsilon_a S (L_\mu L_\nu - \frac{1}{3}\delta_{\mu\nu}) + \bar{\varepsilon}\,\delta_{\mu\nu} \qquad (2)$$

where $\varepsilon_a = \varepsilon_{\parallel} - \varepsilon_{\perp}$ is the anisotropy for perfectly aligned molecules (i.e. $S = 1$ cf. eq.1), $L_{\mu,\nu}$ are cartesian components of the director, the sum $\bar{\varepsilon} = \varepsilon_{\lambda\lambda}/3$ denotes the isotropic part of the tensor and $\delta_{\mu\nu}$ is the Kronecker-delta.

The propagation of light can be described by an inhomogeneous wave equation obtained from Maxwell's equations for dielectric material with no free charges as [12]

$$\Delta \mathbf{E} - \mu_o \varepsilon_o \ddot{\mathbf{E}} = \mu_o \ddot{\mathbf{P}} - \frac{1}{\varepsilon_o} \operatorname{grad} \operatorname{div} \mathbf{P} \qquad (3)$$

where $\mathbf{E}$ is the optical field and $\mathbf{P}$ the induced electric polarization. The dot denotes the time derivative. In linear optics the polarization is given by

$$\mathbf{P} = \varepsilon_o \underline{\underline{\chi}}\, \mathbf{E} = \varepsilon_o (\underline{\underline{\varepsilon}} - \underline{\underline{1}})\, \mathbf{E} \qquad (4)$$

and the electric susceptibility tensor $\underline{\underline{\chi}}$ does not depend on $\mathbf{E}$. In general, solution of the wave equation (3) using the polarization eq. 4 as a source term is rather difficult, if the anisotropic medium is inhomogeneous. This is, however, often the real situation in liquid crystals. Inhomogenities may arise e.g. from spontaneous (or field induced) fluctuations in alignment or they are inherent properties of certain liquid crystal phases like in cholesterics. On the other hand several techniques exist to obtain mono-domains

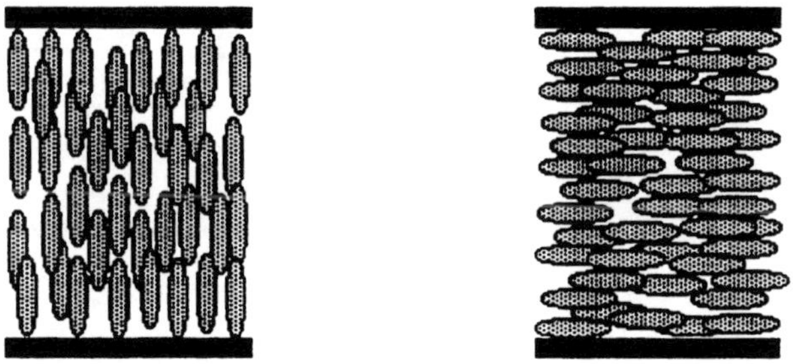

homeotropic planar

Fig. 2: Surface aligned mono-domain nematic liquid crystals

of well-aligned nematic liquid crystals, in which the optical properties can be assumed homogeneous (see Fig. 2). Homeotropic alignment can be achived by coating the inner surfaces of the container with a surfactant (e.g. lecithin). Planar orientation is obtained by unidirectionally rubbing the glass plates. In the homeotropic and planar case an incident plane electromagnetic wave leads to propagation of an ordinary and an extraordinary beam with refractive indices given by

$$n_o = (\varepsilon_\perp)^{1/2} \; ; \; n_e = (\varepsilon_\| \varepsilon_\perp / (\varepsilon_\| \cos^2\Theta + \varepsilon_\perp \sin^2\Theta))^{1/2} \qquad (5)$$

where Θ is the angle between the director and the wave vector. The indices "$\|$" and "$\perp$" refer to coordinate axis parallel and perpendicular to the director, respectively.

For inhomogeneous liquid crystals some interesting results and approximations may be obtained for helical structures. In cholesterics, for instance, the director and hence the optical axis spiral around the helical axis with a pitch p. Solutions of the Maxwell equations may be found in this case analytically for an incident plane wave propagating along the helical axis by transforming the field equation (3) into the director's frame. We will not discuss these solutions and the resulting dispersion relations here in detail but shortly summarize the main results: For a given frequency there are four independent modes with different wavenumbers. Two of these modes describe forward, the other two backward propagating right- and left-handed circular polarized waves, respectively. Three main cases are important for an incident wavelength λ with respect to applications:

i) $\lambda \ll p(n_\| - n_\perp)$ leads to *guided modes*, which means that the normal modes become linearly polarized, propagating as the ordinary and the extraordinary wave. The polarizations follow the director rotation in this case and leave the cholesteric with a phase difference $d(n_\| - n_\perp)$.

ii) $n_\perp p < \lambda < n_\| p$ leads to *Bragg-reflection* for one of the normal modes, i.e. a right-handed helix reflects right circularly polarized light whereas left circular polirizations are transmitted.

iii) $\lambda > (n_\| - n_\perp) p$ leads to *optical activity*, i.e. right- and left-circular polarizations propagate with different phase velocities.

The problem of obligue incidence with respect to the helical axis can be solved numerically e.g. with the 4 x 4-matrix method [13].

Furthermore, approximate solutions of the Maxwell equations can be found in cases where the director changes are small on distances which are large compared to the wavelength. These approximations are in some cases comparable to the geometrical-optics approximation applicable to inhomogeneous isotropic media, e.g. if only the extraordinary refractive index is changed for an extraordinary wave.

The situation may be much more complicated, however, if light-induced changes in the director orientation or the scalar order parameter have to be considered. It will be discussed in the next section how this can happen. It is immediatly clear in this case, that the susceptibility tensor (eqs. 2, 4) is no longer independent of the optical field. Consequently we enter the field of nonlinear optics. Usually these changes are slow compared to light frequencies and result in nonlocal changes of the birefringence. In contrast to isotropic nonlinear material, these changes will not only affect the phase but also the polarization of the incident lightwave, which makes things not easier in general. It will be shown in the following, however, that the above discussed approximations can be used in some particular cases and allows simple interpretations of the nonlinear optical light propagation.

3. BASIC PRINCIPLES AND NONLINEAR EFFECTS WITH CW-LASERS

In nonlinear optics the electric polarization **P** is a complicated nonlinear function of the optical field **E** in general and may be written (in a simplified notation) as

$$\mathbf{P} = \varepsilon_o (\underline{\underline{\chi}}^{(1)} \mathbf{E} + \underline{\underline{\chi}}^{(2)} \mathbf{E}^2 + \underline{\underline{\chi}}^{(3)} \mathbf{E}^3 + ...) \tag{6}$$

where $\underline{\underline{\chi}}^{(n)}$ is the n-th order susceptibility tensor. This equation is equivalent to introduction of a field dependent susceptibility or dielectric tensor

$$\underline{\underline{\varepsilon}} = \underline{\underline{1}} + \underline{\underline{\chi}} = \underline{\underline{1}} + \underline{\underline{\chi}}^{(1)} + \underline{\underline{\chi}}^{(2)} \mathbf{E} + \underline{\underline{\chi}}^{(3)} \mathbf{E}^2 + ... \tag{7}$$

More generally the n-th order susceptibility is a tensor of rank n+1. The first order term contains the linear optical properties like index of refraction and absorption. The second order term describes e.g. the Pockels-effect and is also responsible for second harmonic generation (SHG). The third order susceptibility describes e.g. the Kerr-effect, third harmonic generation and light-induced changes in the refractive index.

In many liquid crystals SHG is not allowed from symmetry properties, i.e. $\underline{\underline{\chi}}^{(2)} = 0$. We will further assume in the following that the light intensities are sufficient to induce nonlinear *macroscopic* polarizations but only a linear response on the individual *molecular* level is considered. Consequently third harmonic generation is also omitted. It is shown in this section that optical torques and absorption of light may induce collective changes in the orientation and alignment of liquid crystals which leads to changes in the dielectric tensor as indicated in eq. 3. Since these effects depend on $\mathbf{E}^2$, the resulting changes in $\underline{\underline{\varepsilon}}$ can be attributed to $\underline{\underline{\chi}}^{(3)}$ and lead to Kerr-like optical nonlinearities.

3.1 Stationary Optical Reorientation

Due to the strong anisotropic susceptibility originating from orientational order, liquid crystal molecules experience torques in the presence of applied fields, which may lead to director reorientation. First experiments on optical reorientation in nematics were performed almost simultaneously by A.S. Zolotko et al. [14], I.C. Khoo et al. [15], B.Y. Zel'dovich et al. [16] and S.D. Durbin et al. [17].

In case of an incident laser beam with intensity I the optical field is given by

$$E = (2\,Z\,I/n)^{1/2} \tag{8}$$

where n is the effective refractive index and $Z = 377\,\Omega$ the vacuum impedance. The optical torque can be written as

$$\mathbf{M}_{op} = \mathbf{P} \times \mathbf{E} = (\varepsilon_o \underline{\underline{\chi}} \mathbf{E}) \times \mathbf{E} \tag{9}$$

This torque tends to align the optical axis and hence the director parallel to the optical field. Note that M_{op} is a quadratic function of the field and the time average is not zero. The resulting rotation or reorientation of the director leads to large changes of the dielectric tensor. It is responsible for extremely large refractive index changes in nematics, which may be in the order of the birefringence $n_a = n_{\parallel} - n_{\perp} = 0.1 \ldots 0.2$. Therefore, the term "Giant Optical Nonlinearity" (GON) is frequently used for this kind of process. As will be shown in the following, intensities needed to induce GONs are only in the order of several hundered Watts per cm^2. Since furthermore this torque-induced nonlinearity is nonresonant and can be achieved in rather thin, nonabsorbing films of some ten microns thickness, it bears a large potential for many applications.

The optical torque is balanced usually by elastic and frictional forces, which arise from director deformation. In nematics there are three fundamental distortions with corresponding elastic constants K_i, which are called "splay" (K_1), "twist" (K_2) and "bend" (K_3). For many practical cases and for simplicity in calculations it is sufficient, however, to work in the one-constant approximation $K = K_1 = K_2 = K_3$, which is not to bad since the three constants differ only slightly in magnitude. The dynamic director reorientation is damped by the rotational viscosity γ_1. The torque balance describing the light induced reorientation can be written in this case as

$$\mathbf{L} \times (\gamma_1 \dot{\mathbf{L}} - K \Delta \mathbf{L} - \varepsilon_o \varepsilon_a' (\mathbf{L} \cdot \mathbf{E}) \mathbf{E}) = 0 \qquad (10)$$

where we have rewritten the optical torque (9) using the expression (2) for the dielectric tensor and $\varepsilon_a' = S\,\varepsilon_a$ is the effective anisotropy depending on the scalar order parameter. Flow-alignment effects are neglected in eq. 10. To discuss this equation, consider a linearly polarized laser beam incident on a homeotropic aligned nematic liquid crystal as depicted in Fig. 3.

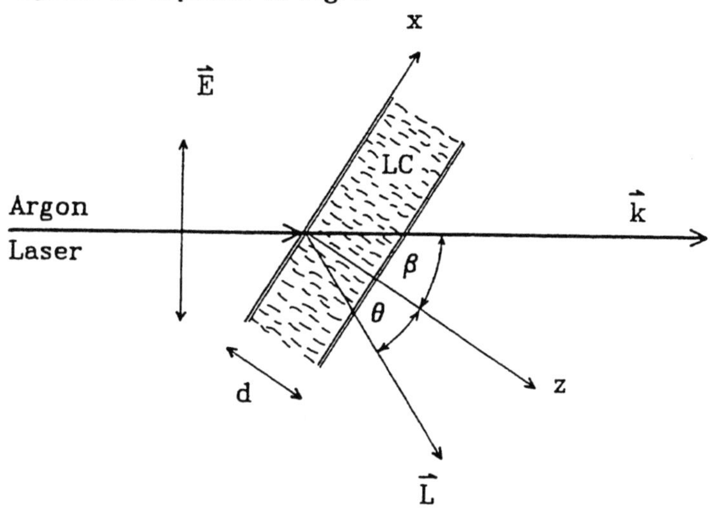

Fig. 3: Schematic arrangement for optical reorientation with a homeotropic aligned nematic liquid crystal. E: Optical field, k: Wave vector, L: Director

The optical field yields in a reorientation of the director **L** in this case, which leads to changes in the extraordinary refractive index but does not change the polarization of the incident wave. The reorientation can be described by an single angle Θ therefore and the torque balance (10) reduces to

$$\gamma_1 \dot{\Theta} - K \Delta \Theta - \frac{1}{2} \varepsilon_o \varepsilon_a' E^2 \sin\{2(\Theta + \beta)\} = 0 \qquad (11)$$

Changes in the extraordinary refractive index established by reorientation are then given by

$$\delta n = n_e(\Theta + \beta) - n_e(\beta) \quad (12)$$

where n_e is given in eq. 5. If we assume e.g. $\beta = 0$ and $E = 0$ then $n_e(0) = n_\perp$. The director may be reoriented to nearly $\Theta = \pi/2$ at sufficient intensities and $n_e(\pi/2) = n_\parallel$. As a result $\delta n = n_\parallel - n_\perp$ which is in the order of 0.1 ... 0.2 as already mentioned.

The strong refractive index changes due to reorientation can be observed experimentally in very impressive self-focussing and self-diffraction effects, which have been studied by several authors [14, 17, 18, 19]. Due to the transverse intensity profile across the laser beam the refractive index is modulated in form of a strong (almost) Gaussian lens, which leads to a far field diffraction pattern consisting of a number of concentric rings (see Fig. 4).

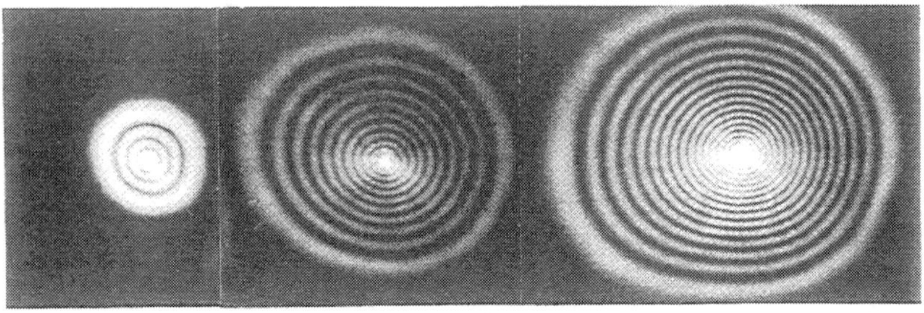

Fig. 4: Self-diffraction patterns obtained in a nematic liquid crystal with different input intensities.

The induced on-axis phase shift is simply given by the number of rings multiplied with 2π in these experiments. Up to 30 rings can be obtained easily with a 100 μm thick liquid crystal. We will compare some of the following theoretical discussions with results obtained in self-diffraction experiments, which have been performed with one of the most frequently used liquid crystal 5CB (pentyl-cyanobiphenyl) in a 100 μm thick homeotropic aligned film.

To calculate the reorientation angle and the resultant refractive index changes one has to solve the (already simplified) nonlinear equation (11), which is not possible in general by analytical means. Therefore, further special cases are treated to give approximate solutions. For a detailed numerical treatment see Ref. 20. First, we will discuss the static case $\partial\Theta/\partial t = 0$ for a plane wave at normal incidence $\beta = 0$ in the small angle limit $\sin\Theta = \Theta$. The linearized eq. 11 then allows to calculate the spatial director field with respect to the "hard" boundary conditions $\Theta(z = 0) = \Theta(z = d) = 0$ to give symmetric solutions like

$$\Theta(z) = \Theta_{2m-1} \sin\{(2m-1)\frac{\pi}{d} z\} \quad (13)$$

where d is the film thickness and m an integer. Asymmetric solutions are possible as well with respect to the boundary conditions but are unstable in general [20]. Inserting (13) into the differential equation gives a threshold condition

$$E_{th}^2 = (2m-1)^2 \frac{K}{\varepsilon_o \varepsilon_a'} \left(\frac{\pi}{d}\right)^2 \quad (14)$$

Below an optical field E_{th} no reorientation occurs. Obviously the first harmonic component m=1 has the lowest threshold and will appear first. This threshold phenomenon results from the fact that the optical torque is zero for normal incidence in the discussed homeotropic arrangement, which is, however, very unstable against any disturbance, in analogy to the Freedericksz transition [2] induced by DC- or low frequency fields. The threshold depends on the elastic constant, the birefringence and the sample thickness in this approximation. With typical values like $K = 7\cdot10^{-12}$ N, $\varepsilon_a' = 0.6$, $d = 100\,\mu m$ and $m = 1$ we obtain $E_{th} = 360$ V/cm which corresponds to a threshold intensity $I_{th} = 265$ W/cm^2. This can be easily realized e.g. with a slightly focussed cw argon-ion laser.

The amplitudes Θ_{2m-1}, however, cannot be calculated in the small angle approximation. For this purpose we return to the nonlinear equation and assume that the spatial director distribution (13) is also approximately valid if the small angle limit is violated. Further the situation $\beta \neq 0$ is assumed and we will concentrate only on the first harmonic m=1 in all further discussions. The reorientation angle at $z = d/2$ is then given by

$$\left(\frac{E}{E_{th}}\right)^2 = \frac{2\Theta_1}{\sin\{2(\Theta_1+\beta)\}} \quad (15)$$

The inverse function of eq. 15 describes the amplitude of reorientation depending on the ratio of laser to threshold intensity and the angle β. The threshold behaviour and the amplitudes near the threshold are included in this simple expression if very small (but finite) values for β are assumed. These small values of β can be considered as a mathematical "simulation" of the director fluctuations, which are necessary to start the reorientation at normal incidence. In Fig. 5 we have plotted the calculated laser induced phase shifts $\delta\Psi = \delta n d$ (using eqs. 12, 13 and 15) as a function of the reduced intensity for different angles β. Note that due to the spatial distribution (13) δn must be averaged across the film. The agreement with experimental data obtained in self-diffraction experiments is fairly good, although some raw approximations have been used. For $\beta = 0$ we can observe the threshold behaviour, which disappears as β is increased.

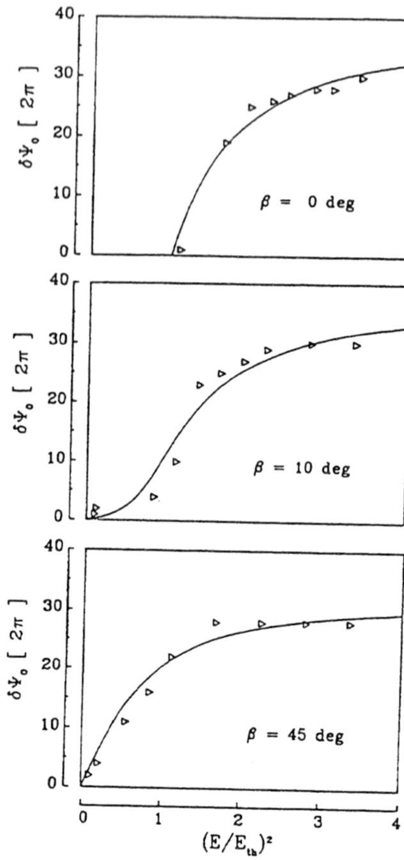

Fig. 5: Number of rings observed in a self-diffraction experiment as a function of reduced intensity. Parameter is the angle of incidence

It should be noted at this point, that the plane wave approximation used so far is, of course, not valid in most experiments. Laser beams are of finite diameter, in general, and additional transverse structures may appear e.g. in two-wave mixing experiments. Due to the diffusive character of the reorientation (cf. eq. 11), any transverse intensity modulation will also influence the reorientation amplitude. In other words, optical reorientation is a nonlocal process and the nonlinear coefficients depend on sample thickness and e.g. the beam diameter. The transverse effects can be considered by a modified threshold intensity and all results discussed so far are still valid if this modified threshold is used. The threshold for finite laser spots can be calculated in most practical cases as [19, 21]

$$E_{th}^2 = \frac{K}{\varepsilon_o \varepsilon_a'} \left(\frac{\pi}{d}\right)^2 (b^2 x^{2(k-1)} + 1) \qquad (16)$$

where $x = \pi w_o / d$ contains the ratio of beam radius w_o to film thickness d, $b = 1.43$ and $k = 0.24$. Eq. 16 shows that reorientation with laser beams of smaller diameter than film thickness leads to much higher thresholds, whereas the influence of finite beams is almost neglectable if $w_o \gg d$.

If experiments are performed with large diameter beams but periodically modulated optical fields, e.g. in grating experiments, the threshold for reorientation with a transverse Fourier component of wave vector q is given by

$$E_{th}^2 = \frac{K}{\varepsilon_o \varepsilon_a'} \left(\left(\frac{\pi}{d}\right)^2 + q^2 \right) \qquad (17)$$

which means that threshold is increasing drastically for gratings with $q \gg \pi/d$ and reorientation is weaker for a given intensity compared to the plane wave case.

3.2 Reorientation Dynamics and Optical Pattern Formation

Next, let us discuss the reorientation dynamics in the plane wave approximation. In the small angle limit an approximate solution (13) of eq. 11 yields (for m = 1)

$$\dot{\Theta}_1 + \frac{1}{\tau} \Theta_1 = \frac{\varepsilon_o \varepsilon_a' \sin 2\beta}{2\gamma_1} E^2 \qquad (18)$$

with $\tau = \dfrac{\gamma_1}{\varepsilon_o \varepsilon_a' (E_{th}^2 - E^2 \cos 2\beta)}$

Eq. 18 describes an exponential relaxation in the case $E^2 = 0$ with a time constant $\tau = \gamma_1 (d/\pi)^2 / K$ which is determined by the viscosity and the elastic constant. The ratio K/γ_1 may be interpreted as a reorientational diffusion coefficient in analogy to general transport phenomena. In the case of nonvanishing intensities the solution of (18) is given by

$$\Theta_1 = \frac{\varepsilon_o \varepsilon_a'}{2\gamma_1} \sin 2\beta \int_{-\infty}^{t} E^2(t') \exp\{-\frac{t-t'}{\tau}\} dt' \qquad (19)$$

which can be integrated for a step-like excitation (i.e. $E^2 = 0$ for

$t<0$ and $E^2 \neq 0$ for $t \geq 0$) to give

$$\Theta_1 = \frac{\varepsilon_0 \varepsilon_a'}{2\gamma_1} \sin 2\beta \; E^2 \tau \left(1 - \exp\left(-\frac{t}{\tau}\right) \right) \qquad (20)$$

Eq. 20 describes an exponential relaxation towards the new equilibrium state with a time constant τ. It should be noted, however, that τ depends on E according to eq. 18. More specific, τ can be positive or negative, depending on wether $E^2 \cos 2\beta$ is smaller or larger than E^2_{th}. If we assume e.g. that $\beta \ll 1$ with an intensity clearly above threshold, then τ is negative and eq. 20 describes an exponential increase as can be seen in Fig.6.

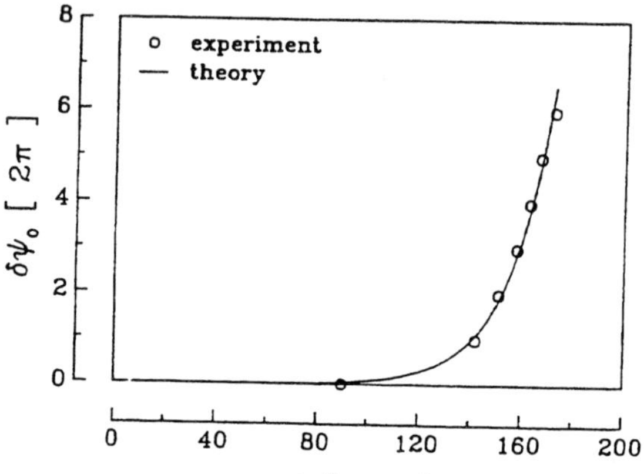

Fig. 6: Number of diffraction rings vs. time for a step-like input intensity started at $t=0$. Experimental conditions are $I = 890\ W/cm^2$, $I_{th} = 520\ W/cm^2$ $d = 120\ \mu m$, $\beta \approx 0$ and $T = 25\,^{\circ}C$.

The experimental points are obtained in a self-diffraction experiment. The new equilibrium reorientation angle is, however, not included within the small angle limit and consequently not described by eq. 20 in this case. Further a " critical slowing down " can be expected by decreasing E^2 towards E^2_{th}, which is typical for all threshold phenomena.

If on the other hand the intensity is drastically increased to $10^9\ W/cm^2$ (e.g. by focussing a 10 ns Q-switched laser pulse of 1 mJ energy to 100 μm spot-size), then τ may be in the range of hundred nanoseconds. This has been verified experimentally by Hsiung et. al. [22] using pulses of 6 ns duration and even with subnanosecond pulses by Eichler and Macdonald [23]. Experiments with short laser pulses and high intensities are discussed later with more detail in the sec. 4 of this paper.

For large β, τ becomes positive and Θ_1 increases like $(1-\exp\{-t/\tau\})$. Reorientation is possible in finite times even with intensities below threshold in this case.

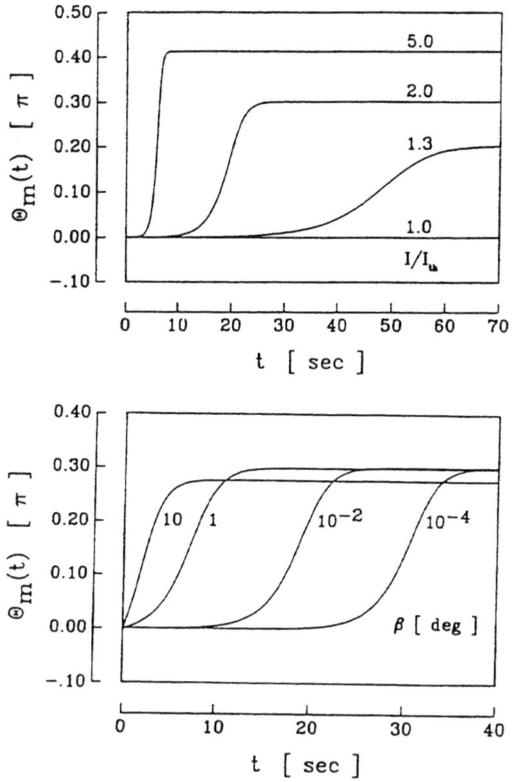

Fig. 7: Dynamics of optical reorientation for a step-like excitation.

A numerical solution of the nonlinear dynamic equation (11) for different intensities and angles β is plotted in Fig. 7. The laser is switched on at $t=0$. For rather small angles β and weak intensities it needs several 10 seconds until a remarkable reorientation can be observed. This time reduces to some seconds if intensity is increased e.g. by a factor of 5 or with larger angles β, respectively.

As mentioned above, the dynamics discussed here in the plane wave approximation are also valid more generally, if the correct thresholds with respect to the real spatial director deformations are used.

Light-induced reorientation has been discussed for homeotropically aligned nematics with incident extraordinary laser beams so far. The results are valid, however, also for extraordinary beams in planar oriented samples. Other cases have been studied as well by several authors, but we may only briefly review some of the obtained results here.

It turns out that optical reorientation in homeotropic cells has much higher thresholds for ordinary polarizations [14], depending on the angle between the director and the wave vector. Reorientation in planar samples with ordinary beams at normal incidence would be also rather difficult because of too high thresholds. Increased thresholds are further observed in many cases where ordinary *and* extraordinary beams propagate simoultaneously within the liquid crystal. This is a result of interference between the two principal polarizations which leads to a periodic torque and reorientation gratings of spatial periode $\Lambda = \lambda / n_a = 2 \ldots 3\,\mu m$ in the direction of beam propagation.

Thresholdless reorientation [24] has been achieved with extraordinary beams at normal incidence in so-called hybridly aligned samples, which means that the container is prepared to provide homeotropic alignment on one surface and planar alignment on the other. This kind of alignment has been used successfully e.g. in nonlinear Fabry-Perot resonators [25] for optical switching and optical bistability. Further, the strong reorientational nonlinearity of hybridly aligned nematics has been applied to experimentally investigate transverse spontaneous optical pattern formation in a simple feeback-mirror arrangement [26] for the first time.

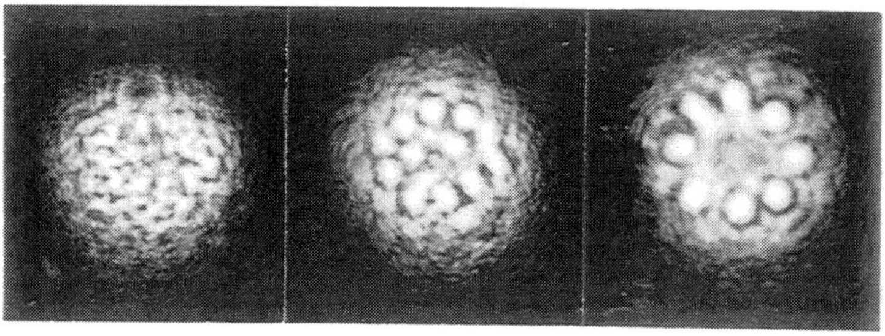

D = 5cm D = 10cm D = 20cm

Fig. 8: Spontaneous optical pattern formation in a nematic liquid crystal for three distances D between the nonlinear film and the feedback mirror

In the experiments, an unfocussed argon-ion laser beam propagates through the nonlinear film to a high reflectivity mirror and back to the film. Examples of the observed patterns are shown in Fig. 8 for different distances between the nonlinear film and the mirror. Intensities were as weak as 10 W/cm^2 in these investigations. The observed spontaneous pattern formation can be explained if it is assumed that the input laser beam contains a small phase modulation after transmitting the liquid crystal film, which then returns back as an amplitude modulated beam. The change from phase to amplitude modulation is a result of light propagation and diffraction effects. The returning transverse amplitude modulation, however, causes refractive index changes and phase modulations in the nonlinear film, thus providing a complex feedback mechanism. The resulting nonlinear dynamic system can lead to the observed self-organization effects and stable pattern formation.

Another interesting experiment [27] has been performed using circularly polarized light normally incident on a homeotropic cell. It was demonstrated that the spin of the lightwave can be transferred to the director motion above threshold and leads to a pressesion of the optical axis around the wave vector. The direction of rotation was reversed for using right or left circular polarizations, respectively.

It must be noted finally, that much less work has been carried out on optical reorientation phenomena in other mesophases. Only a few experiments [28, 29] have been reported concerning smectic phases. Torque-induced distortions of cholesteric phases have been discussed theoretically [30] to offer the possibility for mirrorless optical bistability but have not been observed to our knowledge until now.

3.3 Thermal effects

Optical nonlinearities originating from optical heating have been omitted from our discussion so far, although it is clear that thermal effects may be important in thermotropic liquid crystals. It is well-known that the refractive indices and the birefringence in nematics are strongly temperature dependent as are most other physical properties concerned with orientational order.

On the other hand it is important to note that in many experiments reorientational and thermal effects can be distinguished easily because of different relaxation times and polarization dependences.

In nematic liquid crystals the main thermal nonlinearities result from temperature dependent birefringence and density [22]. The changes of $\varepsilon_{\mu\nu}$ defined in eq. 2 due to a laser induced temperature rise δT are given by

$$\delta\varepsilon_{\mu\nu} = \left(\frac{\partial\varepsilon_{\mu\nu}}{\partial\rho}\frac{\partial\rho}{\partial T} + \frac{\partial\varepsilon_{\mu\nu}}{\partial S}\frac{\partial S}{\partial T} \right)\delta T \qquad (21)$$

Using the Maxwell relation $n = (\varepsilon)^{1/2}$ and the approximation $\partial\varepsilon/\partial\rho = (\varepsilon - 1)/\rho$ we obtain together with eq. 2 for the two principal refractive index changes

$$\delta n_\perp = \frac{\partial n_\perp}{\partial T}\delta T = \left(\frac{n_\perp^2 - 1}{2 n_\perp \rho}\frac{\partial\rho}{\partial T} - \frac{\varepsilon_a}{6 n_\perp}\frac{\partial S}{\partial T} \right)\delta T \qquad (22a)$$

$$\delta n_\perp = \frac{\partial n_\parallel}{\partial T}\delta T = \left(\frac{n_\parallel^2 - 1}{2 n_\parallel \rho}\frac{\partial\rho}{\partial T} + \frac{\varepsilon_a}{3 n_\parallel}\frac{\partial S}{\partial T} \right)\delta T \qquad (22b)$$

A coupling between the density ρ and the order parameter S has been neglected in this calculation of the thermo-optic coefficients ($\partial n/\partial T$).

The temperature dependence of the scalar order parameter can be described for most of the investigated nematics by the empirical expression [4]

$$S = (1 - 0.98\, T_r)^{0.22} \qquad (23)$$

where $T_r = T V^2/(T_{NI} V_{NI}^2)$ is a reduced temperature. V and V_{NI} are the molar volumes at T and the nematic-isotropic phase transition temperature T_{NI}, respectively.

The thermal volume-expansion coefficient is almost independent of temperature not too close to the phase transition and the value [31] for 5CB is e.g. $(\partial\rho/\partial T)/\rho = -8.2 \cdot 10^{-4}\,K^{-1}$.

Numerical values for the thermo-optic coefficients (22a, b) can be calculated by using typical material parameters for 5CB at room temperature like $n_\perp = 1.53$, $n_\parallel = 1.74$, $S = 0.52$, $\varepsilon_a = 1.19$, $\partial S/\partial T = -7.2 \cdot 10^{-3}\,K^{-1}$ to result in $(\partial n_\perp/\partial T) = +5.7 \cdot 10^{-4}\,K^{-1}$ and $(\partial n_\parallel/\partial T) = -2.1 \cdot 10^{-3}\,K^{-1}$ in good agreement with published [7] data. These thermo-optic coefficients are among the largest of all known materials, in particular close to the phase transition where they increase to about $10^{-2}\,K^{-1}$.

It should be noted further that the coefficients parallel and perpendicular are different in sign. This is an interesting feature because both, positive and negative thermo-optic nonlinearities can be realized in the same material by simply changing the optical polarization with respect to the nematic director. The change in sign can be obtained continously with extraordinary beams according to eq. 5 by adjusting the angle between the director and the wave vector. As a consequence, this further implies that the extraordinary thermo-optic coefficient must be zero for a certain angle, which was experimentally found [22] to range around 30 deg in 5CB.

The temperature rise due to absorbed laser intensity can be calculated with the heat-flow equation [29] in principle. Since all spatial gradients in temperature are diffusion limited, the thermal nonlinearity is nonlocal and depends on geometrical factors like beam diameter or sample thickness, very similiar to the reorientational nonlinearity discussed above. The thermal relaxation times are, however, faster and range typically between 10 μs and 10 ms. We will not discuss the thermal diffusion equation here in detail but simply estimate the temperature rise and the resultant refractive index changes for realistic experimental parameters. The steady state temperature rise may be calculated in the relaxation time approximation [29] to

$$\delta T = \frac{\alpha I \tau}{\rho C_p} \tag{24}$$

where α is the absorption coefficient, I the intensity, ρ the mass-density and C_p the heat capacity. τ is the thermal relaxation time depending on the heat diffusion coefficient and geometrical parameters like beam diameter, film thickness or grating wave numbers. If typical values like $\rho = 10^{-3}$ kg/cm^3 and $C_p = 4000\,\mathrm{J\,K^{-1}\,kg^{-1}}$ are used with experimental parameters $I = 1000$ W/cm^2 and $\tau = 1$ ms we obtain with $\alpha = 0.01\,\mathrm{cm}^{-1}$ (5CB "off resonance" at $\lambda = 514$ nm) $\delta T = 2.5 \cdot 10^{-3}$ K which is rather small and may induce only a hardly detectable refractive index change. If, however, $\alpha = 10^5\,\mathrm{cm}^{-1}$ (5CB "at resonance" $\lambda = 280$ nm [32]) then $\delta T = 2.5 \cdot 10^4$ K. Consequently, the absorption must be enhanced in many liquid crystals to investigate the thermal nonlinearity in the visible spectral range e.g. by adding appropriate dyes, which has been pointed out by several authors [33, 34]. The situation is, however, completely different if short laser pulses and much higher intensities are used as will be shown later.

Another possibility to take advantage of the large thermal nonlinearity of liquid crystals is to provide heat production by surface excitation of thin absorbing coatings. This has been successfully demonstrated by Lloyd and Wherret [35], who observed optical bistability with input powers as low as 20 μW using nonlinear Fabry-Perot resonators with one metallic and one dielectric mirror filled with intra-cavity liquid crystals as nonlinear media. Mirrorless optical bistability has been observed [36] in absorbing cholesteric liquid crystals where optical heating has been used to tune the helix-pitch to the Bragg-condition for the incident wavelength.

A new thermo-optical effect has been proposed and demonstrated recently by Macdonald et al. [29] using ferroelectric Sm C* phases. Reorientation of the optical axis has been achieved in 2 μm thick planarly oriented samples as a result of temperature dependent tilt angles and laser heating. The maximum reorientation angle was about 23 deg with intensities of 3000 W/cm^2. Switching times are limited by heat diffusion in these experiments and in the range of 1 ms or somewhat less. Opto-thermally reorientation can be used e.g. as an optical adressable halfwave plate and for opto-optical modulation as can be seen in Fig. 9. It further turns out in these experiments, that optical heating may be useful for investigations of the SmC* - SmA phase transition. The effect has been also investigated dynamically using short laser pulses and the results will be shown in the next part of this paper.

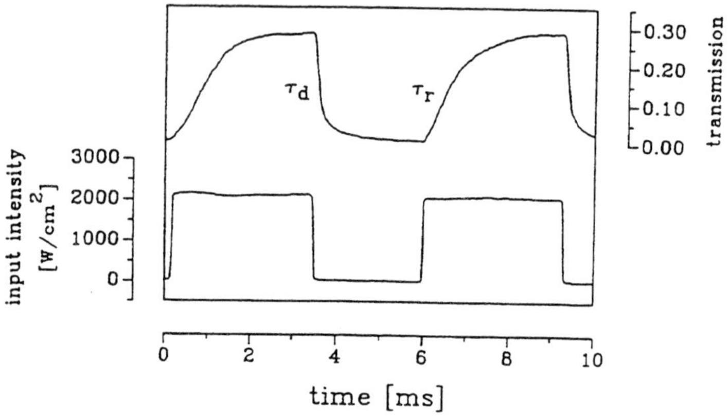

Fig. 9: Opto-optical modulation using a ferroelectric liquid crystal. Upper trace is the modulated, lower trace the modulating intensity.

A further effect in connection with absorbing liquid crystals has been found by Janossy et al. [37]. An anomalously low threshold for reorientation of about 1.5 mW was observed in a commercial guest-host mixture, consisting of a dichroic dye and a homeotropically oriented nematic liquid crystal. This is about 65 times less than for a corresponding transparent nematic film. The underlying mechanism for this threshold reduction is still under investigation and not known yet.

Another optical nonlinear effect which can be neither attributed to director reorientation nor to thermally induced birefringence changes has been reported by Odulov et al. [38] to appear in certain absorbing liquid crystals like MBBA. The observed large birefringence changes were explained to result from photostimulated molecular conformation changes and the observed optical nonlinearity is therefore called *conformational nonlinearity.*

4. NONLINEAR OPTICAL RESPONSE TO SHORT LASER PULSES

In the previous sections we have reviewed some basic fundamentals of liquid crystals and discussed the main Kerr-like optical nonlinearities, which have been observed with low power cw laser radiation. The present section is more specific, dealing with transient nonlinearities in connection with short laser pulses. The motivation for using short excitation pulses is two-fold: First, the dynamics of optical nonlinearities can be investigated with higher temporal resolution. Second, the influence of optical fields on the orientational order of liquid crystals can be studied under extreme conditions e.g. with rather high intensities and strong fields. These aspects are important for both, basic research of (nonlinear) material response to light and applications. The dynamic response of optical nonlinearities determine e.g. the switching-times, which may be realized in opto-optical devices.

As will be shown in the following, the use of short laser pulses offers the unique possibility to study different transient phenomena in liquid crystals. Generally, there is a transfer of energy, linear and angular momentum from the light wave to the liquid crystal during the impulsive excitation, which result in temperature rise, flow and reorientation of the molecules, respectively. Since these processes influence the optical properties of the material, the corresponding relaxation processes can be monitored with a weak probe beam. These advantages have been already pointed out in one of the first experiments using short laser pulses for the investigation of nematic liquid crystals reported by Hsiung et al. [22]. In this reference transient laser heating and optical reorientation in 5CB was observed using Q-switched pulses of 6 ns duration, but no flow effects.

Armitage and Delwart [34] used nanosecond pulses to investigate the dynamics of thermal effects in dye-doped nematic films near the nematic-isotropic phase transition. They observed a critical slowing-down in connection with the increasing thermo-optic coefficients in the vicinity of the transtion temperature and a delay in the optical response. The latter was attributed to the scalar order parameter relaxation time. Transient reorientation and thermal heating in nematic films with nanosecond pulses has been further reported by Khoo et al. [28, 33]. Nonlinear absorption and refractive index changes have been investigated with the z-scan technique by P. Palffy-Muhoray and co-workers [9, 39] using laser pulses of 7ns duration.

In order to achieve even higher time-resolution we demonstrated [23] that reorientation phenomena can be observed in nematic films even if subnanosecond pulses are used. In self-diffraction grating experiments with excitation pulses of 80 picoseconds duration it was shown that molecular reorientation is responsible for the diffraction effects. Time resolved grating experiments [40, 41, 42, 43] show, however, that the dynamic response to short intense ps laser pulses is rather complex in general, because of coupling between density, temperature, flow and reorientation. On the other hand flow effects and density waves (ultrasound) can be excited much easier with picosecond than with nanosecond pulses. In the following we report about recent results of time resolved investigations with picosecond laser pulses, which have been performed in nematic, smectic and cholesteric liquid crystals using the light-induced dynamic grating technique [44].

The wave-mixing arrangement for the experiments is sketched in Fig. 10. Two pump- pulses of 80 ps FWHM at $\lambda_{exc} = 532$ nm obtained from a frequency doubled mode-locked Nd:YAG laser with a symmetric beam splitter are slightly focussed to a spot size of about 500 µm diameter with an intersection angle of $\vartheta = 1$ deg onto the liquid crystal sample. The two beams are linearly polarized, either parallel or perpendicular to each other. The optical fringes resulting from interference of the two beams modulate the optical properties of the liquid crystal film due to optical nonlinearities. The fringe spacing is about $\Lambda = \lambda / \sin \vartheta = 30\,\mu m$ in the following.

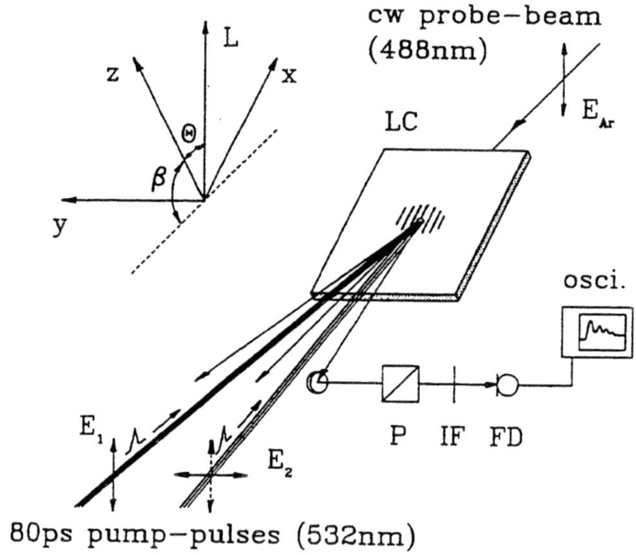

Fig. 10: Experimental wavemixing arrangement with LC, liquid crystal; FD, fotodetector; IF, interference filter (488 nm); P, polarizer

The center of the induced optical grating is monitored by diffraction of a weak cw laser at $\lambda_{pr} = 488$ nm on a spot of 100 μm. The first order diffracted intensity of the probe beam is measured as a function of time with the help of fast fotodetectors. In most experiments fast fotodiodes are used in connection with oscilloscopes and the temporal resolution is between 200 ps and 2 ns. Higher resolutions of several picoseconds are realized by using a streak-camera instead. The diffraction efficiency of a thin phase grating into the first order diffracted beam is given by [44]

$$\frac{I_1}{I_o} = | J_1(\delta\psi_o) |^2 \tag{25}$$

where J_1 is the first order Bessel function (of first kind) and $\delta\psi_o = 2\pi\,\delta n_o\, d/\lambda$ denotes the amplitude of the induced phase grating for a given polarization of the probe beam. I_1 is the diffracted intensity of a probe beam with input intensity I_o. For weak gratings $\delta\psi_o \ll 1$ the approximation $J_1 = \delta\psi_o/2$ holds and the diffracted intensity is proportional to the square of the induced refractive index changes.

4.1 Ultrasound Generation and Flow-Alignment in Nematics

Fig. 11 displays oscilloscope traces of the diffracted probe beam intensity on two different time scales obtained with a homeotropic aligend nematic film. An intensity grating (i.e. parallel polarized pump beams) of 80 ps duration has been used for the excitation. The initial alignment includes a finite angle β with the intersection plane of the two pump beams to overcome the " nematic barrier "

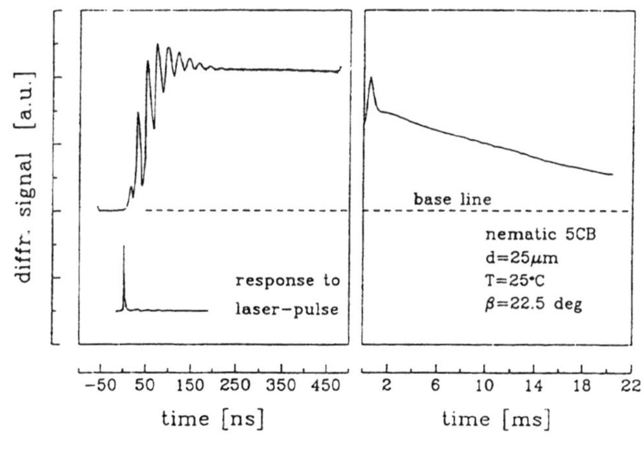

Fig. 11: Diffracted probe-beam intensity vs. time after ps grating excitation with parallel polarized pump-beams. Excitation energy is $W_p = 0.25$ mJ.

which usually occurs in reorientation phenomena at normal incidence.

Obviously, the dynamic response of liquid crystals to short intense laser pulses is rather complex, showing several characteristic relaxation processes and time constants. The grating is built-up on a nanosecond time scale and the diffracted signal is still increasing long after the pump pulse has left the sample. Afterwards the grating washes out and the diffracted signal decays with a double exponential law with two time constants of about 200 µs and 40 ms, respectively. The slow relaxation process can be attributed to collective molecular reorientation effects whereas the faster process is the decay of thermal gratings due to heat diffusion. The observed slow relaxation times exhibit a characteristic temperature dependence shown in Fig. 12 which has been explained [43] with the temperature dependences of the rotational viscosity γ_1 and the elastic constant K, confirming that reorientation has been induced with picosecond laser pulses. The thermal grating relaxation is treated in section 4.3 with more detail.

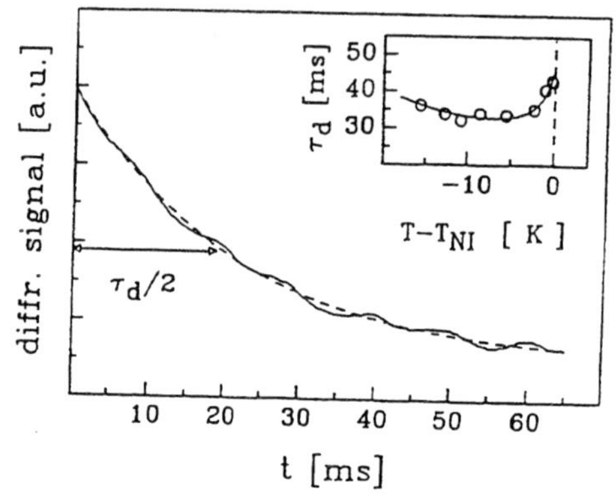

Fig. 12 : Reorientation grating decay after excitation with 80 ps laser pulse. Inset: Evaluated relaxation times vs. reduced temperature.

The strong oscillations during the grating build up are due to laser-induced standing ultrasound waves [45, 46] and diffraction at the resulting density modulations (" forced Brillouin scattering "). The holografic ultrasound generation is explained schematically in Fig. 13. A density modulation with the spatial frequency q of the intensity grating is excited at t = 0 by thermal volume expansion or electrostrictive forces, respectively. The periodic stretching and squeezing of the fluid launches two counterpropagating sound waves if the corresponding frequency

$$\omega = c_s q \qquad (26)$$

is included within the Fourier-spectrum of the excitation pulse. c_s is the speed of sound. As a result, the density modulation and the corresponding phase grating relax in form of a damped standing wave leading to oscillations in the diffracted signal ("acoustic ringing").

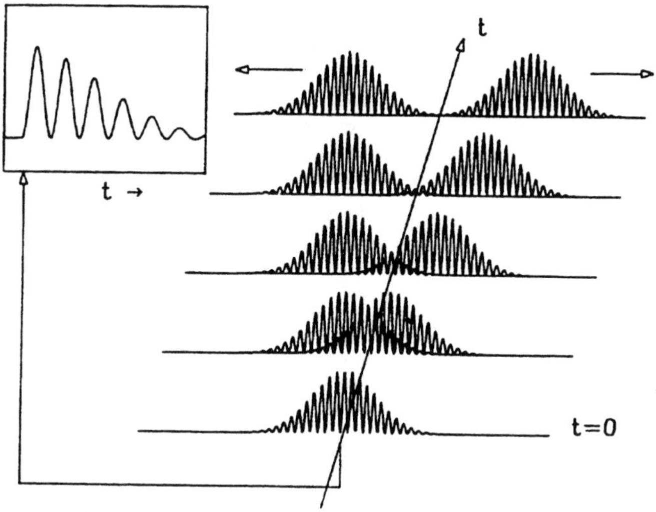

Fig. 13: Schematic representation of holografic ultrasound generation

The numerical value of the frequency given by (26) for $q = (2\pi/30\,\mu m)$ and [46] $c_s = 1500\,m/s$ is $\omega/2\pi = 50$ MHz in agreement with the oscillations in Fig. 11. The holografic sound wave generation is very effective if the pulse duration $\tau_p \ll \omega^{-1}$.

The damping of the acoustic grating signal is given by two mechanisms in general. First, the viscosity η of the fluid leads to a phonon life-time [47]

$$\tau_1 = 2/(\eta q^2). \qquad (27)$$

Second, the two sound wave packets travel out of the excitation region leading to effective damping of the signal which can be estimated with

$$\tau_2 = w_o/c_s \qquad (28)$$

where w_o is the excitation beam radius. The phonon life-time can be observed only if $w_o \gg 2c_s/(\eta q^2)$. The damping of the signal displayed in Fig. 11 is given by the second process eq. 28.

The observed acoustic gratings are a new quality appearing in transient grating experiments with picosecond laser pulses since they are connected with translational motions of the liquid crystal molecules. These motions are not important in most experiments with cw lasers and have been neglected so far. In liquid crystals translational motions are usually coupled to rotation and reorientation of the molecules via anisotropic frictional forces. Shear-induced flow-alignment of liquid crystal films between two moving glass-plates [2] is a well-known example for this phenomenon. The importance of flow-alignment effects in ps grating experiments has been demonstrated by Khoo et al. [42] and by us [41, 43] almost simultaneously. In Ref. 42 reorientation effects due to temperature and density modulations have been observed in an experimental arrangement where *parallel* polarizations of the excitation beams nearly normally incident on a homeotropic liquid crystal did *not allow* optical torque-induced reorientation in principle. In Refs. 41, 43 investigations have been performed with a different geometry which *allows* laser-induced reorientation effects, on one hand, but uses *crossed* excitation beam polarizations to avoid thermal and acoustic gratings, on the other hand. It was shown that the excitation pulse is followed by a delayed reorientation process in this case, which can be explained in terms of a large inertial moment due to flow-alignment, which is not driven by temperature and density modulations but by photo-elastic stresses.

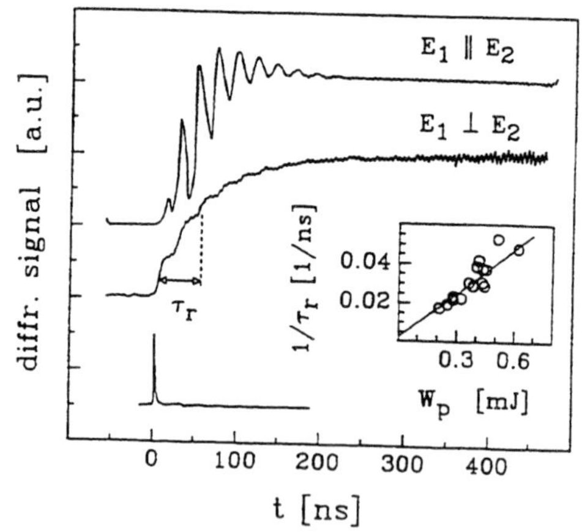

Fig. 14: Diffracted probe-beam signal vs. time with parallel and perpendicular pump-beam polarization. Excitation energy is $W_P = 0.25$ mJ.

Thermal gratings and the resulting acoustic gratings can be suppressed if crossed polarization $E_1 \perp E_2$ of the pump beams are used with moderate excitation energies (below 0.6 mJ in the present experiments) instead of parallel polarizations $E_1 \parallel E_2$ which is shown Fig.14. If $E_1 \parallel E_2$ the two beams interfere to give an intensity grating while in the case $E_1 \perp E_2$ a polarization grating with a much weaker intensity modulation (due to imperfect polarizations of the beams) is formed. As a consequence, polarization-dependent effects like reorientation can be excited periodically with crossed polarizations while intensity-dependent modulations like thermal gratings are suppressed at the same time. Furthermore shear forces and transverse flow effects can be excited with polarization gratings, which result from photoelastic stresses [47].

Since the elasto-optic coefficients relating these stresses to the light field are not known for liquid crystals, the force density **F** corresponding to Maxwell stresses [48]

$$\mathbf{F} = (\mathbf{D}\cdot\nabla)\mathbf{E}^* - \frac{1}{2}\nabla(\mathbf{E}\cdot\mathbf{D}^*) \qquad (29)$$

can be used as a qualitative approximation. **E** is the optical field, **D** is the displacement vector, and the asterik denotes the complex conjugate. The complex notation gives the time-averaged real value of **F**. The main force for a polarization grating with an experimental geometry as depicted in Fig. 10 is given by the x-component $F_x = \frac{1}{2}\varepsilon_o \varepsilon_\perp n_\perp q E^2 \cos(qy)$ and other components are negligible. The director motion can be described by a single reorientation angle Θ in this case, which is obtained by the torque balance eq. 11 in addition to the Navier-Stokes equation [49, 50]

$$\gamma_1 \dot{\Theta} + M_{el} + M_f + M_{op} = 0 \qquad (30a)$$

$$\rho \dot{v}_x - \gamma_s \Delta v_x = F_x \qquad (30b)$$

where $M_{el} = -K\Delta\Theta$ is the elastic torque, $M_f = -\frac{1}{2}(\gamma_1 - \gamma_2 \cos 2\Theta)\, \partial v_x/\partial z$ is a flow coupling term and M_{op} is the optical torque (cf. eq. 9). γ_s and γ_2 are flow viscosities, ρ is the mass density and v_x the flow velocity excited by F_x. The inertial moment (density) connected with pure reorientation phenomena (i.e. no flow) is in the order of $\mu = 10^{-16}$ kg/m and has been ommited from eq. 30a because it would result in reorientation rise time of picoseconds or less, which was not observed in these experiments. The inverse flow alignment process has been also estimated to be of less importance and is neglected here.

We will not discuss the solution of eqs. 30 here in much detail but briefly summarize the main results as given in Ref. 43 : A single equation describing the time dependence of the reorientation grating amplitude in terms of an overdamped oscillator can be obtained by elimination of the flow velocity. The apparent second time derivative is a consequence of the flow elimination and is connected with a large inertial moment $\tilde{\mu} \approx \rho/q\,q'$ of the flowing molecules in the excited regions, where q is the wave number of the excited transverse and $q' = 2\pi/d$ the longitudinal spatial mode. It can be calculated to $\tilde{\mu} = 10^{-8}$ kg/m which is much larger than the inertial moment given above and consequently leads to much slower reorientation rise times. The dynamics of the reorientation grating amplitude according to eqs. 30 for excitation with a delta-pulse is given by

$$\Theta_1 \sim (\exp\{-t/\tau_d\} - \exp\{-t/\tau_r\}) \tag{31}$$

where $\tau_r = \rho/\gamma_s(q^2 + q'^2)$ and $\tau_d = \gamma_1/K(q^2 + q'^2)$ can be interpreted as the rise time and the decay time of the reorientation process.

As with simple reorientation models, the relaxation of the grating is given by the rotational viscosity and the elastic constant of the nematic fluid, whereas the observed rise times are governed by the damping of the induced shear flow and hence the flow viscosity. The observed rise times τ_r are in the order of 20 to 50 ns depending on pump energy as shown in Fig. 14. The energy dependence can be explained assuming a non-Newtonian flow viscosity, which means that the viscosity and thus τ_r^{-1} contains a term which is proportional to the shear rate $\partial v_x/\partial z$. Since v_x is proportional to the pump energy W_p (cf. eq. 29) a linear dependence of the reciprocal rise time on pump energy can be expected, in agreement with the experimental observations shown in Fig. 14. By extrapolating the graph to $W_p \rightarrow 0$ we get (using τ_r as given with eq. 31) $\gamma_s = 0.02$ kg/ms in excellent agreement with published data for small shear rates [51].

It should be noted at the end of this section that longitudinally translational motion of molecules and ultrasound has been also observed even with polarization gratings at higher excitation energies [40]. The origins of these acoustic gratings are, however, still under investigations.

4.2 Nonlinear Absorption and Thermal Gratings

Laser-induced thermal birefringence changes in slightly absorbing nematic liquid crystals have been investigated as important optical nonlinear mechanisms beside the "giant" reorientational nonlinearity in various experiments [22, 28, 33, 34] using short laser pulses. The observed large thermo-optic effects are, however, somewhat contradicted by a rather weak (linear) absorption for the most stable and applicable family of liquid crystals, the cyanobiphenyls, throughout the whole visible and near infrared spectral range as mentioned above. As a consequence, the absorption of these transparent materials has been often enhanced by adding appropriate dyes or by coating the inner walls of the liquid crystal cell with absorbing layers [29, 35] to obtain strong laser heating and the related thermal nonlinearity.

However, enhanced absorption and thermal gratings *without* any additional absorbing material has been also observed in 5CB using short intense laser pulses in the green spectral range, recently. Deeg and Fayer [52] observed two-photon absorption and excited state gratings in wave-mixing experiments with 300 fs laser pulses in isotropic 5CB at $\lambda_{exc} = 575$ nm. It has been pointed out in this work, however, that the two-photon absorption is not sufficient to produce enough heat for the generation of strong thermal gratings, because the relaxation of the first excited singlet is mainly radiative. Nonlinear absorption in 5CB has been also observed [39] using nanosecond pulses in the green spectral range.

We verified experimentally [53] that strong thermal gratings in the nematic and the isotropic phase of 5CB are generated by a three-photon process if short picosecond laser pulses at $\lambda_{exc} = 532$ nm are used. The experiments have been performed in the transient grating arrangement discussed above using a planarly aligned nematic sample and $\beta = 0$.

The polarizations ($E_1 \parallel E_2$) of the pump beams, the grating wave vector and the normal axis of the liquid crystal film are orthogonal to each other and may be labelled as the x-, y- and z-coordinate axis, respectively. The sample can be rotated around the z-axis to provide that the optical field is either parallel or perpendicular to the nematic director. This is important to avoid strong additional reorientation gratings as discussed in the previous section.

The absorbed intensity grating modulates the temperature and hence the birefringence of the fluid (see sec. 3.2) leading to diffraction of the probe beam. The temperature rise δT due to laser heating can be described by the heat flow equation [7, 44]. Solution of this equation in a plane wave approximation neglecting higher Fourier-components of the grating may be written as [53]

$$\delta T = T - T_o = \frac{Q_o \tau_p}{\rho c} \exp\{-t/\tau\} \cos\{q y\} \qquad (32)$$

where the thermal grating relaxation time $\tau = D_{eff}^{-1} q^{-2}$ contains the relevant (effective) heat diffusion coefficient and the grating wave number $q = 2\pi / \Lambda$. The heat production (per time and volume) referring to the amplitude (peaks) of the induced grating is denoted as Q_o, τ_p is the laser pulse duration, ρ the mass density, c the heat capacity and T_o is the initial temperature. In evaluating eq. 31 we have assumed that the heat production is fast compared to the thermal relaxation times and that the main heat diffusion washing out the grating is from the peaks to the nulls of the grating.

The heat diffusion in nematic liquid crystals can be described by a 2nd rank tensor $\underline{\underline{D}} = \underline{\underline{\lambda}}_w \rho^{-1} c^{-1}$ which is given by the heat conductivity tensor λ_w, the mass density ρ and heat capacity c. The effective heat diffusion coefficient D_{eff} is e.g. given by the two principal values $D_\perp$ and $D_\parallel$ in the nematic phase if the temperature gradient (i.e. the grating wave vector) is applied perpendicular or parallel to the planar aligned director, respectively. Typical values for D_{eff} and τ are dicussed below, together with the experimental results.

The heat production in the sample and thus the amplitude of the induced thermal grating at the beginning of the decay is determined by the absorbed intensity of the ps-excitation pulse. In many cases it is sufficient to use expressions [7, 44, 53] like $Q = \alpha I$ where α is the absorption coefficient and I the intensity, which implies, however, that the absorbed intensity is fast and completely dissipated as heat. Following the ideas of Deeg and Fayer [52] strong two-photon absorption (2PA) from the lowest energy band S_0 to the first excited singlet states S_1 can be expected in experiments with green intense laser pulses at $\lambda_{exc} = 532$ nm (see Fig. 15). The 2PA spectra essentially follows the one-photon absorption which peaks at $\lambda = 280$ nm in pure 5CB [32]. The 2PA probability is proportional to the number density of ground state molecules and the photon density, i.e. the intensity. The absorption coefficient can be written in this case as $\alpha_{2PA} = \gamma_2 I$, where γ_2 is a constant. However, since the lifetime of the S_1-state is in the

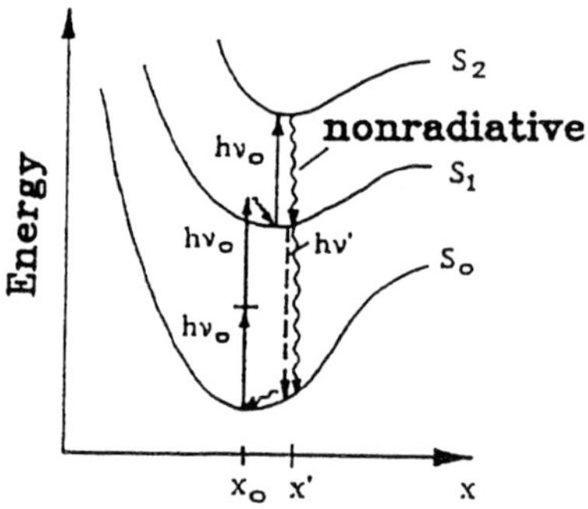

Fig. 15: Simplified energy scheme of 5CB (Ref. 13)

order of one nanosecond or somewhat less [54], there is a strong probability that a third photon of the short picosecond laser pulse is absorbed by an excited state-excited state transition $S_1 \rightarrow S_2$. The one-photon excited-state absorption is very effective if the excitation pulse duration is short compared to the S_1 lifetime, but large compared to the solvent cage reorganization time of 4 ps. The latter is important because the transition is shifted down to the excitation wavelength $\lambda_{exc} = 532\,nm$ by this process [52].

The probability for the one-photon excited state absorption is proportional to the number density of singlet S_1 molecules which is proportional to the intensity absorbed by 2PA in our case, i.e. $\alpha_{12} = k\,\alpha_{2PA}\,I = \gamma_3 I^2$ where $\gamma_3 = k\,\gamma_2$ and k is a constant. The complete three-photon absorption process can then be described by the nonlinear absorption coefficient

$$\alpha = \alpha_{2PA} + \alpha_{12} = \gamma_2 I + \gamma_3 I^2 = \gamma_2 I\,(1 + kI) \qquad (33)$$

The intensity independent $S_0 - S_1$ one-photon absorption has been neglected because it is rather weak for our excitation wavelength as discussed above.

The heat production due to this nonlinear absorption is provided by fast radiationless recombination from the S_2 singlet states mainly and can be written as

$$Q = \alpha_{12} I = \gamma_3 I^3 \qquad (34)$$

The temperature rise and the amplitude of the induced thermal grating can be calculated by using eqs. 34 and 32. As a result, the induced phase grating amplitude $\delta\psi_o$ is expected to be proportional to I^3 or $W_p^{\,3}$, respectively, because $\delta n \sim \delta T$. In Fig. 16 we have plotted $\delta\psi_o$ as a function of the excitation energy on a logarithmic scale, which has been obtained from diffraction efficiency measurements (using eq. 25) with isotropic 5CB displayed in the inset of the same figure. The experimental data fit quite well to a cubic law which is a strong indication that the three-photon absorption process is mainly responsible for the heat production. Laser heating due to 2PA or one-photon absorption would result in a quadradic or linear dependency, respectively.

The coefficients γ_3 have been evaluated from these experiments to $\overline{\gamma}_3 = 3.7 \cdot 10^{-25}\,m^3/W^2$ in the isotropic phase and $(\gamma_3)_\perp = 0.14\,\overline{\gamma}_3$, $(\gamma_3)_\parallel = 1.8\,\overline{\gamma}_3$ in the nematic phase at $T = 25\,^oC$. The nonlinear absorption is anisotropic in the nematic phase with stronger absorption for polarizations parallel to the director. This is not surprising since the involved $S_o - S_1$ transition is polarized along the long molecular axis in cyanobiphenyls.

Nonlinear absorption of course leads to nonlinear transmission of the samples, which has been clearly observed with liquid crystal films of 200 μm thickness.

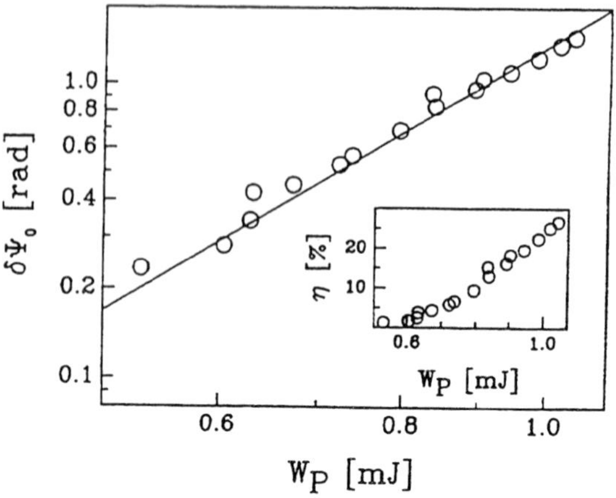

Fig. 16: Evaluated thermal grating modulation-depth vs. excitation energy for isotropic 5CB. The inset shows the measured diffraction efficiencies.

Relaxation of the picosecond induced thermal gratings ranges on a microsecond time scale and exhibit a critical slowing behaviour in the vicinity of the nematic-isotropic phase transition. Fig. 17 shows the evaluated thermal relaxation times which depend on director orientation and temperature. The two branches for τ in the nematic phase correspond to heat diffusion parallel and

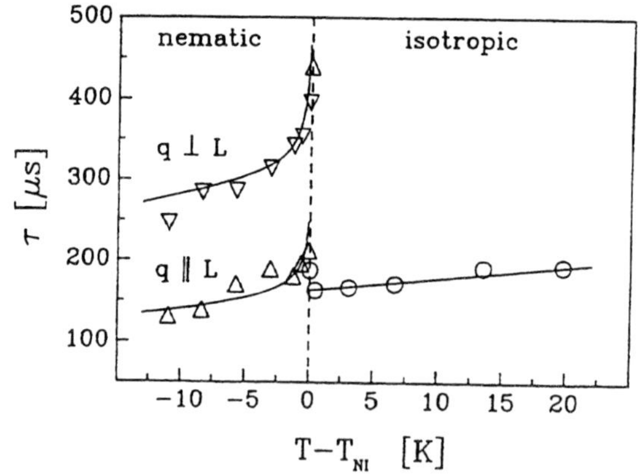

Fig. 17 : Thermal grating relaxation times in 5CB as a function of temperature

perpendicular to the director. Typical values for the relaxation times not too close to the phase transition (e.g. at $|T - T_{NI}| = 10\,K$) are $\tau_\perp = 250\,\mu s$, $\tau_\parallel = 130\,\mu s$ and $\tau_{iso} = 180\,\mu s$ which can be explained with the thermal relaxation time in eq. 32 using $q = 0.2\,\mu m^{-1}$ and heat diffusities like $D_\perp = 0.9 \cdot 10^{-3}\,s^{-1}\,cm^2$, $D_\parallel = 1.8 \cdot 10^{-3}\,s^{-1}\,cm^2$ and $D_{iso} = 1.3 \cdot 10^{-3}\,s^{-1}\,cm^2$ in good agreement with published data [7].

The observed temperature dependence of τ can be explained with the temperature dependence of the heat conductivity tensor and the heat capacity c at the first order nematic-isotropic phase transition [53], whereas changes in the mass density are comparitively small and may be neglected.

4.3 Thermo-Optical Reorientation of Ferroelectric Liquid Crystals

The dynamics of laser-induced tilt angle changes and reorientation effects in ferroelectric liquid crystals discussed in sec. 3.2 have been also investigated using short picosecond laser pulses and the transient grating technique [55]. Due to a rather fast temperature rise induced by impulsive laser heating it is possible to investigate the relaxation of the tilt angle towards the new equilibrium state. This relaxation process is not only of practical interest for the realization of optical switching times but also for fundamental research, since the tilt angle can be used as an order parameter [56] to describe the Sm C*- Sm A phase transition.

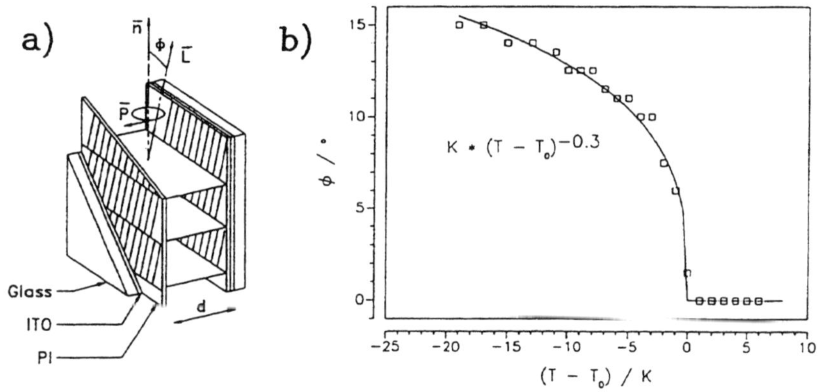

Fig. 18: Surface stabilized Sm C* phase in a bookshelf-like geometry (a) and tilt angle Φ vs. temperature (b)

The experiments have been performed with surface stabilized smectic films in a bookshelf-like geometry as shown in Fig. 18a. The director and hence the optical axis is tilted with respect to the smectic layers including an angle Φ. The tilt angle is temperature dependent as depicted in Fig. 18b. Due to some finite absorption of the laser pulse within the sample the temperature is increased rapidly and the tilt angle follows with a characteristic relaxation time. This reorientation relaxation leads to a delayed increase of the diffracted probe beam as can be seen in Fig. 19. Experimental results obtained within the Sm C* and Sm A phase are shown for comparision. Since the birefringence and film thickness are matched to provide that $n_a d \approx \lambda/2$, periodic reorientation of the optical axis leads to a strong depolarization of the diffracted probe beam. This has been investigated by performing a polarization analysis of the diffracted light. The orthogonally polarized signal is much stronger in the Sm C* phase, indicating optical axis reorientation, whereas no depolarization effects occur in the Sm A phase. The grating diffraction in the Sm A phase is explained by density modulations mainly and no reorientation has been observed. The density relaxation is, however, not connected with acoustic gratings in contrast to the considerations of sec. 2 since the liquid crystal film is rather thin.

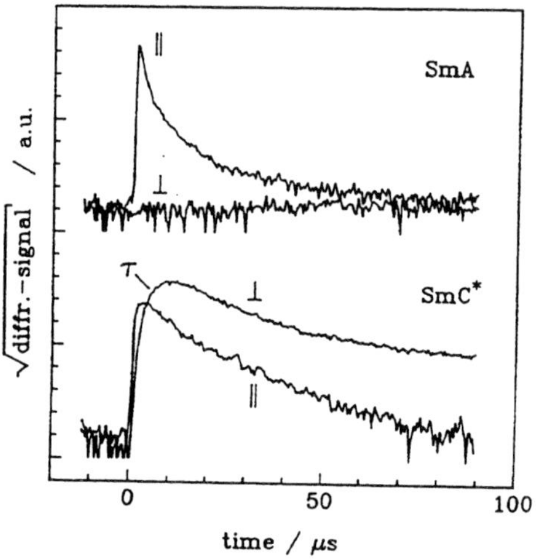

Fig. 19: Square root of the diffracted signal vs. time for Sm C* and Sm A phase of a ferroelectric liquid crystal.

The observed order parameter relaxation times are in the range of 0.6 to 5 μs with the commercial ferroelectric liquid crystal used in these experiments, depending on temperature as shown in Fig. 20. The hyperbolic increase of τ with increasing temperature can be expected from a dynamic model of the phase transition which is, however, beyond the scope of the present article and will not be discussed here.

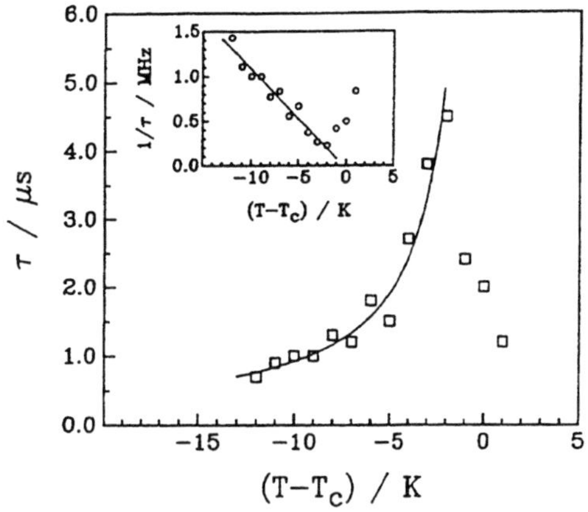

Fig. 20: Tilt angle relaxation times as a function of temperature

The grating decays on a microsecond time scale if low excitation energies ($\leq 100\,\mu J$) are used. Relaxation of the thermal gratings is then determined by surface heat diffusion because the liquid crystal is rather thin compared with the grating periode. The nonexponential decay shown in Fig. 19 can be explained by solving the heat diffusion equation for this particular case.

If, however, the excitation is stronger ($> 100\,\mu J$), a slower component can be observed following the thermal relaxation. The time constants of the slow component are in the range of several milliseconds and can be explained by the relaxation of elastic deformations. Strong distortions of the smectic layers may further lead to permanent gratings which has been also reported by Khoo et al. [42]. These permanent gratings can be erased by carefully heating and subsequent cooling of the whole sample. Eraseable permanent gratings are important e.g. in applications like holografic optical data storages.

4.4 Dye-Doped Cholesteric Liquid Crystals

Opto-thermally induced detuning of the pitch in absorbing cholesteric liquid crystals has been succesfully used for intrinsic (mirrorless) optical bistability [36] and optical switching. The dynamic response of an dye doped cholesteric liquid crystal with an absorption maximum near the selective reflectivity band to a picosecond laser pulse in transient grating experiments [57] is shown in Fig. 21. The film thickness has been about $2\,\mu m$, the excitation wavelength $\lambda_{exc} = 532\,nm$ was close to the absorption peak $\lambda_{max} = 570\,nm$ and the peak reflectivity $\lambda_r = 520\,nm$. The observed

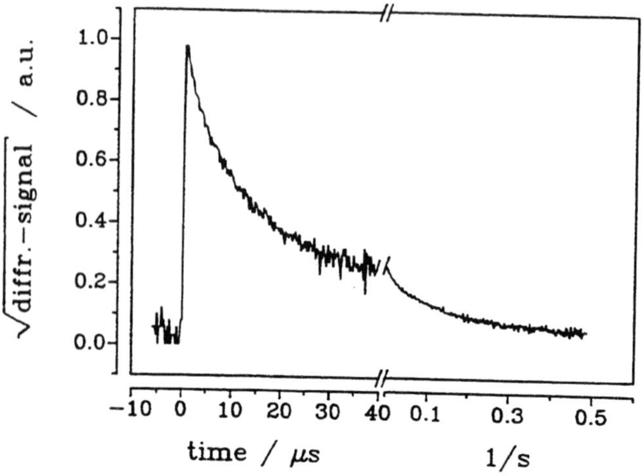

Fig. 21: Square root of the diffracted signal vs. time for an absorbing cholesteric liquid crystal after excitation with 80 ps intensity grating.

relaxation is running on two different time scales mainly. Note that the end of the left curve in Fig. 21 is the beginning of the slower right curve. The faster component exhibits a nonexponential decay on a microsecond scale whereas the slower relaxation can be fitted with a double exponential law using time constants of 30 and 270 ms, respectively.

The microsecond decay is connected with surface heat diffusion out of the thin liquid crystal film and very similiar to the thermal grating relaxation observed in ferroelectric liquid crystals discussed above.

The millisecond component is attributed to elastic deformations of the cholesteric film which relax by surface diffusion, since the sample is much thinner than the induced transversal grating periode. Consequently, the double exponential decay and the two time constants can be appointed to the first ($m = 1$) and third ($m = 3$) longitudinal Fourier-components of the distortion as described in sec. 3.1 of this paper. Since $\tau_m \sim (m\pi/d)^{-2}$ in this case, the observed relaxation times are expected to fulfill the relation $\tau_1 : \tau_3 = 9 : 1$ in good agreement with the experimental observations.

The observed grating rise times are in the order of 200 to 400 ns and may be explained by scalar order parameter relaxation processes or by density relaxations. No decision can be made, however, whether one of these two processes is dominant or not at the present stage of investigations.

5. CONCLUDING REMARKS

The mechanisms of Kerr-like optical nonlinearities in liquid crystals have been reviewed in the present paper. The basic theory of nonlinear effects originating from optical director reorientation and thermo-optic effects have been treated in this article together with a description of related nonlinear processes which can be observed with cw laser radiation. Further, the dynamic nonlinear optical response of liquid crystals to short laser pulses has been discussed with respect to picosecond transient grating experiments.

The motivation for the investigation of nonlinear optical effects in liquid crystals is given by three purposes mainly. First, to study the influence of optical fields on the orientational order of liquid crystals is interesting from a fundamental point of view. Optical fields can be applied with larger spatial and temporal flexibility than static DC or low frequency fields. Further, it was shown that the transient grating technique using short pulses may be used as a powerful tool for the investigation of relaxation processes of various physical phenomena like reorientation, temperature, ultrasound, flow or phase transitions.

Second, unique properties or new kinds of optical nonlinearities in liquid crystals allow the realization and investigation of new nonlinear processes. The nonresonant "giant" reorientational nonlinearity of nematic liquid crystals may be used e.g. for the investigation of strong transverse effects like self-focussing or spontaneous pattern formation in rather thin, nonabsorbing films. Laser-induced optical axis reorientation may further lead to polarization switching effects.

Third, optical nonlinearities in liquid crystals are investigated under the aspects of applications. Nonlinear optical devices like optical bistable switches, opto-optical modulators or phase-conjugating mirrors are important for future optical information technologies or laser physics and have been realized using liquid crystals. In addition to large optical nonlinearities liquid crystals have the advantage that they can be produced in high quality optical films and with large apertures using state of the art techniques, well established in liquid crystal display technologies.

Although most of the investigations have concentrated on the nematic phase up to now, the other mesophases look very promising as well. Examples have been given in this article for ferroelectric and cholesteric liquid crystals. More generally, liquid crystals belong to the class of organic nonlinear optical materials that receive considerable attention at the present time. Strong efforts are made to design new molecules with strong optical nonlinearities which form mesophases or may be useful in liquid crystalline polymers. It should be noted that the interest in these organic materials is also concentrated on electronic nonlinearities [9] because of fast response times (10^{-12} s) and large hyperpolarizabilities, which may open up new aspects and perspectives in the research and application of nonlinear optical effects.

ACKNOWLEDGEMENT

We like to thank C. Beyer (b. Dettmann), J. Schwartz, O. Hess, C. Keilonat, B. Trösken and D. Grebe for collaboration and helpful discussions in the field of nonlinear optics in liquid crystals.

The research programm has been supported by the Deutsche Forschungsgemeinschaft via the Sonderforschungsbereich "Anisotrope Fluide".

REFERENCES

1 Y.R. Shen; *The Principles of Nonlinear Optics*; New York: J. Wiley & Sons 1984

2 P.G. De Gennes; *The Physics of Liquid Crystals*; Oxford: Clarendon Press 1974

3 S. Chandrasekhar; *Liquid Crystals*; Cambridge: University Press 1977

4 W.H. De Jeu; *Physical Properties of Liquid Crystals*; London: Gordon and Breach 1980

5 L.M. Blinov; *Electro-optical and Magneto-optical Properties of Liquid Crystals*; New York: J. Wiley & Sons 1983

6 N.V. Tabiryan, A.V. Sukhov and B.Y. Zel'dovich; "The Orientational Optical Nonlinearity of Liquid Crystals"; *Mol.Cryst.Liq.Cryst.* Vol. 136, pp. 1-140 (1986)

7 I.C. Khoo; "Nonlinear Optics of Liquid Crystals" in *Progress in Optics* Vol. 26; E. Wolf (Ed.); Amsterdam: North-Holland 1988

8 I. Janossy; "Optical Effects in Liquid Crystals " in *Perspectives in Condensed Matter Physics*; L. Miglio (Ed.); Kluwer Academic Publishers 1990

9 P. Palffy-Muhoray; "The Nonlinear Optical Response of Liquid Crystals " in *Liquid Crystals/Applications and Uses* Vol. 1; B. Bahadur (Ed.); Singapore: World Scientific 1990

10 P.S. Pershan; *The Structure of Liquid Crystal Phases* Singapore: World Scientific 1988

11 N.A. Clark and S.T. Lagerwall; *Recent Developments in Condensed Matter Physics*, New York: Plenum Press 1981

12 M. Born and E. Wolf; *Principles of Optics*; Oxford: Pergamon Press 1975

13 D.W. Berreman and T.J. Scheffer; *Mol.Cryst.Liq.Cryst.* Vol.11 p. 395 (1970)

14 A.S. Zolotko et al.; *JETP Lett.*, Vol.32 , p.158 (1980)
15 I.C. Khoo and S.L. Zhuang; *Appl.Phys.Lett.* Vol.37 , p. 3 (1980)
16 B.Y. Zel'dovich, N.V. Tabiryan and Y.S. Chilingarian; *Sov.Phys. JETP* Vol. 54 , p. 32 (1981)
17 S.D. Durbin, S.M. Arakilian and Y.R. Shen; *Phys.Rev.Lett.* Vol. 47, p. 1411 (1981)
18 I.C. Khoo, T.H. Liu and P.Y. Yan; *J.Opt.Soc.Am. B* Vol. 4 , p. 115 (1987)
19 H.J. Eichler, R. Macdonald and C. Dettmann; *Mol.Cryst.Liq.Cryst.* Vol. 174 , p. 153 (1989)
20 L.B. Au, L. Solymar, C. Dettmann, H.J. Eichler, R. Macdonald and J. Schwartz; *Physica A* Vol. 174 , p.94 (1991)
21 L. Csillag et al.; *Mol.Cryst.Liq.Cryst.* Vol. 84 , p. 125 (1982)
22 H. Hsiung, L.P. Shi and Y.R. Shen; *Phys.Rev. A* Vol.30, p.1453 (1984)
23 H.J. Eichler and R. Macdonald; *Proc.Int.Conf.Lasers '88*, R.C. Sze and F.J. Duarte (Ed.), p. 512, McLean: STS Press 1989
24 G. Barbero and F. Simoni; *Appl.Phys.Lett.* Vol. 41 , p. 504 (1982)
25 O. Hess, R. Macdonald and H.J. Eichler; *Opt.Commun.* Vol. 82, p. 526 (1991)
26 R. Macdonald and H.J. Eichler; *Opt.Commun.* Vol.89, p. 289 (1992)
27 E. Santamato et al.; *Phys.Rev.Lett.* Vol. 57 , p. 2423 (1986)
28 I.C. Khoo and R. Normandin; *J.Appl.Phys.* Vol.55 , p. 1416 (1984)
29 R. Macdonald, J. Schwartz and H.J. Eichler; *Int.J.Nonl.Opt.Phys.* Vol. 1, p. 103 (1992)
30 H.G. Winful; *Phys.Rev.Lett.* Vol. 49, p. 1179 (1982)
31 P.P. Karat and N.V. Madhusudana; *Mol.Cryst.Liq.Cryst.* Vol. 36, p. 51 (1976)
32 S.T. Wu, E. Ramos and U. Finkenzeller; *J.Appl.Phys.* Vol. 68, p. 78 (1990)
33 I.C. Khoo and R. Normandin; *Optics Letters* Vol. 9, p. 285 (1984)
34 D. Armitage and S.M. Delwart; *Mol.Cryst.Liq.Cryst.* Vol. 122 , p. 59 (1985)
35 A.D. Lloyd and B.S. Wherrett, *Appl.Phys.Lett.* Vol. 53, p. 460 (1988)
36 R.B. Alaverdan, S.M. Arakelian and Y.S. Chilingarian; *JETP Lett.* Vol. 42, p.451 (1985)
37 I. Janossy, A.D. Lloyd and B.S. Wherrett; *Mol.Cryst.Liq.Cryst.* Vol. 179, p. 1 (1990)
38 S.G. Odulov et al.; Sov.Phys. JETP Vol. 58, p. 1154 (1983)
39 H.J. Yuan, L. Li and P. Palffy-Muhoray; *Electro-optical Materials for Switches, Coatings, Sensor Optics and Detectors;* SPIE Proc. Vol. 1307 (1990)
40 H.J. Eichler and R. Macdonald; *Proc.Int.Conf.Lasers '89* D.G. Harris and T.M. Shay (Ed.), p. 949, McLean: STS Press 1990
41 H.J. Eichler and R. Macdonald; *Mol.Cryst.Liq.Cryst.* Vol. 207, p. 117 (1991)
42 I.C. Khoo et al.; *J.Appl.Phys.* Vol. 69, p. 3853 (1991)
43 H.J. Eichler and R. Macdonald; *Phys.Rev.Lett.* Vol. 67, p. 2666 (1991)

44 H.J. Eichler, D. Günter and D.W. Pohl; *Laser-Induced Dynamic Gratings*; T. Tamir (Ed.), New York: Springer 1986
45 H.J. Eichler and H. Stahl; *J.Appl.Phys.* Vol. 44 , p. 3429 (1973)
46 M.D. Fayer; *IEEE J.Quant.Electron.* Vol. 22 , p. 1437 (1986)
47 W. Kaiser and M. Maier; "Stimulated Rayleigh, Brillouin and Raman Spectroscopy" in *Laser Handbook* Vol. 2 ; F.T. Arecchi and E.O. Schulz-Dubois (Ed.), p.1077; Amsterdam: North Holland 1972
48 J. D. Jackson; *Classical Electrodynamics* ; New York: Wiley & Sons 1975
49 J.L. Ericksen and F.M. Leslie; *Mol.Cryst.Liq.Cryst.* Vol. 7, p. 153 & 407 (1969)
50 S. Hess and I. Pardowitz; *Z.Naturforsch.* Vol. 36a, p. 554 (1981)
51 J. Constant and E.P. Raynes; *Mol.Cryst.Liq.Cryst.* Vol. 62, p. 115 (1980)
52 F.W. Deeg and M.D. Fayer; *J.Chem.Phys.* Vol. 91, p. 2269 (1989)
53 H.J. Eichler, R. Macdonald and B. Trösken; *Mol.Cryst.Liq.Cryst.* (accepted 1992)
54 C. David and D. Baeyens-Volant; *Mol.Cryst.Liq.Cryst.* Vol. 59 , p. 181 (1980)
55 R. Macdonald and D. Grebe (*Preprint*)
56 R. Blinc et al.; *Mol.Cryst.Liq.Cryst.* Vol. 152, p. 503 (1987)
57 A.G. Iljin, G.V. Klimusheva, R. Macdonald, D. Grebe and L.I. Zagainova (*Preprint*)

Studies in Classical
and Quantum Nonlinear Optics

ISBN 1-56072-168-5

SECOND-HARMONIC GENERATION FROM SURFACES OF CENTROSYMMETRIC MEDIA

Kjeld Pedersen
Institute of Physics, University of Aalborg
Pontoppidanstraede 103, DK-9220 Aalborg Øst, Denmark

Abstract

Optical second-harmonic generation from media with bulk centrosymmetry is found to be sensitive to surface conditions on the atomic monolayer level. Over the last ten years this has led to increased interest in second-harmonic generation which has gradually evolved as a new technique for surface investigations. This article presents the phenomenology needed for analysis of experimental results and some microscopic models of the nonlinear response of metals. Experimental techniques are discussed and results for aluminium are presented and compared to results of theoretical investigations.

1 Introduction

Macroscopic symmetry considerations show that electric dipole contributions to optical second-harmonic generation (SHG) are only obtained from materials without centrosymmetry. Materials with bulk centrosymmetry such as metals and semiconductors do, however, show SHG though the conversion efficiency is low. A description of the sources of SHG from these materials requires theories beyond the electric dipole approximation taking spatially nonlocal effects in the description of electromagnetic fields into account. Furthermore, though the materials show bulk centrosymmetry, the symmetry will be lower at the surface and dipole contributions from a thin surface layer will be present. In many cases the contribution from the surface will dominate and optical surface investigations are therefore possible.

The importance of the surface was realized in some of the first investigations of SHG from metals and semiconductors [1]-[4] in the sixties. However, surface science as a discipline has evolved, at least on the experimental side, during the same period as nonlinear optics and the background knowledge about surfaces was not present at that time. Experiments on silver films in air and what is usually called high vacuum ($\sim 10^{-6} - 10^{-8}$ mbar) were performed and no difference was found between samples in air and in vacuum [5]. It is only during the last ten years that the sensitivity of SHG to surface properties on the monolayer scale has been truely realized through experiments on surfaces prepared in ultrahigh vacuum ($\sim 10^{-10}$ mbar). Tom et al. [6] showed that submonolayer sensitivity of SHG for adsorption of O_2, Co, and Na on $Rh(111)$ could be obtained. A clear sensitivity to the structure of the outermost atomic layers of a surface was shown by Heinz et al. [7] who studied the 2 x 1 to 7 x 7 surface reconstruction of $Si(111)$. This type of investigations gave a renewed interest in surface SHG since it opened for the possibility of combining nonlinear optics and surface science, both of them strongly growing fields of research.

The appealing features of SHG as a surface probe are the potentials for fast time resolved measurements and surface investigations in different environments. Most of the conventional surface sensitive probes rely on electron scattering which cannot be used outside ultrahigh vacuum (UHV). With SHG it is possible to follow processes outside the well-controled UHV environment. Also buried interfaces such as solid/solid and solid/liquid interfaces can be studied as long as they are accessible to light. SHG is now gradually getting accepted in the surface science community as a useful probe but much work is still needed in order to take full advantage of the potentials in this new method.

The purpose of the present paper is not to give a detailed review of the whole area of surface SHG. A review with a complete list of references up to 1988 has recently been given by Richmond et al. [8]. Rather, this contribution is meant as an introduction to surface SHG as it is seen from the point of view of an experimentalist. Though some parts of the presentation also apply to semiconductors the main emphasis will be on metals.

As it will be described later real surfaces are often far from the plane, structurally well-ordered, clean surfaces assumed in theories. A macroscopic description of SHG which can help to separate different effects contributing to the nonlinear process is therefore important. After some general considerations about spatial nonlocality in the description of electromagnetic fields, a macroscopic theory representing the surface contribution to SHG by the radiation of a surface dipole layer will be given. Based on a simple free-electron model a microscopic theory for metals will be used to parameterize the nonlinear response and results of more detailed microscopic calculations will be discussed. The presentation of experiments will largely be limited to results obtained for aluminium surfaces in free air and under clean conditions in UHV. Only experiments trying to reveal the fundamental nonlinear process will be discussed. By limiting the presentation in this way a large number of interesting applications of SHG in surface and interface studies are omitted. References for this type of work can be found in the review by Richmond et al. [8].

2 General considerations about sources of SHG

Before any theoretical details are discussed it can be useful to look at some general aspects of SHG from surfaces. Since a major part of the nonlinear interaction takes place near the surface it is important with a proper description of the electromagnetic fields in the surface region. This can only be obtained with a spatially nonlocal theory.

Let us first consider a simple model of the material properties where the dielectric constant changes abruptly from 1 on the vacuum side to a larger value in the material (Fig. 2.1.). With the usual local Fresnel reflection and transmission coefficients this leads to a discontinuity in the normal component of the electric field at the surface. In a nonlocal theory this discontinuity will be removed as sketched in Fig. 2.1.

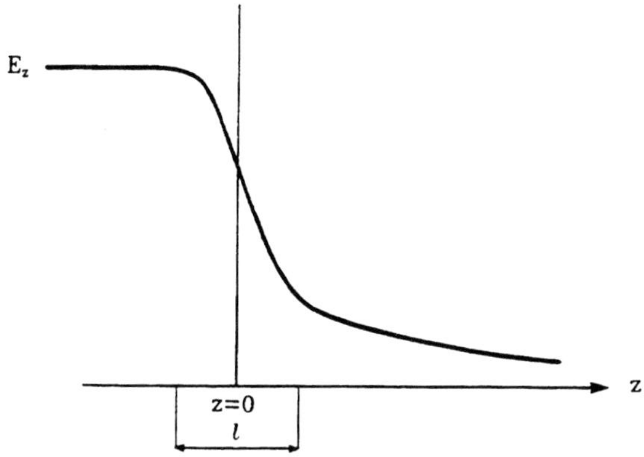

Figure 2.1. *A sketch of the variation of the normal component of the electric field at the surface. In a local approximation the field component changes abruptly at the surface while a nonlocal theory gives a continous variation over a layer much smaller than the wavelength of light.*

The nonlocal relation between the electric field and the linearly induced polarization can be expressed by

$$\vec{P}(\vec{r},\omega) = \int_{-\infty}^{\infty} \overleftrightarrow{\chi}^{(1)}(\vec{r},\vec{r}_1,\omega)\cdot\vec{E}(\vec{r}_1)d^3r_1 \tag{1}$$

where $\overleftrightarrow{\chi}^{(1)}$ is the linear response tensor. This equation relates the induced polarization at space point $\vec{r}$ to the electric field in the surroundings. In the case of a metal the electrons can carry information about the field over distances of the order of the mean free path. Clearly this nonlocal relation is important near the surface where the electric field is strongly varying.

The electronic properties at the surface may differ considerably from the bulk properties and lead to a non-centrosymmetric surface region. As an example the electron density of metals has been described by Lang and Kohn [10] in a free-electron model. Fig. 2.2 shows a sketch of the electron density which varies continously from zero in vacuum over some Fridel oscillations to a constant value in the bulk. The details of the electron density profile at the surface turn out to be important for SHG as it will be discussed in section 4. Analogous to the linear relation in Eq. (1) the nonlocal second order nonlinear interaction can be described by

$$\vec{P}(\vec{r},2\omega) = \int_{-\infty}^{\infty} \overleftrightarrow{\chi}^{(2)}(\vec{r},\vec{r}_1,\vec{r}_2,2\omega) : \vec{E}(\vec{r}_1,\omega)\vec{E}(\vec{r}_2,\omega)d^3r_1d^3r_2 \tag{2}$$

where $\overleftrightarrow{\chi}^{(2)}$ is the nonlinear response tensor. By the same arguments as we used in the linear case above it is clear that the nonlocal character of this equation is particularly

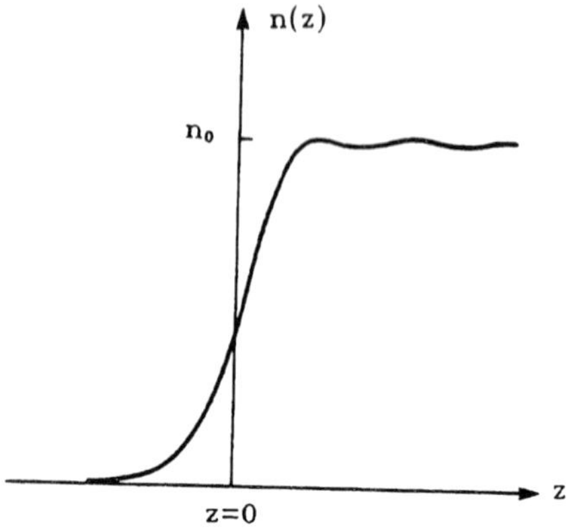

Figure 2.2. *Variation of the density of free electrons near a metal surface. The strong variation in n(z) typically takes place over a length of the order of 10Å.*

important near the surface. The field variations together with the relatively long mean free path of electrons in metals can be considered as a braking of the inversion symmetry. In the nonlocal description of the field-matter interaction even the bulk does not possess centrosymmetry.

The general expression for the nonlinear interaction in Eq. (2) is often approximated by a so-called multipole expansion. This also shows how the dipole approximation comes out of the general nonlocal relation. To obtain the multipole expansion [11] a Taylor expansion of the driving field of the form

$$\vec{E}(\vec{r}_1) = \sum_{n=0}^{\infty} \frac{1}{n!} \left[(\vec{r}_1 - \vec{r}) \cdot \vec{\nabla} \right]^n \vec{E}(\vec{r}) \tag{3}$$

is inserted in Eq. (2). This gives us a nonlinear polarization density of the form

$$P_i(\vec{r}, 2\omega) = \chi_{ijk}^{(L)}(\vec{r}) E_j(\vec{r}, \omega) E_k(\vec{r}, \omega)$$

$$+ \chi_{ijkl}^{(NL)}(\vec{r}) E_j(\vec{r}, \omega) \nabla_k E_l(\vec{r}, \omega)$$

$$+ \cdots \tag{4}$$

The first term on the right hand side gives the local (L) electric dipole contribution and the second term gives the first order nonlocal contribution which is often called near-local (NL) or electric quadrupole and magnetic dipole contribution. The response tensors in Eq. (4) involve spatial averages of the general response tensor in Eq. (2) of the forms

$$\chi_{ijk}^{(L)}(\vec{r}) = \int_{-\infty}^{\infty} \chi_{ijk}^{(2)}(\vec{r}, \vec{r}_1, \vec{r}_2) d^3 r_1 d^3 r_2 \tag{5}$$

and

$$\chi_{ijkl}^{(NL)}(\vec{r}) = \int_{-\infty}^{\infty} \left[\chi_{ijl}^{(2)}(\vec{r},\vec{r}_1,\vec{r}_2)(\vec{r}_2 - \vec{r})_k + \chi_{ilj}^{(2)}(\vec{r},\vec{r}_1,\vec{r}_2)(\vec{r}_1 - \vec{r})_k\right] d^3r_1 d^3r_2 \quad (6)$$

It should be noted that the near-local response tensor in Eq. (6) is a fourth rank tensor. Even higher rank tensors are needed if the expansion is extended to higher orders in nonlocality and the formalism clearly gets complicated if this is needed.

In the bulk of centrosymmetric media the dipole contribution vanishes and only higher order nonlocal terms will lead to SHG. It is usually considered to be a good approximation to represent the bulk nonlinearity by the near-local electric quadrupole and magnetic dipole contribution. At the surface where the centrosymmetry is broken dipole contributions will also be present. Generally, the multipole expansion is not expected to be well suited for describing the nonlinear interaction at the surface. The rapidly varying fields across the surface layer would require several terms in the expansion. An alternative approximation including in principle nonlocality to arbitrary order will be described in the following section.

3 Macroscopic theory

Based on the discussion in the previous section it will be possible to argue for relations between the macroscopic electric field and the induced second-order nonlinear polarization without knowing the detailed microscopic physics of the nonlinear interaction. The reflected SH field will be expressed in terms of nonlinear response tensors. It will be the task of microscopic physics to calculate these response tensors and for experimentalists to measure them. In this sense the macroscopic theory forms the basis for comparing experimental and theoretical results. In this section a useful approximation for the nonlinear response tensors will be treated and after that the resulting SH field generated in reflection will be given.

3.1 Nonlinear response tensors

Detailed descriptions of macroscopic theories have been given by several authors [4, 9, 12, 13, 14]. Here some main points of the description of nonlinear response tensors by Guyot-Sionnest et al. will be given [9]. The material creating the nonlinearity will be divided into bulk and surface regions where the surface region is defined as the layer where the material properties differ from the bulk properties and where the component of the electric field normal to the surface is strongly varying. The extent of this surface layer is expected to be much smaller than the wavelength of light (Fig.3.1).

The wave equation for the electric field at 2ω has the form

$$\left(\frac{1}{4\omega^2\mu_0}\vec{\nabla} \times \vec{\nabla} \times -\epsilon(2\omega)\right)\vec{E}(\vec{r},2\omega) = \vec{P}(\vec{r},2\omega). \quad (7)$$

In this equation a local description of the linear properties is used through the local dielectric constant $\epsilon(2\omega)$. The equation still contains nonlocality, but that is only in the nonlinear polarization on the right hand side which is given by Eq.(2). Inclusion

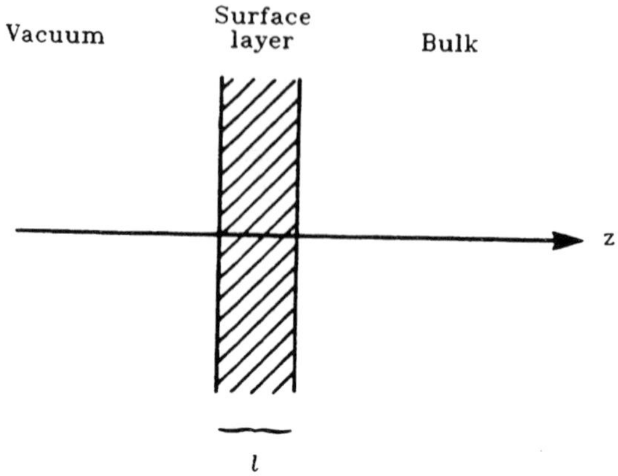

Figure 3.1. *Separation of the material into a bulk and a surface region. The surface region is defined as the layer with strong variations in the electric field and material properties.*

of nonlocality in the linear field problem at 2ω would lead to an integral equation requiring a self-consistent solution [9, 15]. A Green's tensor method can be used to find the field at 2ω. The Green's tensor is the solution to the wave equation with a delta function source:

$$\left(\frac{1}{4\omega^2\mu_0}\vec{\nabla}\times\vec{\nabla}\times-\epsilon(2\omega)\right)\overleftrightarrow{G}(\vec{r},\vec{r}_1)=\delta(\vec{r}-\vec{r}_1)\,\overleftrightarrow{1}\ . \tag{8}$$

In terms of the Green's tensor the SH field resulting from the sources in Eq. (2) is given by

$$\vec{E}(\vec{r},2\omega)=\int_{-\infty}^{\infty}\overleftrightarrow{G}(\vec{r},\vec{r}_1)\cdot\vec{P}(\vec{r}_1,2\omega)\,d^3r_1\,. \tag{9}$$

The form of the Green's tensor will be discussed later.

Due to translational invariance along the surface the variation of the field parallel to the surface will be given by a plane wave factor $exp\,(i2\vec{\kappa}\cdot\vec{R})$ where $\vec{\kappa}$ and $\vec{R}$ are vectors in the surface plane. It is then only necessary to consider equations which are Fourier transformed in the x,y-coordinates and only the z-coordinate needs to be considered in the equations.

The nonlinear polarization is expected to be different in the surface layer and in the bulk because of the broken inversion symmetry and rapid pump field variation near the surface as it was discussed in the previous section. The integral in Eq. (9) is therefore divided into two parts. One part involves the integration over the whole bulk region with a nonlinear polarization described by the near-local representation for the response functions. The other part involves integration over the surface layer with the general nonlocal representation in Eq. (2) for the nonlinear polarization. The boundary conditions for the Green's tensor elements are the same as for the electric field [9], that is $G_{ij}(z, z_1)$ for $j = x, y$ and $\epsilon(z_1)G_{iz}(z, z_1)$ is continuous in z_1 and assumed constant over the surface region. In Eq. (9) the part containing the integration over the surface region is therefore further divided into the parts containing the components of the nonlinear polarization parallel and perpendicular to the surface. Eq. (9) will then take the form

$$E_i(z, 2\omega) = G_{ij}(z, 0)P_{s,j}(2\omega) + \lim_{z_1 \to 0} [\epsilon(z_1, 2\omega)G_{iz}(z, z_1)P_{s,z}(2\omega)]$$

$$+ \int_{-\infty}^{\infty} G_{ik}(z, z_1)P_{B,k}(z_1, 2\omega)\, dz_1 \tag{10}$$

where $j = x, y$ and $k = x, y, z$ and a surface polarization has been defined by

$$P_{s,j}(2\omega) = \begin{cases} \int_s \chi^{(2)}_{jkl}(z, z_1, z_2)E_k(z_1, \omega)E_l(z_2, \omega)\, dz_1 dz_2 dz & j = x, y \\ \int_s \epsilon^{-1}(z, 2\omega)\chi^{(2)}_{jkl}(z, z_1, z_2)E_k(z_1, \omega)E_l(z_2, \omega) dz_1 dz_2 dz & j = z \end{cases} \tag{11}$$

The bulk polarization will be discussed later and we will concentrate on the first two terms in Eq. (10). In Eq. (11) the linear fields are still going to be integrated over the surface layer. The field components parallel to the surface can be considered to be constant in the thin surface region. In the direction normal to the surface the dielectric displacement will be assumed constant. The pump field in Eq. (11) will therefore be represented by

$$E_i(z, \omega) = \begin{cases} E_i(0, \omega) & i = x, y \\ s(z)D_z(0, \omega) & i = z \end{cases} \tag{12}$$

where $s(z)$ is a variable which shall describe the rapid variation of the linear electric field component perpendicular to the surface. Since we are using a local description of the electric fields, the z-dependence of $s(z)$ should be chosen so that the correct variation of the electric field at the surface, known for instance from detailed nonlocal calculations, is reproduced. The form of $s(z)$ is like an inverse macroscopic dielectric constant. However, it is varying over microscopic distances comparable to the interatomic spacing.

By introducing Eq. (12) in Eq. (11) the induced surface polarization density can be expressed by

$$P_{s,i}(2\omega) = \Delta^{(2)}_{ijk}(2\omega)E_j(0, \omega)E_k(0, \omega) \tag{13}$$

where the elements of the response tensor are given by

$$\Delta^{(2)}_{ijk} = \int_s \chi^{(2)}_{ijk}(z, z_1, z_2)\, dz_2 dz_1 dz$$

$$\Delta^{(2)}_{zjk} = \int_s \epsilon^{-1}(z, 2\omega)\, \chi^{(2)}_{zjk}(z, z_1, z_2)\, dz_2 dz_1 dz$$

$$\Delta^{(2)}_{ijz} = \epsilon(\omega) \int_s s(z_2)\, \chi^{(2)}_{ijz}(z, z_1, z_2)\, dz_2 dz_1 dz$$

$$\Delta^{(2)}_{izz} = \epsilon(\omega)^2 \int_s s(z_1)s(z_2)\,\chi^{(2)}_{izz}(z,z_1,z_2)\,dz_2dz_1dz$$

$$\Delta^{(2)}_{zzz} = \epsilon(\omega)^2 \int_s \epsilon^{-1}(z,2\omega)\,s(z_1)\,s(z_2)\chi^{(2)}_{zzz}(z,z_1,z_2)\,dz_1dz_2dz \qquad (14)$$

with $i,j,k \neq z$. Here the dielectric displacement in Eq. (12) has been expressed by $D_z(0,\omega) = \epsilon(\omega)\,E_z(0,\omega)$. Eq. (13) has the form of an electric dipole response and the model is often referred to as a dipole sheet model. The important spatial nonlocality in the nonlinear response has been taken into account by introducing $s(z)$ to describe the proper field variation at the surface. Thus, for each z in the last two indices of $\Delta^{(2)}_{ijk}$ a $s(z)$ appears in the expressions in Eq. (14). Similarly, $\epsilon^{-1}(z,2\omega)$ appears when the first index is a z in order to describe the nonlocality of the field at 2ω. The calculation of the nonlinear field, once the induced polarization is known, is then done in the local limit. The linear field components in Eq. (15) are calculated using local Fresnel transmission coefficients. All the problems with nonlocality have been collected in the nonlinear response tensor $\Delta^{(2)}_{ijk}$ which now describes both the broken inversion symmetry and the field variations at the surface.

In the last term on the right hand side of Eq.(10) describing the bulk contribution there is no dipole contribution for centrosymmetric media. Furthermore, the spatial variations of the electric field over distances comparable to electron mean free paths are expected to be much smaller than at the surface. The near-local approximation is then expected to be valid, and the nonlinear polarization density of the bulk is given by

$$P_{B,i}(\vec{r},2\omega) = \chi^{(NL)}_{ijkl}(\vec{r},2\omega)E_j(\vec{r},\omega)\nabla_k E_l(\vec{r},\omega)\,. \qquad (15)$$

As for the surface contribution the pump field components in Eq. (15) can be calculated by the local Fresnel transmission coefficients.

With the surface and bulk nonlinear polarizations in Eqs. (13) and (15) the general nonlinear response in Eq. (2) has been simplified considerably from the macroscopic point of view since the problem can be treated in terms of local optics. All the problems have been hidden in the nonlinear response tensors and will of course reappear in microscopic calculations of these response functions.

3.2 Reflected SH fields

The goal of the macroscopic theory is to describe the reflected nonlinear electric field. Often we are interested in highly absorbing materials like metals and semiconductors where no transmitted light is available for detection. The reflected SH field can be expressed in terms of the nonlinear response tensors introduced in the previous section. A number of different approaches to this field calculation have been used [9, 16, 17, 18]. The following is based on the Green's function theory developed by Sipe [16], and a detailed treatment of the subject can be found in this reference together with the work by Sipe et al. [12].

In the plane wave approximation the translational symmetry along the surface can be used to describe the source of SHG as a sum of dipole sheet polarizations of the form

$$\vec{P}(\vec{r},2\omega) = \int_{-\infty}^{\infty} \vec{P}(2\vec{\kappa},z_1,2\omega)\,\delta\,(z-z_1)e^{i2\vec{\kappa}\cdot\vec{R}}d^2(2\kappa)\,dz_1\,. \tag{16}$$

The SH field generated by this polarization will be a superposition of the fields $\vec{E}(2\vec{\kappa},z,z_1,2\omega)$ generated by the individual polarization sheets and the total field at point $\vec{r}$ is given by

$$\vec{E}(\vec{r},2\omega) = \int_{-\infty}^{\infty} \vec{E}(2\vec{\kappa},z,z_1,2\omega)e^{i2\vec{\kappa}\cdot\vec{R}}d^2(2\kappa)\,dz_1\,. \tag{17}$$

By solving the Maxwell equations in a homogeneous medium with a polarization sheet source and assuming only one wavevector component $\vec{\kappa}$ in the xy-plane, a Green's tensor can be found [16] in such way that

$$\vec{E}(2\kappa,z) = \int_{-\infty}^{\infty} \overleftrightarrow{G}\,(2\vec{\kappa},z-z_1)\cdot\vec{P}(2\kappa,z_1)\,dz_1 \tag{18}$$

where

$$\begin{aligned}\overleftrightarrow{G}\,(2\kappa,z-z_1) = {} & \frac{2i\mu_0\omega^2}{k_z}\,(\hat{s}\hat{s}+\hat{p}_+\hat{p}_+)e^{ik_z(z-z_1)}\,\Theta\,(z-z_1)\\ & + \frac{2i\mu_0\omega^2}{k_z}\,(\hat{s}\hat{s}+\hat{p}_-\hat{p}_-)e^{-ik_z(z-z_1)}\,\Theta\,(z_1-z)\\ & - \frac{1}{\epsilon}\hat{z}\hat{z}\delta(z-z_1)\,.\end{aligned} \tag{19}$$

The Green's tensor consists of three terms. The first term describes a wave travelling to the right (Fig.3.2), the second term describes a wave travelling to the left, and the last term describes the field at the place of the dipole sheet. The directions of polarization $\hat{p}_+$ and $\hat{p}_-$ for right and left travelling waves are shown on Fig. 3.2.

The reflected field on the vacuum side of a surface ($z < 0$) can be found using the Green's tensor for a homogeneous medium in Eq. (17). First the field generated by bulk sources is considered. Only the part of the Green's tensor describing a wave travelling to the left will contribute. By taking the transmission coefficients t_p and t_s for p- and s-polarized light at the surface into account, the reflected SH field has the form

$$\vec{E}(2\vec{\kappa},z) = \left[\frac{2i\mu_0\omega^2}{k_{z,2}}(\hat{s}\hat{s}t_s+\hat{p}_{-,1}\hat{p}_{-,2}\,t_p)\cdot\int_0^{\infty}\vec{P}_B(2\vec{\kappa},z_1)e^{ik_{z,2}z_1}dz_1\right]e^{-ik_{z,1}z} \tag{20}$$

where the indices 1 and 2 on $\hat{p}_-$ and k_z refer to vacuum and the reflecting medium, respectively. The z-dependence in the integral in Eq. (20) is given by the variation of the pump field which can be calculated from the linear properties of the material. Thus, for a given material the reflected SH field can be expressed explicitly in terms of linear optical properties and the elements of the nonlinear response tensor in Eq. (15).

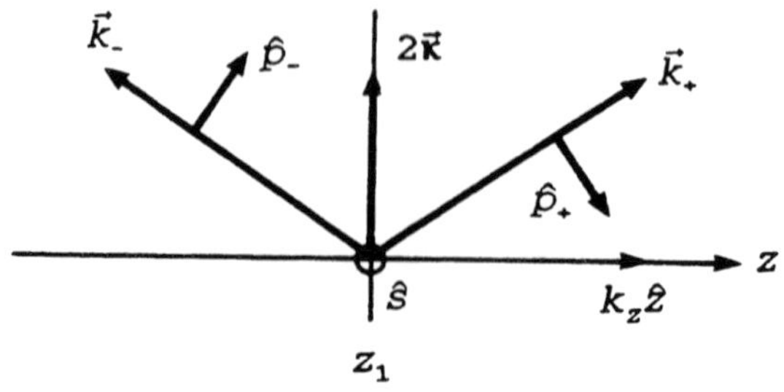

Figure 3.2. *The wavevectors for the nonlinear field generated by a dipole sheet source at z_1 for the wave travelling to the right ($\vec{k}_+$) and the wave travelling to the left ($\vec{k}_-$). The polarization directions $\hat{p}_+$ and $\hat{p}_-$ for the p-polarized components of the field are shown. Only the real parts of the vectors are shown.*

Let us now turn to the SH field generated by the surface dipole sheet source in Eq. (13) placed on top of the surface. The field in vacuum from this source consists of a direct wave travelling to the left and the reflection of the wave travelling to the right. With the reflection coefficients r_p and r_s for p- and s-polarized light the SH field takes the form

$$\vec{E}(2\vec{\kappa}, z) = \frac{2i\mu_0\omega^2}{k_{z,1}}\left[(\hat{s}\hat{s}(1 + r_s) + \hat{p}_{-,1}(\hat{p}_{-,1} + r_p\hat{p}_{+,1})\right] \cdot \vec{P}_s(2\vec{\kappa})e^{-ik_{z,1}z}\,. \tag{21}$$

As for the bulk contribution the field in Eq. (21) can be expressed explicitly by linear optical properties and elements of the nonlinear response tensor, here the surface response tensor in Eq. (13).

The actual tensor elements involved in the description of the reflected SH field depend on the crystal symmetry [12, 14]. For the bulk response the non-vanishing elements of $\chi^{(NL)}_{ijkl}$ can be described generally for a given crystal symmetry in the coordinate system of the crystallographic axes. By evaluating $\chi^{(NL)}_{ijkl}$ for a crystal with cubic symmetry it is found that this tensor can be described by four non-vanishing elements. In a reflection configuration and in the local approximation for the driving field the nonlinear bulk polarization is given by

$$P_{B,i}(2\omega) = \gamma\nabla_i(\vec{E}(\omega)\cdot\vec{E}(\omega)) + \xi E_i(\omega)\nabla_i E_i(\omega) \tag{22}$$

where γ and ξ are elements of the nonlinear response tensor. The first term in Eq. (22) is isotropic while the last term depends on the orientation of the crystal. For a given crystallographic orientation relative to the surface and the plane of incidence the elements of the response tensor can be transformed to a more useful beam coordinate system.

For the surface response tensor in Eq. (13) the elements must be evaluated for each orientation of the surface. The non-vanishing elements of this tensor are determined by the symmetry of the surface layer. Here it is difficult to separate the phenomenology from microscopic physics which is needed in order to determine the extent of the surface layer. For instance, the top layer of an unreconstructed (111) surface has c_{6v} symmetry. However, if also the next layer is considered, the surface has c_{3v} symmetry. Usually c_{3v} symmetry is observed for (111) faces of cubic crystals indicating that more than the top layer is involved in the nonlinear interaction. An example of the need for microscopic considerations in the determination of surface symmetry is the (2 x 1) reconstructed Si(111) [7]. Here the SH signal for a near infrared pump beam (YAG fundamental) shows the two-fold symmetry of the surface layer and not the underlying c_{3v} bulk symmetry since the pump frequency is close to resonance with electronic surface states of the (2 x 1) reconstructed layer. Thus, we can imagine that the observed surface symmetry in some cases will be frequency dependent [19].

Most of the microscopic theories developed so far have considered the non-linear response of an electron gas and neglected the possibility of anisotropic response tensors. Connection between these theories and the macroscopic theory described here is obtained if only the isotropic part of the nonlinear field is retained. In terms of a nonlinear reflection coefficient, here for p-polarized SH light generated from p-polarized pump light, defined by

$$r_p^{(2\omega)} = E_p(2\omega)/(E(\omega))^2 \tag{23}$$

the isotropic part of the reflected SH field is given by

$$r_p^{(2\omega)} = A\left[\delta_{33}\frac{\epsilon(2\omega)}{\epsilon(\omega)}sin^2\theta - \delta_{15}\frac{2g(2\omega)\,g(\omega)}{\epsilon(\omega)}cos^2\theta + \gamma\right] \tag{24}$$

where

$$g(\omega) = \frac{1}{cos\theta}\left[\epsilon(\omega) - sin^2\theta\right]^{\frac{1}{2}} . \tag{25}$$

Here A is a scaling factor involving the angle of incidence θ and the dielectric constant ϵ at the fundamental and SH frequencies. The quantitites δ_{33} and δ_{15} describe the nonlinear polarization perpendicular and parallel to the surface, respectively, while γ describes the bulk response. A tensor element δ_{31} should also appear in Eq. (24), but this term is often neglected. Microscopic theories and experiments aiming at determining γ, δ_{33}, and δ_{15} will be discussed in the following sections.

4 Microscopic theories

A number of different approaches to microscopic calculations of nonlinear optical properties of centrosymmetric media have been developed over the years, starting back in the early days of nonlinear optics in the sixties. General quantum mechanical formalisms were developed by Jha [3] and by Bloembergen and co-workers [4, 20]. Calculations of the bulk nonlinear response could in principle be performed by extending the methods used for linear optical properties using ground states of electronic systems known from bulk band structure calculations. Jha and Warke [21] calculated bulk nonlinear properties by including the effect of the band structure in the linear optical properties and neglecting effects of interband transitions in the nonlinear process. Only recently contributions from interband transitions to bulk nonlinearities of centrosymmetric media have been included in actual numerical calculations [22, 23].

It is generally more complicated to construct microscopic theories for the surface contribution to SHG than for the bulk contribution. The importance of the surface was discussed by Jha in a free-electron model [3]. More detailed discussions of the surface response were started by the work of Rudnick and Stern [24]. They introduced two phenomenological parameters a and b to describe the nonlinear surface polarizability perpendicular and parallel to the surface, respectively. Sipe et al.[25] used a hydrodynamic model to calculate the parameters a and b and found values of the order of unity outside possible surface resonances in the electron gas. Covi and Schaich [26] extended the hydrodynamic model to include a number of steps in the equilibrium electron density at the surface in order to simulate a smooth density profile. The problem of the correct electron density profile at the surface has turned out to be essential for the calculation of the polarizability perpendicular to the surface. Liebsch and Schaich [27] used a density functional theory for a free-electron gas and obtained a parameters an order of magnitude higher than the hydrodynamic result.

In order to introduce the parameterization of surface SHG in terms of the Rudnick and Stern parameters a classical free-electron model will be treated in the following. Later, the polarizability perpendicular to the surface, which is the most interesting, will be discussed.

4.1 Free-electron theories

Let us start by a model where a metal is treated as a gas of non-interacting electrons confined by infinite potential barriers. The velocity of the electrons in this model is often described by a hydrodynamic equation [25,28] (Euler's equation) of the form

$$nm\left[\frac{\partial \vec{V}}{\partial t} + (\vec{V}\cdot\vec{\nabla})\vec{V}\right] = -ne(\vec{E} + \vec{V}\times\vec{B})\,. \tag{26}$$

This equation could also have included a damping term on the right hand side. The general hydrodynamic theory also includes a so-called "quantum pressure" gradient term. In order to keep the treatment as simple as possible for the present purpose these terms are neglected here.

The electromagnetic field applied to the electron gas will induce a modulation of the electron density and the velocity of the electrons. It is assumed that the quantities in Eq. (26) have the forms

$$\begin{aligned} n &= n_0 + \frac{1}{2}\left[n_1 + n_2 + \dots + c.c\right] \\ \vec{V} &= \frac{1}{2}\left[\vec{V}_1 + \vec{V}_2 + \dots + c.c.\right] \\ \vec{E} &= \vec{E}_0^{(NL)} + \frac{1}{2}\left[\vec{E}_1 + \vec{E}_2 + \dots + c.c.\right] \\ \vec{B} &= \vec{B}_0^{(NL)} + \frac{1}{2}\left[\vec{B}_1 + \vec{B}_2 + \dots + c.c.\right] \end{aligned} \tag{27}$$

with time dependences of the form

$$f_n = f_n(n\omega)e^{-in\omega t}\,. \tag{28}$$

The parts of the electric and magnetic fields denoted by $\vec{E}_0^{(NL)}$ and $\vec{B}_0^{(NL)}$ are induced dc-fields generated by nonlinear (NL) interactions in the electron gas. The interesting parts here are the induced current densities at the frequencies ω and 2ω which are given by

$$\vec{J}_1(\omega) = -en_0\vec{V}_1(\omega) \tag{29}$$

and

$$\vec{J}_2(2\omega) = -en_0\vec{V}_2(2\omega) - \frac{1}{2}en_1(\omega)\vec{V}_1(\omega). \tag{30}$$

By inserting Eqs. (27) and (28) in Eq. (26) expressions for the velocities can be found by which the forced parts of the current densities in Eqs. (29) and (30) take the forms

$$\vec{J}_1(\omega) = \sigma(\omega)\vec{E}_1(\omega) \tag{31}$$

and

$$\vec{J}_2(2\omega) = \gamma\vec{\nabla}(\vec{E}_1(\omega)\cdot\vec{E}_1(\omega)) + \beta\vec{E}_1(\omega)(\vec{\nabla}\cdot\vec{E}_1(\omega)) \tag{32}$$

where

$$\sigma(\omega) = \frac{in_0e^2}{m\omega} \tag{33}$$

$$\gamma = -\frac{in_0e^3}{8m^2\omega^3} \tag{34}$$

$$\beta = -\frac{i\epsilon_0 e}{2m\omega} \tag{35}$$

Eqs. (31) and (33) give the well-known linear response of a free-electron gas.

In the bulk a spatially local representation of the driving field is used and the last term in Eq. (32) disappears. We are then left with a bulk contribution which in form is identical to the isotropic part in Eq. (22) and a simple free-electron model of this response function has been obtained. Notice that the induced polarization density is defined in such a way that its time derivative is equal to the current density [28].

Near the surface both terms in Eq. (32) will contribute since the component of the driving field normal to the surface is strongly varying. The nonlinear currents induced parallel and perpendicular to the surface will be considered separately. The component parallel to the surface has the form

$$\vec{J}_{2,\parallel} = \beta\vec{E}_{1,\parallel}\left(\frac{\partial E_{1,z}}{\partial z}\right) \tag{36}$$

where translational invariance of the fields along the surface has been assumed. In a local approximation for the driving field the z-derivative of $E_{1,z}$ acts as a delta function, and we consider the radiation from an effective current sheet at the surface by integrating over the surface region. Since the step hight in $E_{1,z}$ at the surface is $E_{1,z}(0)(\epsilon(\omega)-1)$ and β is independent of the electron density, the integrated current density is given by

$$\vec{J}_{s,\parallel} = \beta(\epsilon(\omega)-1)\vec{E}_{1,\parallel}E_{1,z}(0) \tag{37}$$

where $E_{1,z}(0)$ is the field amplitude just inside the metal. This is consistent with the phenomenological model introduced earlier, and since β is independent of the electron density a well defined approximation for δ_{15} in Eq. (24) can be obtained.

The induced nonlinear current density normal to the surface is given by

$$J_{2,z} = \gamma \frac{\partial}{\partial z}(E_{1,z}^2) + \beta E_{1,z} \frac{\partial E_{1,z}}{\partial z} \tag{38}$$

and involves both terms in Eq. (32). This component of the current density is difficult to evaluate in the free-electron model since γ involves the equilibrium electron density which like $E_{1,z}$ varies discontinously across the surface in this model. Thus, a well defined approximation from the normal component of the surface current cannot be obtained with the present model.

Rudnick and Stern [24] discussed the problems with the surface nonlinearity in a paper in 1971 where they introduced two parameters a and b to describe the surface currents. These parameters are still widely used in the description of surface SHG, and some of the arguments for their introduction will shortly be discussed here.

The rapid variation of the normal component of the electric field as it is sketched in Fig. 2.1 is considered again. The width of the surface layer in the z-direction where this strong variation takes place is called l. Large nonlinearities will occur if the displacement of the electrons by the incident field is of the order of l. The displacement of a free electron in an electric field $E_z exp(-i\omega t)$ along the z-direction is given by

$$z = \frac{e}{m\omega^2} E_z e^{-i\omega t} \tag{39}$$

The field E_0 giving the amplitude of z equal to l is

$$E_0 = \frac{m\omega^2 l}{e} \tag{40}$$

The induced surface current density normal to the surface is now represented as a power series in E/E_0 by

$$J_z = \sigma E_0 \left[\frac{E_z}{E_0} - \frac{1}{2} a \left(\frac{E_z}{E_0} \right)^2 + .. \right] \tag{41}$$

where σ is the linear conductivity. By using Eq. (33) for σ the SH term in Eq. (41) is given by

$$J_z(2\omega) = -\frac{i n_0 e^2 a}{2m\omega E_0} E_z^2 \tag{42}$$

An expression for the effective surface current is obtained by integrating over the surface region simply by multiplying by l:

$$J_{s,z}(2\omega) = -a \frac{i n_0 e^3}{2m^2\omega^3} E_z^2 . \tag{43}$$

In Eq. (43) the expression for E_0 in Eq. (40) has been used.

In the model outlined above the normal component of the nonlinear current is described by the parameter a. All the difficulties with the treatment of the surface problem are collected in this parameter. It should take care of both the nonlocality of the pump field and the variation of the electron density at the surface. For completeness, also the parallel component of the surface current and the bulk current can be corrected by introducing additional parameters:

$$\vec{J}_{s,\parallel}(2\omega) = b\beta\left[\epsilon(\omega) - 1\right] E_{1,\parallel} E_{1,z} \tag{44}$$

$$\vec{J}_B(2\omega) = d\gamma\vec{\nabla}(\vec{E}_1 \cdot \vec{E}_1) . \tag{45}$$

In the work by Rudnick and Stern b was introduced in order to take care of surface roughness. The parameter d can describe band structure effects on the nonlinear response.

Even though these parameters have been introduced by rather simple considereations, they have become the main quantities in both theoretical and experimental investigations of SHG from metals. The form of the SH reflection coefficient is given by Eq. (24). In this equation the a, b, and d parameters are included in δ_{33}, δ_{15}, and γ, respectively.

4.2 Density functional theory

Since the introduction of the phenomenological Rudnick and Stern parameters much of the effort within microscopic physics of SHG has been concentrated on calculations of the a parameter. This is mainly because the induced current normal to the surface which a describes depends strongly on the surface electronic structure and leads to the interesting surface sensitivity of SHG. Furthermore, both theory and experiments show that this often is the dominating contribution to SHG. The most successful calculations of the a parameter have been obtained with density functional theories. An example of the results which can be obtained with this theory is shown in the contribution by Kuchler and Rebentrost [29] in these proceedings.

Lang and Kohn calculated the variation of the electron density near the surface of metals using a density functional theory. Fig. (2.2) shows a sketch of a typical result for $n(z)$ with an electron density corresponding to Al. A number of properties such as surface energy and work function were calculated and good agreement with experimental results were obtained [10, 30].

Liebsch and Schaich used the density functional approach to calculate the nonlinear response to an electric field perpendicular to the surface [27]. This was done by calculating the induced charge density $\delta n(z, 2\omega)$ to second order. Both the calculations of $n(z)$ and the field induced charge densities require self-consistent calculations since the potential which determines n and δn is itself a function of these quantities. Many body effects are included in the calculations through exchange and correlation terms in the potential determining the electron densities.

Once the induced charge density has been found, this can be related to the SH current density and thus to the Rudnick and Stern a parameter by

$$a = \frac{4\omega^4 m^2}{n_0 e^3 E_z^2} \int z\delta n(z, 2\omega) dz . \tag{46}$$

The calculations by Liebsch and Schaich resulted in $|a|$ of the order of 10 to 50 depending on the electron density and excitation frequency. The calculations also showed that the nonlinear current is located in the exponential tail of the equilibrium electron density. Thus the nonlinear response perpendicular to the surface can be considered as nonlinear oscillations of the electron density tail going out into vacuum. From this it is clear that the nonlinear response will depend on the details of this electron tail and that any modification of it, for instance by foreign atoms at the surface, will be seen in SHG.

4.3 Effects of the ion potential

So far the microscopic theories described here have all been restricted to so-called jellium models where the electrons are confined to a certain region by a square-well potential of the ions without any structure from the actual ion positions. The details of the ion potential will affect the electron distribution and are therefore also expected to influence the nonlinear response. One important property of SHG caused by the ion potential is the dependence on crystal orientation both relative to the surface and relative to the plane of incidence and polarization direction of the pump field. There is also the possibility of interband transitions which can appear in the frequency dependence of SHG. Though these effects have been observed experimentally, very few theoretical works have addressed the problem of including the ion potential.

The effects of interband transitions and surface states were formally included in the theory of the SH response by Bower [31]. Detailed calculations have only been performed for the bulk contribution to SHG. Petukhov [22] used a classical Boltzmann approach to calculate the effect of an anisotropic Fermi surface on intraband contributions to SHG from Cu. In another work presented in this book Petukovh and Liebsch [23] have investigated the effect of interband transitions on the nonlinear bulk response of Al.

5 Experiments

Since the first experiments on non-centrosymmetric metals and semiconductors in the sixties a large number of experimental investigations have been performed with the purpose of comparing experimental and theoretical results. The early experiments on metals gave results corresponding to a and b parameters of the order of unity. This agreed quite well with theoretical predictions of that time using very simple models for the surface electron density. It was also found experimentally that no rotational anisotropy was present for a number of metals and for Si and Ge [4]. The conclusion was thererfore that theoretical models neglecting anisotropic contributions and with a simple representation of surface properties gave adequate descriptions of SHG from media with bulk centrosymmetry. These conclusions were based on experiments on samples in free air. During the last ten years experiments on well-characterized surfaces in ultrahigh-vacuum (UHV) have shown that SHG from cubic crystals is very sensitive to surface conditions and often shows anisotropy. Values of the a parameter for clean surfaces are typically found to be higher than 10 but fall upon oxidation to levels comparable to those observed for samples in free air.

The following presentation of experimental results will with a single exception be limited to SHG from aluminium. This is partly done in order to limit the presentation but also because aluminium is one of the best studied materials where good contact to theories has been obtained. Both samples in free air and in UHV will be discussed. For samples in free air the surface is not well-characterized on the atomic scale, and data interpretation is difficult as one might expect from the theoretical considerations in the previous sections. However, this problem should be turned into an advantage of SHG which can be a unique probe of surfaces and interfaces under atmospheric pressure. For clean samples comparisons to microscopic theories become meaningful. Experiments of that kind require UHV and are thus more complicated to perform.

5.1 Experimental technique

Usually harmonic generation requires the use of high laser intensities. For materials with bulk centrosymmetry where the nonlinear interaction takes place in a thin surface layer very small conversion efficiencies are expected. The quantity usually specified in experiments is the nonlinear reflection coefficient defined by $R = I(2\omega)/(I(\omega))^2$. By using the theory of the previous section with $a = 1$ the value of R is found to be of the order of $10^{-20} cm^2/W$. The reflected SH signal can then be evaluated for given laser specifications. One of the commonly used laser types is the Q-switched Nd:YAG laser which delivers high peak power pulses at a low repetition rate, typically 10 to 50 Hz. If we chose 1MW peak power and 10ns pulse width in a 1 cm^2 beam, the reflection coefficient above corresponds to of the order of 300 SH photons generated in each pulse. Even if most of the photons are lost in filtering, this is easily detected with a photomultiplier tube and gated electronics. One could of course try to increase the efficiency by using a higher pump intensity either by increasing the pump power (most YAG lasers deliver more than 1 MW peakpower) or by focusing the beam. However, the possibilities in this direction are limited by the damage threshold of the samples. The best way to increase the efficiency will be to use shorter pulses with higher peak powers but less energy per pulse.

Also mode-locked lasers are now commonly used in surface SHG. These laser systems have high repetition rates, typically 100 MHz, and kW peak powers in picosecond pulses. This generates on average less than one SH photon per pulse and gated photon counting must be used. Here there is usually no damage problems and the beam can be focused on the sample. For instance a synchronously pumped dye laser with an average power of a few hundred mW will give count rates of the order of a few hundred SH photons per second from a metal surface. Even a CW argon-ion laser can be used with a proper detection system though the count rates will be low leading to long data collection times.

Fig. 5.1 shows a sketch of a typical experimental set-up used for surface SHG. The reference arm with a reference sample serves two purposes. Firstly, during data collection it can be used to normalize the data in order to avoid effects of fluctuations in the laser power. Secondly, it can be used to obtain absolute values of the nonlinear reflection coefficient. This makes it possible to compare results obtained under different conditions or with different equipments, and it is important in comparisons with theoretical results.

A proper calibration for absolute measurements requires that the beam profiles are the same on the sample and the reference crystal. The distance from the laser to focusing lenses and from the lenses to the samples should be identical. A reference sample with well-known linear and nonlinear optical properties should be chosen. The

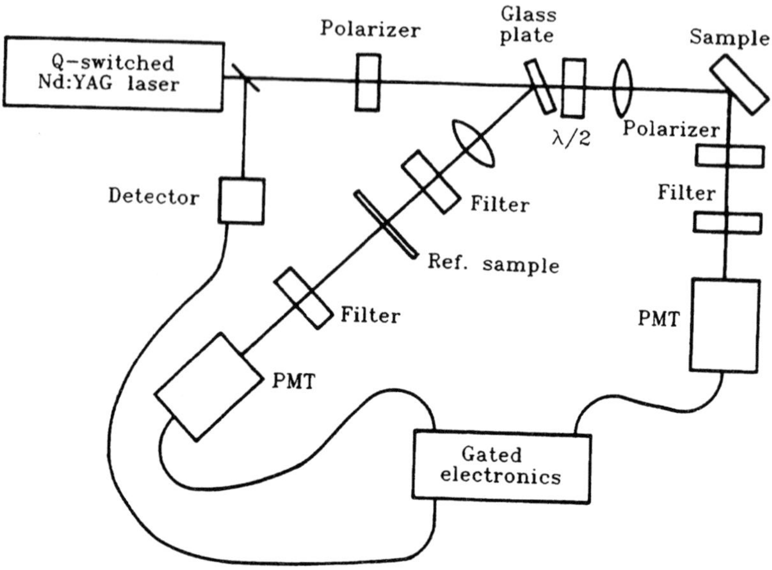

Figure 5.1. *Typical experimental arrangement for surface SHG. The direction of polarization of the pump light is controled by the half-wave plate and the polarization direction of SHG is analyzed with the polarizer on the detector side. The filters consist of coloured glass filters and narrowband bandpass filters or monochromators.*

reference signal should also be of the same order of magnitude as the sample signal in order to avoid inherent nonlinearities in the detection system. Maker fringes from a quartz slab are often used for calibration. Fig 5.2 shows the SH signal obtained from a 1 mm thick y-cut quartz wave plate as a function of the angle of incidence. Here a 1064 nm pump beam polarized in the plane of incidence which contains the x-axis of the crystal has been used. With the nonlinear coefficient d_{11} found in the literature [32] the maximum of the envelope curve can be found [33]. At normal incidence the signal is 10% of the envelope maximum for the wave plate used in these experiments. By using the reflection from the front surface of a glass plate (Fig. 5.1) the quartz signal is reduced by three orders of magnitude to the same level as the sample signal. While the linear properties of quartz are well known over a wide frequency range, the nonlinear coefficient is only specified at 1064 nm. A good estimate for d_{11} in the visible range far from the resonances in the ultra violet is obtained by using Miller's rule [34].

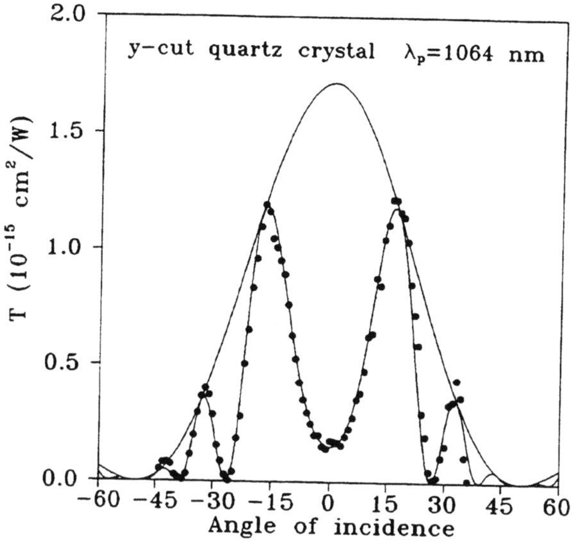

Figure 5.2. *The dependence on angle of incidence for SHG from a y-cut quartz waveplate with the x-axis in the plane of incidence. The curves are obtained by fitting expressions in Ref.[33] to the experimental data.*

5.2 Samples in free air

Silver in the form of evaporated films was the first metal used in SHG experiments [1] and has later been the subject of a large number of investigations. Fig. 5.3 shows the nonlinear reflection coefficient as a function of the angle of incidence for p-polarized pump and SH light and a fundamental wavelength of 1064 nm. The curve is calculated using $a = 1$ and $b = -1$. The value of b has been calculated by a hydrodynamic theory [25] and agrees with measurements of s-polarized SHG where the polarization parallel to the surface is the only source [35].While b depends very little on the conditions of the surface, a is found to be sensitive to the structure of the film. The value $a = 1$ corresponds to the results obtained in the early works on silver films [5] and also to results of more recent works [36, 37]. However, Jiang et al. [35] found that the value of a changes from 7 to -5 upon annealing for a film cold evaporated on a bulk silver substrate. The cold evaporated film consists of silver particles packed together on the substrate. This is known to give local field enhancement of SHG [38]. After annealing the film gets more dense and less rough. This changes the local field at the surface and thus the SH intensity represented by the a parameter.

Since SHG is sensitive to the structure of evaporated films, it can be developed to a probe of such films capable of characterizing them during the evaporation process.

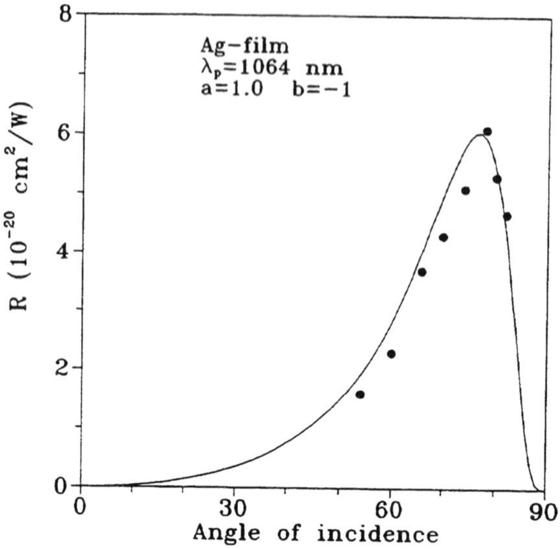

Figure 5.3. *The nonlinear reflection coefficient for p-polarized SH light from a silver film as a function of the angle of incidence of a p-polarized 1064 nm pump beam. The curve is obtained with the Rudnick and Stern parameters a=1 and b=-1.*

However, the values of a obtained for different films are merely a measure of the detected SH intensity and not of differences in the surface electronic structure. To describe the physics behind the a parameters for films a more detailed description of the fields existing locally at the surface is needed. Some work in this direction has been made [38].

Investigations of the dependence on the angle of incidence have been performed on a number of bulk and thin film metals. In the following we will concentrate on bulk aluminium samples in free air. Fig. 5.4 shows the dependence on angle of incidence of p-polarized SHG from Al(111). The wavelength of the p-polarized pump light is 532 nm. Since the sample is a single crystal and well-polished, the complications found with silver films are not expected here. Other complications can appear as it will be shown later. The curve on Fig. (5.4) is again the theory for an isotropic medium with $b = -1$, and here the a parameter is -22.5. It is seen that the dependence on the angle of incidence is well described by this theory.

The same type of dependence on the angle of incidence as in Fig. (5.4) is obtained with the YAG fundamental (1064 nm) as pump wavelength. However, at this wavelength the SH signal is anisotropic when the sample is rotated about a surface normal while no variations were observed with the green pump light. Fig. 5.5 shows the rotational anisotropy for p-polarized pump and SH light. Clearly there is a tendency of three-fold rotational symmetry with three large peaks and three smaller ones in between. This is what one would expect from intrinsic electronic properties of a (111)-

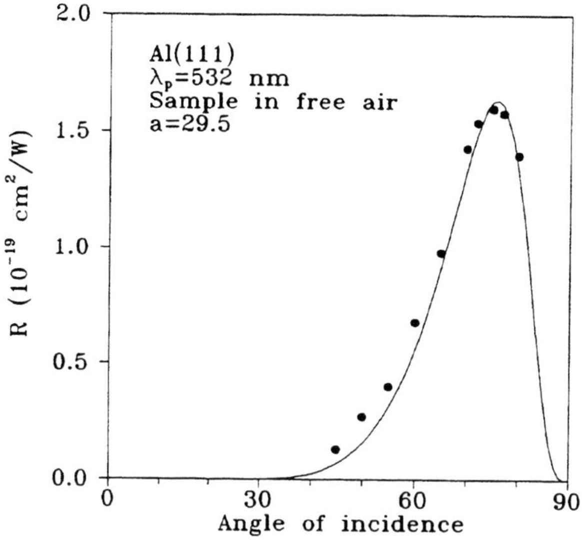

Figure 5.4. *The dependence on the angle of incidence for Al(111) in free air obtained with p-polarized pump and SH light and 532 nm wavelength. The curve is obtained with the parameters a=-22.5 and b=-1.*

oriented surface [12]. Similar results have previously been reported by Govorkov et al. [39]. The symmetry shown in Fig. 5.5 is not a perfect three-fold symmetry and some imperfections in the surface structure must be present. Symmetry considerations show that both bulk and surface contributions to SHG can be present for the (111) surface orientation.

Rotational anisotropy has also been studied for (100)-oriented aluminium crystals and a four-fold symmetry has been obtained as shown in Fig. 5.6. According to the macroscopic theory the rotational anisotropy for this surface orientation should only be due to bulk effects since the nonlinear response of a dipole sheet with four-fold symmetry is isotropic. Detailed investigations of SHG from this surface have been made [40]. By studying the information obtained by varying the angle of incidence and polarization direction of the pump beam as well as the rotational angle of the crystal, some of the elements in the nonlinear response tensor are overdetermined by the experimental data. These investigations show an inconsistency between different ways of determining these tensor elements from the experimental results, and it must be concluded that interpretation of the rotational anisotropy in terms of an intrinsic bulk response is not valid. Other effects not included in the macroscopic theory of a perfectly terminated (100) crystal face must be present and the suspicion immediately falls on the conditions of the surface.

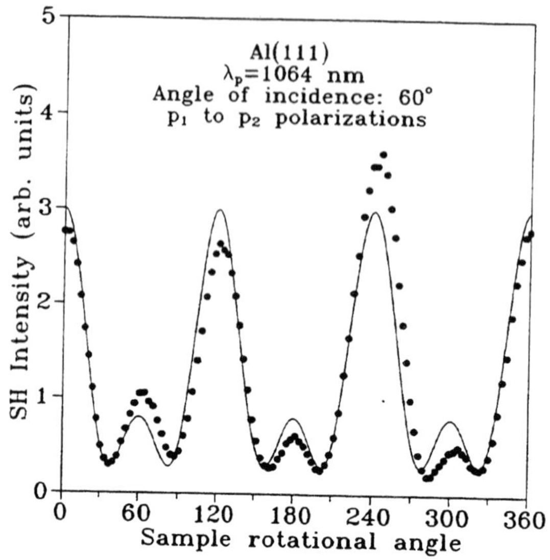

Figure 5.5. *Rotational anisotropy for p-polarized pump and SH light obtained at 60° angle of incidence of a 1064 nm pump beam on Al(111) in free air. The curve represents a three-fold symmetry fitted to the data.*

In order to further investigate the origin of the rotational anisotropy of the (100) -face aluminium the importance of the orientational accuracy of the surface plane relative to the crystallographic axes has been studied. More details about this are presented in the contribution by Jakobsen and Pedersen in these proceedings. Surfaces miscut by a few degrees relative to a (100) plane show a large rotational anisotropy which is well described by a surface polarization with c_{1v} symmetry. It is also found that the anisotropic contribution to SHG increases quadratically with the miscut angle for angles up to 5°. It is concluded that the miscut leads to an ordered structure on the surface. The interface between the metal and the oxide layer cannot be plane in the microscopic sense. Unless the crystal is cut exactly along a low-index plane, there will be steps with a minimum hight of the interatomic spacing on the surface. These steps have been shown to have a pronounced effect on SHG from clean surfaces in UHV [41] and we will return to this later. The same type of effect seems to be present on oxide covered crystals though the oxide creates some disorder in the step structure. A strong enhancement of SHG is observed when the pump field is polarized perpendicular to step edges. The steps are expected to be created during the final mechanical polishing where the surface conditions are like on metal-liquid interfaces and the surface atoms are able to diffuse into the energetically most favourable positions.

Step contributions to SHG are suggested to be responsible for the rotational anisotropy of the Al(100) crystal shown in Fig. 5.6. Though the crystal is well-oriented, residual "orange peel" will result in steps with edges parallel to the crystal-

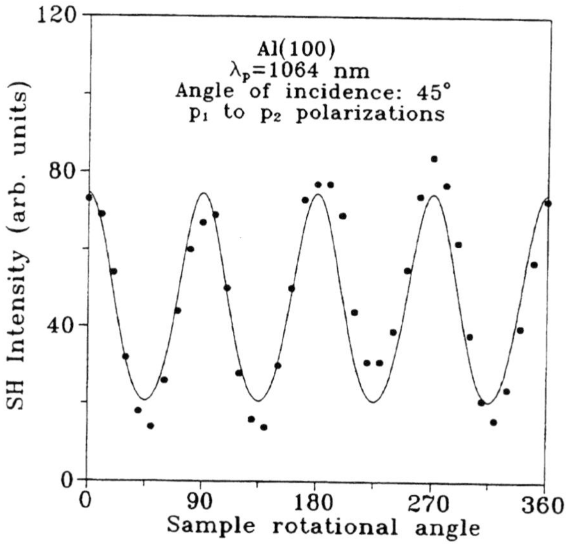

Figure 5.6. *Rotational anisotropy for p-polarized pump and SH light for Al(100) in free air. The angle of incidence at the 1064 nm pump beam is 45°. The curve represents a four-fold symmetry fitted to the data (from Ref.[40]).*

lographic axis. In this way the rotational anisotropy will indicate four-fold symmetry. It is also possible that steps can be involved in the rotational anisotropy of the (111)-oriented surface. However, Petukhov and Liebsch [23] show that the bulk contribution to SHG is at least an order of magnitude larger for the (111) crystal face than for the (100) orientation. It is therefore expected that some part of the anisotropy shown in Fig. 5.5 is caused by intrinsic electronic properties of the crystal while the symmetry imperfections can be due to steps. A direct separation of bulk and surface contributions to the anisotropy is not possible.

The formation of microstructures on the Al/Al_2O_3 interface also affects SHG from polycrystalline materials. In Fig. 5.7 the dependence on the pump beam polarization direction for p-polarized SHG is compared for a bulk polycrystal and a single crystal. For the polycrystal one could expect that the theory for an isotropic medium could be used to describe the results. The dependence on the direction of polarization ϕ (p-polarized pump beam for $\phi = 0$) relative to the plane of incidence is given by [12, 40]

$$I_p(\phi) = A\left[sin^4\phi + (B_R^2 + B_I^2)cos^4\phi + 2B_R\, sin^2\phi\, cos^2\phi\right] \tag{47}$$

where B_R and B_I are the real and imaginary parts of a quantity involving the nonlinear response functions. This equation describes the form of the polarization dependence quite well for the single crystal while it gives a very poor representation of the polycrystal results. The minimum in Fig. 5.7 is narrow which requires the last term

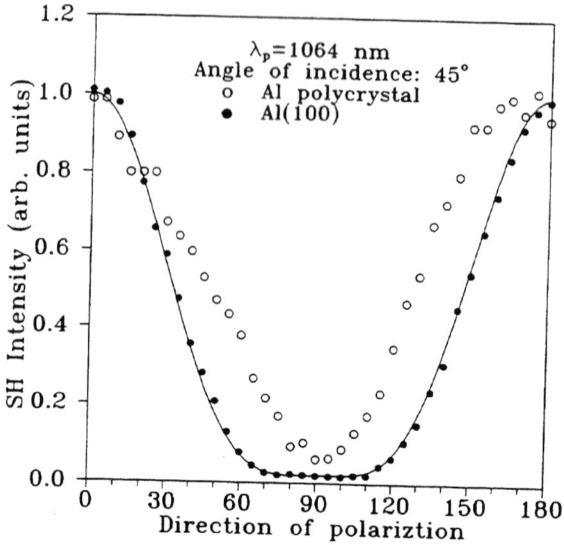

Figure 5.7. *Comparison of the polarization dependence for p-polarized SHG from a polycrystal and a (100) oriented Al crystal obtained with a 1064 nm pump beam at 45° angle of incidence. The data are normalized to the same value for p-polarized pump light ($\phi = 0$) in order to emphasize the differences in polarization dependences for the two materials.*

in Eq. (47) to be large, but B_R is also involved in the term proportional to $cos^4 \phi$ and thereby limited by the ratio between $I(\phi = 0°)$ and $I(\phi = 90°)$.

To explain the results for the polycrystal it must be realized that the sample consists of a large number of microcrystals within the area illuminated by the laser beam. The microcrystals will have different orientations of the crystallographic axes relative to the surface plane and will develop step edges according to the local orientations of the microcrystals. This results in anisotropic SHG from the individual crystals. If the microcrystals are large compared to the wavelength of light (in the present case the size was of the order of 100 μm), their contributions to the total SH signal will add up incoherently and we should add the intensities instead of the fields. This will remove the coupling between the last two terms on the right hand side of Eq. (47), and the observed polarization dependence can be described.

The nonlinear properties of samples in free air are difficult to describe from a microscopic point of view primarily for two reasons. Firstly, the properties of these surfaces are not well-known. Often the dominating part of the SH signal is generated at the interface between the metal and a surface layer created by the contact with the atmosphere. This interface layer cannot be characterized with conventional surface sensitive tools. Secondly, even if the atomic composition of the interface layer was known, the description of nonlinear properties would be considerably more complicated than for a clean surface. As discussed above polycrystals give additional complications because the orientation of the microcrystals at the surface in principle has to be known. On the other hand SHG is unique in the sense that it is a relatively simple surface sensitive probe which can be used under atmospheric pressure. Since it is often possible to give a macroscopic description of SHG from samples in free air, at least some qualitative information about the surfaces can be obtained. The investigations of steps on Si [42] and Al [43] surfaces covered by transparent oxide layers are examples of this. Another important example of investigations of buried interfaces is the liquid/solid interfaces in electrochemical cells where SHG has become a very important tool [8].

5.3 Clean surfaces

In recent years a number of experiments have been designed to investigate the microscopic theories. Such experiments require that the surfaces are clean and structurally well-ordered on the atomic scale. The experiments must be performed on samples in UHV where the pressure of contaminating gases can be kept sufficiently low. In UHV a large number of techniques, many of them based on some kind of electron energy spectroscopy, can be used to characterize the surface. Thus, contrary to samples in free air, the structure and chemical composition of the surface layer can be determined, and realistic comparisons to microscopic theories can be performed.

The technique used to obtain clean surfaces is shortly described in the following. First the crystal is oriented with X-ray diffraction and cut with the desired surface orientation. The sample is then carefully polished in order to obtain a smooth plane surface. It is important that the polishing removes the strained layer created by the cutting and grinding since residual strains will cause the surface to form microcrystals with grain boundaries during annealing. Electropolishing is effective in removing the strained layer but has a tendency of creating curved surfaces with some "orange peel". A combination of electropolishing to remove the strains and mechanical polishing to keep the surface flat can give good results. After preparation the sample is mounted in the vacuum system where it is cleaned by cycles of argon ion sputtering and annealing. It may take a number of cycles to obtain a clean surface since impurities diffuse from the bulk to the surface during annealing. In the experiments presented here the chemical composition of the surface was tested by Auger eletron spectroscopy, and the structure was investigated by Low Energy Electron Diffraction (LEED).

One of the main experimental difficulties in investigations of fundamental nonlinear properties of aluminium surfaces is the strong sensitivity of SHG to microstructures on surfaces. This was discussed in connection with surfaces in free air and turns out to be even more important for clean surfaces. Janz et al. [41] investigated the dependence of SHG on the orientation of the surface plane near a (100) plane for a clean aluminium crystal. The same type of investigation is reported in these proceedings by Jakobsen and Pedersen [43] for a sample with similar geometry but in free air. In both cases it is found that the SH signal increases quadratically with the crystal miscut angle relative to the (100) plane. LEED-patterns for clean miscut samples in UHV show double diffraction spots with a separation indicating a regular array of monoatomic steps on the surface. It is these steps which give rise to large nonlinearities on the surface when the pump light is polarized perpendicular to the step edges.

Electrons near step edges are more loosely bound than electrons at a flat surface. This is known from their higher chemical reactivity tested for instance by oxidation experiments [41]. The loosely bound electrons lead to higher nonlinear polarizability and enhanced SHG efficiency for stepped surfaces. For monoatomic steps the density of steps will increase linearly with the miscut angle leading to quadratic dependence for SHG. The density of steps is limited by interactions between steps which gives a minimum size of the flat terraces by the formation of higher steps for miscut angles above 5°.

The sensitivity to steps must be kept in mind during sample production since misalignments and deviations from plane surfaces will lead to steps. Still it is difficult to avoid steps on surfaces completely. However, the step structure is very sensitive to the sample temperature [43, 44]. As the temperature is increased, larger flat domains separated by higher steps are formed resulting in a lower step density and less SHG. Also, the step contribution to SHG will be highly anisotropic. It is therefore possible to isolate at least the isotropic contribution to SHG originating from intrinsic electronic properties of the flat terraces.

The dependence on the angle of incidence for SHG from clean aluminium crystals in UHV is shown in Fig. 5.8. The results are obtained with 1064 nm pump wavelength and p- to p-polarization configuration. Some effects of steps were observed, and the isotropic contribution from the flat terraces was isolated as described above. Results

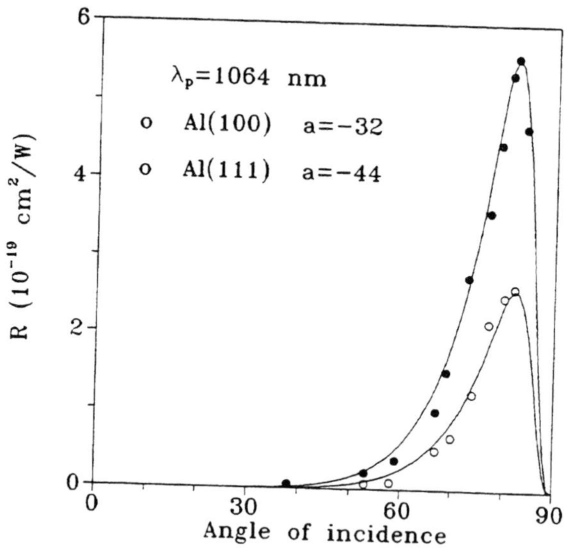

Figure 5.8. *The dependence on angle of incidence for p-polarized SH light generated from clean Al(100) and Al(111) surfaces by a 1064 nm p-polarized pump beam. The curves are obtained with b=-1 and a=-44 for the (111) surface and a=-32 for the (100) surface.*

have been obtained for (111) and (100) oriented surfaces. For both orientations good agreement with respect to the form of the dependence on the angle of incidence is obtained by using $b = -1$ and $a = -44$ for the (111) face and $a = -32$ for the (100) face. This experiment was first done by Murphy et al. [45] for the same two surface orientations. However, in the experimental results, shown in Fig. 5.8, the range of angles of incidence has been extended from a maximum of 75° in ref. [45] to more than 80° in order to get above the peak of the nonlinear reflectivity. Measurements near glancing incidence give special problems with the relatively large beam diameters usually obtained from Q-switched YAG lasers. At 84° angle of incidence the width of the sample in the plane of incidence must be 10 times the laser beam diameter. Due to limitations in sample sizes it is necessary to focus the laser beam which makes it more difficult to compensate for the intensity distribution on the sample in reference measurements. Murphy et al. [45] obtained $a = -37$ and $a = -22$ for the (111) and (100) surface orientations from their experimental data. This is somewhat less than the values used to describe the results in Fig. 5.8. The different results for the a-parameter in the two set of experiments could be caused by differences in surface conditions.

The time-dependent local-density-functional calculations by Liebsch and Schaich [27] give $a = -36 - 9i$ for the YAG fundamental pump frequency. Since no phase information on a is obtained by these experiments, this theoretical magnitude of a agrees with $a = -37$ obtained in ref.[45] for Al(111). It is interesting to notice that the nonlinearity is found to be considerably smaller for the (100) face than for the (111) face both by Murphy et al. [45] and in the present experiments. This type of anisotrophy is not described by the density-functional theory. Murphy et al. ascribed the differences in response of the two surfaces to the potentials felt by the surface electrons. The electrons at the (100) oriented surface are expected to be more tightly bound and thus to give a smaller nonlinear polarization. The surface electron binding shows up in the work function which is 4.41 eV for the (100) face and 4.24 eV for the (111) face.

The difference in response for the two surface orientations of Al is a result of the influence of the crystal potential on the electronic properties of the surface. Another way of detecting effects of the ions is to look for rotational anisotropy in SHG for a given surface orientation. Anisotropic contributions can be seen without rotating the sample by going to normal incidence since the isotropic contribution then has vanished. Fig. 5.9 shows the dependence on the direction of polarization of the pump beam for p-polarized SHG from Al(111) at normal incidence. The curve is a fit of Eq. (47) to the experimental data. A similar polarization dependence for this surface orientation was observed by Murphy et al. [45] at 15° angle of incidence and ascribed to the influence of the band structure. However, it was found that the signal shown in Fig. 5.9 decreased below the detection limit when the sample was heated by a few hundred degrees. The signal in Fig. 5.9 is therefore more likely to be caused by steps than by band structure effects of a flat surface. Clearly SHG from clean Al(111) is dominated by the polarizability perpendicular to the surface while this contribution is reduced for oxide covered samples. Some of the anisotropy shown in Fig. 5.5 could therefore be due to band structure effects. Additional information about rotational anisostropy could be obtained from oxidation experiments on very well prepared Al(111) samples in UHV.

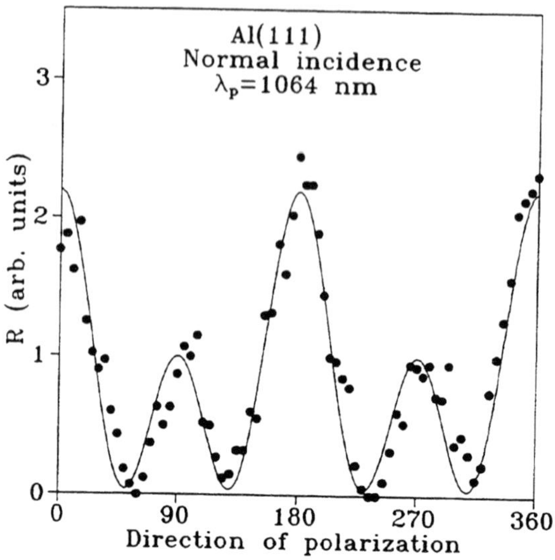

Figure 5.9. *The dependence of p-polarized SHG from Al(111) on the polarization direction of the 1064 nm pump beam at normal incidence. The curve is obtained from Eq.(47).*

In comparisons to microscopic theories the excitation frequency is a very important parameter. Fig. 5.10 shows the dependence on the angle of incidence obtained for an Al(100) crystal with a 532 nm pump beam. Compared to the results in Fig. 5.8 with the YAG fundamental the nonlinear reflection coefficient is higher with green pump light and the a parameter used to describe the data is -40. The results of the local-density-functional calculations [27] is $a = -i50$. As for the infrared pump wavelength the experimental result for the magnitude of a is somewhat lower than the theoretical value for the (100) oriented surface. From a practical experimental point of view it is also important to notice that the maximum of the nonlinear reflection coefficient is shifted down by 5° in angle of incidence for the green pump light relative to the infrared. For large values of a the theory shows that the position of the maximum is determined by the linear optical properties of the crystal. For the green pump light there are no problems with steps on the surface. The polarizability of the steps is found to decrease considerably with excitation frequency from the infrared to the visible. Even for vicinal Al(100) samples miscut by several degrees no effect of effects of steps is observed with green pump light.

The dispersion of SHG from the three low-index surfaces of aluminium, that is Al(100), Al(110), and Al(111), was investigated by Janz et al. [44]. Results for p-polarized SH light generated by p-polarized pump light in the wavelength range from 570 to 850 nm were obtained with a tunable dye laser synchronously pumped by the SH of a CW mode-locked Nd:YAG laser. For the (111) and (110) oriented surfaces

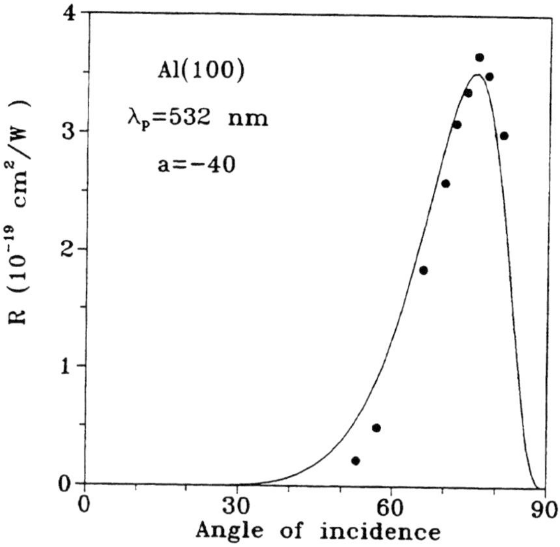

Figure 5.10. *The dependence on angle of incidence for p-polarized SHG from Al(100) obtained with a 532 nm p-polarized pump beam. The curve is obtained with the parameters b=-1 and a=-40.*

the experimental results largely follow the form of the frequency dependence given by the density-functional calculations [27]. However, the absolute values of the nonlinear reflection coefficients are a factor of two lower than predicted by the time-dependent local-density-functional calculations and agree with a calculation where the exchange and correlation effects are neglected. The (100) face was shown to give a lower SH reflection coefficient from the visible region up to 750 nm just as it was found with the YAG fundamental pump wavelength. In the near infrared a considerable increase in R was observed for Al (100). Janz et al. suggested that a surface state located 2.75 eV below the Fermi level at the Γ point [47] could lead to resonances at the SH frequency. More detailed investigations in the wavelength range from 800 nm to 1000 nm are needed in order to confirm this.

External perturbations are often used in the search for information about solid state systems from optical experiments. For instance low-frequency electric fields or strains applied to crystals can change the band structure thereby modifying optical signals. In SHG experiments electric fields applied to surfaces in electrochemical cells have a strong effect on the SH signal [8]. Also magnetic fields can change the nonlinearity of the surface [47]. Controled adsorption of foreign atoms on a surface in UHV is another way of changing the conditions of surface electrons. When aluminium is exposed to oxygen, a chemisorption bonding involving the surface electrons takes place and the polarizability of the surface is reduced. This can help to distinguish between bulk and surface sensitive contributions to SHG since the formation of a transparent oxide layer is expected to have little effect on the bulk contribution.

Janz et al. [48] have investigated oxidation of Al(100), Al(110), and Al(111) with SHG. For all three crystal faces the SH signal decreased by an order of magnitude after oxygen exposure corresponding to adsorption of one monolayer. However, monolayer coverage is obtained after 150 L (1 langmuir is 10^{-6} mm Hg sec.) for the (111) and (110) faces and 400 L for the (100) face. This difference in chemical reactivity is clearly seen in the SH signals which decay faster for the (111) and (110) faces than for the (100) face. The lower reactivity of the (100) face is consistent with the lower SH polarizability observed for this surface. The connection between chemical reactivity and SH polarizability also holds for steps on aluminium surfaces. For SHG from stepped Al(100) a fast decay during the first 100 L of oxygen exposure followed by a more slow decay was observed [41]. Oxygen is known to adsorp preferentially at the step edges where the electrons are loosely bound. As previously mentioned it is also the electrons at steps which give enhanced SHG. Thus the step contribution to SHG decays faster than the contribution from the terraces during oxygen exposure.

6 Conclusion

The discussion of SHG from aluminium shows that good contact with microscopic theories has been obtained. Some variations exist in the experimental results obtained by different groups. These variations are likely to originate from differences in surface conditions especially with respect to surface roughness. The difference in nonlinear response of Al(100) and Al(111) observed in experiments is not described by the density-functional theory. Also the influence of steps and the oxidation of aluminium are not yet described theoretically in detail though some initial attempts have been made [29]. However, the experimental observations which are beyond the present theories at least agree qualitatively with surface properties such as work function and chemical reactivity known from other surface sensitive experiments.

Due to the strong sensitivity of SHG to a variety of surface effects the basic study of surface nonlinearities cannot at least on the experimental side be separated from the surface science aspects of SHG. For example, the sensitivity to steps discussed here makes it necessary to study the effects of these steps in order to isolate nonlinearities of a flat surface. The same holds for adsorption of foreign atoms on surfaces where it is important to know the effect of this on SHG for different adsorbates in order to evaluate results for surfaces supposed to be clean. The large amount of work done in these years where SHG is used to study a wide range of surface effects is therefore also important for the basic understanding of the nonlinear interactions. On the other hand, further developments of SHG for applications in surface science also require more detailed investigations of the basic nonlinear interactions in order to get from qualitative to quantitative information. Theoretical descriptions including band structure effects on surface electrons should be developed and different approaches to calculations of nonlinearities should be investigated. On the experimental side accurate measurements should be performed on surfaces made as ideal as possible and the experiments should aim at nonlinear spectroscopy over a wide frequency range.

References

[1] F. Brown, R.E. Parks, and A.M. Sleeper, "Nonlinear Optical Reflection from a Metallic Boundary", *Phys.Rev.Lett.*, Vol. 14, No. 25, 1965, pp. 1029-1031.

[2] F.Brown and R.E. Parks, "Magnetic-dipole Contribution to Optical Harmonics in Silver", *Phys.Rev.Lett.*, Vol. 16, No. 12, 1966, pp. 507-509.

[3] S.S. Jha, "Second-Order Optical Processes and Harmonic Fields in Solids", *Phys.Rev.*, Vol. 145, No. 2, 1966, pp. 500-506.

[4] N. Bloembergen, R.K. Chang, S.S. Jha, and C.H. Lee, "Optical Second-Harmonic Generation in Reflection from Media with Inversion Symmetry", *Phys.Rev.*, Vol. 174, No. 3, 1968, pp. 813-822.

[5] H. Sonnenberg and H. Heffner, "Experimental Study of Optical Second-Harmonic Generation in Silver", *J.Opt.Soc.Am.*, Vol. 58, No. 2, 1968, pp. 209-212.

[6] H.W.K. Tom, C.M. Mate, X.D. Zhu, J.E. Crowell, T.F. Heinz, G.A. Somorjai, and Y.R. Shen, "Surface Studies by Optical Second-Harmonic Generation: The Adsorption of O_2, Co, and Sodium on the Rh(111) Surface", *Phys.Re.Lett.*, Vol. 52, No. 5, 1984, pp. 348-351.

[7] T.F. Heinz, M.M.T. Loy, and W.A. Thompson, "Study of Si(111) Surface by Optical Second-Harmonic Generation: Reconstruction and Surface Phase Transformation", *Phys.Rev.Lett.*, Vol. 54, No. 1, 1985, pp. 63-66.

[8] G.L. Richmond, J.M. Robinson, and V.L. Shannon, "Second Harmonic Generation Studies of Interfacial Structures and Dynamics", *Prog.Surf.Sci.*, Vol. 28, 1988, pp. 1-70.

[9] P. Guyot-Sionnest, W. Chen and Y.R. Shen, "General Considerations on Optical Second-Harmonic Generation from Surfaces and Interfaces", *Phys.Rev.B.*, Vol. 33, No. 12, 1986, pp. 8254-8263.

[10] N.D. Lang and W. Kohn, "Theory of Metal Surfaces: Charge Density and Surface Energy", *Phys.Rev.B*, Vol. 1, No. 12, 1970, pp. 4555-4568.

[11] O. Keller, "Moment Expansion of the Optical Second-Harmonic Response Tensor", *Phys.Stat.Sol.B*, Vol. 157, 1990, pp. 459-471.

[12] J.E. Sipe, D.J. Moss, and H.M. van Driel, "Phenomenological Theory of Optical Second- and Third-Harmonic Generation fram Cubic Centrosymmetric Crystals", *Phys.Rev.B*, Vol. 35, No. 3, 1987, pp. 1129-1141.

[13] V. Mizrahi and J.E. Sipe, "Phenomenological Treatment of Surface Second-Harmonic Generation", *J.Opt.Soc.Am.B.*, Vol. 5, No. 3, 1988, pp. 660-667.

[14] O.A. Aktsipetrov, I.M. Baranova, and Yu.A. Il'inskii, "Surface Contribution to the Generation of Reflected Second-Harmonic Light for Centrosymmetric Semiconductors", *Sov.Phys.JEPT.*, Vol. 64, No. 1, 1986, pp. 167-173.

[15] O. Keller, "Propagator Study of the Selvedge Field in Linear and Nonlinear Nonlocal Jellium Optics", *Phys.Rev.B.*, Vol. 38, No. 12, 1988, pp. 8041-8060.

[16] J.E. Sipe, "New Green-Function Formalism for Surface Optics", *J.Opt.Soc.Am. B.*, Vol. 4, No. 4, 1987, pp. 481-489.

[17] N. Bloembergen, and P.S. Pershan, "Light Waves at the Boundary of Nonlinear Media", *Phys.Rev.*, Vol. 128, 1962, pp. 606-622.

[18] O. Keller, "Screened Electromagnetic Propagators in Nonlocal Metal Optics", *Phys.Rev.B.*, Vol. 34, No. 6, 1986, pp. 3883-3899; O. Keller, "Tensor-Product Structure of a New Electromagnetic Propagator for Nonlocal Surface Optics of Metals", *Phys.Rev.B.*, Vol. 37, No. 18, 1988, pp. 10588-10607.

[19] Y. Tang, L.J. Simpson, and T.E. Furtak, "Ag(111) Surface Reconstruction Studied by Optical Second-Harmonic Generation", *Phys.Rev.Lett.*, Vol. 67, No. 20, pp. 2814-2817.

[20] N. Bloembergen and Y.R. Shen, "Optical Nonlinearities of a Plasma", *Phys.Rev.*, Vol. 141, No. 1, 1966, pp. 298-305.

[21] S.S. Jha and C.S. Warke, "Interband Contributions to Optical Harmonic Generation at a Metal Surface", *Phys.Rev.*, Vol. 153, No. 3, 1967, pp. 751-759.

[22] A.V. Petukhov, "Source of Strong Anisotropy of Quadrupole-Allowed Second-Order Nonlinear Polarizability of the Noble-Metal Surfaces", *Phys.Rev.B*, Vol. 42, No. 15, 1990, pp. 9387-9390.

[23] A. Petukhov and A. Liebsch, "Volume Interband Contribution to Second-Harmonic Generation from Aluminum", *In these proceedings.*

[24] J. Rudnick and E.A. Stern, "Second-Harmonic Radiation from Metal Surfaces", *Phys.Rev.B.*, Vol. 4, No. 12, 1971, pp. 4274-4290.

[25] J.E. Sipe, V.C.Y. So, M. Fukui, and G.I. Stegeman, "Analysis of Second-Harmonic Generation at a Metal Surface", *Phys.Rev.B*, Vol. 21, No. 10, 1980, pp. 4389-4402.

[26] M. Covi and W.L. Schaich, "Hydrodynamic-Model Calculation of Second-Harmonic Generation at a Metal Surface", *Phys,.Rev.B.*, Vol. 33, No. 6, 1986, pp. 3688-3695.

[27] A. Liebsch and W.L. Schaich, "Second-Harmonic Generation at Simple Metal Surfaces", *Phys.Rev.B.*, Vol. 40, No. 8, 1989, pp. 5401-5410.

[28] J.E. Sipe and G.I. Stegeman, "Nonlinear Optical Response of Metal Surfaces", In: *Surface Polarizations*, V.M. Agranovich and D.L. Mills (Eds.), New York: North-Holland Publishing, 1982, pp. 661-701.

[29] M. Kuchler and F. Rebentrost, "Adsorbate Induced Change of Linear and Nonlinear Response Properties of Simple Metal Surfaces", *In these proceedings.*

[30] N.D. Lang and W. Kohn, "Theory of Metal Surfaces: Work Function", *Phys.Rev.B.*, Vol. 3, No. 4, 1971, pp. 1215-1223.

[31] J.R. Bower, "Effects of Electron Band Structure on Optical Second-Harmonic Generation at Metal Surfaces", *Phys.Rev.B.*, Vol. 14, No. 6, 1976, pp. 2427-2432.

[32] V.G. Dimitriev, G.G. Gurzadyan, and Nikogosyan, "Handbook of Nonlinear Optical Crystals", Berlin: Springer-Verlag, 1991.

[33] J. Jerphagnon and S.K. Kurtz, "Maker Fringes: A Detailed Comparison of Theory and Experiment for Isotropic and Uniaxial Crystals", *J.Appl.Phys.*, Vol. 41, No. 4, 1970, pp. 1667-1681.

[34] Y.R. Shen, "The Principles of Nonlinear Optics", John Wiley & Sons, 1984, p. 37.

[35] H.B. Jiang, L. Li, W.C. Wang, J.B. Zheng, and Z.M. Zhang, "Reflected Second-Harmonic Generation at a Silver Surface", *Phys.Rev.B.*, Vol. 44, No. 3, 1991, pp. 1220-1224.

[36] R. Reinisch, M. Niviere, H. Akhouayri, J.L. Coutaz, D. Maystre, and E. Pic, "Grating Enhanced Second-Harmonic Generation through Electromagnetic Resonances", *Optical Engineering*, Vol. 27, No. 11, 1988, pp. 961-971.

[37] J.C. Quail and H.J. Simon, "Second-Harmonic Generation from Silver and Aluminum Films in Total Internal Reflection", *Phys.Rev.B.*, Vol. 31, No. 8, 1985, pp. 4900-4905.

[38] O.A. Aktsipetrov, E.M. Dubinina, S.S. Elovikov, E.D. Mishina, A.A. Nikulin, N.N. Novikova, and M.S. Strebkov, "The Electromagnetic Mechanism of Surface Enhanced Second-Harmonic Generation and Raman Scattering in Island Films", *Solid State Commun.*, Vol. 70, No. 11, 1989, pp. 1021-1024.

[39] S.V. Govorkov, N.I. Koroteev, I.L. Shumay, and V.V. Yakovlev, "Second-Harmonic Generation in Reflection from Crystaline Metals: The Influence of Structure Disordering", *J.Opt.Soc.Am.B*, Vol. 8, No. 5, 1991, pp. 1023-1027.

[40] K. Pedersen and O. Keller, "Experimental Study of the Phenomenological Parameters Describing Optical Second-Harmonic Generation from Centrosymmetric Metals", *J.Opt.Soc.Am.B.*, Vol. 6, No. 12, 1989, pp. 2412-2419.

[41] S. Janz, D.J. Bottomley, H.M. van Driel, and R.S. Timset, "Influence of Steps on Second-Harmonic Generation from Vicinal Metal Surfaces", *Phys.Rev.Lett.*, Vol. 66, No. 9, 1991, pp. 1201-1204.

[42] C.W. van Hasselt, M.A. Verheijen, and Th. Rasing, "Vicinal Si(111) Surfaces Studied by Optical Second-Harmonic Generation: Step-Induced Anisotropy and Surface-Bulk Discrimination", *Phys.Rev.B.*, Vol. 42, No. 14, 1990, pp. 9263-9266.

[43] C. Jakobsen and K. Pedersen, "Microstructure on Al/Al_203 Interfaces Investigated by Optical Second-Harmonic Generation", *In these proceedings.*

[44] S. Janz, K. Pedersen, and H.M. van Driel, "Dispersion and Anisotropy of the Optical Second-Harmonic Response of Single Crystal Al Surfaces", *Phys.Rev.B.*, Vol. 44, No. 8, 1991, pp. 3943-3953.

[45] R. Murphy, M. Yeganeh, K.J. Song, and E.W. Plummer, "Second-Harmonic Generation from the Surface of a Simple Metal, Al", *Phys.Rev.Lett.*, Vol. 63, No. 3, 1989, pp. 318-321.

[46] G.V. Hansson and S.A. Flodström, "Photoemission from Surface States and Surface Resonances on the (100),(110), and (111) Crystal Faces of Aluminum", *Phys.Rev.B.*, Vol. 18, No. 4, 1978, pp. 1562-1571.

[47] J. Reif, J.C. Zink, C.M. Schneider, and J. Kirschner, "Effects of Surface Magnetism on Optical Second-Harmonic Generation", *Phys.Rev.Lett.*, Vol. 67, No. 20, 1991, pp. 2878-2881.

[48] S.Janz, K. Pedersen, H.M. van Driel, and R.S. Timsit, "Structural Transformations in Adsorbed Oxygen Layers on Al Surfaces Observed Using Optical Second-Harmonic Generation", *J. Vac.Sci. Technol.A.*, Vol. 9, No. 3, 1991, pp. 1506-1510.

Studies in Classical
and Quantum Nonlinear Optics

ISBN 1-56072-168-5

OPTICAL PHASE CONJUGATION AND OPTICAL SIGNAL PROCESSING IN PHOTOREFRACTIVE MATERIALS

Paul Michael Petersen
Department of Physics
Technical University of Denmark

1. INTRODUCTION

Photorefractive media are attractive in nonlinear optics because it is possible to induce changes in the refractive index at very low power levels. Holographic information can be dynamically recorded, read out, and erased by light. The photorefractive effect has been successfully used in many applications, including optical information storage, optical computing(1), optical oscillators(2), optical interconnects(3), neural networks(4), image processing(5), and optical phase conjugation(5).

In this paper, I give an introduction to optical phase conjugation in nonlinear materials and briefly describe the photorefractive effect from the band transport model. In photorefractive materials, phase conjugate waves can be obtained by four-wave mixing. The four-wave mixing process involves the simultaneous interaction of four light waves in the photorefractive medium. In section 4, I discuss optical phase conjugation by four-wave mixing in photorefractive materials and a new approach based on "one-grating nondegenerate four-wave mixing" is introduced. In the classic configuration for four-wave mixing, a probe beam is incident on the photorefractive medium together with two counterpropagating pump beams. In passive self-pumped phase conjugation, however, the pumping beams are built up from scattered light, which is amplified by two-wave mixing in the photorefractive medium. I discuss the linear passive self-pumped phase conjugate oscillator, where the feedback that builds up the two pumping beams is provided by two external mirrors. In section 5, I apply the "one-grating nondegenerate four-wave mixing" theory to the linear photorefractive oscillator. Finally, section 6 deals with the applications of the photorefractive effect, which have been investigated by the Optics Group at the Technical University of Denmark. The applications include fast parallel processing of data from a particle flow, time-average interferometry, and photorefractive interference filters.

2. OPTICAL PHASE CONJUGATION

Optical phase conjugation is a process in nonlinear optics for the generation of a wave whose overall phase and direction of propagation are exactly reversed with respect to that of a wave incident on the medium. If the incident field is given by:

$$E_1(r,t) = \tfrac{1}{2} A_1(r) \exp(-i(\omega t - kz)) + c.c \quad (1)$$

where $A_1(r)$ is the complex amplitude and c.c denotes the complex conjugate, then the phase conjugate of E_1 is defined as:

$$E_2(r,t) = \tfrac{1}{2} A_1^*(r) \exp(-i(\omega t + kz)) + c.c \quad (2)$$

We formally calculate E_2 from Eq.(1) by taking the complex conjugate of the spatial part of $E_1(r,t)$, leaving the temporal part unchanged. The phase conjugate wave E_2 retraces the incident field E_1 and everywhere the wave fronts of E_2 coincide with those of E_1. This auto-retracing property of the phase conjugate wave is important for basic comprehension of how these waves work. One basic property of the phase conjugate wave is illustrated in

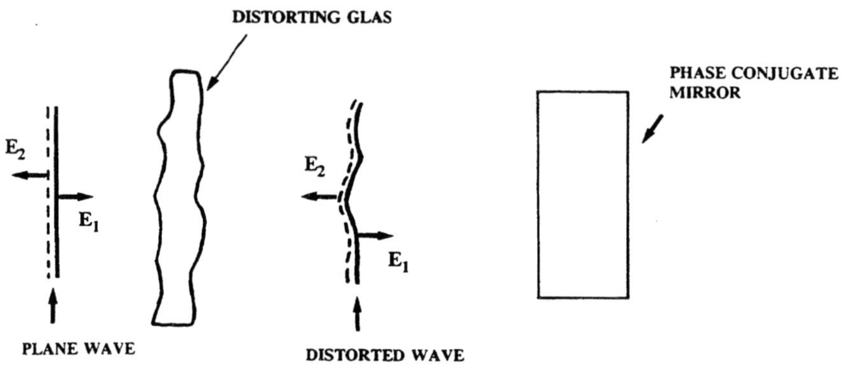

Fig.1 A plane wave E_1 is distorted by a phase-aberrating glass medium. The phase conjugate wave E_2 retraces the original wave E_1 and the distortions are completely removed by returning through the glass medium.

Fig.1. A plane wave is distorted by transmission through a piece of aberrating glass. When the distorted wave is reflected from a phase conjugate mirror, the reflected wavefront is identical to the incident distorted wavefront. When this phase conjugate wave is returned through the piece of glass, the distortions are completely removed. The distortions introduced on the first pass have been completely compensated by returning through the distorter. A mathematical proof of these distortion-correcting properties of the phase conjugate waves can be found in Ref.(6).

In Fig.2(a) we have shown an experimental setup, where an Argon laser beam is incident on a nonlinear photorefractive $BaTiO_3$ crystal. The $BaTiO_3$ crystal generates the optical phase conjugate of the incident laser beam. The phase conjugate beam retraces the incident beam and is displayed on a screen after a reflection in a beam splitter. Just in front of the $BaTiO_3$ crystal we have placed a piece of distorting glass. The aberrated beam is shown together with the phase conjugate beam and the output beam of the laser in

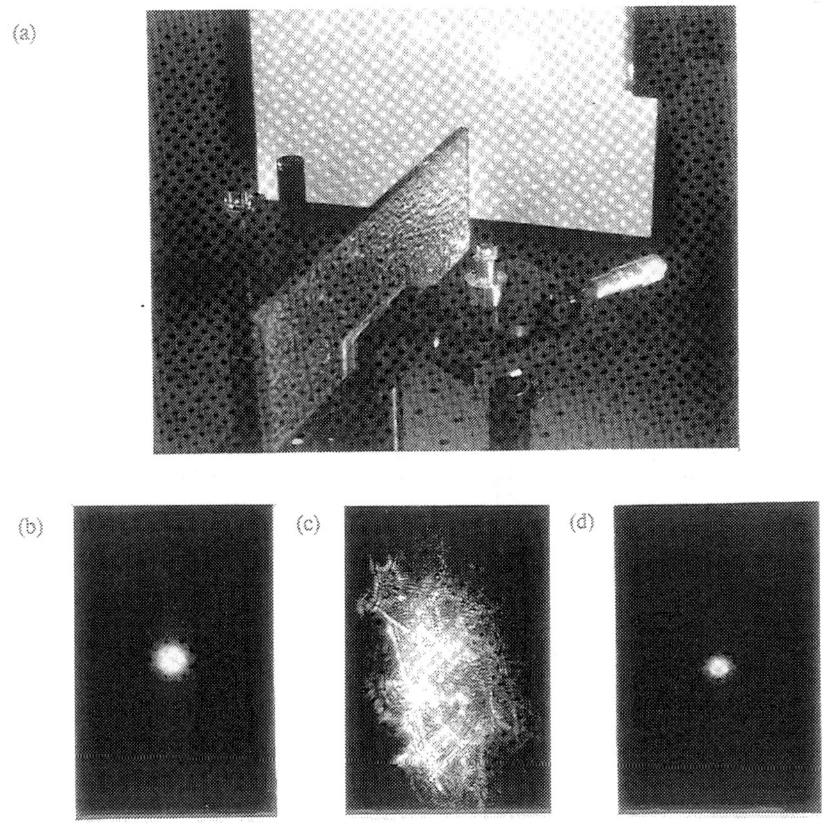

Fig.2 (a) Experimental setup with phase conjugation in a photorefractive $BaTiO_3$ crystal. In this experiment the laser beam is distorted by a severe phase aberrator.
(b) The output laser beam from the Argon laser.
(c) The aberrated beam after the glass aberrator.
(d) The phase conjugate beam after the distortions have been removed.

Fig.2(b-d).The photographs show (b) the output of the laser, (c) the aberrated beam after passage through the glass, and (d) the beam after the distorted beam is phase conjugated and returned through the piece of glass. The aberrations from the glass have been completely removed.

The first observation of optical phase conjugation in nonlinear materials was done by researchers from the Soviet Union[7] in 1972. Their work concerned Stimulated Brillouin Scattering in a CS_2 - filled waveguide. Since this observation, a variety of other nonlinear optical phenomena have been used to observe optical phase conjugation. In Section 4, I describe four-wave mixing, which is today one of the most popular nonlinear mechanisms for generation of phase conjugate waves. But first I outline the nonlinear mechanism of photorefractive materials.

3. THE PHOTOREFRACTIVE EFFECT

In photorefractive materials, a dynamic index grating is induced by a spatial light variation inside the medium. The effect was first observed in 1966[8] where a spatial intensity variation led to a detrimental distortion of the wave front. Since this observation, the photorefractive effect has been observed in many electro-optic crystals such as $LiNbO_3$[9], $Bi_{12}SiO_{20}$[10], $BaTiO_3$[11-13], $KNbO_3$[14], CdS[15], GaAs[16], and InP[16-17]. The effect is attractive for application in optical signal processing and optical phase conjugation, because it is possible to obtain nonlinear refractive index changes at milliwatt power levels. The dynamic index grating is induced by a spatial light variation inside the photorefractive material. The index grating can be erased by a uniform illumination. Without illumination the index grating will decay with a time constant which varies from one millisecond to months, depending on the material. However, for a material like $LiNbO_3$, the grating can be maintained for years if a fixing process[18] is applied to the crystal.

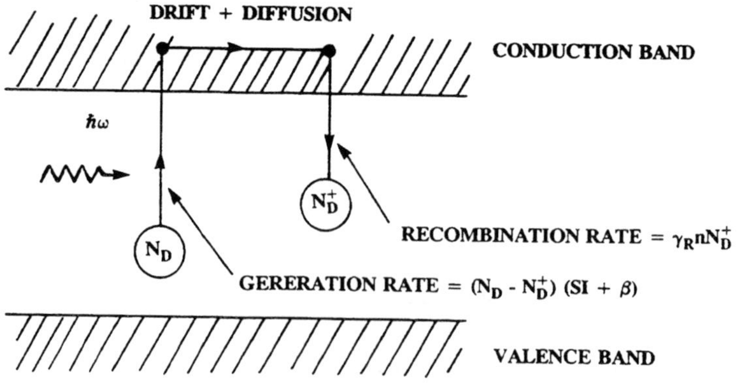

Fig.3 A simple model of the photorefractive effect.

The photorefractive effect is due to photoexcited charge carriers from impurity levels in electro-optic crystals. In this section I use the Kukhtarev[19,20] band transport model to briefly analyze the photorefractive effect. This model is today the most accepted model because so far, it has been able to account for almost all experiments with photorefractive crystals. More comprehensive reviews of the Kukhtarev model can be found in Refs.(21-24). However, it should be mentioned that the Hopping model [13,25] by Feinberg has also been widely used. In Fig.3 a simple band transport model for the photorefractive effect is shown. Incident photons with energy $\hbar\omega$ excite electrons from donor levels into the conduction band, where they are effected by diffusion and drift, which move them away from the illuminated regions until they are trapped in ionized donor levels. The density n of free electrons in the conduction band is governed by:

$$\frac{\partial n}{\partial t} = (N_D - N_D^+)(SI + \beta) - \gamma_R N_D^+ n + \frac{1}{e} \underline{\nabla} \cdot \underline{J} \tag{3}$$

where N_D is the density of donor traps, N_D^+ is the density of ionized donor traps, S is the cross-section of photoionization, I is the light intensity, β is the rate of thermal generation, γ_R is the electron-ionized trap recombination rate, e is the electron charge, and $\underline{J}$ is the current density.

In Eq.(3) $(N_D - N_D^+)(SI + \beta)$ is the rate of electron generation and $\gamma_R N_D^+ n$ is the recombination rate. The current density is given by:

$$\underline{J} = e\mu n\underline{E} + eD\underline{\nabla}n + \underline{J}^f \tag{4}$$

where μ is electron mobility, $\underline{E}$ is the total electric field vector, D is the diffusion constant, and $\underline{J}^f$ is the current density from the photovoltaic effect. The induced space charge field $\underline{E}_{sc}$ is given by:

$$\underline{\nabla} \cdot (\underline{\underline{\varepsilon}} \cdot \underline{E}_{sc}) = e(N_D^+ - N_A - n) \tag{5}$$

where $\underline{\underline{\varepsilon}}$ is the dielectric permittivity tensor and N_A is the immobile acceptor concentration in the photorefractive crystal. Due to the electro-optic effect the space-charge field $\underline{E}_{sc}$ induces a change in the index of refraction given by:

$$\Delta \left(\frac{1}{n^2}\right)_{ij} = r_{ijk} E_k \tag{6}$$

where r_{ijk} is the linear electro-optic tensor.

A space charge field E_{sc} will build up when the photorefractive crystal is illuminated by a light intensity pattern given by:

$$I = I_0 \left(1 + m\cos(\underline{K} \cdot \underline{r})\right) \qquad (7)$$

where m is the modulation depth and $\underline{K}$ is the grating wave vector. The periodic illumination in Eq.(7) generates electrons in the bright regions, where $\underline{K} \cdot \underline{r} = 2\pi p$, p is an integer. The free electrons move from high electron concentration in the bright regions towards low concentration in the dark regions ($\underline{K} \cdot \underline{r} = \pi(2p + 1)$) leaving behind positively charged ionized donors. When the electrons are trapped in the dark regions a space charge field is built up. In the case of an isotropic medium, the field in the photorefractive medium can be calculated from Eq.(3-7). The solution when the photovoltaic effect is neglected is given by[24]:

$$E = E_A + E_G \cos(\underline{K} \cdot \underline{r} + \phi) \qquad (8)$$

where

$$E_G = -mE_s \left[\frac{E_A^2 + E_D^2}{E_A^2 + (E_D + E_S)^2}\right]^{1/2} \qquad (9)$$

$$\tan(\phi) = \left(\frac{E_D}{E_A}\right)\left(1 + \frac{E_D}{E_S} + \frac{E_A^2}{E_D E_S}\right) \qquad (10)$$

In Eqs.(8-10) $E_D = k_B TK/e$ is the diffusion field, $E_S = eN_A/(\epsilon K)$ is the maximum space charge field, E_A is the electric DC field applied to the crystal, ϵ is the permittivity, and k_B is the Boltzmann constant.

4. OPTICAL PHASE CONJUGATION BY FOUR-WAVE MIXING

Optical phase conjugation by four-wave mixing in photorefractive crystals has become an extremely intense field of research during the last few years[26-32]. The four-wave mixing process involves the simultaneous interaction of four light waves in a nonlinear medium. The interference of the light waves produces spatial modulations of the refractive index, and consequently phase gratings are induced in the photorefractive medium. The configuration of the four-wave mixing is shown in Fig.4. Two counterpropagating laser beams A_1 and A_2 impinge on a nonlinear photorefractive medium together with a third probe beam A_4. The interaction of the three beams inside the nonlinear medium leads to the generation of a fourth beam A_3, which is the phase conjugate of the probe beam A_4. The characteristics of the phase conjugate beam are determined by the laws of conservation of energy, momentum, and angular momentum. In general, the four waves induce six gratings[32]. Two of these are transmission gratings, two are reflection gratings, and the other two are reflection gratings which are formed by counterpropagating waves. The analysis of the four-wave mixing, including all six gratings, is complicated and can only be carried out by computer calculation. Fortunately, in many experiments[33] the two transmission

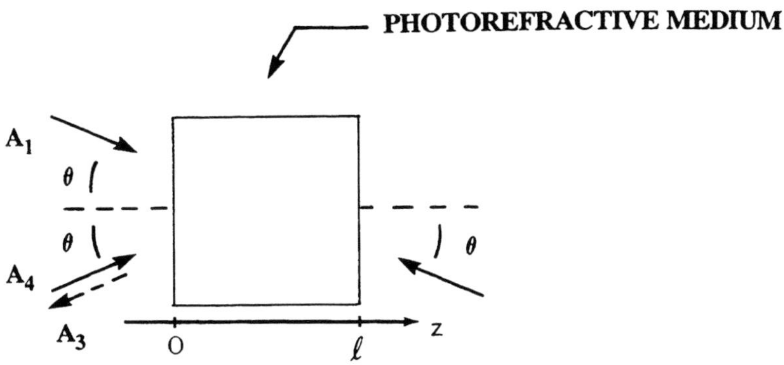

Fig. 4 Configuration for four-wave mixing in a photorefractive medium.

gratings are dominant. Therefore in the commonly used theory[34,35] the reflection gratings are neglected, and the four-wave mixing equations are solved by using simple numerical procedures. The basic equations commonly used[32] to describe the transmission grating nondegenerate four-wave mixing are given by:

$$\frac{dA_1}{dz} = \frac{-\gamma}{I_0} g A_4 - \alpha A_1 \quad , \quad \frac{dA_2^*}{dz} = \frac{-\gamma}{I_0} g A_3^* + \alpha A_2^* \qquad \text{(11a,b)}$$

$$\frac{dA_3}{dz} = \frac{\gamma}{I_0} g A_2 + \alpha A_3 \quad , \quad \frac{dA_4^*}{dz} = \frac{\gamma}{I_0} g A_1^* - \alpha A_4^* \qquad \text{(11c,d)}$$

where $g = A_1A_4^* + A_2^*A_3$ describes the transmission grating, $I_0 = \sum_{i=1}^{4} |A_i|^2$, α is the absorption coefficient, and the complex coupling constant is[35]:

$$\gamma = \frac{\omega r_{eff}\, n_0^3 E_S(E_A + iE_D)}{4c \cos\theta[E_A - \delta\tau(E_D + E_\mu)] + i(E_D + E_S + \delta\tau E_A)} \qquad \text{(11e)}$$

Here $\delta = \omega_1 - \omega_4 = \omega_3 - \omega_2$, τ is the grating saturation recording time, Θ is the angle of incidence of A_4. E_S, E_D, and E_μ are the respective electric fields of maximum space charge, diffusion, and drift, E_A is the externally-applied dc electric field, r_{eff} is the effective electro-optic coefficient, and n_0 is the refractive index of the crystal.

4.1 UNDEPLETED PUMP BEAMS APPROXIMATION

Eqs.(11a-d) describe the exchange of energy and phase between the four waves. In the undepleted pump beam approximation, the equations are solved when $|A_1|^2, |A_2|^2 >> |A_3|^2, |A_4|^2$. The undepleted pump beams approximation is used almost universally in the literature because it leads to a simple mathematical description, which shows some of the most important basic features of the four-wave mixing phase conjugation process. In this approximation, second order products of A_3 and A_4 can be neglected and Eqs.(11) reduce for $\alpha=0$ to:

$$\frac{dA_1}{dz} = 0 \quad , \quad \frac{dA_2^*}{dz} = 0 \tag{12a,b}$$

$$\frac{dA_3}{dz} = \frac{\gamma}{I_0}\,[|A_2|^2 A_3 + (A_1 A_2) A_4^*] \quad , \quad \frac{dA_4^*}{dz} = \frac{\gamma}{I_0}\,[|A_1|^2 A_4^* + (A_1^* A_2^*) A_3] \tag{12c,d}$$

When we apply the boundary conditions $A_3(l) = 0$ (l is the length of the photorefractive medium) and $A_4^*(0)$ known, the phase conjugate beam is obtained from Eq.(12c,d)[36]:

$$A_3(z) = A_4^*(0)\,\frac{A_1/A_2^*}{r_p^{-1}e^{-\gamma l}+1}\,[\exp\,(\gamma(z-l))\,-1] \tag{13}$$

where $r_p = |A_2|^2/|A_1|^2$ is the pump beam ratio. We note from Eq.(13) that A_3 is proportional to the phase conjugate of A_4.

4.2 ONE-GRATING THEORY

In the transmission grating configuration, one of the transmission gratings arises from the interference between beams A_1 and A_4 and the other arises from the interference between A_2 and A_3. The two transmission gratings have the same grating vector, and they add coherently. I recently published[37] another perturbational approach. In my approach I neglect the contribution from the A_2-A_3 interference, and in the following I show that this leads to analytical expressions for the amplitude and phases of all four interacting waves. In section 5 the theory is applied to the linear photorefractive oscillator, and here it is shown that, for a broad parameter range, the one- grating theory is capable of describing the behavior of the four-wave mixing oscillator.

When we neglect the grating from the interference between A_2 and A_3, we have $g = A_1 A_4^*$ in Eqs.(11). Equations (11a) and (11d) now reduce to:

$$\frac{dA_1}{dz} = -\frac{\gamma}{I_0}\,A_1|A_4|^2 \quad , \quad \frac{dA_4^*}{dz} = \frac{\gamma}{I_0}\,A_4^*|A_1|^2 \tag{14a,b}$$

The solutions to Eqs.(14) are[2]

$$A_1(z) = \left[\frac{I_0}{1 + r^2 \exp(2\, Re\, \gamma z)}\right]^{1/2} \exp\left(i\left\{\psi_1(0) - Im\, \gamma z - \frac{Im \gamma}{2\, Re \gamma} \ln\left[\frac{1 + r^{-2} \exp(-2\, Re\, \gamma z)}{1 + r^{-2}}\right]\right\}\right) \tag{15}$$

$$A_4(z) = \left[\frac{I_0}{1 + r^{-2} \exp(-2\, Re\, \gamma z)}\right]^{1/2} \exp\left(i\left\{\psi_4(0) - Im\, \gamma z - \frac{Im\, \gamma}{2\, Re\, \gamma} \ln\left[\frac{1+r^2 \exp(2\, Re\, \gamma z)}{1 + r^2}\right]\right\}\right) \tag{16}$$

where $\psi_1(0)$ and $\psi_4(0)$ are the initial phases of A_1 and A_4 and $r= |A_4(0)|/|A_1(0)|$. The solution for A_3 is now found if we insert the conservation law[38] $A_1A_2+A_3A_4=c$ (c is a constant of integration) into Eq.(11c):

$$\frac{dA_3}{dz} + \frac{\gamma}{I_0} |A_4|^2 A_3 = \frac{\gamma}{I_0} A_4^* c \tag{17}$$

The solution for A_3 is given by:

$$A_3(z) = \exp\left(-\int \frac{\gamma}{I_0} |A_4|^2 dz\right) \left[c\frac{\gamma}{I_0} \int \left(A_4^* \exp \int \frac{\gamma}{I_0} |A_4|^2 dz\right) dz + c_1\right] \tag{18}$$

where c_1 is a complex constant. With the boundary condition $A_3(l) = 0$ and $c = A_1(l)A_2(l)$ and inserting Eq.(16) into Eq.(18) we find that

$$A_3 = A_2(l)\frac{A_4^*(0)}{A_1^*(0)} \frac{\exp\left\{-i\frac{Im\, \gamma}{2\, Re\, \gamma} \ln\left[\frac{1 + r^2 \exp(2\, Re\, \gamma l)}{1 + r^2}\, \frac{1 + r^2 \exp(2\, Re\, \gamma z)}{1 + r^2}\right]\right\}}{[1 + r^2 \exp(2\, Re\, \gamma l)]^{1/2}[1 + r^2 \exp(2\, Re\, \gamma z)]^{1/2}} \times [\exp(\gamma z) - \exp(\gamma l)] \tag{19}$$

The solution for A_2 can be obtained if we insert the conservation law $A_1A_2+A_3A_4=c$ into Eq.(11b):

$$\frac{dA_2}{dz} - \frac{\gamma^*}{I_0} |A_1|^2 A_2 = \frac{-\gamma^*}{I_0} c A_1^* \tag{20}$$

This equation has the same form as Eq.(17), and if we use the same method as in Eq.(18) we obtain:

$$A_2(z) = \frac{A_2(l)\exp\left\{i\dfrac{Im\,\gamma}{2\,Re\,\gamma}\ln\left[\dfrac{1 + r^{-2}\exp(-2\,Re\,\gamma z)}{1 + r^{-2}\exp(-2\,Re\,\gamma l)}\right]\right\}}{[1 + r^{-2}\exp(-2\,Re\,\gamma l)]^{1/2}[1 + r^{-2}\exp(-2\,Re\,\gamma z)]^{1/2}} \tag{21}$$

$$\times \{r^{-2}\exp[-Re\,\gamma(z + l)]\exp[i\,Im\,\gamma(z - l)] + 1\}$$

Equations (15), (16), (19), and (21) are the analytical expressions for all four interacting waves. The expressions are valid when the four-wave interaction is dominated by the grating created by the interference between A_1 and A_4. This is true when either the backward-beam intensity or the phase conjugate reflectivity is small. The theory is valid for complex coupling strengths and can consequently be used when there is nondegenerate interaction. The theory is analytical and can be applied to four-wave mixing experiments with weak backward beams as a simple analytical model instead of a numerical solution[32] of the four coupled-wave equations (11a-d).

MacDonald and Feinberg[39,40] have reported enhancement of the four-wave mixing efficiency in $BaTiO_3$ caused by a frequency shift of the diffracted beams when they are diffracted in a moving grating. MacDonald and Feinberg used the enhancement of the diffraction efficiency to explain why the output frequency of a dye laser spontaneously starts to scan over the entire gain spectrum of the dye when a self-pumped $BaTiO_3$ crystal is used as one of the cavity mirrors[41]. They based their analysis on a numerical integration of Eqs.(11). The four-wave mixing diffraction efficiency versus frequency shift $(\omega_1\text{-}\omega_4)\tau$ from their analysis is given in Fig. 5. With the present analytical theory we can reproduce the numerical curve, if we calculate the diffraction efficiency $D=|A_3(0)|^2/|A_2(l)|^2$ from Eq.(19) with the coupling strength given by $\gamma=\gamma_0\exp(i\phi)/\{1+[(\omega_1\text{-}\omega_4)\tau]^2\}^{1/2}$, where $\phi=\tan^{-1}[(\omega_1\text{-}\omega_4)\tau]$. In Fig.5 the dots correspond to

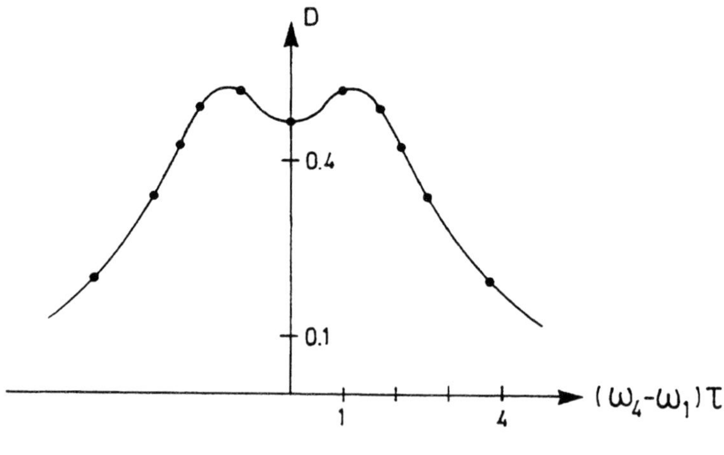

Fig. 5 Diffraction efficiency D versus the frequency shift $\omega_4\text{-}\omega_1$ (in units of the response time τ). The solid curve corresponds to a numerical solution from Ref.(39), and the dots correspond to the analytical results from Eq.(19).

the values of D calculated from Eq.(19) when r=1 and $\gamma_0 l=3.5$[39] are inserted.
In the next section I apply the present analytical theory to a linear photorefractive oscillator. I show that the one-grating theory provides a good description of the oscillator and that it is possible to use the theory even when the backward-beam intensity is not small.

5. THE LINEAR PHOTOREFRACTIVE OSCILLATOR

In Fig. 2 we demonstrated phase conjugation in $BaTiO_3$ with only one beam incident on the $BaTiO_3$ crystal. The two pumping beams are built up from scattered light which is amplified by two-beam coupling inside the crystal. The feedback that builds up the pumping beams is provided by total internal reflection in the $BaTiO_3$ crystal. This principle of phase conjugation is called passive self-pumped phase conjugation. In Fig. 6, the linear passive photorefractive oscillator, where the feedback for photorefractive oscillation is provided by two external mirrors M_1 and M_2, is shown. When a laser beam A_4 is incident upon the photorefractive medium, two oscillation beams A_1 and A_2 are build up in the cavity. The basis of this build-up of oscillations is the two-beam mixing amplification mechanism. The incident beam A_4 produces scattered light from imperfec-

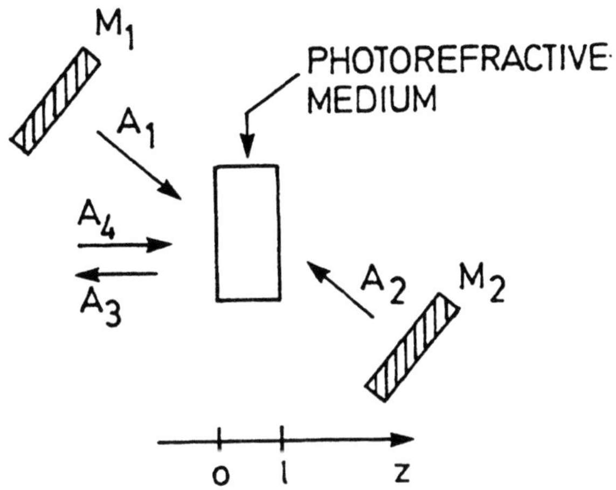

Fig. 6 The linear passive photorefractive oscillator.

tions in the photorefractive crystal. The scattered light interferes with the incident beam itself and, because of beam coupling in the photorefractive medium, the light scattered in the direction of the cavity alignment is amplified until stable oscillations are obtained. The two pumping beams A_1 and A_2 are built up from zero intensity, and therefore the undepleted pump beams approximation from section 4.1 cannot be used. In the following, I apply the one-grating theory from section 4.2 to the linear photorefractive oscillator in

Fig. 6. The boundary conditions for the intensity are given by the mirror reflectivities M_1 and M_2 of the two cavity mirrors:

$$M_1 = \frac{|A_1(0)|^2}{|A_2(0)|^2} \quad , \quad M_2 = \frac{|A_2(l)|^2}{|A_1(l)|^2} \tag{22}$$

The boundary condition for the phases ψ_1 and ψ_2 of the two cavity beams A_1 and A_2 is given by:

$$[\psi_1(l) - \psi_1(0)] + [\psi_2(0) - \psi_2(l)] + 2kL = 2\pi m \tag{23}$$

where m is an integer and L is the cavity length of the empty cavity. Equation (23) gives the condition for constructive interference in the cavity.

The phase-conjugate reflectivity $R = |A_3(0)|^2/|A_4(0)|^2$ of the oscillator in Fig. 6 can be obtained from Eq.(19). The unknown parameter r^2 in this equation can be determined if we insert Eq.(15) and Eq.(21) into the boundary conditions in Eqs.(22). We obtain

$$r^2 = (B + \sqrt{D})/N \tag{24}$$

where $B = -2[M_1M_2 \exp(\mathrm{Re}\ \gamma l) \cos(\mathrm{Im}\ \gamma l) - \exp(2\ \mathrm{Re}\ \gamma l)]$,
$D = B^2 - 4\ (M_1M_2 - 1) \exp(2\ \mathrm{Re}\ \gamma l)\ [M_1M_2 - \exp(2\ \mathrm{Re}\ \gamma l)]$, and
$N = 2[M_1M_2 - \exp(2\ \mathrm{Re}\ \gamma l)] \exp(2\ \mathrm{Re}\gamma l)$.

From Eq.(19) we find the phase-conjugate reflectivity:

$$R = M_2 \frac{1 + \exp(2\ Re\ \gamma l) - 2\exp(Re\ \gamma l)\cos(Im\ \gamma l)}{[1 + r^2 \exp(2\ Re\ \gamma l)]^2} \tag{25}$$

The procedure for calculating the reflectivity R is to calculate r^2 from Eq.(24) and then to obtain R from Eq.(25). Next, we compare the present theory with the previously established exact numerical theory of the linear photorefractive oscillator. In the solution given by Cronin-Golomb et al.[32,35] the phase-conjugate reflectivity of the oscillator is calculated from the following expression:

$$R = \frac{(\Delta + 1)^2\ |T|^2}{M_2|\Delta T + [\Delta^2 + (\Delta+1)^2/M_2]^{1/2}\ |^2} \tag{26}$$

where $\Delta = (I_2(l) - I_1(0) - I_4(0))/I_0$ and $T = \tanh\{(\gamma l/2)\ [\Delta^2 + (\Delta + 1)^2/M_2]^{1/2}\}$. In Eq.(26) Δ has to be found numerically as the zero values of the function:

$$f(\Delta) = M_1M_2 - \left| \frac{T + [\Delta^2 + (\Delta + 1)^2/M_2]^{1/2}}{\Delta T + [\Delta^2 + (\Delta + 1)^2/M_2]^{1/2} + (\Delta + 1)T/M_2} \right|^2 \tag{27}$$

In Fig.7, the phase-conjugate reflectivity R is plotted versus the coupling strength calculated from the simple formula of Eq.(25) when $M_1 = M_2 = 0.5$ and Im $\gamma = 0$. This curve is shown as a solid line. Furthermore, the exact values of the phase conjugate reflectivity calculated from Eq.(26) are shown in Fig.7 as dots. The threshold value of the coupling strength for oscillations in the cavity is, for the one-grating theory, obtained for $r^2 = \infty$. This condition implies that N in Eq.(24) is zero, and therefore the threshold coupling strength $\gamma_t l$ is given by:

$$Re\ \gamma_t\, l = \tfrac{1}{2} \ln (M_1M_2) \tag{28}$$

which is in agreement with the exact theory[2]. Above threshold there is a minor deviation between the two theories. However, at high coupling strengths the one-grating theory again achieves good agreement with the exact theory. In Fig.7 the deviation at Reγl = -8 is less than 0.02%.

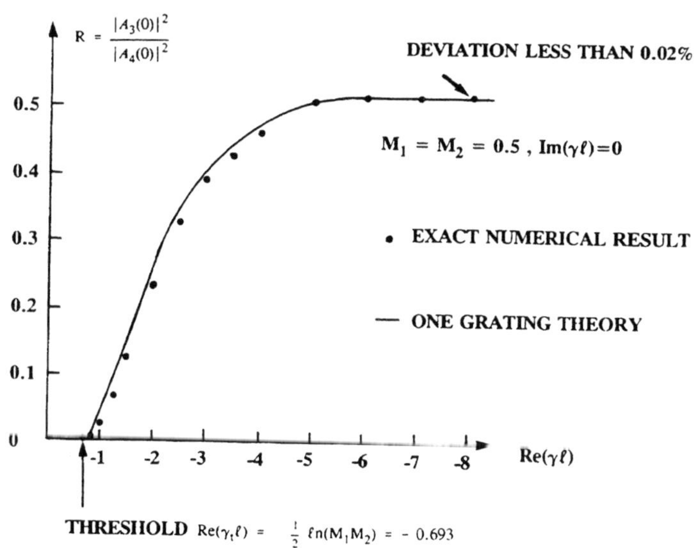

Fig.7 The phase conjugate reflectivity versus coupling strength for the linear photorefractive oscillator. The solid curve is calculated from Eq.(25) and the dots are calculated from the exact theory of Eq.(26). The parameters are: $M_1 = M_2 = 0.5$ and Im $\gamma = 0$.

So far we have based our calculations on the amplitudes of the fields inside the oscillator cavity. However, the phases of the oscillating beams A_1 and A_2 must also fulfill the condition for constructive interference in the cavity. This condition is given in Eq.(23) The phase shift of A_1, arising from the nonlinear interaction inside the photorefractive medium, is obtained from Eq.(15):

$$\psi_1(l) - \psi_1(0) = -Im\gamma l - \frac{Im\gamma l}{2\, Re\, \gamma l} \ln\left[\frac{1 + r^{-2}\exp(-2\, Re\, \gamma l)}{1 + r^{-2}}\right] \tag{29}$$

and from Eq.(21) we find the phase shift for A_2

$$\psi_2(0)-\psi_2(l)=\frac{Im\ \gamma l}{2\ Re\ \gamma l}\ln[\frac{1+r^{-2}}{1+r^{-2}\exp(-2\ Re\ \gamma l)}]+\tan^{-1}\left[\frac{-\sin(Im\ \gamma l)}{r^2\exp\ (Re\ \gamma l)+\cos(Im\ \gamma l)}\right] \tag{30}$$

From Eq.(23) we then find that the cavity detuning $\Delta\psi = 2\pi m - 2kL$ is given by:

$$\Delta\psi = -Im\ \gamma l - \frac{Im\ \gamma l}{Re\ \gamma l}\ln\left[\frac{1 + r^{-2}\exp(-2\ Re\ \gamma l)}{1+ r^{-2}}\right] + \tan^{-1}\left[\frac{-\sin(Im\ \gamma l)}{r^2\exp(Re\ \gamma l) + \cos(Im\ \gamma l)}\right] \tag{31}$$

In Eqs.(23-24) it is assumed that $r^2 \exp[-Re\ \gamma\ (z+l)] \cos[Im\ \gamma\ (z + l)] > 0$. If the frequency ω_4 of the incident beam A_4 is different from the mode frequency of the resonator, the photorefractive grating will spontaneously start to move as shown in Fig.8.

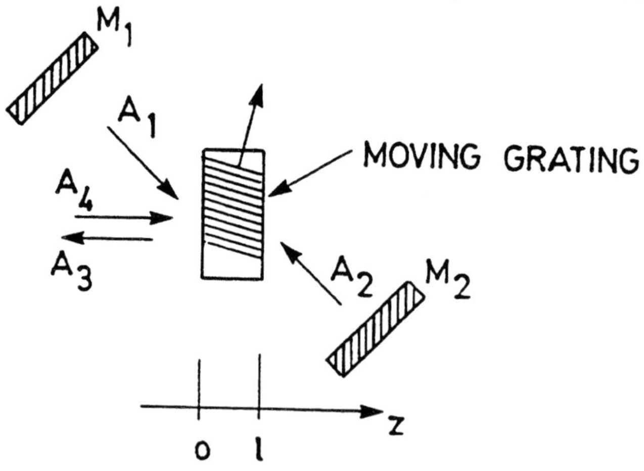

Fig.8 The phase conjugate linear oscillator when the cavity is off resonance. The grating spontaneously starts to move.

The moving grating gives an extra phase shift and compensates for the cavity detuning. When there is no applied field ($E_A = 0$) the coupling constant can, according to Eq.(11e), be written as

$$\gamma = \frac{\gamma_o}{1 + i(\omega_4 - \omega_1)\tau} \tag{32}$$

where γ_0 is the coupling constant without moving grating and with $E_A = 0$. In Fig.7 it is assumed that γ is real. We see now that this is possible only if the frequency of the resonator matches the frequency ω_4 of the incident beam. If there is a mismatch, the relation between the frequency detuning and the cavity detuning can be obtained from Eq.(31). The relationship among the frequency detuning $(\omega_4 - \omega_1)\tau$, the phase conjugate reflectivity R, and the cavity detuning $\Delta\psi$ can be obtained if, in Eq.(24), r^2 is determined from a given frequency detuning. The phase conjugate reflectivity and the cavity detuning $\Delta\psi$ is then obtained from Eq.(25) and Eq.(31), respectively. In Fig.9 the reflectivity R is plotted versus $\Delta\psi$ and compared with the exact numerical calculation by Cronin-Golomb

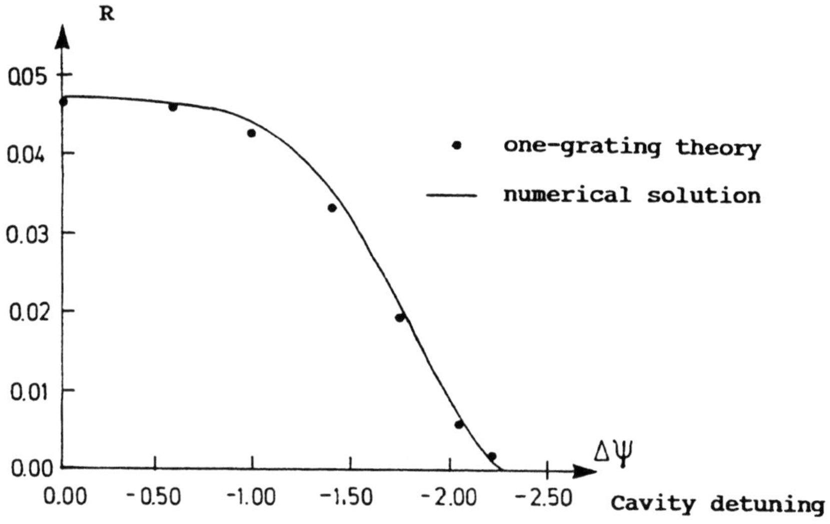

Fig.9 Phase conjugate reflectivity R versus cavity detuning. The one-grating theory is compared with an exact numerical solution. The parameters are $M_1 = 0.9$, $M_2 = 0.05$, and $\gamma_0 l = -5$.

and Yariv[35]. The curve is calculated for $M_1 = 0.9$, $M_2 = 0.05$, and $\gamma_0 l=-5$. From the curve we see that the moving grating is able to compensate for the cavity detuning until $\Delta\psi \approx 1$. When the detuning is further increased, the reflectivity decreases more dramatically. In Fig.9 the exact numerical values of R are a little higher than the analytical reflectivity from the one-grating theory. This is due to the extra contribution from the A_2-A_3 interference which is neglected in the one-grating theory.

6. APPLICATIONS

Photorefractive materials are attractive for optical signal processing, optical interconnections and dynamic interferometry, because they allow parallel recording and readout of holographic information. In this section I describe how the photorefractive effect is used in a flow velocimeter, in time-average interferometry, and in photorefractive interference filters.

6.1 PHOTOREFRACTIVE PARTICLE IMAGE VELOCIMETRY

In photorefractive materials dynamic spatial convolutions and correlations of two-dimensional phase and amplitude images can be obtained by using low-intensity lasers. The spatial convolution and correlation are achieved with classical Fourier transform lenses in two-wave mixing or four-wave mixing experiments[42]. At the Technical University in Denmark, we have developed an autocorrelator[43-46] for flow analysis in particle image velocimetry (PIV). The correlator performs the autocorrelation of a two-dimensional transparency, which is a double exposure photograph of particle flow. The correlator uses

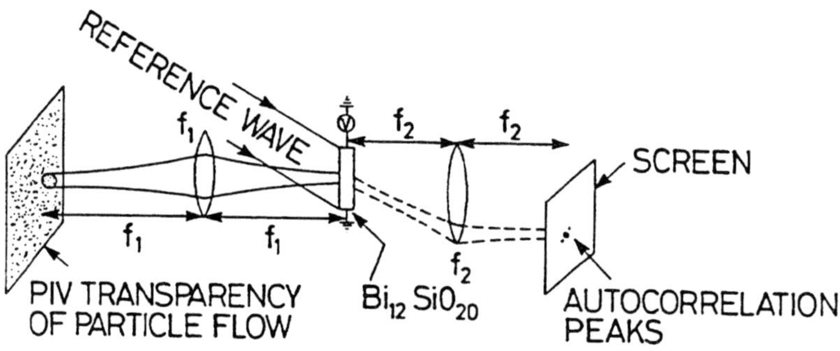

Fig. 10 The BSO autocorrelator.

the two-beam coupling configuration shown in Fig.10. The input is the transparency and the output on the screen is the autocorrelation function. The autocorrelator uses a two step procedure. First, the Fourier transform of a small area of the transparency and a plane reference wave induce a phase grating in the BSO crystal. The reference wave is then switched off and the hologram is read out with the Fourier transformed object beam itself to obtain the autocorrelation on the screen. The result shown in Fig.11(a) consists of two side peaks and a central spot in the case of a uniform particle velocity. The distance between the peaks and their orientation determine the velocity and direction of the flow. If more than one particle velocity is present within the interrogation area, multiple diffraction gratings will be induced in the BSO crystal. In Fig. 11(b) the case of two velocities within the interrogation area is shown.

Figs. 12(a-b) show the PIV measurements[45] from a swirling flow in a cylindrical container with a rotating cover. The cylinder in Fig.12(a) contains a mixture of glycerin and water with pollen as seeding particles. The PIV transparency is made by sweeping a double-pulsed laser beam through a cross section of the cylinder, and reflected light from the seed particles is recorded on a high resolution film. The data processing is performed by the photorefractive optical correlator and the corresponding vector map is shown in

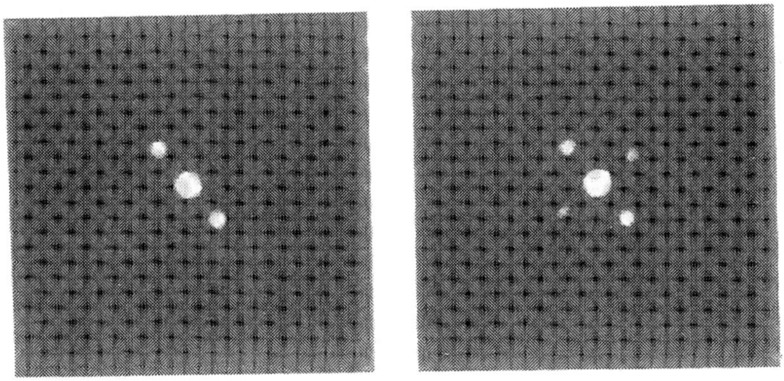

Fig.11 The output from the autocorrelator (a) one velocity in the interrogation area. (b) two velocities within the interrogation area.

Fig.12(b). The autocorrelation function of the input transparency is usually carried out on a digital computer by using two-dimensional fast Fourier transforms. The processing time on a 386 PC with 8 MFLOP array processor plug-in board is about 20s per velocity vector and about 0.5s on a VAX 780 with Numerix array processor. The photorefractive autocorrelator in Fig.10 has the potential for much higher speeds. The processing time for the autocorrelator is given by the response time of the photorefractive BSO crystal, which is typically 5ms[46]. The numerical processing time can, of course, be decreased in faster computers, but so can the photorefractive response time.

The holographic particle image velocimeter in Fig.10 was originally proposed by Coupland and Halliwell[47]. In their analysis, however, they assumed that the recording medium was isotropic. Recently, we included the effect of birefringence and optical activity of the BSO crystal in a numerical model[43]. Our results show that the polarization and the intensity of the autocorrelation peaks are sensitive to a number of parameters, such as voltage applied to the crystal, particle displacement, and polarization of the

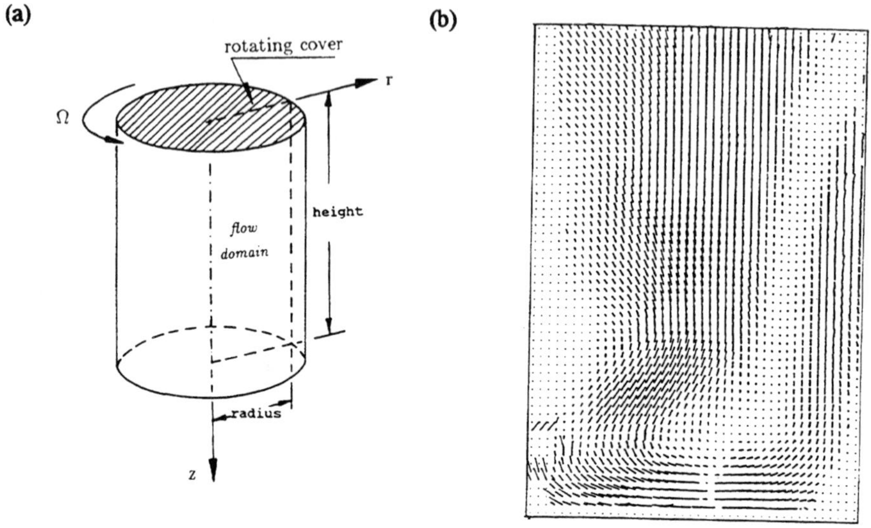

Fig.12 (a) Geometry of cylinder for the particle image velocimetry measurements.
(b) Vector map obtained by the optical correlator.
From Buchhave et al.[45]

incident beams. The polarization characteristics of the photorefractive autocorrelator can be used to enhance the performance of the correlator. Fig.13 shows the intensity profile of the diffracted peaks for three different orientations of a polarizer placed just behind the BSO crystal. The incident beams from an Ar-ion laser at 514.5nm are linearly polarized perpendicular to the plane of incidence. Even though all three peaks correspond to a first-order diffraction in the BSO crystal, their polarization states are different. In Fig.13 the intensity of the center spot decreases as the direction of the linear polarizer at the output is changed from the vertical to the horizontal position. Since the information regarding the velocity distribution is in the two autocorrelation peaks, we can use the polarization states of the correlator output to increase the signal-to-noise ratio of the optical correlator. Fig.14 shows further experimental results for the polarization angle (Fig.14(a)) and the ellipticity (Fig.14(b)) for the diffracted peaks as a function of the direction of the linearly polarized incident readout beam. The polarization angle is defined as the angle between the direction of the major axis of the elliptical polarization and the normal to the plane of incidence. In the ellipticity measurements, the positive values correspond to right-handed polarization whereas the negative values indicate left-handed polarization.

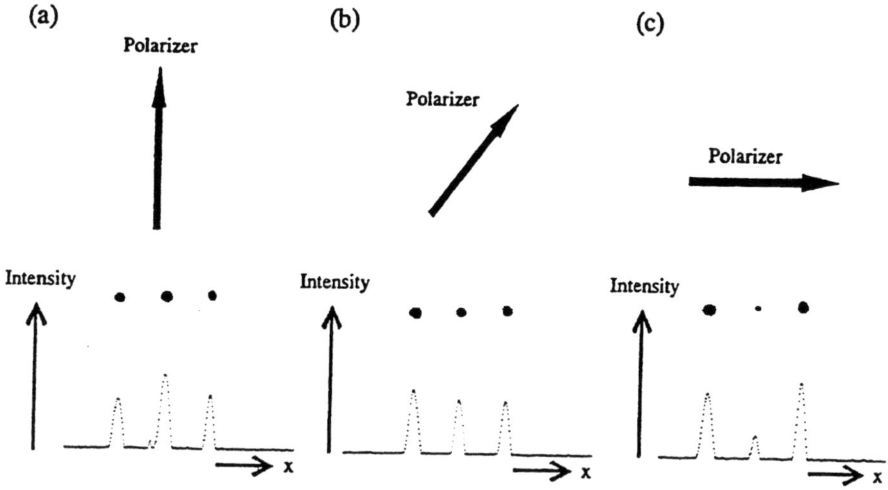

Fig.13 Intensity profiles of the three peaks for three different orientations of a polarizer placed just after the BSO crystal. From Petersen et al.[44]

The results of Fig. 14 show that the polarization of the peaks is sensitive to the polarization of the incident beams. Furthermore, they show that it is possible to find a configuration in which the ellipticity of the center peak is small compared with that of the side peaks. In such a configuration, a polarizer placed after the BSO crystal will increase the signal-to-noise ratio because we remove the central peak.

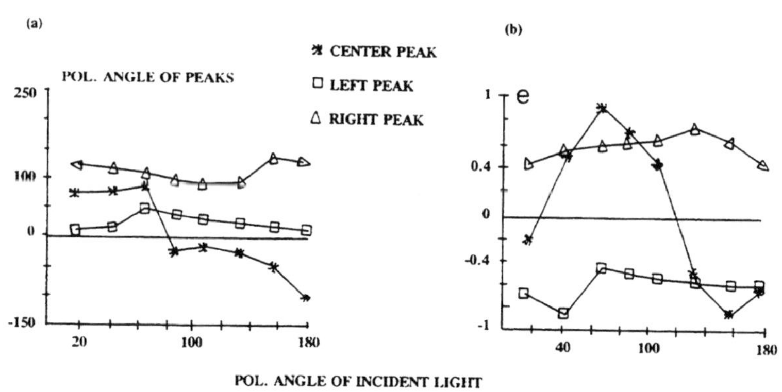

Fig.14 Experimental curves of (a) the polarization angle and (b) the ellipticity e of the 3 diffracted peaks as a function of the polarization angle of the incident readout light. From Petersen et al.[44]

The performance of the photorefractive autocorrelator can be explained from the simple model in Fig.15. The model provides a simple tool for investigating the basic behavior. In the simplest configuration, the PIV-transparency contains only two particles. In this case we can replace the transparency with two pinholes. The two spherical waves that are generated after illumination of the two holes are transformed into two plane waves by the lens (f). These two scattered object waves and the reference beam form gratings in the photorefractive BSO crystal.

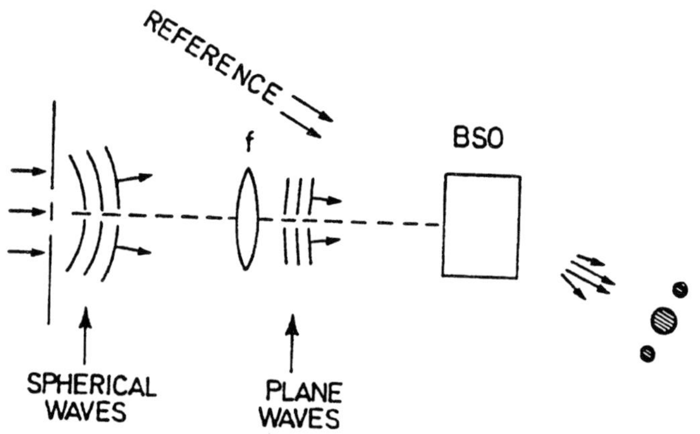

Fig.15 Simple model for the autocorrelation process. From Petersen et al.[44]

There are two gratings G_1 and G_2 caused by the interference of the reference wave with each of the two object waves, and one grating G_3 caused by the interference between the two object waves. The autocorrelation function with the dc spot and the two correlation peaks can be explained as diffraction of the two plane object beams in the two gratings G_1 and G_2. The two autocorrelation peaks correspond to off-Bragg diffraction, when each of the scattered waves diffracts from the grating created by the other scattered wave and the reference beam. The center dc spot consists of two contributions, which correspond to Bragg diffraction in G_1 and G_2. In this analysis of the optical autocorrelator in Fig.10 we have neglected the influence of the grating G_3 and we have furthermore assumed a linear combination of the gratings. The simple interpretation of the autocorrelation, however, can help analyze the photorefractive correlator and can be used for optimizing the parameters of the BSO correlator. In our work[43-44] we have combined the model in Fig.15 with the beam-propagation method[43,48]. Fig.16 shows a typical curve of the diffraction efficiency versus angular Bragg detuning obtained from the beam-propagation method. Since the angular detuning is proportional to the particle displacement, we can obtain the intensity of the autocorrelation peaks from Fig.16. Once the relationship between detuning δ_0 and the particle displacement is established the intensities of the two autocorrelation peaks are given by the diffracted intensities at δ_0 and $-\delta_0$ in Fig.16. The intensity of the center spot can be obtained for zero detuning $\delta = 0$.

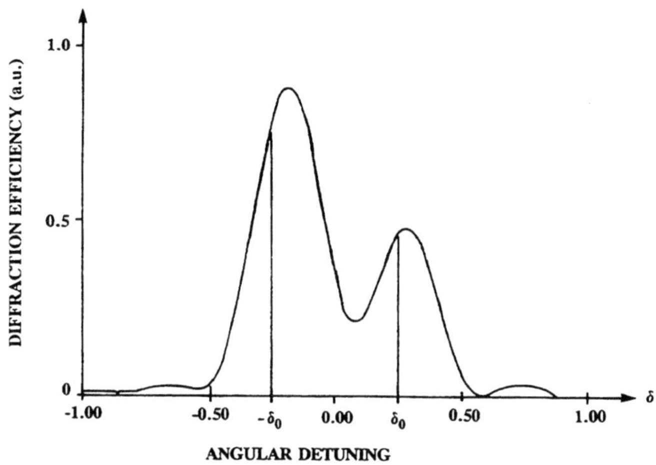

Fig.16 The theoretical diffraction efficiency versus detuning angle from the Bragg incidence. From Petersen et al.[44]

In general, we obtain asymmetric detuning curves and consequently asymmetric autocorrelation functions. This is in contrast to the symmetric detuning curves obtained in the grating theory for isotropic media given by Kogelnik[49]. The complicated polarization and intensity behavior for the diffracted beams in our BSO autocorrelator is due to the optical activity and birefringence in the BSO crystal[43,48].

6.2 PHOTOREFRACTIVE TIME-AVERAGE INTERFEROMETRY

Time-average interferometry of vibrating structures was introduced in 1965 by Powell and Stetson[50]. At that time they used conventional holographic recording techniques which required cumbersome chemical processing. In 1971 Butters and Leendertz[51] introduced electronic speckle pattern interferometry (ESPI) in which the fringes were directly recorded on a video camera. The image quality in ESPI is poor, however, because the interferograms consist of noisy speckle patterns.

Photorefractive materials are attractive in time-average interferometry because they combine high holographic image quality with real-time recording and readout of information. Real-time interferograms of vibrating structures using $Bi_{12}SiO_{20}$ crystals were obtained by Huignard et al[52] and Marrakchi et al[53]. Recently[54] we extended the work in Ref.53 to include a complete amplitude and phase mapping of the vibrating structure using photorefractive BSO crystals. The setup for photorefractive time-average interferometry is shown in Fig.17. The setup is based on the four-wave mixing scheme from section 4. Diffuse light A_4 from a vibrating structure is incident on a BSO crystal, together with two pump beams A_1 and A_2. The interference between the light from each point of the vibrating structure and the forward beam A_1 induces a grating in the crystal.

The phase conjugate interferogram results from diffraction of the backward pump beam A_2 in the recorded gratings. The stationary parts of the vibrating structure induce stationary gratings which lead to maximum light intensity in the phase conjugate image. Higher order fringes correspond to increasing vibration amplitude. In Fig.18(a) the phase conjugate output with a stationary metal plate as object is shown and in Fig.18(b) the phase conjugate output when the metal plate vibrates with a frequency f = 4730 Hz is

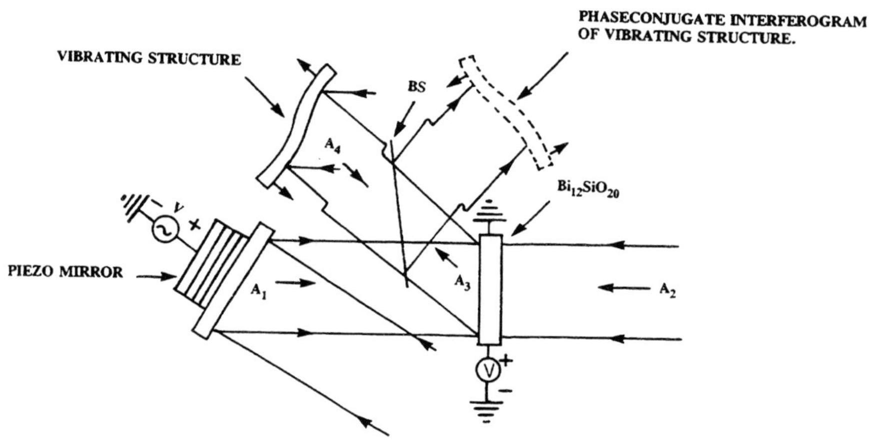

Fig.17 Time-average interferometry in the four-wave mixing geometry

shown. The brightest areas in the conjugate signal corresponds to the stationary vibration states on the plate. If we vibrate the piezo mirror in Fig.17 at the same frequency, amplitude, and phase as parts of the vibrating structure, these parts will lead to a stationary interference pattern, and will give maximum light intensity in the phase conjugate interferograms.

Fig.18(c-e) show the interferogram when the piezo-mirror vibrates with the same frequency as the test structure, and when we increase the amplitude of the piezo-mirror from 222nm to 589nm. In Fig.18(f) we have π-shifted the phase of the piezo-mirror for an amplitude of 589nm.

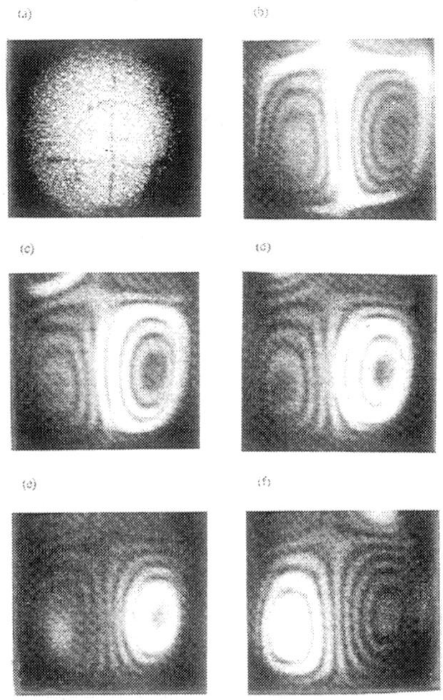

Fig.18 (a) The phase conjugate output with a stationary metal plate as object. (b-f) The phase conjugate interferogram when the metal plate and the piezo-mirror vibrate with a frequency of 4730Hz. The amplitudes of the piezo-mirror are (b) 0nm, (c) 222nm, (d) 406nm, (e) 589nm, and in (f) we have π-shifted the phase of the piezo-mirror for an amplitude of 589nm. From Rohleder et al.[54]

The experimental results in Figs.18 show that we can measure the amplitude and phase anywhere on the vibrating structure by introducing a proper sinusoidal phase shift in the forward pump beam.

6.3 PHOTOREFRACTIVE INTERFERENCE FILTERS

The photorefractive interference filter has only been very briefly investigated in literature[55,56]. The filter has several control parameters for the reflectance and a unique high wavelength selectivity. The configuration for the filter is shown in Fig.19. Two coherent laser beams E_1 and E_2 are incident on a photorefractive medium. Due to the photorefractive effect, these two beams induce a sinusoidal index of refraction of period Λ in the photorefractive medium. A third beam E_3 propagating in the x-direction in Fig.19 will experience layers of alternate high and low index of refraction. This situation is somewhat analogous to the multilayer dielectric mirror[57] with alternating quarter-wave layers of high and low index. In Fig.19 the layers of different index is a half fringe spacing $\Lambda/2$. If this thickness is equal to a quarter-wave for the probe beam E_3, then the reflected waves from all the layers in the filter will interfere constructively at the front surface of the photorefractive medium and a strong reflected signal E^R_3 from the filter can be obtained.

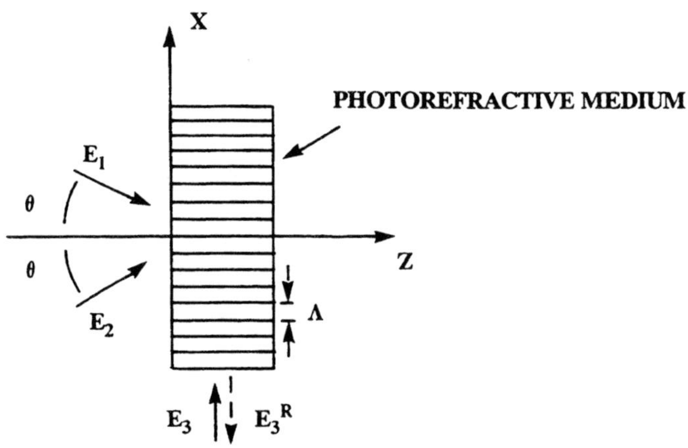

Fig.19 Configuration for the photorefractive interference filter. Two coherent laser beams E_1 and E_2 induce the filter and control the reflection of a third laser beam E_3. E^R_3 is the reflected wave. Θ is the incident angle with respect to the normal. Λ is the fringe spacing.

We assume that the two beams E_1 and E_2 induce an index profile given by:

$$n(x) = n_0 + n_1 \sin(Kx) \tag{33}$$

where $K = 2\pi/\Lambda$ is the grating vector. The reflectance of the filter is given by[57]:

$$R = \frac{|E_3^R|^2}{|E_3|^2} = \frac{\kappa^2 \sinh^2(SL)}{S^2\cosh^2(SL) + (\frac{\Delta k}{2})^2 \sinh^2(SL)} \tag{34}$$

where $\kappa = \pi n_1/\lambda$, $S^2 = \kappa^2 - (\Delta k/2)^2$, $\Delta k = 2\pi(2n_0/\lambda - 1/\Lambda)$, λ is the wavelength of the probe beam E_3, and L is the grating interaction length in the x-direction. For a photorefractive medium the modulation in the refractive index can be obtained from Eq.(6) and Eq.(9) in section 3:

$$n_1 = 2n_0^3 \, r_{eff} \frac{\sqrt{\beta}}{1 + \beta} \left[\frac{E_A^2 + E_D^2}{(1 + E_D/E_s)^2 + E_A^2/E_S^2} \right]^{1/2} \tag{35}$$

where r_{eff} is the effective electro-optic coefficient, and β is the incident intensity ratio $|E_2|^2/|E_1|^2$ of the exciting beams. Eq.(35) is valid for small phase changes.The reflectance R in Eq.(34) has a peak value $R_{max} = \tanh^2(\kappa L)$ at $\Delta k = 0$ and a sharp cutoff with oscillatory sidebands. In Fig.20 we have shown the reflectance versus mistuning ΔkL calculated from Eq.(34) when typical values for $BaTiO_3$: $n_1 = 5\ 10^{-5}$ and L = 5mm are

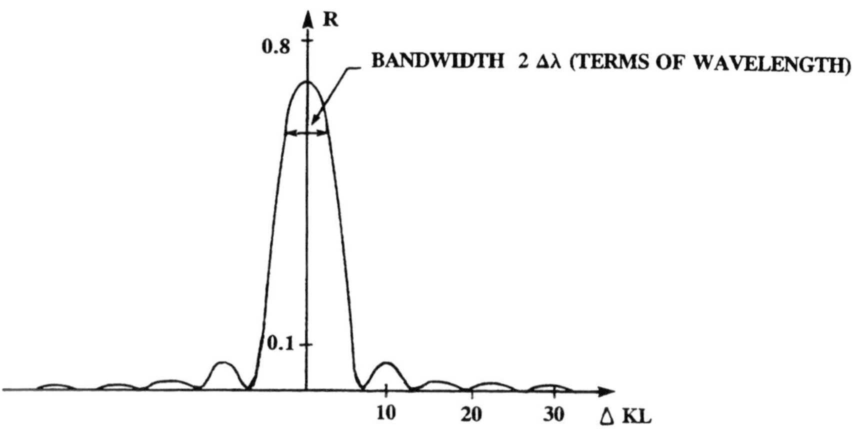

Fig.20 The reflectance versus mistuning for the photorefractive filter calculated from Eq.(34) with typical values for $BaTiO_3$. From Petersen et al.[56]

inserted. The bandwidth of the filter is given by (S must be real):

$$\Delta\lambda/\lambda = 2|n_1/n_0| \tag{36}$$

If we insert $n_1 = 5\ 10^{-5}$ and $n_0 = 2.48$ we obtain $\Delta\lambda/\lambda = 4\ 10^{-5}$. For a probe wavelength $\lambda = 1064$nm we have a wavelength selectivity $\Delta\lambda = 0.43$ Å which is approximately a factor of 100 smaller than for the best conventional dielectric coatings. Furthermore, the center wavelength $\lambda = 2\ \Lambda n_0$ is controlled in real-time by the angle 2Θ between the two exciting beams according to $\Lambda = \lambda_e/(2\sin\Theta)$ where λ_e is the wavelength of the exciting beams. This tuning property is important for many applications. The reflectance can be dynamically controlled by the modulation of the two exciting beams E_1 and E_2.

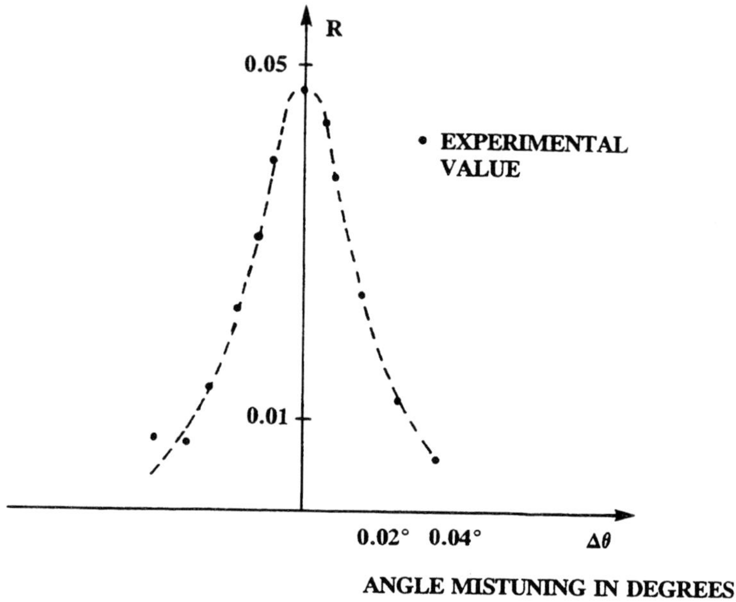

Fig.21 The experimental reflectance R versus angular mistuning of the probe beam E_3. From Petersen et al.[56]

In Fig.21, I have shown the experimental reflectance R versus angular mistuning $\Delta\Theta$ of the HeNe probe beam E_3 for a $BaTiO_3$ crystal with dimensions 5x5x5xmm^3. The experimental reflectance R shows the high angular selectivity of the filter. However, the magnitude of R is lower than expected from Eq.(34). This is probably due to beam fanning in $BaTiO_3$ which is detrimental to the filter performance. Finally, we have neglected absorption losses in Eq.(34).

REFERENCES

1) P. Yeh, A.E. Chiou, J. Hong, P. Beckwith, T. Chang, and M. Khoshnevisan, "Photorefractive nonlinear optics and optical computing", *Opt. Eng.* Vol. 28, 1989, pp. 328-343.

2) S. Kwong, M. Cronin-Golomb, and A. Yariv, "Oscillation with photorefractive gain", *IEEE, J. of Quantum Electron,* Vol. QE-22, 1986, pp. 1508-1523.

3) P. Yeh and C. Gu, "Photorefractive media for optical interconnections", *Int. J. of Nonlinear Optical Physics,* Vol. 1, 1992, pp. 167-201.

4) D. Psaltis, D Brady, X.G. Gu, and S. Lin, "Holography in artificial neural networks", *Nature* (London), Vol. 343, 1990, pp. 325-330.

5) A. Yariv, *Optical Electronics,* New York, Saunders College Publishing, 1991, pp. 608-636.

6) A. Yariv, and R.A. Fisher, In *Optical Phase Conjugation,* Ed. R.A. Fisher, New York, Academic Press, 1983, pp. 1-22.

7) B.Y. Zel'dovich, V.I. Popovichev, V.V. Raguel'sky, and F.S. Faizullov, "Connection between the wave fronts of the reflected and exciting light in stimulated Mandel'Shtam-Brillouin scattering", *Sov. Phys. JETP,* Vol. 15, 1972, pp. 109-113.

8) A. Ashkin, G.D. Boyd, J.M. Dziedzie, R.G. Smith, A.A. Ballman, J.J. Levinstein, and K. Nassau, "Optically induced refractive index inhomogeneities in $LiNbO_3$ and $LiTaO_3$" *Appl. Phys. Lett.*, Vol. 9, 1966, pp. 72-74.

9) R.A. Rupp, R. Sommerfeldt, K.H. Ringhofer, and E. Krätzig, "Space charge field limitation in photorefractive $LiNbO_3$:Fe crystals, *Appl. Phys. B,* Vol. 51, 1990, pp. 364-370.

10) R.A. Mullen, and R.W. Hellwarth, "Optical measurements of the photorefractive parameters of BSO", *J. Appl. Phys.*, Vol. 58, 1985, pp. 40-44

11) M.B. Klein, " Photorefractive properties of $BaTiO_3$" In *Photorefractive Materials and Their Applications I,* Ed. P. Günter, and J.P. Huignard, Berlin: Springer-Verlag, 1988, pp. 195-236.

12) M.C. Gower, "Photoinduced voltages and frequency shifts in a self-pumped phase-conjugating $BaTiO_3$ crystal", *Opt. Lett.*, Vol. 11, 1986, pp. 458-460.

13) J. Feinberg, D. Heiman, A.R. Tanguay, and R.W. Hellwarth, "Photorefractive effects and light-induced charge migration in barium titanate", *J. Appl. Phys.*, Vol. 51, 1980, pp. 1297-1305.

14) P. Günter, "Holography, coherent light amplification and optical phase conjugation with photorefractive materials, *Phys. Rep.*, Vol. 93, 1982, pp. 199-299.

15) A. Ashkin, B. Tell, and J.M. Dziedzic, "Laser induced refractive index inhomogeneities and absorption saturation effects in CdS, *IEEE J. Quantum Electron.*, Vol. QE-3, 1967, pp. 400-406.

16) J.C. Fabre, J.M.C. Jonathan, and G. Roosen, " Photorefractive beam coupling in GaAs and InP generated by nanosecond light pulses", *J. Opt. Soc. Am. B,* Vol. 5, 1988, pp. 1730-1736.

17) R.B. Bylsma, A.M. Glass, D.H. Olson, and M. Cronin-Golomb, "Self-pumped phase conjugation in InP: Fe", *Appl. Phys. Lett.*, Vol. 54, 1989, pp. 1968-1970.

18) W. Phillips, J.J. Amodei, and P.L. Staebler, "Optical holographic storage properties of transition metal doped lithium niobate", *RCA Review*, Vol. 33, 1972, pp. 94-109.

19) V.L. Vinetskii, N.V. Kukhtarev, S. G. Odulov, and M.S. Soskin, " Dynamic self-diffraction of coherent light beams", *Sov. Phys. Usp.*, Vol. 22, 1979, pp. 742-756.

20) N.V. Kukhtarev, V.B. Markov, S.G. Odulov, M.S. Soskin, and V.L. Vinetskii, "Holographic storage in electro-optic crystals. Beam coupling and light amplification", *Ferroelectrics*, Vol. 22, 1979, pp. 961-964.

21) P. Günter, and H.J. Eichler, "Introduction to Photorefractive materials", In *Electro-optic and Photorefractive Materials*, Ed. P. Günter, Berlin, Heidelberg: Springer-Verlag, 1987, pp. 206-228.

22) P. Günter and J. Huignard, "Photorefractive effects and materials", In *Photorefractive materials and Their Applications I*, Ed. P. Günter and J.P. Huignard, Berlin, Heidelberg: Springer-Verlag, 1988, pp. 7-73.

23) L. Solymar, "Theory of volume hologram formation in photorefractive crystals", In *Electro optic and Photorefravtive Materials*, Ed. P. Günter, Berlin, Heidelberg: Springer-Verlag, 1987, pp. 229-245.

24) T.J. Hall, R. Jaura, L.M. Connors, and P.D. Foote, "The photorefractive effect - a review", *Progress in Quantum Electronics*, Vol. 10, 1985, pp. 77-146.

25) S. Ducharme, and J. Feinberg, "Speed of the photorefractive effect in a $BaTiO_3$ crystal", *J. Appl. Phys.*, Vol. 56, 1984, pp. 839-842.

26) B. Fischer, "Theory of self-frequency detuning of oscillations by wave mixing in photorefractive crystals", *Opt. Lett.*, Vol. 11, 1986, pp. 236-238.

27) J. Goltz, C. Denz, H. Klumb, and T. Tschudi, " Four-wave mixing in photorefractive crystals with depleted pumps", *Opt. Lett.*, Vol. 13, pp. 321-323.

28) R.W.Boyd and G.Grynberg, In *Contemporary Nonlinear Optics*, Ed. G.P.Agrawal and R.W.Boyd, Boston, Academic Press, Inc., 1992, pp. 85-118.

29) P.M.Petersen, " Phase mismatching in a perturbational approach to degenerate four-wave mixing", *Optics Comm.*, Vol. 86, 1988, pp. 449-452.

30) H. Rajbenbach, B. Imbert, J.P. Huignard, and S. Mallick, "Near infrared four-wave mixing with gain and self starting oscillators with photorefractive GaAs", *Opt. Lett.*, Vol. 14, 1989, pp. 78-80.

31) T.J. Hall, and A.K. Powell, "Phase conjugations", In *Nonlinear Optics in Solids*, Ed. O. Keller, Berlin, Heidelberg: Springer-Verlag, 1990, pp. 235-264.

32) M. Cronin-Golomb, B. Fisher, J.O. White, and A. Yariv, "Theory and applications of four-wave mixing in photorefractive media", *IEEE J. Quantum Electron.*, Vol. QE-20, 1984, pp. 12-30.

33) B. Fisher, and S. Sternklar, "New optical gyroscope based on the passive phase conjugator", *Appl. Phys. Lett.*, Vol. 47, 1985, pp. 1-3.

34) B. Fisher, S. Sternklar, and S. Weiss, "Photorefractive oscillators", *IEEE J. Quantum Electron*, Vol. QE-25, 1989, pp. 550-569.

35) M. Cronin-Golomb, and A. Yariv, "Plane-wave theory of non-degenerated oscillation in the linear photorefractive passive phase-conjugate mirror", *Opt. Lett.* Vol. 11, 1986, pp. 242-244.

36) S. Kwong, A. Yariv, M. Cronin-Golomb, and B. Fisher, "Phase of phase conjugation and its effect in the double phase conjugate resonator", *J. Opt. Soc. Am. A,* Vol. 3, 1986, pp. 157-160.
37) P.M. Petersen, "Theory of one-grating nondegenerate four-wave mixing and its application to a linear photorefractive oscillator, *J. Opt. Soc. Am. B,* Vol. 8, 1991, pp. 1716-1722.
38) M.R. Belic, "Degenerate four-wave mixing as a Sturm Liouville problem" *Opt. Lett.*, Vol. 12, 1987, pp. 105-107.
39) K.R. MacDonald and J. Feinberg, "Enhanced four-wave mixing by use of frequency shifted optical wave in photorefractive $BaTiO_3$, *Phys. Rev. Lett.*, Vol. 55, 1985, pp.821-824.
40) J. Feinberg, and K.R. MacDonald, "Phase-conjugate mirrors and resonators with photorefractive materials", In *Photorefractive Materials and Their Applications II,* Ed. P. Günter and J.P. Huignard, Vol. 52 of Topics in Applied Physics, Berlin: Springer-Verlag, 1980, pp. 151-203.
41) J. Feinberg, and G.D. Bacher, "Self-scanning of a continuous wave dye laser having a phase-conjugating resonator cavity", *Opt. Lett.*, Vol. 9, 1984, pp. 420-422.
42) J.O. White and A. Yariv, "Real-time image processing via four-wave mixing in a photorefractive medium", *Appl. Phys. Lett.*, Vol 37, 1980, pp. 5-7
43) B. Edvold, P.E. Andersen, P. Buchhave, and P.M. Petersen, "Polarization properties of a photorefractive $By_{12}SiO_{20}$ crystal and their application in an optical correlator", submitted to *IEEE J. Quantum Electron.*
44) P.M. Petersen, B. Edvold, P. Buchhave, P.E. Andersen, and A. Marrakchi, "Photorefractive particle image velocimetry: performance enhancement with silicon oxide crystals", *Opt. Lett.*, Vol. 17, 1992, pp. 619-621.
45) P. Buchhave, M.L. Jakobsen, C.H. Westergaard, and J.N. Sørensen, "PIV: measurement of the early transition of a rotating flow in a closed cylinder", Proceedings from *Fourth Int. Conf. on Laser Anemometry, Advances and Applications,* Cleveland, Ohio, 1991, pp. 45-51.
46) T. Mørck, P.E. Andersen, and C.H. Westergaard, "Processing speed of photorefractive optical correlators in PIV-processing", Proceedings from *Sixth Int. symposium on applications of laser techniques to fluid mechanics*, Lisbon, Portugal, 1992, pp.27.2.1-27.2.6
47) J.M.Coupland and N.A.Halliwell,"Particle image velocimetry: Rapid transparency analysis using optical correlation", *Appl. Optics*, Vol. 27,1988, pp. 1919-1921
48) A. Marrakchi, R.V. Johnson, and A.R. Tanguay, Jr., "Polarization properties of photorefractive diffraction in electrooptic and optically active selenite crystals (Bragg regime)", *J. Opt. Soc. Am. B.*, Vol. 3, 1986, pp. 321-336.
R.V. Johnson, and A.R. Tanguay, Jr., "Optical beam propagation method for birefringent phase grating diffraction", *Opt. Eng.*, Vol. 25, 1986, pp. 235-249.
49) H. Kogelnik, "Coupled wave theory for thick hologram gratings", *Bell Syst. Tech. J.,* Vol. 48, 1969, pp. 2909-2947.
50) R.L. Powell, and K.A. Stetson, " Interferometric vibration analysis by wavefront reconstruction", *J. Opt. Soc. Am.,* Vol. 55, 1965, pp. 1593-1598.

51) J.N. Butters, and J.A. Leendertz, "Speckle pattern and holographic techniques in engineering metrology", *Opt. Laser Technol.*, Vol. 3, 1971, pp. 26-30.
52) J.P. Huignard, J.P. Herriau, and T. Valentin, "Time average holographic interferometry with photoconductive electro optic $Bi_{12}SiO_{20}$ crystals", *Appl. Optics*, Vol. 16, 1977, pp. 2796-2798.
53) A. Marrakchi, J.P. Huignard, and J.P. Herriau, "Application phase conjugation in $Bi_{12}SiO_{20}$ crystals to mode pattern visualization of diffuse vibrating structures", *Optics Comm.*, Vol. 34, 1980, pp. 15-18.
54) H. Rohleder, P.M. Petersen, and A. Marrakchi, " Complete amplitude and phase mapping in photorefractive time-average interferometry", Submitted to *CLEO 93*.
55) P. Mills, and E. Paige, "Holographically formed, highly selective, infrared filter in iron-doped lithium niobate", *Electron. Lett.*, Vol. 21, 1985, pp. 885-886.
56) P.M. Petersen, P.M. Johansen, and T. Skettrup, "Laser-induced interference filters in photorefractive materials", In *Photorefractive Materials, Effects and Devices,* Optical Soc. of America, *Technical Digest Series,* Vol. 14, 1991, pp. 195-198.
57) P. Yeh, *Optical waves in layered media,* J. Wiley and Sons, New York, 1988.

Studies in Classical
and Quantum Nonlinear Optics

ISBN 1-56072-168-5

TOWARDS EXACT QUANTUM THEORY OF STIMULATED RAMAN SCATTERING

Valery I. Rupasov
Landau Institute for Theoretical Physics, Russian Academy of Sciences,
Kosygina str. 2, 117334 Moscow, Russia

Abstract

Recently, it has been shown that the quantum versions of some traditional models used for a description of stimulated Raman scattering (SRS) can be solved exactly without any approximation. In this article we summarize the results which have been obtained for a point-like geometry of a resonance nonlinear medium and discuss the SRS problem for a "long" medium within the framework of the theory of completely integrable quantum systems.

1 Introduction

The modern theory of stimulated Raman scattering (SRS) of light in resonant media is based on investigations of appropriate models of the classical field theory [1, 2] and does not take into account the quantum (spontaneous) nature of the phenomenon. Moreover, there are some questions and problems, for example, the problem of scattered photons statistics, which can be studied only within the framework of the quantum field theory.

Recently, the exact quantum theory of SRS by a system of two-level atoms has been developed [3, 4] for the case of a point-like geometry of media, when the effects of light propagation in a medium are insignificant and may be ignored. Therefore, the developed theory may be adequate in the two cases of a medium geometry:

- the Dicke geometry [5], when all atoms are contained in a small volume of radius $r_0 \ll \lambda$, where λ is a typical wavelength of light;
- a "short" extended medium, when atoms are contained in a pencil-shape sample of length $l \gg \lambda$, but $l \ll l_p$, where l_p is a typical

size of light pulses and, hence, variation of "slow" envelopes of pulses over the length of the sample may be ignored (spatially-homogeneous approximation) [6].

The following main results of the quantum theory have been obtained and are discussed here:

- The critical phenomena in SRS process when the number of atoms M in an atomic system is equal to some critical value $M_{cr} = 2\pi/J$ (where J is the coupling constant) are predicted. If $M < M_{cr}$, the total number of Stokes photons created in the course of scattering N_S is, naturally, proportional to M and the population of an atomic system can be completely inverted. If $M > M_{cr}$, the value of N_S decreases in inverse proportion to M: $N_S \propto M_{cr}^2/M$, and the population of an atomic system can not be completely inverted for any duration and intensity of an incident laser pulse. At the point $M = M_{cr}$ the function $N_S(M)$ is discontinuous: $N_S(M_{cr} - 0) - N_S(M_{cr} + 0) = M/2$.
- The exact solution of the quantum scattering problem allows us to study quantum fluctuations and statistical properties of the scattered field. It was shown that the fluctuations of the number of Stokes photons are very large at the initial stage of scattering, and the intensity of the Stokes pulses undergoes strong random oscillations at an initial stage of SRS.
- In the case of a "long" extended media ($l > l_p$), when the spatially-homogeneous approximation must be abandoned, the SRS problem can be formulated and studied within the framework of the six-vertex model of 2D statistical mechanics [7].

At last, it is necessary to emphasize that all of the quantum models studied in this article should be considered as the quantum versions of the classical models traditionally used in the modern "classical" theory of SRS [1, 2].

2 Formulation of the problem

A laser pulse with a carrier frequency ω_L is incident on a system containing M two-level atoms which have a dipole-forbidden transition $1 \to 2$ with a frequency ω_{12}. The scattered field contains an unshifted component at the frequency ω_L and two shifted (Raman) components: a Stokes component with a frequency $\omega_S = \omega_L - \omega_{12}$ and an anti-Stokes component with a frequency $\omega_{AS} = \omega_L + \omega_{12}$. Furthermore, some of the incident pulse is not scattered and forms a transmitted laser beam.

We assume that all the atoms are in a sphere of radius r_0 which is much smaller than the typical wavelength of the light, λ. The existence of the small parameter $r_0/\lambda \ll 1$ makes it possible to expand the field operators in spherical harmonics [8] and to retain only the electric-dipole harmonic with the angular momentum $j = 1$ in the field-atomic system interaction. If levels 1 and 2 are nondegenerate, a specification of the polarization of the incident light fixes the second quantum number of the dipole photons, the angular momentum projection m. We assume that the incident light is linearly polarized. Choosing a quantization axis for the atoms along the field polarization direction, we then have $m = 0$.

We distinguish three intervals along the frequency axis: a laser interval, a Stokes interval, and an anti-Stokes interval, which are centered on the points $\omega_L, \omega_S,$ andω_{AS}, respectively, and which have width $\Delta \sim \tau_p^{-1}$, where τ_p is the length of the pulses. We use the indices $L, S,$ andAS to specify that the dipole-photon operators $b(\omega) \equiv b(\omega, j = 1, m = 0)$ belong to one of these intervals. In the limit $\Delta \ll \omega_{12}$ all three harmonics are independent; i.e., the field operators with different indices commute with each other. Then, the field operators can be combined into an isotopic triplet

$$b_\sigma = \begin{pmatrix} b_{AS} \\ b_L \\ b_S \end{pmatrix} \tag{2.1}$$

with the commutation relations

$$[b_\sigma(k), b_\mu(q)] = 2\pi\,\delta_{\sigma\mu}\delta(k - q) \tag{2.2}$$

where the isospin variable σ (below we will say simply 'spin variable') takes on three possible values: $\sigma = AS, L, S$.

The system of M two-level atoms is described by the spin operator

$$\mathbf{R} = \sum_{a=1}^{M} \mathbf{r}_a = (R^x, R^y, R^z); \quad R^\pm = R^x \pm iR^y \tag{2.3}$$

with the ordinary commutation rules

$$[R^i, R^j] = e^{ijk}R^k \tag{2.4}$$

where the spin operator $\mathbf{r}_a$ ($r_a = 1/2$) corresponds to the ath atom.

In the terms of the operators $b_\sigma(k)$ the Hamiltonian of the system consisting of the dipole photons and the atoms is

$$H = \int_{-\infty}^{\infty} \frac{\mathrm{d}k}{2\pi} k b_\sigma^+(k) b_\sigma(k) - \int_{-\infty}^{\infty} \frac{\mathrm{d}k\mathrm{d}q}{(2\pi)^2} b_\sigma^+(k)[\sigma_{\sigma\mu}^+ R^- + \sigma_{\sigma\mu}^- R^+] b_\mu(q) \tag{2.5}$$

where a repeated spin index implies a summation, and the 3×3 matrices

$$\sigma^{+}=\begin{bmatrix}0 & J_{AS} & 0\\ 0 & 0 & J_S\\ 0 & 0 & 0\end{bmatrix},\quad \sigma^{-}=\begin{bmatrix}0 & 0 & 0\\ J_{AS} & 0 & 0\\ 0 & J_S & 0\end{bmatrix} \tag{2.6}$$

act in the spin space of the field. The operators representing the total number of photons,

$$N=\int_{-\infty}^{\infty}\frac{\mathrm{d}k}{2\pi}b_{\sigma}^{+}(k)b_{\sigma}(k), \tag{2.7}$$

and the z component of the spin of the field+atoms system,

$$S^z=R^z+\int_{-\infty}^{\infty}\frac{\mathrm{d}k}{2\pi}b_{\sigma}^{+}(k)\sigma_{\sigma\mu}^{z}b_{\mu}(k), \tag{2.8}$$

where

$$\sigma^{z}=\begin{bmatrix}1 & 0 & 0\\ 0 & 0 & 0\\ 0 & 0 & -1\end{bmatrix}, \tag{2.9}$$

commute with the Hamiltonian. Therefore, we have omitted from the Exp. (2.5) the terms $\omega_L N$ and $\omega_{12}S^z$ and are counting the photon frequency from the center of the corresponding interval. Since the size of the incident laser pulse τ_{in} satisfies the inequality

$$\omega_L\tau_{in}\gg 1, \tag{2.10}$$

and this relation is not seriously violated for the components of the scattered field, we extended the integration in the above formulas to infinite limits and ignore the frequency dependence of the coupling constants J_S and J_{AS}.

Introducing the new field operators

$$\epsilon_{\sigma}(x)=\int_{-\infty}^{\infty}\frac{\mathrm{d}k}{2\pi}\,b_{\sigma}(k)\mathrm{e}^{ikx}=\begin{pmatrix}E_{AS}\\ E_L\\ E_S\end{pmatrix} \tag{2.11}$$

with the commutation relations

$$[\epsilon_{\sigma}(x),\epsilon_{\mu}(y)]=\delta_{\sigma\mu}\delta(x-y), \tag{2.12}$$

where x may be treated as a coordinate along any straight line going through the point-like atomic system, one can rewrite the Hamiltonian of the problem as

$$H = -\int_{-\infty}^{\infty} dx\{i\epsilon_\sigma^+(x)\frac{d}{dx}\epsilon_\sigma(x) + \delta(x)\epsilon_\sigma^+(x)[\sigma_{\sigma\mu}^+ R^- + \sigma_{\sigma\mu}^- R^+]\,\epsilon_\mu(x)\} \tag{2.13}$$

Thus, the SRS of light pulses by a point-like system of atoms in the dipole ($r_0/\lambda \ll 1$) resonant ($\omega_L\tau_{in} \gg 1$) approximation is described by an effectively one-dimensional quantum field theory. This type of one-dimensionalization is characteristic of several problems (e.g., the theory of magnetic alloys [9, 10] and the Dicke superradiance [11]) in which a field interacts with a point impurity. In optical problems the role of an impurity is played by an atom or an ensemble of atoms concentrated in a small volume.

In contrast with the theory of magnetic alloys, where the entities of interest are the equilibrium (thermodynamic) properties of a particle+impurity system, the physical formulation of the spontaneous SRS problem corresponds to a many-body scattering problem, in whose initial state there are only laser photons. We will see in the following that the exact solution of this scattering problem can be found without the integrability of the model (2.13).

The model (2.13) can also be used to describe SRS in a "short" medium, when the field propagation effects may be ignored. In this case, the atomic subsystem is again described by the operator **R**, while the variable x takes on the meaning of the coordinate along the sample axis. Consequently, all the results for short medium [3] can be extracted from the results for the Dicke geometry.

In the classical limit and in the limit of continuous description of the nonlinear medium, the equations of motion for the dynamical variables of the field+atoms system completely coincide with the evolution equations used traditionally to describe SRS [1]. This coincidence takes place also in the case when the atomic levels are not depleted in the course of Raman scattering and a nonlinear medium can be described by Bose variables [2]. The corresponding quantum versions of classical models of SRS can be formulated and studied.

3 Scattering problem

3.1 One-particle scattering problem

Before we take up the solution of the general many-body scattering problem, let us examine the one-particle problem, in which a laser photon with a wave vector $\mathbf{k}$ ($|\mathbf{k}| \equiv k = \omega - \omega_L$) is propagating along the z axis and is scattered by m-fold excited system of atoms at the point $z = 0$. The initial state of the field+atoms system is obviously given by

$$|\text{In} >= \left[\frac{(M-m)!}{m!M!}\right]^{1/2} a^+(\mathbf{k})(R^+)^m|0>, \quad <\text{In}|\text{In}>= 1, \qquad (3.1)$$

where

$$a^+(\mathbf{k}) = \int \mathrm{d}z\phi(z)\mathrm{e}^{ikz}a^+(z),$$

and $\phi(z)$ is some arbitrary normalized wave function of the incident photon. The vacuum state is defined as the state in which there are no photons, and all of the atoms are in the ground state:

$$R^-|0>= 0, \quad R^z|0>= -\frac{M}{2}|0>. \qquad (3.2)$$

Using an expansion of the operator $a(\mathbf{k})$ in spherical harmonics with a center at the point $z = 0$, one can write it as the sum of two terms:

$$a^+(\mathbf{k}) = \beta^+(k) + \left(\frac{\sigma_1}{S_0}\right)^{1/2} b_L(k), \qquad (3.3)$$

where the operator $\beta^+(k)$ incorporates all the harmonics with angular momentum $j \geq 2$ which do not interact with the atomic subsystem, $\sigma_1 = 3\pi/\omega_L^2$ is the impact area of the dipole particles [8], and S_0 is the cross-section area of the incident light beam.

The Exp.(3.3) converts our original problem (3.1) into the scattering problem of a dipole particle:

$$|\text{in} >= b_L(k)(R^+)^m|0> \qquad (3.4)$$

By definition its solution is given by

$$|\text{out} >= S|\text{in}> \qquad (3.5)$$

where the S-matrix of scattering of the particle by the impurity is found by solving a one-particle Schrödinger equation

$$(H - E)|\Psi >= 0 \tag{3.6}$$

where E is the eigenenergy of a one-particle state

$$|\Psi >= \int_{-\infty}^{\infty} \mathrm{d}x\, \Phi_{\sigma n}(x)\epsilon_{\sigma}^{+}(x)(R^{+})^{n}|0> \tag{3.7}$$

Then one gets the equation for the one-particle wave function

$$(i\frac{\mathrm{d}}{\mathrm{d}x} - E)\Phi_{\sigma n}(x) = \delta(x)[\sigma_{\sigma\mu}^{+}R_{nm}^{-} + \sigma_{\sigma\mu}^{-}R_{nm}^{+}]\Phi_{\mu m}(x) \tag{3.8}$$

whose solution is given by $(k = E)$

$$\Phi_{\sigma n}(x) = \mathrm{e}^{ikx}\begin{cases} \chi_{\sigma n} & , x < 0 \\ S_{\sigma\mu}^{nm}\chi_{\mu m} & , x > 0 \end{cases} \tag{3.9}$$

where $\chi_{\sigma n}$ is an arbitrary tensor independent on the spatial variable and the scattering matrix is

$$S_{\sigma\mu}^{nm} = \exp\left[i(\sigma_{\sigma\mu}^{+}R_{nm}^{-} + \sigma_{\sigma\mu}^{-}R_{nm}^{+})\right] = \exp\left[i(\sigma^{+}R^{-} + \sigma^{-}R^{+})\right] \tag{3.10}$$

The last equality in (3.10) represents the S-matrix as an operator acting on the spin index of field operators, for example:

$$(\sigma^{+})^{2}b_{L}^{+} = (\sigma^{-})^{2}b_{L}^{+} = 0 \tag{3.11}$$

$$\begin{aligned} \sigma^{+}b_{L}^{+} &= J_{AS}b_{AS}^{+}, \qquad \sigma^{-}b_{L}^{+} = J_{S}b_{S}^{+}, \\ \sigma^{-}b_{AS}^{+} &= J_{AS}b_{L}^{+}, \qquad \sigma^{+}b_{S}^{+} = J_{S}b_{L}^{+}. \end{aligned} \tag{3.12}$$

These and analogous relations follow from the matrix representation of the σ operators (2.6).

Let us rewrite (3.10) as the series

$$S = \sum_{n=0}^{\infty}\frac{i^{n}}{n!}(\sigma^{+}R^{-} + \sigma^{-}R^{+})^{n} \tag{3.13}$$

and consider the square of the operator:

$$(\sigma^{+}R^{-}+\sigma^{-}R^{+})^{2} = (\sigma^{+})^{2}(R^{-})^{2}(\sigma^{-})^{2}+(R^{+})^{2}+\sigma^{+}\sigma^{-}R^{-}R^{+}+\sigma^{-}\sigma^{+}R^{+}R^{-} \tag{3.14}$$

It is easy to see that the initial state (3.4) is an eigenstate of the operator (3.14). Indeed, according to (3.11), (3.12) the first two terms vanish, while the third and forth are equivalent to the operator $(J_S^2 R^- R^+ + J_{AS}^2 R^+ R^-)$. Hence, we have

$$(\sigma^+ R^- + \sigma^- R^+)^{2n} |\text{in}> = (J_S^2 R^- R^+ + J_{AS}^2 R^+ R^-)^n |\text{in}> \qquad (3.15)$$

Thus, acting on a state containing only laser photons, the expression for the S-matrix takes the form:

$$S = A + (\sigma^+ R^- + \sigma^- R^+)\, C, \qquad (3.16)$$

$$A = \cos\,(J_S^2 R^- R^+ + J_{AS}^2 R^+ R^-)^{1/2} \qquad (3.17)$$

$$C = i\,\frac{\sin\,(J_S^2 R^- R^+ + J_{AS}^2 R^+ R^-)^{1/2}}{(J_S^2 R^- R^+ + J_{AS}^2 R^+ R^-)^{1/2}} \qquad (3.18)$$

The operators A and C contain only the spin operators of the impurity and describe the probability amplitude of the elastic and inelastic (Raman) scattering, respectively.

Taking account of the harmonics which do not interact with the impurity, we finally find for the Out-state of the one-particle scattering problem:

$$|\text{Out}> = \left[\frac{(M-m)!}{m!M!}\right]^{1/2} D^+ (R^+)^m |0> \qquad (3.19)$$

where the operator

$$D^+ = a^+(\mathbf{k}) + \left(\frac{\sigma_1}{S_0}\right)^{1/2} \left[(A-1) b_L^+(k) + (\,J_S R^+ b_S^+(k) + J_{AS} R^- b_{AS}^+(k))\, C\right]$$

contains the operators of the transmitted and scattered components of the field.

Now we are able to evaluate the intensities of the scattered light which are found by averaging the corresponding density operator $\hat{I}_\sigma = \epsilon_\sigma^+ \epsilon_\sigma$ (there is no summation over σ) over the Out-state:

$$I_L(t) = 4 I_D(t) \sin^4 [g(m)/2], \qquad (3.20)$$

$$I_S(t) = I_D(t) J_S^2 (m+1)(M-m) \frac{\sin^2 [g(m)]}{g^2(m)}, \qquad (3.21)$$

$$I_{AS}(t) = I_D(t) J_{AS}^2 m (M-m+1) \frac{\sin^2 [g(m)]}{g^2(m)}, \qquad (3.22)$$

where

$$I_D(t) = \frac{\sigma_1}{S_0} I_0(t) \tag{3.23}$$

is the intensity of the dipole component of the incident laser pulse with an arbitrary intensity $I_0(t)$, and the function $g(m)$ is given by

$$g(m) = [J_S^2 (m+1)(M-m) + J_{AS}^2 m(M-m+1)]^{1/2} \tag{3.24}$$

A very important point here is the interference of the transmitted and scattered laser light. From mathematical point of view the interference is taken into account by the nonvanishing commutator of the transmitted and scattered components:

$$[a(\mathbf{k}), b_L^+(k)] = \left(\frac{\sigma_1}{S_0}\right)^{1/2} \tag{3.25}$$

As usual in scattering theory [8], we assume that the incident beam has a finite cross-section S_0 and that the measurements of the scattered light are carried outside the transmitted beam. For this reason, the commutator (3.25) must be taken into account only for operators b_L incorporated into the Out-state, while the operators b_L of measured physical values must be considered as commuting with operators of the transmitted wave $a(\mathbf{k})$.

As we will see below, the region of parameters

$$J_S M, \; J_{AS} M \sim 1 \tag{3.26}$$

is of physical interest. For macroscopically large values of M the terms

$$J_S^2 M, \; J_{AS}^2 M \sim M^{-1} \ll 1 \tag{3.27}$$

are negligibly small and will be omitted below.

If before scattering the atomic system was in the ground state ($m = 0$), one gets from (3.20)-(3.24) the relations:

$$I_L(t) = \frac{1}{4} I_D(t) J_S^4 M^2, \tag{3.28}$$

$$I_S(t) = I_D(t) J_S^2 M \tag{3.29}$$

reproducing the well-known results [12] that the scattering for the unshifted and Stokes components is respectively coherent ($\propto M^2$) and incoherent ($\propto M$).

3.2 Many-body scattering problem

In order to avoid tedious intermediate calculations in the derivation of expectation values of the scattered field it is convenient – using the fact that the S-matrix does not depend on the momentum of the particle – to introduce a discrete description of the field.

In the discrete model the field is described by nodes operators a_j with the commutation rules

$$[a_j, a_l] = \delta_{jl} \tag{3.30}$$

and the initial state of N-particle scattering problem is written as

$$|\mathrm{In}> = \left[\frac{(M-m)!}{m!M!}\right]^{1/2} \left(\prod_{j=1}^{N} a_j^+\right) (R^+)^m |0> \tag{3.31}$$

where the number of nodes, N, is equal to the number of incident laser photons:

$$N = \int_{-\infty}^{\infty} \mathrm{d}t I_0(t) \tag{3.32}$$

Singling out the dipole harmonic in the node operator,

$$a_j^+ = \beta_j^+ + \left(\frac{\sigma_1}{S_0}\right)^{1/2} b_L^+(j), \tag{3.33}$$

we can again reduce the original scattering problem (3.31) to that of solving a set of auxiliary problems containing some number of dipole photons ($N_D = 0 \div N$):

$$|\mathrm{in}> = \left(\prod_{j=1}^{N_D} b_L^+(j)\right) (R^+)^m |0> \tag{3.34}$$

The dipole harmonic of the light is described by a triplet of the fields with the commutation relations

$$[b_\sigma(j), b_\mu^+(l)] = \delta_{\sigma\mu}\delta_{jl} \tag{3.35}$$

and the scattering of jth particle by the impurity is described by the S-matrix

$$S_{j0} = A + (\sigma_j^+ R^- + \sigma_j^- R^+)C, \tag{3.36}$$

where the operators σ_j act in the spin space of jth node.

The out-state of the auxiliary problem is then given by

$$|\text{out}> = \left(\prod_j S_{j0}\right) |\text{in}> \tag{3.37}$$

Taking account of higher harmonics of the field we finally obtain

$$|\text{Out}> = \left[\frac{(M-m)!}{m!M!}\right]^{1/2} \left(\prod_{j=1}^{N} D_j^+\right) (R^+)^m |0>, \tag{3.38}$$

where

$$D_j^+ = a_j^+ + \left(\frac{\sigma_1}{S_0}\right)^{1/2} \left[(A-1)b_L^+(j) + (\, J_S b_S^+(j) R^+ \; + \; J_{AS} b_{AS}^+(j) R^- \,)\, C\right]. \tag{3.39}$$

The expression (3.38) describes a sequential independent scattering of particles by the impurity corresponding to the physical formulation of SRS problem. Therefore, a particle-particle scattering – arising as a result of scattering of different particles by the same impurity and determining the thermodynamic properties of exchange models [9, 10] – has been eliminated from our consideration. Namely this circumstance allowed us to obtain the exact solution of the many-particle scattering problem despite of the fact that the model under consideration is not integrable.

As a result of scattering, the state of the atomic system changes; i.e., the particle with the index $n+1$ is scattered by the impurity whose state arose as a result of the scattering of the preceding n particles. If relaxation processes do not have time to destroy the "memory" in the ensemble of atoms, the scattering amplitude of a particle will depend on its index, and stimulated (or cooperative) Raman scattering occurs.

4 Physical observables

The physical characteristics of the scattered light are determined by averaging the node operators over the Out-state.

First of all the intensity of the scattered components is of our interest. The corresponding expectation values are given as a result of averaging of the intensity operators $\hat{I}_\sigma(n) = b_\sigma^+(n) b_\sigma(n)$ (there is no summation over the spin index σ) and can be written as

$$I_\sigma(n+1) = I_0(n+1)\,\frac{(M-m)!}{m!M!}$$
$$\times \;< (R^-)^m \left(\prod_{j=1}^{n+1} D_j\right) \hat{I}_\sigma(n+1) \left(\prod_{j=1}^{n+1} D_j^+\right) (R^+)^m > \quad (4.1)$$

where $I_0(n)$ is the intensity of the incident light in the node representation. We also used here the unitarity of scattering, i.e., the commutation rule

$$[D_j, D_l^+] = \delta_{jl} \quad (4.2)$$

following from the definition (3.39).

Furthermore, we are calculating here the expectation value of the atomic system population which arises as a result of scattering of n incident photons

$$\mathcal{R}_n = \frac{(M-m)!}{m!M!} < (R^-)^m \left(\prod_{j=1}^{n} D_j\right) R^z \left(\prod_{j=1}^{n} D_j^+\right) (R^+)^m > \quad (4.3)$$

The angle brackets here and below mean averaging over the vacuum state.

To return to a continuous description of the field one has to replace the node index n in the expectation values by the function

$$n(t) = \int_{-\infty}^{t} dt' I_0(t') \quad (4.4)$$

where $I_0(t)$ is an arbitrary intensity of the incident laser light.

4.1 Recurrence-relation method

Let us consider the expectation value

$$Q_{n+1}^m = \frac{(M-m)!}{m!M!} < (R^-)^m \hat{Q}_{n+1} (R^+)^m > \quad (4.5)$$

of the "Heisenberg" operator

$$\hat{Q}_{n+1} = \left(\prod_{j=1}^{n+1} D_j\right) \hat{q}_{n+1} \left(\prod_{j=1}^{n+1} D_j^+\right), \quad (4.6)$$

where $\hat{q}_{n+1}$ is some arbitrary node operator (the "Schrödinger" operator).

Splitting out the most-right operator D_1^+ and the most-left one D_1 in Exp.(4.5) and acting by these operators on the states $(R^+)^m|0>$ and $<0|(R^-)^m$, respectively, one finds the following recurrence relation for the quantities Q_n^m:

$$Q_{n+1}^m = Q_n^m + \frac{\sigma_1}{S_0}\Big\{ - \sin^2[g(m)]\,Q_n^m + J_S^2(m+1)(M-m)\frac{\sin^2[g(m)]}{g^2(m)}Q_n^{m+1} + J_{AS}^2\,m(M-m+1)\frac{\sin^2[g(m)]}{g^2(m)}Q_n^{m-1}\Big\} \tag{4.7}$$

The structure of these relations is determined only by the structure of the D-operators and does not depend on the type of a node operator $\hat{q}$. The type of a node operator determines only the initial condition

$$Q_0(m) = Q_{n+1}^m|_{n=0} \tag{4.8}$$

which is required to solve the recurrence relations. It is obvious that the solution of the one-particle scattering problem can be used as the initial conditions like (4.8).

Let us – as the first approximation of our analysis – treat the expectation values Q_n^m as a function of continuous variables n and m. Then, one can convert the recurrence relations to the differential equation

$$\left[\frac{\partial}{\partial n} - c(m)\frac{\partial}{\partial m}\right] Q(n,m) = 0 \tag{4.9}$$

$$c(m) = \frac{\sigma_1}{S_0}\frac{J_S^2 - J_{AS}^2}{J_S^2 + J_{AS}^2}\sin^2[g(m)] \tag{4.10}$$

with the initial condition

$$Q(n,m)|_{n=0} = Q_0(m) \tag{4.11}$$

Here we have also taken into account (3.26), (3.27).

If before scattering the atomic system was in the ground state, the general solution of the Cauchy problem (4.9)-(4.11) is written as

$$Q(n,m) = Q_0[m(t)] \tag{4.12}$$

where the function $m(t)$ is determined by the equation

$$\int_0^{m(t)} \frac{dm'}{\sin^2[g(m')]} = \frac{\sigma_1}{S_0}\frac{J_S^2 - J_{AS}^2}{J_S^2 + J_{AS}^2} n(t) = \frac{J_S^2 - J_{AS}^2}{J_S^2 + J_{AS}^2}\int_{-\infty}^{t} dt' I_D(t') \tag{4.13}$$

Taking into account the expressions (3.20)-(3.22) as the corresponding initial conditions we finally obtain

$$I_L(t) = 4I_D(t)\sin^4[g(m(t))/2], \tag{4.14}$$

$$I_S(t) = \frac{J_S^2}{J_S^2 + J_{AS}^2} I_D(t)\sin^2[g(m(t))], \tag{4.15}$$

$$I_{AS}(t) = \frac{J_{AS}^2}{J_S^2 + J_{AS}^2} I_D(t)\sin^2[g(m(t))], \tag{4.16}$$

$$\mathcal{R}(t) = -M/2 + m(t) \tag{4.17}$$

4.2 Quantum and correlation effects

The expressions (4.12)-(4.17) for one-point correlation functions in the many-body problem are completely identical to the corresponding expressions for the one-particle problem of the scattering by m-fold excited atomic system. The only manifestation of the many-body nature of the problem is the time dependence of m or, more precisely, the dependence of m on the number of incident photons which have been scattered by the time t.

To estimate the role of many-body effects, including quantum fluctuations, let us evaluate the two-point correlation function

$$Q(n_1, n_2) = < \hat{Q}_{n_1}\hat{Q}_{n_2} >, \tag{4.18}$$

where the Heisenberg operator is given by (4.6). The corresponding recurrence relations for the quantities $Q^m_{n_1,n_2}$ can again be converted into the differential equation for the function of the continuous variables $Q(n_1, n_2, m)$:

$$\frac{\partial Q}{\partial n_1} + \frac{\partial Q}{\partial n_2} - c(m)\frac{\partial Q}{\partial m} = 0 \tag{4.19}$$

with the initial condition

$$Q(n_1, n_2, m)|_{n_2=0} = Q_0(m)Q(n_1, m) \tag{4.20}$$

where the function $Q_0(m)$ is the initial condition in the Cauchy problem for one-point correlation functions (4.11). while the function $Q(n_1, m)$ is its solution. The solution of (4.19), (4.20) is found as the product

$$Q(t_1, t_2) = Q(t_1)Q(t_2) \tag{4.21}$$

of one-point correlations functions.

Thus, the transformation from exact recurrence relations to the differential equation results in the neglect of correlation effects. All of the multipoint correlation functions split up into a product of one-point correlators. The expressions (4.12)-(4.17) describe only average statistical values of physical observables, and information about their fluctuations has been lost in the transformation to differential equations. In order to study correlation and statistical properties of the scattered light we thus need a more careful analysis of the Out-state structure that will be done in what follows.

5 Critical phenomena

The intensity of the scattered field and the population of the atomic system are determined by the function $m(t)$, the equation for which, (4.13), can be rewritten as

$$\frac{\mathrm{d}}{\mathrm{d}t} m(t) = \frac{J_S^2 - J_{AS}^2}{J_S^2 + J_{AS}^2} I_D(t) \sin^2[g(m)] \tag{5.1}$$

with the asymptotic initial condition

$$m(t \to -\infty) \to 0 \tag{5.2}$$

Providing that we confine ourselves here by the physically interesting case $J_S > J_{AS}$, the right side of Eq. (5.1) is positive, so $m(t)$ is a monotonically increasing, bounded $(0 < m(t) < M)$ function.

The total numbers of laser, Stokes, and anti-Stokes scattered photons,

$$N_\sigma = \int_{-\infty}^{\infty} \mathrm{d}t I_\sigma(t)$$

are given by

$$N_L = \frac{J_S^2 + J_{AS}^2}{J_S^2 - J_{AS}^2} \int_0^\mu \mathrm{d}m \, \tan^2[g(m)/2] \tag{5.3}$$

$$N_S = \frac{J_S^2}{J_S^2 - J_{AS}^2}\mu \tag{5.4}$$

$$N_{AS} = \frac{J_{AS}^2}{J_S^2 - J_{AS}^2}\mu \tag{5.5}$$

where $\mu \equiv m(t \to +\infty)$ is the maximum value of the population of the atomic system $m(t)$, which is not always the same as the number of atoms, M.

It can be shown that the time scale of the evolution of $m(t)$ is determined by

$$\tau_s = \frac{2\tau_{\text{in}}}{(J_S^2 - J_{AS}^2)N_D M} \tag{5.6}$$

where τ_{in} and N_D are the duration and the number of the dipole photons in the incident laser pulse. The parameter range $(J_S^2 - J_{AS}^2)N_D M \ll 1$ is thus of no physical interest.

The character of the time evolution of the atomic system population and, hence, the qualitative features of the SRS process are completely determined by the maximum magnitude

$$G \equiv \max[g(m)] = \frac{M}{2}(J_S^2 + J_{AS}^2)^{1/2} \tag{5.7}$$

of the function $g(m)$, and below we discuss all of the three possible regimes of the scattering.

5.1 Pre-critical regime

5.1.1 The given-field limit

Let us first consider the simplest case $G \ll 1$. Then, the function $g(m)$ is also much less than unit for any m that allows us to replace the function $\sin[g(m)]$ by its argument. Then we immediately obtain

$$m(t) = \frac{M}{1 + \exp[-2(t - t_0)/\tau_s]}, \tag{5.8}$$

$$I_L(t) = \frac{I_D}{64}(J_S^2 + J_{AS}^2)^2 M^4 \cosh^{-4}[(t - t_0)/\tau_s], \tag{5.9}$$

$$I_S(t) = \frac{I_D}{4} J_S^2 M^2 \cosh^{-2}[(t - t_0)/\tau_s], \tag{5.10}$$

$$I_{AS}(t) = \frac{I_D}{4} J_{AS}^2 M^2 \cosh^{-2}[(t - t_0)/\tau_s], \tag{5.11}$$

where the intensity of the incident light does not depend on time, $I_D =$ constant and the parameter t_0 determining the shift of the maximum magnitudes $I_\sigma^{\max}$ of the intensities $I_\sigma(t)$ is given by

$$t_0 = (\tau_s/2) \ln M. \tag{5.12}$$

At $J_{AS} = 0$, the expressions (5.8), (5.10) and (5.12) become the same as the results obtained by Rautian and Chernobrod [6] for the case of a "short" medium in the given-field approximation. Thus, the condition $G \ll 1$ should be considered as a quantitative criterion of the validity of the given-field approximation.

Moreover, in the given-field approximation the laser component of the field is eliminated from consideration, and the Hamiltonian of SRS problem is reduced (at $J_{AS} = 0$) to the Hamiltonian of the Dicke superradiance problem [5]. Thus, all the results of the Dicke superradiance theory can be extracted from the SRS theory in the limit $G \ll 1$.

In the course of scattering the atomic system population is completely inverted, i.e., $\mu = M$ and the magnitudes $I_\sigma^{\max}$ are reached at $m = M/2$. The total numbers of Stokes and anti-Stokes photons are proportional to the number of atoms, while the total number of scattered laser photons is proportional to M^3:

$$N_L = \frac{(J_S^2 + J_{AS}^2)^2}{J_S^2 - J_{AS}^2} \frac{M^3}{24} \tag{5.13}$$

At last, it is necessary to stress the extraordinary forth power of the number of atoms in the expression (5.9) for the laser component of the scattered field, $I_L^{\max} \propto M^4$. So, here we deal with a super-coherent scattering process.

5.1.2 Double-humped pulses

The first qualitative change in the structure of the scattered field takes place when $G > \pi/2$. In this case the maximum of the function $\sin[g(m)]$ is reached at the two points $m = m_{1,2}$, where $m_{1.2}$ are found from the condition $g(m) = \pi/2$

$$m_{1,2} = \frac{M}{2}\left[1 \pm \sqrt{1 - (M_0/M)^2}\right], \tag{5.14}$$

and the value M_0 is defined by the condition

$$G(M_0) = \frac{\pi}{2} = \frac{M_0}{2} (J_S^2 + J_{AS}^2)^{1/2} \tag{5.15}$$

For $M > M_0$, the Stokes and anti-Stokes pulses become double-humped, and the distance between humps grows with increasing M. At the moments of time $t_{1,2}$ corresponding to values $m_{1,2}$, the intensities of the Raman components reach their maxima

$$I_S^{\max} = \frac{J_S^2}{J_S^2 + J_{AS}^2} I_D, \tag{5.16}$$

$$I_{AS}^{\max} = \frac{J_{AS}^2}{J_S^2 + J_{AS}^2} I_D \tag{5.17}$$

and do not depend on M.

The population of the atomic system is completely inverted in the course of scattering ($\mu = M$), but the total number of laser photons is the complicated function of M [see (5.3)].

5.2 Post-critical regime

The critical change in the scattering process occurs at $M = M_{\text{cr}}$, where $G(M_{\text{cr}}) = \pi$:

$$M_{\text{cr}} = \frac{2\pi}{(J_S^2 + J_{AS}^2)^{1/2}} = 2M_0 \tag{5.18}$$

For $M > M_{\text{cr}}$ the atomic system population is not completely inverted for the arbitrary intensity and duration of the incident laser pulse. At the point μ determined by the condition $G(\mu) = \pi$:

$$\mu = \frac{M}{2}\left[1 - \sqrt{1 - (M_{\text{cr}}/M)^2}\right] < \frac{M}{2} \tag{5.19}$$

the time evolution of the population and the creation of the Stokes and anti-Stokes photons is asymptotically finished, while the intensity of the scattered laser light reaches the unitary limit $4I_D$.

At the point $M = M_{\text{cr}}$, the function $\mu(M)$ is discontinuous:

$$\mu(M_{\text{cr}} - 0) - \mu(M_{\text{cr}} + 0) = M/2, \tag{5.20}$$

while for $M \gg M_{\text{cr}}$, the quantity $m(M)$ and, along with it, the values N_S and N_{AS} decrease in inverse proportion to M:

$$\mu \sim M_{\text{cr}}^2/M \tag{5.21}$$

The value μ is reached by the function $m(t)$ only asymptotically in the limit $t \to +\infty$ and the vicinity $m > \mu$ is completely forbidden.

Thus, the dynamical process of SRS undergoes the critical change with increasing parameter G: the discontinuous of the function μ (and N_S, N_{AS}) at the point $G = \pi$ and a transition of all of these function from the growth to the fall with increasing G.

6 Probability distribution function

Hitherto the quantum nature of a spontaneous creation of Stokes and anti-Stokes photons has not been taken into account, or, more precisely, we derived the exact recurrence relations which contain full information about the scattering process, but the transition to the differential equation has eliminated quantum fluctuations from the consideration and allowed us to evaluate only average values of physical observables. In order to study quantum-statistical properties of the scattered field we thus need a more careful analysis of the structure of the Out-state.

6.1 Scattering problem for "short" medium

In what follows we confine ourselves by the case of a short nonlinear medium and the resonant Raman scattering ($J_{AS} \ll J_S$). So, an anti-Stokes component is eliminated from our consideration.

In the model the operators of slow envelopes of a laser (L) and Stokes (S) fields are combined into a two-component isotopic spinor

$$\epsilon_\sigma(x) = \begin{pmatrix} E_L(x) \\ E_S(x) \end{pmatrix}, \qquad [\,\epsilon_\sigma(x), \epsilon_\mu^+(y)\,] = \delta_{\sigma\mu}\,\delta(x-y) \tag{6.1}$$

and a short system of M two-level atoms is again described by the spin operator $\mathbf{R}$ (2.3), (2.4).

The model Hamiltonian is

$$H = -\int_{-\infty}^{\infty} \mathrm{d}x\,\{i\epsilon_\sigma^+(x)\frac{\mathrm{d}}{\mathrm{d}x}\epsilon_\sigma(x) + J\delta(x)\,\epsilon_\sigma^+(x)\,[\,\sigma_{\sigma\mu}^+ R^- + \sigma_{\sigma\mu}^- R^+]\,\epsilon_\mu(x)\} \tag{6.2}$$

where $J \equiv J_S$; $\sigma^\pm = \sigma^x \pm i\sigma^y$ are constructed from the ordinary Pauli matrices $\sigma^i = (\sigma^x, \sigma^y, \sigma^z)$, and a summation over a repeated spin index $\sigma = L, S$ is implied. The first term of the Hamiltonian describes a free propagation of both components of the field in the positive direction of x axis (a sample axis) and the second one describes an anisotropic exchange interaction between spins of the field and the impurity located at the point $x = 0$.

Using the discrete description of the field, the initial state of the scattering problem can be written as

$$|\text{in}>=\prod_{j=1}^{N_L} E_L^+(j)|0>, \tag{6.3}$$

where the index $j = 1, \ldots, N_L$ enumerates nodes of a chain (N_L being the number of incident laser photons) and the operator $E_L^+(j)$ with the commutation rules

$$[\epsilon_\sigma(j)\,,\epsilon_\mu^+(k)] = \delta_{jk}\delta_{\sigma\mu}$$

creates a laser photon (a particle with spin "up") in jth node of chain. The vacuum state of the field+impurity system is defined by relations

$$\epsilon_\sigma(j)|0>=0; \quad R^-|0>=0; \quad R^z|0>=-(M/2)|0> \tag{6.4}$$

The scattering of jth laser photon by the impurity is determined by the scattering matrix

$$S_{j0} = A + \sigma_j^- C \tag{6.5}$$

where the operator

$$A = \cos\left(J\sqrt{R^-R^+}\right)$$

corresponds to an elastic channel of the scattering and the operator

$$C = R^+ \frac{i\sin(J\sqrt{R^-R^+})}{\sqrt{R^-R^+}}$$

corresponds to an inelastic (Raman) channel when a laser photon is converted into a Stokes photon (a particle with spin "down"). Thus, the final state of the field+impurity system is given by

$$|\text{out}>=\prod_{j=1}^{N_L}[A\,E_L^+(j) + C\,E_S^+(j)]|0> \tag{6.6}$$

In the initial state (6.3) the number of incident particles is given. For the following, it is convenient to transit from the Fock basis to the basis of coherent states, where only the average number of incident particles $\overline{N}_L$ is given. Then we have:

$$|\text{In}> \;=\; e^{-\overline{N}_L/2} \sum_{N_L=0}^{\infty} \left(\frac{\overline{N}_L^{N_L}}{N_L!}\right)^{1/2} \prod_{j=1}^{N_L} E_L^+(j)|0> \tag{6.7}$$

$$|\text{Out}> \;=\; e^{-\overline{N}_L/2} \sum_{N_L=0}^{\infty} \left(\frac{\overline{N}_L^{N_L}}{N_L!}\right)^{1/2} \prod_{j=1}^{N_L}[A\,E_L^+(j) + CE_S^+(j)]|0> \tag{6.8}$$

As the coherent states basis is complete (and even overcomplete), this transition does not restrict the generality of our consideration. Any initial state of an incident field can be represented as a superposition of the coherent states and then, using the problem solution for an arbitrary vector of the coherent states basis, statistical properties of the scattered field can be derived by the standard way [13].

6.2 Probability distribution function

The product in the Fock out-state (6.6) can be represented as the sum of terms

$$|\mathrm{out}>=\sum_{n=0}^{n=N_L}|\mathrm{n}_S> \tag{6.9}$$

where $|\mathrm{n}_S>$ is the sum all of the states containing n operators E_S^+. The state with no Stokes photons is obviously given by

$$|0_S>=A^{N_L}\prod_{j=1}^{N_L}E_L^+(j)|0> \tag{6.10}$$

and, hence, the probability to create no Stokes photons is

$$\begin{aligned}P_0 &= \mathrm{e}^{-\overline{N}_L}\sum_{N_L=0}^{\infty}\frac{\overline{N}_L^{N_L}}{N_L!}<0_S|0_S>=\mathrm{e}^{-\overline{N}_L}\sum_{N_L=0}^{\infty}\frac{\overline{N}_L^{N_L}}{N_L!}[\cos^2(J\sqrt{M}\,)]^{N_L}\\ &= \exp[-\overline{N}_L\sin^2(J\sqrt{M}\,)]\end{aligned} \tag{6.11}$$

The state with one Stokes photon is given by

$$|1_S>=\sum_{k=1}^{N_L}\left[A^{(N_L-k)}C\,A^{(k-1)}\prod_{j_2=k+1}^{N_L}E_L^+(j_2)\,E_S^+(k)\prod_{j_1=1}^{k-1}E_L^+(j_1)\right]|0> \tag{6.12}$$

and we have for the probability to create one Stokes photon as a result of scattering of $\overline{N}_L$ laser photons:

$$\begin{aligned}P_1 &= \mathrm{e}^{-\overline{N}_L}\sum_{N_L=0}^{\infty}\left(\frac{\overline{N}_L^{N_L}}{N_L!}\right) \\ &\times \left[\sum_{k=1}^{N_L}[\cos^2(J\sqrt{2(M-1)}\,)]^{(N_L-k)}\sin^2(J\sqrt{M})[\cos^2(J\sqrt{M})]^{(k-1)}\right]\end{aligned} \tag{6.13}$$

The structure of this expression is quite obvious. The term

$$\sin^2(J\sqrt{M})$$

is the probability of an inelastic scattering of kth photon, the term

$$\left[\cos^2(J\sqrt{M})\right]^{(k-1)}$$

is the probability of an elastic scattering of the preceding $(k-1)$ photons, and, at last, the term

$$\left[\cos^2(J\sqrt{2(M-1)})\right]^{(N_L-k)}$$

is the probability of an elastic scattering of the following $(N_L - k)$ photons, but now, naturally, by the one-fold excited atomic system.

The summations in (6.13) can be carried out and an explicit expression for P_1 can be obtained.[1] For this purpose, let us change the summation order in (6.13) and rewrite it as

$$\begin{aligned} P_1 &= \overline{N}_L \sin^2(J\sqrt{M})e^{-\overline{N}_L} \\ &\times \sum_{k_1=0}^{\infty}\left[[\overline{N}_L\cos^2(J\sqrt{M})]^{k_1}\sum_{k_2=0}^{\infty}\frac{[\overline{N}_L\cos^2(J\sqrt{2(M-1)})]^{k_2}}{(k_1+k_2+1)!}\right] \end{aligned} \tag{6.14}$$

The sum over the index k_2 defines the special function [14]

$$E_{\alpha,\beta}(z) = \sum_{k=0}^{\infty}\frac{z^k}{\Gamma(\alpha k+\beta)} \tag{6.15}$$

(Γ is the Euler's gamma function), which coincides at the point $\beta = 1$ with the Mittag-Leffler function. Using the integral representation [14]

$$E_{\alpha,\beta}(z) = \int_L \frac{\mathrm{d}\tau}{2\pi i}\frac{\tau^{\alpha-\beta}}{\tau^{\alpha}-z}e^{\tau}, \tag{6.16}$$

[1]The possibility of an explicit calculation of such expressions is the main reason for using the coherent states basis.

where the contour of integration L runs parallel to an imaginary axis in the right half of the complex plane, and carrying out the rest summation over index k_1, we obtain:

$$P_1 = \int_L \frac{d\tau}{2\pi i} \frac{\omega_0}{(\tau + \omega_0)(\tau + \omega_1)} e^{\tau}, \tag{6.17}$$

where

$$\omega_n = \overline{N}_L \sin^2 [J\sqrt{(n+1)(M-n)}] \tag{6.18}$$

The corresponding summation procedure can be carried out in the case of an arbitrary number of Stokes photons. As a result one obtains the following expression for the probability distribution function – the probability to create n Stokes photons as a result of scattering of $\overline{N}_L$ incident laser photons:

$$P_n = \int_L \frac{d\tau}{2\pi i} \frac{\omega_0 \omega_1 \ldots \omega_{n-1}}{(\tau + \omega_0)(\tau + \omega_1)\ldots(\tau + \omega_n)} e^{\tau} \tag{6.19}$$

or, introducing a new variable of integration $z = \tau/\overline{N}_L$,

$$P_n(t) = \int_L \frac{dz}{2\pi i} \frac{\Omega_0 \Omega_1 \ldots \Omega_{n-1}}{(z + \Omega_0)(z + \Omega_1)\ldots(z + \Omega_n)} \exp[\overline{N}_L(t) z] \tag{6.20}$$

$$\Omega_n = \sin^2 [J\sqrt{(n+1)(M-n)}]. \tag{6.21}$$

These expressions are suitable if before scattering the atomic system has been in the ground state. For the following, it is necessary to know the probability distribution function $P_n^{(m)}(t)$ corresponding to the case of an initially m-fold excited system. Repeating the above procedure of calculations we obtain the quite analogous expression:

$$P_n^{(m)}(t) = \int_L \frac{dz}{2\pi i} \frac{\Omega_m \Omega_{m+1} \ldots \Omega_{m+n-1}}{(z + \Omega_m)(z + \Omega_{m+1})\ldots(z + \Omega_{m+n})} \exp[\overline{N}_L(t) z], \tag{6.22}$$

coinciding with (6.20) at $m = 0$.

7 Dicke superradiance

The expression (6.20) can be treated as a Laplace representation of the probability to create the n Stokes photons $P_n(t)$ by the moment of time

t determining by

$$\overline{N}_L(t) = \int_{-\infty}^{t} \mathrm{d}t' I_0(t') \tag{7.1}$$

The Laplace images

$$P_n(z) = \frac{\Omega_0\Omega_1 \ldots \Omega_{n-1}}{(z+\Omega_0)(z+\Omega_1)\ldots(z+\Omega_n)} \tag{7.2}$$

for different n are connected by recurrence relations:

$$z\,P_n(z) = \Omega_{n-1}P_{n-1}(z) - \Omega_n P_n(z), \tag{7.3}$$

and returning to the temporal variable t we obtain a set of evolution equations for the probability distribution function $P_n(t)$:

$$\frac{\mathrm{d}}{\mathrm{d}t}P_n(t) = I_0(t)[\,\Omega_{n-1}P_{n-1}(t) - \Omega_n P_n(t)\,] \tag{7.4}$$

with the initial condition:

$$P_0(t \to -\infty) \to 1, \quad P_{n\neq 0}(t \to -\infty) \to 0 \tag{7.5}$$

Since the z-component of the total spin of the field+atoms system, whose operator

$$S^z = R^z + \int_{-\infty}^{\infty} \mathrm{d}x \epsilon_\sigma^+(x)\sigma_{\sigma\mu}^z\epsilon_\mu(x) \tag{7.6}$$

commutes with the Hamiltonian (6.2), is conserved in the course of scattering, the number of the created Stokes photons n is connected with z-component of the impurity spin $\mathcal{R}$ by the relation $\mathcal{R} = -M/2 + n$. Moreover, the value n may also be treated as a level of excitation (or population) of the atomic system produced by creation of the n Stokes photons. Then, the physical meaning of the equations (7.4) becomes quite obvious: the rate of changing the population of the nth level is determined by the difference between the probabilities of the transition from the $(n-1)$th level to the nth one and from the nth level to the $(n+1)$th one.

Dicke [5] was the first to suggest this type of rate equations to describe a spontaneous decay of an excitation of a point-like system of two-level atoms. Here we derived from the first principals the exact equations of motion for the atomic system population in the case of the Raman scattering for an arbitrary intensity of incident laser pulse. But, it is necessary to stress that these equations are suitable only in the case of a coherent state of an incident field and can be dramatically changed with changing incident field statistics.

To obtain the rate equations for the Dicke superradiance, let us reformulate our scattering problem and consider the problem of scattering of incident Stokes photons (all incident particles have the spin "down") by an atomic system m-fold excited at the initial moment of time. Then, the population of the atomic subsystem is only decreasing in the course of scattering. Assuming that the parameter JM is much less than unit, $JM \ll 1$, the intensity of incident Stokes field I_S^0 does not depend on time and combining I_S^0 and J into a new coupling constant $\gamma = J^2 I_S^0$, we obtain the equations:

$$\frac{\mathrm{d}}{\mathrm{d}t} P_n(t) = \gamma[\,(n+1)(M-n)P_{n+1}(t) - n(M-n+1)P_n(t)\,], \tag{7.7}$$

with the initial condition:

$$P_m(t=0) = 1, \quad P_{n \neq m}(t=0) = 0, \tag{7.8}$$

which have been extensively studied in papers dealing with the Dicke superradiance [15].

8 Correlation functions

The expectation values of the scattered field are determined by averaging the corresponding node operators over the Out-state (6.8). In this section we study the behavior of the average number of Stokes photons

$$< N > \equiv < \mathrm{Out}|\hat{N}|\mathrm{Out} > = \sum_{n=0}^{M-m} n P_n^{(m)}(t), \tag{8.1}$$

created in the course of scattering of $\overline{N}_L$ laser photons and the correlations functions at a single moment of time:

$$\begin{aligned} \mathcal{D}_k(t) &\equiv < \mathrm{Out}|(\hat{N}+1)(\hat{N}+2)\ldots(\hat{N}+k)|\mathrm{Out} > \\ &= \sum_{n=0}^{M-m} (n+1)(n+2)\ldots(n+k) P_n^{(m)}(t) \end{aligned} \tag{8.2}$$

where

$$\hat{N} = \sum_{j=1}^{N_L} E_S^+(j) E_S(j)$$

is the operator of the number of Stokes photons. Here the symbol $< \ldots >$ denotes averaging with the distribution function as a weight one, and the index of the Stokes component is omitted because only fluctuations of Stokes field are studied below.

In this section we consider the case of $G \equiv JM/2 \ll 1$ and, hence, $g(n) \ll 1$ for any n. In this case, one can replace the function Ω_n in (6.22) by its argument $g(n)$ and rewrite it as

$$P_n^{(m)}(t) = \int_L \frac{dz}{2\pi i} \frac{\exp[J^2 \overline{N}_L(t) z]}{(c-1)(1-d)} \frac{(-M+m)_n (m+1)_n (1)_n}{(c)_n (d)_n \, n!} \tag{8.3}$$

where

$$\begin{aligned} c(z) &= \sqrt{z + (\frac{M+1}{2})^2} - \frac{M+1}{2} + m + 2 \\ d(z) &= -\sqrt{z + (\frac{M+1}{2})^2} - \frac{M+1}{2} + m + 2 \end{aligned}$$

and

$$(a)_n = a(a+1)\ldots(a+n-1) = \frac{\Gamma(a+n)}{\Gamma(a)}$$

is the Pochhammer's symbol [16].

Using (8.3) one can represent the norm of Out-state and correlators $\mathcal{D}_k$ as the Laplace integrals

$$\begin{aligned} < \mathrm{Out}|\mathrm{Out} > &= \sum_{n=0}^{M-m} P_n^{(m)} \\ &= \int_L \frac{dz}{2\pi i} \frac{\exp[J^2 \overline{N}_L(t) z]}{(c-1)(1-d)} \, {}_3F_2 \left(\begin{matrix} -M+m, & m+1, & 1 \\ c, & d & \end{matrix} \right) \end{aligned} \tag{8.4}$$

$$\mathcal{D}_k = k! \int_L \frac{dz}{2\pi i} \frac{\exp[J^2 \overline{N}_L(t) z]}{(c-1)(1-d)} \, {}_3F_2 \left(\begin{matrix} -M+m, & m+1, & k+1 \\ c, & d & \end{matrix} \right) \tag{8.5}$$

of the generalized hypergeometric functions [16]

$$ {}_3F_2\left(\begin{matrix} a_1, & a_2, & a_3 \\ b_1, & b_2 & \end{matrix} \middle| \zeta\right) = \sum_{n=0}^{\infty} \frac{(a_1)_n(a_2)_n(a_3)_n}{(b_1)_n(b_2)_n} \frac{\zeta^n}{n!} \tag{8.6} $$

at the point $\zeta = 1$.

In our case of a negative integer parameter $a_1 = -M + m$, the function ${}_3F_2(\zeta)$ is reduced to a polynomial of the $(M - m)$ degree. The possibility of an explicit calculation of the polynomial ${}_3F_2(\zeta = 1)$ is determined by the parameter

$$ \Theta = (a_1 + a_2 + a_3) - (b_1 + b_2). $$

In the expression for the norm of Out-state $\Theta = -1$ and the polynomial in (8.4) belongs to the so-called Saalschütz's polynomials, whose magnitude at the point $\zeta = 1$ is expressed in the explicit form [16]

$$ {}_3F_2\left(\begin{matrix} -n, & a_2, & a_3 \\ b_1, & b_2 & \end{matrix}\right) = \frac{(b_1 - a_2)_n(b_1 - a_3)_n}{(b_1)_n(b_1 - a_2 - a_3)_n} \tag{8.7} $$

In our case

$$ {}_3F_2\left(\begin{matrix} -M + m, & m + 1, & 1 \\ c(z), & d(z) & \end{matrix}\right) = \frac{1}{z}(c - 1)(1 - d) \tag{8.8} $$

and we immediately obtain from (8.4), (8.8)

$$ < \mathrm{Out}|\mathrm{Out} >= \sum_{n=0}^{M-m} P_n^{(m)}(t) = \int_L \frac{dz}{2\pi i} \frac{\exp\left[J^2\overline{N}_L(t)z\right]}{z} = 1 \tag{8.9} $$

Hypergeometric polynomials in the expression (8.5) for correlation functions have the parameter $\Theta = k - 1 \geq 0$, therefore they are not expressed, unfortunately, through elementary functions and Exp.(8.5) cannot be significantly simplified, although numerous integral representations through special functions can be successfully used.

9 Local description of fluctuations

In the course of scattering quantum-statistical properties of the scattered field are significantly changed, therefore, it is more convenient to study a local behavior of the probability distribution function and correlation functions for a period of time much shorter than the duration of Stokes pulse. In other words, in order to study the distribution function properties near some point $n = n_0$ one can use a linear approximation replacing a small piece of a curve $\Omega(n)$ by a tangent line at this point.

At the initial stage of scattering ($n_0 = 0,\ n+m \ll M$) the linear approximations of the functions $\Omega(n)$ and $g(n)$ are the same. Therefore, both regimes of scattering can be studied by the unified way. In other words, at the initial stage quantum-statistical properties of the scattered field are insensitive to critical phenomena. From another side, at the final stage of the evolution of the atomic system population, quantum-statistical properties are different for different regimes of scattering and will be considered separately.

9.1 Initial stage

9.1.1 Fluctuations of the number of photons

Using the linear approximation

$$\Omega(n+m) = \sin^2\left[J\sqrt{(n+m+1)(M-m-n)}\right] \approx J^2M\,(n+m+1) \tag{9.1}$$

for an initial stage of scattering ($n_0 = 0,\ (n+m) \ll M$), we obtain

$$\begin{aligned} P_n^{(m)}(t) &= \int_L \frac{\mathrm{d}z}{2\pi i}\,\frac{(m+1)_n}{(z+m+2)_n}\,\frac{\exp\left[\xi(t)z\right]}{z+m+1} \\ &= \frac{(m+1)_n}{n!}\exp\left(-(m+1)\xi\right)\left[1-\exp\left(-\xi\right)\right]^n \end{aligned} \tag{9.2}$$

where

$$\xi(t) = J^2M\overline{N}_L(t) = J^2M\int_{-\infty}^{t}\mathrm{d}t' I_0(t') \tag{9.3}$$

It is convenient to redefine the correlators (8.2) as

$$\mathcal{D}_k(t) = \sum_{n=0}^{\infty}(n+m+1)(n+m+2)\dots(n+m+k)P_n^{(m)}(t) \tag{9.4}$$

Then, one obtains

$$\mathcal{D}_k(t) = (m+1)_k\int_L \frac{\mathrm{d}z}{2\pi i}\,\frac{\exp\left(\xi z\right)}{z+m+1}F(1, m+k+1; z+m+2; \varsigma = 1)$$

$$= \quad (m+1)_k e^{k\xi}, \tag{9.5}$$

where

$$F(a,b;c;\zeta) = \sum_{n=o}^{\infty} \frac{(a)_n (b)_n}{(c)_n} \frac{\zeta^n}{n!} \tag{9.6}$$

is the Gauss hypergeometric function and we used its expression at the point $\zeta = 1$ through Γ-functions:

$$F(a,b;c;\zeta=1) = \frac{\Gamma(c)\Gamma(c-a-b)}{\Gamma(c-a)\Gamma(c-b)} \tag{9.7}$$

Thus, the average number of Stokes photons grows exponentially

$$< N(t) >= \mathcal{D}_1 - (m+1) = (m+1)\left(e^{\xi(t)} - 1\right) \tag{9.8}$$

and fluctuates very strongly with the relative fluctuation

$$f \equiv \frac{\sqrt{< N^2 > - < N >^2}}{< N >} = \left(\frac{1}{m+1} + \frac{1}{< N >}\right)^{1/2} \tag{9.9}$$

9.1.2 Correlation of fluctuations

To study the behavior of the Stokes component intensity let us now calculate the function of correlation of fluctuations for two different moments of time:

$$\begin{aligned} K(t_1,t_2) &\equiv < \Delta N(t_1)\Delta N(t_2) > \\ &= < N(t_1)N(t_2) > - < N(t_1) >< N(t_2) > \end{aligned} \tag{9.10}$$

where $\Delta N = N - < N >$ is the fluctuation of the number of photons.

If before scattering the atomic system was in the ground state, the first term in the right side of (9.10) can be written as

$$< N(t)N(t+\tau) >= \sum_{m=0}^{\infty}\left[mP_m(\xi(t))\sum_{n=0}^{\infty}(m+n)P_n^{(m)}(\xi(t+\tau)-\xi(t))\right] \tag{9.11}$$

where $\tau > 0$. The distribution function $P_m(\xi(t))$ describes the probability to create m photons from the initial moment of time by the moment of time t, while $P_n^{(m)}(\xi(t+\tau)-\xi(t))$ describes the creation n photons during the period of time τ, taking into account that the atomic system was m-fold excited at the moment of time t. The parameter

$$\xi(t+\tau) - \xi(t) = J^2 M \int_t^{t+\tau} dt' I_0(t')$$

is proportional to the number of incident laser photons scattered during τ.

Using the general expression (9.5), one gets

$$K(t,t+\tau) = e^{\xi(t+\tau)}(e^{\xi(t)}-1) = < N(t) >< N(t+\tau) > + < N(t) > \tag{9.12}$$

As the correlation of fluctuations of the intensities is obviously related with $K(t_1,t_2)$ by

$$K_I(t_1,t_2) \equiv < \Delta I(t_1)\Delta I(t_2) > = \frac{\partial^2}{\partial t_1 \partial t_2}K(t_1,t_2), \tag{9.13}$$

we easily find

$$K_I(t,t+\tau) = < I(t) >< I(t+\tau) > \tag{9.14}$$

and, hence, the autocorrelation function of intensities is

$$Q(t_1,t_2) \equiv \frac{K_I(t_1,t_2)}{< I(t_1) >< I(t_2) >} = 1 \tag{9.15}$$

and does not depend on the time.

In the case of an initially m-fold excited atomic system the above calculations can be repeated and one gets

$$K(t,t+\tau) = (m+1)e^{\xi(t+\tau)}(e^{\xi(t)}-1) \tag{9.16}$$

$$K_I(t,t+\tau) = \frac{1}{m+1} < I(t) >< I(t+\tau) > \tag{9.17}$$

$$Q(t_1,t_2) = \frac{1}{m+1} \tag{9.18}$$

Thus, the initial stage of the scattering process is characterized by extremely large quantum fluctuations of the number of photons and the intensity of the Stokes field. The intensity of the Stokes pulses undergoes random oscillations with a great amplitude and its behavior cannot be described by any classical evolution equations.

It is necessary to emphasize also that the vicinity of large quantum fluctuations is extended to macroscopically large magnitudes of the number of the Stokes photons, because all the results obtained above are valid for macroscopically large $< N >$, provided that $< N > \ll M$, or $< N > \ll n^*$ for the second regime ($JM > 2\pi$) of the SRS.

9.2 Final stage

It is easy to see from (9.9) that the role of quantum fluctuations diminishes with the growth of the level of the atomic system excitation. To study the statistics of Stokes photons at the final stage of scattering one can use a linear approximation of the function $\Omega(n)$, but now only for the pre-critical regime of scattering ($JM < 2\pi$), because for the post-critical one ($JM > 2\pi$) the function $\Omega(n)$ has at the point $n = n^*$ the zero of the second order and cannot be linearized. Thus, two different regime of SRS – having the same statistics at the initial stage – have different quantum statistical properties at the final stage and should be studied separately.

9.2.1 Pre-critical regime

Near the point $n + m = M$ the function $\Omega(n + m)$ can be represented as

$$\Omega(n+m) \approx J^2 M (M - m - n) \tag{9.19}$$

Then, in the vicinity $(M - m - n) \ll M$ the probability distribution function is given by

$$\begin{aligned} P_n^{(m)}(t) &= \int_L \frac{\mathrm{d}z}{2\pi i} \frac{(-M+m)_n}{(1+m-M-z)_n} \frac{\mathrm{e}^{\xi(t)z}}{z+M-m} \\ &= \frac{(M-m)!}{n!(M-m-n)!} \mathrm{e}^{-(M-m)\xi} (\mathrm{e}^{\xi} - 1)^n \end{aligned} \tag{9.20}$$

where m is the initial level of the atomic system excitation. We also have

$$< N(t) > \; = \; (M - m)(1 - \mathrm{e}^{-\xi(t)}) \tag{9.21}$$

$$f \; - \; \frac{1}{\sqrt{< N >}} \sqrt{1 - \frac{< N >}{M - m}} \tag{9.22}$$

Thus, at the final stage of scattering quantum fluctuations are suppressed and here we deal with sub-Poisson statistics.

9.2.2 Post-critical regime

If the parameter $JM > 2\pi$ the function $\Omega(n + m)$ can be represented as the second-order polynomial:

$$\Omega(n+m) \approx J^2[(M/M_{\mathrm{cr}})^2 - 1](n^* - m - n)^2, \tag{9.23}$$

where

$$n^* = \frac{M}{2}[1 - \sqrt{1 - (M_{cr}/M)^2}\,].$$

Then, the probability distribution function is again given by an expression like (5.3):

$$P_n^{(m)}(t) = \int_L \frac{dz}{2\pi i} \frac{\exp[\xi(t)z]}{z + (m - n^*)^2} \frac{(-n^* + m)_n(-n^* + m)_n}{(c)_n\,(d)_n} \qquad (9.24)$$

with the redefined parameters

$$\begin{aligned} c(z) &= \sqrt{-z} - n^* + m + 1 \\ d(z) &= -\sqrt{-z} - n^* + m + 1 \\ \xi(t) &= J^2\overline{N}_L[(M_{cr}/M)^2 - 1] \end{aligned}$$

Thus, both the general case of the pre-critical regime (section 8) and the final stage of the post-critical one are described by the mathematically equivalent expressions (8.3) and (9.24).

10 Long nonlinear medium

We turn now to an analysis of the SRS problem in an extended system of two-level atoms, which has a length $l > l_p$, so the spatially-homogeneous approximation must be abandoned. The Hamiltonian of the problem should be rewritten as

$$\begin{aligned} H = & - \int_{-\infty}^{\infty} dx \left\{ i\epsilon_\sigma^+(x) \frac{d}{dx} \epsilon_\sigma(x) \right. \\ & + \left. J \sum_{a=1}^{M} \delta(x - x_a) \epsilon_\sigma^+(x)[\sigma_{\sigma\mu}^+ r_a^- + \sigma_{\sigma\mu}^- r_a^+] \epsilon_\mu(x) \right\} \qquad (10.1) \end{aligned}$$

where $\{x_a;\, 0 < x_a < l\}$ are coordinates of the atoms along the sample axis.

In the classical limit and in the limit of a continuous description of nonlinear medium, the equations of motion for the dynamical variables of the field+medium system are the same as the equations of the classical theory of SRS, which have been studied [1] by the inverse scattering method [17]. In our approach, the model is obviously to be integrable, since its Hamiltonian is the same – aside from some distinctions unessential for an integrability – as the Hamiltonian of the anisotropic Kondo model, whose integrability was established by Wiegmann [18]. So, we have an opportunity to apply to SRS problem some elegant mathematical techniques developed in the theory of integrable quantum systems [7, 9, 10].

The initial state of the scattering problem is

$$|\,\mathrm{in}>=\prod_{j=1}^{N} E_L(j)|0> \tag{10.2}$$

where again we are using the node description of the field. The scattering of the jth particle by the ath atom is described by the scattering matrix [18, 9]

$$S_{ja}=\frac{1}{2}(1\otimes 1+\sigma_j^z\otimes r_a^z)+\frac{b}{2}(1\otimes 1-\sigma_j^z\otimes r_a^z)+c(\sigma_j^+\otimes r_a^-+\sigma_j^-\otimes r_a^+), \tag{10.3}$$

where

$$b=\cos J; \qquad c=i\sin J, \tag{10.4}$$

and the symbol $\otimes$ means the tensor product of matrices.

It is convenient to treat (10.3) as a 2×2 matrix acting in the spin space of the field with the matrix elements consisting of atom operators $\mathbf{r}_a$:

$$S_{ja}=\begin{pmatrix} 1/2[(1+b)+(1-b)r_a^z] & c\,r_a^- \\ c\,r_a^+ & 1/2[(1+b)-(1-b)r_a^z] \end{pmatrix} \tag{10.5}$$

The scattering of the jth particle by the atomic subsystem is then described by the ordered scalar product of matrices (10.5)

$$\begin{aligned} L_j &= S_{j,M}S_{j,M-1}\dots S_{j,2}S_{j,1}=\begin{pmatrix} A & B \\ C & D \end{pmatrix} \\ &= \frac{1}{2}(1+\sigma_j^z)\otimes A+\frac{1}{2}(1-\sigma_j^z)\otimes D+\sigma_j^+\otimes B+\sigma_j^-\otimes C \end{aligned} \tag{10.6}$$

The out-state of the scattering problem is found as

$$|\,\mathrm{out}>=\prod_{j=1}^{N}[A\,E_L^+(j)+C\,E_S^+(j)\,]|\,0> \tag{10.7}$$

The structure of this expression is quite analogous to that of the out-state in the case of the point-like geometry of nonlinear medium, (3.38), (6.6), but now the operators A and C – in contrast to the corresponding operators in the previous sections – do not form closed algebra. Therefore, having the exact solution of the scattering problem, we are not able to evaluate any physical characteristics of the scattered field.

The scattering problem of our model can be formulated [4] in terms of the six-vertex model of 2D statistical mechanics [7]. In order to construct a closed algebra for the elements of the monodromy matrix L_j, (10.6), one can use one of the main ideas of the theory of exactly solved models of statistical mechanics to generalize the expression for the S-matrix (10.5). Namely, let us consider the S-matrix which depends on some additional spectral parameter λ,

$$S_{ja} = \frac{1}{\cosh 2\eta} \begin{pmatrix} \sin(\lambda + i\eta r_a^z) & (i \sinh 2\eta) r_a^- \\ (i \sinh 2\eta) r_a^+ & \sin(\lambda - i\eta r_a^z) \end{pmatrix}, \tag{10.8}$$

where $\cosh 2\eta = 1/\cos J$. At the point $\lambda = (\pi + 2i\eta)/2$, the generalized S-matrix is the same as the S-matrix of our problem. But now, the elements of the generalized monodromy matrix

$$L_j(\lambda) = S_{j,M}(\lambda) S_{j,M-1}(\lambda) \dots S_{j,2}(\lambda) S_{j,1}(\lambda) = \begin{pmatrix} A(\lambda) & B(\lambda) \\ C(\lambda) & D(\lambda) \end{pmatrix} \tag{10.9}$$

form closed algebra, for example

$$A(\lambda)C(\mu) = \frac{\sin(\lambda - \mu + 2i\eta)}{\sin(\lambda - \mu)} C(\mu) A(\lambda) + \frac{i \sinh 2\eta}{\sin(\mu - \lambda)} C(\lambda) A(\mu), \tag{10.10}$$

and we are able to generalize the out-state as

$$|\,\text{out}; \{\lambda_j\} >= \prod_{j=1}^{N} [A(\lambda_j) E_L(j) + C(\lambda_j) E_S(j)]\, |0>, \tag{10.11}$$

where $\{\lambda_j\}$ is the set of arbitrary different spectral parameters.

Using the recurrence-relation method, one can again derive recurrence relations and a differential equation for physical observables with the coefficient

$$c(m, \{\lambda_j\}) = \frac{\left\langle \prod_{j=1}^{m+1} B(\lambda_j^*) \prod_{j=1}^{m+1} C(\lambda_j) \right\rangle}{\left\langle \prod_{j=1}^{m} B(\lambda_j^*) \prod_{j=1}^{m} C(\lambda_j) \right\rangle} \tag{10.12}$$

depending on the spectral parameters. Recently, Korepin has proposed a general method for evaluating expressions of this type [19], that allows us to hope, carrying out the needed limiting transition $\{\lambda_j\} \rightarrow (\pi + 2i\eta)/2$, to derive closed equations of quantum theory of stimulated Raman scattering in the case when the spatially-homogeneous approximation is inapplicable.

Acknowledgments

I gratefully acknowledge many fruitful discussions with V. M. Agranovich, D. J. Kaup, O. Keller, V. E. Korepin, A. I. Maimistov, F. Persico, V. Tognetti, V. I. Yudson, and V. E. Zakharov.

References

[1] H. Steudel, *Physica*, Vol. 6D, p. 155, 1983;
D. J. Kaup, *Physica*, Vol. 6D, p. 143, 1983;
A. I. Maimistov and E. A. Manykin, *Phys. Lett.*, Vol. A95, p.216, 1983;
A. A. Zabolotskii, *Sov. Phys. JETP*, Vol. 66, p. 47, 1987.

[2] F. Y. F. Chu and A. C. Scott, *Phys. Rev.* Vol. A12, p. 2060, 1975;
K. Drühl, R. G. Wenzel, and J. L. Carlsten, *Phys. Rev. Lett.*, Vol. 51, p. 1171, 1983;
D. J. Kaup, *Physica*, Vol. 19D, p. 125, 1986;
C. R. Menyuk, *Phys. Rev. Lett.*, Vol. 62, p. 2937, 1989;
D. J. Kaup and C. R. Menyuk, *Phys. Rev.*, Vol. A42, p. 2937, 1990.

[3] V. I. Rupasov, *Sov. Phys. JETP Lett*, Vol. 47, p. 73, 1988.

[4] V. I. Rupasov, *Sov. Phys. JETP*, Vol. 67, p. 2003, 1988.

[5] R. H. Dicke, Phys. Rev. **93**, 99 (1954).

[6] S. G. Rautian and B. M. Chernobrod, *Sov. Phys. JETP*, Vol. 45, p. 705, 1977.

[7] R. J. Baxter, Exactly Solved Models in Statistical Mechanics, London: Academic Press, 1982;
H. B. Thaker, *Rev. Mod. Phys.*, Vol. 53, p. 253, 1981.

[8] L. D. Landau and E. M. Lifshitz *Quantum Mechanics: Non-Relativistic Theory*, New York: Pergamon, 1977.

[9] A. M. Tsvelick and P. B. Wiegmann, *Adv. Phys.*, Vol. 32, p. 453, 1983.

[10] N. Andrei, K. Furuya, and J. H. Lowenstein, *Rev. Mod. Phys.*, Vol. 87, p. 331, 1983

[11] V. I. Rupasov and V. I. Yudson, *Sov. Phys. JETP*, Vol. 60, p. 927, 1984.

[12] V. B. Berestetskii, E. M. Lifshitz, and L. P. Pitaevskii, *Relativistic Quantum Theory*, Addison-Wesley, Reading, Mass., 1971.

[13] R. J. Glauber, *Phys. Rev.*, Vol. 131, p. 2796, 1963;
J. R. Klauder and E. C. G. Sudarshan, *Fundamentals of Quantum Optics*, Benjamin, N. Y., 1968;
B. Crosignani, P. di Porto, and M. Bertolotti, *Statistical Properties of Scattered Light*, Academic Press, London 1975.

[14] H. Bateman and A. Erdelyi, *Higher Transcendental Functions*, Vol. 3, McGrow-Hill, N. Y., 1955.

[15] R. Bonifacio, P. Schwendimann, and F. Haake, *Phys. Rev.*, Vol. A4, p. 302, 1971; p. 854 1971;
V. De Giorgio and F. Chielmetti, *Phys. Rev.*, Vol. A4, p. 2415, 1971;
F. Haake and R. J. Glauber, *Phys. Rev.*, Vol. A5, p. 1457, 1972;
M. Gross and S. Haroche, *Phys. Rep.*, Vol. 93, p. 301, 1982.

[16] H. Bateman and A. Erdelyi, *Higher Transcendental Functions*, Vol. 1, McGrow-Hill, N. Y., 1955.

[17] V. E. Zakharov, S. V. Manakov, S. P. Novikov, and L. P. Pitaevskii, *Theory of Solitons: The Inverse Scattering Method*, Consultants Bureau, 1984.

[18] P. B. Wiegmann, *J. Phys.*, Vol. C14, p. 1463, 1981.

[19] V. E. Korepin, *Commun. Math. Phys.*, Vol. 113, p. 177, 1987.

Studies in Classical
and Quantum Nonlinear Optics

ISBN 1-56072-168-5

NON-CLASSICAL FLUCTUATIONS IN SOLIDS AND THEIR OBSERVATION IN QUANTUM OPTICS

Alexander S. Shumovsky
N.N. Bogolubov Laboratory for Nuclear Research
PO Box 79, Moscow 101000, Russia
and Physics Department, Bilkent University
Bilkent 06533, Ankara, Turkey

1. INTRODUCTION

The two main objectives of this paper are on the one hand to show that the "non-classical effects" such as squeezed state may occur in solids at thermal equilibrium, and on the other hand to demonstrate how to observe them with the aid of optical methods.

The first chapter is devoted to the quantum fluctuation effects in condensed matter. The nature of homogeneous and heterophase fluctuations is discussed, and statistical properties of collective excitations are examined.

In the second chapter we turn to the subject of correlation spectroscopy and consider Raman scattering as a possible method of measurement of quantum statistical properties of different collective excitations in solids.

The author wishes to express his gratitude to Organizing Committee of 5th International Topsoe Summer School on Nonlinear Optics and especially to Professor Ole Keller for providing him with the opportunity to give these lectures. Thanks are also due to Prof. A.Andreev, Prof. S.Ciraci, Prof. F.Persico, Prof. R.Pucci, Prof. V.Rupasov, Dr. A.Chizhov, Dr. T.Hakioglu, and Dr. B.Tanatar for useful discussions and collaboration.

2. QUANTUM FLUCTUATIONS IN CONDENSED MATTER

The role of fluctuations in behavior of the macroscopic systems which have been established for the first time by J.W.Gibbs [1] is generally known. There are two main reasons of fluctuations:

i) chaotic thermal processes;

ii) quantum uncertainty relations.

Although both sources of fluctuations act simultaneously it is the custom to believe that the quantum nature of fluctuations is exhibited either at zero temperature or in some special media (quantum liquids [2] and quantum crystals [3]). For example, the critical fluctuations at phase transitions are treated as a classical one within the framework of modern theory based on the renormalization group approach [4].

In the modern theory of phase transitions it is customary to divide all fluctuations into two groups: homogeneous or homophase fluctuations and heterogeneous or heterophase fluctuations (see e.g. [5] [6]). The homophase fluctuations are defined as the thermal excitations of a collective state of system with given symmetry while the homophase fluctuations represent an equilibrium mixture of collective states with different macroscopic symmetry. An example of such a heterophase state is provided by the water between $0C$ and $+4C$ where an equilibrium mixture of liquid phase and fluctuating ice exists [7]. The main reason for heterophase fluctuations is the competition of different microscopic interactions [6], that is they have a quantum nature.

This type of macroscopic fluctuations is very close to that examined in quantum optics in connection with squeezing phenomenon. A simple illustrative example is provided by the famous Bogolubov's theory of superfluidity [8] the main ideas of which can be formulated as the following statements.

1. Below some temperature the ^4He atoms in condensate state with momentum $k = 0$ are weakly coupled with the over-condensate excitations [1]

2. Every operator of atom with $k = 0$ in the interaction Hamiltonian is supposed to be replaced by a C-number according to the rule: $a_0 \propto \sqrt{N_0}$ where N_0 is the mean number of atoms in condensate state; N_0 is supposed to be much greater than the mean number of atoms in over-condensate excitations N_{exc}. Then, instead of real interaction Hamiltonian we obtain an effective Hamiltonian which is a quadratic form with respect to the operators of excited atoms (with $k \neq 0$).

3. This effective Hamiltonian is diagonalized with the aid of Bogolubov Canonical Transformation (BCT). Such a procedure leads to the famous expression for the excitation spectrum with the phonon and roton parts.

All subsequent development of the theory of superfluidity of ^4He is based on that ideas. Their paraphrase is given in a lot of monographs and textbooks on statistical physics and condensed matter physics (e.g. [10], [11]).

[1]The same ideas underlies the semi-phenomenological two-liquid theories of superfluidity [9].

Let us note that substitution of some C-numbers instead of Bose-operators with $k = 0$ corresponds to the averaging of the Hamiltonian with respect to some state of condensate atoms [12], [13]. Thus, the effective Hamiltonian is represented in the following form

$$H_{eff} = \frac{1}{2}\sum_{k\neq 0}[C_k(a_k^+ a_k + a_{-k}^+ a_{-k}) + \lambda < a_0^2 > a_k^+ a_{-k}^+ + \lambda < a_0^{+2} > a_{-k} a_k]$$

$$+\frac{\lambda}{2} < a_0^{+2} a_0^2 > -\mu < a_0^+ a_0 > \qquad (1)$$

where

$$C_k = k^2/2m - \mu + 2\lambda < a_0^+ a_0 > .$$

Here λ is the coupling constant, μ is the chemical potential, k is the momentum of atom, and m is the mass of atom. Brackets $< ... >$ mean the averaging under consideration.

One can see that the Bogolubov's parameterization of condensate state [8]

$$\begin{cases} < a_0^+ a_0 >= N_0 \\ < a_0^2 >= N_0 \\ < a_0^{+2} a_0^2 >= N_0^2 \end{cases} \qquad (2)$$

corresponds to the averaging with respect to the Glauber coherent state [14] with real parameter of coherence α. Actually, in the case of coherent state the averaging of any operator construction $f(a_0^+, a_0)$ is defined in the following way

$$< f(a_0^+, a_0) >=< 0|D^+(\alpha) f(a_0^+, a_0) D(\alpha)|0 >$$

where $|0 >$ is the vacuum state of atoms with $k = 0$ and D is the displacement operator

$$D(\alpha) = \exp(\alpha a_0^+ - \alpha^* a_0).$$

By virtue of the commutation relations

$$\begin{cases} D^+(\alpha) a_0 D(\alpha) = a_0 + \alpha, \\ D^+(\alpha) a_0^+ D(\alpha) = a_0^+ + \alpha^* \end{cases} \qquad (3)$$

we obtain $|\alpha|^2 = N_0$.

The Hamiltonian (1) is diagonalized by the Bogolubov's canonical transformation (BCT) of the form

$$a_k = u_k(b_k + \xi_k b_{-k}^+) \qquad (4)$$

where u and ξ are the complex numbers and

$$\begin{cases} \xi_k = -\eta_k + \sqrt{\eta_k^2 - 1} \\ |u_k|^2(1 - \xi_k^2) = 1 \\ \eta_k = C_k/(\lambda| < a_0^2 > |). \end{cases} \tag{5}$$

In the diagonal representation (4) instead of (1) we get

$$H_{eff} = \frac{1}{2}\sum_{k\neq 0} E_k(b_k^+ b_k + b_{-k}^+ b_{-k}) + \frac{\lambda}{2} < a_0^{+2} a_0^2 > -\mu < a_0^+ a_0 > \\ + \sum_{k\neq 0} |u_k|^2 \xi_k^2 (C_k + \lambda| < a_0^2 > |). \tag{6}$$

where

$$E_k = \sqrt{C_k^2 - \lambda^2| < a_0^2 > |^2}. \tag{7}$$

In the case of Bogolubov's parameterization (2) expression (7) describes the famous spectrum of excitations with phonon and roton parts.

One can see that the eigenstates of Hamiltonian (6) are the Fock number states of quasi-particles described by the BCT (4):

$$b_k^+ b_k \psi_k = n_k \psi_k, n_k = 0, 1, 2, ...$$

In terms of the Bose fields, describing ^4He atoms in over-condensate state, this is a squeezed number state [15]:

$$\psi_k = \frac{b_k^{n_k}}{\sqrt{n_k!}} \Psi_k, \Psi_k = S(\zeta_k)|0>_k |0>_{-k} . \tag{8}$$

Here $|0>_k$ is the vacuum state of corresponding mode

$$b_k|0>_k = 0$$

and $S(\zeta_k)$ is the unitary two-mode squeezing operator [16]

$$S(\zeta_k) = \exp(\zeta_k^* a_k a_{-k} - \zeta_k a_k^+ a_{-k}^+). \tag{9}$$

where ζ_k is a known function of u_k and ξ_k. A state of over-condensate atoms at zero temperature is defined by the wave function

$$\Phi = \prod_{k\neq 0} \psi_k. \tag{10}$$

Let us emphasize here that introduction of squeezed state by Stoler [17] was connected with a single-mode BCT of the type of (4)

We can calculate now the variance of number of atoms in excited (over-condensate) state at zero temperature

$$V(N_{exc}) = < N_{exc}^2 > - < N_{exc} >^2,$$
$$N_{exc} = \frac{1}{2}\sum_{k\neq 0}(b_k^+ b_k + b_{-k}^+ b_{-k}).$$

Employing expressions (4), (8), and (10) then gives

$$V(N_{exc}) = \sum_{k\neq 0} |u_k|^2(|u_k|^2 - 1) \tag{11}$$

where $\forall k \neq 0, |u_k|^2 > 1$. Hence

$$V(N_{exc}) >< N_{exc} >\equiv \sum_{k\neq 0}(|u_k|^2 - 1).$$

In such a case it is customary to say in quantum optics that we have a super-Poisson distribution [18]. In other words, there are strong enough quantum fluctuations of number of excited atoms even at zero temperature. Because of conservation of the total number of atoms it means the existence of corresponding quantum fluctuations of the number of atoms in condensate state. In the case of Bogolubov's parameterization (1.2) the number variance for the atoms in condensate state is

$$V(N_0) = N_0 \equiv |\alpha|^2.$$

Thus, the quantum fluctuation is supposed to be small ($\delta_0 \propto \sqrt{N_0}$).

At the same time, the modern neutron scattering experiments [19], [20] have given the following very important result: at very low temperature N_0 is much smaller than N_{exc} ($N_0 \leq 0.02 N_{exc}$). Thus, we have a discrepancy between the principles of theory and experimental results.

In this connection it should be noted, first of all, that the Bogolubov's parameterization (2) is not the only possibility. Moreover, the coherent state $|\alpha >= D(\alpha)|0>$ is not an eigenstate of a Hermite operator [14]. Thus, an another possibility should be considered.

It is naturally to suppose that the condensate atoms are also in a squeezed state (at zero temperature) [15]:

$$|\zeta_0 >= S(\zeta_0)|0>_0, a_0|0>_0 = 0$$

where

$$S(\zeta_0) = e^{\zeta_0^* a_0^2/2 - \zeta_0 a_0^{+2}/2}. \tag{12}$$

Then, we get a new parameterization of condensate state [21]

$$< a_0^+ a_0 >= \sinh^2|\zeta_0| \equiv N_0,$$
$$< a_0 >= e^{i\theta}\cosh|\zeta_0|\sinh|\zeta_0| \propto N_0,$$
$$< a_0^{+2} a_0^2 >= \sinh^2|\zeta_0|(\cosh^2|\zeta_0| + 2\sinh^2|\zeta_0|) \propto N_0^2 \tag{13}$$

The number variance is given by

$$V(N_0) = 2\cosh^2|\zeta_0|\sinh^2|\zeta_0| \equiv 2N_0(N_0+1).$$

Hence, in that case we have strong quantum fluctuations of number of atoms in condensate state ($\delta_0 \propto N_0$). The probability to have exactly N_0 atoms in the condensate state is

$$P(N_0) = \frac{1}{N_0! \cosh^2|\zeta_0|}(\frac{1}{2}\tanh|\zeta_0|)^{N_0} H_{N_0}(0)$$

where H_m is the Hermite Polynomial of degree m. The squeezing parameter ζ_0 can be chosen in the following way. Because of conservation of the total number of particles we have

$$\sinh^2|\zeta_0| = N - \sum_{k\neq 0} |u_k|^2 \xi_k^2$$

where $|u_k|$ and ξ_k depend on ζ_0 due to the parameterization of condensate state (13). This equation for ζ_0 can be expressed in terms of densities

$$\rho_0 = \rho - \rho_{exc} \tag{14}$$

where $\rho_i = N_i/V$, and V is the volume of the system.

The conception of squeezed state permits us to look at the problem of superfluidity of ^4He from a different point of view. Because of definition (9) the over-condensate state is formed by the groups of atoms (in the momentum space) with any even number and with zero total momentum of a group. Such a group is a boson. Thus, we have some kind of Bose-Einstein condensation of groups of atoms. The motion without friction of that groups can be considered as a reason for superfluidity. This picture has very close analogy with superconductivity of BCS-type [22] and with superfluidity of liquid ^3He [23] where the coupling of particles leads to the formation of bosons (or quasi-bosons) with zero momentum. The contribution of such a condensation can be also measured by the neutron scattering. But in this case it it necessary to decrease the energy of neutrons in comparison with the case of standard condensate because the "sizes" (correlation length) of such a groups are much larger than the atomic size [21].

Another example of squeezed state in condensed matter is provided by the polaritin. In this case we have not a phase transition or heterophase state, but we have strong enough quantum fluctuations.

The retarded interaction of ions in an ionic crystal is realized by the photons generated due to the transversal optical oscillations of lattice. The polaritonon which is a bound state of photon and phonon [24] is described by the Hamiltonian

$$H = \sum_k(\omega_k a_k^+ a_k + \Omega_k c_k^+ c_k) - \sum_k D_k[(a_k^+ - a_k)(c_k + c_{-k}^+) + h.c.]. \tag{15}$$

Here a^+ and a are the photon operators, c^+ and c are the phonon operators, and D is the coupling parameter.This Hamiltonian is a bilinear form with respect to two types of bosons.

In general case, if we have a bilinear boson Hamiltonian

$$H = \sum_{j,k} A_{jk} x_j^+ x_k + \sum_{j,k} (B_{jk} x_j x_k + B_{jk}^* x_k^+ x_j^+) \tag{16}$$

where

$$[x_j, x_k^+] = \delta_{jk}$$

and $A_{jk} = A_{kj}, B_{jk} = B_{kj}^*$, it can be diagonalized with the aid of general BCT of the form

$$X = UY \tag{17}$$

where X is the formal column from the operators x and x^+, and Y is the corresponding column from new (quasi-particle) boson operators y and y^+. The complex matrix U obey the condition

$$\det U = 1.$$

Their elements u and v realize the BCT of the form

$$x_k = \sum_m (u_{km} y_m + v_{km}^* y_m^+). \tag{18}$$

In the diagonal representation the Hamiltonian (16) has the form

$$H = E_0 + \sum_m E_m y_m^+ y_m, , , E_0 = -\sum_{k,l} E_l |v_{kl}|^2 \tag{19}$$

where E_m, u_{km}, and v_{km} are the eigenvalues and eigenfunctions of the following equations

$$\begin{cases} E_m u_{km} = \sum_l (B_{kl} u_{lm} + 2A_{kl} v_{lm}) \\ -E_m v_{km} = \sum_l (B_{kl}^* v_{lm} + 2A_{kl}^* u_{lm}) \end{cases} \tag{20}$$

with the additional conditions

$$\begin{gathered} \sum_k (u_{kl} u_{km}^* - v_{kl} v_{km}^*) = \delta_{lm} \\ \sum_l (u_{kl} u_{ml}^* - v_{ml} v_{kl}^*) = \delta_{km} \\ \sum_k (u_{kl} v_{km} - u_{km} v_{kl}) = 0 \\ \sum_l (u_{kl} v_{ml} - u_{ml} v_{kl}) = 0. \end{gathered}$$

Any thermodynamical equilibrium state of quasi-particles described by the operators y and y^+ is characterized by the density matrix

$$\varrho = e^{-\beta H} / Tr e^{-\beta H} \tag{21}$$

where H is the Hamiltonian in the diagonal representation (19) and β is the inverse temperature. Referred to the initial fields x, x^+ this state (21) should be considered as a squeezed chaotic state [25] while the squeezed number state [25] is an eigenstate of Hamiltonian (19).

A simple illustrative example of the above statement is provided by the special case of Hamiltonian (15) of the form [26]

$$H = \omega a^+ a + \Omega b^+ b + ig(a^+ + a)(b - b^+). \tag{22}$$

In this case the parameters of BCT (18)

$$c_j = u_{j1} a + u_{j2} b + v_{j1} a^+ + v_{j2} b^+$$

are defined by the expressions

$$\begin{aligned} u_{j1} &= \tfrac{E_j+\omega}{2\omega}\Phi_{j1} \\ v_{j1} &= \tfrac{E_j-\omega}{2\omega}\Phi_{j1} \\ u_{j2} &= \tfrac{(E_j+\Omega)(E_j^2-\omega^2)+4\omega g^2}{2E_j(E_j^2-\omega^2)}\Phi_{j2} \\ v_{j2} &= \tfrac{(E_j-\Omega)(E_j^2-\omega^2)+4\omega g^2}{2E_j(E_j^2-\omega^2)}\Phi_{j2} \end{aligned}$$

where

$$E_j^2 = \frac{\omega^2 + \Omega^2}{2} - \frac{(-1)^j}{2}\sqrt{(\omega^2 - \Omega^2)^2 + 16g^2\omega\Omega}$$

and

$$|\Phi_{j1}|^2 = (-1)^{j+1}\frac{\omega(E_j^2 - \Omega)}{E_j(E_1^2 - E_2^2)},$$
$$|\Phi_{j2}|^2 = (-1)^{j+1}\frac{\Omega(E_j^2 - \omega)}{E_j(E_1^2 - E_2^2)}.$$

In the case under consideration for the vacuum state of Bose-fields c we get the following representation

$$|0>_c = \sum_{i,j=0}^{\infty} \lambda_{ij} |i>_a |j>_b . \tag{23}$$

Here $|j>_x$ is the Fock number state of corresponding field x. This state (23) can be considered as a two-mode squeezed state more general than (8), (9). The squeezing operator has the form [27]

$$S = \exp(-\frac{\Delta_4}{2\Delta_1}a^{+2} - \frac{\Delta_2}{\delta_1}a^+ b^+ - \frac{\Delta_3}{2\Delta_1}b^{+2})$$

and

$$|0>_c = S|0>_a |0>_b .$$

Here

$$\begin{cases} \Delta_1 = u_{11}u_{22} - u_{12}u_{21} \neq 0 \\ \Delta_2 = u_{11}v_{21} - u_{21}v_{11} \\ \Delta_3 = u_{11}v_{22} - u_{21}v_{12} \\ \Delta_4 = u_{22}v_{11} - u_{12}v_{21} \\ \Delta_5 = u_{22}v_{12} - u_{12}v_{22}. \end{cases}$$

Using now the diagonal representation of Hamiltonian (22)

$$H = \sum_{j=1}^{2} E_j(c_j^+ c_j + 1/2)$$

we can determine the density matrix (21) and calculate any temperature dependent expectation value. In analogy to quantum optics let us calculate the second order degree of coherence [18] for the phonon subsystem:

$$G_b^{(2)} \equiv \frac{< b^{+2}b^2 >}{< b^+b >^2}. \tag{24}$$

This correlation function is usually used to describe the type of statistics. For example, in the case of Poisson distribution of particles $G^{(2)} = 1$ while in the case of Gaussian distribution which corresponds to a thermal equilibrium state of an ideal Bose-gas $G^{(2)} = 2$.

The straightforward calculations of the averages in (24) lead to the following expressions [28]

$$\begin{cases} < b^+b >= \mu_1 n_1 + \mu_2 n_2 + \eta \\ < b^{+2}b^2 >= 2 < b^+b >^2 + (2n_1+1)^2\nu_1 + (2n_2+1)^2\nu_2 + 2\gamma(2n_1+1)(2n_2+1) \end{cases}$$

where

$$\begin{cases} \mu_i = |u_{i2}|^2 + |v_{i2}|^2, \\ \nu_i = |u_{i2}|^2 |v_{i2}|^2, \\ \eta = |v_{12}|^2 + |v_{22}|^2, \\ \gamma = Re(u_{12}u_{22}^* v_{12}^* v_{22}), \\ n_i = (e^{\beta E_i} - 1)^{-1}. \end{cases}$$

In Fig.1 the degree of coherence (24) is plotted as a function of temperature and "detuning" parameter $\Delta = (\omega - \Omega)/(\omega + \Omega)$.

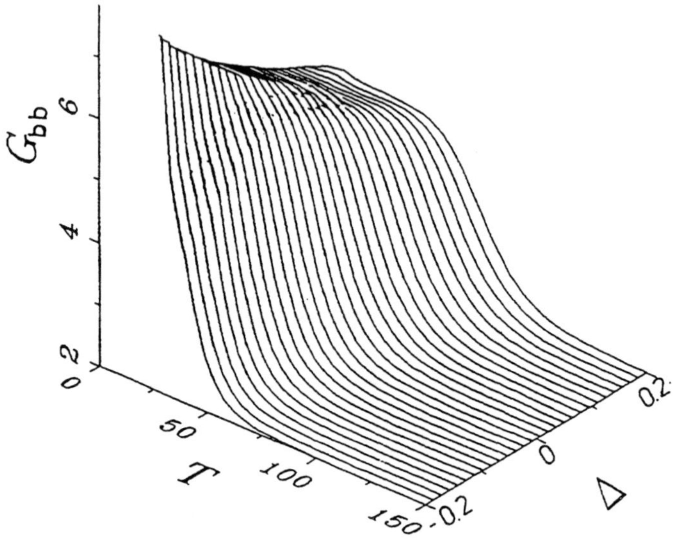

Fig. 1. Phonon degree of coherence (24) versus temperature (in Kelvin) and detuning Δ for the photon frequency $\omega = 5 \cdot 10^{13} sec^{-1}$ and the coupling constant $g = 5 \cdot 10^{-15} erg$.

One can see that at any parameters there is a super Gaussian probability distribution (i.e., more wide than the Gaussian probability distribution: $G^{(2)} > 2$). It means that the phonon–photon interaction in an ionic crystal leads to more strong quantum fluctuations of the number of phonons in comparison with the pure harmonic lattice vibrations. The reason for that fluctuations is the squeezed chaotic state of the system. Increase of temperature leads to the fuzzifying of quantum fluctuations due to the reinforcing of thermal fluctuations in comparison with quantum fluctuations. But even at $50K$ where are large enough phonon density $< b^+ b > \propto 1$ the coherence rate $G^{(2)}$ is twice more than in the case of Gaussian distribution.

Let us consider now the so-called Kobayashi model of ferroelectrics of KDP-type

[29]

$$H = \omega a^+ a - \epsilon \sum_f \sigma_f^x - \frac{1}{2} \sum_{f,f'} J_{ff'} \sigma_f^z \sigma_{f'}^z + \frac{K}{\sqrt{N}} \sum_f \sigma_f^z (a^+ + a). \tag{25}$$

Here operators a^+, a describe the optical phonon degree of freedom, ϵ is the integral of tunneling of a proton in the hydrogen bond, J is the parameter of dipole interaction between the protons of different $O - H - O$ bonds, K is the parameter of dipole-phonon coupling, σ is the corresponding component of Pauli operator, and N is the number of lattice sites ($N \to \infty$).

Using the trial Hamiltonian method [30] it is not hard to show that equilibrium expectation values obey the relations

$$\begin{cases} < a^+ a >= N^{-1} \frac{K^2}{\omega^2} < (\sigma^z)^2 > \\ < (a^+ a)^2 >= N^{-2} \frac{K^4}{\omega^4} < (\sigma^z)^4 > . \end{cases} \tag{26}$$

Here $\sigma = \sum_{f=1}^{N} \sigma_f$. Thus, the variance of phonon number is defined by the expression

$$V(a^+ a) = -\frac{4NK\theta}{\omega^4} \frac{\partial^2 F}{\partial J^2} \tag{27}$$

where θ is the temperature,

$$\begin{cases} F = -\theta \ln 2 \cosh \frac{\lambda}{\theta} + \frac{1}{2}(J + 2\frac{K^2}{\omega})C^2, \\ \lambda = \sqrt{\epsilon^2 + (J + 2\frac{K^2}{\omega})^2 C^2}, \\ J = \frac{1}{N} \sum_{f,f'} J_{ff'} \end{cases}$$

and parameter C is defined by the mean-field equation

$$\sqrt{\epsilon^2 + (J + 2K^2/\omega)^2} = (J + 2K^2\omega) \tanh \frac{\sqrt{\epsilon^2 + (J + 2K^2/\omega)^2 C^2}}{\theta}$$

below the critical temperature

$$\theta_c = \frac{\epsilon}{\tanh^{-1}(\omega\epsilon/2K^2)}.$$

One can see that there is also strong fluctuation of the number of phonons $\delta \propto N^{3/2}$ at $\theta \to 0$ and $\delta \to \infty$ at $\theta \to \theta_c$, and also

$$\forall \theta V(a^+ a) \geq < a^+ a > . \tag{28}$$

The same situation can take place in ferromagnet with spin-phonon interaction, polaron, and in some models of High-T_c superconductors [31].

It is well known that a number of quasi-particles in solids are bosons or "almost" bosons (excitons, magnones, Cooper pairs etc.). Taking into account the above obtained results, it is naturally to expect that due to microscopic interactions the strong (super-Gaussian) fluctuations can take place also for that quasi-particles.

In the next chapter we discuss how to observe such a fluctuations.

2. RAMAN CORRELATION SPECTROSCOPY

The Raman scattering is known to be an important method of condensed matter spectroscopy (e.g., see [32]). It is a special case of optical parametric process with three-wave interaction where one of interacting waves corresponds to a vibration field of medium (see Fig. 2).

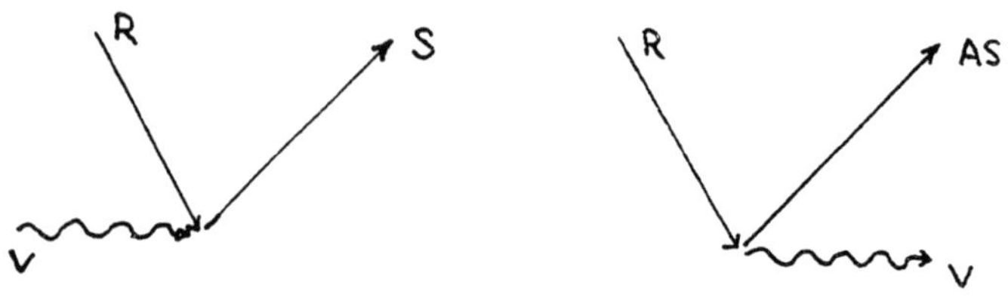

Fig. 2. A schematic illustration of the Raman process. Straight arrows correspond to photons while wavy arrows conform to a vibration mode of a Raman–active medium.

The standard measurements of frequency and angular distribution of scattered light in the Raman process give us an important information about the linear properties of vibration field. In order to measure high order correlation functions of scattered light the Raman correlation spectroscopy can be used [33]. For example, the strong enough energy fluctuations of Stokes photons have been observed in this way [34], [35] (see also [36]). At the same time the different non-classical properties of scattered light such as squeezing, anticorrelations between the Stokes spectral components, collective population trapping and collective jumps have been examined in the theory (e.g., see [37]).

We should establish relations between statistical properties of scattered light and distribution function of vibration mode in the Raman scattering process. For this aim, let us consider a simple model proposed by Walls [38]:

$$H = \sum_j \omega_j a_j^+ a_j + \gamma_s a_v^+ a_s^+ a_r + \gamma_{as} a_{as}^+ a_v a_r + h.c. \tag{29}$$

where $j = r, s, as, v$ for the Rayleigh, Stokes, anti-Stokes, and vibration modes respectively. Here a_j are the Bose operators and γ_i are the coupling constants. In the case of a resonance process the frequencies obey the energy conservation laws

$$\begin{cases} \omega_r = \omega_s + \omega_v \\ \omega_{as} = \omega_r + \omega_v. \end{cases} \tag{30}$$

In addition, the Hamiltonian (29) has two conservation laws expressing the Manley–Rowe relations in nonlinear optics [32]:

$$\begin{cases} a_r^+ a_r + a_s^+ a_s + a_{as}^+ a_{as} = N \\ a_v^+ a_v - a_s^+ a_s + a_{as}^+ a_{as} = N. \end{cases} \tag{31}$$

The standard method to describe the dynamics of model with the Hamiltonian (2.1) consists in the consideration of a "classical pumping field" when operators a_r^+ and a_r are replaced by a C-numbers [39] (see also b19). Then, instead of (29) we get some bilinear effective Hamiltonian which can be solved exactly with the aid of BCT (18).

In the case of completely quantum description the short-time approximation is used [18]. The exact Heisenberg equations in the model under consideration have the following form

$$\begin{aligned} i\tfrac{d}{dt}a_r &= \omega_r a_r + \gamma_s a_s a_v + \gamma_{as} a_v^+ a_{as} \\ i\tfrac{d}{dt}a_s &= \omega_s a_s + \gamma_s a_v^+ a_r \\ i\tfrac{d}{dt}a_{as} &= \omega_{as} a_{as} + \gamma_{as} a_v a_r \\ i\tfrac{d}{dt}a_v &= \omega_v a_v + \gamma_s a_s^+ a_r + \gamma_{as} a_r^+ a_{as}. \end{aligned} \tag{32}$$

The formal solution of the first equation in (2.4) has the form

$$\begin{aligned} a_r(t) = a_r(0)e^{-i\omega_r t} - i\gamma_s \int_0^t e^{-i\omega_r(t-\tau)} a_s(\tau)a_v(\tau)d\tau \\ -i\gamma_{as}\int_0^t e^{-i\omega_r(t-\tau)} a_v^+(\tau)a_{as}(\tau)d\tau. \end{aligned}$$

The formal expressions of the same type can be constructed for all other operators. Now that equations can be solved approximately in powers of t. For example, in the zero approximation we get

$$a_j(t) = a_j(0)e^{-i\omega_j t}.$$

The second approximation have been obtained by iteration in the paper [40].

In order to examine the fluctuations (variances) of the field it is necessary to know the third approximation [31] which was examined in [28]. Let us suppose for simplicity that Stokes and anti-Stokes modes are initially in the vacuum states. Then we get

$$\begin{aligned} a_s(t) = \gamma_s e^{-i\omega_s t}\{ita_v^+(0)a_r(0) + \frac{t^2}{2}\gamma_{as}a_{as}^+(0)a_r^2(0) \\ -i\frac{t^3}{6}a_v^+(0)a_r(0)[n_v(\gamma_s^2 + \gamma_{as}^2) - n_r(\gamma_s^2 - \gamma_{as}^2) \\ +2\gamma_s^2 - \gamma_{as}]\} \end{aligned} \tag{33}$$

and

$$a_{as}(t) = \gamma_{as}e^{-i\omega_{as}t}\{ia_v(0)a_r(0) - \frac{t^2}{2}\gamma_s a_s^+(0)a_r^2(0)$$
$$-i\frac{t^3}{6}[n_v(\gamma_s^2 + \gamma_{as}^2) - n_r(\gamma_s^2 - \gamma_{as}^2) + 2\gamma_r^2$$
$$+\gamma_{as}^2]a_v(0)a_r(0)\} \qquad (34)$$

where $n_j \equiv a_j^+(0)a_j(0)$. On making the further assumption that the density matrix of the system under consideration at $t = 0$ is defined by

$$\rho(0) = \prod_j \rho_j(0)$$

and taking into account the expressions (33) and (34) together with the Hermitian conjugate expressions we can calculate the time-dependent average photon numbers in the Stokes and anti-Stokes modes:

$$I_s(t) = I_r(0)\gamma_s^2 t^2\{(1 + I_v(0))$$
$$-t^2(\frac{1}{3}[(\gamma_{as}^2 - \gamma_s^2)(I_r(0) - 1) + \gamma_s^2](< n_v > +1)$$
$$+\frac{1}{3}(\gamma_s^2 + \gamma_{as}^2)(< n_v > + < n_v^2 >) - \frac{1}{4}\gamma_{as}(I_r(0) - 1))\},$$

$$I_{as}(t) = I_r(0)\gamma_{as}^2 t^2\{< n_v >$$
$$-t^2(\frac{1}{3}[(\gamma_{as}^2 - \gamma_s^2)(I_r(0) - 1) + \gamma_s^2] < n_v >$$
$$+\frac{1}{3}(\gamma_{as}^2 + \gamma_s^2)] < n_v^2 > -\frac{1}{4}\gamma_s^2(I_r(0) - 1))\}.$$

Here $I_j \equiv < a_j^+ a_j >$. Solving these expressions with respect to $< n_v >$ and $< n_v^2 >$, we find the first two moments of distribution of phonons at $t = 0$:

$$< n_v >= \frac{3}{I_r(0)t^2(\gamma_s^2 + \gamma_{as}^2)}\{\frac{I_{as}(t)}{\gamma_{as}^2 t^2} - \frac{I_s(t)}{\gamma_s^2 t^2}$$
$$+I_r(0) - t^2(\frac{1}{12}(\gamma_{as}^2 - \gamma_s^2)(I_r(0) - 1) + \frac{\gamma_s^2}{3}I_r(0))\} \qquad (35)$$

and

$$< n_v^2 >= \frac{3}{I_r(0)t^2(\gamma_s^2 + \gamma_{as}^2)}\{-\frac{I_a(t)}{\gamma_{as}^2 t^2} + I_r(0) < n_v >$$
$$-t^2 I_r(0)(\frac{1}{3}[(\gamma_{as}^2 - \gamma_s^2)(I_r(0) - 1) + \gamma_s^2] < n_v >$$
$$-\frac{\gamma_s^2}{4}(I_r(0) - 1))\}. \qquad (36)$$

It follows from the expressions (35) and (36) that the variance of the number of phonons

$$V(a_v^+ a_v) \equiv < n_v^2 > - < n_v >^2$$

is determined by the mean energies of pumping $I_r(0)$, Stokes $I_s(t)$, and anti-Stokes $I_{as}(t)$ fields in the short-time approximation. In principle, this linear characteristics of the photons can be measured experimentally. But it is necessary to take into account the condition of validity of that approximation: $\max \gamma_j t \ll 1$. For standard Brillouin scattering [41] in gives the following estimation

$$t \preceq 1nsec.$$

Now let us turn our attention to exact steady-state solution of dynamical problem with the Hamiltonian (29). Employing the conservation laws (31) then gives the following exact eigenfunction of the system under consideration

$$|\psi^{(n_r,n_v)} >= \sum_{j=0}^{n_r} \sum_{k=0}^{m_j} \lambda_{jk}^{(n_r,n_v)} |n_r - j - k >_r |j >_s |k >_{as} |n_v + j - k >_v \tag{37}$$

where $m_j = \min\{n_r - j, n_v + j\}$ and n_i denotes the eigenvalue of corresponding operator in exp. (31). The coefficients λ in (31) obey the recursion relations

$$\begin{aligned} x^{(n_r,n_v)} &= \gamma_s \lambda_{j-1,k}^{(n_r,n_v)} \sqrt{(n_r - j - k + 1)j(n_v + j - k)} \\ &+\gamma_s \lambda_{j+1,k}^{(n_r,n_v)} \sqrt{(n_r - j - k)(j+1)(_{nv} + j - k + 1)} \\ &+\gamma_{as} \lambda_{j,k-1}^{(n_r,n_v)} \sqrt{(n_r - j - k + 1)(k(n_v + j - k + 1)} \\ &+\gamma_{as}^{(n_r,n_v)} \lambda_{j,k+1}^{(n_r,n_v)} \sqrt{(n_r - j - k)(k+1)(n_v + j - k)}. \end{aligned} \tag{38}$$

Here

$$x^{(n_r,n_v)} = E^{(n_r,n_v)} - \omega_r n_r - \omega_v n_v$$

and $E^{(\dots)}$ is the eigenvalue of Hamiltonian (29) with given n_r and n_v.

Unfortunately, the analytic solution of recursion relations (29) is unknown in general case. We therefore restrict ourselves to the case $\gamma_{as} = 0$ (we neglect the anti-Stokes component generation). From the physical point of view, it is possible if the initial number of Rayleigh photons is small enough [32].

The model problem with Stokes mode generation has been solved exactly by Carusotto [42]. However, we consider here an another more descriptive method of exact solution which was proposed in paper [43] (see also [31], [44]).

Let us put $\gamma_{as} = 0$ in the previous relations. Then, instead of the recursion relation (38) we get

$$\begin{aligned} &\lambda_{j+1}^{(n_r,n_v)} \sqrt{(n_r - j)(j+1)(n_v + j + 1)} \\ &= x^{(n_r,n_v)} \lambda_j^{(n_r,n_v)} - \lambda_{j-1}^{(n_r,n_v)} \sqrt{(n_r - j + 1)j(n_v + j)}. \end{aligned} \tag{39}$$

In view of our restriction, we drop here index k describing the anti-Stokes photons, as well as the summation over k in exp. (37). This recursion relation (39) can be represented in the following way

$$P_{j+1}^{(n_r,n_v)} = xP_j^{(n_r,n_v)}(x) - q_j^{(n_r,n_v)}P_{j-1}^{(n_r,n_v)}(x) \tag{40}$$

where $P(x)$ are some new orthogonal polynomials connected with the coefficients of wave function (37) by the relations

$$P_j^{(n_r,n_v)}(x) = \xi_j^{(n_r,n_v)}\lambda_j^{(n_r,n_v)}$$
$$\xi_{j+1}^{(n_r,n_v)} = \xi_j^{(n_r,n_v)}\sqrt{(n_r-j)j(n_v+j+1)}$$

and $q_j^{(n_r,n_v)} = (n_r-j+1)j(n_v+j)$. Introducing the generating function

$$F(x,t) \equiv \sum_{k=0}^{\infty} P_k(x)\frac{t^k}{k!}$$

one can define these polynomials by the differential equation

$$xF(x,t) = \frac{\partial F}{\partial t} + t(n_r\frac{\partial}{\partial t} - t\frac{\partial^2}{\partial t^2})(tF). \tag{41}$$

It is also useful to express these polynomials in terms of the Bernoulli polynomials $B_l^{j+1}(n_r;n_v)$. We have

$$P_j^{(n_r,n_v)}(x) = \sum_{l=0}^{j} \beta_l^j(n_r;n_v)B_l^{j+l}(n_r;n_v)x^l \tag{42}$$

where

$$\beta_l^{j+1}(n_r;n_v)(j-l) = -jq_j^{(n_r,n_v)}\beta_l^{j-1}(n_r;n_v)$$
$$\beta_l^{j+1}(n_r;n_v)(n_r-j)l = j\beta_{l-1}^j(n_r;n_v).$$

In terms of polynomials P the equation for eigenvalues E (or for x) of the system under consideration takes the form

$$P_{n_r+1}^{(n_r,n_v)}(x) = 0. \tag{43}$$

For any fixed n_r this equation has n_r+1 real roots. The value $x^{(n_r,n_v)} = 0$ is the root of equation (43) for even n_r only. The roots $\{x_i^{(n_r,n_v)}\}$ at any fixed n_r are ordered symmetrically with respect to zero. It should be emphasized that the number n_v occurs as a parameter in our equations and, in general, can have any integer value.

With the help of the above relations, we obtain for the coefficients of the eigenfunctions the following expression

$$\lambda_j^{(n_r,n_v)}(x_i) = \lambda_0^{(n_r,n_v)}\frac{P_j^{(n_r,n_v)}}{\Phi_j^{(n_r,n_v)}} \tag{44}$$

where x_i $(i = 1, 2, ..., n_r + 1)$ is any solution of equation (43) and

$$\Phi_j^{(n_r,n_v)} = \sqrt{(j!)^3 \binom{n_r}{j} \binom{n_v + j}{j}}$$

The coefficient λ_0 is determined from the normalization condition.

Using the Hadamard criterion [45], it is not hard to show that the maximum root has the following asymptotic behavior

$$x_{\max} \propto \begin{cases} n_r^{3/2} & n_r \to \infty, n_v \ll n_r \\ n_r n_v^{1/2} & n_v \to \infty, n_v \gg n_r. \end{cases} \tag{45}$$

In view of the symmetry of roots of the equation (43) it follows from the definition of eigenvalues $E^{(n_r,n_v)}$ that for any n_v and γ_s it is possible to find a value $n_r^{(0)}$ such that for any $n_r > n_r^{(0)}$ there is a negative branch of spectrum $E^{(n_r,n_v)}(x_{\min})$ in the system with Hamiltonian (29) at $\gamma_{as} = 0$. In other words, we have instability of spectrum of the eigenvalues in the system under consideration.

There can be a few reasons for such an instability from the physical point of view. First of all, we have supposed the small number of initial photons in the Rayleigh mode. In the opposite case the contribution of anti-Stokes process must be taken into consideration. Some other reasons have been considered in paper [31].

We now turn to the description of dynamical properties of the system under consideration. The time dependent wave function is represented by

$$|\psi(t)> = \sum_{n_r,n_v=0}^{\infty} e^{-i(n_r\omega_r + n_v\omega_v)t} \sum_{l=1}^{n_r+1} C_l^{(n_r,n_v)} e^{-i\gamma_s x_l^{(n_r,n_v)} t}$$

$$\times \sum_{j=0}^{n_r} \lambda_j^{(n_r,n_v)}(x_l^{(n_r,n_v)}) |n_r - j>_r |j>_s |n_v + j>_v . \tag{46}$$

The coefficients C are determined by the initial conditions which we choose in the form

$$\sum_{l=1}^{n_r+1} C_l^{(n_r,n_v)} \lambda_j^{(n_r,n_v)}(x_l^{(n_r,n_v)}) = \begin{cases} p_{n_r} f_{n_v} & \text{if j=0} \\ 0 & \text{otherwise} \end{cases} \tag{47}$$

For simplicity, we suppose the initial state of the Rayleigh mode to be either a number or a coherent state with the probability distribution

$$p_{n_r} = \begin{cases} \delta_{n_r n_r'} \\ \frac{\alpha^{n_r}}{\sqrt{n_r!}} e^{-|\alpha|^2/2}. \end{cases}$$

In order to simulate the uncorrelated and correlated states of the vibration mode we choose the probability distribution f_{n_v} in the following way:

$$f_{n_v} = \begin{cases} \delta_{n_v n_v'} & \text{uncorrelated} \\ \frac{1}{\sqrt{n_v! \mu}} (\frac{\nu}{2\mu})^{n_v/2} H_{n_v}(0) & \text{correlated} \end{cases}$$

In other words, we consider here either number, or squeezed vacuum state of the phonons.

The results of our calculations are represented by the Figs. 4 and 5 for the Rayleigh mode initially in a number state. One can see a qualitative difference between the scattering process on the uncorrelated and correlated phonons in this situation.

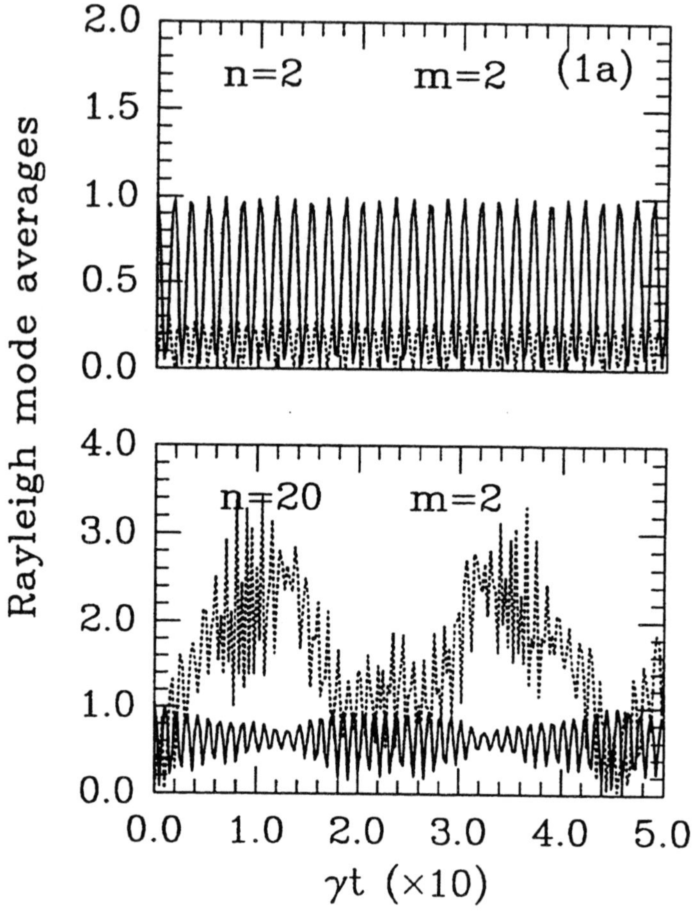

Fig. 4. The time dependence of the normalized mean number of photons (solid line) and variance (dotted line) for the Rayleigh mode and uncorrelated photons.

In the first case the increase of number of initial Rayleigh photons leads to the periodical renewal of a sub-Poissonian statistics of scattered light in the Rayleigh mode while in the second case that increase implies the strong super-Poisson distribution of scattered phonons.

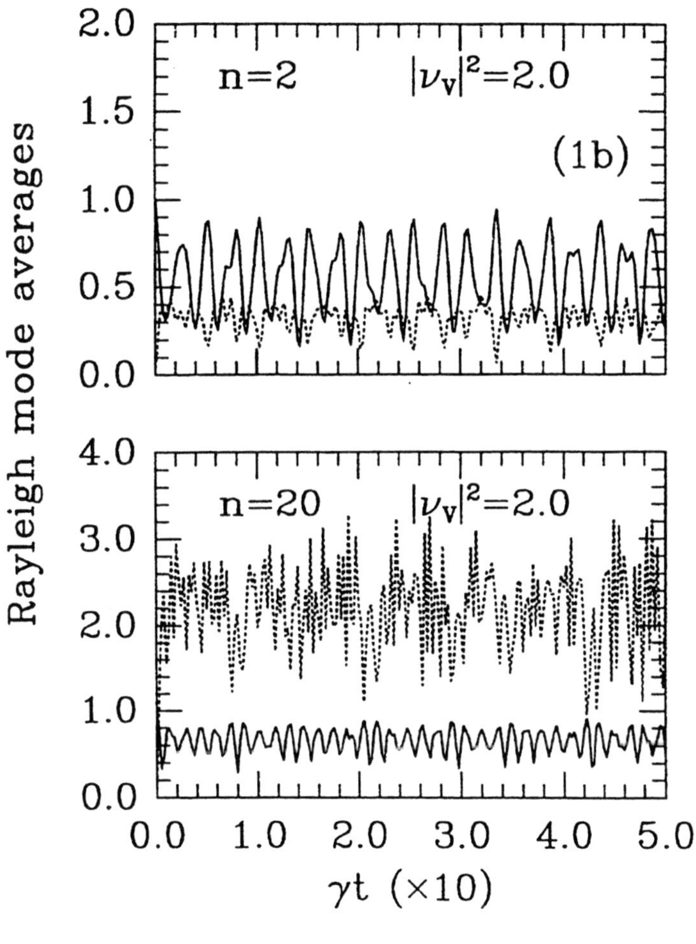

Fig. 5. Same as Fig. 4 for the correlated phonons.

This qualitative difference opens the possibility of experimental observation of the correlated phonons. Qualitatively the same behavior is obtained also for the Stokes mode (Fig. 6).

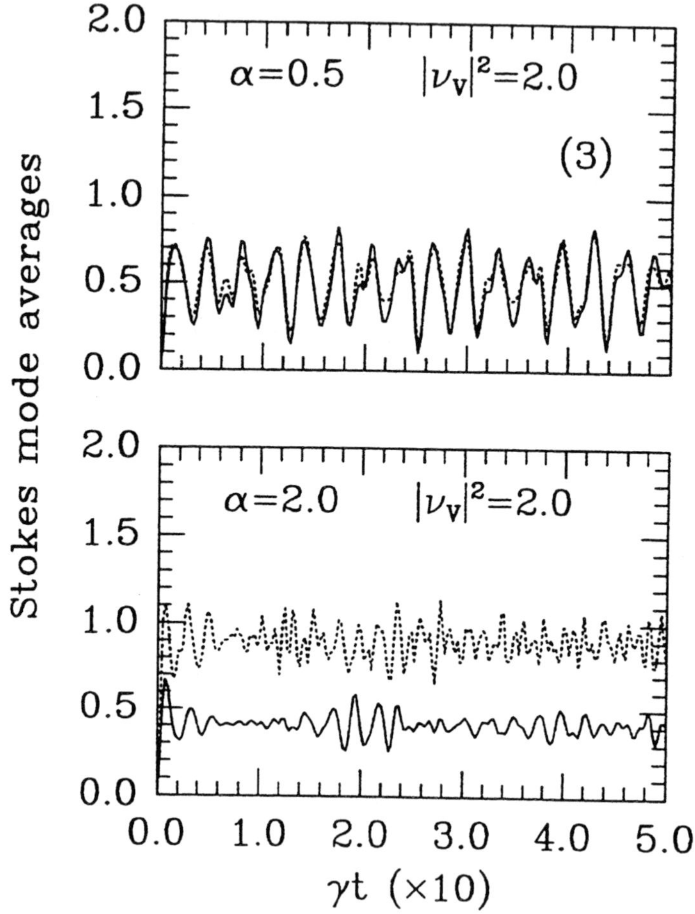

Fig. 6. The time dependence of the normalized mean number of photons (solid line) and variance (dotted line) for the Stokes mode.

In the case of initial coherent state of the Rayleigh photons for both initial states of phonons we have almost the same behavior of scattered light (Fig. 7). The same result can be observed also in the case of initial squeezed state of the Rayleigh mode [44].

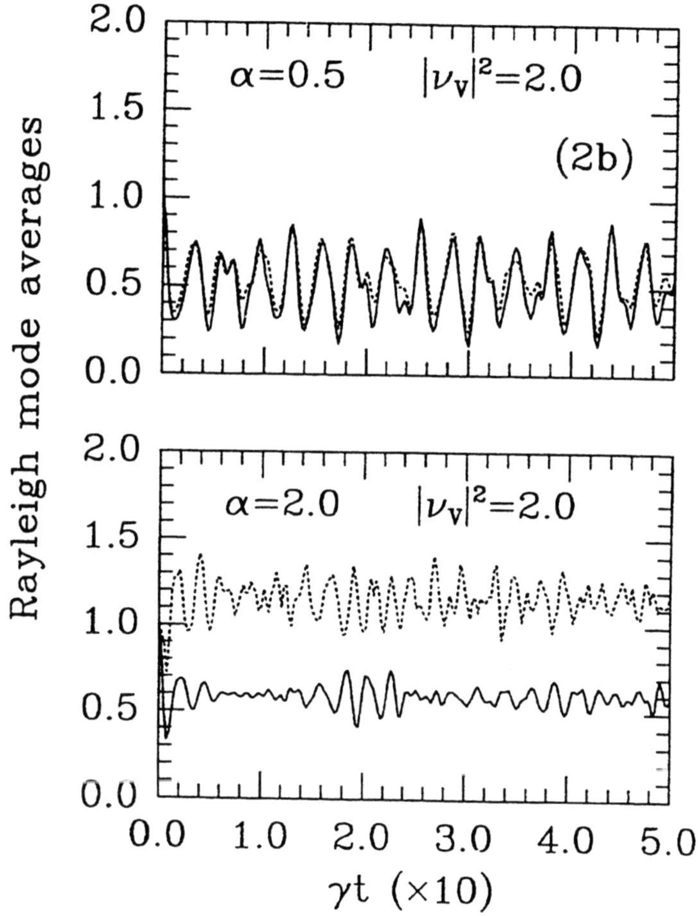

Fig. 7. The time dependence of normalized mean number of photons (solid line) and variances (dotted line) for the Rayleigh mode initially in a coherent state.

One can see also that there is some kind of collapses and revivals of quantum oscillations which is known from the theory of two- and three-level atoms interacting with an electromagnetic field [45], [46]. It is not surprised because due to the definition of exact eigenfunction any time-dependent expectation value contains the sum of oscillating terms. Initially the system is in a definite state and all oscillations are correlated with each other. At $t > 0$ because of the different oscillation frequencies the different terms have the significant phase shifts and therefore the averaging leads to the collapse of oscillations. After some time the phase shifts become divisible by 2π or almost 2π. Then the quantum oscillations are partially restored (revivals).

3. SUMMARY

Let us formulate now the main results of the present work.

1). The strong enough fluctuations of the mean number of collective excitations of the Bose-type can take place in condensed matters because of correlations initiated by the microscopic interactions. Measurements of such a fluctuations can give us an important information about the distribution function of that excitations, in other words, about the parameters of microscopic interactions.

2). These fluctuations can be measured by the Raman scattering process both on the short times and for the steady-state regime. In the second case, one must use the source generating a number (or sub-Poissonian) state of light. In principle, it is possible now.

Undoubtedly, there are a number of problems which should be solved (the process with anti-Stokes generation, the influence of real pumping, the role of detuning from the exact resonance and so on). Thus, in our opinion, there is an interesting field for the future research.

References

[1] J.W.Gibbs. The Collected Works. Longmans, Green. New York. 1931.

[2] Z.M.Galasiewicz. Superfluidity and Quantum Liquids. Pergamon Press. Oxford. 1970.

[3] D.I.Pushkarov. Defectons in Crystals. World Scientific. Singapore. 1989.

[4] K.G.Wilson. Rev. Mod. Phys. 55 (1983) 583.

[5] M.Fisher. Nature of Critical Points. Colorado University Press. Boulder. 1965.

[6] V.I.Yukalov and A.S.Shumovsky. Lectures on Phase Transitions. World Scientific. Singapore. 1990.

[7] C.A.Angell. Water: A Comprehensive Treatise. V. 7. Plenum Press. New York. 1982.

[8] N.N.Bogolubov. J. Phys. USSR 11 (1947) 23.

[9] R.P.Feinman. Phys. Rev. 94 (1954) 262.

[10] J.M.Ziman. Elements of Advanced Quantum Theory. Cambridge University Press. Cambridge. 1969.

[11] M.Umezava, H.Matsumoto and M.Tachiki. Thermo Fluid Dynamics and Condensed State. North-Holland. Amsterdam. 1982.

[12] N.M.Hugenholtz. Prog. Phys. 28 (1965) 201.

[13] G.G.Emch. Algebraic Methods in Statistical Mechanics and Quantum Field Theory. Wiley. New York. 1972.

[14] R.J.Glauber. Phys. Rev. 131 (1963) 2766.

[15] A.S.Shumovsky. In 9th Gen. Conf. of Condensed Matter Div. of European Phys. Soc. Eds. J.Lahuerte, M.Le Bellac and F.Raimond. V. 13A, p. 223. EPS. Nica. 1989.

[16] R.Loudon and P.L.Knight. J. Mod. Optics 34 (1987) 709.

[17] D.Stoler. Phys. Rev. D4 (1971) 1925 and 2309.

[18] J.Perina. Quantum Statistics of Linear and Nonlinear Optical Phenomena. D.Reidel Pub. Co. Dordrecht. 1984.

[19] T.R.Sosnick, W.M.Snow, P.E.Sokol, and R.N.Silver. Europhys. Lett. 9 (1989) 707.

[20] I.V.Bogoyavlenskii, L.V.Karnatsevich, J.A.Kozlov, and A.V.Puchkov. Physica B 176 (1992) 151.

[21] A.S.Shumovsky. Comm. JINR, E17-29-355. Dubna. 1992.

[22] A.A.Abrikosov. Fundamentals of the Theory of Metals. North-Holland. Amsterdam. 1987.

[23] D.D.Osheroff, R.C.Richardson, and D.M.Lee. Phys. Rev. Lett. 28 (1072) 885.

[24] V.M.Agranovich and V.L.Ginzburg. Spatial Dispersion in Crystal Optics and the Theory of Excitons. Interscience. New York. 1965.

[25] A.S.Shumovsky. Theor. Math. Phys. 89 (1991) 438 (in Russian).

[26] A.V.Chizhov, R.G.Nazmitdinov, and A.S.Shumovsky. Quantum Optics 3 (1991) 1.

[27] O.I.Mokhov. On the Squeezed Vacuum States Corresponding to the General Two-Mode Bogolubov Transformation. Subm. to J. Quantum Nonlin. Phenomena.

[28] A.V.Chizhov, B.B.Govorkov Jr., and A.S.Shumovsky. To be published in J. Mod. Phys. B (1993).

[29] H.Kobayashi. J. Phys. Soc. Jap. 24 (1968) 497.

[30] N.N.Bogolubov Jr., B.I.Sadovnikov, and A.S.Shumovsky. Mathematical Methods of Statistical Mechanics of Model Systems. CRC Press. Boca Ration. (1993).

[31] A.S.Shumovsky. In Modern Nonlinear Optics. Ed. M.Evans. Wiley. New York. 1993.

[32] Y.R.Shen. The Principles of Nonlinear Optics. Wiley. New York. 1984.

[33] B.Crosignani, P.Di Porto, and M.Bertolotti. Statistical Properties of Scattered Light. Academic Press. New York. 1975.

[34] I.A.Walmsley and M.G.Raymer. Phys. Rev. A 33 (1986) 382.

[35] M.D.Duncan, R.Mahon, L.L.Tankersley, and J.Reintjes. J. Opt. Soc. Am. B7 (1990) 1336.

[36] J.Mostowski and M.G.Raymer. In Contemporary Nonlinear Optics. Eds. P.Agarwal and R.W.Boyd. Academic Press. New York. 1992.

[37] A.S.Shumovsky and Tran Quang. In Interaction of Electromagnetic Field with Condensed Matter. Eds. N.N.Bogolubov, A.S.Shumovsky, and V.I.Yukalov. World Scientific. Singapore. 1990.

[38] D.F.Walls. Z. Phys. 237 (1970) 224.

[39] F.Persico and G.Vetri. Phys. Rev. A 12 (1975) 2083.

[40] P.Szlachetka, S.Kielich, J.Perina, and V.Perinova. J. Phys. A 12 (1979) 1921.

[41] Light Scattering in Semiconductor Structures and Superlatices. Eds. D.J.Lockwood and J.F.Young. Plenum Press. New York. 1991.

[42] S,Carusotto. Phys. Rev. A 40 (1989) 1848.

[43] Yu.Orlov, I.Pavlotsky, A.Shumovsky, V.Suslin, and V.Vedenyapin. To be published in Int. J. Mod. Phys. B.

[44] A.S.Shumovsky and B.Tanatar. To be published in Phys. Rev. A.

[45] H.I.Yoo and J.H.Eberly. Phys. Rep. 118 (1985) 239.

[46] Fam Le Kien and A.S.Shumovsky. Int. J. Mod. Phys. B 5 (1991) 2287.

Studies in Classical
and Quantum Nonlinear Optics

ISBN 1-56072-168-5

NONLINEAR OPTICAL RESONATORS AND INTEGRATED DEVICES

Guy Vitrant and R. Reinisch
Laboratoire d'Electromagnétisme de Microondes et Optoélectronique
(LEMO) - ENSERG
23, rue des Martyrs
BP 257
38016 Grenoble, France

1. INTRODUCTION

Nonlinear (NL) optical planar resonators have been attracting a lot of attention for some decades now, because of their potential applications in optical information processing and the wealth of physical phenomena they display. Indeed, a host of interesting types of behavior have been observed, depending on the configuration under study (Fabry-Pérot or ring resonators, distributed couplers, multilayered structures, etc. ..) and on the range of parameters considered (finesse, angle of incidence, finite beam width, type of nonlinearity (absorptive, dispersive, saturating), absorption, material response time, etc. ...). These exciting effects range from bi- and/or multistability, self-pulsing and chaos [1,2], spatial soliton formation [3,4], symmetry breaking, etc. ... The description of most of these phenomena requires a numerical approach. Nevertheless, simple equations have been proposed to model a wide class of nonlinear planar resonators. Lugiato and coworkers [5] initiated a mean field treatment of ring as well as Fabry-Pérot type resonators leading to a kind of approach much simpler than the set of original counterpropagating equations because of the particular mean field assumption along the longitudinal axis of the cavity. The same type of equations has been derived from a pole analysis to describe the field within the nonlinear optical (NLO) resonator by Haelterman et al [6], leading to the possibility of performing a modal analysis of the response of these NL devices. In the first part of this lecture, the basic equations describing nonlinear optical resonators within the so-called "Modal Theory" are derived. We will insist on the importance of tranverse effects due to the finite width of the incident beam and show that the existence of a transverse dimension by itself strongly determines the nonlinear dynamical behavior. The very interesting question of the formation and evolution of spatial patterns arising spontaneously in nonlinear optical systems is covered in more details in the lecture by W. Firth [7] and will not be further investigated here.

The distributed coupler appears as a key device which allows for a synchrounous excitation of a guided wave in a thin film, based either on prism coupler technic or grating coupler one. They are found to be equivalent to a nonlinear Fabry-Pérot type resonator, excited under oblique incidence. In addition distributed couplers are well-known in the field of nonlinear integrated devices since they are at the basis of the nonlinear M-line thin film characterisation technic.

2. MODAL APPROACH OF NONLINEAR PLANAR RESONATORS.

2.1 Stationary equations of an ideal Fabry-Pérot (FP) resonator

2.1.a linear Fabry-Pérot. Let us here consider an ideal Fabry-Pérot devices as depicted on Fig. 1. A cavity of thickness L, is bounded by two mirrors whose coefficients of reflexion and transmission in amplitude are respectively R_f and R_b, and coefficients of transmission are T_f and T_b. Let $r=(R_fR_b)^{1/2}$, $t=(T_fT_b)^{1/2}$ and F_c be the finesse of the cavity with $F_c=\pi\sqrt{r}/(1-r)$. The cavity is filled with a nonlinear material of refractive index $n= n_o+ n_2 I$ where I is the local light intensity and n_2 the nonlinear coefficient. The x-axis is parallel to the mirrors of the resonator and thus perpendicular to the direction of propagation of the incident beam at normal incidence. So, it is called the transverse direction and the y-axis is called the longitudinal axis (Fig. 1). The complex amplitude of the incident beam on the input interface is denoted $A_i(x)$ without any assumption on its x-dependence. Let $A_r(x)$ and $A_t(x)$ denote respectively the reflected and transmitted field amplitudes. A time dependence $e^{-j\omega t}$ is assumed throughout this section in which we deal with the stationary problem. $\lambda=2\pi c/\omega$ denotes the wavelength in vacuum.

Let us first consider the case of the associated linear resonator ($n_2=0$) illuminated with a plane wave defined by $A_i(x)=A_ie^{j\beta x}$ with $\beta=2\pi/\lambda\sin(\theta_i)$ where θ_i and θ_t are respectively the input and output angles, and θ the angle inside the resonator. We recall the expression of the linear transmission of the resonator:

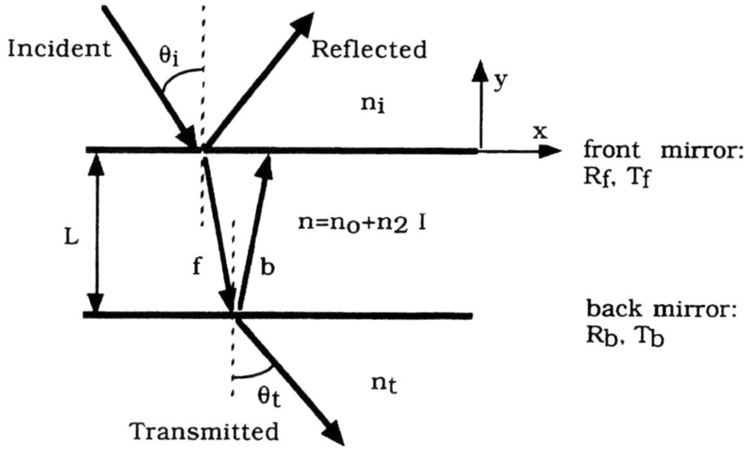

Figure 1: Schematic representation of an "ideal" NL Fabry-Pérot resonator under oblique incidence illumination.

$$T = \frac{te^{j\phi/2}}{1 - re^{j\phi}}$$

$$\text{where } \phi(\beta^2) = 2L\sqrt{\left(\frac{2\pi}{\lambda}n_o\right)^2 - \beta^2} \qquad (1)$$

is the round trip phase change. The complex poles $\beta_m{}^2$ of the function $T(\beta^2)$ are defined by $1\text{-}re^{j\phi}=0$ and correspond to the eigenmodes of the resonator. Provided that the cavity finesse is high enough, it is possible to expand the transmission of the resonator in the vicinity of a given pole, leading to the general lorentzian expression for the transmission:

$$T = \frac{j\xi}{\beta_m{}^2 - \beta^2} = \frac{j\xi}{\rho_m{}^2 - \beta^2 + j\kappa_m{}^2} \qquad (2)$$

Notice that it is $\beta_m{}^2$ and not β_m which enters eq. (2). Besides, the real and imaginary parts of the pole $\beta_m{}^2$, respectively $\rho_m{}^2$ and $\kappa_m{}^2$, are introduced and will be seen to play an important role in the following.

Let us consider now a finite width beam under the averaged incidence θ_i, caracterised by a complex amplitude $A_i(x)e^{j\beta x}$ where $A_i(x)$ is a slowly-varying function of x. Applying a spatial Fourier transform to eq. (2) leads to the spatial differential equation governing $A_i(x)$:

$$\frac{d^2A_t}{dx^2} + 2j\beta\frac{dA_t}{dx} + \left[\rho_m{}^2 - \beta^2\right]A_t + j\kappa_m{}^2A_t = j\xi A_i(x) \tag{3}$$

2.1.b Nonlinear Fabry-Pérot. In the linear case, for a given mode $\beta_m{}^2$, the field inside the resonator obeys a Schrödinger equation:

$$\frac{d^2E(y)}{dy^2} + k^2E(y) = \beta_m{}^2E(y) \quad \text{with} \quad k = \frac{\omega}{c}n \tag{4}$$

In the nonlinear case, the refractive index depends on the light intensity, and k^2 writes $k^2=k_o{}^2+\Delta k^2(|E(y)|)$ and become a function of y. In fact, assuming a weak nonlinear regime, Δk^2 can be seen as a perturbation of the unperturbed equation (4). A classical stationary perturbation method leads to an approximate expression of the perturbed nonlinear eigenvalue $\beta_{m,nl}{}^2$:

$$\beta_{m,nl}{}^2 = \beta_m{}^2 + \frac{\int_{-\infty}^{+\infty}\Delta k^2(y)|E_m(y)|^2\,dy}{\int_{-\infty}^{+\infty}|E_m(y)|^2\,dy} \tag{5}$$

This equation (5) can be easily written in a well-known form when dealing with waveguides as obtained by Stegeman et al [8] in an early work. In the case of a pure Kerr nonlinearity, the nonlinear wave vector $\beta_{m,nl}{}^2$ can be written as $\beta_m{}^2+\eta\zeta^2|A_t|^2$ because the transmitted field A_t is in fact proportional to the intracavity field. η is the sign of the non linearity (+1 for self-focusing and -1 for self-defocusing) and ζ^2 is related to the overlap integral (5). Since the eigenmodes of the ideal planar resonator are actually plane waves, we have, so far implicitely restricted our calculations to plane waves; the generalization of eq. (5) to the case of finite-width beams, for which $\beta_{m,nl}{}^2$ becomes a function of x is not straightforward *a priori*. Nevertheless, for sake of simplicity we shall merely posit, by analogy with the nonlinear prism coupler theory of Carter and Chen [9], that the linear plane wave result [eq. (4)], when the output amplitude A_t is a function of x is still valid in such a way that the substitution of $\beta_m{}^2$ to $\beta_{m,nl}{}^2$ from eq. (5) in eq. (3) provides directly the equation of the nonlinear FP resonator. This equation writes for an ideal Kerr type nonlinearity:

$$\frac{d^2A_t}{dx^2} + 2j\beta\frac{dA_t}{dx} + \left[\rho_m{}^2 - \beta^2 + \eta\zeta^2|A_t|^2\right]A_t + j\kappa_m{}^2A_t = j\xi A_i(x) \tag{6}$$

There are several arguments in favor of this simple NL treatment.

First the basic nonlinear equation (6) can be seen as a result of a coupled mode theory in a nonlinear structure including transverse effects along x [10]. Indeed the y dependence of the field in the resonator, at a position x, can be expanded on the eigenmodes of the resonator. In the vicinity of a given resonance, #m, the sum reduces to two modes only: the two modes m^+ and m^- which are forward and backward propagating along the x axis and have the same y-dependence. The contribution of the other ones are negligible provided that they are not significantly excited, in other words that they are "far enough" from the mode #m. This is indeed possible if the cavity finesse is high enough. Under that assumption the y-field profile of the nonlinear response is constant and independent of the nonlinearity and this justifies eq. (6). Thus the nonlinearity leads to a modification of the wavevector β_m^2 only.

Secondly the validity of the modal equation (6) cand be tested numerically by comparison with a bidimensional calculation. This bidimensionnal calculation consists of solving together the equations for the forward- and backward- propagating waves in the usual slowly varying amplitude approximation:

$$\begin{cases} -2\,jk_y \dfrac{\partial E_f}{\partial y} + 2\,j\beta \dfrac{\partial E_f}{\partial x} + \dfrac{\partial^2 E_f}{\partial x^2} = A_{NL}\left(|E_f|^2 + 2|E_b|^2\right)E_f \\ +2\,jk_y \dfrac{\partial E_b}{\partial y} + 2\,j\beta \dfrac{\partial E_b}{\partial x} + \dfrac{\partial^2 E_b}{\partial x^2} = A_{NL}\left(|E_b|^2 + 2|E_f|^2\right)E_b \end{cases} \tag{7}$$

where:

$$\begin{cases} k_y = \sqrt{\left(\dfrac{\omega}{c} n_o\right)^2 - (\beta_o)^2} \\ A_{NL} = -2\dfrac{\omega^2}{c^2} n_o n_2 \end{cases} \tag{8}$$

together with the boundary conditions on the two mirrors:

$$\begin{cases} E_f(x,0) = R_f E_b(x,0) + T_f A_i(x) \\ E_b(x, y = -L) = R_b E_f(x, y = -L) \end{cases} \tag{9}$$

Here, "b" and "f" subscripts stand for backward and forward.

2.1.c Normalized equation. It is convenient to introduce in eq. (6) dimensionless parameters which have a nice physical interpretation and leads to a simple "universal" equation for nonlinear optical resonators [11]. These are:

$$\begin{cases} X = \kappa_m x \\ f = \dfrac{\zeta}{\kappa_m} A_t, \quad g = \xi \dfrac{\zeta}{\kappa_m{}^3} A_i \\ \Delta = \dfrac{\beta^2 - \rho_m{}^2}{\kappa_m{}^2} \\ \alpha = \dfrac{\beta}{\kappa_m} \end{cases} \tag{10}$$

The number of significant parameters is greatly reduced by this treatment. According to eqs. (10) the x-axis is scaled by $1/\kappa_m$, and the transmitted and incident amplitudes are scaled with respect to the nonlinear coefficient. Δ measures the detuning in terms of the line width, $(\kappa_m)^2$, as α does for the angle of incidence. With the reduced quantities defined in eqs. (10), the modal equation (6) takes the simple form:

$$\frac{d^2f}{dX^2} + 2j\alpha\frac{df}{dX} + \left[j - \Delta + \eta|f|^2\right]f = jg(X) \tag{11}$$

which shows that any nonlinear Fabry-Pérot resonator in the single mode approximation is characterized by only three reduced parameters: the detuning Δ, the angle of incidence α and the sign of the nonlinearity η $(= \pm 1)$.

2.1.d Numerical comparison between modal and bidimensional theories. The results of the modal and bidimensionnal theories are shown on Fig. 2a&b in the special case of normal incidence. In addition, the good agreement between the two theories has been checked for every angle of incidence. The influence of the finesse of the cavity as well as of the incident beam width on the transmission behavior predicted by these two theories will now be investigated. A gaussian incident beam defined by $A_i(x)=A_i\exp(-x^2/w^2)$ is assumed (the reduced waist is $W=\kappa_m w$).

The influence of the cavity finesse is seen on figure 2a on which we have plotted the response of the resonator, represented by the transmitted amplitude at the center of the incident beam, $f(X=0)$, versus the input amplitude, g. The nonlinearity is positive ($\eta=+1$). The curves (a-c) are obtained with the bidimensional calculation for Fabry-Pérot resonators whose reflection coefficients are respectively 0.6, 0.9 and 0.95, leading to finesse cavities respectively equal to 3, 7, 30. The corresponding reduced quantities are the same for these three curves: $\Delta=2.6$, $W=4$, $\alpha=0$. Thus, the modal curves associated to the three cases (a-c) are identical, and appear on the plot (d) of fig. 2a. Figure 2a shows that the agreement is already good for a finesse F_c as low as 7 and that it is all the more good as F_c is high. In fact, this is not surprising since the transmission peaks of the resonator (in variable β^2) are F_c times thinner than they are separated. Let us recall that the origin of optical bistability lies in the self-retuning of the resonator on one of its β^2 mode. Thus, in the range of β^2 corresponding to the bistable loop (approximately Δ times the resonance width around the peak), the transmission of the resonator is well described by a lorentzian function: this is the basic approximation of the modal theory. As a consequence, the error due to the modal approach can be evaluated in the following way: it is the difference between the approximate expression of the transmission (Eq. 2) and the rigorous one (Eq. 1) in the range of interest of β^2.

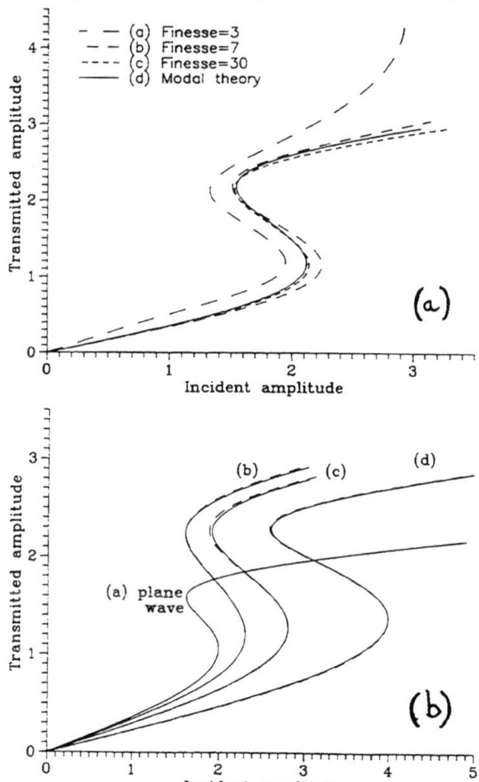

Figures **2**.
a) comparison between the bidimensional calculation for various finesses (curves a-c) to the modal calculation (curve d). The agreement is already good for finesses as high as 7.
b) comparison for narrow incident beams and a finesse of 30. The beam radius (at intensity 1/e) are 1.26, 0.63 and 0.3 times the cavity thickness for respectively curves b, c and d. In every case, the result of the bidimensional calculation (solid curve) coincides with the one deduced from the modal theory (dashed curve). (after ref. [11]).

One can thus predict that the modal theory is valid even for highly focused incident beam, provided the extension of the beam in the β^2 domain does not allow the excitation of other resonances than the closest one, characterized by $(\gamma_m)^2$. This result is demonstrated on figure 2b. For this example, we consider the resonator of the curve (c) of figure 2a, but now approximately ten times thicker. It is approximately 5 λ thick and its finesse is F_c = 30.6. The plane-wave response is plotted in (a) and the curves (b) to (d) correspond to the responses at the center of the incident beam for three decreasing values of the beam waist, X_o. Every curves (b) to (d) include in fact two plots: the full line one corresponds to the bidimensional calculation whereas the dotted line one has been obtained with the modal theory. In this way, one can verify that the agreement is always very good whatever the value of the beam width x_o is, even when x_o is smaller than L. We also notice that the plane wave response is never recovered, even for "large" incident beams.

As predicted in Ref. [11], the modal theory is only valid provided the overlap between the different resonances can be neglected. Surprisingly, from fig. 2a, this requirement is already fulfilled for finesses as low as a few units. In addition, the incident beam waist can be chosen smaller than the resonator thickness. As a consequence, the unified modal theory applies to a wide class of NL optical resonators. Thus the other curves of this paper have been calculated with the modal theory.

The domain of validity of the modal theory being now established, it can be used for studying extensively the influence of the different parameters on the reponse of the NL optical resonators. The normalised form of the modal equation is here very helpfull since it shows that any nonlinear resonator can be described by only three parameters already stated in section 2.1.c, eq. (10). This kind of approach has been used for the study of nonlinear optical resonators by several authors, namely Firth et al [12] and Balkarei et al [13].

2.1.e Generalisation to multilayered structures.The basic equation (6) has been derived in the stationary case for "ideal" nonlinear Fabry-Pérot. This equation obviously holds for every kind of nonlinear structures, the only condition is that a resonance associated to a pole of the transmission function might be isolated from the others [10]. This is in paticular the case for multilayers GaAs Fabry-Pérot type structures [10,14], a structure of particular technological interest [15,16]. Let us consider for example the case of the structure depicted on Fig. 3. Such a device works in reflection and corresponds to a recently published stucture [15,16]. This kind of nonlinear structure has been studied by Olin and coworkers [17,18] by bidimensional models which requires heavy time-consuming computations. Our approach has the advantage of being much more faster due to the elimination of the longitudional y-variable in the modal equation.

The front and back mirrors are assumed linear and are constituted by, respectively, 7 and 14 periods of quarter-wave layers of

AlAs/$Al_{0.1}Ga_{0.9}As$, such that their reflectivities in intensity are R_F=91.7% and R_B=97.6%. The refractive indices of AlAs, GaAs and $Al_{0.1}Ga_{0.9}As$ are respectively chosen equal to 2.956, 3.660 and 3.551 at the wavelength in vacuum of 840 nm. The active medium exhibits a negative nonlinearity and consists of 130 periods of alternating (~10 nm) GaAs and $Al_{0.3}Ga_{0.7}As$ layers. We have checked that it is possible to replace it by a single layer with an averaged value of the permittivity, whose complex refractive index n_o is 3.537+j 6.7 10^{-4}. The layers of the mirrors are quarter-wave ones at a wavelength λ=838 nm, and the thickness of the central layer is chosen equal to 2.7258m.

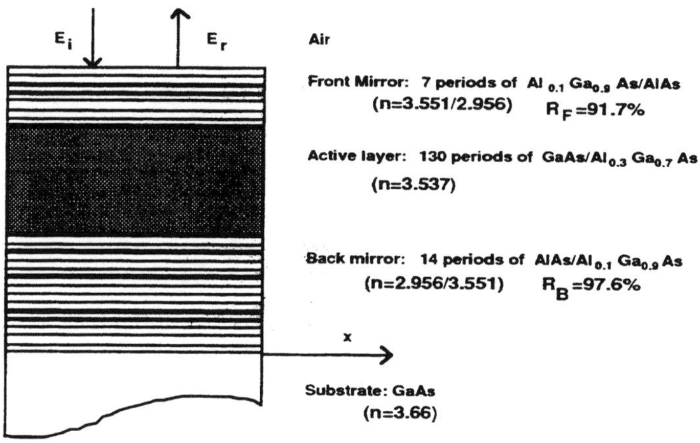

Figure 3: Geometry of the nonlinear multiple-quantum-well Fabry-Pérot-type structure of $Al_{1-x}Ga_xAs$/GaAs/AlAs.

One of the important points is to determine the reflected field of the NL device (Fig. 3). This question will be important also for distributed couplers in waveguides in section 3.2.

In the linear plane wave case, there exists a linear relation between the reflected field $E_r(\beta)$, the transmitted one which is here the intracavity field $E_t(\beta)$, and the incident one $E_i(\beta)$ which may be written as [10]:

$$E_r(\beta) = pE_i(\beta) + qE_t(\beta) \quad (12)$$

in which p and q are two constant complex coefficients nearly independent of β in the vicinity of the resonance. The coefficients p and q entering eq. (12) only depend on the features of the linear transverse field map. This kind of expression has already been derived by Ulrich [19] for a linear prism coupler, a configuration which is in fact very similar to the Faby-Pérot considered herre. Thus, in the case of a finite-beam-width excitation, the reflected field spatial distribution, $E_r(x)$ is related to the transmitted one, $E_t(x)$ by the same relation:

$$E_r(x) = pE_i(x) + qE_t(x) \quad (13)$$

In the nonlinear case, since p and q are nearly independent of β in the β-range considered here, i.e. "close" to a resonance, they do not depend of the nonlinearity [10], and equation (13) holds in the nonlinear case too: the relation between the reflected, transmitted and incident fields remains linear.

A linear analysis of the multilayered stucture depicted Fig. 3 allows for the determination of all of the parameters involved in the modal equation (6). The numerical values are given in table 1.

parameter	value
$(\beta_m)^2$	$1.8901\ 10^{12} + j\ 0.61878\ 10^{12}\ m^{-2}$
ξ	$0.1805\ 10^{12}\ m^{-2}$
ζ^2	$3.701\ 10^{15}\ V^{-2}$
p	$-0.9927 + j\ 0.00055$
q	$3.586 + j\ 0.0157$

Table 1 - numerical values of the parameters for the modal theory.

A comparison between the exact and modal theories can easily be performed in the nonlinear plane-wave case. Indeed, the plane-wave approximation allow for the determination of the exact nonlinear response of any nonlinear stratified structure [20]. In ref. [20], Danckaert et al consider in a detailed way the question of the validity of the commonly used Slowly Varying Enveloppe Approximation (SVEA) and derive an important criterium. It is shown that SVEA is no longer valid when dealing with thin films as already pointed out by Chen and Mills [21].

It is seen in Fig. 4 that the agreement between the modal reflectivity obtained from eqs. (6,13) and the exact reflectivity is excellent. Hence, we can use the tested coefficient (table 1) to study transverse effects in the nonlinear structure (Fig. 3) illuminated by an incident Gaussian beam.

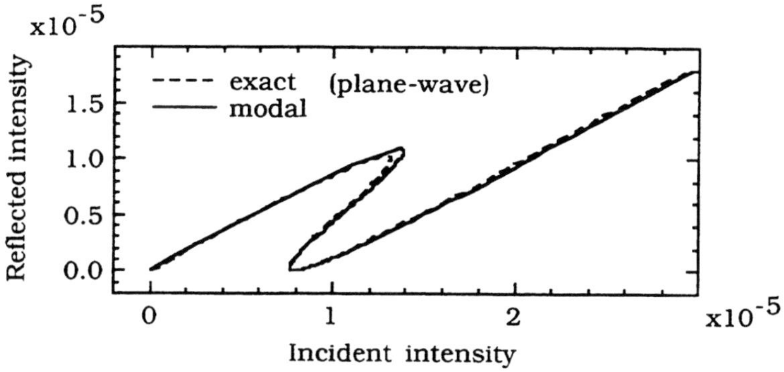

Figure 4: Comparison between the exact and modal theories in the plane-wave case. Normal incidence.

The influence of the incident beam width w (w is here the full width at half maximum) on the response of the NL Fabry-Pérot resonator is analyzed at normal incidence. Figure 5 shows the bistability curves in reflection for Gaussian beams whose waists are w=3, 6 and 12 μm. The total reflected power integrated across the beam section is plotted versus the total incident power for the different cases. It is seen that, for a given detuning, there exists a minimum waist for bistability to exist as pointed out by Olin et al [17].

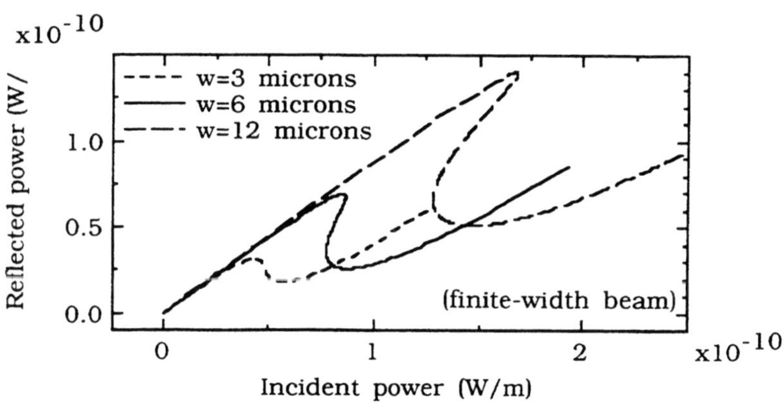

Figure 5: response of the device for three incident beam widths w (w is the full width at half intensity).

To summarize [6,10], the basic modal equation holds for nonlinear resonators as various as the ones schematically drawn on Fig. 6. It includes "ideal" Fabry-Pérot and multilayered structure in bulk or integrated configuration using Bragg reflectors (see [22] for example) already described, but also distributed couplers as these used in guided wave optics. Such distributed couplers based on the use of gratings or prisms will be described with more details later in this chapter.

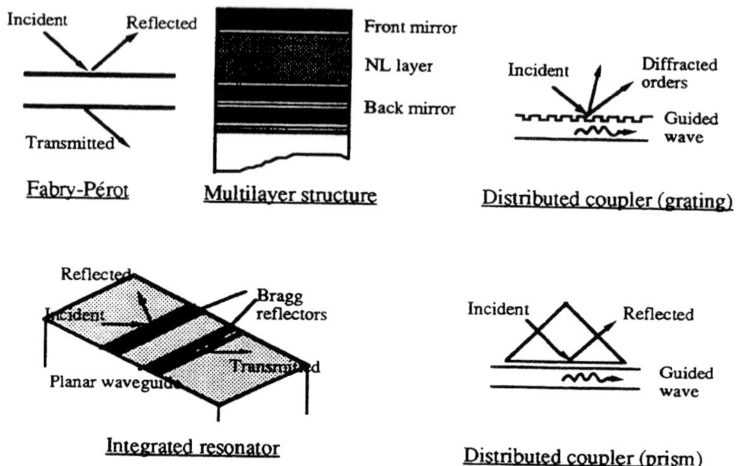

Figure 6: examples of NL optical resonators for which the modal theory applies.

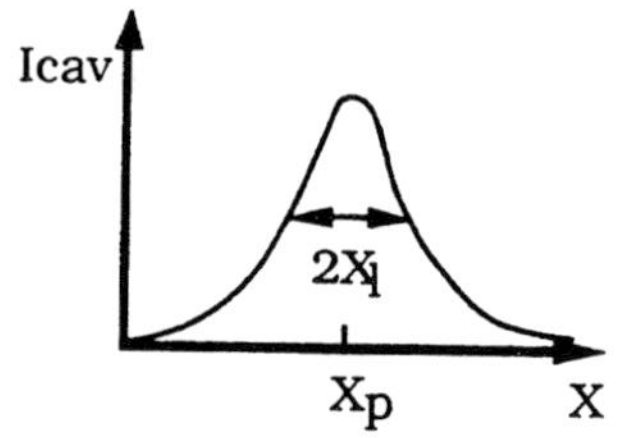

$$I_{cav} = \frac{C}{(X - X_p)^2 + X_l^2}$$

position width

$$X_{pNL} = X_p + C_{NL} I_{cav}$$

Figure 7: Main feature of the modal treatment: an isolated resonance leads to a response function which is a lorentzian function whose position depend on the nonlinear term.

The basic feature of these resonators is the following: they all exhibit an isolated resonance leading to an intracavity amplitude which is a lorentzian function of a parameter X which acts on the eigenmode of the resonator, whatever X is. In the pure nonlinear phase effect, the position of the resonance depends on the cavity intensity, I_{cav} through a nonlinear coefficient C_{NL}. On the contrary, the width of the resonance, X_l, would depend on I_{cav} only through a nonlinear absorption phenomenon since it induces an intensity-induced damping of the resonator and an increase of X_l with I_{cav}.

2.2 Dynamical study of nonlinear multilayer structures.

2.2.a General dynamical equations. The general principles of the method presented in the previous section, in the stationary case, apply in a similar way to the dynamical case [23]. In the dynamical case, the pulsation ω of the light is no more accurately defined and we have to take into account the fact that β_m is a function of ω:

$$\beta_m^2(\omega) = \left.\frac{\partial \beta_m^2}{\partial \omega^2}\right|_{\omega_o} \left(\omega^2 - \omega_o^2\right) + \beta_{mo}^2 \tag{14}$$

where β_{mo}^2 is the pole at the central pulsation ω_o. Applying a Fourier transform in the time domain in a similar way as the one did in the space domain, one obtains a spatio temporal equation:

$$2j\omega_o \frac{d\beta_m^2}{d\omega^2} \cdot \frac{\partial A_t}{\partial t} + \frac{\partial^2 A_t}{\partial x^2} + 2j\beta \frac{\partial A_t}{\partial x} + \left[\rho_m^2 - \beta^2 + \eta\zeta^2 |A_t|^2\right] A_t$$
$$+ j\kappa_m^2 A_t = j\xi A_i(x,t) \tag{15}$$

which describes the space and time evolution of the transmitted field in the resonator. Equation (15) results from a slowly varying enveloppe approxiation for $A_t(x,t)$ which does not include the varying phase term $\exp(j\omega_o t - j\beta x)$. Again the normalisation given by eq. (10) together with:

$$\tau = c_t t = \frac{\kappa_m^2}{2\omega_o} \frac{1}{\left| \left.\frac{d\beta_m^2}{d\omega^2}\right|_{\omega_o} \right|} t \tag{16}$$

leads to a very simple reduced dynamical equation:

$$j\frac{\partial f(X,\tau)}{\partial \tau} + \frac{\partial^2 f}{\partial X^2} + 2j\alpha \frac{\partial f}{\partial X} + \left[j - \Delta + \eta |f|^2\right] f = jg(X,\tau) \tag{17}$$

provided the dispersion coefficient $d\beta_m^2/d\omega^2$ is real. An inspection of eq. (15) allows to derive some physically interesting quantity. Considering the plane-wave case i.e. $d/dx=0$ and $d^2/dx^2=0$, and assuming that the resonator is linear and tuned on the resonance, i.e. that $\rho_m=\beta$, shows that $\tau_c = 1/c_t$ where c_t is defined by eq. (16) corresponds to the cavity build-up time. The reduced time τ is thus given in units of the cavity build-up. Let us point out that the nonlinear coefficient ζ^2 is complex in the general case, the real part is related to nonlinear dispersive effect and the imaginary part to nonlinear absorption.

2.2.b Stability analysis of the plane-wave nonlinear stationary solution. The stability of the nonlinear stationary solution is important when considering the study of nonlinear systems based on Kerr type nonlinearities. Indeed, stationary solutions like bistable loops may be of no pratical interest when exhibiting instabilities. The existence of such instabilities, self-pulsing and chaos has been predicted in nonlinear optical resonators more than ten years ago by Ikeda [24], Winful and Cooperman [1], Lugiato et al [2]. Other interesting dynamical phenomena has been predicted such as critical slowing down close to turning points of bistable-like stationary characteristics and noncritical slowing down [25] close to unstable solution for instance. These analytical results have been confirmed experimentally, for a variety of bistable systems, including optical ones, both with an absorptive [26] as well as with a dispersive nonlinearity [27].

The stability analysis of the plane-wave solution can be extended to a more realistic situation taking into account a non instantaneous nonlinearity, described by a phenomenological Debye model. Let is the nonlinearity response time in cavity build-up time units be T_d. We have now two equations to describe the system, one for the field (modal equation) and one for the material which appears here through the nonlinear term U [28]:

$$\begin{cases} i\dfrac{df}{d\tau} = -(i - \eta(\Delta - U))f + g \\ T_d\dfrac{dU}{d\tau} = -U + |f|^2 \end{cases} \tag{18}$$

In order to perform a stability analysis, we consider small variations of the nonlinear system around the stationary state: f_o, $U_o = |f_o|^2$. The two real variables $\delta|f|^2 = |f|^2 - |f_o|^2$ and $\delta U = U - U_o$ represent the variations from the stationary states. Introducing these variations, assumed to be small, into eq. (18), it is easy to uncouple the obtained set of two equations and to derive a third order differential equation for δU:

$$\begin{aligned} T_d\frac{d^3\delta U}{dt^3} + (1 + 2T_d)\frac{d^2\delta U}{dt^2} + \left[2 + T_d\left(1 + (\Delta - U_o)^2\right)\right]\frac{d\delta U}{dt} \\ = -\left\{3U_o^2 - 4\Delta U_o + 1 + \Delta^2\right\}\delta U \end{aligned} \tag{19}$$

Let us now turn to the question of the stability analysis of the solutions of eq. (18). This is most conveniently done considering eq. (19) dealing with the variation of the nonlinear term δU and its characteristic equation for the exponent λ governing the time response of δF and δU ($\delta U=\mu e^{\lambda t}$):

$$T_d\lambda^3 + (1+2T_d)\lambda^2 + \left[2 + T_d\left(1+(\Delta - U_o)^2\right)\right]\lambda + R = 0$$

$$\text{with } R = 3U_o^2 - 4\Delta U_o + 1 + \Delta^2 = \frac{dg^2}{dU_o} \tag{20}$$

The nature of the solution of eqs. (18) is determined [29] by the signs of the real parts of the roots of the characteristic equation (20). Applying the Routh-Hurwitz criterium, one obtains that the stationary solutions are unstable when one of the two following conditions is satisfied:

$$\begin{cases} 1) & R < 0 \\ 2) & \dfrac{1+2T_d}{T_d}\left[2 + T_d\left(1+(\Delta - U_o)^2\right)\right] < R \end{cases} \tag{21}$$

The first one corresponds to the negative slope part of the (U_o,S^2) stationary response curve and is well known. On the contrary, the second one can only be satisfied on the positive slope portions, and corresponds to a pair of complex conjugate roots crossing the imaginary axis (i.e. Hopf bifurcation)[30]. It is worth noting that the existence of the unstable solutions derived from the second inequality is directly related to the three-dimensional nature of the problem. The second inequality can be solved analytically in order to determine the unstable domain in the U_o variable. Instabilities only exist provided:

$$D = \Delta^2 + 4\left(1 - T_d^2\right)\left(1 + \frac{1}{T_d}\right) > 0 \tag{22}$$

where only the detuning Δ and the material response timeT_d are involved. The inequality (22) is very important in terms of material and device requirements since it determines the position of the instability domain. Indeed, this inequality directly leads to the following result: for a given detuning Δ, instabilities only exist if T_d is smaller than a maximum value, T_{dmax}, which is a fonction of Δ as plotted Fig. 8a.

Some important characteristics of the unstable domain are visualised on Fig. 8b. Indeed the boundaries separating the stable and unstable domains are plotted in the (U_o,T_d) plane, for several values of Δ. One can deduce some interesting results regarding the shape of the curve $U_o(T_d)$. Indeed, the system is unstable if $0<T_d<1$ whatever Δ. But the minimum cavity field U_o which can lead to unstabilities diverges when T_d decreases to zero. As a consequence, for a given driving field

S related to U_o through the stationary relation $U_o(g^2)$, the resonator is unstable in a window of the material response time $T_{dmin}<T_d<T_{dmax}$. It is of practical interest to note that the stability property of the stationary solution is sensitive to the material response time as this quantity is close to zero: this could be used to measure the material response time T_d.

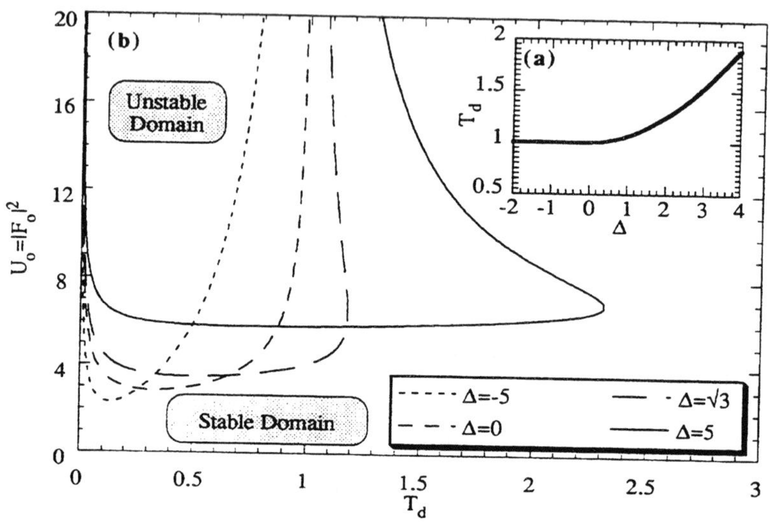

Figure 8: (a) The minimum value of T_d allowing for the existence of instabilities is plotted versus the detuning Δ. (b) The limit between the stable and the unstable domains is characterized by a critical intracavity intensity $U_o=|F_o|^2$ which is plotted versus the response time, T_d, of the nonlinear material, assumed to obey a Debye equation. The detuning Δ is considered as a parameter. Instabilities are due to the existence of two time scales in the system and are found to occur for a bounded domain of T_d. The time unit is the cavity build-up time.

This seems in contradiction with numerical results of Ikeda [24] and Abraham and Firth [31] who showed that instabilities occurs on the lower branch of the bistable curve $U_o(g^2)$ considering a full model without the single mode approximation. As a consequence we can deduce that this type of instability appearing on the lower branch arises from mode competition when several modes are simultaneously excited. This situation is excluded in our single mode description which applies to high finesse resonators only. We have seen here that unstable regions located on the upper branch can exist in the single mode regime, at least in certain region of parameter space depending on the cavity detuning and the material response time. This result confirms in a general feature the one obtained by Lugiato et al for a similar system [2].

Let us note that Ikeda instabilities which occur for Debye time shorter than the round trip time are not predicted by the dynamical modal theory. Indeed, the one mode approximation implies a limitation of the bandwidth of the order of several times the resonance width. In the time domain this means that the modal theory does not allow to describe phenomena at the round trip time scale. This corresponds to a Debye relaxation time between the round trip time and the build up time of the cavity. A new theoretical approach for the study of the influence of diffraction on Ikeda instability in the Kerr-tpe nonlinear ring cavity has been proposed by Haelterman [32] in the mainframe of the modal theory using some extensions.

2.2.c Nonlinear dynamics of the plane-wave solution. The matter of this subsection as already be considered by P. Mandel [25] who proposed a simple so-called generic equation (eq.(3) of Ref.23) in order to describe the switching dynamics of a bistable system. This model was verified for a variety of experimental systems [26,27] (that can be adequately described by a one dimensional (1D) model) and the analytical results obtained from the generic equation have been shown to be in good agreement with the experimental results. Here, we will generalise Mandel's results to the three dimensional case, and thereby we will show under whose assumptions such a generic equation can be derived. A quantitative comparison between exact solutions of the modal equation and the analytical predictions will be given. We will also be able to discuss the dependence on the Debye time of important dynamical phenomena like critical and noncritical slowing down. We will also show that generic equations can be derived from the modal theory in two limiting cases, namely i) $T_d=0$ and ii) $T_d>>1$.

i) The case $T_d=0$ corresponds to an instantaneously responding nonlinear medium. The analysis here follows directly from eq.(18) (or from eq.(19) by putting $T_d=0$). Therefrom, the following equation on the intensity $|f|^2$ (for small variations with respect to the stationary state) can be derived:

$$\frac{d^2|f|^2}{dt^2}+2\frac{d|f|^2}{dt}=g^2-\left[1+\left(\Delta-f^2\right)^2\right]|f|^2 \tag{23}$$

In the right hand side (RHS) of eq. (23), one recognises the expression of the stationary response curve, and therefore this member can be interpreted as a measure of the deviation of the output intensity from its steady state value, for a given input field S. The existence of the second order derivative in the LHS of eq. (23) is fundamentally due to the fact that the basic eq.(18) is complex, i.e. that the phase space has two dimensions (phase and amplitude, or equivalently, the parts of the field in phase and out of phase with the driving field). This will be the case for every dispersive nonlinear driven oscillator. Therefore, eq.(18) is much more general than the context of nonlinear optical resonators in which it has been derived here, and it can be written down for any two-dimensional nonlinear dispersive system.

If, moreover, slowing-down occurs (i.e. in a dynamical regime close to the unstable branch or close to the switching point), $|f|^2$ is a slowly varying function of time, and a kind of slowly varying intensity approximation in the time domain applies. The second and third derivatives in the left hand-side of eq.(19) can then be neglected, yielding an equation with is very similar to Mandel's generic equation.

ii) We now consider the limit $T_d>>1$, where the nonlinear material response is much slower than the cavity build up process. In this case, the time derivative of U turns to be small (of order T_d^{-1}). The time response of the device is then governed by the material response time, and all the higher derivatives appearing in eq.(19) can be neglected, yielding:

$$T_d\left[1+(\Delta-U_o)^2\right]\frac{dU}{dt} = g^2-\left[1+(\Delta-U)^2\right]U \tag{24}$$

Eqs. (24 & 23 without the second derivative in the LHS) have the same structure and can therefore be handled along the same lines. They still do not present analytical solutions, due to the appearance of the third order term in the RHS. However, since they have been derived from a local expansion around an arbitrary fixed point, we introduce little extra errors by further developing the RHS around an arbitrary fixed point (I_i*,I*), e.g. one of the switching points. This allows to apply the general results obtained by Mandel and coworkers [25,33]. Scaling laws, e.g. a pulse area scaling law, critical and noncritical slowing down are thus generic properties of bistable systems, whatever the system's phase space dimension may be. A quantitative comparison between the analytical results and exact solutions of the modal equation from which they were derived was made, yielding a good agreement between both.

Another surprising nonlinear phenomenon, which can be seen as the contrary of the slowing down and which is called speeding up phenomenon has been recently predicted analytically [28]. It can be checked by numerically solving directly the set of eqs.(18) for suitably chosen initial conditions. The situation considered here corresponds to a detuning $\Delta=3>\sqrt{3}$ so that the stationary response curve shown in Fig. 9a is indeed unstable. The hatched region shows the portion of the curve on the positive slope part which can be unstable, depending on the value of the material reponse time. The nonlinear system is left free to evolve under a constant excitation starting from an initial condition (on $|f|^2$) which is close to the stationary solution. The characteristic time evolution of the system, T_c, is plotted as a function of the medium's response time T_d on Fig. 9b, for point A of Fig. 9a which can be unstable. Indeed T_c is larger than zero when $0.26<T_d<1.2$. In the Debye time domain $3<T_d<15$ we observe a very surprising effect since the characteristic time of the nonlinear resonator decreases when the material response time increases: moreover, for $T_d>4$, the characteristic time of the system remains much shorter than the response time of the nonlinearity. This corresponds to a speeding up phenomenon.

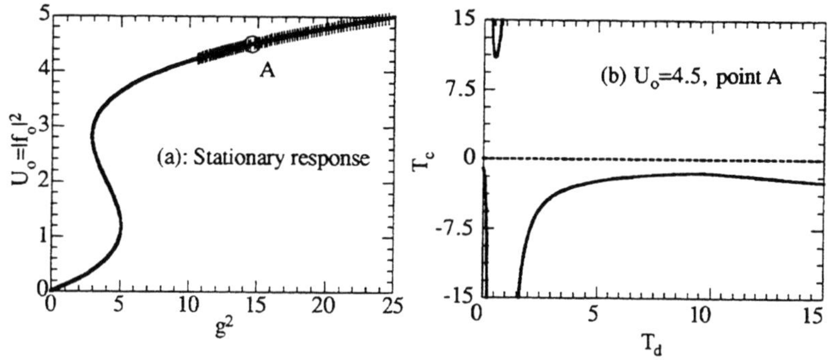

Figure 9: (a) The steady-state response is plotted as the intracavity intensity, U_o, versus the incident intensity, g^2. Here Δ is chosen equal to 3 in order to have bistability. (b) The characteristic response time is plotted versus the material response time, T_d, for point A of (a). Notice that the system becomes unstable for $0.3<T_d<1.54$ and that speeding up appears for $T_d>5$: the characteristic time is still about five times *smaller* than the Debye time.

A direct example of the speeding-up phenomenon is shown on Fig. 10 where the cavity intensity, $|f|^2$ is plotted versus time for conditions which correspond to point A on Fig.9a. For $T_d=1$, according to Fig.9b, the system is unstable. Indeed, on Fig.10, one sees that the system starts to oscillate and that the oscillation amplitude is growing at a rate which corresponds to a characteristic time of the order of 10. For $T_d=1.68$, the system is stable (see Fig. 9b). Indeed, on Fig. 10, the system is seen to return to its stationary state after oscillations which are damped with a characteristic time also of the order of 10. However, for higher value, e.g. $T_d=10$, the system is seen to return faster to its equilibrium than it would do since it has a characteristic time of the order of 1. This is a surprising property of eqs.(18) which could probably be observed experimentally.

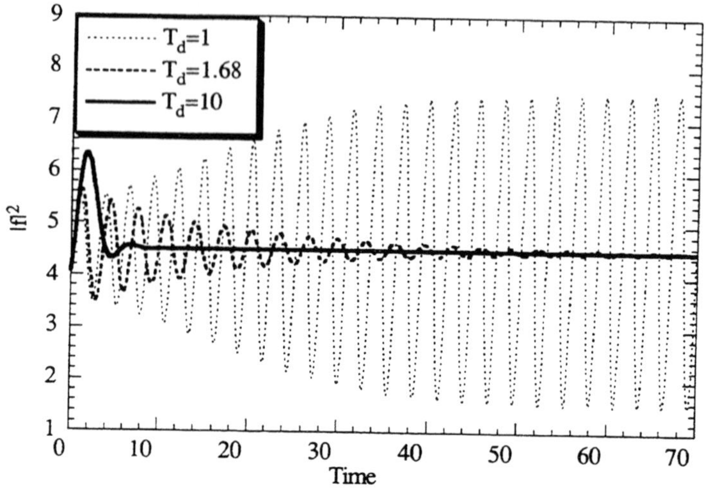

Figure **10**: Nonlinear dynamics in the conditions of point A of figure 9a. The cavity intensity is plotted against time for three different material response times. The system is unstable for Td=1.0 and stable for Td=1.68 and 10. In the unstable domain a self-pulsing behaviour is observed. The response for Td=10 clearly demonstrates the speeding up phenomenon since the system returns faster to equilibrium than for Td=1.68.

As shown by the curve "Td=1" on Fig.10, in the unstable domain, the system settles in a self-pulsing regime, a behaviour typical for a Hopf-bifurcation. The occurrence of self-pulsing here needs to be ascribed to the two timescales appearing in the equations, and not to mode competition since the theory was developed here in the single mode approximation. Contrary to Lugiato et al [2] for a very similar set of equations, and to a Yefimov & Shkerdin [34] for another set of equations describing the same physical system, the general equations of single mode nonlinear resonator with non instantaneous material reponse time does not exhibit a bifurcation cascade, eventually leading to chaotic behaviour. All the periodic solutions obtained are themselves found to be stable.

Distributed feedback gratings (DFBG) are another kind of nonlinear resonators which are not described by the modal theory here presented in details. Indeed they are used closed to the border of the stop band where several modes are always simultaneously excited. Thus, the single mode approximation is no more valid and a DFBG structure is very different from a Fabry-Pérot resonator[22]. On the contrary, interference filters belongs to the FP-like device family because they exhibit an isolated resonance. In addition, it is worth noting that DFBGs are less interesting than FPs in terms of applications because the required irradiance for nonlinear effects is much higher due to the fact that the border of the stop band is less sharp than the resonance in an associated interference filter [22]. However, DFBGs can exhibit the same kind of dynamical behavior than nonlinear FPS.

Winful and Cooperman [1] have shown that the DFBG can exhibit self pulsing and chaos when driven by a cw input provided the material characteristic ime T_d is smaller than the transit time defined by $T=nL/c$ (L is the length of the structure). Self-pulsing is very interesting for practical applications since it realizes a fast all-optical clock function. More recently, Cauwenberghs et al [35] have shown that the behavior depends also on the coupling constant Γ between the forward and backward propagating waves in the DFB. For low Γ, self-pulsing is not present and optical bistability can be obtained. The transition between the Bragg reflection domain inside the stopband to transparency outside is related to the development of a spatial soliton-like standing wave pattern. When the coupling is strong enough, this transition is very sensitive to the nonlinearity, leading to self-pulsing. In this case, the relaxation of the material works as a delay line which appears to enhance instabilities and brings the system to a chaotic regime. Cauwenberghs et al [35] considered NL planar multilayered structures which they treated as a stack of homogeneous NL layers. In each layer, using the slowly varying envelope approximation (SVEA), the nonlinear dynamical wave equation reduces to a set of coupled first order partial equations in the amplitudes of forward and backward propagating waves, in a similar manner as shown by Firth [36] for the stability analysis of NL Fabry-Pérot resonators. A more detailed study of periodic structures is given in [37,38].

2.2.d Modulational instability of the plane wave solution. A great number of papers have been devoted to transverse effects in nonlinear structures in recent years, as shown in the recent review paper by Abraham and Firth [39].

One of the important results is the existence of spatial solitons for a positive Kerr type nonlinearity [40]. The existence of spatial solitons in resonators has been established analytically [6]. This solution corresponds to the exact balance of diffraction by the self-induced refractive index gradient (self-focusing). This means that a spatial soliton builds up in the cavity and that the resonance is perfectly verified across the entire beam width in such a way that no light is reflected by the FP resonator, exactly as in the case of a linear FP resonator illuminated by a plane wave at resonance. Such solitonlike solutions in bistable devices have been theoretically investigated by Moloney [4] for the particular case of a nonlinear ring cavity. To date, only the observation of spatial solitons in planar waveguides, without any cavity, has been reported [41,42]. However, the same transverse profiles are predicted in both cases: with and without cavities [6].

Besides "Time domain transverse effects", i.e. a non-permanent regime should also affect significantly the nonlinear response and lead to behavior unexpected from the stationary study [43,44]. Indeed, the emergence of regular or random spatial patterns from a gaussian or plane wave excitation depends very sensitively on the particular mechanism involved in the system [45]. In particular, the spontaneous emergence of spatial solitary waves in NL FP resonators or ring cavities pumped by gaussian beams has been described [46,47] and interpreted by Lugiato and coworkers in terms of dissipative structure formation [5,48].

A novel instability of the modulation type which is closely related to what occurs in FP type resonators, has recently been predicted for counterpropagating beams in a Kerr medium [49-51]. This effect is specific of counterpropagating waves and does not occur in ring cavities. Kaplan predicted [52] the possible use of such counterpropagating beams for making bistable devices, studied by Leberre et al [53].

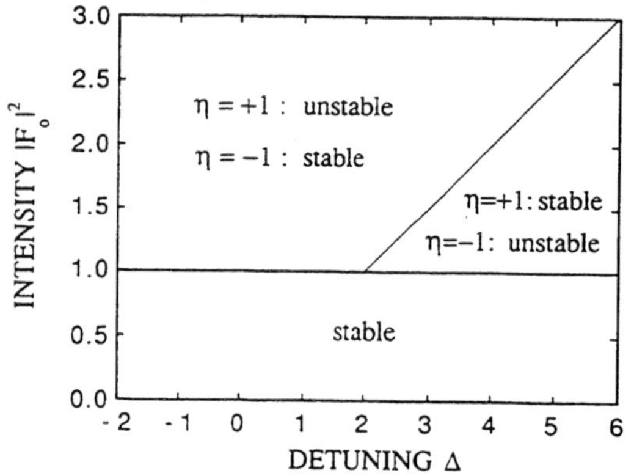

Figure 11: Domains of spatial modulational instability with a plane-wave excitation.

In order to analyze the stability of the stationary solution, taking into account the transverse dimension, we have to introduce in eq. (17) the decomposition $f=f_0+\delta f$ with the following ansatz [54]:

$$\begin{pmatrix} \delta f \\ \delta f^* \end{pmatrix} = e^{\lambda\tau} \cos(kx) \begin{pmatrix} \delta f_k \\ \delta f_k^* \end{pmatrix} \qquad (25)$$

Equation (25) introduces the possibility for the amplitude of a Fourier component of wavevector k to analyse the temporal behavior of the spatial spectrum of these solutions. When looking for the plane wave solution, only the case k=0 is considered. The $k\neq0$ type of instability is called modulational instability.

The corresponding domain of instability in the plane $(|f_0|^2, \Delta)$ is shown on Fig.11 in the case of an instantaneous nonlinearity $(T_d=0)$ and an incident plane wave. The plane-wave instability can be checked numerically by integrating directly the modal equation (17) including the transverse terms. Figures 12a&b present the evolution of the transmitted intensity profile for an input plane wave that verifies the instability condition presented on Fig. 11. The plane wave response instability has been seeded by adding a small periodic perturbation of amplitude $|f_0|/100$. We observe the grouwth of a periodic stripe structure: in the case of normal incidence (Fig. 12a) the pattern is a stationary one, as in the case of slightly oblique incidence, the structure has a translation motion at the speed 2α where α is related to the angle of incidence (Fig. 12b).

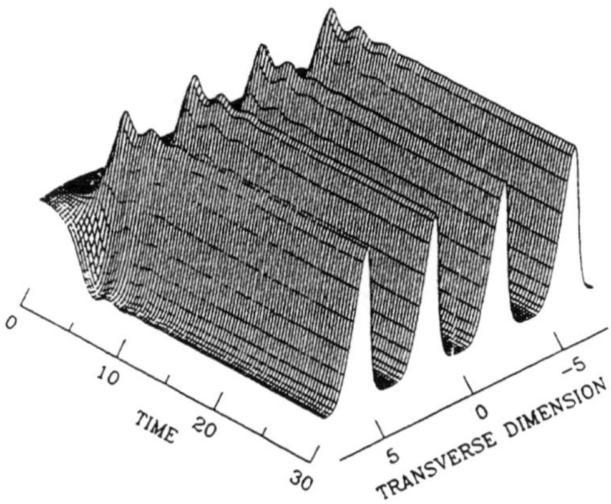

(a) under normal incidence, $\alpha=0$

Figure 12: Evolution of the transmitted beam amplitude profile in the case of plane wave excitation with parameters corresponding to the unstable region plotted Fig. 11: $|f_0|^2=2$, $\Delta=1$. (after ref. [54]).

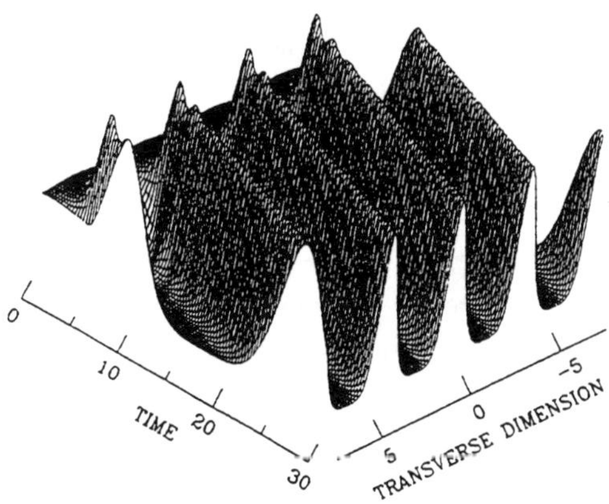

(b) under oblique incidence, $\alpha=0.1$

2.2.e<u>Influence of the angle of incidence: The difference between Fabry-Pérot resonators and integrated distributed couplers:</u> The angle of incidence θ [11] is expected to play an important role. One of the main results is that optical bistability only exists under normal or nearly normal incidence and disappears when the resonator is tilted [11]. The limit between the bistable domain and the switching one is characterized by a threshold in angle (called θ_t). In addition, when the angle of incidence is increased, the response tends to the solution of

Carter and Chen's equation (eq.(3) of ref. [9]) which describes the nonlinear prism coupler (see Ref. [55]). Indeed, a tilted Fabry-Pérot resonator does not include any kind of feedback along the x direction and can be seen as a propagative wave device very similar to a prism coupler which, also, does not exhibit any bistability when dealing with a locally NL material.

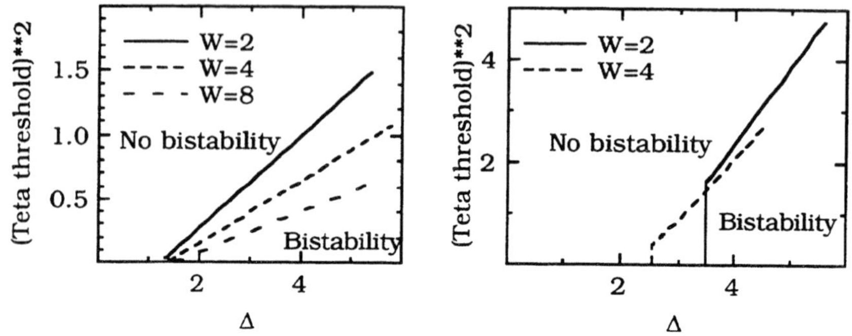

(a) (b)

Figure 13 Influence of the angle of incidence: threshold for bistability plotted versus cavity detuning: (a) positive nonlinearity, (b) negative nonlinearity. Every curve corresponds to a specific value of the gaussian beam width, W.

Because the number of parameters describing a Fabry-Pérot resonator is small, it is possible to derive general results concerning θ_t. The essential information is summarized on figs. 13a and 13b respectively for positive and negative nonlinearities. The normalized value of the angle of incidence squared, i.e. $(\omega n_o/c \sin(\theta_t))^2/\kappa_m{}^2$, is plotted versus the detuning Δ for different values of the gaussian beam width W. On Figs. 13, $\theta_t{}^2$ is seen to be almost proportional to Δ in every case, provided Δ is larger than a threshold Δ_t (W). For a given $\Delta>\Delta_t(W)$, bistability exists in the domain limited by the Δ axis and by the threshold straight line corresponding to W. For $\Delta<\Delta_t$ (W), bistability never exists. For $\eta=+1$ (fig. 13a), $\theta_t=0$ when $\Delta=\Delta_t(W)$. This means that the bistable angular domain becomes smaller as Δ approaches $\Delta_t(W)$. In addition, it decreases when the beam waist W increases. In fact, although this does not clearly appear on figure 13a, the threshold curve tends to a limiting curve when W becomes infinitely large. Let us now notice that the same features are obtained in the case $\eta=-1$ (fig. 13b). Nevertheless, one specific result is to be pointed out. First, $\theta_t \neq 0$ when $\Delta=\Delta_t(W)$. This means that the minimum width of the angular domain for which bistability occurs is different from 0. This is due to nonlinear enhanced diffraction, due to the self-defocusing phenomenon, which facilitates the excitation of the two counter-propagating modes. This idea also explains why the threshold value of the incidence angle decreases when W increases: the larger W, the narrower the width of the associated spatial frequency spectrum, and the less efficient the excitation of the two counter-propagating modes, $\pm \beta_m$.

The practical conclusion of this general study is the order of magnitude of the maximum angle of incidence for which bistability is still possible. Roughly speaking, for a positive nonlinearity, the limiting angle is of the order of the angular resonance width (related to $\kappa_m{}^2$). Due to self-defocusing, it is almost 3 or 4 times larger for a negative nonlinearity. The disappearance of bistability under oblique incidence is mathematically due to the fact that the second derivative in eq.(17) becomes negligible. In fact this second derivative term accounts for the simultaneous excitation of two counterpropagating waves, a phenomenon which appears to be responsible for optical bistability. Let us point out, finally, that the disappearance of optical bistability in a tilted FP is only due to transverse effects and is not predicted by a plane-wave calculation which introduces an effective longitudinal feedback. Indeed, even under oblique incidence, the plane wave assumption implicitly links the behaviors at the different points along x, which is not the case in the concrete situation: roughly speaking, the light propagates along a zig-zag path between the two mirrors.

2.3.Dynamical spatio-temporal behavior.

A modal dynamical approach has been used recently for nonlinear resonators. Balkarei et al studied nonlinear resonators [13] in semiconductor. Recently Mc Donald and Firth [56] showed that it is possible to take advantage of the spatial soliton-like spontaneous formation due to modulational instabilities for making memories. Indeed, starting with a transverse pattern in which coupled spatial solitons are produced in a situation similar to the one of Fig. 12, McDonald and Firth have shown that the pattern remains stable if one or more of the spatial peaks (related to solitons) is removed. This allow the use of a spatial for making memory (through the presence or the absence of a soliton at a given position) and more generally for all-optical processing, as visualized on Fig. 14.

In two transverse dimensional systems the question of spatial transverse pattern formation is even more interesting. Indeed stable or not self-organised patterns can be produced in nonlinear passive resonators [7,12] as well as in lasers. There is an analogy between the question of spatial transverse instabilities in nonlinear resonators and temporal soliton instability in synchronously-pumped ring resonators. This equivalence can be seen directly on the equation describing such resonators [57]. Instabilities shown recently by Haelterman et al [57] in ring resonators are very similar with those presented here.

In particular, the influence of the angle of incidence has been recently demonstrated experimentally in a nonlinear Fabry-Pérot filled with carbon disulfide illuminated by 200ps long pulses at 532nm [58]. The reflectivity of the mirrors (r=.785) gives a finesse close to 13 and the thickness is e =0.75mm. The incident beam focused onto the input face of the Fabry-Pérot has a waist of 1.33mm in the x dimension. The time evolution of the transverse profile of the output pulse has been observed using a streak camera, while detuning was monitored.

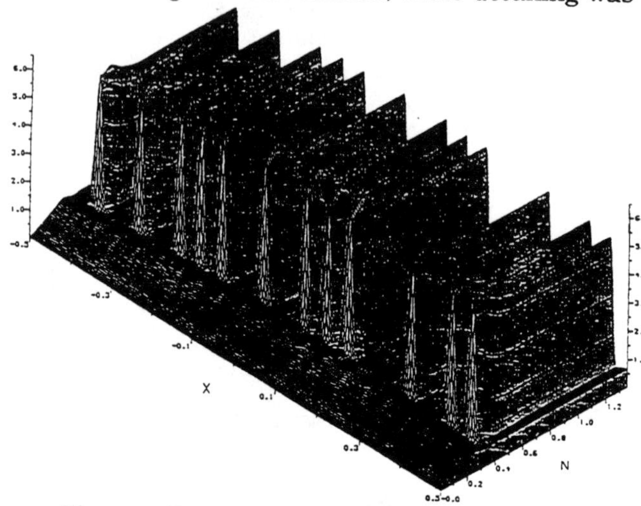

Figure 14: transverse pattern encoding in non linear resonator (after McDonald and Firth[56])

The results are presented for two incident angles: for $\theta=0°$ on Fig. 15 and for $\theta=12°$ on Fig. 16 (which corresponds to an intracavity angle of the order of 7°). Indeed, as predicted in general by a stationary analysis presented in the section 2.2.d, optical bistability, which can exist at normal incidence, disappears when the resonator is tilted. The value of the critical angle depends on the detuning, beam width and other parameters, but its order of magnitude is given by $\theta_c=(\Delta\theta^2)^{1/2}$ where $\Delta\theta^2$ is the resonance line-width in the θ^2 variable. With the experimental parameters one gets $\theta_c\approx 1°$.

The results of the calculations are given on Figs. 15a and 16a, respectively, for $\theta=0°$ and 12°. They have to be compared to the corresponding streak camera pictures given, respectively, on Figs. 15b and 16b. In these figures, the horizontal axis represents time and the vertical one, the transverse space coordinate. The experimental pictures contain two spots corresponding, at the bottom, to the incident pulse and, at the top, to the transmitted one. For both, the darker the image, the higher the light intensity. The theoretical figures correspond to contour lines of the computed output spatio-temporal pulse. It means that the darker the drawing the higher the slope, and that peaks in intensity appear rather white. In both cases, the experimental detuning of the resonator is about -50°. Let us compare qualitatively the different behaviour regimes:

- Under normal incidence, the calculation gives a self-pulsing behaviour which appears on Fig. 15a as portions of rings in the space-time plane. The result is an output pulse whose duration is a bit longer than the incident one, due to an energy storage in the cavity. This storage is only possible at high intensity when the resonator is retuned by the nonlinearity. Self-focusing is responsible for a spatial confinement of light. This simulation is confirmed by the experimental result on Fig. 15b. In addition, spontaneous symetry breaking is visible on Fig. 15a.

(a) theory (b) experiment

Figure 15: Spatio-temporal response under normal incidence (time and space are respectively plotted horizontally and vertically).

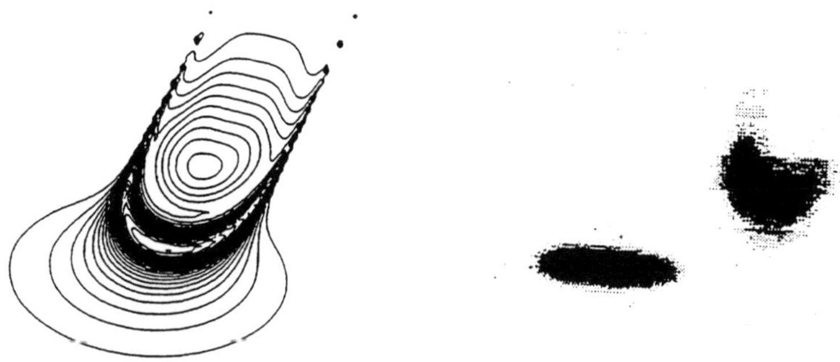

(a) theory (b) experiment

Figure 16: Same as Fig. 15 but under oblique incidence (7° inside).

- Under oblique incidence (Fig. 16), the behavior is clearly strongly different. The self-focusing phenomenon cannot affect the response because cumulative effects are broken by the spatial shift occuring at each round trip. One can see on both figures the typical shape in "croissant" which shows that the resonator is retuned by the nonlinearity: when the resonator is tuned, it works as a delay line both in time and space. In order to understand the meaning of the oblique lines let us consider a tuned linear resonator with a Dirac's excitation in space and time: some light will flow out of the resonator at every round trip, but every time a bit farther and a bit later. The relation between the spatial and temporal delays is linear and gives the oblique straight lines appearing on Figs. 16a and 16b.

The qualitative agreement between theoretical and experimental results is good. This confirms the theoretical predictions and, among them, the fact that the angle of incidence is a very important parameter when dealing with a nonlinear Fabry-Pérot resonator. It is not possible here to speak of bistability because the pulse duration (200ps) is not large enough as compared to the build-up time of resonator (34ps). Nevertheless, a dynamical analysis which takes into account transverse effects describes qualitatively what happens experimentally.

3. NONLINEAR INTEGRATED OPTICAL RESONATORS

3.1 Distributed couplers: A usefull tool for nonlinear material caracterisation.

3.1.1 <u>The use of distributed couplers for thin film study</u>: Prism and grating couplers are commonly used for the excitation of a guided wave into a planar waveguide. The angle of incidence of the input beam is chosen in such a way that the phase matching condition between the incident beam and the guided wave is achieved (see Fig. 6). In the case of a grating coupler, phase-matching is achieved through one diffracted evanescent order. For coupling the light into a waveguide with a good efficiency it is necessary to use half a coupler i.e. extending from $-\infty$ to 0. Indeed the optical tunnel effect across the gap in the prism coupler as well as the diffraction in the grating coupler are responsible for radiation losses of the guided mode which is called a leaky mode in these conditions[59,60]. The electromagnetic leakage is necessary to allow for the excitation process when an incident wave is present. The leakage associated to the prism or the grating is supressed for $x>0$ in order to "close" the optical device which becomes a waveguide. Under this condition the coupling efficiency can be as high as 80% theoretically [60-62]. The prism and grating coupler look therefore very similar, the only difference lies in the way to achieve the phase-matching condition.

For thin film caracterisation a full coupler is generally used as drawn on Fig. 6 instead of half one since there is no particular interest propagation [63]. On the other hand, it is interesting to look at the intensities of the various reflected and transmitted beams when the angle of incidence varies. Indeed when the phase-matching condition is realized (at resonance), the situation changes strongly when compared to the off resonance case. This is visible on the reflectivity/transmitivity response curve plotted versus the angle of incidence.

Here the result is not at all the same for prism and gratings couplers. Let us consider non absorbing waveguides. In the case of the prism coupler, there is only one emergent beam from the structure (reflected one) and the energy balance condition implies that the reflectivity of an incident plane wave is equal to one in modulus. However, the phase of the reflected beam varies rapidly of π across the resonance. Considering now a finite width incident beam which can be expanded in spatial Fourier components, the reflected beam from the structure can be determined as a superposition of the reflection of the different plane wave components [60]. Therefore the intensity reflected beam profile observed at infinity, in the focal plane of a lens, is exactly the incident intensity beam profile and that no special feature is observed. However, the rapid variation of the reflected plane wave phase with the angle of incidence produces an important Goos-Haenchen shift (see [64] for example) which distords the reflected profile of a finite width beam in the near field limit. Thus the interpretation of the observed M-lines in linear prism coupler experiments requires a carefull analysis.

The grating coupler presents special features which avoid the difficulties associated with the prism coupler. Other practical advantages makes the grating coupler quite popular. Indeed as pointed out by Golubenko and coworkers[65], the grating coupler present visible M-lines with prefect non absorbing materials, visible at infinity with a high contrast. This is due to the fact that there are essentially two emergent beams. In usual waveguide grating couplers, the diffraction efficiency is low for orders other than the resonant ones. For first order grating couplers the 0-order propagating reflected and transmitted beams are the only ones which are important. Nevertheless the existence of two emergent beams instead of one introduces a new feature: the energy exchange between the reflected and transmitted beams. A typical curve is shown on Fig.17: off resonance the reflectivity is low but, on resonance, the reflectivity reaches one and the transmission is nearly 0 for plane waves. Such a result is in contrast with the corresponding one of the prism coupler.

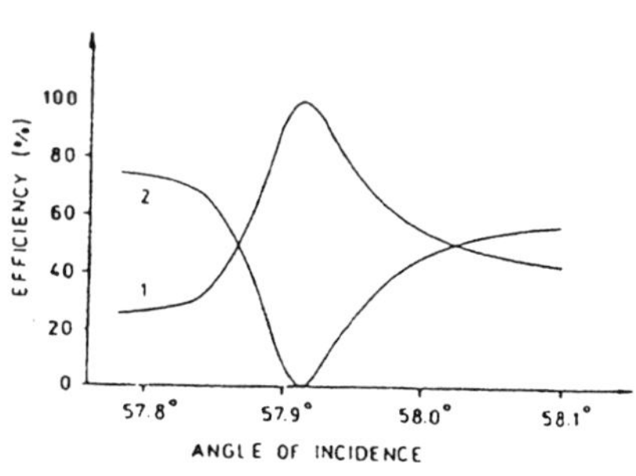

Figure 17: Typical reponse curve of a linear grating coupler excited by a plane wave. Intensity of the zeroth diffraction reflected, curve 1, and transmitted, curve 2, orders as a function of the angle of incidence in the vicinity of a waveguide mode excitation. (After Golubenko et al [65])

3.1.2 nonlinear distributed couplers: Nonlinear distributed couplers have been used for nonlinear measurements in organics by Chen and Carter [66] in an early work. It has been used more recently by several authors (see refs. [67-70] for example). Basically the effect is that a variation of the incident intensity induces a variation of the index of the material constituting the waveguide and so a shift of the position of the M-line associated to the resonant excitation of the NL guided mode. This shift allows for a determination of the magnitude and of the sign of the nonlinearity. A typical curve is plotted in Fig. 18 corresponding to a polymer under ps pulsed excitation: It is seen that the nonlinear resonance curve is not only shifted but that its shape is also changed.

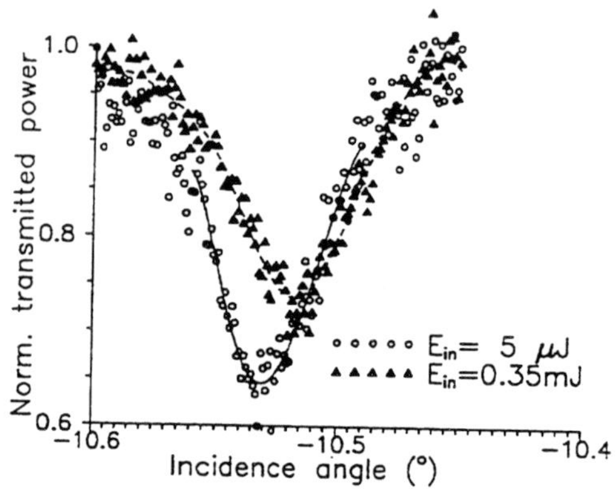

Figure 18: A typical nonlinear change in the M-line associated with the excitation of a guided wave. The M-line is not only shifted but its shape is also changed, a typical feature of local nonlinearities. (After Marques et al [67])

The change in the shape of the high irradiance M-line with respect to the low irradiance one is an interesting feature. Indeed locality of the nonlinearity has been predicted to play an important role in nonlinear distributed couplers [71-72]. Indeed, depending on the initial detuning (Δ), optical limiting and switching has been predicted theoretically [9] and observed experimentally [73-75]. However, the distributed coupler has been found to be bistable only in the plane wave approximation [71,72,76]. Under finite width beam excitation, as already explained in section 2.2.e, the coupler can not be bistable because of the lack of optical feedback in the structure. This is not the case for the Fabry-Pérot resonator under normal or closely normal incidence.

When nonlocality is considered [72], the material introduces a feedback along x in the structure. This feedback can lead to optical bistability provided its x-extension is long enough. The transition from a pure Kerr-law response without bistability but only switching to bistability is shown in Fig. 19. As the diffusion length, D, in coupling length unit increases, the slope of the steep portion increases until a switch is obtained at a critical value of D, named D_{th}. For $D>D_{th}$ bistability occurs. The variation of $1/D_{th}$ with the wave vector detuning $\Delta\beta$ from the optimum low-power coupling condition is shown in Fig. 20.

We find that $1/D_{th}$ increases linearly with the detuning. The important point is that D_{th} is of the order of the coupling length which lies generally in the range 100µm-1mm. This means that the nonlinear distributed coupler requires a strongly nonlocal nonlinearity to be able to exhibit bistability.

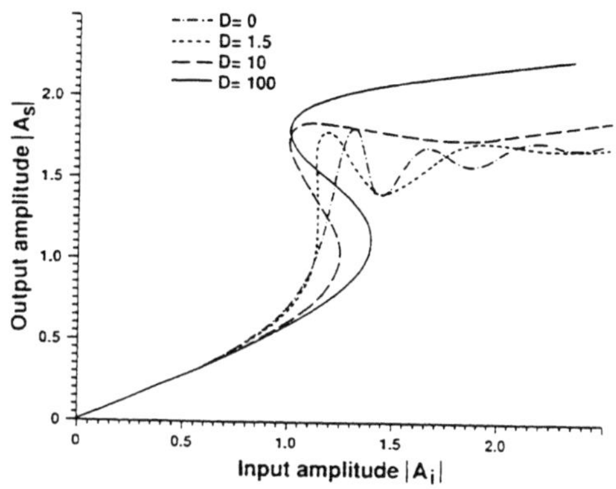

Figure 19: The output excited guided wave amplitude $|A_s|$ is plotted versus the input amplitude $|A_i|$ in normalized units. The diffusion length which scales the nonlinear nonlocality, D, is expressed in coupling length unit. The normalized detuning $\Delta=1.8$, normalized beam waist, $w_o=10$. [72]

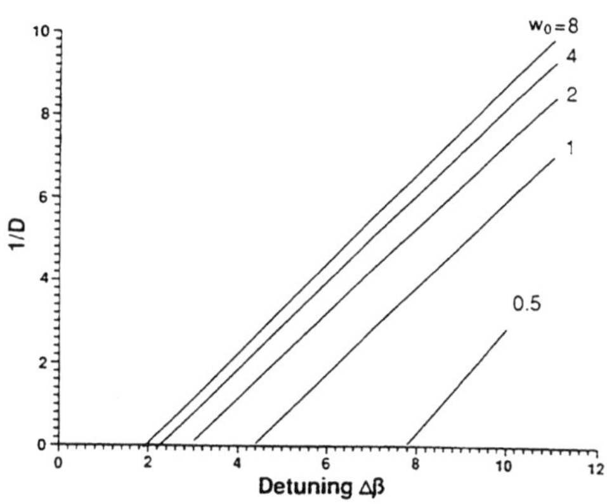

Figure 20: The inverse of the minimum diffusion length, D_{th}, is plotted versus the normalized detuning, $\Delta\beta$, for various values of the beam spot size, w_o. [72]

The theoretical analysis underlying the results presented in Fig. 19-20 has been tested in a beam scaning experiment[77]. A beam scanner is a distributed output coupler similar to a distributed input coupler except that it is used to outcouple a guided wave. In a pump-probe configuration, the angle of emergence of the outcoupled weak guided wave (probe) will depend on the nonlinear effect produced by a pump guided wave. In the case where the pump is also the probe, spatial nonlinear focusing may occur [78]. The comparison between the theoretical analysis and the beam scanning experiment is shown in Fig. 21. A good agreement can be noticed. Here the nonlinearity has a thermal origine which is highly nonlocal.

It is worth noting that the deformation of the nonlinear M-line (as observed in Fig. 18) is a signature of a local nonlinear effect. Indeed a completely delocalized nonlinear effect, i.e. homogeneous, is equivalent to a variation of the refractive index which is known to induce a shift of the resonance only. However a detailed interpretation of the shape of the resonance is far from being completely achieved since several effects may distord the NL M-line. In particular, the nonlinear absorption, which is described by the imaginary part of n_2, induces a change in absorption when the incident power varies. The absorption of the material has an influence on the contrast of the M-line and it should be possible to deduce it from the nonlinear M-line spectroscopy technique. That is rather more important that nonlinear absorption plays an important role in nonlinear integrated devices [79](see also [55]). Indeed, the maximum nonlinear phase shift which can be produced in any device is limited by the ratio $Im(n_2)/Re(n_2)$.

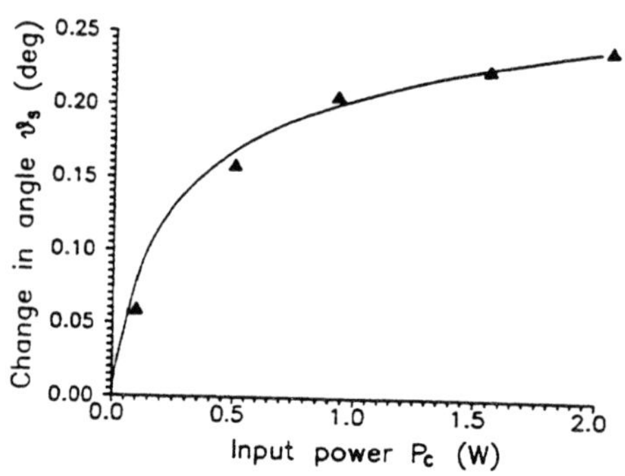

Figure 21: Graph of the change in the out-coupling angle versus incident power P_C and comparison with experimental data from a ZnS slab waveguide on fused quartz. The nonlinearity is here of thermal origine (after ref [77].

The modal theory for prism coupler, an essential tool for the interpretation of the above cited results, can be extended to nonlinear grating couplers. It is the aim of the next section to present recent progress in that matter.

3.2 Nonlinear modal theory of grating couplers.

The well-known theory which allows deriving the optical response of a nonlinear grating coupler (NLGC) not only for the guided wave but also for any radiated diffracted orders consist in solving the Maxwell equations together with the expression of the nonlinear polarization [80]. But this theory has the two following drawbacks: it is computer time consuming and it does not allow to deal with transverse effects since it was developed for plane-wave excitation only. On the other hand, several formalisms have been published [81,82] which allow studying transverse effects in NLGC but, unfortunately, are only concerned with guided waves. However, recent extensions of the modal theory to NLGC [83] allow to study the radiated diffracted orders also. the approach is based on the simultaneous use of the rigorous theory of diffraction in linear optics [84] and of the coupled mode formalism [85] and applies under the following conditions:

(i) it is possible to isolate a resonance

(ii) the angle of incidence remains close to the resonance.

Within the range of validity of points (i) and (ii), the transverse field map of a NLGC and of the associated linear one are identical. This is due to the fact that the Kerr nonlinearity induces an index variation of the guiding layer which remains small compared to the index step at the boundaries of this layer: we are in the weak nonlinear regime. Under this condition it is possible to show, using the coupled modes theory [85] that the linear relation between the amplitude of the nth radiated diffracted order in reflection or transmission F_n, the incident amplitude A_i and the excited guided wave amplitude A_{gw} still holds in the nonlinear regime[83]:

$$F_n(x) = p_n A_i(x) + q_n A_{gw}(x) \tag{26}$$

This equation is a generalization of eq. (13) to nonlinear grating couplers. The amplitude A_{gw} obeys a nonlinear equation which is equation (15) since A_{gw} is proportional to A_t. In addition, because the guided wave regime correponds necessary to oblique incidence, the second derivative in x can be neglected [11]. Numerical results are here presented to demonstrate the validity of the approximations used in this treatment. The situation corresponds to the structure depicted in Fig. 22 and the results are presented in Figs. 23. The agreement between exact [80] and approximate [83] theories is evident for the guided wave amplitude A_{gw} (Fig. 23 a) but also for the 0-order diffracted reflected (Fig. 23b).

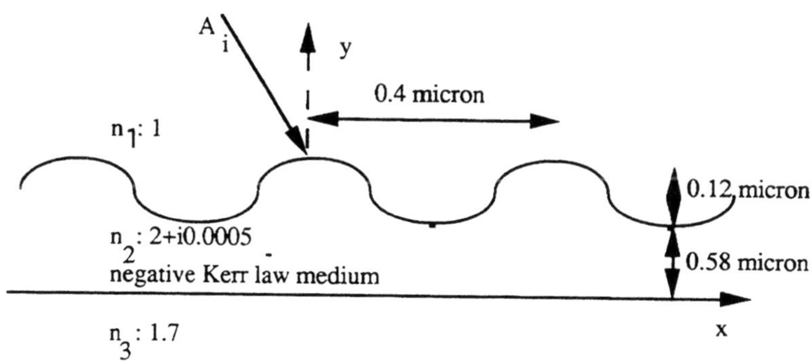

Figure **22**: Structure under study for Figs. 23.

Finally, it is important to note that only the nonlinearity is taken perturbatively into account in the modal theory. In particular, the grating is not described perturbatively but exactly through the use of the rigorous linear theory of diffraction. This is a very important point since perturbative calculations of grating couplers in the resonant domain fail even for small groove depths.

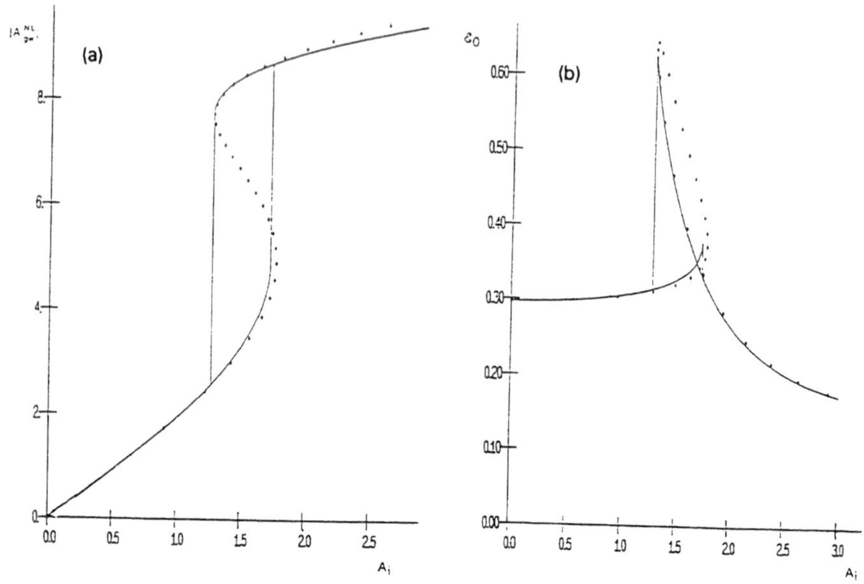

Figure 23:(a) $|A_{gw}|$ is plotted as a function of A_i. Full line: exact theory of ref. [80], dotted line: modal theory of ref. [83].

(b) same as (a) but for the reflectivity in the 0th reflected diffracted order.

4. CONCLUSION

Nonlinear resonators have been studied for almost 30 years since the early prediction of bistability in seeded laser by Lascher in1964 [86] and in passive nonlinear resonators by Seidel in 1969 [87]. They have received a lot of interest because of potential applications in all-optical signal processing. In addition, they present a great interest in more fundamental research because of exotic dynamic and spatial behavior they can exhibit: self-pulsing, chaos, spatial soliton formation, spontaneous modulational instability. Nonlinear resonators can be studied numerically by solving directly coupled nonlinear wave equations. However this treatment, although sometimes necessary in the strong nonlinear regime, is generally unnecessary for high finesse resonators. In this case a longitudinal single mode approximation applies which leads to simple equations. It is then possible to perform some analytical studies: although the basic equations are simple, their solutions are very rich and far from being completely understood. In addition all the nonlinear optical resonators appear as being described with a small number of parameters. This modal theory constitutes a powerful tool for the study of a wide class of nonlinear resonators.

Besides, transverse effects are very important in nonlinear system because the superposition theorem does not apply. One of the very important and unexpected result is that a nonlinear Fabry-Pérot type resonator is not bistable under oblique incidence. This result also applies to the case of distributed couplers as the ones used in planar integrated optics for in-coupling the light.

5. REFERENCES

1 H.G. Winful and G.D Cooperman, *Self-Pulsing and Chaos in Distributed Feedback Bistable Optical Devices*, Appl. Phys. Lett., Vol. 40, 1981, pp. 298-300.

2 L.A. Lugiato, .M. Narducci, D.K. Bandy and C.A. Pennise, *Self-Pulsing and Chaos in a Mean-Field Model of Optical Bistability*, Optics Comm. Vol. 43, 1982, pp. 281-286.

3 H.G. Winful, R. Zamir and S. Feldman, Appl. Phys. Lett., Vol. 58, 1991, p. 1001.

4 J.V. Moloney, *Two dimensional transverse solitary waves as asymptotic states of the field in a bistable optical resonators*, IEEE J. Quantum. Electron., Vol. 21, 1985, p.1393.

5 L.A. Lugiato and C. Oldano, *Stationary spatial patterns in passive optical systems*, Phys. Rev. A, Vol. 37, 1988, p. 3896.

6 M. Haelterman, G. Vitrant and R. Reinisch, *Transverse Effects in Nonlinear Planar Resonators. I. Modal Theory*, J. Opt. Soc. Am. *B*, Vol. 7, 1990, pp. 1309-1318.

7 W.J. Firth, this book 1993.

8 G.I. Stegeman, C. Liao and H.G. Winful, *Distributed Feedback Bistability in Channel Waveguides*, OBII, C.M. Bowden, H.M. Gibbs and S.L. Mc Call eds., Plenum Press, New York, 1984, p. 389.

9 G.M. Carter and Y.J. Chen, *Nonlinear Optical Coupling between Radiation and Confined Modes*, Appl. Phys. Lett. Vol. 42, 1983, p. 643 .

10 R. Reinisch, G. Vitrant and M. Haelterman *Coupled-mode theory of diffraction-induced transverse effects in nonlinear optical resonators*, Phys. Rev. B, Vol. 44, No. 15, 1991, pp. 7870-7878.

11 G. Vitrant, M. Haelterman and R. Reinisch, *Transverse Effects in Nonlinear Planar Resonators II: Modal Analysis for Normal and Oblique Incidence*, J. Opt. Soc. Am., Vol. B7, 1990, pp. 1319-1327.

12 W.J. Firth, A.J. Scroggie, G.S. McDonald and L.A. Lugiato, *Hexagonal Patterns in Optical Bistability*, Phys. Rev. A, Vol. 46, 1992, pp. 3609-3612.

13 Y.I. Balkarei, M.G. Evtikhov, J.V. Moloney and Y.A. Rzhanov, *Self-Focusing Effects in a Bistable Semiconductor Interferometer*, J. Opt. Soc. Am. B, vol. 7, 1990, pp. 1298-1302.

14 G. Vitrant, B. Vögele and R. Reinisch, *Study of Transverse Effects in Nonlinear Fabry-Pérot Resonators*, Annales de Physique, colloque n°1, Vol. 16, pp. 25-32 (1991).

15 J.L. Oudar, B. Sfez and R. Kuszelewicz, *Bistable Optical Switching in GaAs Multiple Quantum Well Epitaxial Etalons*, SPIE, Vol. 1361, 1990, pp. 490-498.

16 B.G. Sfez, J.L. Oudar, J.C. Michel, R. Kuszelewicz and R. Azoulay, *High Contrast multiple Quantum Well Optical Bistable device with Integrated Bragg Reflectors*, Appl. Phys. Lett., Vol. 57, 1990, p.324.

17 U. Olin and O. Sahlén *Theory for the Temperature rise in Optical Bistable GaAs/AlGaAs Etalons including Diffraction, Heat Conduction and Carrier Diffusion*, Optics Letters, Vol. 14, 1989, p. 566.

18 U. Olin, *Effects of Diffraction and Diffusion in Dispersive Optical Bistability*, J. Opt. Soc. Am. Vol. B7, 1990, p. 35.

19 R. Ulrich *Theory of the Prism-Film Coupler by the Plane-Wave Analysis*, J. Opt. Soc. Am, Vol. 60, 1970, p. 1337.

20 J. Danckaert, K. Fobelets, I. Veretennicoff, G. Vitrant and R. Reinisch, *Dispersive Optical Bistability in Stratified Structures*, Phys. Rev. B, Vol. 15, 1991, pp. 8214-8225.

21 W. Chen and D.L. Mills, Phys. Rev. B, *Optical Response of Nonlinear Multilayer Structures: Bilayers and Superlattices*, Vol. 36, 1987, pp. 6269-6278.

22 G. Vitrant, *Third Order Nonlinear Integrated Optical Resonators* In *Guided Wave Nonlinear Optics* D.B. Ostrowski and R. Reinisch (Eds.) Kluwer Academic Publishers (1992) pp. 285-303.

23 M. Haelterman and G. Vitrant, *Dynamics and Transverse Effects in Nonlinear Planar Resonators: a Unified Modal Theory* in OSA Proceedings on *Nonlinear Dynamics in Optical Systems*, N.B. Abraham, E.M. Garmire & P. Mandel, eds., Vol. 7 1991, p.210.

24 K. Ikeda, *MultiplValued Stationary State and its Instability of th Transmitted light by a ring cavity system*, Optics Comm., Vol. 30, 1979, pp. 257-261.

25 P. Mandel, *Scaling Properties of Switching Pulses*, Optics Comm., Vol. 55, 1985, pp. 293-296.

26 B. Macke, B. Segard and J. Zemmouri, *Optical Bistability and Instability*, Optics Comm., Vol. 60, 1986, p. 323.

27 J.Y. Bigot, A. Daunois and P. Mandel *Phys. Lett. A*, Vol. 123, 1987, p. 123.

28 J. Danckaert, G. Vitrant, R. Reinisch and M. Georgiou *Nonlinear Dynamics in Single Mode optical Resonators* submittted for publication in Phys. Rev. A.

29 H. Haken, *Synergetics-An introduction* Springer Verlag, Berlin 1977.

30 P. Bergé and Y. Pomeau *Le Chaos, théorie et expériences* Eyrolles eds., Paris, 1988.

31 E. Abraham and W.J. Firth, *Periodic Oscillations and Chaos in a Fabry-Pérot Cavity containing a nonlinearity of finite response time*, Optica Acta, Vol. 30, 1983, p. 1541.

32 M. Haelterman, *Period-doubling Bifurcations and Modulational instability in the Nonlinear ring Cavity: an Analytical Study*, Optics letters, Vol 17, 1992, pp. 792-794.

33 P. Mandel, *Switching Times in Optical Bistability: Analytical Results* in *Optical Bistability and Optical Computing*, H.Y. Zhang and K.K. Lee Eds., World Scientific, Singapore, 1988, pp. 118-135.

34 A.A. Yefimov and G.N. Shkerdin Sov. J. of Communication Technology and Electronics, Vol. 36, 1991, p. 59.

35 G. Cauwenberghs, H. Thienpont and I. Veretennicoff, *Switching dynamics and transient response of Kerr nonlinear optical heterostructures* in *OSA Proceedings on Nonlinear Dynamics in Optical Systems*, N.B. Abraham, E.M. Garmire & P. Mandel, eds., Vol. 7 1991, pp. 138-140.

36 W.J. Firth *Stability of nonlinear Fabry-pérot resonators*, Optics Comun., Vol. 39, 1981, pp. 343-346.

37 A. Yariv and P. Yeh *Optical waves in crystals*, 1983 J. Wiley&Sons.

38 J. Sipe *Gap solitons*, In *Guided Wave Nonlinear Optics* D.B. Ostrowski and R. Reinisch (Eds.) Kluwer Academic Publishers (1992) pp. 305-318.

39 Abraham, N.B. and Firth, W.J. *Overview of transverse effects in nonlinear-optical systems*, J. Opt. Soc. Am., Vol B 7, 1990, pp. 951-962.

40 A. Barthelemy, S. Maneuf and C. Froehly *Propagation Soliton et autoconfinement de faisceaux laser par non-linéarité optique de Kerr*, Opt. Comm., vol. 55, 1985, pp. 201-206.

41 Aitchinson, J.S., Weiner, A.M., Silberberg, Y., Oliver, M.K., Jackel, J.L., Leaird, D.E., Vogel, E.M. and Smith, P.W.E., *Observation of spatial optical solitons in a nonlinear gleass waveguide*, Optics Letters, Vol. 15, 1990, pp. 471-473.

42 De la Fuente, R., Barthelemy, A. and Froehly, C., *Spatial-soliton-induced guided waves in a homogeneous nonlinear Kerr medium* Optics Letters, Vol. 16, 1991, pp. 793-795.

43 N.N. Rosanov and V.E. Semenov, *The kinetics of the hysteresis change of the beam profile in nonlinear interferometers*, Opt. Commun., Vol. 38, 1981, p. 435.

44 J.V. Moloney and H.M. Gibbs, *Role of diffractive coupling and self-focusing or defocusing in the dynamical switching of a bistable optical cavity*, Phys. Rev. Lett., Vol. 48, 1982, p. 1607.

45 J.V. Moloney, H. Adachihara, R. Indik, C. Lizarraga, R. Northcutt, D.W. McLaughlin and A.C. Newell, *Modulational-induced optical pattern formation in a passive optical-feedback system*, J. Opt. Soc. Am., Vol. 7, 1990, pp. 1039-1044.

46 N.N. Rosanov and G.V. Khodova, *Diffractive autosolitons in nonlinear interferometers*, J. Opt. Soc. Am., Vol. B7, 1990, p. 1057.

47 H. Adachihara, D.W. McLaughlin, J.V. Moloney and A.C. Newell, *Solitary waves as fixed points of infinite-dimensional maps for an optical bistable ring cavity: Analysis*, J. Math. Phys., Vol. 29, 1988, pp. 63-85.

48 L.A. Lugiato and R. Lefever, *Spatial dissipative structures in passive optical systems*, Phys. Rev. Lett., Vol. 58, 1987, p. 2209.

49 W.J. Firth, A. Fitzgerald and C. Paré, *Transverse instabilities due to counterpropagation in Kerr media*, J. Opt. Soc. Am., Vol. B7, 1990, pp. 1087-1097.

50 G.P. Agrawal, *Transverse modulation instability of copropagating optical beams in nonlinear Kerr media*, J. Opt. Soc. Am., Vol. B7, 1990, pp. 1072-1078.

51 C.T. Law and A.E. Kaplan, *Instabilities and amplification of counterpropagating waves in a Kerr nonlinear medium*, J. Opt. Soc. Am., Vol. B8, 1991, pp. 58-67.

52 A.E. Kaplan, *Optical bistability that is due to mutual self-action of counterpropagating beams of light*, Optics Letters, Vol. 6, 1981, pp. 360-362.

53 M. LeBerre, E. Ressayre, A. Tallet, K. Tai and H.M. Gibbs, *Transverse optical bistabilities and instabilities*, IEEE, J. Quantum Electron., Vol. QE21, 1985, pp. 1404-1416.

54 M. Haelterman and G. Vitrant *Drift Instability and Spatiotemporal Dissipative Structures in a Nonlinear Fabry-Pérot Resonator Under Oblique Incidence*, J. Opt. Soc. Am., Vol. B9, 1992, pp. 1563-1570.

55 G. Assanto, *Third order nonlinear integrated devices*, in *Guided Wave Nonlinear Optics* D.B. Ostrowski and R. Reinisch (Eds.) Kluwer Academic Publishers (1992) pp. 257-284.

56 G.S. McDonald and W.J. Firth, *Spatial solitary-wave optical memory'* J. Opt. Soc. Am., Vol. B7, 1990, pp.1328-1335.

57 M. Haelterman, S. Trillo and S. Wabnitz, *Additive-Modulation-Instability Ring Laser in the Normal Dispersion Regime of a Fiber*, Optics Letters, Vol. 17, 1992, pp. 745-747.

58 G. Vitrant, M. Haelterman, G. Martinelli, A. Barthelemy and C. Froehly, *Dynamical Analysis of a Nonlinear Fabry-Perot Resonator Under Normal and Oblique Incidence: Comparison with Experimental Results*, Inst. Phys. Conf. No115, IOP Publishing Ltd, 1991, pp. 149-152.

59 G.I. Stegeman, D. Sarid, J.J. Burge and D.G. Hall, *Scattering of Guided Waves By Surface Periodic Gratings for Arbitrary Angles of Incidence: Perturbation Field Theory and Implications to Normal-Mode Analysis*, J. Opt. Soc. Am., Vol. 71, 1981, pp. 1497-1507.

60 R. Petit and M. Cadilhac, *Théorie Electromagnétique du Coupleur a Prisme*, J. optics (Paris), Vol. 8, 1977, pp. 41-49.

61 R. Ulrich *Theory Of Prism-Film Coupler by Plane-Wave Analysis*, J. Opt. Soc. Am., Vol. 60, 1970, pp. 1337-1350.

62 R. Ulrich *Efficiency of Optical-Grating Couplers*, J. Opt. Soc. Am., Vol. 63, 1973, pp. 1419-1431.

63 J.S. Wei and W.D. Westwood, *A New Method for Determining Thin-Film Refractive Index and Thixkness Using Guided Optical Waves*, Appl. Phys. Lett., Vol. 32, 1978, pp. 819-821.

64 D.L. Lee, *Electromagnetic Principles of Integrated Optics*, John Wiley & Sons, 1986.

65 G.A. Golubenko, A.S. Svakhin, V.A. Sychugov, A.V. Tischenko, E. Popov and L. Mashev, *Diffraction Characteritics of Planar Corrugated Waveguides*, Optical and Quantum Electronics, Vol. 18, 1986, pp. 123-128.

66 Y.J. Chen and G.M. Carter, *Measurement of Third Order Nonlinear Susceptibilities by Surface Plasmons*, Appl. Phys. Lett., Vol. 41, 1982, pp. 307-309.

67 M.B. Marques, G. Assanto, G.I. Stegeman, G.R. Möhlmann, E.W.P. Erdhuisen and W.H.G. Horsthuis *Large, Nonresonant, Intensity Dependent Refractive Index of 4-dialkylamino-4-nitro-diphenil-*

polyene side chain Polymers in Waveguides, Appl. Phys. Lett., Vol. 58, 1991, pp. 2613-2615.

68 N. Bartuch, A. Bräuer, P. Dannberg, H.H. Hörhold and D. Raabe, *Measurement of high nonresonant third-order nonlinearity in MP-PPV waveguides*, Int. Journal of Optoelectronics, Vol. 5, 1991.

70 M. Kull and J.L. Coutaz, *Two-Wavelength Nonlinear Prism Coupling to Planar Waveguides in Semiconductor Doped Glasses*, Springer Series in Wave Phenomena, Vol. 9 *Nonlinear Optics in Solids, ed. O. Keller, 1990, pp. 185-188.*

71 G.I. Stegeman, G. Assanto, R. Zanoni, C.T. Seaton, E. Garmire, A.A. Maradudin, R. Reinisch and G. Vitrant, *Bistability and Switching in Nonlinear Prism Coupling*, Appl. Phys. Lett., Vol. 52, 1988, pp. 869-871

72 G. Vitrant, R. Reinisch, J.C. Paumier, G. Assanto and G.I. Stegeman, *Nonlinear Prism Coupling with Nonlocality*, Optics Letters, Vol. 14, 1989, pp. 898-900.

73 G. Vitrant and P. Arlot, *Demonstration of Optical Bstability with a Nonlinear Prism Coupler*, J. Appl. Phys., Vol. 61, 1987, pp. 4744-4748.

74 R.M. Fortenberry, G. Assanto, R. Moshrefzadeh, C.T. Seaton and G.I. Stegeman, *Pulsed Excitation of Nonlinear Distributed Coupling into Zinc Oxide Optical Guides*, J. Opt. Soc. Am., Vol. B5, 1988, pp. 425-431.

75 H. Sauer, N. Paraire, A. Koster and S. Laval, *Optimization of a Slicon-On-Sapphire Waveguide Device for Optical Bistable Operation*, J. Opt. Soc. Am., Vol. B5, 1988, pp.443-451.

76 G. Vitrant, P. Arlot and R. Reinisch, Procedings of the Fourth International Symposium on Optical and Optoelectronic Applied Science and Enginering, 1987, Vol. 800, p. 169.

77 G. Assanto, G.I. Stegeman and G. Vitrant, *Analysis of All-Optical Integrated Beam Scanning Devices*, Appl. Phys. Lett., Vol. 54, 1989, pp. 1854-1856.

78 P. Arlot and G. Vitrant, *Theoretical Study of a Nonlinear Prism Output Coupler*, Appl. Phys. Lett., Vol. 50, 1987, p. 650.

79 V. Mizrahi, K.W. DeLong, G.I. Stegeman, M.A. Saifi and M.J. Andrejco, Optics Lett., *Two-Photon Absorption as a Limitation to All-Optical Switching*, Vol. 14, 1989, pp. 1140-1142.

80 P. Vincent, N. Paraire, M. Nevière, A. Koster and R. Reinisch, *Gratings in Nonlinear Optics and Optical Bistability*, J. Opt. Soc. Am., Vol. B2, 1985, pp. 1106-1116.

81 G.I. Stegeman, E.M. Wright, N. Finlayson, R. Zanoni and C.T. Seaton, *Third Order Nonlinear Integrated Optics*, J. of Lightwave Tech., Vol. 6, 1988, pp. 953-970, and references cited herein.

82 G. Assanto, M.B. Marques and G.I. Stegeman, *Grating Coupling of Light Pulses into Third-Order Nonlinear Waveguides*, J. Opt. Soc. Am., Vol. B8, 1991, pp. 553-561.

83 R. Reinisch, M. Nevière, P. Vincent and G. Vitrant, *Radiated Diffracted Orders in Kerr-Type Grating Couplers*, Optics Comm., Vol. 91, 1992, pp. 51-56.

84 R. Petit ed., *Electromagnetic Theory of Gratings*, Springer Berlin, 1980.

85 C. Vassalo, *Théorie des guides d'ondes électromagnétiques*, CNET-ENST, Eyrolles, Paris 1985. H. Kogelnik, *Theory of dielectric waveguides*, in: Integrated Optics, ed. T. Tamir, Springer, Berlin, 1975.
86 G.J. Lasher, *Analysis of a Proposed Bistable InjectionLaser*, Solid State Electronics, Vol. 7, 1964, pp. 707-716.
87 H. Seidel, *Bistable Optical Circuit Using Saturable Absorber Within a Resonator Cavity*, U.S. Patent 3 610 731, 1969.

Studies in Classical
and Quantum Nonlinear Optics

ISBN 1-56072-168-5

OPTICAL ECHO PHENOMENA IN DISORDERED SOLIDS

Vladimir I. Yudson
Institute of Spectroscopy, Russian Academy of Sciences
Troitzk 142092, Moscow, Russia

Abstract

The influence of disorder on the echo channel of optical four-wave mixing has been studied. In the stationary regime ("optical phase conjugation") the nonlinearly backscattered light is shown to possess coherent properties similar to those in the case of ordered media. It has been found that the "optical echo" (the pulse regime of mixing) from disordered semiconductors is drastically influenced by elastic scattering of electrons and holes. The echo signal reveals a sharp angular dependence and its dependence on the delay time changes from exponential to power law with increasing delay time. This is in contrast to the case of ordered semiconductors where in the absence of dissipation the echo signal does not decay up to a rather long delay time determined by a geometry of pumping. The survival of both the phase conjugation and the optical echo phenomena in the presence of disorder is due to a special class of time reversed trajectories and can be considered as a manifestation of "weak localization effects".

1 Introduction

Among a variety of nonlinear optical processes optical echo phenomena are of particular interest. Their distinguishing feature is that an amplitude $\mathcal{E}_s$ of the echo signal is determined by a phase conjugated amplitude $\mathcal{E}_i^*$ of incident light. We consider a typical nonlinear optical process of four-wave mixing. It is described by the nonlinear polarization

$$P_{nl} = \chi^{(3)} E_3 E_2 E_i, \tag{1.1}$$

where $\chi^{(3)}$ is the nonlinear optical susceptibility of the third order. The electromagnetic field E_i of the incident (probe) wave is implied to be small compared to fields E_2 and E_3 of the pump waves. The electric field of each wave is represented in the form of a plane wave

$$E = \mathcal{E}(\mathbf{r},t)\exp\left[i(\mathbf{kr}-\omega t)\right] + \mathcal{E}^*(\mathbf{r},t)\exp\left[-i(\mathbf{kr}-\omega t)\right], \tag{1.2}$$

with an amplitude $\mathcal{E}(\mathbf{r},t)$ slowly varying with space and time. Hereafter we concentrate on the following channel of the mixing process (1.1):

$$\mathcal{E}_s \sim \mathcal{P}_{nl} = \chi^{(3)}\mathcal{E}_3\mathcal{E}_2\mathcal{E}_i^*. \tag{1.3}$$

The frequencies and wave vectors of the waves involved in the process (1.3) are evidently connected by the relations:

$$\omega_s = \omega_2 + \omega_3 - \omega_i. \tag{1.4}$$

$$\mathbf{k}_s = \mathbf{k}_2 + \mathbf{k}_3 - \mathbf{k}_i. \tag{1.5}$$

As well known, the complex conjugate Schrödinger equation describes formally a time reversal evolution of a system:

$$i\frac{\partial \Psi}{\partial t} = H\Psi \qquad \longrightarrow \qquad -i\frac{\partial \Psi^*}{\partial t} = H^*\Psi^*. \tag{1.6}$$

Thus, as the signal field (1.3) is proportional to the complex conjugate amplitude of the incident field, it can be interpreted as the time reversal of the incident field.

In the stationary regime, the process (1.3) is usually referred to as "optical phase conjugation" [1]-[3]. And when the incident and pump fields are short pulses with some time delay, one deals with the "optical echo" [3] in the true sense. A peculiar feature of both phenomena is a coherent reproduction of the incident field.

Up to now a nonlinear medium has been implied to be "ordered" in a sense that light waves (excitations of the medium) are not subjected to scattering caused by random structure imperfections, etc. The purpose of this paper is to analyze whether the optical echo phenomena survive in the presence of a static disorder. For optical phase conjugation this question was studied recently and a part of results was reported at the previous School [4]. Their summary and some new results are presented in the next section. And the major part of this paper is devoted to the optical echo from disordered semiconductors.

2 Optical phase conjugation in disordered media

2.1 Peak of nonlinear backscattering

The problem of phase conjugation in disordered media has been posed recently [5],[6]. Elastic scattering of light caused by static disorder results in diffusive-like propagation of light inside the medium. The phase matching condition (1.5) is no longer valid on a macroscopic scale of

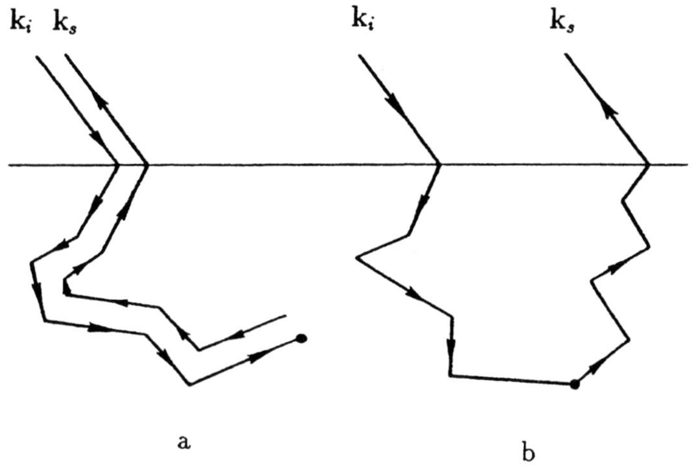

Figure 1: Generation of conjugated light in a disordered medium: (a) the contribution to the peak of nonlinear backscattering: (the trajectory of generated light is the time reversal of the trajectory of incident light); (b) the contribution to the diffuse background (the trajectories of generated and incident light are independent).

the sample size L. But for the case of a weak disorder (when the light wavelength $\lambda \ll l$, l being the mean free path for elastic scattering) the condition (1.5) is still valid locally - with the accuracy of the order $1/l$ - at an elementary act of four-wave mixing in a region of size l.

It has been shown in (1.5) that even in the case $l \ll L$, a peak of conjugated light may occur in the direction opposite to the direction of incidence of the probe wave. The ratio of peak intensity of this backscattered signal wave to the intensity of the diffuse background at the same frequency ω_s may be of the order $l/\lambda \gg 1$. This is in contrast to a well-known case of linear backscattering (see, e.g. [7]) where this ratio does not exceed 2. The angular width $\Delta\theta_{nl} \sim \lambda/L$ of the nonlinearly backscattered peak is considerably smaller than that ($\Delta\theta_l \sim \lambda/l$) for the linear backscattering. All these peculiarities of optical phase conjugation are due to the fact that the contribution to the backscattered peak is made by scattering events shown in Fig.1a, while those shown in Fig.1b contribute to the diffuse background. Contributions of all the trajectories of the type (a) with a fixed point of nonlinear interaction are added coherently due to compensation of random phases at exactly reversed paths. And contributions of the trajectories of the type (b) add

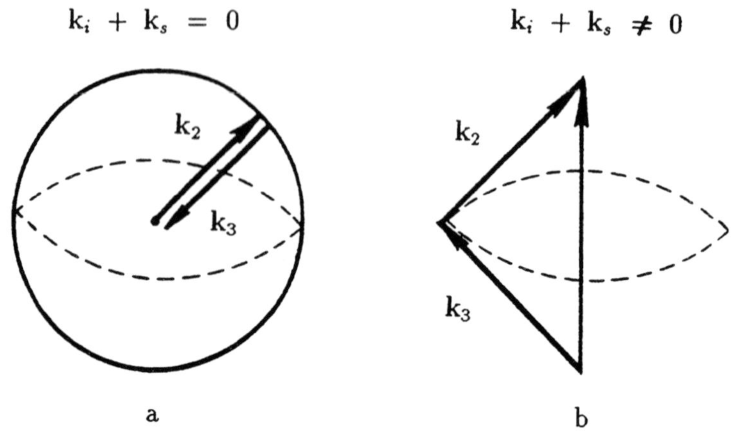

Figure 2: Phase space of the pump wave vectors $\mathbf{k}_2$ and $\mathbf{k}_3$ determined by the phase matching condition: (a) for the backscattering ($\mathbf{k}_i + \mathbf{k}_s = 0$); (b) for the diffuse scattering ($\mathbf{k}_i + \mathbf{k}_s \neq 0$).

incoherently as there is no compensation of random phases. The volume of phase space of momenta $\mathbf{k}_2$ and $\mathbf{k}_3$ of the pump waves restricted by (1.5) is much greater for the case (a) of counterpropagating incident and generated waves than for the case (b), see Fig.2. This is the reason of a relatively high peak intensity.

2.2 Correlation properties of nonlinearly scattered light

Further analysis [8] has been devoted to the question whether phase conjugated light generated by a disordered nonlinear medium possesses the same striking coherent properties as it does in the case of an ordered medium, e.g. a restoration of an incident wave front. To answer this question it is necessary to examine correlations of amplitudes $\mathcal{E}_{s\alpha}$ ($\alpha = 1, 2, \ldots$) of several outgoing (signal) waves corresponding to different incoming (probe) waves with amplitudes $\mathcal{E}_{i\alpha}$ and wave vectors $\mathbf{k}_{i\alpha}$ starting at a point A of a source (see Fig.3).

The correlation function $\langle \mathcal{E}_{s\alpha} \mathcal{E}^*_{s\beta} \rangle$ may be represented as $f_{\alpha\beta} \mathcal{E}^*_{i\alpha} \mathcal{E}_{i\beta}$, where the factor $f_{\alpha,\beta}$ is shown [8] to be practically constant for a rather large difference of the angles of incidence of the waves $|\theta_\alpha - \theta_\beta| \sim 1$. This angle difference exceeds considerably the angular width $\Delta\theta_{nl}$ of the

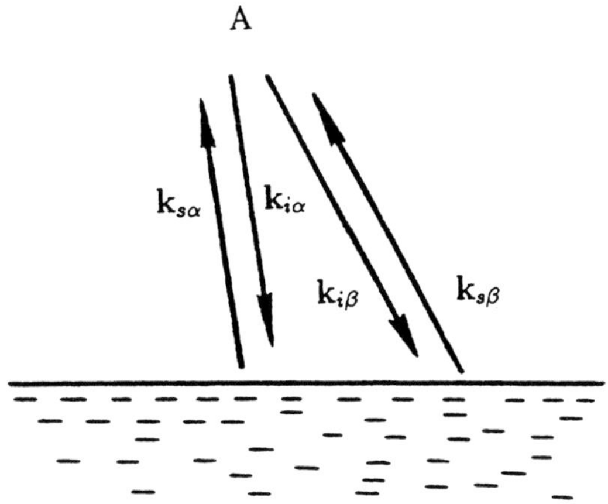

Figure 3: The conjugated image of a source A.

nonlinearly backscattered peak for each of the waves. As a consequence, the disorder-averaged intensity I_s of the signal (conjugated) field at the point A

$$\langle I_s \rangle = \sum_{\alpha\beta} \langle \mathcal{E}_{s\alpha} \mathcal{E}^*_{s\beta} \rangle \approx f \sum_{\alpha\beta} \mathcal{E}^*_{i\alpha} \mathcal{E}_{i\beta} = I_i \tag{2.1}$$

does reproduce variations of the intensity I_i of the source field, as it takes place for phase conjugation in an ordered medium.

Phase conjugation plays a crucial role for the coherence of backscattered waves. In the case of usual linear backscattering of light from disordered media there is no correlation of different backscattered waves if the difference in the angles of incidence of incoming waves exceeds the angular width $\Delta\theta_l \ll 1$ of the linear backscattering peak.

The survival of coherent properties of backscattered light at optical phase conjugation in a disordered medium allows a simple qualitative explanation. Remind that the coherent reflection of conjugated light is provided by the processes shown in Fig.1a with opposite wave vectors $\mathbf{k}_2 = -\mathbf{k}_3$ of pump waves involved in the mixing. Using the Fourier transformation of the product $\mathcal{E}_2(\mathbf{r})\mathcal{E}_3(\mathbf{r})$ in the l.h.s. of (1.3)

$$\mathcal{E}_2(\mathbf{r})\mathcal{E}_3(\mathbf{r}) = \sum_{k_2,k_3} \mathcal{E}_2(\mathbf{k}_2)\mathcal{E}_3(\mathbf{k}_3) \exp\left[i(\mathbf{k}_2 + \mathbf{k}_3)\mathbf{r}\right], \tag{2.2}$$

and separating in the sum (2.2) the coherent part with $\mathbf{k}_2 + \mathbf{k}_3 = 0$ we obtain for the signal field (1.3)

$$\mathcal{E}_s \sim C\mathcal{E}^*_i. \tag{2.3}$$

Here

$$C = \sum_k \mathcal{E}_2(\mathbf{k})\mathcal{E}_3(-\mathbf{k}) \tag{2.4}$$

is the random complex constant determined by statistics of the pump field in the disordered medium. It is important that this constant is the same for all the incoming probe waves and its random phase cancels out from the correlation function (2.1). Thus, in the case of disordered media the nonlinearly backscattered conjugated light possesses the following coherent properties: the correlation functions containing nonequal numbers of $\mathcal{E}_s$ and $\mathcal{E}_s^*$ vanish in the leading order while the correlation functions with equal numbers of $\mathcal{E}_s$ and $\mathcal{E}_s^*$ are proportional to those for the incoming probe light.

We see that the coherent nature of the phase conjugation process survives in the presence of disorder. In the next section we will consider an influence of disorder on optical echo phenomena.

3 Optical echo from disordered semiconductors

A striking phenomenon of spin echo is known for more than 40 years. A formal equivalence between the spin 1/2 and a typical model of resonance optics - the model of a "two-level atom" - underlies the phenomenon of optical echo from isolated atoms, molecules, impurity centers in solid matrices, see, e.g. [3].

For inhomogeneously broadened optical transitions in the ensemble of two-level systems (TLS) the initial polarization created by a laser pulse decays fast on the time scale $(\Delta\omega)^{-1}$, where $\Delta\omega$ is an inhomogeneous linewidth. In the standard configuration of photon echo experiment the sample is excited by the second laser pulse at a time $T \gg (\Delta\omega)^{-1}$. An echo signal is emitted at the time $2T$. The peak amplitude of the echo signal decays as $S(2T) \sim \exp[-2T/T_d]$, where the dephasing time T_d describes a decay of coherence of an individual TLS. Such decay may be caused by interaction with phonons or by other irreversible processes.

Recently, photon echo from band-to-band transitions in a semiconductor has been observed [9]. Despite the fact that electronic excitations in ordered semiconductors are not localized but extend over the whole sample, the concept of TLS is still applicable to understand the echo phenomenon. Indeed, due to quasimomenta conservation, the light field induces transitions between electron states in the valence band (v,$\mathbf{p}$) and in the conduction band (c, $\mathbf{p}$) with coinciding "momenta" $\mathbf{p}$ (neglecting a small momentum of light). Thus, an ordered semiconductor may be considered as an ensemble of independent TLS with transition frequencies determined by momenta $\mathbf{p}$.

The echo pulse decay is a result of dephasing processes breaking the picture of independent TLS. Such dephasing in semiconductors may be due to the carrier-carrier interaction and scattering of carriers caused by phonons or a static disorder. Estimation of the Coulomb interaction between carriers shows [10] that the corresponding rate of dephasing decreases with the carrier density and, therefore, can be rather small for not too strong pumping.

The role of a static disorder in optical-dephasing experiments was studied in [11]-[13]. The presence of a static random potential causes scattering of carriers from the given state $\mathbf{p}$ to states with other momenta and this makes the concept of independent TLS meaningless. The corresponding time τ of the mean free path of carriers plays the role of the dephasing time, at least, at a not too long delay time T.

This conclusion may seem paradoxical as one can consider a set of exact eigenstates $\{(v;\mu),(c;\nu)\}$ of electrons and holes for the given realization of random potential. In the absence of dissipative processes these states do not decay at all. And then, what about a decay of the echo signal?

The point is that in disordered semiconductors the photon field induces transitions generally between all the (v,μ) and (c,ν) states in contrast to the case of ordered semiconductors, where only $(v,\mathbf{p})$ and $(c,\mathbf{p})$ states are coupled by the light field. Thus, the set of exact eigenstates of a disordered semiconductor can no longer be treated as an ensemble of independent TLS. The dipole transition matrix elements $\mu_{\mu\nu}$ between the states (v,μ) and (c,ν) are now random quantities depending on the disorder. Randomness of the transition matrix elements results effectively in the decay of the echo signal.

This picture is valid for a "weak" disorder. In the case of a "strong" disorder the carriers are localized (the Anderson localization) and the light field couples again only pairs of states restoring the concept of independent TLS.

We see that as a non-dissipative time-reversal process, elastic scattering allows for the nondecaying echo-signal in both limiting cases of ordered and strongly disordered media [14]. The influence of the Anderson localization on the echo signal was studied recently [13] for specific models (the model of an infinite hole effective mass and the model of an ordered conduction band). For these models the echo signal can straightforwardly be expressed in terms of the known correlation functions (density-density and current-current, respectively) of the localization theory (see, e.g., [15],[16]). The purpose of this paper is to consider a more general and realistic model of a semiconductor with both electrons and holes scattered by disorder. We study in detail the effect of time-reversed successions of scattering events on the decay of the optical echo. It is shown that in the region of long delay time $T \gg \tau$ the echo signal $S(2\tau)$ decays according to a slow power law in contrast to the fast exponential decay at $T \sim \tau$.

3.1 The model and basic equations

We consider a simple model of a disordered two-band semiconductor described by the Hamiltonian of non-interacting electrons and holes:

$$\begin{aligned} H \equiv H_e + H_h &= \int \Psi_e^+(\mathbf{r})[\epsilon_e(\hat{\mathbf{p}}) + U_e(\mathbf{r})]\Psi_e(\mathbf{r})d\mathbf{r} \\ &+ \int \Psi_h^+(\mathbf{r})[\epsilon_h(\hat{\mathbf{p}}) + U_h(\mathbf{r})]\Psi_h(\mathbf{r})d\mathbf{r}, \end{aligned} \qquad (3.1)$$

where $\epsilon_{e(h)}(\hat{\mathbf{p}})$ are kinetic energy operators of electrons (holes); $U_{e(h)}(\mathbf{r})$ are effective random potentials, and $\Psi^+_{e(h)}$, $\Psi_{e(h)}$ are creation and annihilation field operators obeying the usual fermion anticommutation relations:

$$\{\Psi^+_{e(h)}(\mathbf{r}), \Psi_{e(h)}(\mathbf{r}')\} = \delta(\mathbf{r}-\mathbf{r}'); \quad \{\Psi_e^+, \Psi_h\} = \{\Psi_{e,h}, \Psi_{e,h}\} = 0 \qquad (3.2)$$

The random potentials $U_{e,h}$ are assumed to be the Gaussian variables with zero average and correlation functions

$$\langle U_{e(h)}(\mathbf{r})U_{e(h)}(\mathbf{r}')\rangle = \gamma_{e(h)}\delta(\mathbf{r}-\mathbf{r}'). \qquad (3.3)$$

For simplicity, we neglect a possible correlation between U_e and U_h. Even present, this correlation can be shown to be insignificant everywhere except a small exotic range of parameters.

Interband optical transitions are described by the interaction Hamiltonian

$$H_{int} = -\mu \int [\Psi_e^+(\mathbf{r})\Psi_h^+(\mathbf{r})E(\mathbf{r},t) + h.c.]d\mathbf{r}, \qquad (3.4)$$

where $E(\mathbf{r},t)$ is a classical electromagnetic field and μ is the dipole transition matrix element. Here, for simplicity, we restrict ourselves to a scalar field and spinless carriers, though the approach can be straightforwardly extended to take into account polarization and spin effects.

As we are interested in the echo channel (1.3) of the four-wave optical mixing process, we should extract the corresponding term from the general expression for the time-dependent polarization in the presence of external fields

$$P_{nl}(\mathbf{r},t) = \langle 0|\hat{S}^{-1}(t)\hat{p}(\mathbf{r},t)\hat{S}(t)|0\rangle, \qquad (3.5)$$

where

$$\hat{p}(\mathbf{r},t) = \mu\Psi_h(\mathbf{r},t)\Psi_e(\mathbf{r},t), \qquad (3.6)$$

the evolution operator $\hat{S}(t)$ is given by the time-ordered exponent

$$\hat{S}(t) = \mathrm{T}\exp\left[-i\int_{-\infty}^{t} H_{int}(t')dt'\right] \qquad (3.7)$$

and quantum averaging over the ground state $|0\rangle$ is performed. Operators in (3.5-3.7) are taken in the interaction representation with respect to the Hamiltonian (3.1).

Expanding (3.7) we obtain the following expression for the nonlinear polarization responsible for the echo signal:

$$\mathcal{P}(\mathbf{r},t) = -\mu \int_{-\infty}^{t} dt_i dt_2 dt_3 d\mathbf{r}_i d\mathbf{r}_2 d\mathbf{r}_3 \exp\left[i\sum_\alpha \eta_\alpha(\mathbf{k}_\alpha \mathbf{r}_\alpha - \omega_\alpha t_\alpha)\right]$$
$$\mathcal{E}_3(t_3)\mathcal{E}_2(t_2)\mathcal{E}_i^*(t_i)[\langle 0|\hat{p}(\mathbf{r}_i,t_i)\hat{p}^+(\mathbf{r}_2,t_2)|0\rangle\langle 0|\hat{p}(\mathbf{r},t)\hat{p}^+(\mathbf{r}_3,t_3)|0\rangle -$$
$$-\frac{1}{2}\langle 0|\hat{p}(\mathbf{r}_i,t_i)\hat{p}(\mathbf{r},t)\hat{p}^+(\mathbf{r}_2,t_2)\hat{p}^+(\mathbf{r}_3,t_3)|0\rangle], \qquad (3.8)$$

where $\eta_{2,3} = 1; \eta_i = -1$. In the derivation of (3.8) the light field has been assumed to be a superposition of three plane pulses with carrying wave vectors and frequencies $\mathbf{k}_\alpha$ and ω_α ($\alpha = i, 2, 3$), respectively. The initial pulse "i" is followed by the pump pulse (split into two waves "2" and "3") with a time delay T, so that the amplitudes $\mathcal{E}_\alpha(t_\alpha)$ are nonzero only at $t_i \approx 0, t_2 \approx T, t_3 \approx T$. The frequencies of the pump waves "2" and "3" coincide: $\omega_2 = \omega_3 \equiv \omega_p$.

It is convenient to express the correlation functions in (3.8) through the retarded and advanced Green functions of electrons and holes [17]:

$$G^{(R)}_{e(h)}(\mathbf{r},t;\mathbf{r}',t') = -i\theta(t-t')\langle 0|\Psi_{e(h)}(\mathbf{r},t)\Psi^+_{e(h)}(\mathbf{r}',t')|0\rangle, \qquad (3.9)$$

$$G^{(A)}_{e(h)}(\mathbf{r},t;\mathbf{r}',t') = i\theta(t'-t)\langle 0|\Psi_{e(h)}(\mathbf{r},t)\Psi^+_{e(h)}(\mathbf{r}',t')|0\rangle. \qquad (3.10)$$

Using (3.2) we find for the square brackets in (3.8):

$$[\ldots] = G_h^{(A)}(\mathbf{r}_i,t_i;\mathbf{r}_3,t_3)G_h^{(R)}(\mathbf{r},t;\mathbf{r}_2,t_2)G_e^{(A)}(\mathbf{r}_i,t_i;\mathbf{r}_2,t_2)G_e^{(R)}(\mathbf{r},t;\mathbf{r}_3,t_3). \qquad (3.11)$$

Changing time variables $t, t_{2,3} \to t+2T, t_{2,3}+T$, we obtain an expression for the echo polarization (an inessential phase factor has been omitted):

$$\mathcal{P}(\mathbf{r},2T+t) = \mu^4 \int dt_i dt_2 dt_3 d\mathbf{r}_i d\mathbf{r}_2 d\mathbf{r}_3 \exp\left[i\sum_\alpha \eta_\alpha(\mathbf{k}_\alpha \mathbf{r}_\alpha - \omega_\alpha t_\alpha)\right]$$
$$\mathcal{E}_3(t_3)\mathcal{E}_2(t_2)\mathcal{E}_i^*(t_i)G_h^{(A)}(\mathbf{r}_i,t_i;\mathbf{r}_3,t_3+T)G_h^{(R)}(\mathbf{r},t+2T;\mathbf{r}_2,t_2+T)$$
$$G_e^{(A)}(\mathbf{r}_i,t_i;\mathbf{r}_2,t_2+T)G_e^{(R)}(\mathbf{r},t+2T;\mathbf{r}_3,t_3+T), \qquad (3.12)$$

where we have also changed the notations in such a way that the amplitudes $\mathcal{E}_\alpha(t_\alpha)$ differ from zero only in the time range $|t_\alpha| \le \Delta t_p$ (Δt_p is the pulse duration).

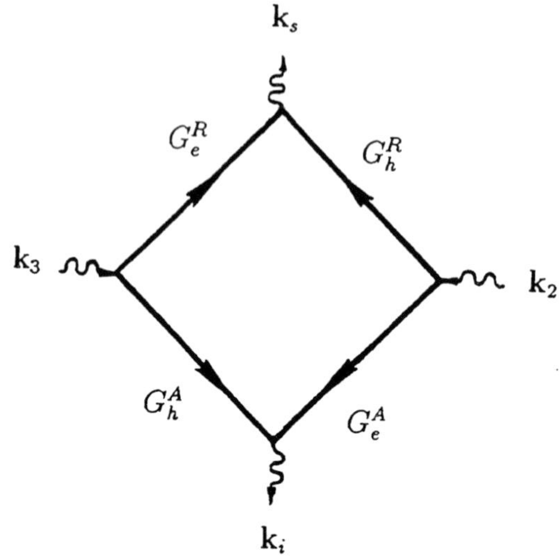

Figure 4: The diagram for the echo response. The solid lines correspond to the (nonaveraged) electron and hole Green functions.

Eq.(3.12) is the basic point of the following analysis. It is represented diagrammatically in Fig.4, where the solid lines correspond to the Green functions (3.9,3.10).

3.2 Optical echo from an ordered semiconductor

In order to gain an idea of the echo phenomenon in semiconductors we consider first the situation in the absence of disorder. In this case the Green functions (3.9,3.10) are represented as

$$G_{e,h}^{(0)R(A)}(\mathbf{r},t;\mathbf{r}',t') = \frac{1}{V}\sum_p \int \frac{dE}{2\pi} \exp\{i[\mathbf{p}(\mathbf{r}-\mathbf{r}') - E(t-t')]\} G_{e,h}^{(0)R(A)}(\mathbf{p},E), \tag{3.13}$$

$$G_{e,h}^{(0)R(A)}(\mathbf{p},E) = [E - \epsilon_{e,h}(\mathbf{p}) \pm i\delta]^{-1} \tag{3.14}$$

with $\delta \to +0$, V being a sample volume and the upper (lower) sign corresponding to the retarded (advanced) function. Integration over E in (3.13) gives

$$G_{e,h}^{(0)R(A)}(\mathbf{p},t-t') = \mp i\theta[\pm(t-t')] \exp[-i\epsilon_{e,h}(\mathbf{p})(t-t')]. \tag{3.15}$$

Using (3.15) we obtain for (3.12)

$$\mathcal{P}(\mathbf{r}, 2T+t) = \mu^4 \exp(i\mathbf{k}_s\mathbf{r}) \int dt_i dt_2 dt_3 \exp[-i\sum_\alpha \eta_\alpha \omega_\alpha t_\alpha)]$$
$$\mathcal{E}_3(t_3)\mathcal{E}_2(t_2)\mathcal{E}_i^*(t_i)$$
$$\frac{1}{V}\sum_p \exp[-i\epsilon_e(\mathbf{p})(T+t-t_3) + i\epsilon_e(\mathbf{p}-\mathbf{k}_3+\mathbf{k}_i)(T+t_2-t_i)]$$
$$\exp[-i\epsilon_h(-\mathbf{p}+\mathbf{k}_s)(T+t-t_2) + i\epsilon_h(-\mathbf{p}+\mathbf{k}_3)(T+t_3-t_i)], \qquad (3.16)$$

where $\mathbf{k}_s$ is the wave vector (1.5) of the generated polarization.

The momenta $\mathbf{p}$ contributing to the sum in (3.16) are typically of the order of an inverse lattice spacing a^{-1} and, therefore, are much greater than the wave vectors of light $\mathbf{k}_\alpha$. Assuming that the length of ballistic path of carriers during the time interval T is much smaller than the light wavelength λ

$$\frac{vT}{\lambda} \ll 1 \qquad (3.17)$$

(v is a typical velocity of carriers), we can neglect all the light wave vectors under the integral (3.16). Then the terms proportional to T cancel in the exponent and the carrier energies enter (3.15) as a sum

$$\epsilon_e(\mathbf{p}) + \epsilon_h(\mathbf{p}) \equiv \xi_p. \qquad (3.18)$$

After integrations over t_α we obtain

$$\mathcal{P}(\mathbf{r}, 2T+t) = \mu^4 \exp(i\mathbf{k}_s\mathbf{r}) \frac{1}{V}\sum_p \exp(-i\xi_p t)$$
$$\tilde{\mathcal{E}}_3(\xi_p - \omega_3)\tilde{\mathcal{E}}_2(\xi_p - \omega_2)\tilde{\mathcal{E}}_i^*(\xi_p - \omega_i), \qquad (3.19)$$

where $\tilde{\mathcal{E}}_\alpha(\omega)$ is the Fourier transform of $\tilde{\mathcal{E}}_\alpha(t)$ and differs from zero only at frequencies within the pulse spectral width $\Delta\omega_p$, the latter being small compared with light frequencies $\omega_\alpha \approx \omega_p$. Thus, the contribution to (3.19) is given by a domain with

$$\xi_p \approx \omega_p; \qquad (3.20)$$

this is nothing but an approximate energy conservation law in light absorption (emission) processes with creation (annihilation) of the electron-hole pairs with the momenta p and $-p$.

It is convenient to replace the summation over p by integration over an energy variable ξ:

$$\frac{1}{V}\sum_p \mathcal{F}(\xi_p) \to \int \nu(\xi)\mathcal{F}(\xi)d\xi \approx \nu(\omega_p)\int \mathcal{F}(\xi)d\xi \qquad (3.21)$$

with $\nu(\xi)$ being an analogue of a single-particle density of states. For example, taking the energy dispersion laws in the effective mass approximation

$$\epsilon_{e,h}(\mathbf{p}) = \frac{1}{2}E_G + \frac{p^2}{2m_{e,h}} \tag{3.22}$$

(where E_G is the energy gap of the semiconductor) we have

$$\nu(\xi) = \frac{1}{\pi^2}\sqrt{\frac{\xi - E_G}{2m}}. \tag{3.23}$$

Here m is the reduced effective mass of electron-hole pairs

$$\frac{1}{m} = \frac{1}{m_e} + \frac{1}{m_h}. \tag{3.24}$$

Using (3.21) we obtain the spectrum of the echo pulse

$$\tilde{\mathcal{P}}(\omega) = \mu^4\nu(\omega_p)\tilde{\mathcal{E}}_3(\omega - \omega_3)\tilde{\mathcal{E}}_2(\omega - \omega_2)\tilde{\mathcal{E}}_i^*(\omega - \omega_i) \tag{3.25}$$

and the echo signal amplitude $\mathcal{P}(2T)$:

$$\mathcal{P}(2T) = \mu^4\nu(\omega_p)\int \tilde{\mathcal{E}}_3(\xi - \omega_3)\tilde{\mathcal{E}}_2(\xi - \omega_2)\tilde{\mathcal{E}}_i^*(\xi - \omega_i)d\xi. \tag{3.26}$$

The magnitudes of (3.25,3.26) depend crucially on a spectral overlap of the exciting pulses.

In the considered approximation (3.17), the magnitude of the echo signal does not depend on the delay time T (in the absence of dissipative processes). This is in accordance with the treatment of a semiconductor as an ensemble of independent TLS. However, this treatment is no longer valid for the case when (3.17) does not hold. In the remaining part of the subsection we consider this case in more detail.

When $vT/\lambda \geq 1$, we cannot neglect the wave vectors of light. Taking them into account we have now a T-dependent factor under the sum in (3.16):

$$\exp(-iT\mathbf{pW}), \tag{3.27}$$

where

$$\mathbf{W} = \frac{\mathbf{k}_3 - \mathbf{k}_i}{m_e} - \frac{\mathbf{k}_2 - \mathbf{k}_i}{m_h}. \tag{3.28}$$

Averaging over the directions of $\mathbf{p}$ gives for (3.27) $\sin(TpW)/(TpW)$, which brings an additional factor

$$\frac{\sin\left[TW\sqrt{2m(\xi - E_G)}\right]}{TW\sqrt{2m(\xi - E_G)}} \tag{3.29}$$

under the integral (3.26). The variable ξ takes values within the spectral interval $\Delta\omega_p \sim (\Delta t_p)^{-1}$ around ω_p. Taking a characteristic momentum $p = \sqrt{2m(\xi - E_G)}$ of the order of inverse spacing a^{-1}, a variation of the sine argument in (3.29) can be estimated at $\sim (a/\lambda)(T/\Delta t_p)$. In the case

$$(a/\lambda)(T/\Delta t_p) \leq 1 \tag{3.30}$$

this variation may be neglected and we obtain the decay law of the echo signal (3.26) from an ordered semiconductor:

$$\mathcal{P}(2T) \propto \frac{\sin\left[TW\sqrt{2m(\omega_p - E_G)}\right]}{TW\sqrt{2m(\omega_p - E_G)}} \tag{3.31}$$

Note that besides a decay at

$$T > T_W \equiv \frac{1}{W\sqrt{2m(\omega_p - E_G)}} \sim \frac{\lambda}{v} \tag{3.32}$$

the expression (3.31) possesses the "Friedel-like" oscillation.

With an increase of the delay time T we come to the opposite case

$$(a/\lambda)(T/\Delta t_p) \geq 1 \tag{3.33}$$

when the sine in (3.29) oscillates fast while ξ runs through the spectral interval $\Delta\omega_p$. It results in a faster (exponential for smooth spectral functions $\tilde{\mathcal{E}}_\alpha(\xi - \omega_\alpha)$) decay of the echo signal. Assuming, for example, the Lorentzian shape of the spectral functions

$$\tilde{\mathcal{E}}_\alpha(\xi - \omega_\alpha) \propto \frac{1}{(\xi - \omega_\alpha)^2 + (\Delta\omega_p)^2} \tag{3.34}$$

we obtain an additional to (3.27) factor

$$\exp\left(-T/T_p\right), \tag{3.35}$$

where

$$T_p = \frac{1}{W\Delta\omega_p}\sqrt{\frac{2(\omega_p - E_G)}{m}} \tag{3.36}$$

is of the order $\Delta t_p(\lambda/a)$.

The established decay (3.31,3.35) of the echo signal from an ordered semiconductor in the absence of any dissipative processes indicates that the treatment of a semiconductor as an ensemble of independent two-level systems $\{(v;p),(c;p)\}$ is of limited validity. Indeed, this treatment neglects light wave vectors. Taking them into account one finds that successive light pulses couple different "two-level systems" (except the special excitation configuration with a small W (3.28)).

Thus, light induced coupling of TLS can be a reason of echo signal decay in ordered semiconductors at a long delay time. Note that this type of decay is rather sensitive to the excitation configuration.

In the following subsection we consider the echo phenomenon in the presence of disorder.

3.3 Optical echo from disordered semiconductors

The path of carriers corresponding to the times T_W and T_p is of the order of wavelength λ and $\lambda(\omega_p \Delta t_p)$, respectively. In disordered semiconductors these lengths may be considerably greater than the mean-free path l due to disorder induced elastic scattering of carriers. Here we consider just this situation.

The expression (3.12) for the echo response should be averaged over all the realizations of the random disorder (3.3). This averaging is performed by means of a standard diagrammatic technique [17] developed for the description of electrons in disordered metals. We restrict our consideration to the case of a weak disorder when the averaged Green function is given by (3.13) with (3.14) replaced by

$$G^{R(A)}_{e,h}(\mathbf{p}, E) = [E - \epsilon_{e,h}(\mathbf{p}) \pm i/(2\tau_{e,h})]^{-1}, \tag{3.37}$$

where the mean free time $\tau_{e,h}$ of electrons (holes) is given by

$$\frac{1}{\tau_{e,h}} = 2\pi\nu_{e,h}(E)\gamma_{e,h}; \tag{3.38}$$

$\nu_{e,h}(E)$ is the electron (hole) density of states and $\gamma_{e,h}$ is determined by (3.3). The criterion for the weakness of disorder is given by an inequality

$$\tau_{e,h}\sqrt{m(E - E_G)} \gg 1 \tag{3.39}$$

In the time representation the averaged Green function has the following form:

$$G^{R(A)}_{e,h}(\mathbf{p}, t - t') = \mp i\theta[\pm(t - t')] \exp\left[-i\epsilon_{e,h}(\mathbf{p})(t - t') - \frac{|t - t'|}{2\tau_{e,h}}\right], \tag{3.40}$$

where an additional (as compared to (3.15)) factor describes a disorder induced decay of phase coherence.

A more complicated task is to average the product (3.12) of the Green functions. As known from the theory of electrons in disordered metals,

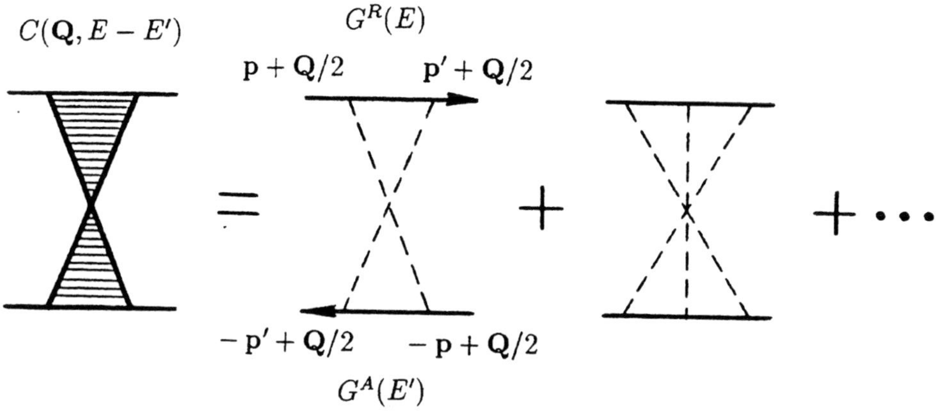

Figure 5: The "Cooperon" : a series of diagrams with "maximally crossed" disorder (dashed) lines. The solid lines correspond to the averaged Green functions.

the leading contribution to the conductivity (a linear response function) can be calculated in the Drude approximation, when the average product is replaced by the product of averaged Green functions. However, it does not hold true for the considered echo response! In the Drude approximation the echo signal $\mathcal{P}(2T)$ (3.12) would be evidently proportional to

$$\exp\left[-\left(\frac{1}{\tau_e} + \frac{1}{\tau_h}\right)T\right], \tag{3.41}$$

(see Fig.4) and would be negligible at a long delay time

$$T \gg \tau_{e,h} \tag{3.42}$$

The point is that the Drude approximation ignores totally important correlations between time-reversal trajectories of carriers. For coinciding successions of scattering events the random phases of the retarded G^R and advanced $G^A = (G^R)^*$ Green functions cancel in the product $G^R G^A$ (with mutually time-reversal in- and out-states), see Fig.5. The solid and dashed lines in Fig.5 correspond to the averaged Green functions (3.37) and the correlation functions of disorder (3.3). The series of diagrams in Fig.5 with maximally crossed disorder lines defines a famous "Cooperon" propagator in the weak localization theory (see, e.g., [15], [16]). It is responsible for an enhanced backscattering of light from disordered media, it provides a weak localization corrections to the Drude conductivity of disordered metals, etc. Because of Drude term vanishing at $T \gg \tau_{e,h}$,

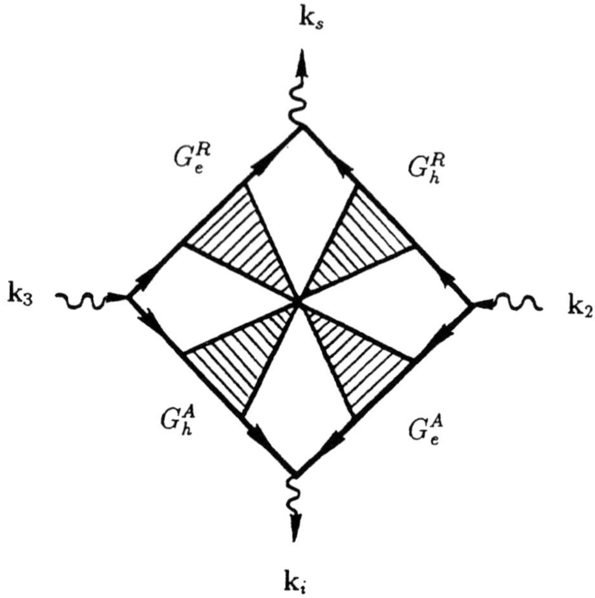

Figure 6: The diagramm making the leading contribution to the echo signal.

in the considered echo response the Cooperon contribution becomes a leading one.

Neglecting correlations scattering of electrons and holes we average the electron and hole Green function separately. In the leading approximation in the weak disorder parameter (3.39) the averaged echo polarization (3.12) is determined by a two-Cooperon diagram shown in Fig.6.

An expression for the Cooperon is obtained by a standard summation of the series in Fig.5 :

$$\gamma_{e,h} C_{e,h}(\mathbf{Q}, E - E') = \frac{\gamma_{e,h}}{\tau_{e,h}} \frac{1}{-i(E - E') + D_{e,h}\mathbf{Q}^2}, \tag{3.43}$$

where it is implied that $(E - E')\tau_{e,h} \ll 1; |\mathbf{Q}| v_{e,h}\tau_{e,h} \ll 1 \quad (v_{e,h} = p/m_{e,h})$, and the diffusion coefficients are given by

$$D_{e,h} = \frac{v_{e,h}^2 \tau_{e,h}}{3}. \tag{3.44}$$

Using (3.43) we arrive at the following representation for (3.12) corresponding to Fig.6 :

$$\mathcal{P}(\mathbf{k}_s, 2T) = \mu^4 \int \prod_{\alpha=i,2,3} dt_\alpha \prod_{\beta=e,h} \frac{dE_\beta d\Omega_\beta}{(2\pi)^2} \frac{d\mathbf{p}_\beta d\mathbf{p}'_\beta d\mathbf{Q}_\beta}{(2\pi)^{9/2}}$$
$$\mathcal{E}_3(t_3)\mathcal{E}_2(t_2)\mathcal{E}_i^*(t_i) \exp\left[-i(E_e + E_h - \omega_p)(t_i - t_2 - t_3)\right]$$
$$\frac{\gamma_\beta}{\tau_\beta} \frac{1}{-i\Omega_\beta + D_\beta Q_\beta^2} \exp\{-i[\Omega_\beta(T - t_i/2) + (\Omega_e - \Omega_h)(t_2 - t_3)]\}$$
$$\delta(\mathbf{p}'_e + \mathbf{p}'_h - \mathbf{k}_i - \frac{\mathbf{Q}_e + \mathbf{Q}_h}{2})\delta(\mathbf{p}_e - \mathbf{p}_h - \frac{\mathbf{k}_3 - \mathbf{k}_2}{2})\delta(\mathbf{Q}_e + \mathbf{Q}_h - \mathbf{k}_3 - \mathbf{k}_2)$$
$$G_\beta^R(\mathbf{p}_\beta + \frac{\mathbf{Q}_\beta}{2}, E_\beta + \frac{\Omega_\beta}{2})G_\beta^R(\mathbf{p}'_\beta + \frac{\mathbf{Q}_\beta}{2}, E_\beta + \frac{\Omega_\beta}{2})$$
$$G_\beta^A(-\mathbf{p}_\beta + \frac{\mathbf{Q}_\beta}{2}, E_\beta - \frac{\Omega_\beta}{2})G_\beta^A(-\mathbf{p}'_\beta + \frac{\mathbf{Q}_\beta}{2}, E_\beta - \frac{\Omega_\beta}{2}), \tag{3.45}$$

where, for simplicity, we set $\omega_\alpha = \omega_p, \alpha = i, 2, 3$. This rather cumbersome expression includes integrations over "fast" $(\mathbf{p}, E)$ and "slow" $(\mathbf{Q}, \Omega)$ variables. Neglecting $\mathbf{Q}, \mathbf{k}_\alpha$ and Ω in the Green functions arguments we obtain for (3.45):

$$\mathcal{P}(\mathbf{k}_s, 2T) = \mu^4 \int \prod_{\alpha=i,2,3} dt_\alpha \prod_{\beta=e,h} \frac{dE_\beta}{2\pi} \frac{d\mathbf{Q}_\beta}{(2\pi)^{3/2}} F^2(E_e, E_h) \frac{\gamma_\beta}{\tau_\beta}$$
$$\mathcal{E}_3(t_3)\mathcal{E}_2(t_2)\mathcal{E}_i^*(t_i) \exp\left[-i(E_e + E_h - \omega_p)(t_i - t_2 - t_3)\right]$$
$$\exp\left[-(D_e Q_e^2 + D_h Q_h^2)T\right] \delta(\mathbf{Q}_e + \mathbf{Q}_h - \mathbf{k}_3 - \mathbf{k}_2), \tag{3.46}$$

where the poles integrations over $\Omega_{e,h}$ have been performed (and we have neglected small terms Ωt_α in comparison with ΩT). The factor $F^2(E_e, E_h)$ in (3.46) results from integrations over fast momenta $\mathbf{p}, \mathbf{p}'$:

$$F(E_e, E_h) = \int \frac{d\mathbf{p}}{(2\pi)^3} \frac{1}{[E_e - \epsilon_e(p)]^2 + \frac{1}{4\tau_e^2(E_e)}} \frac{1}{[E_h - \epsilon_h(p)]^2 + \frac{1}{4\tau_h^2(E_h)}} \tag{3.47}$$

This expression depends primarily on a combination

$$m_e(E_e - \frac{E_G}{2}) - m_h(E_h - \frac{E_G}{2}) \equiv M\epsilon \tag{3.48}$$

(where $M = m_e + m_h$) and has a sharp maximum at $\epsilon = 0$. On the other hand, a variable $E \equiv E_e + E_h$ in (3.47) is evidently restricted to a narrow spectral region near ω_p. These two conditions determine the values of $E_{e,h}$

$$E_e \approx \frac{m_h}{M}\omega_p + \frac{m_e - m_h}{2M}E_G \tag{3.49}$$

$$E_h \approx \frac{m_h}{M}\omega_p + \frac{m_h - m_e}{2M}E_G \qquad (3.50)$$

entering the definitions of $\tau_{e,h}$, $D_{e,h}$, etc. Calculating (3.47) and performing the remaining integrations in (3.46) we find eventually

$$\mathcal{P}(2T) = \frac{\mu^4}{M(\frac{1}{m_h\tau_e} + \frac{1}{m_e\tau_h})} \int \tilde{\mathcal{E}}_3(E-\omega_p)\tilde{\mathcal{E}}_2(E-\omega_p)\tilde{\mathcal{E}}_i^*(E-\omega_p)\frac{dE}{4\pi}$$
$$\frac{1}{[\pi T(D_e+D_h)]^{3/2}} \exp\left[-\frac{(\mathbf{k}_2+\mathbf{k}_3)^2 D_e D_h}{D_e+D_h}T\right]. \qquad (3.51)$$

The equation (3.51) expresses the central result of this subsection. For arbitrary directions of light wave vectors $\mathbf{k}_2$ and $\mathbf{k}_3$ the signal decays exponentially with the decay time

$$T_{\Delta k} = \frac{D_e + D_h}{D_e D_h |\mathbf{k}_2 + \mathbf{k}_3|^2} \sim \tau\frac{\lambda}{l}. \qquad (3.52)$$

However, for counterpropagating pump waves of the second pulse $\mathbf{k}_2 + \mathbf{k}_3 = 0$, and the echo polarization (3.51) decays in a slow power law:

$$\mathcal{P}(2T) \sim T^{-3/2} \qquad (3.53)$$

It allows a qualitative interpretation. The echo signal survives due to compensation of random phases along time-reversed trajectories of carriers. The time-reversed trajectories corresponding to the diagram in Fig.6 are shown in Fig.7. Note, that the distance between the points "2" and "3", "i" and "f" is of the order of the mean free path l which is much shorter than a typical size $\sqrt{DT}$ of the diffusion trajectories. For integration goes over all the positions of the "merged" points (2,3), one can consider the electron and hole diffusion trajectories as parts of a closed trajectory starting and ending in the merged points (i,f). The factor $T^{-3/2}$ arises as a probability of returning to the starting point in the time T for diffusion motion. To avoid the dephasing caused by the external momentum $\mathbf{k}_2 + \mathbf{k}_3$ entering the merged points (2,3) the size $\sqrt{DT}$ of the diffusion trajectory should not exceed $|\mathbf{k}_2 + \mathbf{k}_3|^{-1}$. Note that namely the parameter $|\mathbf{k}_2 + \mathbf{k}_3|\sqrt{DT}$ enters the exponent in (3.51). Thus, the echo response (3.51) is rather sensitive to the geometry of excitation: to avoid exponential attenuation the angle between $\mathbf{k}_2$ and $\mathbf{k}_3$ should not exceed $\lambda/\sqrt{DT}$.

Of course, for the investigated case of a weak disorder allowing a regular theoretical treatment, the echo signal (3.51) is considerably smaller than that (3.26) for an ordered semiconductor. However, it demonstrates a significance of time-reversed trajectories for survival of the echo response at a long delay time.

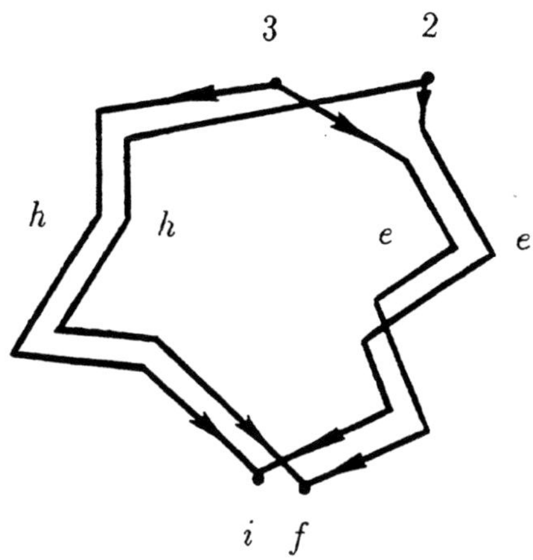

Figure 7: The trajectories of the carriers corresponding to the diagram in Fig.6.

4 Conclusions

We have studied the influence of disorder on four-wave optical mixing with the phase conjugation of a probe wave. Both stationary and pulse regimes corresponding to "optical phase conjugation" and "optical echo" phenomena have been considered. Both phenomena have a lot of common features. They have been shown to survive in the presence of disorder despite phase randomization caused by disorder induced scattering. The survival is due to a special class of time-reversed trajectories, along which random phases are compensated. In the theory of electrons in disordered media these trajectories correspond to the particle-particle diffusion propagator ("Cooperon") and are responsible for weak localization corrections to conductivity, etc. In the considered optical phenomena these weak localization effects are not only corrections but make the main contributions to the signal magnitude. The reason is that upon the neglect of correlation effects (the "Drude approximation") one obtains only an exponentially small contribution to the signal in the considered range of parameters. One should have in mind, however, that the magnitude of the signal provided by the weak localization effects is small compared to that in ordered media. Note also that even in disordered media the phase matching condition plays an important role for both the optical phase conjugation and the optical echo. As a consequence, both phenomena are very sensitive to the geometry of experiments.

Examining correlations of the backscattered waves corresponding to different incoming waves in the optical phase conjugation in disordered media we see that the backscattered waves possess a "phase memory" even for large differences between the incident angles of the incoming waves (in contrast to the case of linear backscattering). Thus, the optical phase conjugation in disordered media possesses the coherent properties similar to those in the case of ordered media.

The optical echo from weakly disordered semiconductors decays with the increasing delay time T. The decay law changes from exponential (at a short T) to power (at a longer T) law as a consequence of the weak localization effects. This is in contrast to the case of ordered semiconductors where, in the absence of dissipation, the echo signal does not decay up to a rather long delay time determined by the geometry of pumping.

Acknowledgements
The author is thankful to V.M.Agranovich, V.E.Kravtsov and P.Reineker for helpful discussions and cooperation in the works reviewed in the section 2. I wish to express my gratitude to O.Keller and all the participants of the Aalborg Summer School for warm and creative scientific atmosphere and valuable remarks.

References

[1] Y.R.Shen, The Principles of Nonlinear Optics (Wiley, New York, 1984).

[2] B.Ya.Zeldovich, N.F.Pilipetski and V.V.Shkunov, Phase Conjugation of Light (Nauka, Moscow, 1985) (in Russian).

[3] L.Allen, J.H.Eberly, Optical Resonance and Two-Level Atoms (Wiley-Interscience Publ., New York, 1975).

[4] V.E.Kravtsov, V.M.Agranovich and V.I.Yudson, in: Nonlinear Optics in Solids, ed. by O.Keller (Springer, New York, Berlin, 1990).

[5] E.Hanamura, Phys.Rev. **B 39** (1989) 1152.

[6] V.E.Kravtsov, V.I.Yudson, and V.M.Agranovich, Phys.Rev. B **41**, (1990) 2794.

[7] S.John, Comments Cond.Mat.Phys. **14** (1988) 193.

[8] V.I.Yudson and P.Reineker, Phys.Rev. **B 45** (1992) 2073.

[9] P.C.Becker, H.L.Fragnito, C.H.Brito Gruz, R.L Fork, J.E.Cunnigan, J.E.Henry, and C.V.Shank, Phys.Rev.Lett. **61**, (1988) 1647.

[10] V.L.Gurevich, M.I.Muradov, and D.A.Parshin, Europhys.Lett. **12**, (1990) 375.

[11] L.Root and J.L.Skinner, J.Chem.Phys. **81** (1984) 5310.

[12] J.Hegarty and M.D.Sturge, J.Opt.Soc.Am. **B2** (1985) 1143.

[13] Ch.Lonsky, P.Thomas, and A.Weller, Phys.Rev.Lett. **63** (1990) 652.

[14] G.Noll, U.Siegner, S.G.Shevel, and E.O.Göbel, Phys.Rev.Lett. **64** (1990) 792.

[15] B.L.Altshuler, A.G.Aronov, D.E.Khmelnitskii, and A.I.Larkin, in "Quantum Theory of Solids", ed. by I.M.Lifshits (Mir Publishers, Moscow, 1982).

[16] P.A.Lee, T.V.Ramakrishnan, Rev.Mod.Phys. **57**(1985) 287.

[17] A.A.Abrikosov, L.P.Gor'kov and I.E.Dzialoshinskii, Methods of Quantum Field Theory in Statistical Physics (Prentice-Hall, New York 1963).

Studies in Classical
and Quantum Nonlinear Optics

ISBN 1-56072-168-5

SELF-LOCALIZATION PHENOMENON AND INFLUENCE OF RAMAN SELF-SCATTERING ON THE DYNAMICS OF FEMTOSECOND SOLITONS IN A DUAL-CORE OPTICAL-FIBER NONLINEAR COUPLER

F.Kh. Abdullaev, R.M. Abrarov, S.A. Darmanyan, and V.I. Goncharov
Thermal Physics Department of the Uzbek Academy of Sciences
Ul. Katartal 28, Chilanzar, Tashkent 700135, Uzbekistan

1 INTRODUCTION

During recent years much attention has been paid to the creation of optical devices providing, with the help of ultrashort optical pulses, an ultrafast information exchange between various channels of a system. The dual-core optical-fiber nonlinear coupler (DCOFNC) and birefringent fibers are the basic elements for the creation of such devices.

After Jensen [1] and Maier [2] have shown a possibility of the existence of switching, bistability and other phenomena for continuous waves in a system of nonlinear coupled fibers many works have appeared, where these problems are studied. A review of recent results in this area of investigations including analyses of chaotic phenomena can be found in ref. [3].

New phenomena arise when considering the ultrashort pulse dynamics in such systems. In a paper [4] the soliton propagation in the DCOFNC has been numerically investigated. Varying the ratio between the coupling constant of the fibers (ϵ) and the square of the soliton amplitude, the authors have shown that at small values of this ratio, the wave is localized in one fiber, and with an increase of the ratio, a strong energy exchange occurs between the fibers. A similar phenomenon has also been found in ref. [5]. The analytical and numerical calculations made in refs. [6,7] demonstrate the existence of soliton coupled states, repulsive regimes between solitons, as well as all-optical switching of solitons.

The analytical results obtained so far for solitons in DCOFNC's are based on a very specific choice of the initial conditions and parameters, namely the one where the solitons in each of the cores have the same amplitude. In contrast, the numerical results cover a wide class of initial conditions and coupling constant values. In this work, we report analytical and numerical results which describe the soliton dynamics in DCOFNC's in various regimes.

In section 2, applying the perturbation theory for solitons in the adiabatic approximation and a computer simulation, we consider a case where one soliton is injected into one of the fibers, the field being absent in the other fiber. Then, we study a case where solitons are injected in each fiber with very different initial amplitudes. In both cases we observed the phenomenon of field self-localization in one fiber.

Section 3 deals with the influence of the stimulated Raman scattering (SRS) effect upon the dynamics of solitons in nonlinear couplers. The necessity of taking into account the SRS is caused by the importance of the phase relation between the solitons in DCOFNC operation. In the case of solitons, SRS results in a Soliton Self Frequency Shift (SSFS). This shift is proportional to ν^4 as it has been shown in ref. [8].

Let us briefly introduce the basic equations describing the nonlinear evolution of the electromagnetic fields in tunnel-coupled single-mode fibers. Thus these take the form

$$i\frac{\partial q_n}{\partial x} + \frac{1}{2}\frac{\partial^2 q_n}{\partial \tau^2} + \mid q_n \mid^2 q_n = \epsilon q_m + C_1 q_n \frac{\partial \mid q_n \mid^2}{\partial \tau} .$$

$$n, m = 1, 2 , \qquad n \neq m . \tag{1}$$

Here, x, τ, and q are the normalized length along the fiber, the travelling time and the amplitude, respectively, ϵ is the linear coupling coefficient between the fibers, and C_1 is the SRS constant. In fused quartz optical fibers $C_1 \simeq 10,6fs/\tau_0$, τ_0 being the full width at half-maximum power [8,9].

When $\epsilon = C_1 = 0$, the system of equations in (1) decouples into two unperturbed nonlinear Schrődinger (NLS) equations with a soliton solution

$$q_n = 2\nu_n sech[2\nu_n(\tau - \xi_n)]exp[2i\mu_n(\tau - \xi_n) + i\delta_n] . \tag{2}$$

Below we shall study the behaviour of the system of equations in (1) using various initial conditions.

2 SOLITON LOCALIZATION PHENOMENON

Let us consider the case where a soliton (Eq. (2)) is injected into the first fiber and the field is absent in the second one and let us ignore the SRS effect (i.e. $C_1 = 0$).

Supposing that in the evolution process when $\epsilon \ll 1$, only a little amount of energy will be transmitted into the second fiber, one can ignore the nonlinear effect in this fiber. Applying the adiabatic perturbation theory developed in [10,11] one then obtains the following expression for the soliton amplitude

$$\nu_x = \frac{\epsilon^2}{2\nu\sqrt{\pi}} Re\left\{(i-1)\int_{-\infty}^{+\infty} \frac{zdz}{sinh(z)} \int_0^{2\nu^2 x} \frac{dy}{\sqrt{y}}\, exp\left[i\left(\frac{z^2}{4y} - y\right)\right]\right\} . \tag{3}$$

To find an analytical expression for the function $\nu(x)$ let us consider small propagation distances i.e. $2\nu_0^2 x \ll 1$. The analysis of Eq. (3) shows that for this case one can obtain the following expression

$$\nu(x) = \nu_0 - \frac{\epsilon^2}{\nu_0^3} sin^2(\nu_0^2 x) , \tag{4}$$

where $\nu_0 \equiv \nu(x = 0)$. The same result can be obtained by ignoring the dispersive term in the second equation of the systems in (1), since the expression in Eq. (4) is valid for distances less than the dispersive length.

The results show that at small ϵ's the wave field remains localized in the first waveguide. Computer simulations show that the adiabatic theory describes the value of the soliton amplitude oscillation period well, but the soliton shape is modulated by the large scale oscillations not described by Eq. (4). To describe these it is necessary to take into account the influence of the continuum spectrum.

2.1 Case of two solitons of unequal amplitudes

Let us consider the case where solitons are injected into each fiber, and let us suppose that the soliton amplitudes are very different, i.e. $\nu_2/\nu_1 \ll 1$. Applying the adiabatic perturbation theory for solitons one can obtain the following system of equations for the soliton parameters

$$\nu_2 \approx \nu_{02}\left[1 - \frac{\pi\epsilon}{\nu_{01}^2} sin(\nu_{01}^2 x) sin(\psi_0 - \nu_{01}^2 x)\right] , \tag{5}$$

$$\nu_1 \approx A - \nu_2 \ , A = \nu_{01} + \nu_{02} = const \ , \tag{6}$$

$$\Delta\xi \approx \Delta\xi_0 \left[1 - \frac{\pi\epsilon}{\nu_{01}^2} sin(\nu_{01}^2 x) sin(\psi_0 - \nu_{01}^2 x)\right] , \tag{7}$$

where $\Delta\xi = \xi_2 - \xi_1$, $\psi = \delta_2 - \delta_1$, and ν_{0n} and ψ_0 are the initial values of the amplitude and phase difference (we choose here the following initial conditions: $\nu_{0n} = 0$, $\xi_{01} - \xi_{02} \ll 1$).

It also follows from the solution of the system of equations in (5)-(7) that the soliton with the large amplitude moves practically with constant group velocity, and that the small amplitude soliton oscillates near the center of the large amplitude soliton. This means a presence of the soliton coupled state in this case. Here, like in the case of Section 2, the field is localized in the first fiber and the solitons exchange only an insignificant part of the total energy.

3 THE SSFS INFLUENCE ON THE SOLITON DYNAMICS IN THE DCOFNC

Let us consider the case where solitons with equal amplitudes are injected in each fiber. Performing the same actions as in the previous sections and taking into account that the operator of perturbation has the form

$$R_{mn} = -i\epsilon q_m - iC_1 q_n \frac{\partial \mid q_n \mid^2}{\partial \tau} \tag{8}$$

one realizes that the influence of SRS results in a velocity change of each soliton propagation separately as well as in a velocity change of the center of mass, i.e.

$$\frac{d}{dx}(\xi_1 + \xi_2) = 2(\mu_1 + \mu_2) \ ,$$

$$\frac{d}{dx}(\mu_1 + \mu_2) = \frac{8}{15}C_1(2\nu)^4 \ . \tag{9}$$

This velocity change which can influence also other soliton parameters, in particular leads to an x-dependent phase difference x, even for the case where the initial phase difference ϕ_0 equals 0 or π. In turn, this results in energy exchange between solitons, but over small distances we can assume that $\psi \approx \psi_0 = \delta_{20} - \delta_{10} = const$. Then the further analysis for the initial values of soliton phase difference $\phi_0 = 0, \pi$, as it already has been shown in [6], leads to a closed equation for the variable $r = 2\nu(\xi_2 - \xi_1)$, namely,

$$r_{xx} - 16\nu^2\epsilon\frac{rcosh(r) - sinh(r)}{sinh^2(r)}cos(\psi_0)\beta = (ln\beta)_x r_x \ , \tag{10}$$

where $\nu = (\nu_1 + \nu_2)/2$, $\beta = (1 - \frac{\epsilon\pi^2 r}{12\nu^2 sinh(r)}cos(\psi_0))$. Here we have taken into account the terms proportional to ϵ^2, which as will be shown below, can play an essential role. The analysis of the equation above is rather simple in the region of small $r(r \ll 1)$. Thus, in this case Eq. (10) is reduced to

$$r_{xx} - k^2 r = 0 \ , \tag{11}$$

where $k^2 = \frac{16}{3}\nu^2\epsilon cos\psi_0(1 - \frac{\epsilon\pi^2}{12\nu^2}cos\psi_0)$. It follows from Eq. (11) that when $\epsilon cos\psi_0 < 0$, the r oscillates with a period $L = 2\pi/k$. When $\frac{12\nu^2}{\pi^2} = \epsilon_c > \epsilon cos\psi_0 > 0$ there exists a regime of repulsion, which for $|\epsilon| > \epsilon_c$ is changed to a regime of attraction. These results are in good agreement with the numerical results given in [7] (see Figs. 7-10 in [7]).

Simulations of the pulse propagation in the fiber were obtained from a numerical solution of Eq. (1) using an improved explicit scheme [12]. The numerical calculations were made for the case of equal initial amplitudes and initial phase difference equal to 0 and π for various values of the constants ϵ and C_1. The results of the computer simulation show that in the case of pulses of duration $\tau_0 > 1ps$, the influence of the SRS effect is weak and mainly results in a change of the center of mass velocity of the interacting solitons. The SRS process begins to influence essentially upon the pulse dynamics at a duration of the order of some hundreds of femtoseconds. As seen from Fig. 1a the SSFS effect results when $C_1 = 0.05(\tau_0 \approx 200fs)$ in a quite considerable change of the pattern of soliton interaction in comparison with the case where this effect is not taken into account (Fig. 1b). Already at a length of $x \approx 5$, the coupled state is broken and a strong energy redistribution takes place between the solitons (Fig. 1a) leading to field localization in one waveguide. A similar effect in the absence of the SRS has been observed in numerical experiments in [9] for much larger distances $x \approx 60$ (see Fig. 4 of [9]). Therefore, the SSFS effect results in an abrupt increase of the soliton coupled state instability. Another pattern of interaction is revealed in the case of larger coupling between the waveguides. In this case a stable motion of a coupled pair of solitons with initial amplitudes$A_{1,2} = 1$ is observed (Fig. 2a). Such a motion is followed by a periodic exhchange of a little amount of energy (15%) between the solitons. In this case the increase of $C_1(0 < C_1 < 0.4)$ influences only the velocity of the center of mass motion of the coupled pair, without changing the character of the soliton interaction. But the amplitude increase leads to a strong energy redistribution among the solitons and to a coupled state decay.

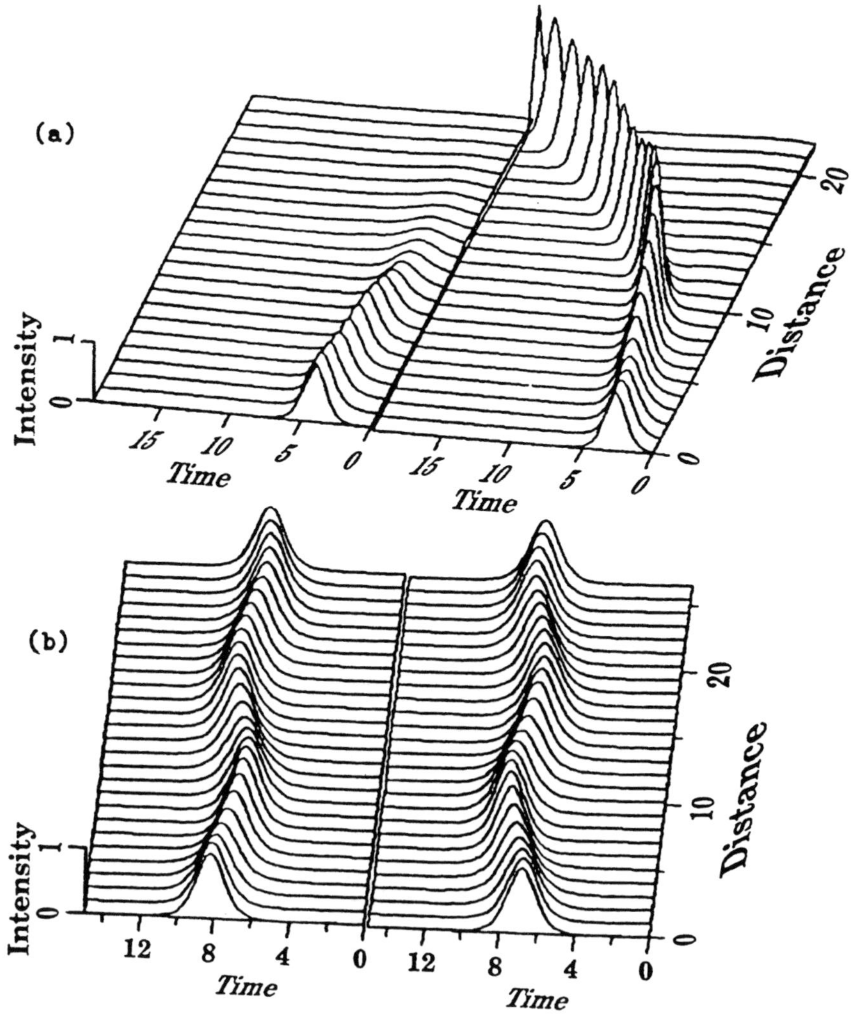

Figure 1. *Soliton evolution with* $A_{1,2} = 1$, $\psi_0 = 0$, $\epsilon = -0.1$, *initial distance between solitons* $\Delta = 1.4$.: *(a) - with account of SRC* ($C_1 = 0.05$), *(b) - without account of SRC* ($C_1 = 0$).

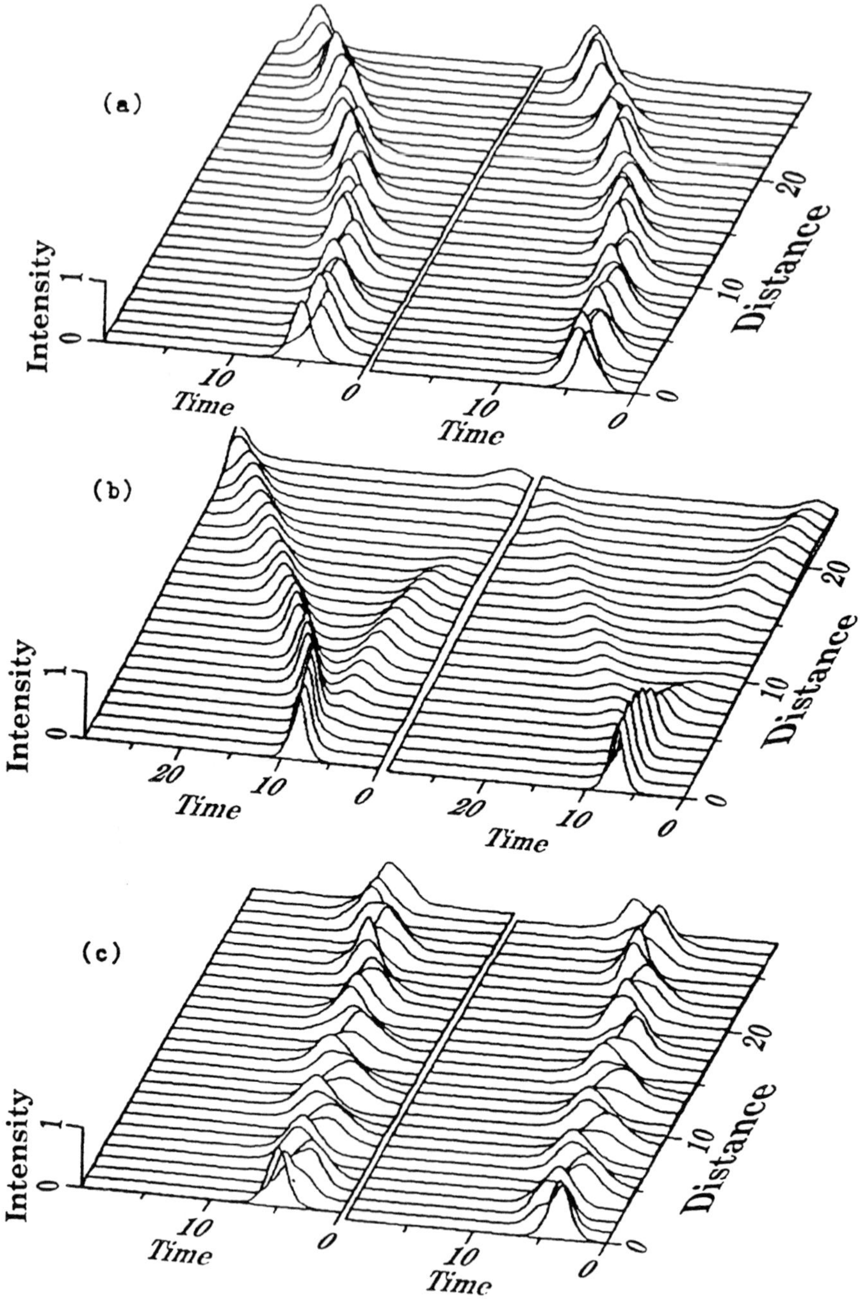

Figure 2. *As in Fig. 1: (a) -* $\epsilon = -0.8$ *and* $C_1 = 0.1$*; (b) -* $\psi_0 = \pi$, $\epsilon = -0.2$, $C_1 = 0.1$*; (c) - the same as in (b) with* $\epsilon = -0.8$.

In the case of weak coupling ($|\epsilon| < \epsilon_C$) and $\psi_0 = \pi$, a soliton repulsion regime occurs (Fig. 2b). As seen for increasing distance between the pulses there exist independent processes of energy exchange between the channels of propagation. When $|\epsilon| > \epsilon_C a$ the character of the soliton interaction is changed for the reverse one and a coupled state stable to the SSFS effect, is formed (Fig. 2c). If now one increases the initial amplitudes ($A_{1,2} = 2$), then the interaction regime again becomes of the repulsive character, since in this case the inequality $|\epsilon| < \epsilon_c$ is again fulfilled. Thus, one can include that also for both psec pulses and fsec pulses there is a threshold value of coupling, ϵ_C. When $|\epsilon| > \epsilon_C$ the behaviour of solitons with different durations is identical - a stable coupled state for $\psi_0 = 0$ as well as for $\psi_0 = \pi$ is formed. If $|\epsilon| < \epsilon_C$, then in the case of femtosecond solitons with equal initial phases, the coupled state is instable. The shorter the durations and the larger the initial amplitudes of the optical pulses, the more strong is the energy redistribution resulting in a decay of the coupled pair.

References

[1] A.A. Maier, *Kvantovaya Electron.*, (Moscow), Vol. 9, 1982, p. 2296 (in Russian) [*Sov.J.Quantum Electron.*, Vol. 2, 1982, p. 1490].

[2] S.M. Jensen, *IEEE J.Quant.Electr.*, Vol. QE-18 No. 10, 1982, p. 1580.

[3] S. Trillo and S. Wabnitz, *J.Opt.Soc.Am.*, Vol. B5, 1988, p. 483.

[4] L.M. Andrushko, K.S. Karplyuk and S.V. Ostrovsky, *Radiotech.i Elektron.*, Vol. 32, 1987, p. 427 (in Russian).

[5] S. Trillo, S. Wabnitz, E.M. Wright and G.I. Stegeman, *Opt.Lett.*, Vol. 13, 1988, p. 672.

[6] F.Kh. Abdullaev, R.M. Abrarov and S.A. Darmanyan, *Opt.Lett.*, Vol. 14, 1989, p. 131.

[7] E.M. Wright, G.I. Stegeman and S. Wabnitz, *Phys.Rev.*, Vol. B40 No. 2, 1989, p. 4455.

[8] J. Gordon, *Opt.Lett.*, Vol. 11 No. 10, 1986, p. 662.

[9] M.N. Islam, G. Sucha, I. Bar-Joseph, M. Wegener, J.P. Gordon and D.S. Chemla, *J.Opt.Soc.Am.*, Vol. B6, 1989, p. 1149.

[10] V.I. Karpman and E.M. Maslow, *JETPh*, Vol. 73, 1977, p. 573.

[11] D.J. Kaup and A.C. Newell, *Proc.Roy.Soc.*, Vol. A361, 1987, p. 413.

[12] T.A. Petrova and A.D. Yunakovsky, *Preprint, No. 68, USSR Academy of Sciences, Inst. of Appl.Phys.*, Gorkiy, 1983.

Studies in Classical
and Quantum Nonlinear Optics

ISBN 1-56072-168-5

PHOTOINDUCED PHENOMENA IN RB: KTP WAVEGUIDES

K.S. Buritskii, V.A. Chernykh, E.M. Dianov,
V.P. Gapontsev, V.A. Maslov, E.A. Shcherbakov*
General Physics Institute, Russian Academy of Sciences
38 Vavilov St., Moscow, 117942, Russia
*R&D Center IRE-Polus, Vvedenskogo Sq. 1,
Fryazino, Moscow, 141120, Russia

1 Introduction

Potassium titanyl phosphate (KTP) crystal is of particularly significance in the field of nonlinear optics because of its superior properties: high nonlinear and electro-optical coefficients, high optical damage resistance, mechanical stability, wide phase-matching range for second harmonic generation (SHG) etc [1]. KTP has found increasing commercial application for frequency doublers and optical parametric oscillators, where a crystal is influenced by high peak power light pulses. However, various photoinduced phenomena, such as damage of crystal surface [2], bulk darkening [3], the refractive index change due to photorefraction [4], may occur concurrently with nonlinear frequency conversion depending on the experimental conditions (peak power, pulse duration, pump and SH power ratio, crystal quality). The mechanism of these phenomena in KTP is practically unknown; there are only some assumptions about its origin. The investigation of photoinduced effects is very important due to numerous practical applications of KTP.

The goal of this paper was to study some photoinduced phenomena in a KTP crystal while high power light pulses propagating through the channel and coupled waveguides fabricated in KTP [5]. Photoinduced phenomena (for example, photoinduced heat) begin to influence the waveguide properties at moderate light power levels in cw regime. It was reported [6] about the influence of the thermal nonlinearity on SHG efficiency in the blue region ($\lambda_p = 0.86\ \mu m, P_\omega = 100\ mW, P_{2\omega} = 4\ mW$). On the other hand, KTP is very interesting as a material with the high third-order nonresonant nonlinearity $\chi^{(3)}$. The influence of light self-action due to photorefractive, thermooptical, photoelastic etc effects should be excluded when the "fast" third-order nonlinearities caused by nonlinear polarizabilities of electron bonds are studied [7].

2 Experimental

A schematic diagram of the experimental setup is depicted in Fig.1. The pump source was a mode-locked and Q-switched Nd:YAG laser delivering 150 ps pulse trains at 1.064 μm at a 500 Hz repetition rate. The optical pulses from the laser were coupled first to a single-mode anisotropic polarization maintaining fiber (at the fiber output the polarization degree was not less than 10 dB). The light was focused on the fiber input end-face by a thin lens with f=1 cm. The beam was first collimated at the fiber output with a $20^{\times}$ microobjective, then directed through a glass plate and coupled into a waveguide by a $20^{\times}$ microobjective. The average light power was measured by using glass plates at the fiber and the waveguide outputs. The light power launched into the waveguide was changed by a relative displacement of the focusing lens and the input facet of the fiber. Such a method takes advantage of the fixed position of the fiber output with respect to the waveguide input. In addition, the light power distribution had gaussian shape on the fiber output facet, what led to a high coupling efficiency. Photoinduced phenomena were observed from the peak power level of 1 kW on the input facet of the waveguide. This value corresponded to the peak intensity of the order of 10 GW/cm^2 and to the average power level of approximately 1 mW.

KTP crystals used for waveguide fabrication were grown by the flux method. Samples were cut perpendicular to the Z axis in slabs approximately 6x10x1 mm^3 in size and then polished on all faces. The channel waveguides, 6 μm in width, and coupled waveguides of the same width with a 6 μm gap between the waveguides were fabricated by ion exchange in a molten mixture of rubidium and barium nitrates with 97/3 Ba $(NO_3)_2/RbNO_3$ molar ratio. The ion-exchange conditions were chosen to obtain the single-mode waveguides in the needed polarization at 1.06 μm.

3 Results and discussion

3.1 Transmission of high-peak-power light pulses through the channel waveguide

Evidently, when a channel waveguide is excited in pulse regime at a low power level, the dependence of the output light power on the input light power remains linear and the pulse envelope does not change its shape. As our experiments showed, when increasing the input peak power a saturation of the $P_{out}(P_{in})$ dependence and flattening of the upper part of the pulse envelope were observed, which testifies to the development of third-order nonlinear processes in the crystal. While further increasing input power the optical breakdown of the crystal happened.

If one looks at breakdown patterns, one can distinguish two kinds of them. The first one is analogous to the thermooptical breakdown, which occurs in cw regime. This is the case when input facets heat up to the melting point leading to crater formation. It should be mentioned that the average input power in the pulse regime did not exceed 10 mW. Evidently, it was insufficient for crystal melting. We should admit that irradiation of the crystal with high peak power optical pulses could cause the formation of the color centers, which induced the increase of the optical loss. Such an avalanche-type increase of absorption coefficient gives the thermooptical increase

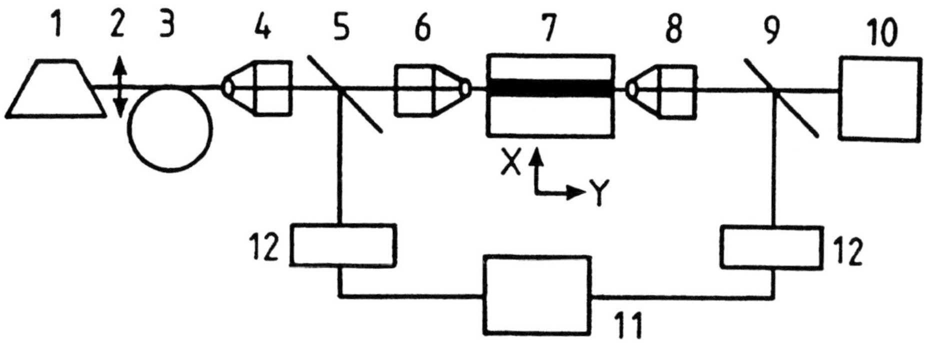

Fig. 1. Experimental setup: 1-laser, 2-lens, 3-single-mode fiber, 4,6,8-microobjectives, 5,9-beamsplitters, 7-waveguide, 10-spectroanalyzer, 11-voltmeter, 12-photodiode.

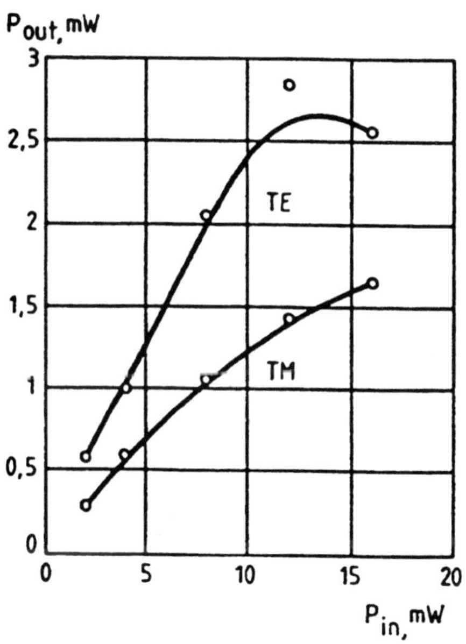

Fig. 2. Average output power vs average input power for TE and TM modes.

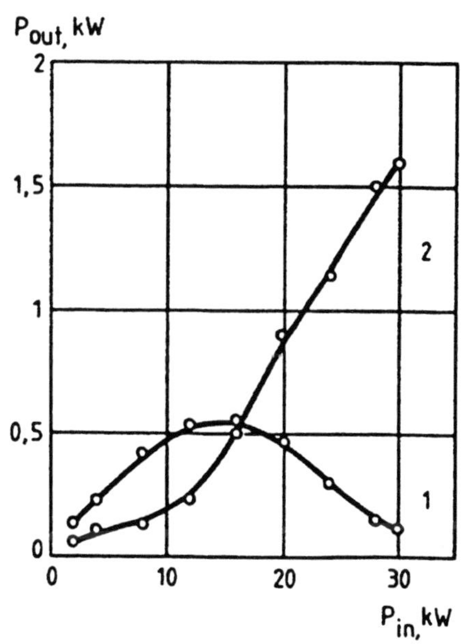

Fig. 3. Power redistribution between 1 and 2 channels of the directional coupler vs peak input power.

of the refractive index and leads to self-focusing of the light beam. This is a typical case of the laser induced thermal breakdown [8].

The second case corresponds to a shorter (up to 100 ps) pulse. The thermal crater is absent, but one can see a crack on the crystal surface. This crack can be formed both on the input and rear endfaces. The destruction region is located within the waveguide, that suggests that mechanical destruction was caused by inner strains occured. The possible origins of strains are the local crystal heat owing to the light self-focusing, the excitation of stimulated Brillouin scattering and the gradual accumulation of elastic strains due to color centers formation.

We measured the dependence of the average output power P_{out} versus the input power P_{in} at power levels lower than the breakdown threshold (Fig. 2) with approximately 1 min exposure time at each point. The exposure time was chosen in accordance with the time needed to reach the steady state. In the linear region full transmission, defined as the dependence slope, is larger for TE-mode than that in case of TM-mode (due to its power loss) and therefore transmission nonlinearity occurs at a lower power level. The transmission does not restore at room temperature, which indicates stable color center formation in the waveguide region. Flattening of the pulse envelope caused, in our opinion, by two-photon absorption of high-peak-power pulses has been also observed.

Apart from amplitude effects, we also observed the change of the output radiation spectrum at the waveguide output when the Z-polarized wave was excited. The analysis of the spectrum obtained revealed that Raman scattering occured in the waveguide. Comparing the position of Raman components with that being published [9] we easily identified the ν_1 and ν_5 lines with frequencies 690 cm^{-1} and 270 cm^{-1}. However, the components with 420 cm^{-1} and 530 cm^{-1} frequencies are ascribed as a rule to X-polarized pumping wave and their origin still remains unknown.

3.2 Transmission of high-peak-power light pulses through coupled waveguides

We have also investigated the passing of subnanosecond light pulses (TM-polarization) through the Rb:KTP directional coupler. The nonlinear directional coupler is one of the possible arrangements for all-optical switching. It consists of two channel optical waveguides placed in close proximity. In this geometry the tunneling optical field provides coupling between the waveguides. In our preceding paper [7] we observed strong reshaping of the output pulse envelope (for TE-polarization) in both channels, when light power was increased. This fact was explained by the presence of the "fast" third-order nonlinearity with the response time of less than 100 ps.

The value of the nonlinear coefficient n_2 and the response time of the nonlinearity depend on the specific nonlinearity mechanism. In this paper we report an observation of slow redistribution of the light power between two channels of the directional coupler with characteristic time of the order of 1 min, when TM-polarized waveguide mode is excited. The waveguides were single-mode for TM-polarization. The measured dependencies of the output power in each of the coupled waveguides on the input power are displayed on Fig. 3. The exposure time at each point corresponds to the time needed for the establishment of the steady state and is about 5 min. Probably the origin of the slow redistribution of the output power is the crystal heat due to photoinduced loss in the waveguide. When the crystal is heated, the change of the refractive index is determined by the thermooptic coefficient $dn_e/dT = 3 \times 10^{-5}$ 1/grad at 1.06 μm [10]. To break the coupling between the waveguides, the value of the refractive index change should satisfy the condition $k\Delta n^* L \cong 2\pi$, which corresponds to $\Delta n^* \cong 10^{-4}$ and therefore to the temperature change of about 3°C.

4 Conclusions

We have observed nonlinear transmission of high-peak-power subnanosecond light pulses through channel and coupled waveguides based on KTP crystal. The photoinduced effects arise when the peak power in the waveguide approaches the value of 1 kW, which corresponds to the power density of the order of 10 GW/cm^2. Apparently these photoinduced phenomena should be taken into consideration when KTP-based integrated nonlinear and electrooptical devices are created. Different kinds of the catastrophic optical breakdown have been registered and possible physical mechanisms of crystal destruction have been discussed. Besides the Raman scattering in a Rb:KTP waveguide has been also observed and the Raman spectrum has been partially interpreted.

Acknowledgements

The authors gratefully acknowledge Dr. P.S. Fisher for helpful assisstance in the experiment. This research was supported in part by AT&T.

References

[1] J.D. Bierlein, H. Vanherzeele, "Potassium titanyl phosphate: properties and new applications", *J.Opt.Soc.Am.B.*, Vol. 6, No.4, 1989, pp. 622-633.

[2] F. Ahmed, "Laser damage threshold of $KTiOPO_4$", *Appl.Opt.*, Vol. 28, No. 1, 1989, pp. 119-125.

[3] J.C. Jacco, D.R. Rockafellow, E.A. Teppo, "Bulk darkening threshold of flux-grown $KTiOPO_4$", *Opt.Lett.*, Vol. 16, No. 17, 1991, pp. 1307-1309.

[4] J.K. Tyminski, "Photorefractive damage in KTP used as second-harmonic generator", *J.Appl.Phys.*, Vol. 70, No. 10, 1991, pp. 5570-5575.

[5] J.D. Bierlein, A. Ferretti, L.H. Brixner, W.Y. Hsu, "Fabrication and characterization of optical waveguides in $KTiOPO_4$", *Appl.Phys.Lett.*, Vol. 50, No. 20, 1987, pp. 1216-1219.

[6] M.J. Jongerius, R.R. Drenten, R.B.J. Droste, "Blue second-harmonic generation in segmented $LiNbO_3$ and KTP waveguides", *Tech.Dig. Conference on Laser and Electro-Optics*, OSA, Washington Dc, 1992, paper CThA2.

[7] K.S. Buritskii, E.M. Dianov, V.A. Maslov, V.A. Chernykh, E.A. Shcherbakov, "Nonlinear directional coupler based on Rb:KTP waveguides", *Appl.Phys.B.*, Vol. 54, No. 3, 1992, pp. 167-170.

[8] Y.R. Shen, *The Principles of Nonlinear Optics*, New York: John Wiley &Sons, Inc., 1984, 10.

[9] G.H. Watson, "Polarized Raman spectra of $KTiOAsO_4$ and isomorphic nonlinear-optical crystals", *J. Raman Spectr.*, Vol. 22, 1991, pp. 705-713.

[10] D.J. Gettemy, W.G. Harker, G. Lindholm and N.P. Barnes, "Some optical properties of KTP, $LiIO_3$ and $LiNbO_3$", *IEEE J. Quantum Electron.*, Vol. 24, No. 11, 1988, pp. 2231-2237.

Studies in Classical
and Quantum Nonlinear Optics

ISBN 1-56072-168-5

GREEN LIGHT GENERATION FROM A NEODYMIUM FIBRE LASER

K.S. Buritskii, V.P. Gapontsev, E.M. Dianov, V.A. Maslov, I.E. Samartsev*, V.A. Chernykh and E.A. Shcherbakov*
General Physics Institute, Russian Academy of Science,
Vavilov Str. 38, Moscow 117942, Russia
*R&D Center IRE-Polus, Vvedenskogo Sq. 1,
Frayazino, Moscow 141120, Russia

1 Introduction

Recently, much attention has been attracted to the problem of creating compact blue-green light sources. Particularly, the frequency doubling of semiconductor lasers has been investigated for this purpose. However, rare-earth doped fibre lasers are an interesting alternative to diode lasers due to their tunability, high quality output beam etc [1]. Such lasers are compatible with channel KTP-based waveguides, which are very promising for nonlinear frequency conversion from the near-IR to green and blue. The phase-matched type II interaction or quasi-phase-matched type I interaction can be used depending on the fundamental wavelength [2], [3].

This paper is concerned with the phase-matched (type II) nonlinear frequency conversion of Nd^{3+} - doped fibre laser radiation in the channel Rb:Ba:KTP-waveguide.

2 Fibre laser

A tunable fibre laser used in our experiment consisted of a Nd^{3+}-doped fibre pumped by a laser diode ($\lambda = 0.8\ \mu m$), directional coupler and diffraction grating (Figure 1). A two-core germanium-containing silica fibre with Nd^{3+}-doped inner core allowed us to use a multimode laser diode for pumping. The active fibre possessed the following parameters: length-5 m, activator concentration-600 ppm, active and pumping core diameters-6.5 μm and 60 μm respectively. The maximum pumping power in the active fibre was 300 mW.

The wavelength of the laser radiation could be changed in the range of 1.05-1.13 μm (Figure 2) by the grating rotation. The output power reached 14 mW and the linewidth did not exceed 0.2 nm.

3 Waveguide fabrication

Channel waveguide was fabricated on the z-cut flux grown KTP substrate to be directed along the Y-axis by the ion exchange from the mixed $RbNO_3/Ba(NO_3)_2$ melt at 340^0C [4]. The process duration was 15 min.

The waveguide obtained (length L=1 cm, width W=7 μm) was single-mode for the pumping radiation with the wavelength of $\simeq$ 1.1 μm. The single-mode regime was maintained for both TE- and TM- polarizations. The value of the optical loss was estimated to be 1 dB/cm for pumping and 2 dB/cm for SH radiation.

4 Numerical estimation

The second-order nonlinear interactions in waveguides have been treated by several authors [5]-[6]. We assume that purely transverse-mode fields propagate in the channel guide. The field components for the propagating mode are described by:

$$E_\omega(x,y,z,t) = \frac{1}{2}\left\{A_\omega(z)E_\omega(x,y)exp[i(\omega t - \beta_\omega z)] + c.c.\right\} ,$$

$$E_{2\omega}(x,y,z.t) = \frac{1}{2}\left\{A_{2\omega}(z)E_{2\omega}(x,y)exp[i(2\omega t - \beta_{2\omega} z)] + c.c.\right\} ,$$

where $\beta = 2\pi n^*/\lambda, n^*$ is the effective refractive index, $A_{\omega,2\omega}$ are the corresponding amplitudes and $E_{\omega,2\omega}(x,y)$ are the mode field distributions. In the phase-matching case ($\beta_\omega = \beta_{2\omega}$) the SHG efficiency η is given by the following expressions, if the pump depletion and optical loss are ignored:

$$\eta = P_{2\omega}/P_\omega = KP_\omega L^2 \tag{1}$$

$$K = \frac{8\pi^2}{c\epsilon_0\lambda_\omega^2(n_\omega^*)^2 n_{2\omega}^*} d_{eff}^2 J^2 \tag{2}$$

$$J = \frac{\int\int_{-\infty}^{\infty} E_{2\omega}(x,y)E_\omega^2(x,y)dxdy}{\left[\int\int_{-\infty}^{\infty} E_{2\omega}^2(x,y)dxdy\right]^{1/2}\int\int_{-\infty}^{\infty} E_\omega^2(x,y)dxdy} \tag{3}$$

where ϵ_0 is the vacuum permittivity, c is the vacuum velocity of light, $P_{\omega,2\omega}$ is the light power and J is the fields overlap integral.

The type II phase-matching (TE_{00}^ω + $TM_{00}^\omega \Rightarrow TE_{00}^{2\omega}$) was obtained in the Rb:Ba: KTP channel waveguide with the nonlinear coefficient $d_{eff} = d_{15} = 6.1$ pm/V. The maximum refractive index change ($\Delta n \simeq 0.01$) and depth ($d \simeq 3\mu m$) [4] for a Rb:Ba:KTP waveguide are close to those of the Ti-diffused waveguides in $LiNbO_3$. For this reason, the value of the overlap integral J was estimated by using the results for Ti:$LiNbO_3$ waveguides. The optimum value of the overlap integral equals $1/J^2 \simeq 100\ \mu m^{-2}$ [5]. Substituting these data in (1)-(2) we have estimated the figure of merit K$\simeq$ 16% $W^{-1}cm^{-2}$ in a Rb:Ba:KTP waveguide. Calculations of the K with the data of the nonlinear coefficient $d_{15} = 1.91$ pm/V, reported in [6], gives K=1.6% $W^{-1}cm^{-2}$.

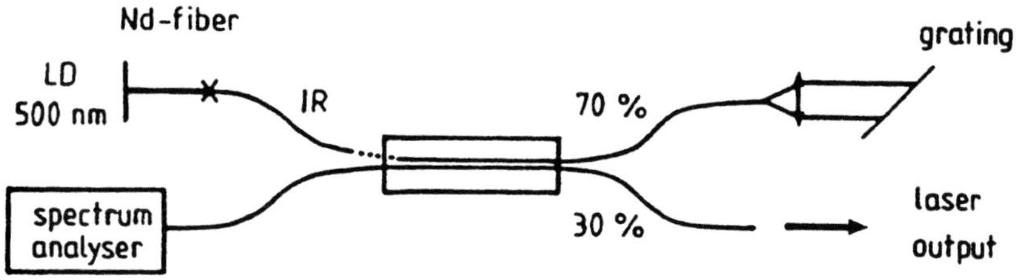

Fig. 1. Schematic of the neodymium doped fibre laser.

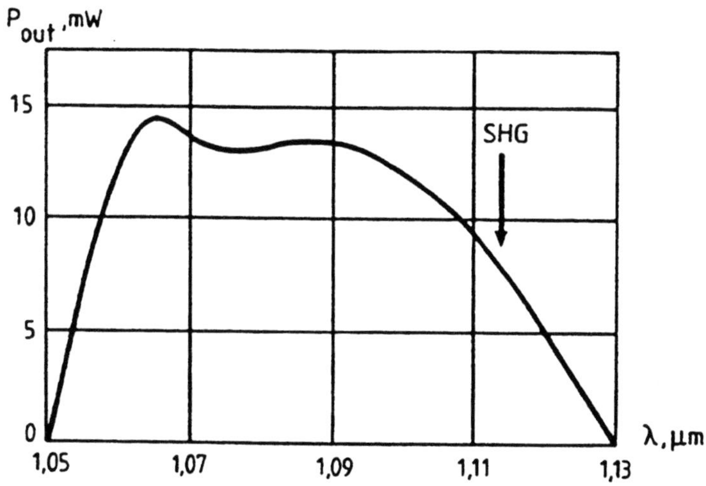

Fig. 2. Fibre laser tuning curve.

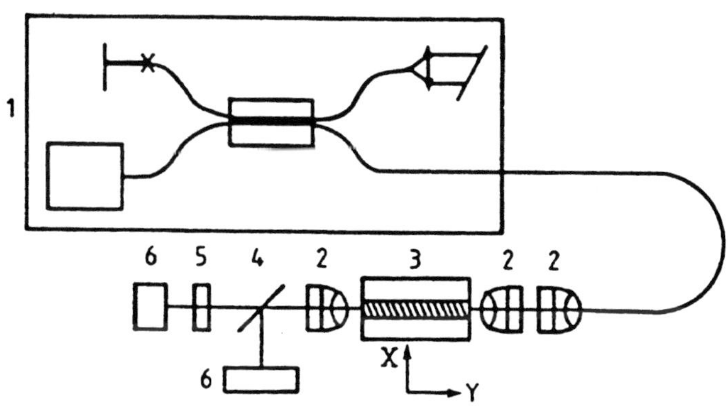

Fig. 3. Schematic diagram of the experimental setup: 1-fibre laser, 2-20$^{\times}$ microobjectives, 3-channel waveguide, 4-beamsplitter, 5-IR-cut filter, 6-photodiodes.

5 Experiment

The fibre laser radiation was coupled to the channel waveguide through the waveguide polished endface by a $20^{\times}$ microobjective with a 30% efficiency. At the output endface the SH radiation was detected after collimation (Figure 3).

Type II phase-matching condition has been achieved for pumping wavelength $\lambda_\omega = 1.112\ \mu m$ at room temperature (T=22^{0}c). At this wavelength, an output power of 7 mW of the fibre laser radiation resulted in a 1.5 mW power at the waveguide output total in TE- and TM-mode.

The maximum SHG efficiency obtained was 0.023%. Consequently, the corresponding normalized value was K=15% $W^{-1}cm^{-2}$. The comparison with the theoretical values emphasizes the importance of the precise measurement of the nonlinear coefficients of the KTP.

6 Summary

We have been the first to demonstrate the SHG of the Nd^{3+}-doped fibre laser radiation in the channel Rb:Ba:KTP waveguide. The normalized SHG efficiency obtained was K=15% $W^{-1}cm^{-2}$.

Acknowledgements

This work was supported in part by AT&T.

References

[1] V.P. Gapontsev, I.E. Samartsev, A.I. Zayatz, R.R. Loryan, "A high power, CW-operated fiber lasers", *Techn. Dig. of the Advanced Solid State Lasers Top. Meet.*, Feb. 17-19, 1992, Santa Fe, NM, p. 176.

[2] J.D. Bierlein, H. Vanherzeele, "Potassium titanyl phosphate: properties and new applications", *JOSA B*, Vol. 6, No. 4, 1989, p. 1216.

[3] C.J. Van der Poel, J.D. Bierlein, J.B. Brown, "Efficient type I blue second-harmonic generation in periodically segmented $KTiOPO_4$ waveguides", *Appl.Phys.Lett.*, Vol. 57, No. 20, 1990, p. 2074.

[4] K.S. Buritskii, E.M. Dianov, Yu.M. Grjaznov, N.G. Dobjakova, V.A. Maslov, V.A. Chernykh, E.A. Shcherbakov, "Characterization of Rb:KTP channel waveguides", *Sov. Lightwave Commun.*, Vol. 1. No. 2, 1991, p. 107.

[5] R. Regener, W. Sohler, "Efficient second-harmonic generation in Ti:$LiNbO_3$ channel waveguide resonators", *JOSA B*, Vol. 5, No. 2, 1988, p. 267.

[6] H. Vanherzeele, J.D. Bierlein, "Magnitude of the nonlinear-optical coefficients of $KTiOPO_4$" *Opt.Lett.*, Vol. 17, No. 14, 1992, p. 982.

Studies in Classical
and Quantum Nonlinear Optics

ISBN 1-56072-168-5

QUANTUM INTERFERENCES OF EXCITONS IN CdSe.

J. Erland, K.-H. Pantke, B.S. Razbirin[1],
V.G. Lyssenko[2] and J.M. Hvam
Fysisk Institut, Odense Universitet,
Campusvej 55, DK-5230 Odense M, Denmark

The technique of nonlinear quantum beat spectroscopy is applied to the semiconductor CdSe. From the observation of quantum beats in the time integrated and spectrally resolved self-diffracted signal in a nonlinear four-wave mixing experiment, we extract information about energy splittings and dephasing times of bound excitons. The technique can be applied to any semiconductor system with a strong, resonantly enhanced, third order susceptibility.

1. INTRODUCTION

In atomic and molecular high resolution spectroscopy, linear quantum beat spectroscopy (QBS) has been developed since the early sixties[1]. Quantum beats are observed as a modulation of the resonance fluorescence coming from an atom or a molecule, when two or more levels of the same parity are coherently excited by a short light pulse. The coherent superposition of states has cross terms causing a modulation of the fluorescence rate with frequency $\nu = \Delta E/h$, where ΔE is the level splitting and h is Planck's constant. These modulations, or quantum beats, can be observed for as long as the excited states remain coherent, i.e. within the dephasing time T_2 of the transitions. The dephasing times for transitions in atoms or molecules are usually long, due to the weak interactions in the gas phase allowing for observations of quantum beats in the MHz regime. Because of the strong dephasing interactions in solids it has only recently been possible to extend QBS to the GHz regime[2,3,4].

In a nonlinear four-wave mixing experiment, quantum beats are observed as a modulation of the self-diffracted signal due to the interference of the different frequency components of the third order polarization. This technique has been applied to quantum beat studies of the excitonic polarization in GaAs/AlGaAs multiple quantum wells[5,6,7]. However, in most of these experiments no spectral analysis of the emitted signal was performed.

In this work the technique of nonlinear quantum beat spectroscopy (NQBS)[8,9] is applied to study impurity bound excitons in bulk CdSe crystals, in the THz regime, because of the strong (bound) excitonic enhancement of the nonlinear polarization. In NQBS the self-

---

[1] Permanent address: A.F. Ioffe Physical-Technical Institute, 194021 St. Petersburg, Politekhnicheskaya 26, Russia.

[2] Permanent address: Institute of Microelectronics Technology and Superpure Materials, Chernogolovka, Moscow District 142432, Russia.

diffracted signal recorded in a time-domain four-wave mixing experiment[10] is spectrally resolved and by a Fourier transformation[11] the (quantum) beat spectrum for each frequency component in the emitted signal is obtained. This beat spectrum reveals information about the energy splitting of the different beating levels from which a consistent level diagram is established. A line width analysis reveals the corresponding dephasing times.

In this article the technique of NQBS is addressed in more detail. The experimental equipment and setup as well as the CdSe sample is described. Results from NQBS is presented and discussed. Finally the relation between the time-domain and the frequency-domain in the experiments is shown to be consistent.

2. THE TECHNIQUE OF NQBS

The time-domain four-wave mixing in NQBS is performed in a two beam configuration with incoming pulses with wavevectors $\mathbf{k}_1$ and $\mathbf{k}_2$, as illustrated in Fig. 1. The nonlinear signal, which is a photon echo in the case of inhomogeneously broadened transitions[12], is detected time integrated in the direction $2\mathbf{k}_2$-$\mathbf{k}_1$ as function of the time delay τ between the two pulses, predominantly for $\tau>0$, i.e. when pulse #1 arrives before pulse #2. By spectrally resolving the nonlinear signal the strong resonance enhancement through the third order susceptibility $X^{(3)}$ becomes evident for the excited levels giving the possibility to identify the close-lying states involved in the nonlinear mixing. As each spectral component of the signal is detected in the time-domain, a Fourier transformation reveals a beat spectrum for each spectral component, where only beats with frequency $1/T_2 < \nu < 1/\tau_L$ can be observed i.e. limited by the fastest dephasing time T_2 of the beating levels and the coherence time τ_L of the laser. A line width analysis of lines in the beat spectrum gives the relation between the full width at half maximum ν_{FWHM} and the dephasing time T_2. The beat signal decays as $v(\tau) = e^{-\tau/T_2}$, $\tau>0$, and in the experiment $v^2(\tau)$ is detected. Fourier transforming $v^2(\tau)$, the spectrum $V(\nu)$ is obtained:

$$V(\nu) = \int_0^\infty v^2(\tau) e^{-i2\pi\nu\tau} d\tau = \frac{1}{\frac{2}{T_2} + i2\pi\nu} \tag{1}$$

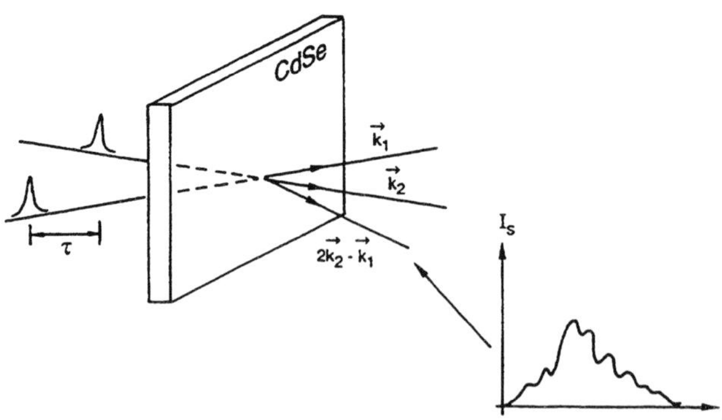

Fig. 1. Sketch of experimental configuration.

The relevant relation is obtained from:

$$\frac{|V(\nu_{HWHM})|}{|V(0)|} = \frac{1}{2} \tag{2}$$

giving

$$\nu_{FWHM} = 2\nu_{HWHM} = \frac{1}{T_2}\cdot\frac{\sqrt{12}}{\pi} \approx \frac{1}{T_2} \tag{3}$$

i.e. the fastest dephasing time T_2 of the beating levels is obtained as the inverse of the full width at half maximum ν_{FWHM}.

3. EXPERIMENTAL SETUP AND THE CdSe SAMPLE

The NQBS experiments are realized with pulses generated by a mode-locked Argon-ion laser, which pumps synchronously a tunable dye laser (DCM) with a repetition rate of 82 MHz. Although the pulse duration is about 7 ps, a higher time resolution is obtained due to a shorter coherence time of the pulses important for coherent spectroscopy[13]. Noticing that the incoherent part of the pulses does not contribute significantly to the spectral width of the pulses, a coherence time of less than 500 fs is found from a spectral width of about 4 meV assuming a Gaussian pulse-shape. This gives the opportunity to extend NQBS to the THz regime. The diffracted signal is dispersed in a spectrometer and finally detected by an OMA-system. The thin (≈ 10 μm) CdSe sample of good optical quality is immersed in a liquid Helium cryostat and all experiments are performed at 2K. In Fig. 2 is shown the luminescence spectrum in the region of the bound excitons after band-to-band excitation. Two pronounced resonances I_1 and I_2 are seen, due to excitons bound to neutral acceptors (A°,X) and neutral

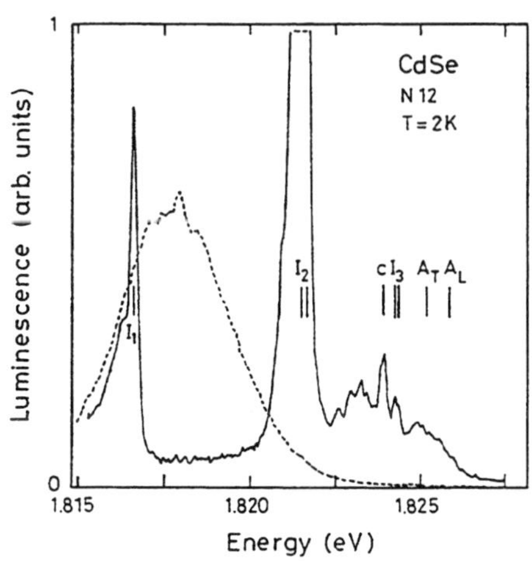

Fig. 2. Luminescence spectrum of CdSe sample (solid curve) and spectral profile (dashed) of the laser pulse in the NQBS.

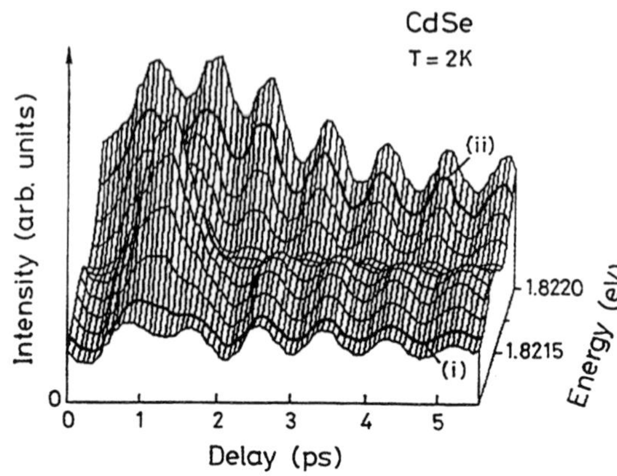

Fig. 3. Correlation traces with quantum beats for different detection energies.

donors (D^o,X), respectively. The I_2-line has a doublet structure I_2' and I_2'', which could hardly be resolved in the luminescence spectrum[14]. More shallow resonances are also present as indicated by the small luminescence band between the I_2-line and the free exciton A_T. In Fig. 2 is also shown the broad laser line covering the resonances to be investigated.

4. EXPERIMENTAL RESULTS AND DISCUSSION

Exciting near the I_1-line (see Fig. 2) and detecting the diffracted signal around the I_2-resonance, the oscillations in Fig. 3 on a picosecond timescale are observed. It is clear that going through the I_2 resonance, the oscillations change character. There is an approximate phase shift of π, which is to be expected for the polarization when going through a strong resonance, but there is also a more fundamental change in the irregular, but pronounced, beat signal. In order to get information about the involved frequencies these data have been Fourier analyzed.

The resulting frequency spectra of two selected traces (i) and (ii) are shown in Fig. 4. Detecting just below the I_2 line (trace (i)), the beat spectrum is as shown in the left part of Fig. 4. The dominating structure in the spectrum is the I_2-I_1 peak showing a fine structure due to the splitting of the I_2 line. The I_2'-I_2'' beat frequency can further be observed as a small shoulder near the zero frequency peak. Additionally, three pronounced beats I_2-a, I_2-b and I_2-c can be observed, where again the I_2-a beat clearly shows a doublet structure due to the splitting of the I_2 line. The levels denoted a, b and c are not clearly identified in luminescence and absorption spectra (Fig. 2). Comparing the intensities of the doublet structure in the I_2-I_1 peak and the I_2-a peak, showing a mirror image, we can conclude that the energy of the a-level is higher than the energy of the I_2- level, and we assume the same for the b and c levels. By detecting the nonlinear beat signal in this spectral range (trace (ii)), indeed we observe a pronounced resonance enhancement of the nonlinear signal, and the corresponding beat spectrum in the right part of Fig. 4 shows a large peak at the a-I_1 beat frequency. Making use of the I_2-b and I_2-c frequencies in the left part, we may identify two shoulders at lower fre-

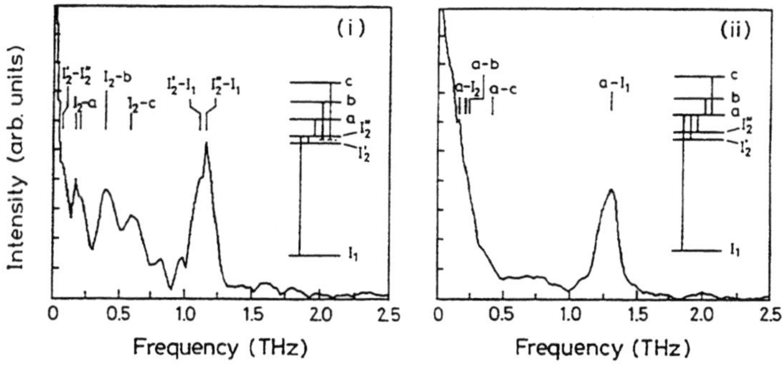

Fig. **4**. Fourier transforms of the quantum beat signals in curve (i) and (ii) of Fig. 3. The insets show the inferred levels.

quencies as a-b and a-c beats and thereby establish a complete (tentative) level diagram as shown in the insets of Fig. 4. The energy positions of the lines are calculated using the I_1-line as a reference and compared with literature values[15] in Table 1. It should be mentioned that the origins of the lines a, b and c are not known.

A line width analysis of the presented Fourier spectra, that contains information about dephasing times of the involved resonances, has been performed using Eq. (3). Our group has previously determined the dephasing time of the I_1-bound exciton to be extremely long, i.e. 1 ns for low intensity and temperature.[16] From the present analysis, we infer a dephasing time of 8 ps for the I_2-bound exciton and somewhat lower values for the more shallow lines, as shown also in Table 1. All these dephasing times are shortened with increasing excitation density.

Table 1. Position of exciton lines and their dephasing times from the present work and from literature.

Line	Pos. (eV)	Pos.[15] (eV)	T_2 (ps)
I_1	1.8169	1.8169	1000[16]
I_2'	1.8217		
I_2''	1.8218	1.8218	8
a	1.8223	1.8224	5
b	1.8234	1.8236	5
c	1.8241	1.8247	4

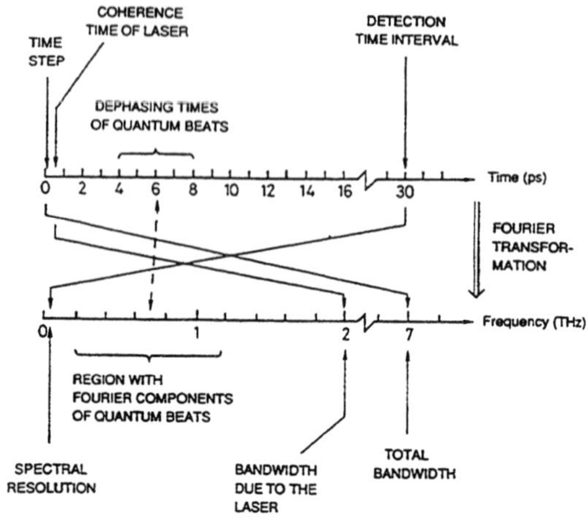

Fig. 5. Characteristic times and frequencies in the presented NQBS experiments.

To discuss the consistency between the time-domain and the frequency-domain the characteristic times and frequencies of the experiments are shown in Fig. 5. From a detection time interval of about 30 ps the spectral resolution in the Fourier spectra is found to be of about 50 GHz, well below the region with Fourier components of the quantum beats. The total bandwidth of 7 THz, as found from a time step of about 70 fs, is reduced to about 2 THz due to the coherence time of the laser. From the line width analysis of the quantum beat lines dephasing times well below 30 ps are inferred, and therefore clearly resolved. The analysis in Fig. 5. thus shows the consistency between the time-domain and the frequency-domain in the presented NQBS experiments.

5. CONCLUSIONS

In this work the technique of nonlinear quantum beat spectroscopy has been applied to a bulk CdSe crystal at low temperature. The quantum interferences of different bound exciton states below the free exciton result in a strong modulation of the four-wave mixing signal. The corresponding beat spectra reveal the energy splittings of the bound excitons, as observed in a luminescence experiment, and from these a level diagram was established. From a line width analysis the corresponding dephasing times was found to be between 4 ps and 8 ps and decreasing, when the bound exciton states become more shallow.

ACKNOWLEDGEMENT

We are grateful to Dr. D.K. Nelson for measuring the luminescence spectrum of sample N12. The work was supported by the Danish Natural Science Research Council.

REFERENCES

1. S. Haroche, "Quantum Beats and Time-Resolved Fluorescence Spectroscopy", In: *High Resolution Spectroscopy* (K. Shimoda Ed.), Heidelberg: Springer Verlag, 1976, pp. 253-313.
2. V. Langer, H. Stolz, and W. von der Osten, "Observation of Quantum Beats in the Resonance Fluorescence of Free Excitons", *Phys. Rev. Lett.* Vol. 64, No. 8, 1990, pp. 854-857.
3. H. Stolz, V. Langer, E. Schreiber, S. Permogorov, and W. von der Osten, "Picosecond Quantum-Beat Spectroscopy of Bound Excitons in CdS", *Phys.Rev. Lett.* Vol. 67, No. 6, 1991, pp. 679-682.
4. W.A.J.A. van der Poel, A.L.G.J. Severens, and C.T. Foxon, "Quantum Beats in the Exciton Emission of Type II GaAs/AlAs Quantum Wells", *Opt. Commum.* Vol. 76, No. 2, 1990, pp. 116-120.
5. K. Leo, T.C. Damen, J. Shah, and K. Köhler, "Quantum Beats of Free and Bound Excitons in GaAs/Al_xGa_{1-x}As Quantum Wells", *Phys. Rev. B* Vol. 42, No. 17, 1990, pp. 11359-11361.
6. B.F. Feuerbacher, J. Kuhl, R. Eccleston, and K. Ploog, "Quantum Beats of Light and Heavy Hole Excitons in a Quantum Well", *Solid State Commun.* Vol. 74, No. 12, 1990, pp. 1279-1283.
7. D. Oberhauser, K.-H. Pantke, J.M. Hvam, K.-H. Schlaad, G. Weimann, and C. Klingshirn, "Laser Induced Gratings in GaAs/AlGaAs Multiple Quantum Wells", *Proc. Int. Conf. Laser '91*, San Diego, 1991 (in press).
8. K.-H. Pantke, B.S. Razbirin, V.G. Lyssenko, J. Erland, and J.M. Hvam, "Nonlinear Quantum Beat Spectroscopy in CdSe", *Proc. 21st. Int. Conf. Phys. Semicond.*, Beijing, China, 1992, Singapore: World Scientific Publishing Co. (in press).
9. K.-H. Pantke, P. Schillak, J. Erland, V.G. Lyssenko, B.S. Razbirin, and J.M. Hvam, "Nonlinear Quantum Beats of Excitons in CdSe", *phys. stat. sol. (b)* Vol. 173, No. 1, 1992, pp. 91-98.
10. C. Dörnfeld and J.M. Hvam, "Optical Nonlinearities and Phase Coherence in CdSe Studied by Transient Four-Wave Mixing", *IEEE J. Quantum Electron.* Vol. QE-25, No. 5, 1989, pp. 904-912.
11. Y.C. Chen, K. Chiang, and S.R. Hartmann, "Spectroscopic and Relaxation Character of the $^3P_0$-$^3H_4$ Transition in LaF_3Pr^{3+}", *Phys. Rev. B* Vol. 21, No. 1, 1980, pp. 40-47.
12. J.M. Hvam, "Transient Nonlinear Optics in Semiconductors", In: *Proc. Int. Summer School Nonlinear Optics in Solids*, Aalborg, Denmark, 1989, (O. Keller Ed.), Heidelberg: Springer Verlag, 1990, pp. 213-234.
13. N. Morita and T. Yajima, "Ultrahigh - Time Resolution Coherent Transient Spectroscopy with Incoherent Light", *Phys. Rev. A* Vol. 30, No. 5, 1984, pp. 2525-2536.
14. Private communication with Dr. D.K. Nelson, A.F. Ioffe Physical-Technical Institute, St. Petersburg.
15. E. F. Gross, B. S. Razbirin, V. P. Fedorov, and Yu. P. Naumov, "Bound Exciton Complexes in CdSe Crystals", *phys. stat. sol.* Vol. 30, No. 2, 1968, pp. 485-494.
16. H. Schwab, V.G. Lyssenko, and J.M. Hvam, "Spontaneous Photon Echo from Bound Excitons in CdSe", *Phys. Rev. B* Vol. 44, No. 8,1991, pp. 3999-4001.

Studies in Classical
and Quantum Nonlinear Optics

ISBN 1-56072-168-5

HIGH POWER Nd GLASS LASER WITH INTRACAVITY NONLINEAR PLASMA ELEMENTS

A.I. Ershov
General Physics Institute, Russian Academy of Sciences
Vavilov str. 38, GSP-1, 117942, Moscow, Russia

It is known that nonlinear optical properties of intracavity plasma in laser beam can be used to control laser operation itself [1]. In the past, the operation of laser with intracavity reflecting plasma, so-called laser with plasma mirror, was obtained for gas lasers - CO_2 [2], and iodine photodissociation [3] lasers, and for solid state Nd glass laser [4]. Fig. 1 shows two possible schemes of Nd glass laser with intracavity plasma. In both these configurations carbon target is used. In the simplest scheme, laser consists of several active Nd glass rods placed between a normal dielectric mirror and carbon target in the focus of a lens. The generation arises due to the reflection of laser radiation from plasma and target surface. Thus, on the one side, plasma is created by laser radiation, and on the other side, plasma itself influences the lasing. In another scheme, one of the active rods is placed inside an unstable resonator, created by dielectric mirrors 1 and 3, so there is a three-mirror laser configuration which consists of two dielectric mirrors and one plasma mirror.

In such a system, there are several possible modes of operation [5,6]: (i) quasi-continuous operation in time scale compared to millisecond pumping pulse of flash-lamps; (ii) generation of high-power nanosecond pulse consisted of several dozens of picosecond pulses with energy per each picosecond pulse about 1 J, (can be treated as Q-switch and mode locking operation), and (iii) generation of several thousand subnano-

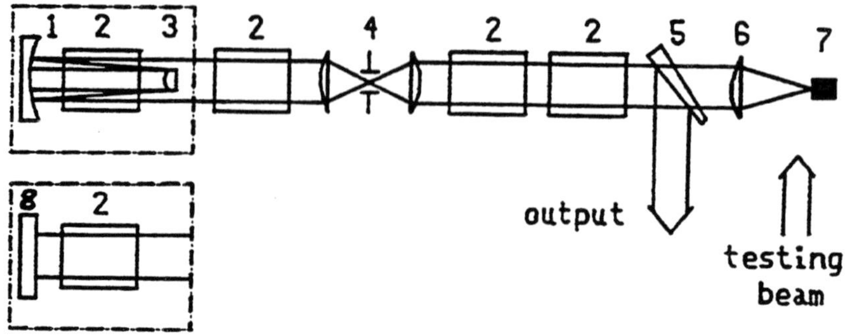

Figure 1. Experimental setup of the laser with intracavity plasma. 1,3,8 - dielectric mirrors, 2 - active elements, 4 - spatial filter, 5 - wedge, 6 - focusing lens, 7 - carbon target.

second pulses during the pumping pulse (quasi-cw mode locking without Q-switch). In previous experiments, mode (ii) was observed in the three-mirror configuration, while mode (iii) - in the two-mirror configuration. Quasi-continuous lasing to some extend could be observed in both configurations with intensity of the order of (10 - 100) MW/cm^2 on the target surface.

When Nd glass laser radiation with intensity in this range incidents on the carbon target, its reflection from the target becomes weaker as the intensity increases. In this range plasma of carbon target is quite rare and there is no reflection from the plasma but only the scattering from the target surface. Decreasing of reflection is connected with the increasing of absorption of laser radiation in plasma. At higher intensities plasma becomes quite dense, and we can expect not only reflection from the target surface, but also reflection from the plasma itself which can increase when increasing the intensity. When the radiation, reflected from plasma, comes back to laser, a positive feedback arises which leads to very sharp increasing of generated laser power. On the other side, if intensity on the target does not exceed a critical value of ~100 MW/cm^2, reflection from the target could be considered for negative feed-back in the laser which tends to smooth out its free-running relaxational oscillations.

In the present paper, the diagnostics of intracavity plasma in quasi-continuous mode has been made, and the results obtained have been used in the study of the lasing itself.

The experiments have been carried with both two- and three-mirror configurations. Four neodymium glass rods with diameter 45 mm and length 680 mm have been used as amplifying stages. The value of total small-signal gain of the active medium in amplifying stages exceeded 10^5 and thus ensured small reflection from plasma ($r \sim 10^{-4}$) to be essential for the lasing. Special measures were made to suppress stray lasing without plasma mirror: active rods were cut out at the angle of 85° to the optical axis; all surfaces of optic elements were tilted relative to the axis, and anti-reflection coating of the optical elements was used.

PLASMA DIAGNOSTICS

Plasma has been tested by employing the interferometric technique. Two separate methods were used: double exposure holographic interferometry and nonlinear dispersion interferometry.

Double exposure holographic interferometry was carried out at the second harmonic of an auxiliary Q-switch YAG: Nd^{3+} laser with pulse duration of 10 ns. Main laser radiation was focused by a f=70 cm lens into 0.6 mm diameter spot. Moment of Q-switch laser generation was chosen at the appropriate part of the first half of the main laser millisecond pulse when burning crater at the target was not very large. Fig.2 shows holographic interferograms at different times from the beginning of the millisecond pulse, corresponded to different intensities on the target. One can see the development of plasma flume with a space scale of the order of several centimeters. But electron density in plasma becomes lower very quickly at large distances from the target surface, so in these regions plasma is very rare and does not interact with laser beam. On the other side, more dense plasma near the target has a very short relaxation time (as compared to the millisecond pulse), so in this region a quasi-stationary plasma flow exists with parameters determined by the laser intensity only. In quasi-cw mode, when radiation intensity on the target lies within the range 10 - 100 MW/cm^2 near-target carbon plasma could be weakly ionized (degree of first ionization is less than 1). Refractive index of the plasma then will be determined by the following equation:

$$n - 1 = -C_e N_e + C_a N_a + C_i N_i, \tag{1}$$

where the three terms in the right part are connected with electrons (e), neutral atoms (a), and ions (i), N and C mean

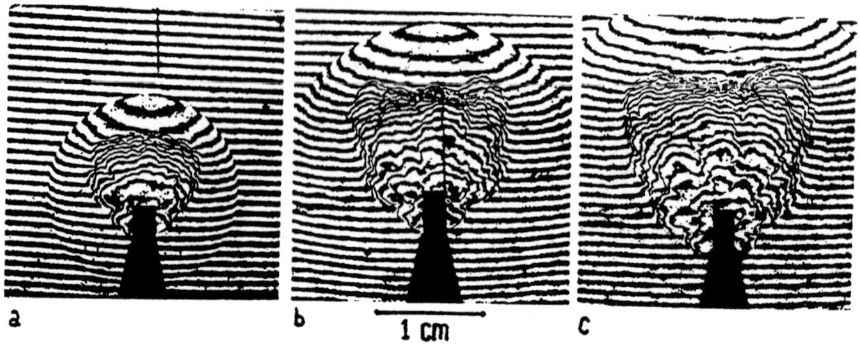

Figure 2. Holographic interferograms of carbon plasma at the moments 20(a), 30(b), and 40(c) μs from the beginning of the millisecond pulse.

concentration of appropriate particles and refraction per one particle. When $N_e = N_i << N_a$, as in our case, the last term can be neglected. Eq.(1) shows that knowing only the distribution of refractive index n is not sufficient to define the concentration of particles in the plasma.

In the present work, the distribution of electron density near the target has been defined by means of nonlinear dispersion interferometry sensitive only to wavelength-dependent part in the refractive index. The term in Eq.(1), corresponding to electrons, is the only such a part, and is determined by:

$$C_e = 2e^2/mw^2, \tag{2}$$

whereas the refraction of neutral atoms and ions far from resonances can be considered approximately as independent of wavelength. In Eq.(2), m is the electron mass, w denotes the light frequency. Fig.3 shows the scheme of the interferometer. Intracavity plasma of carbon target was probed by the first and second harmonics of YAG: Nd^{3+} laser passing through the plasma along the same path. After passing plasma, the first harmonic wave was doubled in a nonlinear crystal CDA, so that at the output of the interferometer we had two second harmonic waves. As a result of their interference, an interferometric pattern was created, with fringe shifts determined by the difference of refractive index at the first and second harmonics of probe radiation.

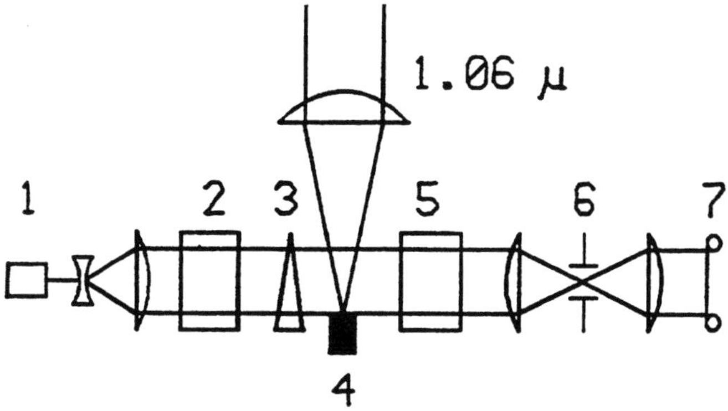

Figure 3. Nonlinear dispersion interferometer. 1 - YAG:Nd^{3+} laser, 2 - *KDP* doubler, 3 - wedge, 4 - carbon target, 5 - *CDA* doubler, 6 - spatial filter, 7 - photo film.

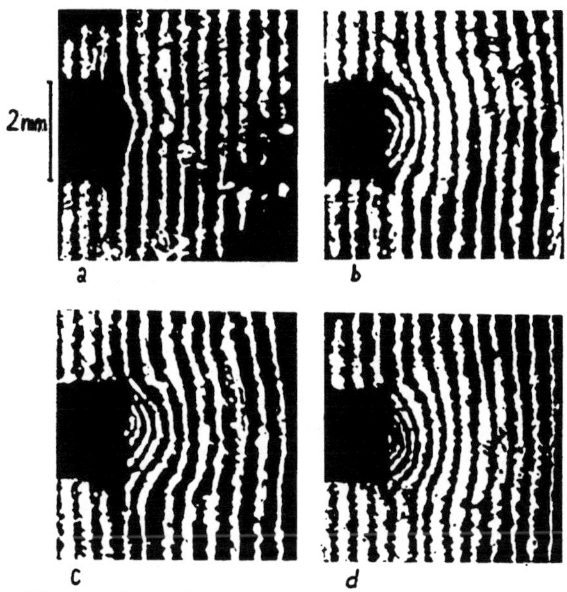

Figure 4. Dispersion interferograms for *I*: 30(a), 60(b), 87(c), and 100(d) MW/cm^2.

Fig.4 shows the dispersive interferograms of plasma near the target for different intensities in the range 30 - 100 MW/cm^2. The highest value of intensity is close to the threshold of radiation reflection from plasma itself and thus

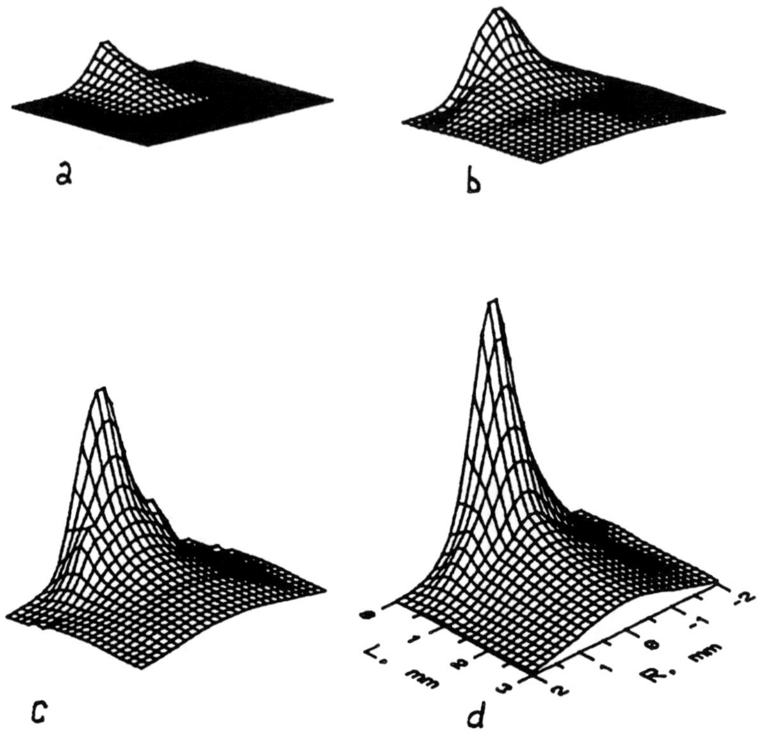

Figure 5. Electron density near the target for I: 30(a), 60(b), 87(c), and 100(d) MW/cm^2.

to the boundary of quasi-continuous lasing mode. Distributions of electron density obtained from dispersive interferometry are presented in Fig.5. One can see proportional increasing of the size of a dense part of plasma and value of its electron density near the target with the increasing of intensity.

Dispersive interferograms make it possible to define electron density in plasma for different intensities. Fig.6 shows the dependencies of electron density near the target on intensity within the range of quasi-cw lasing. The electron density, which gives start to the appearance of sharp reflection from plasma is $5 \cdot 10^{18}$ cm^{-3} that is only 1/200 part of critical electron density at 1.06 μ. But it is important to point out that the reflection itself appears in much more

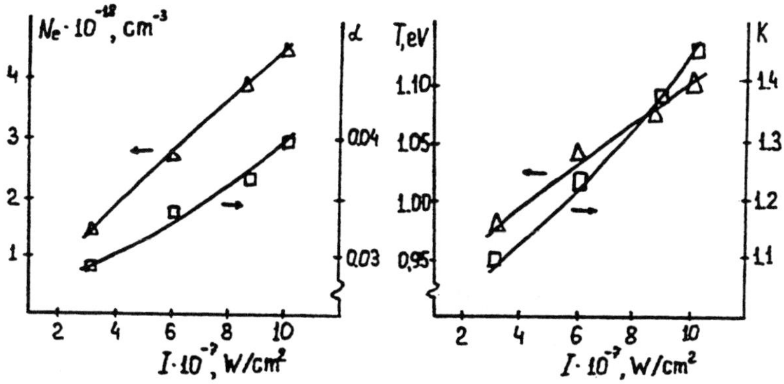

Figure 6. Electron density N, degree of ionization α, and temperature T near the target laser as well as total absorption of laser light by plasma K.

dense plasma and these values of laser intensity and electron density are the highest stable values, above which the sharp increasing of electron density and generated laser intensity occurs.

Using the results of holographic interferometry now makes it possible to define other interesting plasma parameters. Fig.6 shows curves for electron density, degree of ionization, and temperature as well as total absorption of laser light by plasma. For absorption calculations, the following formula has been used [7].

$$\mu_w = 2.12\, N_e^2 / (T^2 w^2) \qquad (3)$$

where μ_w is the absorption coefficient (in cm^{-1}) for light frequency w.

QUASI-CW LASING

The results, obtained by using interferometric diagnostics have be used in investigating the operation of the Nd glass laser with intracavity plasma. Fig.7 shows typical oscillograms of smooth quasi-continuous operation obtained. The maximum energy was up to 500J in 1ms pulse. Main part of the energy could be removed from the resonator with an additional mirror placed before the focusing lens, and only a small part was spent for creating plasma, cf. Fig.1. Better

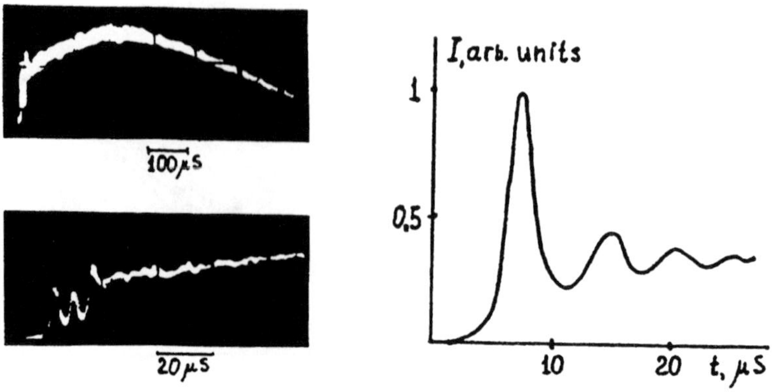

Figure 7. Oscillograms or the output pulses (energy W = 200 J) (a,b) and calculated dependence mirror $I(t)$ (c).

quality of the lasing was obtained in the three-mirror configuration, where first amplifying stage placed inside the additional unstable resonator was pumped below its own generation threshold. One can see that free-running relaxational oscillations are smoothed out during first several pikes. It was obtained that this smoothing out is due to the nonlinear absorption in plasma. Quasi-cw operation of the laser can be described by the following equations:

$$\begin{cases} \frac{1}{c}\frac{\partial I_+}{\partial t} = -\frac{\partial I_+}{\partial z} + \sigma n I_+ + \frac{h\nu n}{\tau}\Omega \\ \frac{1}{c}\frac{\partial I_-}{\partial t} = -\frac{\partial I_-}{\partial z} + \sigma n I_- + \frac{h\nu n}{\tau}\Omega \\ \frac{dn}{dt} = W - \frac{\sigma}{h\nu} n (I_+ + I_-) - \frac{n}{\tau_o} \end{cases} \quad (4)$$

where $I_{+,-}(z,t)$ is the intensity of light waves propagating in two opposite directions in the laser, n represents the inversion of population, $h\nu$ = 1.16 eV is the photon quantum energy, $\sigma = 2.5\cdot10^{-20}$ cm^2 is the cross section of stimulated emission, $\tau = 2\cdot10^{-4}$ s is the spontaneous lifetime, $W = 1.6\cdot10^{21}$ cm^{-3} is the pumping power, and $\Omega = 10^{-6}$ is the spontaneous emission angle, determined by the configuration of the laser. These equations were solved with the boundary conditions $I_-(0,t) = I_+(0,t)$, $I_-(l,t) = r \cdot I_+(l,t)$, where $r = R \cdot K^{-1}(I_+)$ is the reflection from the target, calculated with account of the absorption K in plasma. This model better fits the two-mirror configuration because it does not consider the unstable resonator but can be used for approximate description of the three-mirror scheme as well. The calculated curve $I(t)$ presented in Fig. 7 shows good agreement with the experimental one and thus indicates that nonlinear absorption in plasma does smooth out pikes-like quasi-cw lasing.

REFERENCES

[1] A.N.Malkov, A.M.Prokhorov, V.B.Fedorov, and I.V.Fomenkov, 'Laser radiation - target interaction in laser-target system', **Proceedings of General Physics Institute**, Vol.10, 1988, pp.127-156 in Russian.

[2] M.Piche, and P.A.Belanger, 'Mode-locking by intracavity plasma. Pt 1: Linear model', **Canad. J. Phys.**, Vol.59, 1981, pp.130-137; 'Mode-locking by intracavity plasma. Pt 2: Experimental results', **Canad. J. Phys.**, Vol.60, 1981, pp.77-87.

[3] J.Schmiedberger, and M.Vrbova, 'Iodine photodissociation laser with plasma mirror', **Laser and Particle Beams**, Vol.4, parts 3&4, 1986, pp.427-433.

[4] V.A.Batanov, D.A.Dementjev, A.N.Malkov, et all, 'Neodymium laser with reflected plasmo-optic shutter', **Sov. Phys. JETP** Vol.77, 1979, pp.2186-2199, in Russian. Trans. in English.

[5] A.I.Ershov, V.B.Fedorov, and I.V.Fomenkov, 'Emission spectrum and limits of quasi-cw operation of a neodymium laser with an intracavity plasma formed from a carbon target', *Sov.J.Quantum Electron,* Vol.17, 1990, pp.988-993 in Russian, Vol.20(8), 1990, pp.905-909 in English.

[6] P.Grjibek, B.Kubichek, and M.Vrbova, 'Mode locking operation of neodymium laser with plasma mirror', *Sov.J.Quantum Electron,* Vol.10, 1983, pp.2391-2393, in Russian. Trans. in English.

[7] J.P.Raizer, *Lazernaja iskra i rasprostranenie razrjadov (Laser spark and physics of discharges)*, Moscow, Nauka, 1974, 308 pp. in Russian.

Studies in Classical
and Quantum Nonlinear Optics

ISBN 1-56072-168-5

OPTICAL DISPERSION RELATIONS AND COMPLEX REFRACTIVE INDEX FOR ATOMS IN SQUEEZED VACUA

S.S. Hassan, R.K. Bullough‡ and H.A. Batarfi‡*

* Ain Shams University, Faculty of Science,
Mathematics Department, Cairo, Egypt

‡ Department of Mathematics, UMIST, PO Box 88, Sackville Street,
Manchester, M60 1QD, U.K.

1. INTRODUCTION

Canonical quantisation of the electromagnetic (e.m.) field in the fashion of Dirac [1] leads to the conclusion that an electric field component $\mathcal{E}(\chi)$ with phase χ and its quadrature field component $\mathcal{E}(\chi + \frac{1}{2}\pi)$ have fluctuations $\Delta\mathcal{E}(\chi)$ and $\Delta\mathcal{E}(\chi + \frac{1}{2}\pi)$ which satisfy Heisenberg's uncertainty principle $\Delta\mathcal{E}(\chi)\ \Delta\mathcal{E}(\chi + \frac{1}{2}\pi) \geq \frac{1}{2}\hbar$. The field is then said to be *squeezed*, if, for some choice of χ, one of $\Delta\mathcal{E}(\chi)$ and $\Delta\mathcal{E}(\chi + \frac{1}{2}\pi)$ is less than $(\frac{1}{2}\hbar)^{\frac{1}{2}}$ [2-6]. The field is a squeezed *vacuum* field if the means $\overline{\mathcal{E}}(\chi)$ and $\overline{\mathcal{E}}(\chi + \frac{1}{2}\pi)$ both vanish. Squeezed vacuum fields are evidently excited states of the quantised e.m. field as compared with the true vacuum state of that field.

The possibility of generating squeezed vacuum fields in the laboratory [2,3,7-11] raises the question of how particular physical systems may respond to such fields in ways which are characteristic of that squeezing. Theories of resonance fluorescence from a single atom [12] and from several cooperating atoms [3,13,14] indicate that both of these are possible examples, as are the optical bistability of two-level atoms inside an optical cavity [15] or without cavity feedback [16].

In this short paper we are concerned to investigate the effect of a broad band, multimode, squeezed vacuum field on a linear dielectric medium. In particular we investigate effects of the squeezed vacuum field on the refractive index of that dielectric. We are concerned in general to compute dispersion relations for an extended slab-like dielectric made up of many atoms bathed in the squeezed vacuum and forming a Fabry-Perot cavity. The system is excited by a single mode harmonic electric field of arbitrary strength. In the absence of any squeezing of the vacuum field the system excited by a strong enough harmonic field is nonlinear and bistable (multistable [17,18]). But to describe that optical bistability it becomes necessary to resort [17,18] to extensive numerical computations. In this paper we leave such computations for the corresponding case of the squeezed vacuum to later work and focus attention on the linear regime achieved for weak enough coherent driving fields and which can be handled analytically for incident fields of small enough aperture. We are then concerned to compute the effect of the squeezed vacuum on the familiar, linear, complex refractive index. We call this quantity m in this paper.

2. BLOCH EQUATIONS OF MOTION

For an extended system of N_a 2-level atoms coupled to the quantised radiation field the Hamiltonian in dipole approximation can be taken in the form [19]

$$H = \sum_{k,\lambda} \hbar\omega_k\, a^{\dagger}_{k,\lambda}\, a_{k,\lambda} + \tfrac{1}{2}\hbar\omega_o \int_V \sigma_z(x,t)dx - \int_V p(x,t).e(x,t)dx \tag{1}$$

where the operator densites are

$$\sigma_z(\mathbf{x},t) = \sum_{i=1}^{N_a} \sigma_z^{\,i}(t)\delta(\mathbf{x}-\mathbf{x}_i) ,$$

$$p(\mathbf{x},t) \equiv p\sigma_x(\mathbf{x},t) = p\sum_{i=1}^{N_a} \sigma_x^{\,i}(t)\delta(\mathbf{x}-\mathbf{x}_i). \tag{2}$$

The $(\sigma_x^i, \sigma_y^i, \sigma_z^i) \equiv \vec{\sigma}^{\,i}$ are three spin-$\frac{1}{2}$ Pauli matrices for each atom labelled i and their su(2) Lie algebra means that the operator densities satisfy the commutation relations

$$[\sigma_x(\mathbf{x},t), \sigma_y(\mathbf{x}',t)] = 2i\sigma_z(\mathbf{x},t)\delta(\mathbf{x}-\mathbf{x}') \tag{3}$$

with cyclic permutations of x,y,z. The field operator $\mathbf{e}(\mathbf{x},t)$ is given by

$$\mathbf{e}(\mathbf{x},t) = i\sum_{\mathbf{k},\lambda} \mathbf{g}_{\mathbf{k},\lambda} (a_{\mathbf{k},\lambda}(t)\, e^{i\mathbf{k}.\mathbf{x}} - a^{\dagger}_{\mathbf{k},\lambda}(t)\, e^{-i\mathbf{k}.\mathbf{x}}) \tag{4}$$

where the vectors $\mathbf{g}_{\mathbf{k},\lambda} = \sqrt{\dfrac{2\pi\, \hbar\omega_k}{V_o}}\; \hat{\epsilon}_{\mathbf{k},\lambda}$. So $\hat{\epsilon}_{\mathbf{k},\lambda}$ is a unit polarisation vector for the mode $\mathbf{k}$ with polarisation index $\lambda = 1,2$ and $V_o \longrightarrow \infty$ is the quantisation volume : V_o contains V the macroscopic region occupied by the atoms here taken to be in a slab-like geometry and the ith atom lies at the site $\mathbf{x}_i \in V$. The $a_{\mathbf{k},\lambda}$, $a_{\mathbf{k},\lambda}^{\dagger}$ satisfy

$$[a_{\mathbf{k},\lambda}, a^{\dagger}_{\mathbf{k}',\lambda'}] = \delta_{\mathbf{k},\mathbf{k}'}\delta_{\lambda,\lambda'}. \tag{5}$$

The operator field $\mathbf{e}(\mathbf{x}_i,t)$ at $\mathbf{x}_i$ drives the atom sited at $\mathbf{x}_i$. Following Ref [19] this field has positive and negative frequency parts $\mathbf{e}^{\pm}(\mathbf{x}_i,t)$ and each of these is expressible as the sum of three parts

$$\mathbf{e}^{\pm}(\mathbf{x}_i,t) = \mathbf{e}_o^{\pm}(\mathbf{x}_i,t) + \mathbf{e}_{self}^{\pm}(\mathbf{x}_i,t) + \mathbf{e}_{int}^{\pm}(\mathbf{x}_i,t) \tag{6}$$

where

$$\mathbf{e}_o^{\pm}(\mathbf{x}_i,t) = \pm\, i \sum_{\mathbf{k},\lambda} \mathbf{g}_{\mathbf{k},\lambda} \begin{Bmatrix} a_{\mathbf{k},\lambda}(0) \\ a^{\dagger}_{\mathbf{k},\lambda}(0) \end{Bmatrix} e^{\mp i(\omega_k t - \mathbf{k}.\mathbf{x}_i)} \tag{7}$$

are the *free* field operators, and the $\mathbf{e}^{\pm}_{self}$ are reaction field operators describing the back reaction of the atom i on itself. These reaction fields are given by [19-21]

$$\mathbf{e}_{self}^{\pm}(\mathbf{x}_i,t) = \pm\, i\, \frac{2}{3}\left(\frac{\omega_o}{c}\right)^3 p\, \sigma_{\pm}^{\,i}(t) \pm \frac{2}{3\pi}\left(\frac{\omega_o}{c}\right)^3 p\, \ln\left(\frac{\omega_c}{\omega_o}\right)\left(\sigma_+^{\,i}(t) - \sigma_-^{\,i}(t)\right) \tag{8}$$

with $\sigma_{\pm} = \frac{1}{2}(\sigma_x \pm i\sigma_y)$ while ω_c is a cut-off frequency $\approx$ the Compton frequency. The remaining operator terms $\mathbf{e}^{\pm}_{int}(\mathbf{x}_i,t)$ in (6) are the interatomic field operators which act on atom i from all of the other atoms j $\neq$i. Upto neglect of certain 'slow' retardation effects Ref. [19] puts

$$e^{\pm}_{Int}(x_l,t) = np \int_{V-v} dx' \; \bar{\sigma}_{\pm}(x',t) \; \hat{u}\hat{u} : \begin{cases} F(x,x';\omega) \\ F^*(x,x';\omega) \end{cases} \tag{9}$$

where $n = \langle \sum_{l=1}^{N_a} \delta(x-x_l) \rangle_{Avr}$ is a mean number density of atoms obtained after an ensemble average over the sites x_l in V, the $\bar{\sigma}_{\pm}$ are ensemble averaged *operators* scaled [19] to a single atom, and $F(x,x';\omega) \equiv (\nabla\nabla + \omega^2c^{-2}U)\{\exp(i\omega c^{-1}|x-x'|)\}|x-x'|^{-1}$ is a c-number photon propagator (in which U is the unit tensor). The frequency ω is the frequency of a harmonic driving field which is still to be extracted from the expectation value of the free field operators, equations (7) : $\omega \simeq \omega_0$ the atomic frequency which appears in the Hamiltonian, equation (1). As explained in Ref. [19], $p = p\hat{u}$, taken in a fixed direction $\hat{u}$ for simplicity. More generally $\hat{u}$ should be properly averaged over all of its directions. This average is a part of the more general ensemble average.

The field operator (9) does not depend on the atomic sites x_l and these sites are eliminated by the ensemble average. This ensemble average over the x_l and more generally over all orientations of atom i at site x_l is a complicated process involving all possible inter-atomic scattering processes even in linear dielectric theory in the case when the vacuum field is the ordinary vacuum field [22-24], and (9) is only a leading order contribution corresponding to a smooth distribution of each of the x_l inside V without any correlations between different x_l [22-24]. At this level of approximation, and as explained in Refs. [22-24], the behaviour of $F(x,x';\omega)$ as $x' \longrightarrow x$ inside V means that a small sphere v about x must be removed from the region of integration and the integrals then become conditionally convergent : v has the vanishingly small radius a which $\longrightarrow 0$ after the integration, and experts will recognize (and cf.[22-24]) that this is the origin of the Lorentz field of dielectric theory investigated also by C.M. Bowden [25] at this meeting. Interatomic correlation then corrects this Lorentz field appropriately, and the choice of a *sphere* for v assumes that this gives the best approximation at this (zeroth) order. Note that F is now embodying Coulomb dipole-dipole interaction terms which were omitted from the original Hamiltonian (1).

Exactly as in Ref. [19] the Hamiltonian, equation (1), means that Heisenberg's operator equations of motion for the atoms are the operator Bloch equations

$$\dot{\sigma}_+(x,t) - i\omega_o\sigma_+(x,t) = ip\hbar^{-1}e^-(x,t)\,\sigma_z(x,t)$$

$$\dot{\sigma}_z(x,t) = -2\,p\hbar^{-1}.(e^-(x,t)\,\sigma_y(x,t) + h.c)$$

$$\dot{\sigma}_-(x,t) = (\dot{\sigma}_+(x,t))^{\dagger}. \tag{10}$$

As in Ref. [19] these operator equations have still to be ensemble averaged over the x_l and be transformed to c-number equations by taking expectation values $r_{\pm,3}(x,t) \equiv n^{-1}\langle \sigma_{\pm,z}(x,t)\rangle$. Unlike in the case of the ordinary vacuum treated in Ref. [19], these two steps must now be taken in the proper order viz. expectation values at fixed atomic site configurations $\{x_l\}$, then ensemble average over the x_l, essentially a Born-Oppenheimer procedure. If we use equations (6) - (9), taking into account that (9) as written contains a premature ensemble average which introduces the atomic number density n and the operators $\bar{\sigma}_{\pm}$, and if we assume that the actual initial state of the field is a coherent state field of frequency ω plus the broad band squeezed vacuum field, then, on tracing over this free field state as well as over the initial states of the atoms, we arrive at the c-number Bloch equations in a frame rotating at the frequency ω which are

$$\dot{r}_+(\mathbf{x},t) + (\gamma(1+2N) + i\delta)\, r_+(\mathbf{x},t) + \gamma M(\mathbf{x}) r_-(\mathbf{x},t)$$

$$= 2i(E^+(\mathbf{x},t) - \Omega e^{-ik_oz})\, r_3(\mathbf{x},t)$$

$$\dot{r}_3(\mathbf{x},t) + 2\gamma(1+2N) r_3(\mathbf{x},t) + \gamma = -i(E^+(\mathbf{x},t) - \Omega e^{-ik_oz}) r_-(\mathbf{x},t) + \text{c.c.},$$

$$\dot{r}_-(\mathbf{x},t) = (\dot{r}_+(\mathbf{x},t))^*, \tag{11}$$

at each point $\mathbf{x}$ inside V. The $r_{\pm}(\mathbf{x},t)$, $r_3(\mathbf{x},t)$ are now in the rotating frame. The analogous calculation is carried out in Ref.[19] and the symbols are, with n the ensemble averaged number density defined below (9),

$$E^+(\mathbf{x},t) = np^2\hbar^{-1} \int_{V-v} d\mathbf{x}'\; r_+(\mathbf{x}',t)\; \hat{u}\hat{u} : \mathbf{F}\,(\mathbf{x},\mathbf{x}';\omega), \tag{12}$$

while Ω is the Rabi frequency associated with the coherent driving field; $\delta \equiv \omega - \omega_0$ is the detuning of the frequency ω of the single mode coherent field from the atomic frequency ω_0, 2γ is the A-coefficient, and N is the mean number of photons in the modes of the broad band squeezed vacuum field. Evidently (12), which is an expectation value, appeals to the fact that, evaluated at site $\mathbf{x}_i$ of atom i, the field operator $e^{\pm}_{int}(\mathbf{x}_i,t)$ which becomes (9) is driven by all of the other atoms except i, and more detail is in [19].

3. THE SQUEEZED VACUUM

In equations (11) the essentially new feature compared with Ref.[19] is the complex parameter $M(\mathbf{x})$ arising from terms $M_{\mathbf{k},\lambda} = \langle a_{\mathbf{k},\lambda} a_{\mathbf{k},\lambda}\rangle$ evaluated for the squeezed vacuum field at the frequency ω : these terms are non-vanishing in the squeezed vacuum and satisfy for each $(\mathbf{k},\lambda)$(compare Refs.[26-28]) $|M_{\mathbf{k},\lambda}|^2 \leq N_{\mathbf{k},\lambda}(N_{\mathbf{k},\lambda}+1)$. For minimum uncertainty $|M_{\mathbf{k},\lambda}|^2 = N_{\mathbf{k},\lambda}(N_{\mathbf{k},\lambda}+1)$. In the experiments of Refs. [7-9] the squeezed vacuum is generated by an optical parametric oscillator and there is significant pairwise correlation between modes of the squeezed vacuum. Relevant analyses in these terms are in e.g. Ref.[28] for $N_a = 1$ atoms and in Ref. [29] for $N_a = 2$ atoms. If one takes the case of N_a *Rydberg* atoms [13,14,26,27] interacting at microwave frequencies one can assume the correlation between modes or one might assume the modes of the squeezed vacuum field are squeezed independently. In this latter case non-stationary field properties arise resulting in time dependent and squeezing dependent renormalisations of the parameters M [13,14,26,27]. A simpler view is in Ref. [15]. Rather than assuming as in Refs. [13,14,26,27] that the atoms are bathed in a squeezed vacuum with pairwise mode correlation and time independent phase shifts, one simply assumes [15] (and cf. Ref.[28]) that a light beam in a squeezed vacuum state is *injected* into the dielectric medium forming the cavity at a slight tilt to the cavity axis. But in this case the cavity should be a single-mode cavity such as a ring cavity, and in order to eliminate a phase mismatch, the atomic system should be an atomic beam of macroscopic length and small radius as suggested in Refs. 6,8 of Ref. [15].

In this paper we assume the atoms *are* bathed in the squeezed vacuum. and consider two cases : one in which the squeezed vacuum has axial symmetry and direction only along the slab axis (this case proves to be very much as was assumed in Ref. [15]) and one in which the squeezed vacuum has axial symmetry and direction along an axis *orthogonal* to the slab axis. At this instant we have not worked out all of the details of the theory for the first case and only the second case is handled analytically in this paper. This second case itself raises problems, since either the atoms must form an atomic beam of small radius as in Ref.[15], or the exciting field of frequency ω which is assumed to travel along the direction of the slab axis has to be of very small aperture ; this very small aperture (radius ~ $\frac{1}{2}c\omega^{-1}$) raises the problem of significant diffraction effects which we shall not handle here.

We do not give the details of the calculations which introduce both radiative shifts and the damping terms in the A-coefficient 2γ which appear in equations (11) : the self-field operators (8) shift the atomic frequencies ω_0 to their (ordinary) vacuum shifted values and introduce the terms in γ (but not those in $2N\gamma$) into (11). The squeezed vacuum in the trace over its states then introduces the damping terms in $2N\gamma$ and introduces further radiative shifts depending on this mean photon number N of the squeezed vacuum. Somewhat more complicated are the corresponding renormalisations of the squeezing parameter $M(\mathbf{x})$ which arise and we must report on these elsewhere. Moreover we shall ignore any correlations between the modes of the squeezed vacuum and treat these modes as squeezed independently. Such omissions are unavoidable in this short and first report on the effect of squeezing on an extended dielectric.

Indeed compared with the N_a Rydberg atom problem of Refs. [13,14,26,27] we already have in the equations (11) a striking new feature. For Rydberg atoms and therefore *microwave* wavelengths actual experiments have shown [30] that the Dicke model is a good, physically applicable, model. This model places all of the N_a 2-level atoms at the same single site $\mathbf{x}_i = \mathbf{x}_0$ (say) and no ensemble average over the $\mathbf{x}_i$ is evidently required. For the shorter wavelengths considered in this paper the atoms must be assumed to be distributed inside V and all of the resulting spatially dependent effects must be taken into account. We find in particular that, as compared with the Rydberg atom problem, $M \rightarrow M(\mathbf{x})$ and depends on the point $\mathbf{x}$ inside V : in particular M gains a phase ϕ which depends on $\mathbf{x}$.

A principle effect of the extended system is to introduce the retarded inter-atomic field (9) which becomes the c-number field (12). In the Rydberg atom Dicke model analysis, since $\mathbf{x}_i = \mathbf{x}_0$ for all labels i, this field becomes in effect a collective N_a-atom self-field (one proves the radiative shifts are not collective however [26,27]). In the extended system the retarded inter-atomic fields create the nonlinear or linearised refractive indices $m(\omega)$ [17,18,31,32] (and see below).

The effect of the squeezed vacuum is an effect on the modes of the free field operators $\mathbf{e}_0^{\pm}(\mathbf{x},t)$, equation (7). In the operator equations (10) these act only at single sites $\mathbf{x}_i$ so that the effect of the squeezed vacuum is to introduce a contribution to the atomic self-field in addition to that described by equations (8). There is a physically significant field-matter decorrelation approximation made here, as there was in the simpler unsqueezed problem [19]. But otherwise, traced over the squeezed vacuum states the free-field operator introduces the squeezing parameter N as written in (11) (together with the further radiative shifts mentioned) *providing* the squeezed modes have spherical symmetry lying, without correlations, in some annulus with $k = |\mathbf{k}|$ in $k_{min} \le k \le k_{max}$ such that $k = k_0$, $k_0 \equiv \omega c^{-1}$, lies in this annulus : by the assumption of spherical symmetry $N_{\mathbf{k},\lambda} \equiv \langle a^{\dagger}_{\mathbf{k},\lambda}\, a_{\mathbf{k},\lambda}\rangle = N_k$ and depends only on $k = |\mathbf{k}|$ for each mode of the squeezed vacuum, and $N \equiv N_{k_0}$.

However in this isotropic case $M(\mathbf{x}) = M\, j_0(k_0 r)$ where $j_0(x) \equiv (\sin x)x^{-1}$, since each mode $(\mathbf{k},\lambda)$ of the squeezed vacuum introduces at each point $\mathbf{x}$ the squeezing parameters $\langle a_{\mathbf{k},\lambda}\, a_{\mathbf{k},\lambda}\rangle\, e^{2i\mathbf{k}.\mathbf{x}}$ and their adjoints in $a^{\dagger}_{\mathbf{k},\lambda}$: here $|\mathbf{k}| = k_0$ and $r = |\mathbf{x}|$ and this result assumes an origin for $\mathbf{x}$ at some arbitrary point inside the whole quantisation region $V_0 \to \infty$.

The analytical solution of equations (11) in these spherically symmetric terms, nonlinear or linearised, seems impractible at present. For an actual experiment it seems likely that the squeezed vacuum should in any case take on axial symmetry. As a specimen calculation for this paper we therefore consider the two such cases axial symmetry along the z-axis the axis of the slab-like region V and axial symmetry along the x-axis orthogonal to z. By assumption, modes lie parallel to z or to x in each case. This means that the photon numbers N of the squeezed vacuum also have the same properties and $N \to \frac{1}{3}N$ associated with the directions z or x only : there is then a complication over the combination $1+2N$ multiplying γ in equations (11). To avoid this we shall assume in this paper that the remaining modes of the free field which are not squeezed are nevertheless filled with the photon occupation number N. With these assumptions about the squeezed vacuum, the equations (11) are as written, but $M(\mathbf{x}) = M \cos 2k_0 z$ or $M(\mathbf{x}) = M \cos 2k_0 x$ for the z-directed or x-directed squeezed vacua respectively.

In Refs. [17,18] we have solved the nonlinear problem (11) for a harmonic driving field when $M \equiv 0$. We make use of the optical extinction theorem [17,18,31-33]. It will become evident that for the z-directed squeezed vacuum problem with $M(\mathbf{x}) = M \cos(2k_0 z)$, in which $k_0 = \omega_0 c^{-1}$ is the wave number of the free external field incident on V, the action of this extinction theorem becomes substantially more complicated than its usual form. Therefore in this short paper we consider further only the case of x-directed squeezed vacua. We hope to report results for the z-directed squeezed vacua elsewhere, again utilising the extinction theorem in this connection.

Dielectric theory for finite regions V becomes complicated by real, and observable, diffraction effects. For simplicity it is usual to assume that V is not only slab-like but is translationally invariant in all directions orthogonal to the slab-axis. This translational invariance is incompatible with $M(\mathbf{x}) = M \cos(2k_0 x)$! For a simplification for present purposes we therefore assume that the incident harmonic field is normally incident along the axis of z and is of very small aperture - indeed of radius $< \frac{1}{2} k_0^{-1}$. Evidently this *will* introduce severe diffraction problems which we do not attempt to solve here. Otherwise this small aperture means $M(\mathbf{x})$ is effectively constant and we can then *formally* adopt transverse translational invariance into the theory. Note that these considerations actually address both the diffraction and phase-matching problems which would arise in practice in the system considered in Ref. [15].

4. SOLUTION FOR THE NONLINEAR AND LINEAR REFRACTIVE INDICES

To solve equations (11) with $M(\mathbf{x}) \equiv M$, a complex number independent of $\mathbf{x}$, we can now follow our previous work [17,18,31] on the equivalent problem in the normal vacuum and which has $M \equiv 0$ and make an ansatz for the slowly varying envelopes of the dipole densities, $r_{\pm}(\mathbf{x},t)$, at frequency ω. In the steady state these envelopes will vary (slowly) only in space and for transverse translational invariance, then only in the direction of propagation, also taken to be the z-direction. Thus we assume [17,18,31] that $r_{\pm}(\mathbf{x},t) \equiv r_{\pm}(\mathbf{x},\omega)$, and that

$$r_{+}(\mathbf{x},\omega) = f(z)\, e^{-imk_0 z} + \Lambda b(z)\, e^{imk_0 z} = (r_{-}(\mathbf{x},\omega))^{*} \tag{13}$$

where f,b vary slowly in z on the wave-length scale k_0^{-1} while $r_3(\mathbf{x},t) \equiv r_3(\mathbf{x},\omega) \equiv r_3(z)$ does likewise : the wave number $k_0 = \omega c^{-1}$. The ω-dependent quantity Λ is a reflexion coefficient and the complex number m is thus a nonlinear generalisation of the complex refractive index [17,18,22-24,31] in the case when the coherent single mode field incident on the slab formed by the atoms is normally incident on the planar surface z = 0 of that slab : the axis of the slab lies along z and as mentioned the second surface of the slab will be z = L. The ansatz (13) thus describes forward and backward going dipole waves at harmonic frequency ω appropriate to the Fabry-Perot cavity of width L formed by the atoms.

With (13), the expression for E^{+}, equation (12), yields to first order in the derivatives f′, b′ of the slowly varying envelopes f,b [17,18,31]

$$2i\, E^{+} r_3(z) = \frac{i8\pi n p^2 \hbar^{-1}}{3(m^2-1)} \left[\left(\frac{m^2+2}{3} f - \frac{2imf'}{(m^2-1)k_0} \right) e^{-imk_0 z} + \left(\frac{m^2+2}{3} b + \frac{2imb'}{(m^2-1)k_0} \right) e^{imk_0 z} \right] r_3(z) + (\text{terms in } e^{\pm ik_0 z}) . \tag{14}$$

We can put this expression for $E^{+} r_3$ into the first of the Bloch equations, equations (11), and (despite the non-linearity of this problem as it actually is at this stage) still appeal to the 'optical extinction theorem' due originally to Ewald and Oseen for linear systems. The 'theorem' is described in Refs. [22-24] (particularly Ref. [22]) and the references to these papers and also in Refs. [31,32] in particular, and cf. also [33]. In essence this theorem says that scattering from dipole sources distributed uniformly over the whole of the surface of the dielectric V exactly annihilates the incident free field at all points inside V, replacing waves $e^{\mp ik_0 z}$ of that free field inside V by dipole waves with a new wave number $m\, k_0$ $(m \neq 1)$. In the nonlinear case these dipole waves are the ansatz (13). Exactly as in the linear case, annihilation of the free field leads to two conditions, one connecting the amplitudes of the dipole waves with the amplitude of the free field, and the second one relating the amplitudes of the forward and backward going dipole waves in terms of the reflexion coefficient Λ - now explicitly expressed in terms of the parameter m. In the nonlinear case there then remains as a third condition a pair of *differential* equations in f and b. Technically with (14) into (11) one simply compares coefficients of waves $e^{\pm ik_0 z}$ and waves $e^{\pm imk_0 z}$ to see extinction of the incident free field, an expression for the reflexion coefficient Λ in terms of $m(\omega)$, and finally the two first order differential equations

$$(\gamma(1+2N)+i\delta)\, f = \frac{i8\pi np^2\hbar^{-1}}{(m^2-1)}\, r_3(z) \left(\frac{m^2+2}{3} f - \frac{2im}{(m^2-1)k_0} f' \right) - \gamma M b^* e^{-i(m^*-m)k_0 z},$$

$$(\gamma(1+2N)+i\delta)\, b = \frac{i8\pi np^2\hbar^{-1}}{(m^2-1)}\, r_3(z) \left(\frac{m^2+2}{3} b + \frac{2im}{(m^2-1)k_0} b' \right) - \gamma M f^* e^{i(m^*-m)k_0 z}. \quad (15)$$

One can now see that this step, use of the optical extinction theorem with M = constant, becomes complicated for the z-directed vacuum for which $M(x) = M \cos 2k_0 z$. Even so, and as in the nonlinear problem for the ordinary vacuum handled in Refs. [17,18,31], equations (15) have an explicit z-dependence involving the inversion density $r_3(z)$. In the linear limit for the attenuator, $r_3(z) = -\frac{1}{2}$ independent of z, and f and b can be taken as constants so that f′ and b′ both vanish. In the nonlinear problem with the squeezed vacuum which leads to (15) we can still follow the argument of Refs. [17,18,31] and choose the *parameter* m, introduced as such at equation (13), self-consistently so that the usual Fabry-Perot action of the slab-like cavity can be described in terms of it. Explicit expressions for this Fabry-Perot action are found through the extinction of the incident fields. The *self-consistent* procedure is to choose a "mean inversion" R_3 such that $R_3^{-1} = L^{-1}\int_0^L r_3^{-1}(z)\, dz$, where L is the slab width. One can actually choose $f(0) = b(0) = f(L) = b(L) = P_0$, amplitudes of the dipole waves; then m must satisfy

$$\eta f = -\gamma M b^* e^{-i(m^*-m)k_0 z}$$
$$\eta b = -\gamma M f^* e^{i(m^*-m)k_0 z} \quad (16)$$

where
$$\eta \equiv \gamma(1+2N) + i\delta - \frac{i8\pi np^2\hbar^{-1}}{m^2-1} \cdot \frac{m^2+2}{3} R_3$$

when
$$f'(z) = \beta(z)\, f, \quad b'(z) = -\beta(z)\, b(z) \quad (17)$$

with

$$\beta(z) \equiv \frac{k_0(m^2-1)(m^2+2)}{6im} \left(1 - \frac{R_3}{r_3(z)} \right). \quad (18)$$

This self-consistent argument for the nonlinear dielectric in the ordinary vacuum is sketched in [31] and given in complete detail in [17,18]. Notice that in the ordinary vacuum M = 0 and from $\eta = 0$ with N = 0 one obtains the usual (linear) expression for m as a dispersion relation in the detuning δ : the nonlinearity is thus concealed in the self-consistent route which chooses the "mean inversion" R_3 self-consistently. However to analyse the results further in the case of the ordinary vacuum one must compute to see optical bistable (multistable) behaviour [17,18]. The same situation obtains now in the case of the squeezed vacuum.

One can see immediately that (16) means that (take the c.c. of the second equation and multiply the two equations)

$$\left|1+2N + i\Delta - i\,\frac{8\pi np^2\hbar^{-1}}{3\gamma}\,\frac{m^2+2}{m^2-1}\,R_3\right|^2 = |M|^2 \tag{19}$$

(where $\Delta \equiv \delta\gamma^{-1}$). This is a nonlinear dispersion relation for the parameter m which includes effects of the squeezed vacuum through N and M and the nonlinearity through R_3. We must defer analysis of the optical bistability in the squeezed vacuum embodied in (19), since the numerical work is still to be done. However, Refs. [17,18] show exactly how this was done in the case of the ordinary vacuum and the same numerical methods can be used for the squeezed vacuum. We shall report results later. Here we shall simply *linearise* the problem by assuming a sufficiently weak incident field and not consider optical bistability in the squeezed vacuum any further. For such a weak incident field r_3 is a constant involving saturation only from the squeezed vacuum — so $r_3 \simeq -\frac{1}{2}(1+2N)^{-1} \equiv R_3$. The generalised dispersion relation with the effect of the squeezed vacuum included thus takes the form, from (19), that

$$[1+2N + i\Delta + i\alpha\,\frac{m^2+2}{3(m^2-1)}\cdot\frac{1}{1+2N}\,] = |M|\,e^{i\phi} \tag{20}$$

where $\alpha \equiv 4\pi np^2\hbar^{-1}\gamma^{-1}$. The phase ϕ is made up of the phase $\phi_{sq.vac}$ of the squeezing parameter M and an arbitrary phase θ_0 :$\phi = \phi_{sq.vac} - \theta_0$. This phase θ_0 from (19) can be tracked back through (16) to the phase of the dipole field (13) and thus back, via the extinction theorem, to the phase of the incident field. The phase ϕ is thus a relative phase which can be varied by choosing the phase of the squeezed vacuum $\phi_{sq.vac}$. We can thus sensibly analyse the generalised dispersion relation (20) for different values of the phase ϕ. Notice that ϕ has not involved any z-dependence because of our assumption that the squeezed vacuum is x-directed and that the incident field has the very small aperture of radius $\frac{1}{2}k_0^{-1}$.

5. ANOMALOUS DISPERSION AND ABSORPTION

If we put m = X-iY (X,Y both real) then (20) is the two equations

$$\begin{aligned}&1+2N - \alpha(1+2N)^{-1}\,2XY[(X^2-Y^2-1)^2 + 4X^2Y^2]^{-1} = |M|\cos\phi\,,\\ &\Delta + \tfrac{1}{3}\alpha(1+2N)^{-1}[(X^2-Y^2-1)(X^2-Y^2+2) + 4X^2Y^2].\\ &\quad.[(X^2-Y^2-1)^2 + 4X^2Y^2]^{-1} = |M|\sin\phi\,.\end{aligned} \tag{21}$$

Some unrewarding algebra then leads to the following results:

$$X = -DY^{-1},\ Y = \{(D^2+\tfrac{1}{4}(1+CD)^2)^{\frac{1}{2}} - \tfrac{1}{2}(1+CD)\}^{\frac{1}{2}} \tag{22}$$

where $$C = 2\Gamma^{-1}(B-\tfrac{1}{3}\alpha/(1+2N))$$

and $$D = \tfrac{1}{2}\ \Gamma\alpha(1+2N)^{-1}[B-\tfrac{1}{3}\alpha/(1+2N))^2 + \Gamma^2]^{-1} \qquad (23)$$

in which $\Gamma \equiv |M| \cos\phi -(1+2N)$, $B = |M|\sin\phi - \Delta$. For $N = M = 0$, the case of the ordinary vacuum, D reduces to the familiar expression $D = -\tfrac{1}{2}\alpha[(\Delta+\tfrac{1}{3}\alpha)^2 + 1]^{-1}$ and $X = -DY^{-1}$: D is thus a Lorentzian in the detuning δ of total width 2γ and centred at the shifted resonance frequency $\omega_0 - \tfrac{1}{3}\gamma\alpha$:Y itself can be evaluated and X calculated and this shows the usual anomalous dispersion. These features continue in the presence of the squeezed vacuum for which $N,M \neq 0$. Results are shown in the five figures Figs. (1a,b) (2a,b) and (3).

Notice that even when $N = M = 0$, $D \neq 0$ because $\gamma \neq 0$. The A-coefficient 2γ enters the theory via $\mathbf{e}_{self}$ and radiation reaction. This is a measure of spontaneous emission describing optical scattering by the separate atoms originally (i.e. before ensemble average) occupying distinct sites $\mathbf{x}_l$. In the case of the ordinary vacuum, $N = M = 0$, correlation at all orders between such sites leads [22-24] to expressions for the total scattering cross-section [24] involving pair correlation functions and 'internal field factors' [33] depending on both X and Y. The equations (22) with (23) evidently involve only the *self* correlation as modified by the effects of the squeezed vacuum. The missing correlation terms are, in effect, eliminated at the step at (9) which defines $\dot{\mathbf{e}}^{\pm}_{lnt}(\mathbf{x},t)$ only to 'leading order' in these correlations - namely for the smooth dielectric [31]. Evidently the squeezed vacuum modifies the total scattering cross-section in a rather complicated way but the details of these modifications are not yet worked out. Still it is evident that one can expect to see *observable* changes both in this scattering cross-section and the corresponding differential scattering cross-sections as modified by a squeezed vacuum and we must hope for actual experiments on these lines. Here we report only on the changes in $Y \equiv Im(m)$, essentially the absorption spectrum due to external scattering without interatomic correlation as monitored by the transmitted light (an approximation to the total scattering cross-section), and $X \equiv Re(m)$ the dispersion and anomalous dispersion. The Figs. (1a,b) for X have α =

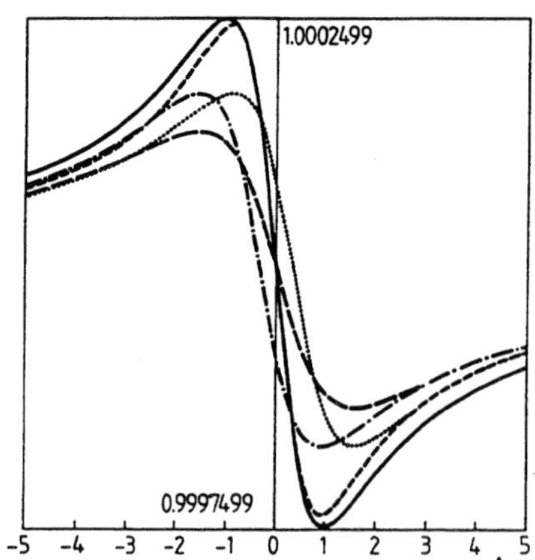

Fig. (1a): Anomalous dispersion, plot of $X \equiv Re(m)$ against scaled detuning Δ for $N = 0$, normal vacuum and $N = 0.1$, with $\phi = 0, \tfrac{1}{2}\pi, -\tfrac{1}{2}\pi$ and π for the squeezed vacuum; $\alpha = 10^{-3}$.

10^{-3} and $|M| = (N(N+1))^{\frac{1}{2}}$. Fig.(1a) shows (full line) X against Δ for N = 0 (the normal vacuum), and four other curves for N = 0.1 for which $\phi = 0$, $\frac{1}{2}\pi$, $-\frac{1}{2}\pi$, π respectively (the lines 'dash', 'dot', 'dash-dot' and 'large dash' respectively in Fig. (1a)). Since $\alpha = 10^{-3}$ is small, X is very close to unity where $\Delta \equiv \delta\gamma^{-1} = 0$ the case of the ordinary vacuum (N = M = 0). Otherwise these dispersion curves are significantly changed by the non-zero parameters N, M of the squeezed vacuum and particularly by change in the phase ϕ, associated as explained with the phase of M. As might be expected most substantial quantitative changes occur for $\phi = \pi$ (large dashes in Fig(1a)). But $\phi = +\frac{1}{2}\pi$ and $\phi = -\frac{1}{2}\pi$ differ in the way they push the whole curve to the right ($\phi = \frac{1}{2}\pi$) and to the left ($\phi = -\frac{1}{2}\pi$). Fig.(1b) with N = 1.0 confirms these several features.

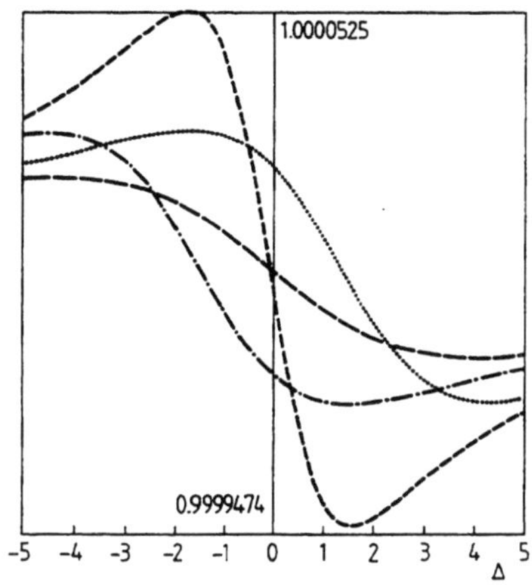

Fig. (1b): Anomalous dispersion for N = 1.0 and $\phi = 0, \frac{1}{2}\pi, -\frac{1}{2}\pi, \pi$; $\alpha = 10^{-3}$

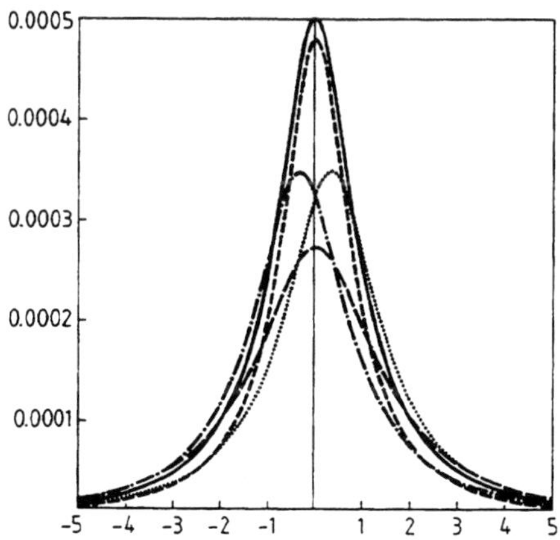

Fig. (2a): Absorption spectrum, plot of Y = Im{m} against scaled detuning Δ, for N = 0, and for N = 0.1 with $\phi = 0, \frac{1}{2}\pi, -\frac{1}{2}\pi$ and π; $\alpha = 10^{-3}$.

The Figs. (2a,b) also have $\alpha = 10^{-3}$ and $|M| = (N(N+1))^{\frac{1}{2}}$. Here the dependence on the phase ϕ is very marked: for N = 0.1, Fig. (2a) the peak height is almost halved for $\phi = \pi$, and the peak maxima are shifted to right or left for $\phi = +\frac{1}{2}\pi$ or $\phi = -\frac{1}{2}\pi$ and peak heights reduce to about 2/3; peak shapes and widths are changed to more or to less degree in all

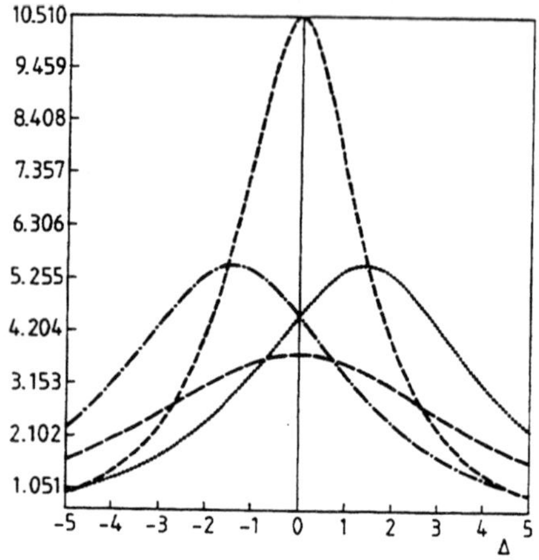

Fig. (2b): Absorption spectrum for N = 1.0 and $\phi = 0, \frac{1}{2}\pi, -\frac{1}{2}\pi, \pi$; $\alpha = 10^{-3}$.

cases compared with N = 0. For N = 1.0, Fig. (2b), the maximum peak height, which was 0.0005 for N = 0, is reduced to 0.0001 for $\phi = 0$ and this becomes as small as 0.000038 for N = 1.0 and $\phi = \pi$. Again peaks shift to left or right for $\phi = -\frac{1}{2}\pi$ or $+\frac{1}{2}\pi$. All of these effects induced by the squeezed vacuum would seem to be measurable. Note however that $\alpha = 10^{-3}$ means number densities $n \leq 10^{11}$ atoms cc^{-1}. The Fig. (3) compares with Fig. (2b). It has N = 1 but α is increased from 10^{-3} to $\alpha = 5$: the effects of the squeezed vacuum remain very plain but the Lorentz field shifts are also now very apparent.

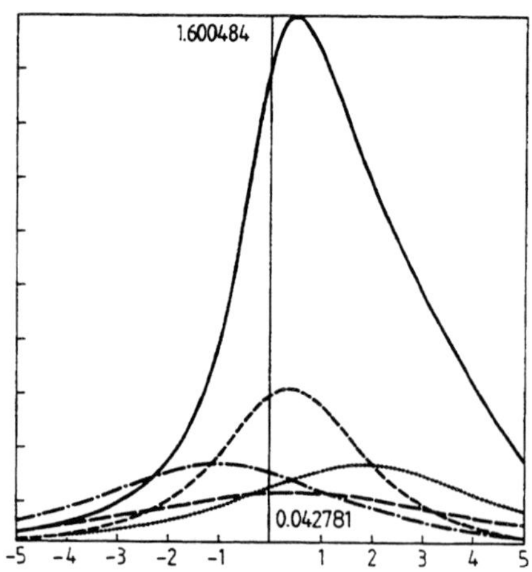

Fig. 3: Absorption spectrum for N = 0, and for N = 1.0 with $\phi = 0, \frac{1}{2}\pi, -\frac{1}{2}\pi, \pi$; $\alpha = 5$.

Finally we quote here the analytic expressions for the peak maxima of Y against Δ and of X against Δ:

$$\Delta_i = - d_i - \tfrac{1}{3}\,\alpha_1 + |M|\,\sin\phi\ ; \qquad i = 1,2,3 \qquad (24)$$

and Δ_2 gives the position of the maximum in Y while Δ_1 and Δ_3 give respectively the position of the maximum and of the minimum in X : $\alpha_1 \equiv \alpha(1+2N)^{-1}$ and the d_i are roots of $4d^3 + 3\alpha_1 d^2 - 4\Gamma^2 d - \alpha_1\Gamma^2 = 0$ where $\Gamma = |M|\cos\phi - (1+2N)$: d_1, d_2 and d_3 can be put in the forms

$$d_1 = A_3 + A_2 - \frac{1}{4}\alpha_1$$

$$d_2 = - \frac{1}{2} A_3 - \frac{1}{2}A_2 - \frac{1}{4}\alpha_1 - i\,\frac{\sqrt{3}}{2}\,(A_3 - A_2)$$

$$d_3 = - \frac{1}{2} A_3 - \frac{1}{2}A_2 - \frac{1}{4}\alpha_1 + i\,\frac{\sqrt{3}}{2}\,(A_3 - A_2)\ ,$$

where $A_1 = i\sqrt{3}\,\Gamma\,[256\,\Gamma^4 + 144\alpha_1^{\,2}\Gamma^2 + 27\alpha_1^{\,4}]^{\frac{1}{2}}$

$$A_2 = - \left[\frac{1}{64}\,\alpha_1^{\,3} + \frac{1}{144}\,A_1\right]^{\frac{1}{3}}$$

$$A_3 = - \left[\frac{1}{64}\,\alpha_1^{\,3} - \frac{1}{144}\,A_1\right]^{\frac{1}{3}}\ . \qquad (25)$$

By choosing the cube-roots all of the d_i are real as they must be. These analytical results confirm the values displayed in the five figures Figs. (1a,b), (2a,b), (3) and show again that even the small squeezing parameter N = 0.1 can induce substantial and observable changes particularly as the phase ϕ changes from $\phi = 0$ to $\phi = \pi$.

REFERENCES

1. P.A.M. Dirac, 'The quantum theory of the emission and absorption of radiation', *Proceedings of the Royal Society of London, Series A*, Vol. 114, (1927), p.243.
2. R.K.Bullough, 'Squeezed and Non-classical Light' *Nature*, Vol. 333 No. 6174, (1988), pp.601-602.
3. 'Squeezed and Non-classical Light', (P. Tombesi and E.R. Pike Eds.,) New York, Plenum Press, NATO ASI Series B: Physics Vol. 190 (1989).
4. Articles in Special Issue on 'Squeezed Light', *Journal of Modern Optics*, Vol. 34 Nos. 6,7 (1987), (R. Loudon and P.L. Knight Eds.).
5. Articles in Special Issue 'Squeezed states of the Electromagnetic Field', *Journal of the Optical Society of America*, Series B, Vol.4 No.10 (1987) (H.J. Kimble and D.F. Walls Eds.)
6. K. Zaheer and M.S. Zubairy in 'Advances in Atomic, Molecular and Optical Physics' (D. Bates and B. Bederson Eds.) New York, Academic Press, Vol. 38 (1990), p.143.
7. R.E. Slusher, L.W. Holberg, B. Yurke, J.C. Mertz and J.F Valley, *Physical Review Letters*, Vol. 55 (1985), p.2409.
8. R.M. Shelby, M.D. Levinson, S.M. Perlmutter, R.G. de Voe and D.F. Walls, *Physical Review Letters*, Vol. 57 (1986), p.691.
9. L.A. Wu, H.J.Kimble, J.L.Hall and H. Wu, *Physical Review Letters*, Vol. 57 (1986), p.2520.
10. M.G. Raizen, L.A. Orozco, M. Xiao, T.L. Boyd and H.J. Kimble, *Physical Review Letters*, Vol. 59 (1987), p.198.

11. L.A. Orozco, M.G. Raizen, M. Xiao, R.J. Brecha and H.J. Kimble, *Journal of the Optical Society of America*, Series B, Vol. (1987), p.1490.
12. H.J. Carmichael, A.S.Lane and D.F. Walls, *Physical Review Letters*, Vol. 58 (1987), p.2539.
13. M.R.B. Wahiddin, S.S. Hassan, J. Timonen and R.K. Bullough, 'Transient and spectral properties of several Rydberg atoms in a squeezed vacuum' in "Coherence and Quantum Optics VI" (J.H. Eberly, L.Mandel and E. Wolf Eds.) New York, Plenum Press (1990) pp. 1195-99.
14. M.R.B. Wahiddin, S.S. Hassan, R.R. Puri, J. Timonen and R.K. Bullough, 'Collective optical phenomena for several or many Rydberg atoms in a squeezed vacuum' in "Proceedings of the European Conference on Optics, Optical Systems and Applications, Rome, 1990, Institute of Physics Conference Series No. 115, Section 1 (1990), pp.85-90.
15. P. Galatola, L.A. Lugiato, M.G. Porrera and P. Tombesi, *Optics Communications*, Vol. 81 (1991), p.175.
16. S. Singh, J. Rai, C.M. Bowden and A. Postan, *Physical Review, Series A*, Vol. 45 (1992), p.5162.
17. M.N.R. Ibrahim, M.Sc. Thesis, University of Manchester (1989).
18. R.K. Bullough, S.S. Hassan, M.N.R. Ibrahim, N. Nayak and G.P.Hildred, 'Analytical and numerical results in a fundamental non-linear refractive index theory of optical multistability in a Fabry-Perot cavity'. To be published.
19. S.S. Hassan and R.K.Bullough, 'The Driven Dicke Model and its Macroscopic Extension : Bistability or Bifurcation' in "Optical Bistability" (C.M. Bowden, M.Ciftan and H.R. Robl Eds.) New York, Plenum Press (1981), pp. 367-404.
20. R. Saunders, R.K. Bullough and F. Ahmad, *Journal of Physics A : Mathematics and General Physics*, Vol. 8 (1975), p.759.
21. See also J.R. Ackerhalt and J.H. Eberly, *Physical Review, Series D*, Vol. 10 (1974), p.3350.
22. F. Hynne and R.K. Bullough, *Philosophical Transactions of the Royal Society of London, Series A*, Vol. 312 (1984), pp. 251-293.
23. F. Hynne and R.K. Bullough, *Philosophical Transactions of the Royal Society of London, Series A*, Vol. 321 (1987), pp. 305-360.
24. F. Hynne and R.K. Bullough, *Philosophical Transactions of the Royal Society of London, Series A*, Vol. 330 (1990), pp. 253-313.
25. C.M. Bowden 'Formulation of generalised Maxwell-Bloch equations for optically dense media' and 'New dipole-dipole interaction effects in quantum and nonlinear optics'. Two lectures at this meeting.
26. M.R.B. Wahiddin, Ph.D. Thesis, University of Manchester (1989).
27. S.S. Hassan, M.R. Wahiddin, R.K. Bullough and R. Saunders, 'Atoms in a broad band squeezed vacuum : I Master equation for the Dicke model'.To be published.
28. Z. Ficek and and P.D. Drummond, 'Three level atom in a broad-band squeezed vacuum field. I. General theory', *Physical Review, Series A*, Vol. 43, No.11 (1991), pp.6247-6257.
29. R.R. Puri, Amitabh Joshi and R.K. Bullough, 'The two atoms Dicke model in a squeezed vacuum: the fluorescence spectrum including dipole-dipole interactions', *International Journal of Modern Physics, Series B*, Vol.5, No.19 (1991), pp.3115-3125.
30. R.K. Bullough, Two lectures on Cavity Quantum Electrodynamics at this meeting.
31. R.K. Bullough and S.S. Hassan, 'The optical extinction theorem in the nonlinear theory of optical multistability' in Proceedings of the Society of Photo-optical Instrumentation Engineers, Bellingham, Washington, Vol. 369 (1982), p.257.
32. R.K. Bullough and F. Hynne, 'Ewald's optical extinction theorem' in "P.P. Ewald and his dynamical theory of X-ray diffraction"

(D.W.J. Cruickshank, J.J. Juretschke and N. Kato Eds.) Oxford, Oxford University Press (1992), pp 98-110.

33. For relev a nt material see also O. Keller, 'Self-consistent field calculations in the nonlinear optics of mesoscopic systems' and 'Nonlocal electrodynamics of an atom in front of a normal or phase conjugating mirror', Lectures at this meeting and their references.

Note in proof By mistake we used $|M| = \frac{1}{2}(N(N+1))^{\frac{1}{2}}$ instead of twice this value for all of the Figs. (1a,b),(2a,b) and 3 of this paper. Thus, as displayed, all of the Figures have $|M| < (N(N+1))^{\frac{1}{2}}$ and the squeezing is not maximised. We have since computed them again with $|M| = (N(N+1))^{\frac{1}{2}}$ and find significant changes of detail but no essentially qualitative changes. We will report these new computations elsewhere.

Studies in Classical
and Quantum Nonlinear Optics

ISBN 1-56072-168-5

MICROSTRUCTURE ON Al/Al_2O_3 INTERFACES INVESTIGATED BY OPTICAL SECOND-HARMONIC GENERATION

C. Jakobsen and K. Pedersen
Institute of Physics, University of Aalborg
Pontoppidanstraede 103, DK-9220 Aalborg Øst, Denmark

Abstract

Microstructure on the interface between (100) oriented aluminium crystals and their native oxide layers is studied by optical second-harmonic generation (SHG). The second harmonic (SH) signal from mechanically polished crystals which are miscut a few degrees toward a [100] direction varies strongly when the crystals are rotated about a surface normal. By comparing this rotational anisotropy for samples in free air to SHG from monoatomic steps on vicinal $Al(100)$ in ultra-high vacuum (UHV) it is shown that a well ordered structure on the atomic scale is formed on the interface between the crystaline aluminium and the amorphous oxide layer.

1 INTRODUCTION

In the bulk of media with centrosymmetry electric dipole contributions to SHG are absent. To the lowest order in spatial nonlocality only electric quadrupole and magnetic dipole contributions of the form [1]

$$P_{i,b}^{(2\omega)} = \Gamma_{ijkl} E_j^{(\omega)} \nabla_k E_l^{(\omega)} \tag{1}$$

give rise to bulk SHG. At a surface or an interface the centrosymmetry is broken and electric dipole contributions to SHG are present. This together with the effect of spatial nonlocality to arbitrary order is represented by the polarization density of a dipole sheet at the surface given by [2]

$$P_{i,s}^{(2\omega)} = \chi_{s,ijk}^{(2)} E_j^{(\omega)} E_k^{(\omega)} \tag{2}$$

The symmetry of the surface or interface layer is reflected in the elements of the third rank tensor $\overleftrightarrow{\chi}_s^{(2)}$. For a surface with c_{4v} symmetry as one would expect from an unreconstructed (100) face of a cubic crystal SHG due to the surface contributions is isotropic [3]. For lower symmetry new elements in $\overleftrightarrow{\chi}_s^{(2)}$ appear and lead to anisotropy in SHG.

In the present work the surface sensitivity of SHG is used to study structural properties of aluminium crystals which have been cut with the surface oriented a few degrees off a (100) face. An interesting aspect of SHG is the possibility of investigating buried interfaces if they are accessible to light [4]. The oxide layer on aluminium is transparent over a wide spectral range and is not expected to contribute to SHG. This allows us to investigate the interface between the metal and the oxide layer.

2 EXPERIMENTAL RESULTS AND DISCUSSION

Single crystals of aluminium were cut with the surface a few degrees away from a (100) plane. The miscut is specified by the angle α between the surface normal and the [100] direction in a (100) plane. The surfaces were carefully mechanically polished in order to get through the strained layer left by the cutting and grinding processes and to obtain a smooth mirrorlike surface.

The samples were mounted in an experimental setup where they could be rotated about their surface normals and the angle of incidence could be varied. A Q-switched Nd:YAG laser from which the fundamental wavelenght of 1064 *nm* and its SH of 532 *nm* were available was used as pump source. The reflected light from the sample was passed through coloured glass filters and interference filters in order to isolate the SH radiation which was detected by a photomultiplier tube connected to gated electronics.

Fig.1 shows the rotational anisotropy of the SH signal obtained with the 1064 *nm* pump wavelenght and 45° angle of incidence on a (100) aluminium crystal miscut by 5°. The results are given in terms of a nonlinear reflection coefficient defined by $R = I(2\omega)/(I(\omega))^2$. Both the p_1 to p_2 and s_1 to s_2 polarization configurations show strong anisotropy when the YAG fundamental is used as pump wavelenght. With the YAG SH as pump wavelenght no significant rotational anisotropy was detected.

In order to describe the results on Fig.1 we have investigated the response of the form given by Eq. (2) when the interface has c_{1v} symmetry. The symmetry reduction from c_{4v} of the (100) face to c_{1v} is caused by the miscut of the sample since height variations across the surface are restricted to steps of integer atomic distances (Fig.2).

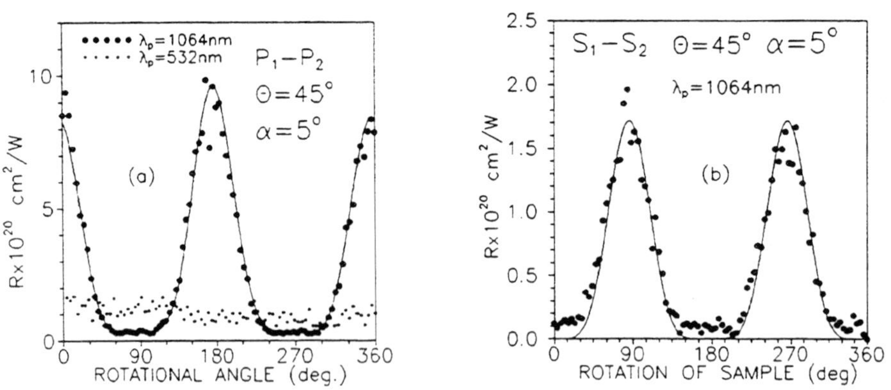

Figure 1: SHG as a function of the rotation angle ϕ obtained from a 5° miscut sample in free air for (a) p_1 to p_2 and (b) s_1 to s_2 polarizations.

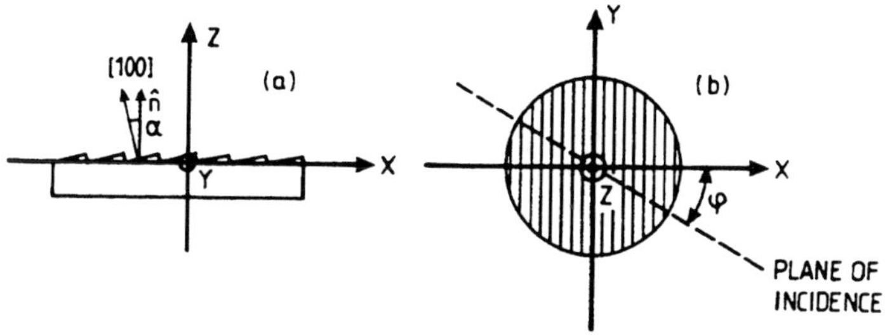

Figure 2: A sketch of the sample geometry seen from (a) the mirror plane and (b) a top view.

The response tensor in Eq. (2) will then contain 10 non zero elements. However, if it is assumed that the SH response is determined by the $\chi_{s,111}$ element describing the nonlinear polarization perpendicular to the steps generated by a field in the same direction the SH intensities for p_1 to p_2 and s_1 to s_2 polarizations are given by

$$I_{p_1 \to p_2} = \mid a_{p,p} + b_{pp}\chi^{(2)}_{s,111} cos^3\phi \mid^2 \tag{3}$$

$$I_{s_1 \to s_2} = \mid b_{s,s}\chi^{(2)}_{s,111} sin^3\phi \mid^2 \tag{4}$$

where ϕ is the angle between the plane of incidence and the direction perpendicular to the steps. The curves on Fig.1 are obtained by fitting the parameters a and b in Eqs. (3) and (4) to the experimental results. It is seen that the experiments are represented quite well by this model. For $\phi = 90°$ we should get the response of a surface without steps. The signal is enhanced by a factor of 30 by the steps for the p_1 to p_2 polarization configuration.

Steps on atomically clean aluminium surfaces have recently been shown to enhance SHG strongly [5]. In order to compare our results for oxide covered samples to results for clean well characterized surfaces a sample miscut by 4° was mounted in a UHV system. The surface was cleaned by cycles of argon ion sputtering and annealing to 400°C. The presence of steps on the clean surface was revealed by Low Energy Electron Diffraction (LEED) where the pattern showed streaks which at low temperatures tended to form double spots.

Fig.3 shows the rotational anisotropy of the clean sample for the p_1 to p_2 and s_1 to s_2 polarization configurations.

The results are largely represented by a $cos^6\phi$ for the p_1 to p_2 case and a $sin^6\phi$ for the s_1 to s_2 case as expected for c_{1v} symmetry. By comparison to Fig.1 it is seen that the value of the nonlinear reflection coefficient goes up when the surface is cleaned as one would expect [6]. The maximum signal ($\phi = 0°$) for the p_1 to p_2 case increases relatively more than the signal at $\phi = 90°$. This shows that the oxide layer has a stronger effect on the steps than on the terraces.

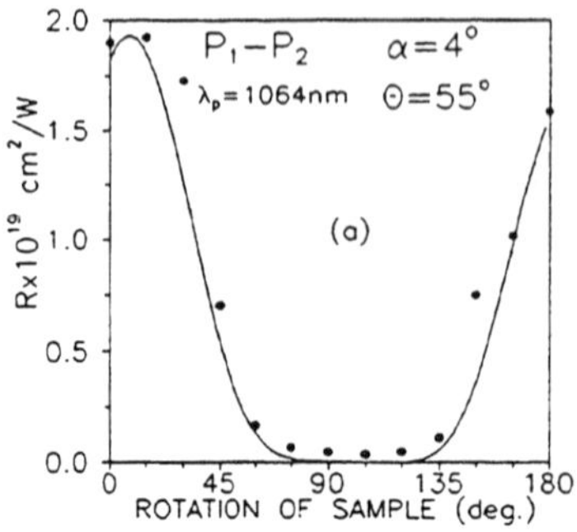

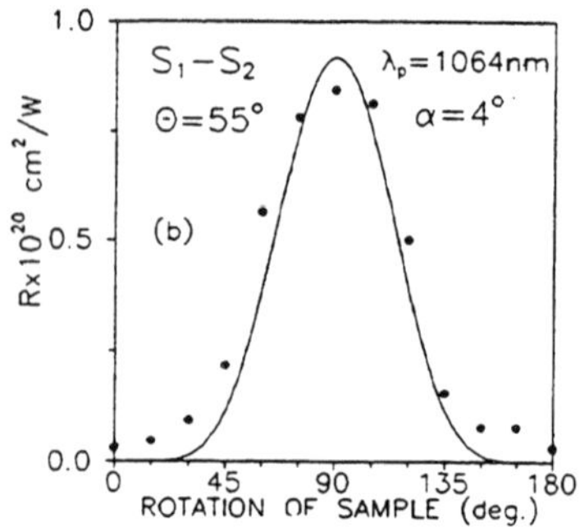

Figure 3: SHG as a function of the rotation angle ϕ obtained from a clean 4° *miscut sample in UHV for (a) p_1 to p_2 and (b) s_1 to s_1 polarizations.*

The interpretation of the signal enhancement in terms of atomic steps is supported by the observed temperature dependence of SHG from stepped surfaces. Fig.4 shows the variation of the SH signal as the sample temperature is changed from $-130°C$ to $450°C$. The signal enhancement decreases linearly with the temperature up to $450°C$. Above this temperature the rotational anisotropy has disappeared. This temperature dependence has previously been observed by Janz [7] and explained by the formation of larger flat domaines at higher temperatures which reduces the density of steps and thus the density of SH sources.

The result clearly show that the reduction of the surface or interface symmetry created by miscutting the samples strongly enhances the SH signals. In order to study the dependence on the miscut angle a sample with a cylindrical surface was prepared. The angle α varied continuously from $-10°$ to $10°$ across the sample (see Fig.5). The sample was mechanically polished and studied in free air.

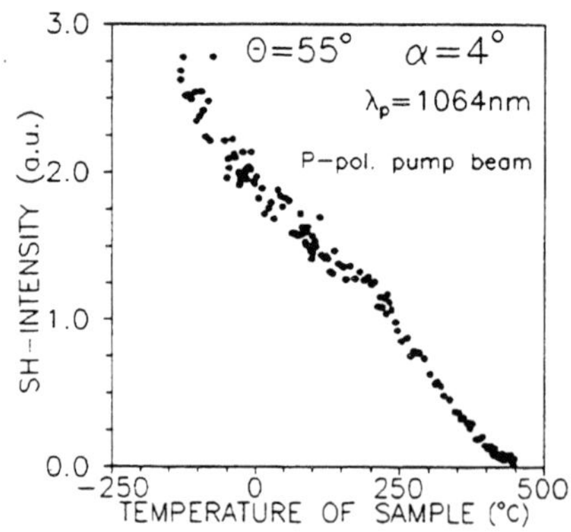

Figure 4: Temperature dependence of the step enhanced SHG from a clean sample in UHV.

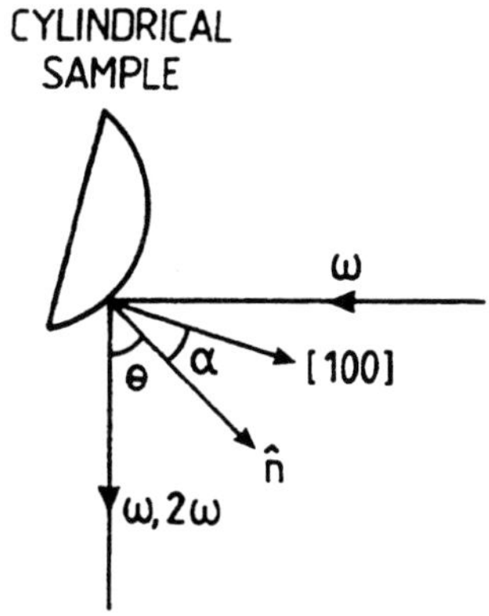

Figure 5: The geometry of the cylindrical sample with varying miscut angle.

With a p-polarized pump beam the p-polarized SHG was measured as a function of α by scanning the laser beam across the sample.

The result in Fig.6 shows that the SH signal grows quadratically with α when $|\alpha|$ is less than 5° and thus that the density of SH sources is proportional to α for small miscut angles. These results in Fig.6 have a strong resemblance to the results obtained by Janz et al. [5] with a 820 nm pump beam on a similar sample cleaned in UHV. They found from LEED experiments that monoatomic steps with a density proportional to α existed up to approximately 5° miscut. For higher miscut angles the steps start to interact and form larger terraces separated by height differences of more than one atomic layer. It is suggested that the same mechanisms are responsible for the present results obtained on oxidized samples.

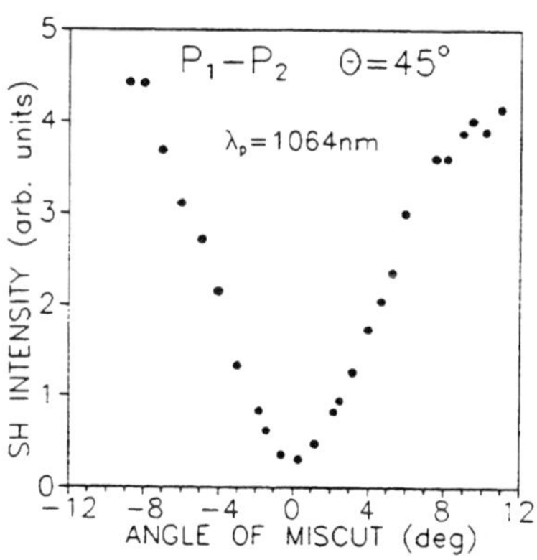

Figure 6: The dependence of SHG on the miscut angle showing a quadratic behaviour when $|\alpha|$ is less than 5°.

3 CONCLUSION

We have found that SHG form $Al(100)/Al_2O_3$ interfaces are enhanced when the $Al(100)$ crystal is miscut by a few degrees. By comparing results with respect to rotational anisotropy and frequency dependence for mechanically polished samples in free air to results for clean samples in UHV, we are lead to suggest that atomic steps are responsible for the enhancement of SHG. Also the dependence on the miscut angle for oxidized samples shows the same behaviour as for clean samples where the steps can be characterized by LEED.

The experiments indicate that a step structure with a high degree of order is formed between the crystaline aluminium and the amorphous oxide layer when the samples are carefully mechanically polished. The step density is proportional to the miscut angle up to 5° indicating a fairly regular step spacing. At higher miscut angles the SHG enhancement saturates which is taken as a signature of the creation of less but higher steps.

References

[1] N.Bloembergen, R.K. Chang, S.S. Jha, and C.H. Lee, "Optical Second-Harmonic Generation in Reflection from Media with Inversion Symmetry", Phys.Rev. **174,** pp. 813-822 (1968).

[2] P. Guyot-Sionnest, W. Chen, and Y.R. Shen, "General Considerations on Optical Second-Harmonic Generation from Surfaces and Interfaces", Phys.Rev. **B33,** pp. 8254-8263 (1986).

[3] J.E. Sipe, D.J. Moss, and H.M. van Driel, "Phenomenological Theory of Optical Second- and Third-Harmonic Generation from Cubic Centrosymmetric Crystals", Phys.Rev. **B35,** pp. 1129-1141 (1987).

[4] G.L. Richmond, J.M. Robinson, and V.L. Shannon, "Second-Harmonic Generation Studies of Interfacial Structure and Dynamics", Prog.Surf.Sci. **28** pp. 1-70 (1988).

[5] S. Janz, D.J. Bottomley, H.M. van Driel, and R.S. Timsit, "Influence of Steps on Second-Harmonic Generation from Vicinal Metal Surfaces", Phys.Rev.Lett. **66** pp. 1201-1204 (1991).

[6] S. Janz, K. Pedersen, H.M. van Driel, and R.S. Timsit, "Structural Transformations in Adsorbed Oxygen Layers on Al Surfaces Observed Using Optical Second-Harmonic Generation", J.Vac.Sci.Technol. **A9,** pp. 1506-1510 (1991).

[7] S. Janz, "Second-Harmonic Generation from Aluminum Surfaces", Ph.D.Thesis, University of Toronto (1991).

Studies in Classical
and Quantum Nonlinear Optics

ISBN 1-56072-168-5

ADSORBATE-INDUCED CHANGE OF LINEAR AND NONLINEAR RESPONSE PROPERTIES AT SIMPLE METAL SURFACES

M. Kuchler and F. Rebentrost
Max-Planck-Institut für Quantenoptik
8046 Garching near Munich, Germany

1. INTRODUCTION

In the last years great efforts of research have gone into the field of nonlinear optics and in particular into second harmonic generation (SHG) at surfaces. One reason for this is the interest in the fundamental physical processes of the nonlinear response which also is dependent on ground state and linear response properties. Therefore by this one has a rather rigorous test of present understanding of surface electronic structure in general. The SHG process is sensitive e.g. to the electronic properties including free and bound electrons, surface states, interband transitions, adsorbates etc.. On the other hand nonlinear optical methods have many important applications in surface science. SHG is inherently sensitive to the interfacial region between centrosymmetric media since it is dipole forbidden in the bulk but allowed at a surface where the symmetry is broken. Therefore it can e.g. give information on buried surfaces. Furthermore SHG has the potential to resolve dynamical processes like phase transitions, desorption and adsorption processes and surface reaction kinetics on a subpicosecond time level.

First theoretical considerations based on a macroscopic point of view were already given in the original works by Bloembergen et al. [1] and later reviewed by Richmond [2] and Shen [3]. In contrast to the macroscopic description the microscopic theory is still on a rather low level. An exception are s-p bonded simple metals where a free electron gas model works rather well and the understanding of the fundamental processes has made good progress. One of the pioneering works is by Rudnick and Stern [4] who investigated the quantum mechanical second-order response of a free electron gas. They realized that the SHG process is mainly governed by three nonlinear currents: a bulk current of magnetic dipole character extending about one skindepth into the metal and the surface-parallel and -perpendicular currents which both have electric dipole and quadrupole sources and are localized within a few angstroms at the surface. There are important contributions to the surface currents from the discontinuity (symmetry breaking) at the surface and from field gradients caused by the screening of the external field. This extremely simplified model assumes a non-interacting electron gas and a sharp truncation of the electron density at the surface. The first of these simplifications was later resolved by Sipe and coworkers [5] by a hydrodynamic model which considers the collective motion of electrons but still assumes a sharp cut-off of the electron distribution at the surface. The results are however very similar to the free electron gas model. Recently Weber and Liebsch [6] used a jellium model description for the metal substrate and treated the density response within density functional theory (DFT) and the local density approximation (LDA). They found a great enhancement of the perpendicular nonlinear current with origin mainly in the exponentially decreasing tail of the electron density in the vacuum. This shows the need for a full quantum mechanical calculation of the detailed microscopic response properties to the perpendicular electric field component of the incident light. The LDF treatment of the nonlinear response will now be discussed in more detail.

2. DENSITY FUNCTIONAL THEORY

The most sophisticated calculations for response properties of metal surfaces up to now have been done with density functional theory. DFT has been used in calculations of the electronic ground state as well as of the linear and nonlinear response. The central idea (for a review see e.g. [7]) of DFT is that the ground state energy is a unique functional of the ground state charge density $n(\vec{r})$

$$E[n(\vec{r})] = \int v_{eff}(\vec{r})n(\vec{r})d^3r + T[n(\vec{r})]$$

where $T[n(\vec{r})]$ is the kinetic energy of the noninteracting electron gas and

$$v_{eff}(\vec{r}) = v(\vec{r}) + \frac{1}{2}\int \frac{n(\vec{r}')}{|\vec{r}-\vec{r}'|}d^3r' + v_{xc}[n(\vec{r});\vec{r}]$$

with the positive ion potential and other external potentials $v(\vec{r})$, the Coulomb potential of the electrons and the many-body interactions between the electrons $v_{xc}[n(\vec{r});r]$. By this the N-particle many-body problem is formally reduced to the solution of N independent one-particle Kohn-Sham equations [8]

$$\left(-\frac{1}{2}\nabla^2 + v_{eff}[n(\vec{r});\vec{r}]\right)\Psi_i = E_i\Psi_i$$

and

$$n(\vec{r}) = \sum_{occ,i} |\Psi_i(\vec{r})|^2.$$

These equations are in principle exact and have to be solved selfconsistently by an iterative procedure. The physics of the problem is now of course hidden in the exchange correlation potential $v_{xc}[n(\vec{r});\vec{r}]$ which depends on the electron density at every point of the metal and makes the equations still numerically intractable. The approximation one usually makes is to assume a 'slowly varying density' which leads to the *local density approximation* where $v_{xc}[n(\vec{r})]$ depends only on the density at point $\vec{r}$. Several parametrisations for $v_{xc}[n(\vec{r})]$ are given in the literature [9, 10]. This approximation gets worse at large density gradients and low electron densities both of which are the case at metal surfaces. LDA might thus be a rather poor approximation for metal surfaces. The calculations of Lang and Kohn [11, 12] however gave good results for work functions and surface energies showing that despite of this the method works quite well for ground state properties.

For the response properties a new problem arises since DFT in the present form is essentially a ground state theory and has no formal justification for its application to exited states. Nevertheless Zangwill and Soven [13] used time-dependent density functional theory (TDFT) to calculate the third harmonic response of rare gas atoms. Later Liebsch [14] treated the response properties of jellium metal surfaces. In both cases the theoretical results are in rather good agreement with experiment showing that DFT can be used also with some success for response properties. For the application to metal surfaces further simplifications are possible. First the parallel surface and the bulk currents depend only on linear response properties (i.e. on the linear dielectric constants). Thus only for the perpendicular surface current a detailed calculation is necessary. The small extension of this current allows to set up a non-retarded theory which implies that the incident electric field $\vec{E}$ has infinite wavelength and is perpendicular to the surface $\vec{E}(\vec{r},t) = E_0\hat{z}e^{-i\omega t}$ and can always be obtained from a potential of the form $\Phi(z,t) = -zE_0e^{-i\omega t}$. It is then sufficient to calculate the density response for homogenous electric fields rather than the full density and current response for general external fields. First and second-order perturbation theory for a non-interacting electron gas can be used to obtain the linear and nonlinear density response. Note however that the independent particle wave functions and energies are required to evaluate the response functions. For the DFT case these have no formal resemblance to the real independent particle wave functions and energy levels and the main justification comes from comparison of calculated results with experiment.

3. SHG FROM A JELLIUM SURFACE

A simpler approach [6] to the response properties is possible in the low frequency limit ($\omega \lesssim 0.1\omega_p$) where ω_p is the plasma frequency of the metal [6]. The electrons then react instantaneously to the field and an electrostatic calculation is expected to be sufficient. The second-order density response $n_{2\omega}(z)$ to an external field E_0 perpendicular to the surface is then related to the static response $n_2(z)$ by expanding the the electron density in powers of E_0

$$n(z,t) = n_{DC}(z) + n_\omega(z) sin\omega t + n_{2\omega}(z) cos2\omega t + \dots$$

and

$$n_{2\omega}(z) = -\frac{1}{2}E_0^2 n_2(z)$$

The perpendicular surface current is obtained from $n_2(z)$ by the continuity equation

$$\frac{\partial}{\partial z} j_\perp(z,t) + \frac{\partial}{\partial t} en(z,t) = 0.$$

The Rudnick-Stern parameter a describes phenomenologically [4] the integrated per-

Theory		$r_s = 2.0$	$r_s = 3.0$	$r_s = 4.0$
LDA	[14]	-28.4	-12.9	-8.6
LDA-RPA	[14]	-22.1	-10.0	-6.6
ext. Thomas-Fermi	[15]	-34.7	-14.9	-6.2
infinite barrier-RPA	[14]	-0.8	-1.0	-1.2
hydrodynamic model	[16]	-0.2	-0.2	-0.2

Table 1: Rudnick-Stern parameter $a(\omega = 0)$ for various bulk densities given by the Wigner-Seitz radius r_s calculated by several theoretical models (see text).

pendicular surface current and is given by n_2 as

$$a = -4n_b(4\pi e)^2 \int_{-\infty}^{\infty} n_2(z)zdz$$

where n_b is the average electron density in the bulk. From this expressions for the emitted SHG intensity follow, e.g. for the case that both excitation and detection are for p-polarized light

$$I_{2\omega}(\theta) \sim \sin^2\theta \cos^4\theta|1 - 2b + 0.5a\sin^2\theta|^2$$

Here θ is the angle of incidence of the pump light and $b = -1$ is the corresponding parameter describing the surface-parallel component.

In table 1 a comparison of the values of the $a-$parameter for several bulk densities (given by the appropriate Wigner-Seitz radius) obtained with various models is presented. All values are calculated in the low frequency limit. The top row shows the results from a full DFT-LDA type calculation. The LDA-RPA results would be obtained by ignoring the exchange correlation part of the potential equal and retaining only the Coulomb interaction. The 'extended Thomas-Fermi' type calculation is essentially a hydrodynamic model in which the exponential tail of the electron distribution in the vacuum is also accounted for. Finally the table gives the results of calculations where the surface was modelled by a step-like potential or electron density. Here the results are an order of magnitude smaller. The LDA and LDA-RPA calculations are however in rather good agreement for a large frequency regime [14]. A deviation of about 20% in the low frequency limit of the LDA and LDA-RPA shows that LDA seems to overestimate the SH response but the fundamental physics is certainly contained in it. The results of the 'extended Thomas-Fermi' type calculation are also surprisingly close to the LDA calculations. It is therefore obviously much more important to describe in some (may be approximate) way the Coulomb screening of the external electric field in the tail of the electron distribution outside the metal rather than having a good parametrisation for the exchange correlation energy.

An example of a DFT-LDA-type calculation in the electrostatic limit is shown in fig. 1 for a jellium metal with a bulk electron density corresponding to Aluminum. Note especially that the dominant contributions to the linear response as well as to the nonlinear response are located in the exponential fall-off of the electron density into the vacuum. Clearly a surface step profile cannot account for this effect. One expects therefore that adsorbates with an equilibrium distance typically in this region will have a pronounced effect on the second-order response.

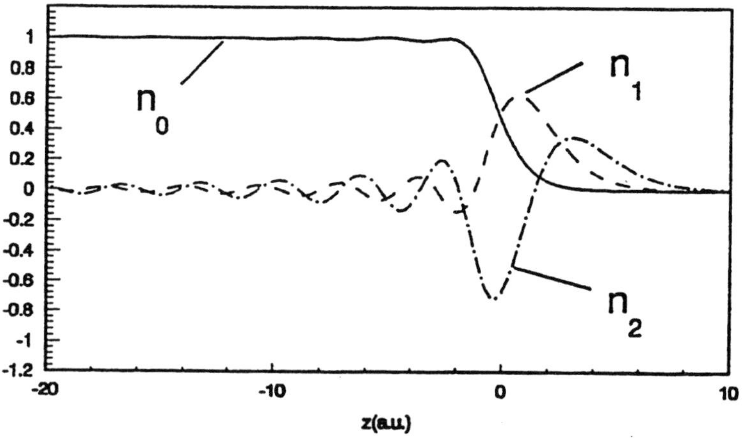

Figure 1: Ground state density $n_0(z)/n_b$, and the linear $n_1(z)$ and nonlinear $n_2(z)$ densities induced at a metal surface by a static uniform electric field (scaled units).

4. INFLUENCE OF ADSORBATES ON SHG FROM SIMPLE METALS

The above discussion has shown that the mechanisms for linear and nonlinear optical response at clean free-electron like metal surfaces are fairly well understood by a jellium model of the metal [6, 14]. Up to now however much less is known about the influence of adsorbates on the response properties. Some specific investigations considered alkali-adlayer/metal systems [17]. In the present calculations we introduce an extension of the atom+jellium model of chemisorption by Lang and Williams [18] to study the influence of a single adatom on the response properties of simple metals. The external static field perpendicular to the surface is taken into account fully selfconsistently. Because of the effective screening properties of a metal the adatom is treated as a localized perturbation and the Kohn-Sham equations of DFT are cast into the Lippmann-Schwinger scattering-theoretic form

$$\psi_{E_0}^{MA}(\vec{r}) = \psi_{E_0}^{M} + \int d^3r' G_{E_0}^{M}(\vec{r},\vec{r}')\, \delta v_{eff,E_0}(\vec{r}')\, \psi_{E_0}^{MA}(\vec{r}'),$$

where $\psi_{E_0}^{M}$ and $\psi_{E_0}^{MA}$ are wavefunctions of the metal and the metal-adsorbate system, respectively, and δv_{eff} denotes the difference of the effective one-electron potentials between the combined metal adatom system and the bare metal. $G_{E_0}^{M}$ is the Green function for the bare metal in the presence of an external static homogenous field E_0 perpendicular to the surface. The resulting densities and related quantities are then written in terms of the bare metal contribution and the change of the system due to the presence of the adatom, i.e. $n^{MA}(\vec{r}) = n^M(z)+\delta n(\vec{r})$. The change in the $a-$parameter [4] resulting from adsorption of independent adatoms is formally expressed in terms of the coverage Θ (number of adatoms/unit area) and the single-atom contribution to the second-order polarizability $\delta\mu_2 = -\int \delta n_2(\vec{r})\vec{r}d^3r$

$$a^{MA}(\Theta) \;=\; a^M \;+\; \delta a(\Theta) \;=\; 4n_b(4\pi e)^2\left(-\int_{-\infty}^{\infty} z n_2^M(z)dz \;+\; \Theta\delta\mu_2\right).$$

A linear dependence of the $a-$parameter on coverage is valid at low coverages as long as adsorbate-adsorbate interactions can be neglected [3].

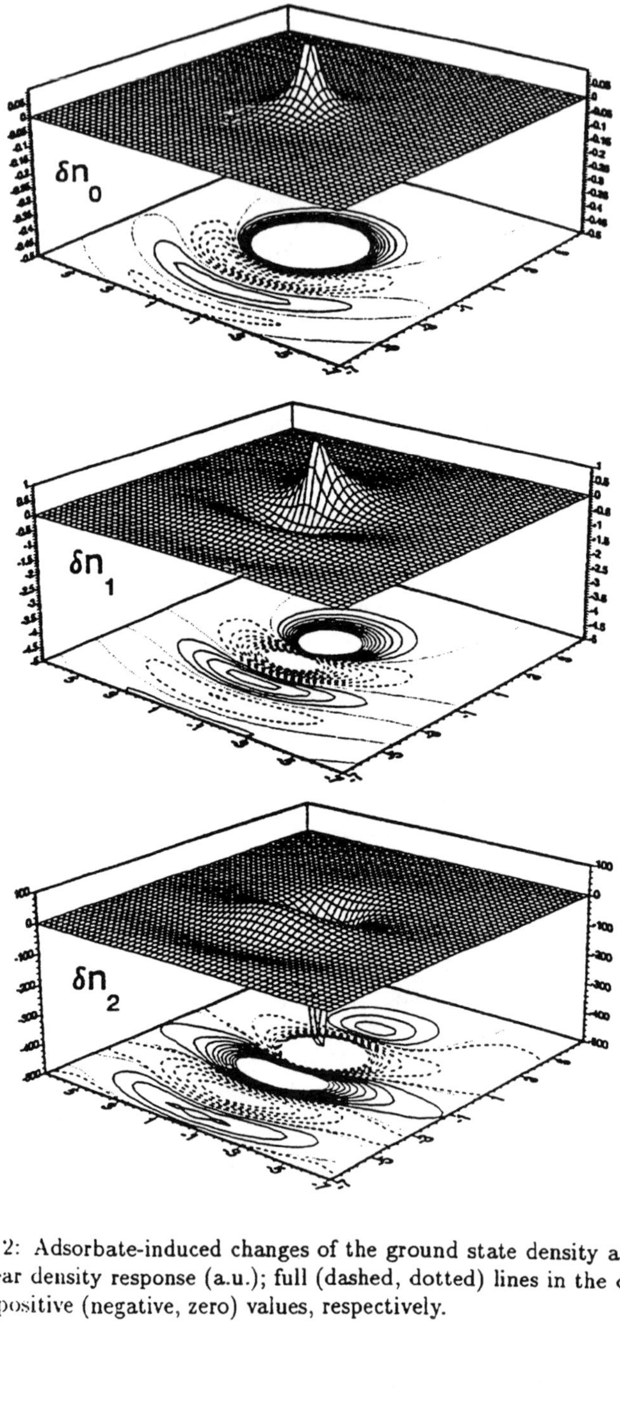

Figure 2: Adsorbate-induced changes of the ground state density and the linear and nonlinear density response (a.u.); full (dashed, dotted) lines in the contour plots represent positive (negative, zero) values, respectively.

As an example for the results one can obtain from such DFT calculations of the $a-$parameter we show in fig. 2 the density response of hydrogen adsorbed on a substrate with $r_s = 2.07\ a_B$ (~Al) in zeroth, first and second order in E_0 in a region of 7 a_B around the adatom. At a coverage of $\Theta = 0.004\ a.u.$ the $a-$parameter changes by $\delta a = 3.3$. This implies a reduction of the SHG-signal by a factor of $I_{2\omega}^{MA}/I_{2\omega}^{M} = 0.85$ for the case of p-polarization and an angle of incidence of 60^0. This reduction takes place because hydrogen adsorbed on an aluminum substrate acts primarily as an acceptor and tends to reduce the polarizability of electrons. Thus the present method allows to investigate in a systematic and semiquantitative manner the relation between electronic response properties and nonlinear optics at adsorbate-covered metal surfaces.

REFERENCES

1. N. Bloembergen, R.K. Chang, S.S. Jha, and C.H. Lee, Phys. Rev. **174**, 1968, 813
2. G.L. Richmond, J.M. Robinson, and V.L. Shannon, Prog. Surf. Sci., **28**, 1987, 1
3. Y.R. Shen, *The principles of nonlinear optics*, New York: Wiley, 1984
4. J. Rudnick, and E.A. Stern, Phys. Rev. B4, 1971, 4274
5. J.E. Sipe, V.C.Y. So, M. Fukui, and G.I. Stegeman, Phys. Rev. **B21**, 1980, 4389
6. M. Weber, and A. Liebsch, Phys. Rev. **B35**, 1987, 7411; **B36**, 1987, 6411
7. S. Lundqvist, and N.H. March (eds.), *Theory of the inhomogenous electron gas*, New York: Plenum, 1983
8. W. Kohn, and L.J. Sham, Phys. Rev. **140A**, 1965, 1133
9. E.P. Wigner, Phys. Rev. **46**, 1934, 1002
10. L. Hedin, and B.I. Lundqvist, J. Phys. C4, 1971, 2064
11. N.D. Lang, and H.D.Kohn, Phys. Rev. B3, 1971, 1215
12. N.D. Lang, and H.D. Kohn, Phys. Rev. **B1**, 1970, 4555
13. A. Zangwil, and P. Soven, Phys. Rev. **A21**, 1980, 1561
14. A. Liebsch, and W.L. Schaich, Phys. Rev. **B40**, 1989, 5401
15. A. Chizmeshya, and E. Zaremba, Phys. Rev. **B37**, 1988, 2805
16. M. Corvi, and W.L. Schaich, Phys. Rev. **B33**, 1986, 3688
17. H. Ishida, and A. Liebsch, Phys. Rev. **B42**, 1990, 5505
18. N.D. Lang, and A.R. Williams, Phys. Rev. **B18**, 1978, 616

Studies in Classical
and Quantum Nonlinear Optics

ISBN 1-56072-168-5

SELECTIVE LINEAR AND NONLINEAR SPECTROSCOPY OF IMPURITY CENTRES IN LOW-TEMPERATURE MATRICES

Peeter Kukk
Institute of Physics, Estonian Academy of Sciences
Riia 142, EE2400 Tartu, Estonia

1. INTRODUCTION

Inhomogeneous broadening of spectral lines in the optical spectra of impurity molecules in low-temperature solids is caused by variation of the electronic and Raman transition frequencies of different molecules due to the microscopic nonequivalence of the host sorrounding every particular molecule [1]. During the last two decades a number of linear and nonlinear steady-state spectroscopic methods for eliminating (suppressing) the static inhomogeneous broadening in optical spectra have been introduced into everyday experimental practice. Among them laser-induced fluorescence line-narrowing (FLN) [2], two-photon spectroscopy [3], hole burning [4], nonlinear optical Kerr effect (RIKE) [5], Rayleigh-type optical mixing spectroscopy [6], coherent anti-Stokes and Stokes Raman scattering (CARS and CSRS) [7], have attracted considerable attention.

The influence of the Doppler broadening on the measured steady-state nonlinear optical spectra in gaseous media has been investigated in several papers [8,9]. In a similar way, assuming that the local changes of different transition frequencies of molecules are positively or negatively correlated, the two-photon spectroscopy [3] and CARS (CSRS) [10] experiments in low-temperature solids have been theoretically interpreted. However, the influence of the local crystal field on the transition frequencies of impurities is not characterizable with one parameter and in every case the correlation of transition frequencies needs additional investigations [11].

Within the present work different theoretical calculations concerning the homogeneous/inhomogeneous broadening of spectral lines in the linear and nonlinear optical spectra were summarised and corresponding experimental investigations were carried out by FLN, CSRS, CARS, CARS excitation and

CARS saturation methods. It is shown that according to theoretical predictions the widths of the spectral lines in these experimentally measured spectra are determined by the inhomogeneous variation of the Raman transition frequencies.

2. MODEL CALCULATIONS

For simple comparision of different theoretical calculations in the present work the impurity molecule in a wide-gap dielectric matrix is approximated by only four energy levels: ground state, first excited electronic state and one vibrational state in both the ground and the excited electronic state. A static inhomogeneous distribution of transition frequencies in this particular case may be described by a three-parameter function $N(\omega_{00'},\Omega,\Omega')$. N represents the number (concentration) of molecules with particular frequencies of a pure electronic transition, $\omega_{00'}$, and of the Raman transitions, Ω and Ω', in the ground and the excited electronic state, respectively. An experimental or a theoretical determining of $N(\omega_{00'},\Omega,\Omega')$ for impurity molecules in solid state samples is rather a complicated problem. Up to date with the help of two spectroscopic methods - laser-induced fluorescence line narrowing and hole burning, only two-parameter distribution functions, $N(\omega_{00'},\Omega)$ and $N(\omega_{00'},\Omega')$, have mainly been investigated. It has been established that in many cases the frequencies of pure electronic and Raman lines in molecular impurity spectra are partly correlated and partly uncorrelated [12]. The former property is disclosed from the dependence of the Raman transition frequency on that of the electronic transition energy and *vice versa*. The uncorrelated part determines the residual widths of vibrational lines, or in other words, the achievable spectral resolution.

Here the theoretical results will be presented, where the local field induced shifts of the electronic and the Raman transition frequencies of impurity molecules were assumed to be as not correlated with each other. This corresponds to the following factorization: $N(\omega_{00'},\Omega,\Omega')=N(\omega_{00'})N(\Omega)N(\Omega')$. The model calculations made with these assumptions allow one to study interesting effects of the inhomogeneous broadening of various transitions but excludes some of the site-selective properties of the linear as well as the nonlinear processes studied earlier [13,14]. Data about the reflection of the homogeneous/ inhomogeneous broadening of the spectral lines in optical spectra measured with the help of more frequently used linear and nonlinear spectroscopic methods is collected in the Table 1 [1,3,4,6,7,15]. Here, H or INH indicate respectively the homogeneous or inhomogeneous (and homogeneous) broadening of the corresponding spectral line. In the latter case, E (R) specifies that the broadening is caused by the inhomogeneous distribution of the electronic (Raman) transition frequency of the molecule. Data in the Table may be summarised in in three main points:

Table 1.

Method	Transition frequency	$\omega_{00'}$	Ω	$\omega_{00'}+\Omega'$/ [Ω']
Absorption, visible		INH-E		INH-E
infrared			INH-R	
Laser-induced fluorescence line narrowing		H	INH-R	
Excitation of flourescence		H		H
Spontaneous Raman scattering			INH-R	
Raman excitation profile		INH-E		INH-E
Hole-burning, visible		H	INH-R	
infrared			H	
Two-photon spectroscopy		INH-E	INH-R	INH-R
Coherent anti-Stokes Raman scattering		H	INH-R	INH-E
		ω_1- ω_2~0	ω_1-ω_2~Ω	ω_1-ω_2~Ω
CARS excitation profile		INH-E		INH-E
CARS saturation		H	INH-R	[INH-R]
Raman-induced Kerr effect		H	INH-R	H
		ω_1-ω_2~0	ω_1-ω_2~Ω	ω_1-ω_2~0

1. When $\omega_1 = \omega_{00'}$ and the luminescence spectrum is detected at the same frequency, or in nonlinear spectroscopy, ω_2 is scanned over w_1 , then, due to the selective detection of population changes induced by narrow-line laser excitation, a narrow line with the width determined by the homogeneous width of the spectral line corresponding to the pure electronic transition of the impurity molecule is observable in the spectrum.
2. When the frequency of the exciting laser ω_1 is fixed on the frequency of the pure electronic transition $\omega_{00'}$ and the luminescence (or Raman) spectrum is measured with high selectivity, or in nonlinear spectroscopy, the frequency of the other laser ω_2 (ω_1-ω_2=Ω or Ω') is scanned, then the spectral resolution is determined by the sum of the homogeneous and inhomogeneous widths of the vibrational lines.
3. The widths of the excitation profiles (characterize the intensity dependence of the vibrational lines on the exciting frequency) are determined by the sum of the widths of the homogeneous and inhomogeneous distributions of the electronic transition frequencies.

A typical value of the line broadening, caused by inhomogeneous distribution of the Raman transition frequencies, is in the range of 0.1-4.0 cm^{-1} , which is of the same order as the homogeneous linewidth of the Raman lines. This pecularity allows one to measure relatively narrow-line Raman spectra of impurity molecules in the ground (but not in the excited) electronic state with almost the same resolution with the help of nonresonant or resonant spontaneous Raman, CARS, CSRS, RIKE or FLN techniques. High-resolution vibrational spectra of the excited electronic state can be measured with the help of hole-burning spectroscopy or by measuring spontaneous Raman, CARS, CSRS, or RIKE spectra generated from the excited electronic state of the molecule populated by an intensive laser beam .

3. EXPERIMENTAL RESULTS AND THEIR INTERPRETATION

Two different matrices for tetracene molecules have been used: crystalline anthracene and glassy polystyrene. The inhomogeneous width of the pure electronic line of the lowest-frequency singlet-singlet transition of tetracene at 4.2 K is 2-4 cm^{-1} or 300-340 cm^{-1} in the anthracene or polystyrene respectively. The optical lifetime-determined homogeneous width of 0-0' line of tetracene impurity at low temperatures is of the order of 0.001 cm^{-1}.

A FLN and a resonant CARS spectrum of tetracene in anthracene in the 1150-1250 cm^{-1} spectral region at 1.6 K are presented in Fig.1. There is no essential difference between the linear and the nonlinear spectrum with respect to the spectral resolution and the signal to noise ratio in this sample. Relative intensities of different spectral lines in the two spectra are different due to the differences in the transition probabilities in the linear and the nonlinear spectrum.

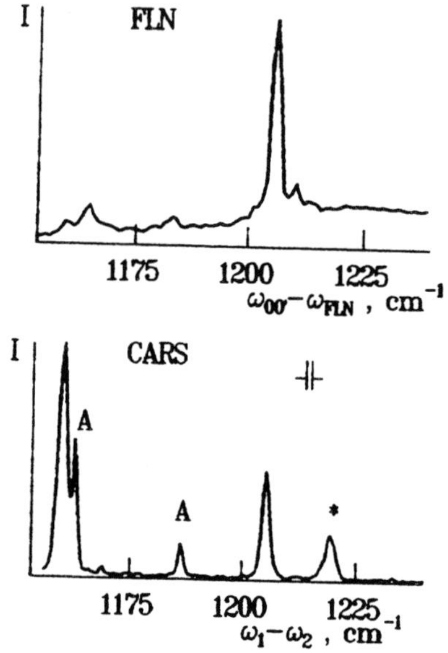

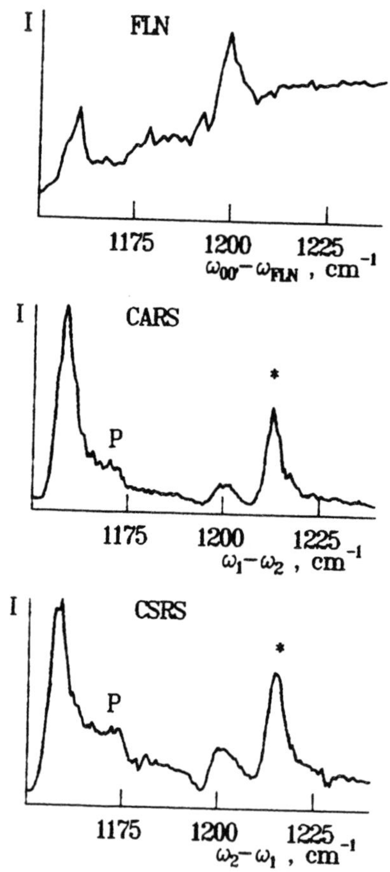

Fig.1. FLN and CARS spectra of tetracene in an anthracene monocrystal at 1.6 K. The asterisk (*) indicates the lines corresponding to the vibrational transitions in the excited electronic state of tetracene.

Fig.2. FLN, CARS and CSRS spectra of tetracene in polystyrene at 4.2 K. Letter (P) indicates the line corresponding to the nonresonant CARS of polystyrene vibrations.

Through the nonresonant CARS process, also anthracene matrix vibrations (1163 and 1186 cm^{-1} , indicated by the letter A) give their contribution to the CARS spectrum.. Two vibrations with almost equal frequencies give their contribution to the 1160 cm^{-1} CARS line: one in the ground and the other one, in the excited electronic state of tetracene. A CARS line at ω_1-ω_2= 1206 cm^{-1} corresponds to the ground-state vibration of tetracene. The line at 1220 cm^{-1} consists of two parts: one is due to the CARS on a vibration of Ω'=1220cm^{-1} in the populated excited electronic state and the other, due to the weak resonance at the CARS frequency [16]. The lines, corresponding to the Raman transitions in the excited electronic state, appear in the CARS spectrum due to the saturation of the electronic transition with intensive excitation and generating a CARS process from the populated excited electronic state. The last line is absent in the FLN spectra, as the spectral lines at vibrational frequencies of the excited electronic state are not reflected in the ordinary luminescence spectra.

LIF, resonant CARS and CSRS spectra of tetracene in polystyrene are compared in Fig.2. The widths of the spectral lines in all the three spectra exceed the spectral resolution by several times. Within the experimental error, the linewidths in the CARS and CSRS spectra do not depend on temperature (4.2 - 70 K). This indicates that at low temperature the observed lines are broadened due to the inhomogeneous distribution of the Raman transition frequencies. The ratio of the electronic and the Raman inhomogeneous broadening, $\Delta\omega_{00'}/\Delta\Omega$, estimated from these spectra, is of the order of 60. CARS and CSRS spectra are quite similar and the lines corresponding to vibrational transitions in the ground (1200 cm^{-1}) and the excited (1215cm^{-1}) electronic state have almost equal widths in both spectra. This agrees with the theoretical expectations cited above, but disagrees with the model, where a strict correlation between electronic and vibrational frequencies was assumed [10]. In the latter case the CARS and CSRS linewidths corresponding to the excited electronic state vibrations should be equal, but in the case of the ground electronic state, different. The parameters determined from the experiment give the width of the CSRS (1200 cm^{-1}) line, $\delta<1.5$ cm^{-1} , which is much less than the 4 - 5 cm^{-1} observed in the experiment.

4. CONCLUSION

In this paper, different steady-state linear and nonlinear spectra of molecular impurities at low temperatures have been compared. The experimental investigations allow a conclusion that the residual widths of vibrational lines are determined by the sum of homogeneous and inhomogeneous broadening of the Raman lines. This experimental result coincides with the theoretical assumpion that for the impurity molecules in low-temperature hosts the inhomogeneous distributions of the frequencies of the electronic and the vibrational transition may be regarded as uncorrelated with each other.

REFERENCES

1.*Zero-phonon Lines and Spectral Hole Burning in Spectroscopy and Photochemistry.* O.Sild and K.Haller (Ed.), Berlin, Heidelberg: Springer Verlag, 1988, 183 pp.

2.A.Szabo, *"Laser-induced fluorescence line narrowing in ruby"*, Phys. Rev. Lett., Vol.25, No.4, 1970, pp.924-926.

3.B.Dick and R.M.Hochstrasser, *"Non-linear spectroscopic studies using sum- and difference frequency generation"*, Chem. Phys., Vol.91, No.1, 1984, pp.1-11.

4.*Persistent Spectral Hole-Burning: Science and Application.* W.E.Moerner (Ed.), Berlin, Heidelberg: Springer Verlag, 1988, 315 pp.

5.J.H.Lee, J.J.Song M.A.F.Scarparo, and M.D.Levenson, *"Coherent population oscillations and hole burning observed in Sm^{2+} CaF_2 using polarization spectroscopy"*, Optics Lett., Vol.5, No.4, 1980, pp.196-198.

6.T.Yajima and H.Souma, *"Stydy of ultra-fast relaxation processes by resonant Rayleigh-type optical mixing. I Theory"*, Phys. Rev., Vol.A17, No.1, 1978, pp.309-323.

7.P.L.Decola, J.R.Andrews, R.M.Hochstrasser, and H.P.Tromsdorff, *"Simultaneous narrow-line Raman spectra of ground and excited electronic states"*, J. Chem. Phys., Vol.73, No.9, 1980, pp.4695-4696.

8.S.A.J.Druet, J.-P.E.Taran, and Ch.J.Borde, *"Line shape and Doppler broadening in resonant CARS and related nonlinear processes through a diagrammatic approach"* J. de Physique, Vol.40, No.9, 1979, pp.819-840.

9.J.-L.Oudar and Y.R.Shen, *"Nonlinear spectroscopy by multiresonant four-wave mixing"*, Phys. Rev., Vol.A22, No.3, 1980, pp.1141-1158.

10.T.-C.Chang, C.K.Johnson, and G.J.Small, *"Multiresonant four-wave miwing spectroscopy of pentacene in naphthalene"*,J. Phys. Chem., Vol.89, 1985, pp.2984-2992.

11.J.Kikas and M.Ratsep, *"Correlation of transition frequencies in impurity spectra of inhomogeneous solids"*, phys. stat. sol. (b), Vol.112, No.2, 1982, pp.409-415.

12.L.A.Rebane, A.A.Gorokhovskii, and J.V.Kikas, *"Low temperature spectroscopy of organic molecules in solids by photochemical hole burning"*, Appl. Phys., Vol. B29, No.2, 1982, pp.235-250.

13.A.P.Marchetti, W.C.McColgin, and J.H.Eberly, *"Inhomogeneous broadening and excited-vibrational-state lifetimes in low-temperature organic mixed crystals"*, Phys.Rev.Lett., Vol. 35, No.4, 1975, pp.387-390.

14.J.K.Steehler and J.C.Wright, *"Multiply enhanced site selective CARS spectroscopy"*, J. Chem. Phys., Vol.83, No.7, 1985, pp.3188-3199.

15.P.Kukk and A.Freiberg, *"Steady-state third-order nonlinear spectroscopy in inhomogeneous media"*, Proc. Estonian Acad. Sci. Phys. Math., Vol.30, No.4, 1981, pp.357-362.

16.P.Kukk and A.Freiberg, *"Resonant coherent anti-Stokes and Stokes Raman scattering spectra of molecular impurities at low temperatures"*, J. Chem. Phys., Vol.98, No.1, 1992, pp.118-127.

Studies in Classical
and Quantum Nonlinear Optics

ISBN 1-56072-168-5

OPTICAL SECOND-HARMONIC GENERATION FROM NIOBIUM QUANTUM WELLS

A. Liu and O. Keller
Institute of Physics, University of Aalborg
Pontoppidanstraede 103, DK-9220 Aalborg Øst, Denmark

1. INTRODUCTION

Optical second-harmonic generation (SHG) has recently attracted particular interest as a tool capable of yielding valuable information about the structure and electronic excitations of surface systems.The high surface sensitivity of this purely optical technique has been demonstrated by the recent experiments on Cs-covered Cu(111) [1] and Rb-covered Ag(110) surfaces [2]. Due to the alkali-metal adsorption, an oscillatory increase of the SHG intensity was observed when increasing the alkali-metal coverage from zero up to a few of monolayers. With a further increase of the coverage, the SHG intensity decreases oscillatorily and approaches the background value which results mainly from the corresponding bulk-metal contributions [1,2]. To obtain a theoretical understanding of the SHG enhancement and the Friedel-type oscillations as a function of the alkali-metal coverage, the so-called time-dependent density-functional approach was used [2]. No satisfactory explanation was obtained. Since the spectra of these metal on metal systems possess discrete quantum states localized near the surface [3,4], it seems important to evaluate the intersubband contributions to the second-harmonic signal in order to better understand the physical mechanisms behind the experimental findings of Refs.[1] and [2].

In this work we discuss optical SHG from a quantum well with special emphasis on the significance of the intersubband transitions. Within the framework of the random-phase-approximation the nonlocal linear and nonlinear conductivity response tensors are analyzed. In the case where only two bound states of the quantum well are taken into account in the analysis, we obtain the exact solutions to the first-harmonic (FH) and second-harmonic (SH) local fields by using the well known Green's function technique [5]. For a niobium quantum well deposited on a quartz substrate, we calculated the efficiency of the optical SHG. Our numerical results show that the SH energy reflection spectrum exhibits three resonance peaks in the vicinity of the energy separation between the two sub-levels. The two resonances stem from the FH and SH local-field enhancement inside the quantum well and the third is due to the level resonance of the nonlinear conductivity response.

2. SELFCONSISTENT CALCULATIONS OF THE FH AND SH LOCAL FIELDS

Let us consider a quantum well system consisting of a superthin niobium film of thickness d deposited on a quartz substrate, and let us assume that the system exhibits infinitesimal translational invariance parallel to the plane of the film. In a Cartesian xyz-coordinate system, the vacuum/film interface coincides with the xy-plane, and the boundaries of the film are located at $z = 0$ and $z = d$. The quartz substrate occupies the space $z > d$ (see Fig. 1(a)).

Due to the assumed translational invariance of our system parallel to the well (xy-plane) all vector-field components (F) appearing in the analysis have the generic

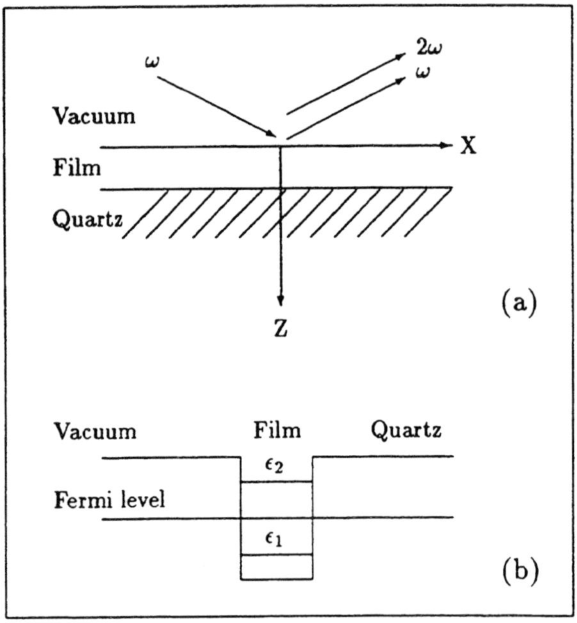

Fig. 1. (a) Schematic illustration showing the vacuum-niobium film-quartz system and the Cartesian xyz-coordinate. It is also depicted that the incident light of frequency ω generates a reflected SH (2ω) signal from the system. (b) Schematic illustration of a niobium quantum well with two bound states (energies ϵ_2 and ϵ_1), with indication also of the Fermi level of the system.

form

$$F(\vec{r},t) = F(z;\vec{q_{\|}},\omega)e^{i(\vec{q_{\|}}\cdot\vec{r}-\omega t)} , \tag{1}$$

where ω and $\vec{q_{\|}} = (q_{\|},0,0)$ denote the angular frequency and the wave vector projection along the quantum well (QW) of the incident electromagnetic wave. Space points are denoted by $\vec{r}$ and t is the time. Using the electromagnetic scattering-theory formalism [6], the local electric field $\vec{E}_1(z;q_{\|},\omega)$ at the fundamental frequency (ω) has to be determined from the integral equation

$$\vec{E}_1(z;q_{\|},\omega) = \vec{E}_0(z;q_{\|},\omega) - i\mu_0\omega \int\int_{QW} \overleftrightarrow{G}(z,z';q_{\|},\omega)\cdot \overleftrightarrow{\sigma}(z',z'';q_{\|},\omega) \cdot\vec{E}_1(z'';q_{\|},\omega)dz''dz' . \tag{2}$$

In the equation above $\vec{E}_0$ is the so-called external field consisting of the incident field (pump field) plus the field reflected from the vacuum/substrate surface in the absence of the quantum well. The tensor $\overleftrightarrow{G}$ is the electromagnetic propagator (known as the Green's function) of the vacuum/ substrate system [5]. Finally, the tensor $\overleftrightarrow{\sigma}$ is the appropriate linear nonlocal conductivity response function of the quantum well.

Within the framework of the parametric approximation [7], the local electric field $\vec{E}_2(z;2q_{\|},2\omega)$ at the SH frequency (2ω) is governed by

$$\vec{E}_2(z;2q_{\|},2\omega) = \vec{E}_B(z;2q_{\|},2\omega) - 2i\mu_0\omega \int\int_{QW} \overleftrightarrow{G}(z,z';2q_{\|},2\omega) \cdot \overleftrightarrow{\sigma}(z',z'';2q_{\|},2\omega)\cdot \vec{E}_2(z'';2q_{\|},2\omega)dz''dz' , \tag{3}$$

where

$$\vec{E}_B(z;2q_{\|},2\omega) = -2i\mu_0\omega \int\int\int_{QW} \overleftrightarrow{G}(z,z';2q_{\|},2\omega)\cdot \overleftrightarrow{\Sigma}(z',z'',z''';2q_{\|},2\omega) : \vec{E}_1(z'';q_{\|},\omega)\vec{E}_1(z''';q_{\|},\omega)dz'''dz''dz' , \tag{4}$$

is the so-called background field for the analysis of SHG. The background field is induced by the FH local field $\vec{E}_1$ inside the quantum well. In Eq.(4), $\overleftrightarrow{\Sigma}$ is the second-order nonlinear nonlocal conductivity response function of the quantum well.

In the present work we restrict our attention to the study of the optical paramagnetic response of the quantum well, and analyze the case where the quantum well has only two bound states, one above and the other below the Fermi level (see Fig. 1(b)). Within the random-phase-approximation and in the long-wavelength ($q_{\|} \rightarrow 0$) and low-temperature ($T \rightarrow 0$) limits, the nonlocal linear conductivity tensor $\overleftrightarrow{\sigma}$ takes a diagonalized form, and the diagonal elements are given by [8]

$$\sigma_{xx}(z,z';\omega) = \sigma_{yy}(z,z';\omega) = \frac{ie^2}{\pi\hbar^2\omega}\frac{\epsilon_{21}(\epsilon_F-\epsilon_1)^2}{[\hbar(\omega+i/\tau)]^2-\epsilon_{21}^2}\phi(z)\phi(z') , \tag{5}$$

$$\sigma_{zz}(z,z';\omega) = \frac{ie^2}{2\pi m\omega}\frac{\epsilon_{21}(\epsilon_F-\epsilon_1)}{[\hbar(\omega+i/\tau)]^2-\epsilon_{21}^2}\Phi(z)\Phi(z') , \tag{6}$$

with

$$\phi(u) = \Psi_1(u)\Psi_2(u), \qquad u = z, z' \tag{7}$$

$$\Phi(u) = \Psi_1(u)\frac{d\Psi_2(u)}{du} - \Psi_2(u)\frac{d\Psi_1(u)}{du} \quad . \qquad u = z, z' \tag{8}$$

In the equations above $\Psi_n(z)(n = 1, 2)$ denote the (real) wave functions of the two energy eigenstates, ϵ_{21} is the energy separation between the upper (ϵ_2) and lower level (ϵ_1), ϵ_F is the Fermi energy of the system, τ is the relaxation time of the electrons, and finally e and m are the charge and mass of the electrons, respectively.

Within the so-called electric dipole approximation [7] the nonvanishing element of the nonlinear conductivity tensor $\overleftrightarrow{\Sigma}$ is given by

$$\begin{aligned}\Sigma_{zzz}(z, z', z''; 2\omega) &= \frac{ie^3\epsilon_{21}}{\pi\hbar^2\omega}(\epsilon_F - \epsilon_1)\frac{P_{1\to 2}\Psi_1(z)\Psi_2(z)}{[\hbar(\omega + i/\tau)]^2 - \epsilon_{21}^2}\delta(z' - z)\delta(z'' - z') \\ &\equiv \Sigma_{zzz}(z; 2\omega)\delta(z' - z)\delta(z'' - z') \ ,\end{aligned} \tag{9}$$

where

$$P_{1\to 2} = \int_{QW} \Psi_1(z)z\Psi_2(z)dz \tag{10}$$

is the matrix element associated with an electronic dipole transition from the state ϵ_1 to the state ϵ_2. The background field $\vec{E}_B$ thus takes the following form:

$$\vec{E}_B(z) = -2i\mu_0\omega \int_{QW} \overleftrightarrow{G}(z, z') \cdot \Sigma_{zzz}(z')[E_{1x}(z')E_{1z}(z')\vec{e}_x + E_{1z}^2(z')\vec{e}_z]dz', \tag{11}$$

where $\vec{e}_x$ and $\vec{e}_z$ are unit vectors along the x- and z-directions. For brevity, we have in Eq.(11) left out the superfluous reference to ω from our notation.

Due to the fact that the z- and z'-dependences of the linear nonlocal conductivity tensor takes a separable form, it is realized that exact solutions for the local electric fields $\vec{E}_1$ and $\vec{E}_2$ can be obtained. Since the local fields for the FH and SH frequencies satisfy the same type of selfconsistent integral equations, this enables us to obtain easily the SH local field once the FH local field has been calculated. For p-polarized light, one can obtain, by combining Eq.(2) with Eqs.(5) and (6), the following results for the FH local field:

$$E_x(z) = E_{0x}(z) + \beta_x F_{xx}(z) + \beta_z F_{xz}(z) \ , \tag{12}$$

$$E_z(z) = E_{0z}(z) + \beta_x F_{zx}(z) + \beta_z F_{zz}(z) \ , \tag{13}$$

where

$$F_{xx}(z) = \frac{\mu_0 e^2}{\pi\hbar^2}\frac{\epsilon_{21}(\epsilon_F - \epsilon_1)^2}{[\hbar(\omega + i/\tau)]^2 - \epsilon_{21}^2}\int_{QW} G_{xx}(z, z')\phi(z')dz' \ , \tag{14}$$

$$F_{xz}(z) = \frac{\mu_0 e^2}{2\pi m}\frac{\epsilon_{21}(\epsilon_F - \epsilon_1)}{[\hbar(\omega + i/\tau)]^2 - \epsilon_{21}^2}\int_{QW} G_{xz}(z, z')\Phi(z')dz' \ , \tag{15}$$

$$F_{zx}(z) = \frac{\mu_0 e^2}{\pi\hbar^2} \frac{\epsilon_{21}(\epsilon_F - \epsilon_1)^2}{[\hbar(\omega + i/\tau)]^2 - \epsilon_{21}^2} \int_{QW} G_{zx}(z, z')\phi(z')dz' \ , \tag{16}$$

$$F_{zz}(z) = \frac{\mu_0 e^2}{2\pi m} \frac{\epsilon_{21}(\epsilon_F - \epsilon_1)}{[\hbar(\omega + i/\tau)]^2 - \epsilon_{21}^2} \int_{QW} G_{zz}(z, z')\Phi(z')dz' \ . \tag{17}$$

The two constants β_x and β_z which are related to the external field $\vec{E}_0$ are given by

$$\beta_x = \frac{[\int_{QW} E_{0x}\phi dz][1 - \int_{QW} F_{zz}\Phi dz] + [\int_{QW} E_{0z}\Phi dz][\int_{QW} F_{xz}\phi dz]}{[1 - \int_{QW} F_{xx}\phi dz][1 - \int_{QW} F_{zz}\Phi dz] - [\int_{QW} F_{xz}\phi dz][\int_{QW} F_{zx}\Phi dz]} \ , \tag{18}$$

$$\beta_z = \frac{[\int_{QW} E_{0z}\Phi dz][1 - \int_{QW} F_{xx}\phi dz] + [\int_{QW} E_{0x}\phi dz][\int_{QW} F_{zx}\Phi dz]}{[1 - \int_{QW} F_{xx}\phi dz][1 - \int_{QW} F_{zz}\Phi dz] - [\int_{QW} F_{xz}\phi dz][\int_{QW} F_{zx}\Phi dz]} \ . \tag{19}$$

By replacing the external field $\vec{E}_0$ by the background field $\vec{E}_B$ in Eqs.(12), (13), (18) and (19) and by substituting $\vec{q}$ and ω with $2\vec{q}$ and 2ω in Eqs.(14)-(17), one can readily obtain the SH local field inside the quantum well. The detailed calculations of the local-field distributions across the quantum well will be presented in our forthcoming paper [8].

3. NUMERICAL CALCULATIONS OF THE SH ENERGY REFLECTION COEFFICIENT

To calculate the efficiency of optical SHG from a quantum well we take as a starting point Eq.(3). By letting the observation point be located outside the quantum well on the vacuum side ($z < 0$), and by inserting into Eqs.(3) and (4) the appropriate expressions for the electromagnetic propagator [5], it is readily realized that the second-harmonic field in vacuum takes the form $\vec{E}_2^{out} exp(-2iq_\perp^0 z)$, where $q_\perp^0$ is the perpendicular component of the wave vector of the incident light. The amplitude $\vec{E}_2^{out}$ essentially is determined by the integral of the SH local field and the background field over the quantum well. The SH energy reflection coefficient is defined as

$$R_{2p} = \frac{2}{\epsilon_0 c_0} \left| \frac{E_2^{out}}{(E_1^{in})^2} \right|^2 \ , \tag{20}$$

where E_1^{in} denotes the amplitude of the incident field, and ϵ_0 and c_0 are the vacuum permittivity and the speed of light in vacuum, respectively.

In our numerical calculations we have used the parameters characterizing a symmetric niobium quantum well deposited on a quartz substrate. The energy eigenvalues of the bound states of the quantum well are calculated by using a finite-barrier model [9]. For a niobium quantum well of thickness $d = 10$Å, the energy separation obtained between the two sub-bands near the Fermi level (one just above and the other just below) is $\epsilon_{21} = 2.435eV$, and the energy difference between the Fermi level and the lower level ($\epsilon_F - \epsilon_1$) is $0.7eV$. In Fig. 2 is shown the SH energy reflection coefficient R_{2p} as a function of the normalized photon energy for the fundamental frequency. An angle of incidence of 70° was used. In our calculations two different relaxation times, namely, $\tau = 3.3 \times 10^{-15}s$ and $1.3 \times 10^{-14}s$ were used. It appears from Fig. 2 that the

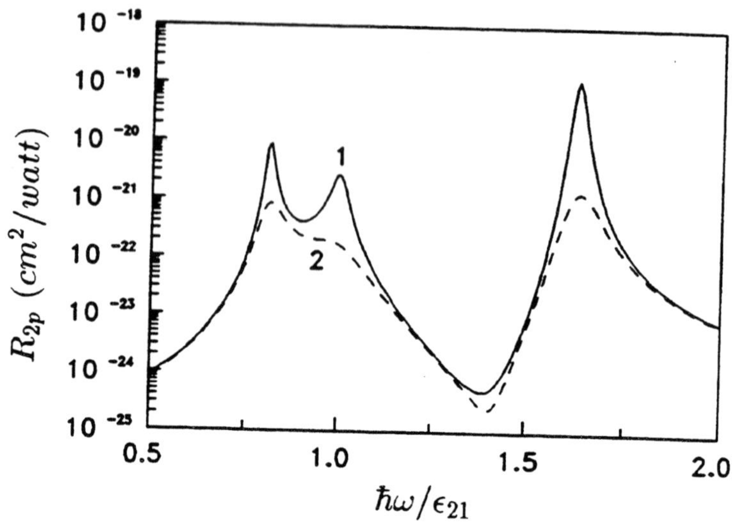

Fig. 2. SH energy reflection coefficient R_{2p} as a function of the normalized photon energy of the fundamental frequency, $\hbar\omega/\epsilon_{21}$, for two different relaxation times: $\tau = 1.3 \times 10^{-14}s$ (curve 1) and $3.3 \times 10^{-15}s$ (curve 2).

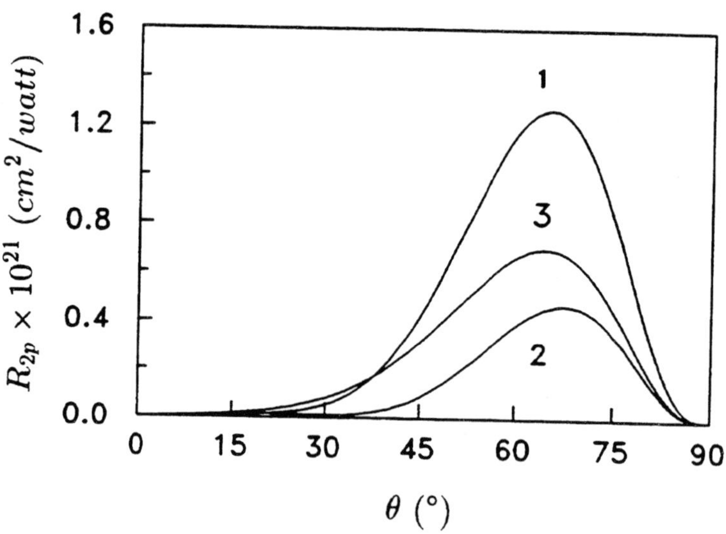

Fig. 3. SH energy reflection coefficient R_{2p} as a function of the angle of incidence θ for three different frequencies: $\hbar\omega = 1.65eV$ (curve 1), $0.85eV$ (curve 2) and $0.80eV$ (curve 3).

SH energy reflection coefficient spectrum exhibits three resonance peaks in the vicinity of the energy separation between the two levels. The one located around the electronic energy spacing ϵ_{21} is easily recognized as the result of an enhancement of the second-order conductivity (see Eq.(9)). The enhancement of the linear conductivity does not lead to a resonance of the FH local field in the presence case [8] and thus does not result in a resonance in the SH energy reflection spectrum. The remaining two resonance peaks stem from the local-field resonances. The one appearing at the higher frequency is due to the FH local-field resonance, and the one located at the lower frequency is due to the SH local-field resonance. By a comparison of curves 1 and 2 in Fig. 2, one can see that the magnitudes of the three resonance peaks are strongly dependent upon the relaxation times of the electrons. The peak values decrease with increasing electron collision frequency. In the higher damping case (see curve 2 in Fig. 2), the resonance peak near the electronic energy separation almost disappears, but the two other resonances are still clearly seen.

In Fig. 3 is presented the dependence of the SH energy reflection coefficient on the angle of incidence for three different photon energies, i.e. $\hbar\omega = 0.80\epsilon_{21}, 0.85\epsilon_{21}$ and $1.65\epsilon_{21}$. A relaxation time of $3.3 \times 10^{-15} s$ was employed. One can see from Fig. 3 that the SH reflection coefficient is very small at near-normal incidence and at the large angles of incidence near 90°. A maximum of the energy reflection coefficient occurs around 70°. It appears from Fig. 3 that the change of the fundamental frequency gives only a slight effect on the peak location in the angular spectrum.

4. SUMMARY

In the present paper we have discussed the second-harmonic generation of light from a single quantum well within the framework of the random-phase-approximation. Limiting ourselves to the study of the electric dipole response from the quantum well and taking into account two sub-bands only, i.e. one above and the other below the Fermi level, we calculate, for the p-polarization, the conversion factor for the reflected SHG from a niobium quantum well deposited on a quartz substrate. Our numerical results show that the SH energy reflection spectrum exhibits three resonance peaks in the vicinity of the energy separation between the two levels. These stem from the local-field resonances at the first- and second-harmonic frequencies inside the quantum well and from the nonlinear response-function resonance.

REFERENCES

1. S. Å. Lindgren, and L. Walldén, "Interband transitions, surface-barrier photoabsorption, and charge-screening waves observed via optical second-harmonic generation of Cs-covered Cu(111)", *Phys. Rev. B*, Vol. **45**, No. 11, 1992, pp.6345-6347.

2. K. J. Song, D. Heskett, H. L. Dai, A. Liebsch, and E. W. Plummer, "Dynamical screening at a metal surface probed by second-harmonic generation", *Phys. Rev. Lett.* , Vol. **61**, No. 12, 1988, pp.1380-1383.

3. M. A. Mueller, T. Miller, and T.-C. Chiang, "Determination of the bulk band structure of Ag in Ag/Cu(111) quantum-well systems", *Phys. Rev. B*, Vol. **41**, No. 8, 1990, pp.5214-5220.

4. S. Å. Lindgren, and L. Walldén, "Electron-energy-band determination by photoemission from overlayer states", *Phys. Rev. Lett.*, Vol. **61**, No. 25, 1988, pp.2894-2897.

5. O. Keller, "Tensor-product structure of a new electromagnetic propagator for nonlocal surface optics of metals", *Phys. Rev. B*, Vol. **37**, No. 18, 1988, pp.10588-10606.

6. O. Keller, "Nonlocal electrodynamics of an atom in front of a normal or phase-conjugating mirror", *J. Quantum Nonlinear Phenomena*, to be published.

7. O. Keller, "Random-phase-approximation study of the response function describing optical second-harmonic generation from a metal selvedge", *Phys. Rev. B*, Vol. **33**, No. 2, 1986, pp.990-1009.

8. O. Keller, and A. Liu, to be published.

9. O. Keller, and A. Liu, "Optical diamagnetic response of a niobium quantum well", *Phys. Lett. A*, Vol. **167**, No. 3, 1992, pp.301-305.

Studies in Classical
and Quantum Nonlinear Optics

ISBN 1-56072-168-5

LINEAR AND NONLINEAR OPTICAL EFFECTS OF GUEST MOLECULES IN MOLECULAR SIEVE CRYSTALS

F. Marlow[1] *and J.Caro*
Zentrum für Heterogene Katalyse,
Rudower Chaussee 5, D-O-1199 Berlin, Germany

ABSTRACT

Molecular sieves are widely used in shape-selective catalysis and adsorption. Increasingly, "new applications" of modified zeolite-like molecular sieves as advanced materials are reported, e.g. in optical data storage and in nonlinear optics, as semi-conductors or sensors. A promising way to create new material properties is the modification of molecular sieves by the adsorption of organic guest molecules. Organic molecules can be incorporated in a highly ordered manner into the pore system of molecular sieves. For molecular sieves with anisotropic channel geometry, this leads to anisotropic optical properties. Furthermore, dipolar organic guests form directed dipole chains inside the molecular sieve and, therefore, the guest-host complex exhibits strong second order nonlinearities. Modified molecular sieves are a new class of nonlinear optical materials tunable by their guest molecules.

---

[1] joined project with the Bundesanstalt für Materialforschung und -prüfung, Dept. 10.14

1. INTRODUCTION

1.1 Why Molecular Sieves?

The interconnected pores and cavities of zeolite-like molecular sieves form a highly regular micropore system with a void volume of up to 0.3 $cm^3 g^{-1}$. The access to the pore system is controlled by channels with diameters between 3 and 13 Å. More than two hundred different synthetic and natural molecular sieve structures are known. Due to their chemical composition, mainly as alumophosphates or alumosilicates, zeolite-like molecular sieves show no resonances in the visible region. Therefore, one can use them as an inert matrix and incorporate various optically interesting guests, e.g. molecules, clusters or polymers, into their pore system. The resulting guest-host structure is determined by the pore geometry. Consequently, this composite is highly ordered. Ordered systems have promising properties for optical applications, especially in nonlinear optics.

In 1988 second harmonic generation (SHG) of molecular sieves loaded with organic molecules was found [1] indicating that adsorbed molecules do not only form an ordered array, but in addition there is a breaking of inversion symmetry in this system. Otherwise the SHG would be forbidden. Consequently, molecular sieve based guest-host complexes exhibit all the properties one needs to design new nonlinear optical materials: Tunable electronic resonances (by the guest molecules) and an ordered and directed arrangement of these chromophors (variable by the molecular sieve).

Although there is a variety of nonlinear optical materials, non of them allows to build up optically integrated circuits comparable to electronically ones [2]. So, there is a need to find new materials for optical information processing. Nanometer-scaled molecular sieve based composites are a good candidate, let us try to examine his aptitudes.

1.2 Guests in Molecular Sieves

In the last few years a variety of techniques ranging from the inclusion of bulky dye molecules during the molecular sieve crystallization [3] to post-synthesis modifications like intracrystalline polymerization [4] or the building up of ship-in-the-bottle like complexes in zeolite cavities [5] have been reported. In our case, we have introduced the dye molecules into the molecular sieve pores via a gas phase loading process. The evacuated molecular sieve crystals were brought into contact with para-nitroaniline (PNA) (6h, 150°C) and dimethylaminobenzonitril (DMABN) (6h, 100°C). The molecular sieves have been synthesized hydrothermally at 150°C-180°C: AlPO-5 [6] with pores of 7.3 Å and ZSM-5 [7] with 5.5 Å.

2. LINEAR OPTICS

2.1 Anisotropic Absorption of Light

Zeolite crystals loaded with PNA show the same bright yellow color as polycrystalline PNA powder. This color results from the absorption of natural light by PNA in the blue region arising from the near UV resonances. However, if the AlPO-5 crystals are observed in transmission of polarized light, the crystals oriented with their length axes in the direction of the electric field oscillation are dark and the others are light (Fig. 1a). Turning the plane of polarization, other crystals become dark (Fig. 1b). Obviously, the PNA molecules preferably absorb light oscillating parallel to their molecular axes. The transition dipole moment of the lowest electronic transition in PNA lies along this axis. Consequently, if the plane of polarization is parallel to the length direction of the hexagonal columns, the PNA molecules aligned by the one-dimensional pores - extending in the crystal's length direction - absorb light and the crystal looks dark.
Fig. 1 shows also that the two endings of the crystals do not exhibit a pronounced anisotropy in light absorption indicating a lower PNA concentration near the endings of the channels. Recently, the anisotropy of light absorption has been verified for PNA in ZSM-5 and in VPI-5 [8].

Figure 1 Microscopic picture of AlPO-5 crystals loaded with PNA in transmission of polarized light [8]. The different polarization planes are indicated by an arrow. The correspondence between crystal habit and channel arrangement is given schematically.

2.2 Birefringence and Fluorescence

The anisotropy of the linear optical properties has been verified in studies of the index of refraction [8] and in fluorescence measurements. The birefringence of the crystals depends strongly on the adsorbed molecules. Typical values for the difference in the index of refraction $\Delta n = n^e - n^o$ are $\Delta n=0.08$ for PNA adsorption, $\Delta n=0.03$ for DMABN and $\Delta n<0.001$ for water. These values refer to AlPO-5 at 633 nm.
Fluorescence of guest molecules in molecular sieves is a sensitive tool to detect the orientation of the molecules inside the pores and their interactions with the surrounding pore surfaces. For DMABN in AlPO-5, the fluorescence light is strongly polarized. The fluorescence spectra are dependent on co-adsorbed water and interactions with the zeolite matrix.

2.3 Raman Scattering [9]

Which are the structures formed by the molecules inside the pores? Are they solid-, liquid- or gas-like? How perfect is the alignment of the guests by the molecular sieve pores? Raman spectroscopy is a powerful tool to answer these questions. The different physical states of PNA are well investigated by Raman spectroscopy [10]. Therefore, the PNA seems to be a suitable probe molecule to trace the constraints of molecules inside molecular sieve pores.
An assemblage of directed PNA molecules is expected to show polarization dependent effects in Raman scattering. PNA has only one relevant electronic transition in the visible region. For free molecules, this transition is polarized parallel to the molecular axis. If the pores align the molecules perfectly, there will be only transitions parallel to the pores. This means that Raman scattering in such assemblage can only be excited with light polarized in the direction of the pores. Consequently, the scattered radiation must be fully polarized in the same way. The occurrence of other Raman signals has to be regarded as a proof for deviations from a perfect alignment of the incorporated molecules.
Our experimental findings are in complete accordance with these expectations. The Raman spectrum of the PNA-loaded AlPO-5 consists of a very intensive broad background of the molecular sieve matrix and some pronounced bands arising from the PNA guest molecules. Up to some small contributions to other components, these adsorbate bands are only present if the electric field of the exciting light is parallel to the length direction of the crystals. The scattered radiation is

found with the same polarization. Our measuring accuracy implies an alignment better than ± 16°.
From comparing the band positions [9] it follows that the Raman bands of PNA in the zeolite correspond neither to these of PNA in solutions nor to those of molten PNA. The bands are similar to those of solid PNA. This indicates that there are hydrogen bonds between the neighboring molecules in the PNA dipole chain like in the solid. However, in the solid different chains are antiparallel [10], whereas in the molecular sieve the PNA chains have the same direction (see the next chapter). Summarizing, our model of PNA in the molecular sieve pores is that of an ensemble of directed one-dimensional solids.

3. NONLINEAR OPTICS

Optical second harmonic generation (SHG) is a sensitive tool to detect the breaking of inversion symmetry. Therefore, SHG can be used to investigate the structure of interfaces, thin films and other directed systems. In a pioneering paper, Cox et al. [1] reported the fundamental observation that a random powder assemblage of aluminophosphate molecular sieves containing PNA generates the second harmonic.
In our polarization dependent investigations, an individual crystal loaded with PNA or DMABN was positioned in the beam of a ps-Nd:YAG laser [11] as shown in Fig. 2.

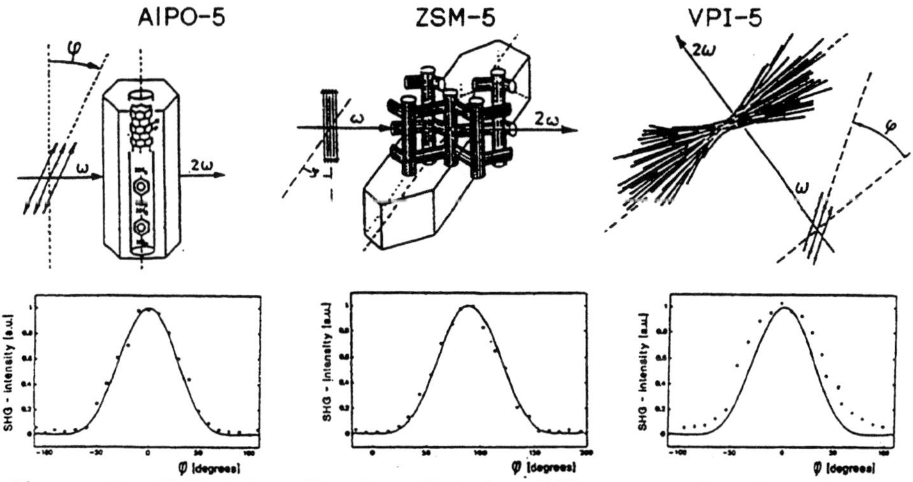

Figure 2 SHG intensity for PNA in different molecular sieves as a function of the polarization direction of the incident light.

a) PNA in AlPO-5, ZSM-5 and VPI-5. When the polarization plane of the incident laser light is varied, the detected SHG passes a maximum at Φ=0° for PNA in AlPO-5 (Fig.2). As expected the measured values are well fitted by $\cos^4\Phi$. Since not any SHG was detected at Φ=90°, the molecular axis of PNA must coincide with the length direction of the crystal (z-axis). This finding corresponds with the inner pore system of AlPO-5, namely that the unidirectional and unidimensional channels extend in z-direction of the hexagonal prisms (cf. Fig. 1).
SHG prooves that the loaded crystals are not centrosymetric on a microscopic scale. That means that different PNA dipole chains in the molecular sieve must have the same preferred direction. The formation of dipole chains is energetically understandable, but the same preferred direction seems to be the result of the PNA uptake mechanism. The preferred direction must exist for regions of several light wavelengths. Macroscopically it is possible that there are several regions of different preferred directions (e.g. the upper and the lower half of the 100 μm crystals). The SHG of these regions does not cancel each other.

The result for PNA in ZSM-5 is quite similar. This means for the interpretation that only PNA molecules in the straight channels contribute to the SHG. Very likely, the molecules in the sinusoidal channels do not form directed dipole chains.

For PNA in VPI-5 maximum SHG is observed if the light is polarized in length direction of the needle-like individual crystals. This indicates that the unidimensional pore system containing the PNA dipole chains extends in this direction. Because of the polycrystalline nature of the VPI-5 crystallite, deviations from the $\cos^4\Phi$ behaviour are observed, especially the non-zero value of the SHG intensity at the minimum of the curve.

b) DMABN in AlPO-5 [12]. For this system two components of SHG are found. One is polarized parallel to the z-axis and exhibits its maximum for excitation with the same polarization. The second component, however, is polarized perpendicular to the z-axis and shows the maximum efficiency for mixed excitation (Φ =45°). Obviously, different components of the hyperpolarization tensor contribute to the resulting SHG. This is understandable since there are more than one excited state in DMABN which are of relevance for the visible region. The corresponding transition dipole matrix elements have different directions.

4. CONCLUSIONS

The guest-host systems investigated have a high variability. By choosing appropriate guests one can try to achieve SHG or other nonlinear optical effects, e.g. sum frequency generation, third harmonic generation or electrooptical effects.

The guest molecules are very well aligned and ordered by the molecular sieve pores. The intermolecular interactions in the formed dipole chains are comparable to those in solids.

ACKNOWLEDGEMENT

The Authors thank S. Dähne, W. Hill, K. Hoffmann, J. Kornatowski, R. Vetter, L. Werner for their support and contributions to this work. This work is supported by the DECHEMA.

REFERENCES

1. S.D. Cox, T.E. Gier, G.D. Stucky and J. Bierlein, *J. Am. Chem. Soc.*, Vol. 110, 1988, 2986 pp.
2. R. Dorn, D. Baums, P. Kersten and R. Regener, *Adv. Mater.*, Vol. 4, 1992, 464 pp.
3. R. Hoppe, G. Schulz-Ekloff, D. Wörle, M. Ehrl and C. Bräuchle, In: Zeolite Chemistry and Catalysis, P.A. Jacobs et al. (Eds.), Amsterdam: Elsevier, *Stud. Surf. Sci. Catal.*, Vol. 69, 1991, 199 pp.
4. S.D. Cox, G.D. Stucky, *J. Phys. Chem.*, Vol. 95, 1991, 710 pp.
5. J. Weitkamp, In: Catalysis and Adsorption by Zeolites, G. Öhlmann et al. (Eds.), Amsterdam: Elsevier, *Stud. Surf. Sci. Catal.*, Vol. 65, 1991, 21 pp.
6. G. Finger, J. Richter-Mendau, M. Bülow and J. Kornatowski, *Zeolites*, Vol. 11, 1991, 443 pp.
7. J. Kornatowski, *Zeolites*, Vol. 8, 1988, 77 pp.
8. F. Marlow, J. Caro, *Zeolites*, Vol. 12, 1992, 433 pp.
9. F. Marlow, W. Hill and J. Caro, *J. Raman Spec.*, subm.
10. M. Harrand, *J. Raman Spec.*, Vol. 4, 1975, 53 pp.; M. Harrand, *J. Raman Spec.*, Vol 8, 1979, 161 pp.
11. L. Werner, J. Caro, G. Finger and J. Kornatowski, *Zeolites*, Vol. 12, 1992, 658 pp.
12. F. Marlow and J. Caro, in prep.

Studies in Classical
and Quantum Nonlinear Optics

ISBN 1-56072-168-5

POLARIZATION SWITCHING OF AN ANISOTROPIC INTERFERENCE FILTER

Krassimir Panajotov Panajotov
Institute of Solid State Physics
72 Tzarigradsko chaussee blvd. 1784 Sofia, Bulgaria

Among the potential advantages of optical bistable devices to the commercial computer building electrical triggers there is one more: to distinguish the different logical states by the polarization of light having a constant intensity. Treating this problem we have shown that a hybrid bistable device on the basis of an electrically driven interference filter with a central layer of lithium niobate can be driven by the light polarization [1], [2].

For illustration in Fig. 1 the experimentally obtained bistable behaviour [1] of such a filter is shown when only the azimuth angle β_0 of the linearly polarized incident light is changed, and in Fig. 2 the switching between three stable states is induced when both the azimuth angle and the ellipticity of the polarization of the input light are changed [2]. The reason for this behaviour is a polarization splitting of the transmittance characteristic of the filter due to the presence of the anizotropic material.

In analogous all optical bistable devices with anisotropic nonlinear material will show a similar behaviour. In order to construct such a device we have produced interference filters with a thermally nonlinear $ZnSe$ layer and mica as a central layer. The high reflecting mirrors of the filters are deposited by vacuum thermal evaporation on both sides of the mica sheets. The filters structure is $(HL)^4\ MH\ (LH)^4$, where H and L denote quarterwave layers of ZnS and Na_3AlF_6 with refractive indexes $n_H = 2.3$ and $n_L = 1.35; H'$ is the $ZnSe$ layer with refractive index $n'_H = 2.5$, absorption 1000 cm^{-1} and thickness $t'_H = \lambda_0/n'_H$, and M denotes the mica. The wavelength of deposition of these layers is $\lambda = 0.635\ \mu m$. At the end the mica filters are glued and pressed between glass plates.

The experimental setup for the angular measurement of the filters is shown in Fig. 3. Laser light ($\lambda = 0.6328\ \mu m$) with vertical (s) polarization goes through the filter F, which could be rotated both in the plane of incidence at an angle Θ and in the plane parallel to its surface at an angle χ. After that the light goes through a polarizer P and falls on a detector D. The signal from the detector is then led to the y-input of a $x - y$ recorder. A signal proportional to the angle of incidence Θ is led into the x-input.

In the next four figures (Figs. 4-7) the experimental angular dependencies of the filter transmittance for the s and p-polarization are shown for different filter orientations, i.e. for four different values of χ. It appears that the filter transmittance depends on its orientation. The initial value of χ (Fig. 4) is chosen so that mainly s-

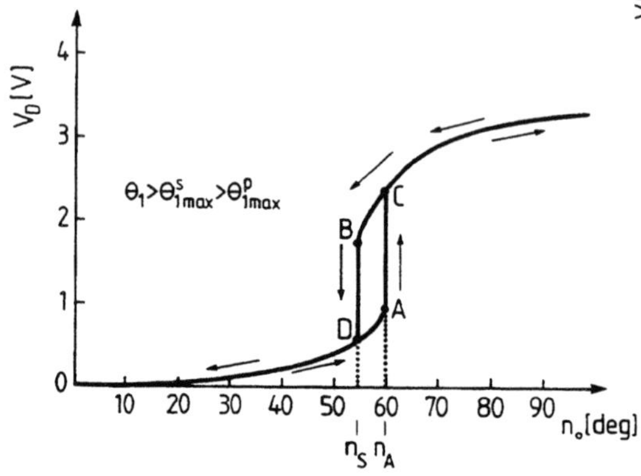

Figure 1

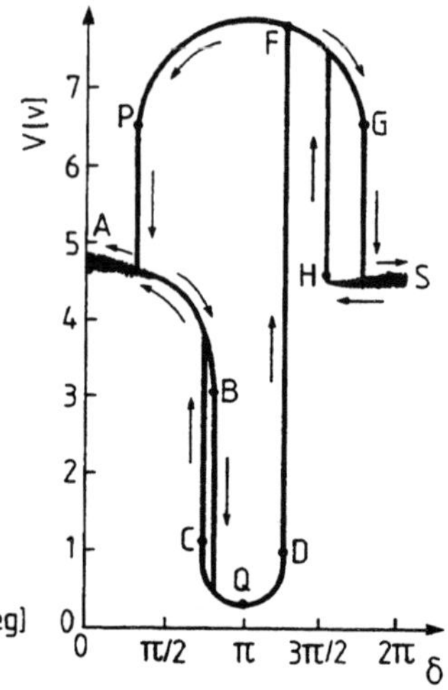

Figure 2

polarized light is transmitted through the filter (with s-polarization of the incident light). A little amount of p-polarized light is also observed at its output (dashed line). There are six well depicted maxima for the p-polarized transmitted light: three of them appear almost at the same position as the maxima for the s-polarization and the other three at smaller angles of incidence ($\Theta_{p1'}, \Theta_{p3}$ and Θ_{p5}). The three strong maxima for s-polarization (Θ_{s2}, Θ_{s4} and Θ_{s6}) correspond to the different interference orders m-1 and m-2. The presence of the very small amount of p-polarized light means that two of mica main coordinate system axes lay almost in the plane of incidence.

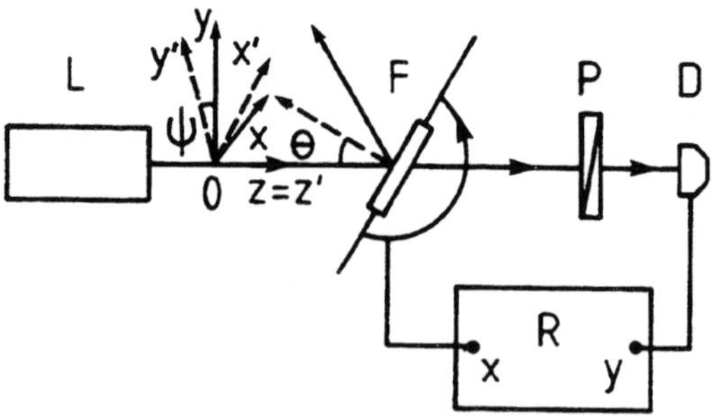

Figure 3

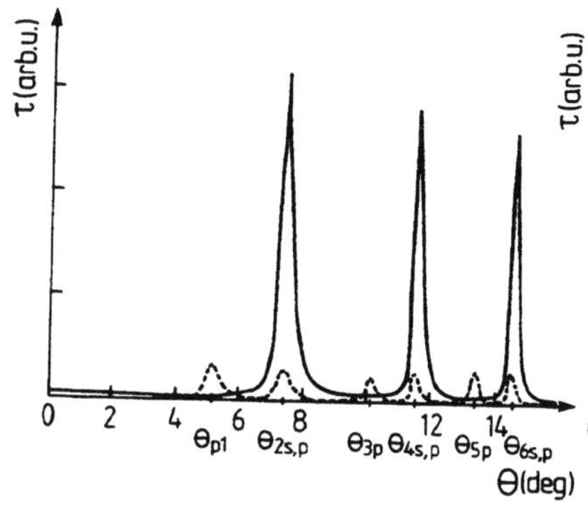

Figure 4

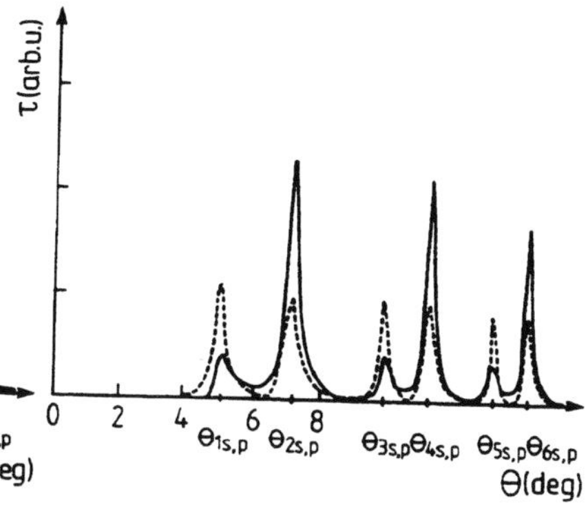

Figure 5

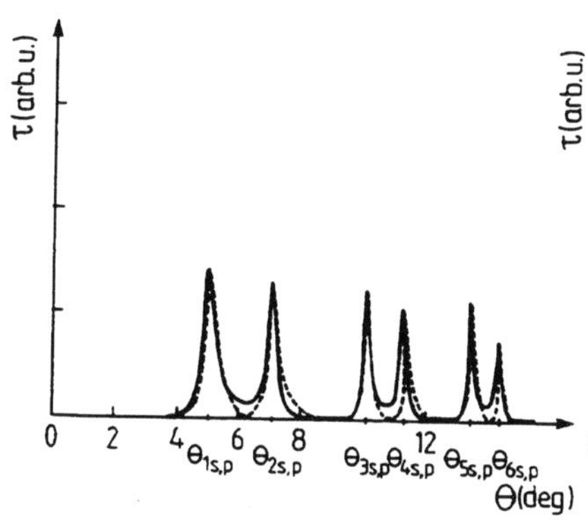

Figure 6

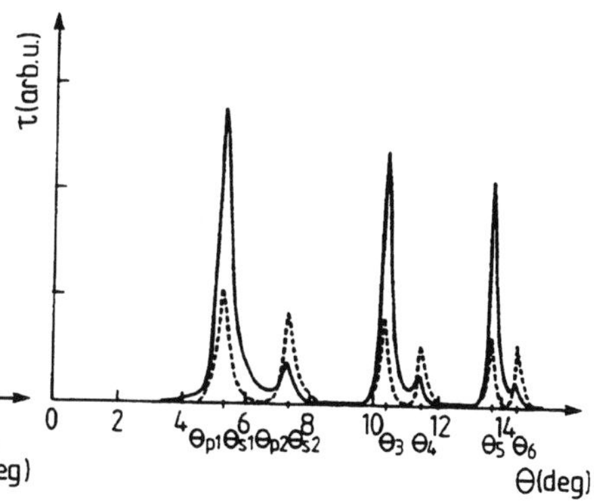

Figure 7

Increasing the angle χ (Fig. 5) the strong s-maxima decrease and three other appear at smaller angles. At the same time the values of the p-maxima increase, remaining almost equal. At $\chi = 30°$ (Fig. 6) the light is approximately equally split into couples of maxima and the amount of p and s-polarized light is also approximately equal. At $\chi = 45°$ (Fig. 6) the s-maxima are moved almost to the initial place of the p-polarization maxima. The existence of an angular splitting for each polarization means that the mica main coordinate system does not yet coincide with the plane of incidence.

The experimental results demonstrate the polarization sensitivity of the mica interference filter to its space orientation, a fact which has not been obtained for the lithium niobate filter. Here one should point out that this space orientation could be changed not only by rotating the filter until it is parallel to the surface plane (as it was done), but also by changing the plane of incidence for the fixed filter. This will

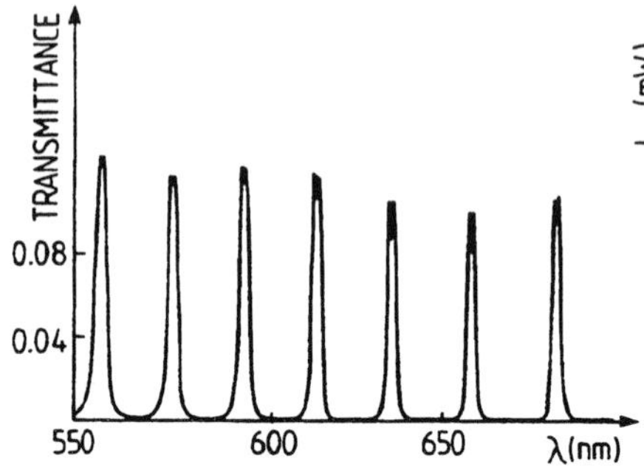

Figure 8

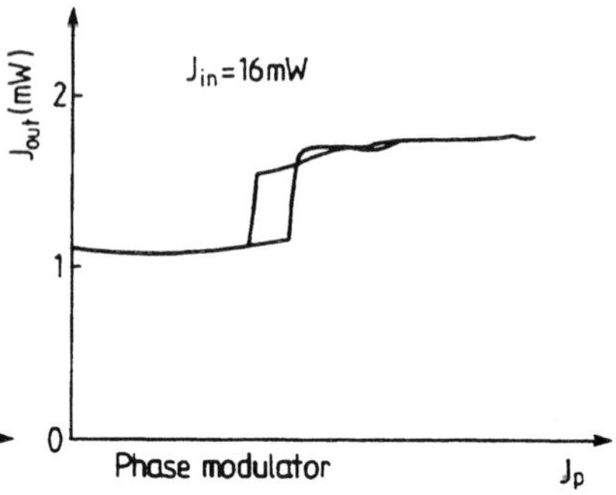

Figure 9

give a new possibility to control and drive bistable devices on the basis of such a filter. The experiments for achieving a bistable behaviour, however, have not led to a positive result. The reason stems from the significantly decreased angular halfwidth of the filter, compared to the same filters, but not containing mica as the central layer. This does not allow strong light focusing in order to increase the light power density. It is possible to achieve power densities, sufficient for switching of these filters, by decreasing the thickness of the mica or by preparing filters for other spectral regions, where more powerful light sources are available.

This led us to the preparation of interference filters with the same construction but for wavelength around 590 nm and with a thinner ($4 - 5\ \mu m$) slide of mica. On fig. 8 the filter transmittance for unpolarized light is shown for the spectral region 550-700 nm where well depicted polarization splitting can be seen. The preliminary experimental result showing polarization switching is shown in Fig. 9. The switching is achieved by changing only the polarization of the input light from the DYE laser having a wavelength 575 nm and an intensity of about 16 mW.

Finally some remarks on the theory used. We have applied the 4x4 matrix method [3], which is a generalization for an anisotropic multilayer structure of the 2x2 Abeles method [4]. The mica is in the monoclinic singony, i.e. its index ellipsoid has three different axes along its main Cartesian coordinate system K. As the mica splits at {100} for the permittivity tensor in the coordinate system K fixed to the plane of incidence xOz with z-axis along the direction of light propagation, we have:

$$\epsilon = \begin{pmatrix} \epsilon_1\, cos^2\psi + \epsilon_2\, sin^2\psi & sin\psi cos\psi\,(\epsilon_2 - \epsilon_1) & 0 \\ sin\psi cos\psi\,(\epsilon_2 - \epsilon_1) & \epsilon_2\, cos^2\psi + \epsilon_1 sin^2\psi & 0 \\ 0 & 0 & \epsilon_3 \end{pmatrix} \tag{1}$$

where ψ is the angle between the x axis of the index ellipsoid and the plane of incid-

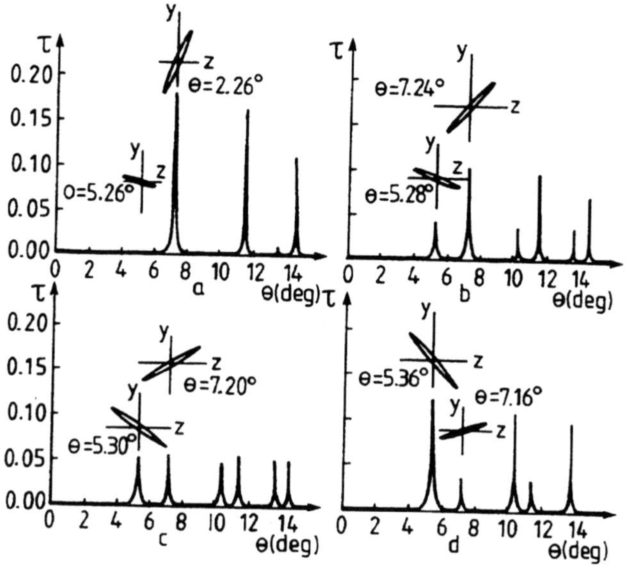

Figure 10

ence. The light propagation through such an anizotropic multilayer obeys the system of differntional equations:

$$\frac{d\phi_i}{dz} = \Delta_{ij}\phi j \; ; \qquad i,j = 1,4 \tag{2}$$

where

$$\phi = \begin{pmatrix} E_x \\ H_y \\ E_y \\ \div H_x \end{pmatrix} \; ; \; \Delta = \begin{pmatrix} 0 & 1-\eta^2/\epsilon_{33} & 0 & 0 \\ \epsilon_{11} & 0 & \epsilon_{12} & 0 \\ 0 & 0 & 0 & 1 \\ \epsilon_{21} & 0 & \epsilon_{22} & 0 \end{pmatrix} ,$$

$\eta^2 = 1 - n_0^2 sin^2\Theta$, and ϵ_{ij} is defined by (1). To solve (2) we find its eigenvalues λ_i from the biquadratic equation:

$$\lambda_i^4 - \left[\epsilon_{22} + \epsilon_{11}(1-\eta^2/\epsilon_{33})\right]\lambda_i^2 + (1-\eta^2/\epsilon_{33})\left[\epsilon_{11}\epsilon_{33} - \epsilon_{12}\epsilon_{21}\right] = 0 \,. \tag{3}$$

The expressions for the corresponding eigenvectors ϕ_{ji} are:

$$\phi_{1j} = \Delta_{12}/\lambda_j, \; \phi_{2j} = 1, \; \phi_{3j} = \Delta_{12}\phi_{2j}/\lambda_j, \; \phi_{41} = \Delta_{41}\Delta_{12}/(\lambda_j^2 - \Delta_{43}) \,,$$

for $\Delta_{12}\Delta_{21} - \lambda^2 = 0$ and

$$\phi_{1j} = \Delta_{12}\phi_{2j}/\lambda_j, \; \phi_{2j} = \Delta_{23}/(\lambda_j^2 - \Delta_{12}\Delta_{21}), \; \phi_{3j} = 1/\lambda_j, \; \phi_{4j} = 1 \,,$$

for $\Delta_{12}\Delta_{21} - \lambda^2 \neq 0$.

The final 4×4 characteristic matrix L_{ij} is obtained from:

$$L_{ij} = \phi_{ik}\delta_{kl}e^{-\lambda_l h}\phi_{lj}^{-1}\,, \tag{4}$$

where h is the thickness of the mica. Combining this result with the well known 2x2 characteristic matrix of an isotropic multilayer it is possible to calculate the transmitance, absorption and reflection of the interference filters with a mica central layer. We apply this method to determine the mica refractive indexes $n_1 = 1.588$, $n_2 = 1.5825$, $n_3 = 1.562$ for $\lambda = 0.6328\ \mu m$ and thickness $h = 40.315\ \mu m$ by fitting the theoretical angular dependencies of the filter transmittance to the experimental ones. The calculated angular characteristics of the filter are shown in Fig. 10 for different angles ψ (ψ differs from χ with a constant of about 15 degree). There is a good coincidence of these results with the experimental ones. The output light polarization at the first two angles of maximal transmittance is shown on the figure and it is seen that the changes of its azimuth with ψ are stronger than that of its ellipticity.

In order to obtain numerically assesment for the bistable behaviour of the filter we applied the equation of heat conductivity in a first order approximation [5]:

$$T - T_0 = I_{in}A(T)/(2k\rho\sqrt{\pi}), \tag{5}$$

where k is the substrate heat conductivity ($k = 4mW/C/cm$ for mica), T and T_0 are the temperatures in the center of, and outside the laser beam, and $A(T)$ is the filter absorption, calculated via the above described formalism. The absorption depends indirectly on the temperature because of the change of the $ZnSe$ refractive index through the thermooptic effect:

$$n_{ZnSe}(T) = n_{ZnSe}(T_0) + (T - T_0)(dn/dT)\,. \tag{6}$$

Assuming that only $ZnSe$, having a thermooptic coefficient $dn/dT = 1.1 \times 10^{-4}$, absorbs the light we need for up-switching an intensity of 20 mW. This is in good agreement with the results of Fig. 9. In turn, this gives a switching power density of $1kW/cm^2$ for a laser beam radius of $\rho = 26\ \mu m$.

In conclusion we would like to point out that a bandpass filter with mica central layer splits the transmitted light into two mutually perpendicular polarizations even for small angles of incidence. This polarization angular splitting could be controlled by filter rotation into a plane parallel to its surface. Moreover, even for s-polarized incident light at its output the other linear polarization (p) is observed at certain angles of incidence. This results in a polarization optical switching of these filters tuned by the angle of incidence and the filter orientation.

References

[1] G. Zartov, K. Panajotov, and R. Peyeva, "Optical Bistability Polarization Dependencies with an Interference Modulator", *Appl.Phys.B*, Vol. 39, No. 1, 1986, pp. 111-115.

[2] G.D. Zartov, K.P. Panajotov, and R.A. Peyeva, "Hybrid bistable device driven by light polarization", *Opt.Commun.*, Vol. 61, No. 6, 1987, pp. 6-10.

[3] R. Azzam and N. Bashara, *Ellipsometry and polarized light*, Amsterdam: North-Holland Publishing Company, 1977, pp. 388-400.

[4] M. Born and E. Wolf, *Principles of Optics*, Oxford: Pergamon Press, 1968, pp. 66-81.

[5] I. Janossy, M.R. Taghizadeh, J.G.H. Mathew and S.D. Smith, "Thermally Induced Optical Bistability in Thin-Film Devices". *IEEE J. Quant. Electr.* Vol. QE-21, No. 9, 1985, pp. 1447-1452.

Studies in Classical
and Quantum Nonlinear Optics

ISBN 1-56072-168-5

OPTICAL RESPONSE OF A SEMICONDUCTOR MULTIFILM CONFIGURATION UNDER PULSED BEAM EXCITATION

T. Peschel, U. Peschel, F. Lederer
Friedrich-Schiller-Universität Jena
Physikalisch-Astronomische Fakultät
O-6900 Jena, Germany

1. INTRODUCTION

Within the last years there has been a significant amount of interest in the investigation of the optical response of planar resonators made from semiconducting materials as GaAlAs and CdSSe. These structures lead to new all-optical devices characterized by a high parallelism of data processsing (more than 10^5 pixel/cm^2) and short switching times (as low as 1ps). Because of the small dimensions of each pixel a plane wave approximation is not valid and a more careful description of the light propagation in the film structure is required. We have to solve Maxwell's equations for an inhomogenous wave propagation in space and time. Typical beam parameters under investigation are diameters greater than $10\mu m$ and durations between 1ps and 1ns. In addition we have to find a model which describes the interaction of light and matter in the investigated frequency domain satisfactorily and connects it consistently with the field equations. To reduce absorption losses we restrict ourselves to photon energies below the gap energy of the semiconductor.

2. OPTICS

Fig. 1 shows the structure investigated, which consists of two Bragg-reflectors and an active film in the middle. We use a method published in [1] to describe its optical response. For simplicity we restrict ouselves to TE-polarized fields, the equations for TM-polarization are similar. The transmission response function $\tau(\beta,\omega)$ of such a structure shows well defined single peaks on certain angles of incidence. Every one of them corresponds to a complex pole of τ at β_0, where $\mathcal{R}e\left(\beta_0^2\right)$ determines the position and $\mathcal{I}m\left(\beta_0^2\right)$ the width of the resonance. In the vicinity of a resonance $\tau(\beta,\omega)$ can be approximated by:

$$\tau(\beta,\omega) \approx \frac{\hat{\tau}}{\beta_0^2(\omega)-\beta^2} \tag{1}$$

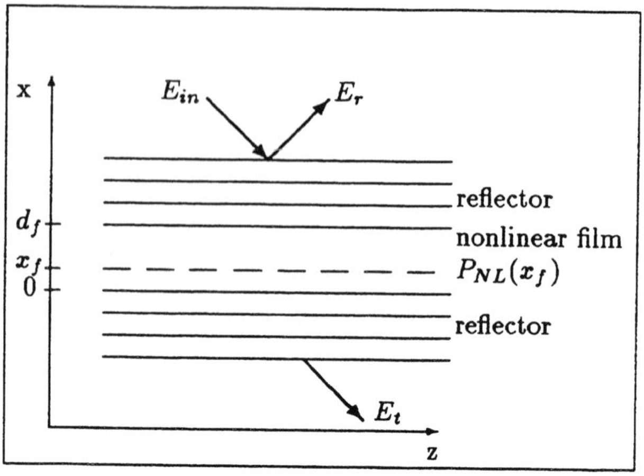

Fig.1 Structure under investigation

If we have weak nonlinearities the influence of the polarization in the film can be expressed using the Green's function of the linear structure.

$$E_t(\beta,\omega) = \tau(\beta,\omega)E_{in}(\beta,\omega) - \mu_0\omega^2 \int_0^{d_f} G_t(x_f;\beta,\omega)P_{NL}(x_f;\beta,\omega)\,dx_f \tag{2}$$

At resonance the Green's function can be split into the transmission response function τ already discussed multiplied by a term g_t which is nonresonant because it depends on the layers above the excitation only.

$$G_t(x_f;\beta,\omega) = \tau(\beta,\omega)g_t(x_f;\beta,\omega) \tag{3}$$

To study processes with a limited temporal duration we expand $\beta_0^2(\omega)$ into a Taylor series around the central frequency ω_0 of the incident beam. Carrying out an inverse Fourier transformation we return to an equation in real space,

$$\left[\Delta + i\left.\frac{\partial\beta_0^2(\omega)}{\partial\omega}\right|_{\omega_0}\frac{\partial}{\partial t} + \beta_0^2\right]E_t(z,t) \tag{4}$$

$$+\mu_0\omega_0^2\hat{\tau}\int_0^{d_f} g_t(x_f;\beta_0,\omega_0)P_{NL}(x_f;z,t)\,dx_f = \hat{\tau}E_{in}(z,t)$$

where Δ is the Laplacian which acts in y and z directions parallel to the films. P_{NL} can be expressed in the following way by the transmitted field

$$P_{NL}(x_f,z,t) \approx \epsilon_0\chi(x_f,z,t)f(x_f,\beta_0,\omega_0)E_t(z,t) \tag{5}$$

where $f(x_f, \beta_0, \omega_0)E_t(z,t)$ denotes the optical field in the nonlinear film. Its shape $f(x_f, \beta_0, \omega_0)$ is determined by the subsystem below the excitation only and hence is nonresonant. Inserting eq.5 into eq.4 we obtain:

$$\left[i \left. \frac{\partial \beta_0^2(\omega)}{\partial \omega} \right|_{\omega_0} \frac{\partial}{\partial t} + \Delta + \beta_0^2 + k_0^2 \tilde{\chi}(z,t) \right] E_t(z,t) = \hat{\tau} E_{in}(z,t) \tag{6}$$

$$\tilde{\chi}(z,t) = \hat{\tau} \int_0^{df} g_t(x_f; \beta_0, \omega_0) f(x_f, \beta_0, \omega_0) \chi(x_f; z, t)\, dx_f \tag{7}$$

Note that $\chi(x_f; z, t)$ is driven by the field in the nonlinear film and hence depends on the transmitted field E_t.

If the incident beam is a stationary plane wave with propagation vector β parallel to the films eq.6 simplifies to an algebraic one.

$$E_t = \frac{\hat{\tau}}{\beta_0^2 - \beta^2 + k_0^2 \tilde{\chi}} \quad E_{in} \tag{8}$$

where $\tilde{\chi}$ is a function of E_t. If eq.8 has more than one solution E_t for one fixed E_{in} we find bistability.

The investigation of incident beams of finite duration and diameter is much more involved. The bistable region is reduced by shortening the beam dimensions and vanishes completely if the beam width approaches values of about $(\mathcal{I}m\beta_0^2)^{-0.5}$. In this case the spatial frequency domain of the incident beam becomes broader than that of the resonance and the optical field is not influenced by any shift of the position of the resonance.

The limitation for the switching time follows from the damping constant of eq.6 which is given by $(\mathcal{I}m\,(\beta_0^2))^{-1} \frac{\partial}{\partial \omega} \beta_0^2$. This relaxation time corresponds to the photon life time in our structure.

3. MATTER

To describe the nonlinear interaction of light and matter in semiconductors we have to consider one- and two-photon processes simultaneously. The susceptibility caused by two-photon processes χ_{TP} is proportional to the intensity of the optical field and acts instantaneously. Its real part crosses zero at a photon energy of approximately 2/3 of the gap energy and thus the relation between absorptive and refractive effects changes considerably within the gap. We calculated the coefficient of the two-photon susceptibility χ_3 following the treatment of M. Sheik-Bahae e.a. [2].

It is more difficult to include single-photon processes into the simulations because of their more complex dependence on carrier densities N_c and temperatures T_c. We first computed a set of single-photon susceptibilities χ_{SP} for different carrier densities and temperatures including saturation, coulomb effects, screening and gap shrinkage corresponding to [3]. The actual value of the susceptibility was then determined by a spline interpolation during the calculation.

If we study the behaviour of the single-photon susceptibility we find that the imaginary part falls exponentially within the gap but the real part decreases only like $1/\omega$. This means that carriers generated by any process (e.g. by multi-photon absorption) influence the refractive index in the whole frequency domain.

Because of the large influence of excited carriers we have to take into account their dynamics that is generation, recombination, diffusion, heating and temperature relaxation,

$$\frac{\partial}{\partial t}N_c = -\frac{N_c}{T_R}+\frac{4\epsilon_0}{\hbar}\mathcal{I}m\left(\chi_{SP}+\frac{1}{2}\chi_{TP}\right)\mid E\mid^2-\frac{\mu}{3e}\Delta U_{exc}+D\Delta N_c \quad (9)$$

$$\frac{\partial}{\partial t}U_{exc} = -\frac{U_{exc}}{T_w}+\frac{\epsilon_0}{\hbar}\mathcal{I}m(\chi_3)\mid E\mid^4(2\hbar\omega-E_G-3K_BT_{lattice}) \quad (10)$$

where μ and D are the ambipolar mobility and diffusion constant respectively. In addition to the well known equation for the carrier density we use a second one which describes the evolution of their inner energy. The reason is that carriers generated by multi-photon processes have a big excess of inner energy U_{exc} or a much higher temperature than that of the lattice. This excess energy relaxes relatively fast with a time constant T_w of about 1ps.

Recombination is to slow to play any role in the time domain under investigation ($T_R \approx 3.5ns$). Diffusion influences the pulse evolution much more. We can define a diffusion time t_D as a function of a characteristic length b.

$$t_D = \frac{b^2}{4\pi^2 D} \quad (11)$$

For directions perpendicular to the film b corresponds to one half of the wavelength in the material and the resulting t_D is smaller than 1ps, i.e. the carrier distribution is nearly uniform and the integal in eq.7 simplifies considerably. For directions parallel to the film we find the opposite situation, where b is determined by the beam width and hence t_D is usually greater than 1ns, i.e. diffusion plays only a minor role.

4. RESULTS FOR AL-GA-AS-STRUCTURES

We present the simulations of the pulse evolution for two different excitation energies. We restricted our numerical calculations to one dimension z parallel to the films. The semiconductor structure investigated was optimized for every wavelength. It consists of twenty five $\lambda/4$ double layers of GaAlAs on every side and one $\lambda/2$ GaAs film in the middle, the substrate was GaAs. We have a resonance at an angle of incidence of 60° with a width of 0.03^o.

The first calculations (see Fig.2) were made for a photon energy slightly above half the gap energy ($\lambda = 1.75\mu m$). At this frequency the real part of the Kerr coefficient χ_3 is positive and the ratio between real and imaginary part is about 170. Because of the focusing character of the Kerr-type nonlinearity we choose an angle of incidence 0.03^o above that of resonance. In the beginning the transmitted field develops like we expected. It switches to the upper state (with a maximum optical field in the central film of about $7 \times 10^7 \frac{V}{m}$) when the incident field reaches its maximum. But with growing intensity the carrier density in the central film generated by two-photon absorption increases very rapidly and causes a negative shift of the refractive index opposite to that of the two-photon susceptibility.

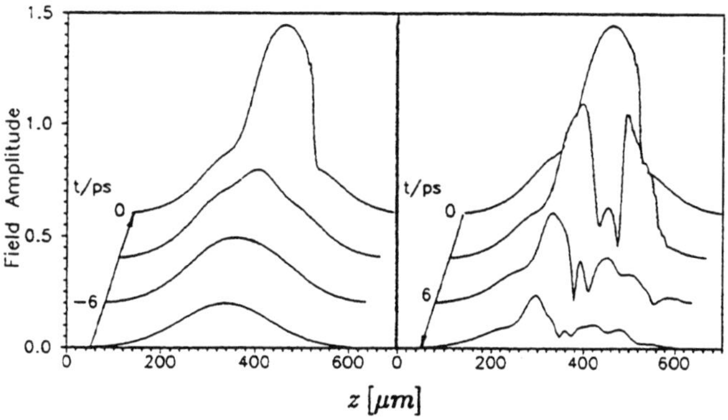

Fig.2 Evolution of the transmitted field for $\lambda = 1.75\mu m$ (maximum field:$7 \times 10^4 \frac{V}{m}$)

These two processes are in competition, but at the end the single-photon processes succeed and drive the system back to a state far from resonance. The result is a break down of the output field beginning at the center of the beam. We can state that even relatively small amounts of two-photon absorption change the evolution of the output field considerably and reduce the stability of the switched-on state.

Let us now consider the situation for a photon energy 50 meV below the gap ($\lambda = 0.92\mu m$) (see Fig. 3). Now all possible changes in the refractive index are negative and so we choose an angle of incidence which is slightly smaller than that of the resonance (-0.03^o). When carrier generation starts the resonance is shifted towards the angle of incidence and the system increases its transmission. This process goes on and the carrier concentration soon exceeds even the saturation value, i.e. electron-hole pairs are generated by two-photon absorption but annihilated in a one-photon process. The growing carrier density causes the system to cross the resonance and to leave it again. Like in the first example the peak of the pulse which corresponds to the switched-on state splits into two parts. In our simulations the field travels from the left to the right because of the oblique angle of incidence. It falls from the switched-on state at the left peak, oscillates some times and reaches the switched-off state. Because of the relatively uniform carrier distribution between the peaks the oscillations are well pronounced. The total field strength needed for nonlinear operation is about 5 times smaller than in the first example. In both cases a large amount of carriers is generated, i.e. it will take the device at least the recombination time to reach the initial state.

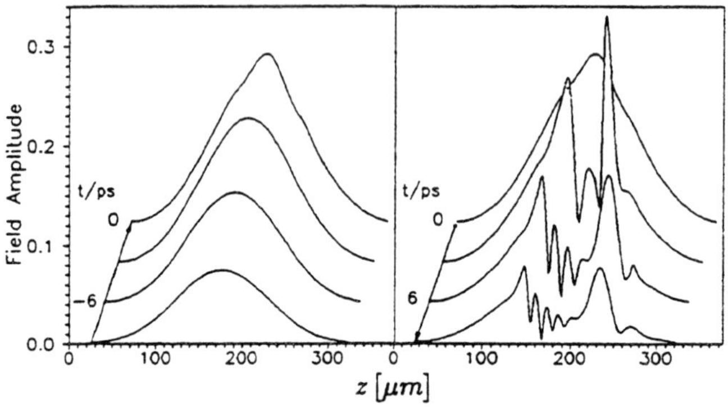

Fig.3 Evolution of the transmitted field for $\lambda = 0.92\mu m$ (maximum field:$1.4 \times 10^4 \frac{V}{m}$)

5. SUMMARY

We simulated the evolution of finite light pulses in a resonant semiconductor multifilm configuration. The results show that stationary plane wave approximations can't be used to describe the pulse evolution and that the interaction of one- and two-photon processes influences the optical properties of the semiconductor considerably. It is impossible to neglect the changes in the refractive index caused by excited carriers even for photon energies close to half the gap energy. The carrier recombination seems to be a serious limitation for the repetition rates of all-optical switching devices.

REFERENCES

1. T.Peschel, and F.Lederer,'Optical Response of nonlinear planar resonators', to be published in *Phys.Rev. B*, 1992
2. M.Sheik-Bahae, D.C. Hutchings, D.J. Hagan and, E.W. Van Stryland, 'Dispersion of Bound Electronic Nonlinear Refraction in Solids', *IEEE J. Quantum Electron.*, Vol.27, No. 6, 1991, pp. 1296-1309
3. L.Banyai and, S.W.Koch,'A simple Theory for the Effects of Plasma Screening on the Optical Spectra of Highly Excited Semiconductors', *Z.Phys.B*, Vol.63, No. 3, 1986, pp. 283-291

Studies in Classical
and Quantum Nonlinear Optics

ISBN 1-56072-168-5

VOLUME INTERBAND CONTRIBUTION TO SECOND-HARMONIC GENERATION FROM ALUMINUM

*A.V. Petukhov**, *A. Liebsch*
Institut für Festkörperforschung, Forschungszentrum Jülich, Germany
Permanent Address: Institute of Crystallography,
Russian Acad. of Sci., Leninsky pr., 59, Moscow 117333, Russia

1. INTRODUCTION

Second-harmonic generation (SHG) during reflection of light from a centrosymmetric medium is a rapidly developing field of research [1] which is making important contributions to the understanding of the electronic and structural properties of surfaces. Most theoretical and experimental studies of SHG were carried out on metallic systems. The theoretical studies of these systems were mainly based on the jellium model [1, 2] which neglects the influence of the crystal potential. The anisotropy of SHG for different crystal faces was observed for several metals [1, 3, 4]. A detailed understanding of the anisotropic contributions to the bulk or surface harmonic polarization is still lacking. Apart from phenomenological studies rather few attempts [5] have been made to investigate the SHG anisotropy in metals using a microscopic theory.

The purpose of the present work is to provide a microscopic evaluation of the SHG bulk anisotropy for Al which has been extensively studied experimentally. As a result of the band gap of about 1.5 eV, the interband contribution to the dielectric function $\epsilon(\omega)$ is significant at relatively low frequencies. Thus, at typical laser frequencies ($\hbar\omega = 1.17$ eV or $\hbar\omega = 2.34$ eV) electronic transitions in this spectral range should noticeably affect also the nonlinear optical properties of Al.

2. FORMALISM

Within the one–electron approximation [6] the nonlinear bulk current can be expressed as:

$$\mathbf{j}_2 = -\frac{e}{2\pi m\omega}(\mathbf{E}_1 \cdot \mathbf{E}_1)\left[\overleftrightarrow{\epsilon}(2\mathbf{q}, 2\omega) - \overleftrightarrow{1}\right] \cdot \mathbf{q}$$

$$-\frac{e}{4\pi m\omega}\mathbf{E}_1 \left\{\mathbf{q} \cdot \left[\overleftrightarrow{\epsilon}(\mathbf{q}, \omega) - \overleftrightarrow{1}\right] \cdot \mathbf{E}_1\right\} \tag{1}$$

$$-(e^3/m^3\omega^2) \sum_{\mathbf{k}bb'b''} \mathbf{M}_1(\mathbf{M}_2 \cdot \mathbf{E}_1)(\mathbf{M}_3 \cdot \mathbf{E}_1) N(b''\mathbf{k} + 2\mathbf{q}, b'\mathbf{k} + \mathbf{q}, b\mathbf{k}, \omega, \omega),$$

where the matrix elements and the resonance coefficient are given by

$$\mathbf{M}_i = (b^{(n)}\mathbf{k} + n\mathbf{q} \mid \mathbf{p} + \hbar\mathbf{k} + \frac{n+m}{2}\hbar\mathbf{q} \mid b^{(m)}\mathbf{k} + m\mathbf{q}), \tag{1a}$$

$$N(b''\mathbf{k}'', b'\mathbf{k}', b\mathbf{k}, \omega, \omega) = \frac{1}{E_{b''\mathbf{k}''} - E_{b\mathbf{k}} - 2\hbar\omega - 2i\epsilon}$$

$$\times \left[\frac{f_0(E_{b''\mathbf{k}''}) - f_0(E_{b'\mathbf{k}'})}{E_{b''\mathbf{k}''} - E_{b'\mathbf{k}'} - \hbar\omega - i\epsilon} - \frac{f_0(E_{b'\mathbf{k}'}) - f_0(E_{b\mathbf{k}})}{E_{b'\mathbf{k}'} - E_{b\mathbf{k}} - \hbar\omega - i\epsilon}\right], \tag{1b}$$

(n, m)=(0,2) if $i = 1$; (n, m)=(2,1) if $i = 2$; and (n, m)=(1,0) if $i = 3$. $b^{(0)} \equiv b, b^{(1)} \equiv b', b^{(2)} \equiv b''$ are band indices.

In the case of small nonlocality the nonlinear current can be expanded in powers of q. In cubic crystals the terms of zero'th order in q cancel due to the inversion symmetry. The quadrupole–allowed nonlinear current at the harmonic frequency is determined by [1]

$$\mathbf{j}_2 = \chi_1 \mathbf{q}(E_1)^2 + 2\chi_2 \mathbf{E}_1(\mathbf{q} \cdot \mathbf{E}_1) + \zeta \sum_i \mathbf{e}_i q_i (E_{1i})^2, \tag{2}$$

where $\mathbf{e}_i$ is the unit vector along the ith axis of the natural coordinate system and the parameter ζ determines the anisotropic properties of $\mathbf{j}_2$. As long as the pump frequency ω is below the bulk plasma frequency ω_p, the linear field far from the surface is purely transverse, i.e. $\mathbf{q} \cdot \mathbf{E}_1 = 0$. Thus the second term in Eq.(2) does not contribute to $\mathbf{j}_2$. Only this case will be studied in the present paper.

To linear order in q the local values for the dielectric function can be substituted in the first and second terms of Eq.(1). These terms contribute only to the isotropic polarizabilities χ_1 and χ_2. The last term in Eq.(1) may be decomposed in the following manner:

$$\sum = \sum\nolimits_0(b'' = b' = b) + \sum\nolimits_1(b'' = b' \neq b) + \sum\nolimits_2(b'' \neq b' = b)$$

$$+ \sum\nolimits_3(b'' = b \neq b') + \sum\nolimits_4(b'' \neq b' \neq b \neq b''). \qquad (3)$$

The first term accounts for the intraband contribution, the next three terms arise from two–band transitions, and the last term from transitions between three different bands. To linear order in q, the intraband term vanishes because $N(b\mathbf{k} + 2\mathbf{q}, b\mathbf{k} + \mathbf{q}, b\mathbf{k}, \omega, \omega) = O(q^2)$. Thus, *within the one–electron approximation developed in Ref.6, the anisotropy of the second–order volume response is entirely due to real or virtual interband transitions. Pure intraband transitions do not contribute to the SHG anisotropy.*

For the calculation of the electronic wave functions and energy bands we use the nearly–free–electron approximation. This approximation has been successfully applied in studies of the linear optical properties of simple metals [7]. For Al the most important Fourier components of the crystal potential are the components V_{002} along the principal axes of the natural coordinate system [7]. In the present work these components are taken into account separately and transitions between three different bands [see last term in Eq.(3)] are neglected.

To first order in q the most interesting terms arise from the expansion of the resonant coefficient N. The expansion of the numerators gives rise to terms which are nonzero only if one of energy bands crosses the Fermi level. Moreover, these terms become very important when $\hbar\omega$ is far below the onset of real interband transitions. Other terms which arise from the expansion of the denominators are of order $1/c^2$ and become important near the resonance conditions. As a result of these terms *the resonance effects in nonlocal nonlinear optics can become very large.*

3. RESULTS

In Fig.1 we show the frequency dependence of the real and imaginary parts of the anisotropic and isotropic polarizabilities, ζ and χ_1. Both functions are given in units of the isotropic second–order polarizability, χ_1^{free}, of the free–electron gas with the same electron density n and relaxation time τ. As a result of the normalization to χ_1^{free}, the singular behavior at small ω is avoided. In these calculations the relaxation time is taken to be $0.8 \cdot 10^{-14}$ s corresponding to room temperature.

The potential component V_{002} is seen to strongly affect the nonlinear polarizabilities in a wide range of ω down to the adiabatic limit. When the fundamental or harmonic frequencies coincide with the onset of the direct interband transitions between parallel bands ($2V_{002} = 1.52$ eV) both ζ and χ_1 reveal resonance-like behavior. But there is an essential difference between the resonances of these polarizabilities: the contributions to ζ involve terms of order $1/\epsilon^2$ whereas the resonances in χ_1 are only of order $1/\epsilon$. Thus, the ratio ζ/χ_1 also exhibits resonant-like behavior at these frequencies.

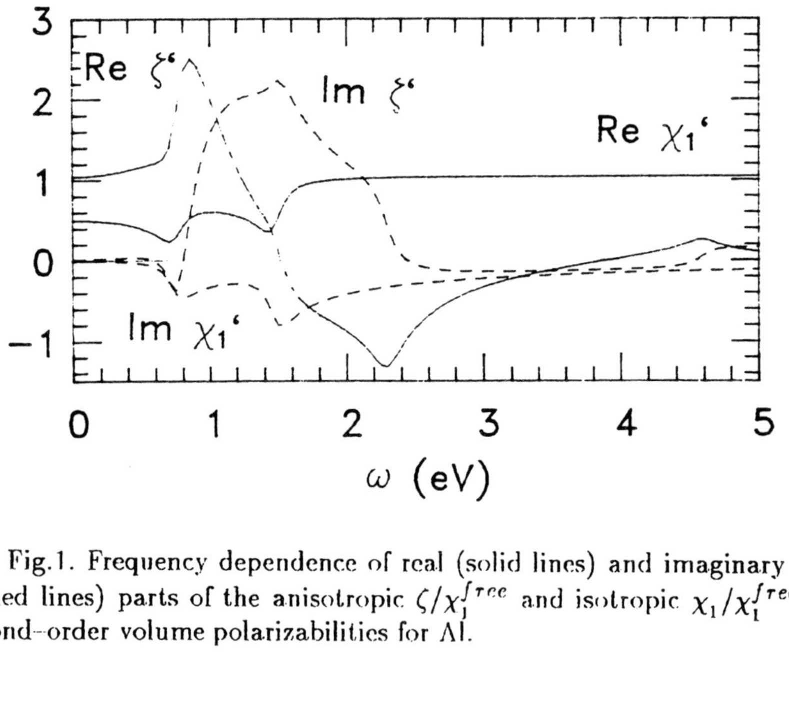

Fig.1. Frequency dependence of real (solid lines) and imaginary (dashed lines) parts of the anisotropic ζ/χ_1^{free} and isotropic χ_1/χ_1^{free} second-order volume polarizabilities for Al.

Additional resonances in ζ arise when $\hbar\omega$ or $2\hbar\omega$ coincide with 4.6 eV, which corresponds to the onset of normal interband transitions from the lowest band to states above the Fermi level along the [002] directions. These spectral features are due to the terms of order q in the expansion of the resonant coefficient N. Since these terms do not contribute to χ_1, the isotropic parameter does not show resonances at these frequencies. Thus, we can conclude, that in Al the crystal potential affects the isotropic second-order polarizability and gives rise to strong anisotropy. The anisotropy parameter ζ is found to be more sensitive to the crystal potential than the isotropic component.

It is interesting to compare the results of the present calculations with the Boltzmann equation approach studied previously [5]. We have recalculated ζ and χ_1 within the present two-band model using the formalism of Ref.[5]. In the adiabatic limit $\omega \to 0$ both approaches give exactly the same values. This result is not surprising because just at these frequences the Boltzmann equation approach should be valid because the interband transitions has been neglected in [5]. However, it is interesting to point out that the strong anisotropy of the nonlinear current at low frequencies within the quantum-mechanical model is related to a combination of intraband and virtual interband contributions. On the other hand, within the Boltzmann equation approach the same result is interpreted in terms of pure intraband transitions.

The results for the second-harmonic radiation field will be presented in terms of the dimensionless parameter d which characterizes the relative value of the bulk contribution to SHG [8]. In the jellium model one has $d = 1$. In real Al the crystal potential affects both the dielectric function $\epsilon(2\omega)$ and the nonlinear polarizabilities χ_1 and ζ. Thus, the effective value of d depends on the angle of incidence θ and the angle of azimuthal rotation, ψ, of the sample about the surface normal. Neglecting terms of order $1/\epsilon(\omega)$ we have for Al(111) at 1.17 eV:

$$d^{(111)} \approx (1.2 + i0.2) + (2.9 - i1.9) \cos 3\psi / \sin \theta,$$

where at $\psi = 0$ the projection of $\mathbf{q}$ onto the surface plane is assumed to point in the positive $[11\bar{2}]$ direction. For Al(100) we find:

$$d^{(100)} \approx (1.45 + i0.65) + (0.15 + i0.3) \cos 4\psi$$

and for Al(110):

$$d^{(110)} \approx (1.3 + i0.35) + (0.2 + i0.4) \cos 2\psi + (0.12 + i0.25) \cos 4\psi,$$

where ψ is the angle between the plane of incidence and the (001) crystallographic plane.

4. CONCLUSION

The crystal potential of Al is shown to strongly affect the quadrupole-allowed second-order polarizability in a wide range of frequencies down to the adiabatic limit. As a result, a strong dependence of the intensity of the bulk contribution on the mutual orientation of the laboratory and natural coordinate systems is found. The comparison of our results with experimental data [3, 4] shows that for clean Al in ultra-high vacuum the surface contribution to the SHG anisotropy plays the principal role. On the other hand, the bulk anisotropy becomes important on Al surfaces where the normal surface polarization is reduced due to adsorption or oxidation.

REFERENCES

1. G. L. Richmond, J. M. Robinson, and V. L. Shannon, "Second harmonic generation studies of interfacial structure and dynamics", *Prog. Surf. Sci.*, Vol. 28, 1988, pp.1–70.

2. A. Liebsch and W. L. Schaich, "Second–harmonic generation at simple metal surfaces", *Phys. Rev. B*, Vol. 40, No. 8, pp. 5401–5410.

3. K. Pedersen, O. Keller, "Experimental study of the phenomenological parameters describing optical second-harmonic generation in centrosymmetric metals", *J. Opt. Soc. Am. B*, Vol. 6, No. 12, 1989, pp.2412–2419.

4. S. Janz, K. Pedersen, and H. M. van Driel, "Dispersion and anisotropy of the optical second–harmonic response of single–crystal Al surfaces", *Phys. Rev. B*, Vol. 44, No. 8, 1991, pp. 3943–3954.

5. A. V. Petukhov, "Source of a strong anisotropy of quadrupole–allowed second–order nonlinear polarizability of the noble-metal surfaces", *Phys. Rev. B*, Vol. 42, No. 15, 1990, pp.9387–9390.

6. S. S. Jha and C. S. Warke, "Interband contribution to optical harmonic generation at a metal surface", *Phys. Rev.*, Vol. 153, No. 3, 1967, pp.751–759.

7. N. W. Ashcroft and K. Sturm, "Interband absorption and the optical properties of polyvalent metals", *Phys. Rev. B*, Vol. 3, No. 6, 1971, pp.1898–1910.

8. P. Guyot–Sionnest, A. Tadjeddine, and A. Liebsch, "Electronic distribution and nonlinear optical response at the metal-electrolyte interface", *Phys. Rev. Lett.*, Vol. 64, No. 14, 1990, pp. 1678–1681.

Studies in Classical
and Quantum Nonlinear Optics

ISBN 1-56072-168-5

NUMERICAL ANALYSIS OF THE FIFTH HARMONIC GENERATION EFFICIENCY IN SOME ORGANIC LIQUIDS

Ion M. Popescu, Niculae N. Puscas,
Paul E. Sterian and Dorin Irimescu
Polytechnic Institute of Bucharest,
Physics Department

1 INTRODUCTION

In the last years many theoretical and experimental studies have been concerned with the generation of higher order harmonics in organic solvents.

Theoretical models used for the calculation of the different order harmonic intensities include: the equivalent internal field model /1/, the optical wave (plane /1/, Gaussian /1/) propagation model in isotropic nonlinear media and so on. This paper presents a theoretical analysis of the fifth harmonic generation in organic liquids which can result both from the direct processes which implies the fifth order nonlinearities and from different step processes which implies the second, third and fifth order nonlinearities, based on the coupled amplitude equations.

2 THE COUPLED AMPLITUDE EQUATION MODEL

For high intensities of the incident electric field the nonlinear polarization of an isotropic medium may be expressed in a series of ascending powers of the electric field:

$$P^{NL} = \chi^{(2)}EE + \chi^{(3)}EEE + \chi^{(4)}EEEE + \chi^{(5)}EEEEE \ . \tag{1}$$

It may be considered that in such a medium the fifth harmonic results from two simultaneous processes /2,3/:

1) The direct process involving the fifth-order nonlinearity:

$$\omega + \omega + \omega + \omega + \omega = 5\omega$$

with

$$k_1 + k_1 + k_1 + k_1 + k_1 = k_5 + \Delta_{15} \ . \tag{2}$$

2) The following step processes involving the second, third and fifth-order nonlinearity:

$$\begin{cases} \omega + \omega = 2\omega \\ \omega + \omega + \omega + 2\omega = 5\omega \end{cases} \tag{3}$$

with

$$\begin{cases} k_1 + k_1 = k_2 + \Delta_{12} \\ k_1 + k_1 + k_1 + k_2 = k_5 + \Delta_{125} \ ; \end{cases}$$

$$\begin{cases} \omega + \omega = 2\omega \\ \omega + 2\omega + 2\omega = 5\omega \end{cases} \tag{4}$$

with

$$\begin{cases} k_1 + k_1 = k_2 + \Delta_{12} \\ k_1 + k_2 + k_2 = k_5 + \Delta_{125} \ ; \end{cases}$$

$$\begin{cases} \omega + \omega = 2\omega \\ \omega + \omega + \omega = 3\omega \\ 2\omega + 3\omega = 5\omega \end{cases} \tag{5}$$

with

$$\begin{cases} k_1 + k_1 = k_2 + \Delta_{12} \\ k_1 + k_1 + k_1 = k_3 + \Delta_{13} \\ k_2 + k_3 = k_5 + \Delta_{235} \ ; \end{cases} \tag{6}$$

with

$$\begin{cases} \omega + \omega = 2\omega \\ \omega + 2\omega = 3\omega \\ 2\omega + 3\omega = 5\omega \end{cases}$$

with

$$\begin{cases} k_1 + k_1 = k_2 + \Delta_{12} \\ k_1 + k_2 = k_3 + \Delta_{13} \\ k_2 + k_3 = k_5 = \Delta_{235} \ ; \end{cases}$$

$$\begin{cases} \omega + \omega + \omega = 3\omega \\ \omega + \omega + 3\omega = 5\omega \end{cases} \tag{7}$$

with

$$\begin{cases} k_1 + k_1 + k_1 = k_3 + \Delta_{13} \\ k_1 + k_1 + k_3 = k_5 + \Delta_{135} \ ; \end{cases}$$

$$\begin{cases} \omega + \omega + \omega + \omega = 4\omega \\ \omega + 4\omega = 5\omega \end{cases} \tag{8}$$

with

$$\begin{cases} k_1 + k_1 + k_1 + k_1 = k_4 + \Delta_{14} \\ k_1 + k_4 = k_5 + \Delta_{145} \ ; \end{cases}$$

$$\begin{cases} \omega + \omega = 2\omega \\ 2\omega + 2\omega = 4\omega \\ \omega + 4\omega = 5\omega \end{cases} \tag{9}$$

with

$$\begin{cases} k_1 + k_1 = k_2 + \Delta_{12} \\ k_2 + k_2 = k_4 + \Delta_{24} \\ k_1 + k_4 = k_5 + \Delta_{145} \ ; \end{cases}$$

$$\begin{cases} \omega + \omega + \omega = 3\omega \\ \omega + 3\omega = 4\omega \\ \omega + 4\omega = 5\omega \end{cases}$$

with

$$\begin{cases} k_1 + k_1 + k_1 = k_3 + \Delta_{13} \\ k_1 + k_3 = k_4 + \Delta_{134} \\ k_1 + k_4 = k_5 + \Delta_{145} \ . \end{cases} \tag{10}$$

The nonlinear polarization expressed by (1) acts like a source in the Maxwell's equations and for a lossless nonlinear medium one has obtained for the direct process (2) and the step process (3)-(10) the differential equations for the coupled normalized amplitudes (a_1, a_2, a_3, a_4, a_5) of the interacting waves /2,3/.

3 THE NUMERICAL ANALYSIS OF THE HARMONIC GENERATION EFFICIENCY

Based on the nonlinear interacting processes, a study of the coupled amplitude equations /2,3/ which describe the fifth harmonic generation in organic liquids was made. This study emphasized the competition between different processes which determine the different order harmonics in the fifth harmonic generation and allows the evaluation of the conversion efficiencies. We considered the following initial conditions for the normalized amplitudes and phases: $a_1(0) = 1; a_{2,3,4,5} = 10^{-5}; \Psi(0) = \Phi(0) = 0$. The dependence of the different order normalized amplitudes (fundamental a_1, third a_3 and fifth a_5) versus the normalized coordinate $\tilde{z}$ for the process (2), and the step process (7), which emphasize the highest conversion efficiencies for the fifth harmonic generation in benzene is presented in Fig. 1 in the case where there is no wave vector mismatch for the third as well as for the fifth order harmonic generation process. The best conversion efficiency ($9.918.10^{-1}$) was obtained for a normalized distance $\tilde{z} = 4$ using for irradiation a laser having $\lambda = 1,06\mu m$ and a power of about $10^{11} W/cm^2$. For the same process, a good conversion efficiency is obtained when the wave vector mismatch $\tilde{\Delta}_{15} = 0$ and $\tilde{\Delta}_{13} = -9,74 \times 10^{-2}$ (Fig. 2) for $\tilde{z} = 3,4$. Good conversion efficiencies were obtained in the cases of cyclohexane ($9,225 \times 10^{-1}$) (Fig. 3) and n hexane (Fig. 4) ($2,881 \times 10^{-1}$) for $\tilde{z} = 8$ respectively. The lowest conversion efficiencies were obtained in the cases of carbon tetracholoride ($3,683 \times 10^{-6}$) (Fig. 5) and chloroform (Fig. 6) ($5,486 \times 10^{-6}$) for $\tilde{z} = 0,1$ and $\tilde{z} = 0,11$ respectively.

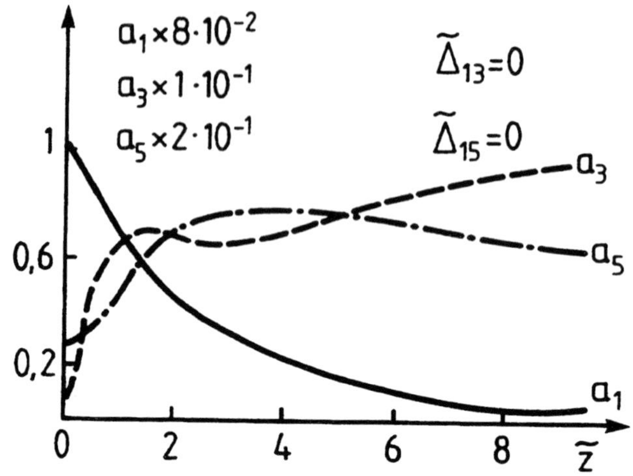

Figure 1. *Dependence of the fundamental,* a_1(——), *the third harmonic,* a_3(– – –), *and the fifth harmonic,* a_5(– · – ·), *in benzene for the processes (2)-(7) on the normalized distance* $\tilde{z}$ *for the following values of the normalized wave vector mismatch:* $\tilde{\Delta}_{13}=\tilde{\Delta}_{15}=0$.

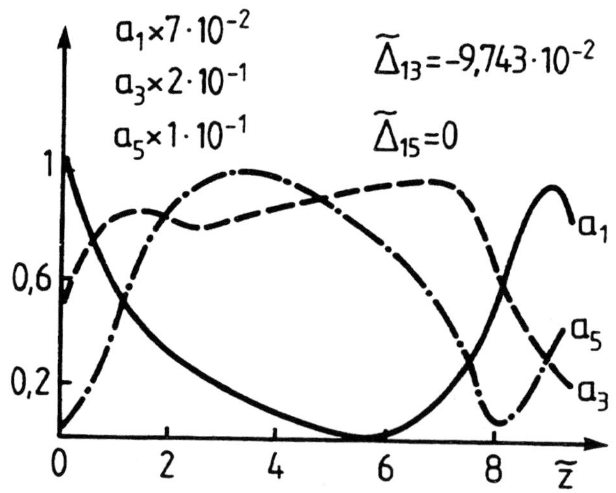

Figure 2. *Dependence of the fundamental,* a_1(——), *the third harmonic,* a_3(– – –), *and the fifth harmonic,* a_5(– · – ·), *in benzene for the processes (2)-(7) on the normalized distance* $\tilde{z}$ *for the following values of the normalized wave vector mismatach:* $\tilde{\Delta}_{15}=0; \tilde{\Delta}_{13}=-9,74\times10^{-2}$.

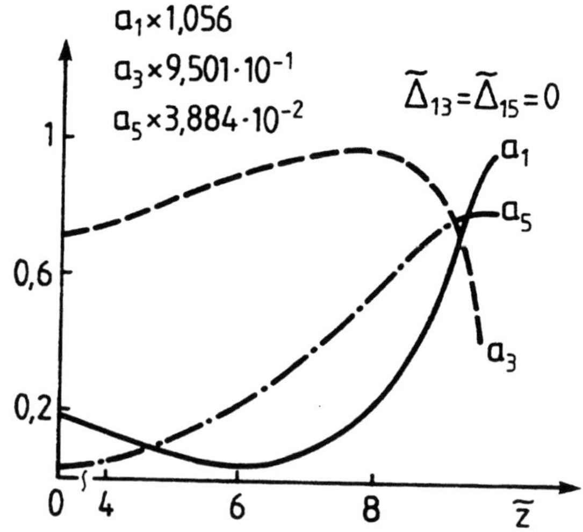

Figure 3. *Dependence of the fundamental,* a_1(——), *the third harmonic,* a_3(– – –), *and the fifth harmonic,* a_5(– · – ·), *in cyclohexane for the processes (2)-(7) on the normalized distance* $\tilde{z}$ *for the following values of the normalized wave vector mismatch:* $\tilde{\Delta}_{13} = \tilde{\Delta}_{15} = 0$.

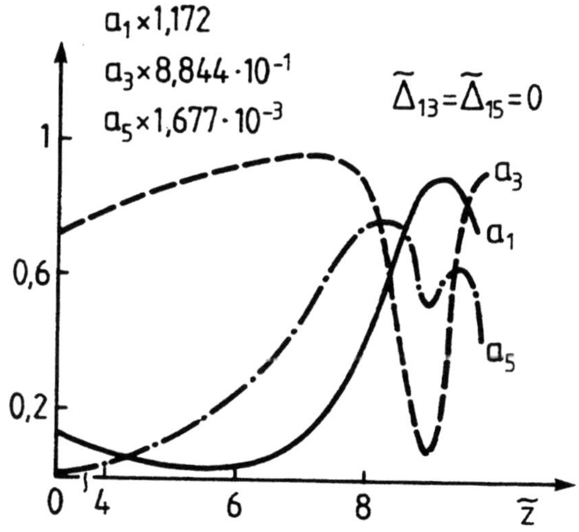

Figure 4. *Dependence of the fundamental,* a_1(——), *the third harmonic,* a_3(– – –), *and the fifth harmonic,* a_5(– · – ·), *in hexane on the normalized distance* $\tilde{z}$ *for the following values of the normalized wave vector mismatch:* $\tilde{\Delta}_{13} = \tilde{\Delta}_{15} = 0$.

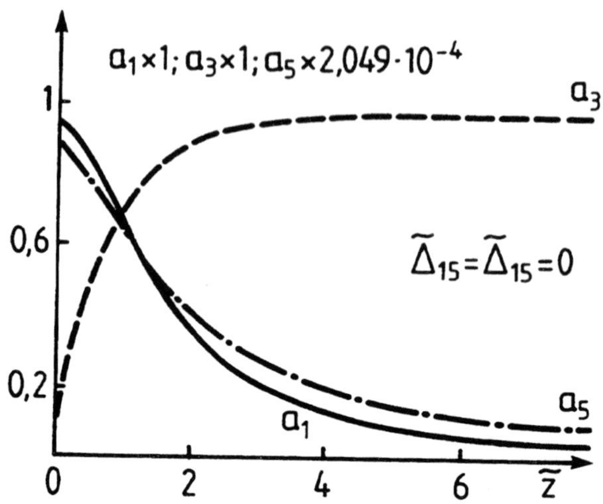

Figure 5. *Dependence of the fundamental, a_1(——), the third harmonic, a_3(– – –), and the fifth harmonic a_5(– · – ·) in carbon tetrachloride for the processes (2)-(7) on the normalized distance $\tilde{z}$ for the following values of the normalized wave vector mismatch:* $\tilde{\Delta}_{13}=\tilde{\Delta}_{15}=0$.

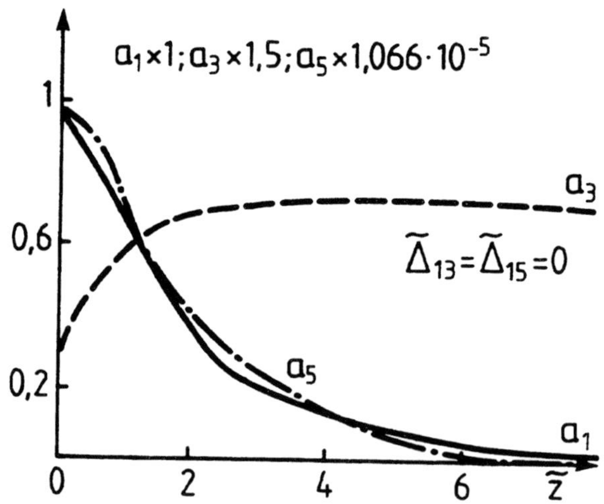

Figure 6. *Dependence of the fundamental, a_1(——), the third harmonic, a_3(– – –), and the fifth harmonic, a_5(– · – ·), in chloroform for the processes (2)-(7) on the normalized distance $\tilde{z}$ for the following values of the normalized wave vector mismatch:* $\tilde{\Delta}_{15}=0$; $\tilde{\Delta}_{13}=0$.

4 CONCLUSIONS

Taking into account the dependence of the higher order harmonics on the normalized distance $\tilde{z}$, one may obtain information concerning the efficiencies of the interacting processes which determine harmonic generations (fifth in our case) and also the competition between different processes.

References

[1] F. Kaisar and J. Messier, "Third harmonic generation in liquids", *Phys.Rev. A.*, Vol. 32 No. 4, 1985, pp. 2352-2363.

[2] J.V. Tomov, M.C. Richardson, "Fifth-harmonic generation in isotropic media", *IEEE J. Quantum Electrron.*, Vol. QE12 No. 9, 1976, pp. 521-532.

[3] I.M. Popescu, N.N. Puscas, P.E. Sterian, D. Irimescu and A.Gh. Podoleanu, "Theoretical study of the fifth harmonic generation in organic liquids", *SPIE*, Vol. 1361, Physical Concepts of Materials for Novel Optoelectronic Device Applications I, 1990, pp. 1041-1047.

Studies in Classical
and Quantum Nonlinear Optics

ISBN 1-56072-168-5

THE PARAMETRICAL INTERACTION OF LIGHT WITH A TWO-COMPONENT NON-LINEAR MEDIUM

N.L. Romashin
Institute of Radioengineering and Electronics,
Russian Academy of Sciences, Russia

1. STATEMENT OF THE PROBLEM

The possibility of efficient amplification of the electromagnetic radiation when interacting with the two-component non-linear medium has been studied in the paper. Two relativistic electron beams have been taken as the medium. One of the beams, namely the pumping beam, is density modulated with the pumping frequency ω_i. In the first approach the interaction process on such system can be considered as Compton scattering of Lengmure waves of pumping beams by the another working relativistic electron beam. However, due to the non-linearity inherent in both beams this process becomes more complex and the scattering by the pumping beam also takes place. The system under consideration can be treated as a new type of free electrons laser. The condition for resonance parametric interaction of electromagnetic radiation of signal wave having the frequency ω_s and the longitudinal wave number h_s with two electron beams take the form:

$$\omega_i - h_i v_{z0} \approx \omega_s - h_{s0} v_{z0} \qquad (1)$$

where h_i is the longitudinal wave number one of the Lengmure waves of pumping beam; v_{z0} is the longitudinal velosity of the working electron beam. ,From this we can obtain the condition for resonance longitudinal interaction of the working beam with the combinational wave. This wave has got the longitudinal wave number $h_0 = h_{s0} - h_i$ and the frequency $\omega = \omega_s - \omega_i$

$$v_{z0} \approx \omega / h_0 \qquad (2)$$

The advantage of this interaction mechanism is the possibility of its application for amplification and generation of powerful electromagnetic radiation including the light at the expense of the consigerable Doppler frequency upconversion

$$\frac{\omega_s}{\omega_i} \approx \frac{1 + |v_{z0}/v_i|}{1 - v_{z0}/v_{s0}} \qquad (3)$$

where $v_{s0} = \omega_s / h_{s0}$, $v_i = \omega_i / h_i$ signal and pump wave phase velocities respectively. It follows from this that the system under consideration allows to get a considerable additional gain in the frequency up conversion and in the frequency ω_s due to using the slow pumping beam.

The following model has been considered. Two relativistic electron beams move in the opposite directions in a waveguide with radius a in the longitudinal magnetic field H_0: the pumping beam with radius b^i and electron velocities v_z^i and v_φ^i, the working beam with radius b and electron velocities v_z and v_φ. E_{0k} (TM) and H_{0k} (TE) axially symmetric modes of the flat waveguide were chosen as signal waves. The longitudinal modulation with respect to the charge density of the pumping beam at the pumping frequency ω_i was assumed to be given in the form $\rho_i \exp(-i\omega_i t + i h_i z)$. With the aim to obtain a high frequency up conversion (3) the following ratio of the working beam and the pumping beam velocities was chosen $v_{z0} \sim c$, $v_{z0} \gg v^i_{z0}$.

2. BASIC EQUATIONS

The self-consistent problem of interaction in the system under consideration has been solved analytically. To solve this problem relativistic three dimensions equations of the electrons' motion and the exitation theory [1] describing the fields excited in the waveguide by the current wave in electron beam have been used. The interaction of the working electron beam with the combinational wave in case the condition of syschronizm (2) holds is described by the equation of the longitudinal motion:

$$\begin{aligned}\frac{dv_z}{dt} &= \frac{e}{m_0\gamma\gamma_0^2}(E_z - \mu_0 v_{\varphi 0} H_r) - \frac{1}{2}\left(\frac{e\mu_0}{m_0\gamma}\right)^2 \left[\left(\gamma_0^{-2} - \left(\frac{v_{z0}v_{\varphi 0}}{c^2}\right)\right)\right.\\ &\times \mathrm{Re}\, ihA_{\varphi s}A^*_{\varphi i} + \mu_0 v_{\varphi 0}\left(\left(\frac{v_{z0}}{c}\right)^2 - \gamma_0^{-2}\right)\mathrm{Re}(A^*_{\varphi i}E_{zs} + A_{\varphi s}E^*_{zi}) \qquad (4)\\ &+\mu_0 v_{z0}\mathrm{Re}\, i\left(\frac{E_{rs}}{\omega_i - v_{z0}h_i}\left(\frac{\omega_i}{h_i}\varepsilon_0 E^*_{ri} - v_{\varphi 0}H^*_{zi} + \omega_H A^*_{\varphi i}/\gamma\right)\right.\\ &\left.\left. - \frac{E^*_{ri}}{\omega_s - v_{z0}h_s}\left(\frac{\omega_s}{h_s}\varepsilon_0 E_{rs} - v_{\varphi 0}H^*_{zs} + \omega_H A_{\varphi s}/\gamma\right)\right)\right],\end{aligned}$$

where $v_z = v_{z0} + \tilde{v}_z$, $v_\varphi = v_{\varphi 0} + \tilde{v}_\varphi$, $\gamma_0 = (1 - (v_{z0}/c))^{-1/2}$, $\gamma = (1 - ((v_{z0}^2 + v_{\varphi 0}^2)/c^2))^{-1/2}$, $\omega_H = e\mu_0 H_0/m_0$: $A_{\varphi s}$, $A_{\varphi i}$ are azimuthal components of the vector potentials of TE waves at frequencies ω_s, ω_i; H_{zs}, H_{zi} are magnetic field longitudinal components of TE waves at frequencies ω_s, ω_i; E_{zs}, E_{zi}, E_{rs}, E_{ri} are electric field longitudinal and radial components of TM waves at frequencies ω_s, ω_i.

In the equation (4) the term in square brackets defines the longitudinal force the combinational wave effect the beam with.

High frequency fields effecting the working and pumping beam electrons are excited in the waveguide by high frequency currents of the beams:

$$\vec{j} = \rho\vec{v}, \vec{j}^i = \rho^i\vec{v}^i \qquad (5)$$

where $\rho = \rho_0 + \mathrm{Re}(\tilde{\rho}_s + \tilde{\rho}_i + \tilde{\rho})$, $\tilde{\rho}_{s,i} = \exp(-i\omega_{s,i}t + ih_{s,i}z)$, ρ is the charge density of the working beam; the superscript i corresponds to quantities determined for the pumping beam.

With the help of the excitation theory the following expressions have been obtained for high frequency fields

$$\vec{E} = \mathbf{H}_1\vec{j} + \mathbf{H}_1^i\vec{j}^i, \quad \vec{H} = \mathbf{H}_2\vec{j} + \mathbf{H}_2^i\vec{j}^i \tag{6}$$

These fields are excited in the system by the waves of the working and pumping beam currents at frequencies ω_i, ω, ω_s.

Combinational currents components excite the electromagnetic radiation being amplified at the frequency ω_s. It has been found that currents of working $j_{\varphi s}$, j_{rs} and pumping j^i_{rs}, j^i_{zs} beams give the maximum contribution into the excitation of the field being amplified at the frequency ω_s.

The working beam current $j_{\varphi s}$ excites H_{0k} modes. The interaction process in this case is the scattering of the pumping beam waves by the working beam. Currents j_{rs}, j^i_{rs}, j^i_{zs} excite E_{0k} modes. In this case an additional interaction mechanism appears due to the scattering of the working beam wave with the frequency ω by the pumping beam. The rotating pumping beam excites simultaneously TE and TM pumping waves.

3. CHARACTERISTIC EQUATION

For more detailed analysis of the interaction process characteristic equations for the interaction with signal H_{0k} and E_{0k} modes $(k = 1, 2, 3, \ldots)$ separatly have been obtained. Special features of interaction in each of these cases are more clearly seen at the condition of resonance interaction with only one wave with k number. Then these equations are reduced to algebraic ones. For H_{0k} mode at the frequency ω_s with the longitudinal wave number $h_{kM} = h_{s0}$ at the condition of synchronism we have:

$$(h - (h_{kM} - h_i))(h + (h_{kM} + h_i))[(h - h_e)^2 - \Gamma h_{pr}^2] = -\frac{h_e^2}{2a^2\gamma^3}\frac{v^i_{\varphi 0}}{v^i_{z0}}\frac{c^2}{v_{z0}v^i_{z0}}\frac{J_{z0}}{J_e}\left(\frac{J^i_{z0}}{J_e}\right)^2 \rho_i^2 \left(\frac{J_e e}{m\varepsilon_0 2\pi c^3}\right)^3 R_{kM} \tag{7}$$

$$R_{kM} = \frac{J_1^2(K_k b)}{J_0^2(\mu_{0k})}\left[\frac{v_{\varphi 0}}{c}\frac{p_i^2}{h_i h_e}\left(H^i_{1zzi} + \left(\frac{c}{v_{z0}}\right)^2 H_{1zri}\right.\right. \times \left.\left(1 + K_k B\frac{J_0(K_k b)}{J_1(K_k b)}\right)\right)\left(H^i_{2\varphi ri} - \frac{v_{\varphi 0}}{v^i_{\varphi 0}}\left(\frac{p_i}{h_i}\right)^2 H^i_{1zzi}\right) \left. + \left(\gamma_0^{-2} - \left(\frac{v_{z0}v_{\varphi 0}}{c^2}\right)^2\right)\frac{v^i_{\varphi 0}c}{v^2_{z0}}(H^i_{2\varphi ri})^2\right], \tag{8}$$

where $h = h_s - h_i$ is the unknown longitudinal wave number of electron waves at the frequency ω_i, $K_k = \mu_{0k}/a$, $J_1(\mu_{0k}) = 0$, $\rho_i^2 = \rho_i^i \rho_i^{i*}/(\rho_0^i)^2$, $h_{pr}^2 = \omega_p^2/(v_{z0}^2 \gamma \gamma_0^2)$, $h_e = \omega/v_{z0}$, J_{z0}, J_{z0} are longitudinal currents of the working and pumping beams respectively; Γ is a coefficient defining the effect of beam's own high frequency fields at the frequency ω; J_e =17kA H_{1zzi}^i, H_{1zri}^i are elements of the tensor $\mathbf{H}_{1i}^i$, describing the effect of the TM wave electric fields of pumping beam with longitudinal current on the working beam, $H_{2\varphi ri}^i$ is an element of the tensor $\mathbf{H}_{2i}^i$, describing the effect of the TE wave magnetic fields of pumping beam with asimutal current on the working beam; $p_i = \sqrt{h_i^2 - K_i^2}$, $K_i = \omega_i/c$.

This equation can have complex roots when the amplification of the electromagnetic radiation takes place. The efficiency of this amplification depends on the magnitude of the right-hand side characteristic equation.

A special feature of the two-beam combinational interaction is the simultaneous interaction of TM pumping wave electric fields and TE pumping wave magnetic fields in rotating working beam $v_{\varphi 0} \neq 0$. As a result term proportional to $1/\gamma^3\gamma_0$ appears in the right-hand side of characteristic equation. In usial one-beam systems the dependence of the right-hand side of characteristic equation on relativistic factor has the form $1/\gamma^5$ [2]. Such efficiency increase is especially appreciable for ultrarelativistic beams with $\gamma > 10$.

For E_{0k} mode at the frequency ω_s with the longitudinal wave number $h_{kE} = h_{s0}$ and synchronism condition we have

$$(h - (h_{KE} - h_i))(h + (h_{KE} + h_i))[(h - h_e)^2 - \Gamma h_{pr}^2]$$
$$= -\frac{h_e h_{ke}}{2a^2\gamma^2}\left(\frac{c}{v_{z0}}\right)^3 \left(\frac{c}{v_{z0}^i}\right)^2 \frac{J_{z0}}{J_e}\left(\frac{J_{z0}^i}{J_e}\right)^2 \left(\frac{J_e e}{m\varepsilon_0 2\pi c^3}\right)^3 \rho_i^2 R_{KE} \qquad (9)$$

Special feartures of the interaction with the E_{0k} mode become the most distinction in case of rectilinear beams $v_{\varphi 0} = 0$, $v_{\varphi 0}^i = 0$. In this case we have:

$$R_{KE} = \frac{J_1^2(g_k b)}{J_1^2(\nu_{0k})}\frac{p_i}{h_i} H_{1zri}^i \left[\frac{p_e}{h_e\gamma_i} H_{1zr}^i \frac{J_1(g_k b^i)}{J_1(g_k b)} - \frac{p_i}{h_i\gamma} H_{1zri}^i\right] \qquad (10)$$

where $g_k = \nu_{0k}/a$, $J_0(\nu_{0k}) = 0$, $p_e = \sqrt{h_e^2 - k^2}$, H_{1zr}^i is the element of the tensor $\mathbf{H}_1^i$, describing the effect of the working beam on the pumping beam at the frequency ω. The first term in the expression (10) appears due to the scattering wave with frequency ω of working beam by pumping beam and excitation of the signal wave by the combinational current of the pumping beam j^i_{rs}. Dependence of the first term on the relativistic factor has the form $1/\gamma$ when $b > b^i$ and it doesn't depend on γ when $b < b^i$. This is responsible for the appearance of terms proportional to $1/\gamma^3$, $(b > b^i)$, $1/\gamma^2$ $(b^i > b)$ in the right-hand side of the equation (9). For the relativistic beam it means an extremely considerable increase of the interaction efficiency.

Thus, in comparison with the conventional scattering in one-component medium (electron beam) this system has got the following adventages: a much greater ratio of the amplified radiation frequency and the pumping frequency (possibly by several orders of the magnitude) and a greater amplification coefficient (by several orders depending on the beam parameters). It may be assumed that if newly developed electron beam accelerators be used in this system, it will be possible to extend to a much more shorter wave band.

REFERENCES

[1] V.A.Solntsev, N.L.Romashin "Contribution to the problem of constructing different forms of the theory of excitation of periodic waveguides", Radio engineering & electronic physics, Vol.28, No.9, 1983, pp. 97-102.

[2] D.A.Deacon, L.R.Elias, J.M.Madey, G.J.Ramian, H.A.Schwettman, and T.I.Smith "First operation of free-electron laser", Phis. Rev. Letts., Vol.38, No.16, 1977, pp.892-894.

Studies in Classical
and Quantum Nonlinear Optics

ISBN 1-56072-168-5

OPTICAL FREQUENCY DOUBLING IN ALUMINIUM NITRIDE THIN FILMS

J. F. Roux, J.L. Coutaz*, D. Blanc#, A. Cachard# and J. C. Pommier#*

* LEMO-ENSERG, BP 257, 38016 Grenoble Cedex, France

TSI, Université de St Etienne, 23 rue du Dr Michelon, 42023 St Etienne Cedex 2, France

1. INTRODUCTION

Aluminium nitride (AlN) is a piezoelectric crystal belonging to the crystal class 6mm, with physical features close to that of zinc oxyde or zinc sulfide. As the latters exhibit good linear and nonlinear optical properties, it is foreseeable that AlN is also a good material for nonlinear optics. To our knowledge, the first estimation of the second order nonlinear efficiency of AlN was given by Levine [1] using a bond charge model: the Miller parameter of AlN was found to be of the same order of magnitude than that of ZnO or GaN.

However, it is almost impossible to grow huge crystals in order to perform optical experiments with bulk AlN. Nowadays, good quality thin films of AlN are fabricated [2,3] and therefore nonlinear experiments can be performed on these layers. We concentrate here our study on second harmonic generation (SHG) in ALN films. We have already reported the electro-optic coefficients for such films [4], which were determined using integrated optics techniques. The r_{ij} coefficients are of the same order of magnitude than those of thin films of ZnO [5].

The first report of SHG in AlN was published by Fujii *et al.* [6]. These authors measured $d_{33}=23.2 \cdot d_{11,quartz}$ for sputtered films of AlN. In addition, AlN exhibits a large energy gap (E_g~6 eV), which makes it transparent down to 0.19 μm. These two qualities, high nonlinearity and wide domain of transparency, make it a very good candidate for nonlinear thin film material. It is the aim of this paper to investigate in details the nonlinear optical properties of AlN layers.

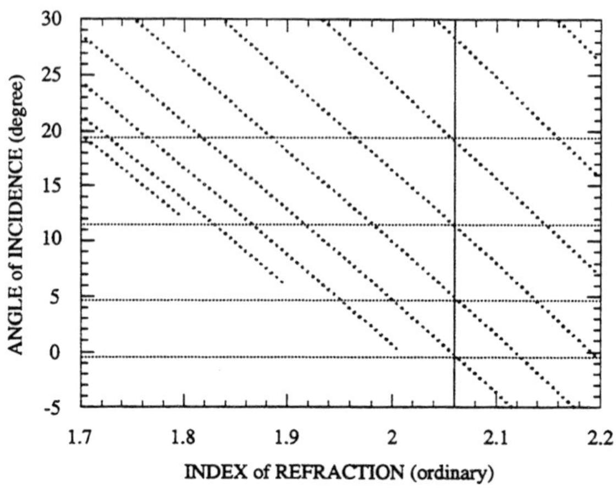

Figure 1: Dispersion curve of the AlN waveguide at λ=1.064 µm in TE polarization. The film thickness is 2.95 µm. The dashed lines correspond to the calculated dispersion curves of the different modes. The horizontal straight lines are the measured angular position of the m-lines. The vertical line indicates the value for which the calculated and measured m-lines angles cross for the same index of refraction of AlN. The precision of the determination of the index is much better than 0.01.

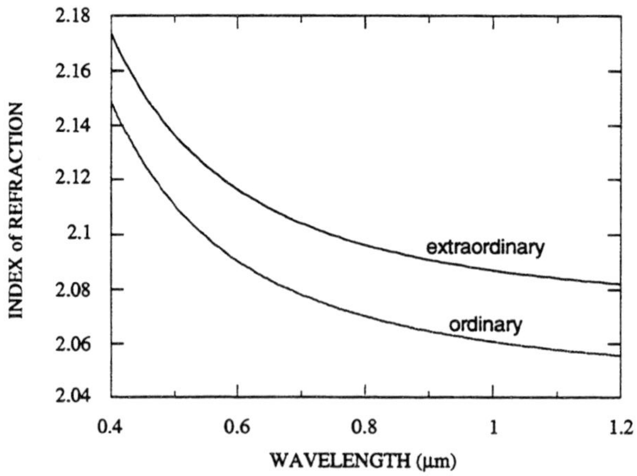

Figure 2: Dispersion of the index of refraction of AlN. The ordinary and extraordinary indices obey the Cauchy 's law :

$$n_o=2.044+0.0167/\lambda^2$$
$$n_e=2.071+0.0164/\lambda^2$$

4. SECOND HARMONIC GENERATION

AlN is known to belong to the point group 6mm. As we are working in the transparency range of this material, we can apply the Kleinman symmetry relations, thus only two non-linear coefficients remain : d_{31} and d_{33}. As the C-axis of the AlN layer is normal to the surface, we can measure the d_{31} (d_{33}) coefficient using a TE (TM) polarized pump beam and measuring the transmitted TM SH beam. The effective nonlinear coefficient d_{eff} is :

for the TE-TM experiment : $d_{eff} = d_{31} \sin(\theta_1)$

for the TM-TM experiment : $d_{eff} = \sin(\theta_1)(3d_{31} \cos^2(\theta_1)+d_{33} \sin^2(\theta_1))$ (2)

where, in the plane of incidence, θ_1 is the angle between the Poynting vector and the optical axis inside the material.

The experimental set-up is shown in figure 3. The Nd-YAG laser supplies pulses of 10 ns duration at 1.064 µm, with a repetition rate of 10 Hz. The SH signal passes through infrared filters (Schott KG5) and interference filter at 0.532 µm, and then it is detected by a photomultiplier tube. A small part of the pump beam is reflected by a glass slide towards a KDP crystal which generates a reference SH signal. Both signals are integrated by a boxcar and a computer records the ratio between them. The samples are placed on a motorized rotation support. For the pump polarization dependence experiments, the polarization of the laser is rotated with a $\lambda/2$ plate.

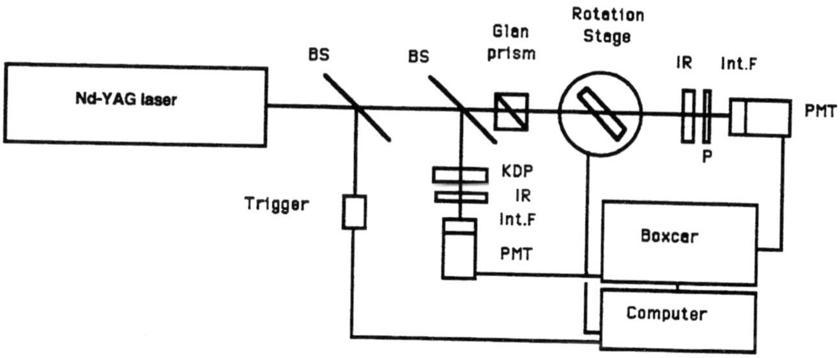

Figure 3: Experimental set up. BS stands for beam spliter, PMT for photomultiplier tube, Int. F for interference filter, IR for KG5 glass filter and P for polarizer.

In order to determine the nonlinear coefficients of our films (deposited at 600°C), we measure the transmitted SH signal generated by the sample. Results recorded for TE-TM and TM-TM interactions are shown in figures 4 and 5.

The absolute value of d_{ij} is obtained by comparison with the SH intensity generated by a quartz slide in a Maker fringes experiment performed in the same conditions as for AlN ($d_{11,quartz} = 0.30$ pm/V [8]). To derive d_{ij} from both measures (AlN and quartz), we take into account [9] the transmission coefficients at the different interfaces of the device, as well as the dispersion of the material (in particular, the dispersion of AlN measured previously).

The theoretical curve for the TM-TM interaction gives the value $|d_{33}|=1.85 \pm 0.15$ pm/V, and the TE-TM one's gives $|d_{31}|=0.04$ pm/V if we set the ratio d_{33}/d_{31} positive. The error range on d_{33} is mainly due to the accuracy of the refractive index measurement. These values are much weaker than those previously reported [6]. This tends to restrain the prospect of using AlN as an efficient frequency doubler. But we only report here results incoming from bulk investigation; because of the high refractive index of AlN, this configuration does not lead to reach angles θ_1 large enough (see equations 2) to improve the SHG merit of AlN as it could be done in a waveguide configuration.

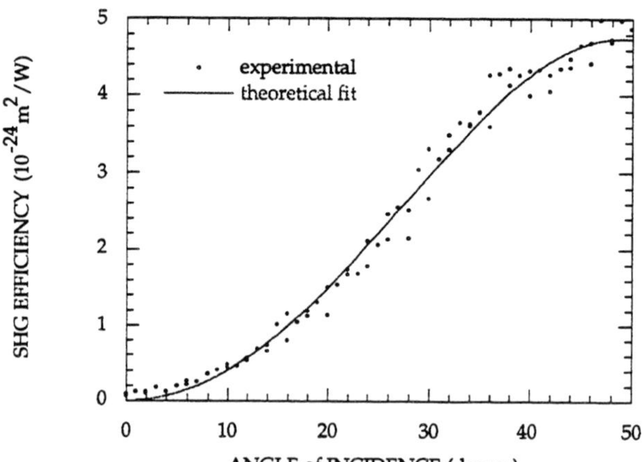

Figure 4 : Experimental and theoretical results of SHG efficiency (pump TE, SH beam TM).

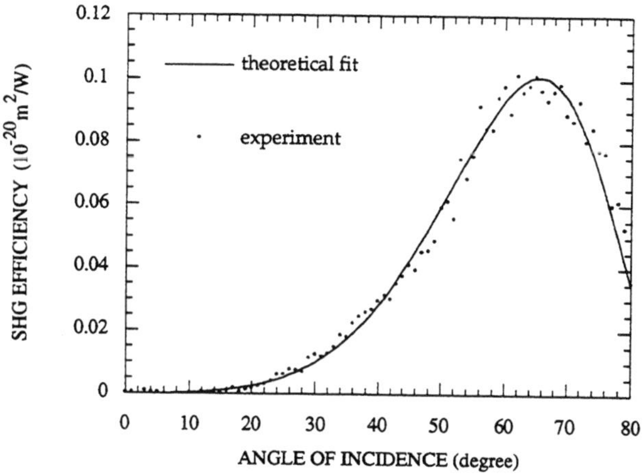

Figure 5: Experimental and theoretical results of SHG efficiency (pump TM, SH beam TM).

5. SECOND HARMONIC GENERATION IN INHOMOGENEOUS SAMPLES

As mentionned before, some layers were grown at low temperature (T<150°C).We record the SHG signal from these layers at normal incidence. Figure 6 shows the SHG signal versus the polarization of the pump beam. Assuming that the crystal belongs to the crystal class 6mm and that the C-axis is perpendicular to the AlN film, the second harmonic signal should be zero (see equations 2). The unexpected nonlinear behaviour of these layers can be modelized by setting the optical axis of the material not perpendicular to the film surface. This model has been confirmed by a x-ray diffraction experiment at grazing incidence in order to probe the first hundreth of nanometers of the film. The x-ray diffraction patterns point out that the layer is mainly homogeneous except in the vicinity of the AlN surface where the optical axis direction becomes parallel to the film.

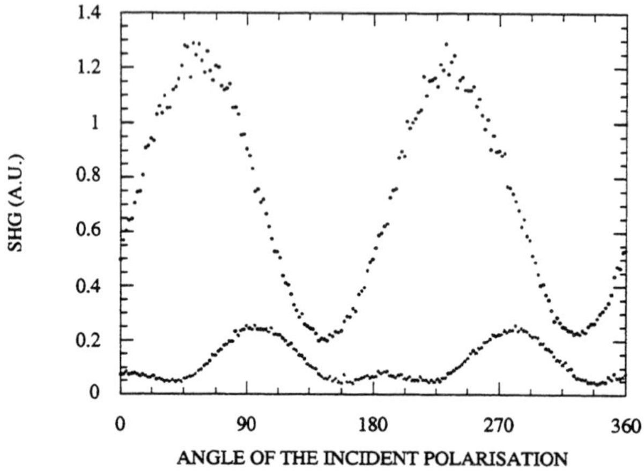

Figure 6: SHG from inhomogeneous layer of AlN. The signal is analysed in two perpendicular directions corresponding to the edges of the sample.

6. CONCLUSION

We report new values, comparable to those of urea, for the non-linear coefficients of sputtered AlN. In addition, the dispersion relation of the refraction index is given, showing a possible phase matching for SHG at 1.355 μm. Therefore, large harmonic conversion can be expected from AlN layers, especially in waveguide configuration.

Layers grown at low temperature exhibit orientation inhomogeneity, which was pointed out from SHG measurements. This confirms that SHG can be used with benefit to investigate the crystalline symmetry of materials.

REFERENCES

1 B. F. Levine, Phys. Rev. Lett. **25,** 440 (1970)

2 A. Cachard, R. Fillit, I. Kadad and J.C. Pommier,Vacuum **41,** 1151 (1990)

3 T. Shiosaki, M. Adachi and A. Kawabata, Thin Solid Films **96,** 129 (1982)

4 P. Gräupner, J. C. Pommier, A. Cachard and J. L. Coutaz, J. Appl. Phys. **71,** 4136 (1992)

5 D. Persegol, E. Pic and J. Plantier, J. Appl. Phys. **62,** 2563 (1987)

6 Y. Fujii, S. Yoshida, S. Misawa, S. Maekawa and T. Sakudo, Appl. Phys. Lett. **31**, 815 (1977)

7 W. P. Chen and J. M. Chen, J. Opt. Soc. Am. **71**, 189 (1981)

8 D. A. Roberts, IEEE J. Quant. Electron. **28**, 2057 (1992)

9 J. Jerphagnon and S. K. Kurtz, J. Appl. Phys. **41**, 1667 (1970);
S. K. Kurtz, in *Laser Handbook*, pp 923-974, edited by F.T. Arecchi and E.O. Schulz-Dubois, North Holland, Amsterdam (1972)

Studies in Classical
and Quantum Nonlinear Optics

ISBN 1-56072-168-5

COMPETING SUBLEVEL COHERENCE EFFECTS ON NEARLY RESONANT PHASE-CONJUGATION IN THREE-LEVEL GAS MEDIA

C. Schmidt-Iglesias and R. Corbalan
Departament de Física, Universitat Autònoma de Barcelona
08193, Bellaterra, Spain

1. INTRODUCTION

Sublevel coherence under efficient transverse optical pumping (TOP) conditions has been recently pointed out as a distinct mechanism for phase-conjugation (PC) through backward degenerate four wave mixing (DFWM), that can be easily controlled by an external magnetic field [1-3]. In case of homogeneously broadened systems [1,2], coherent population trapping [4] in the ground state sublevels has been ascribed to the origin of a pronounced level-crossing dip, ocurring at pump intensities much lower than those required for saturation of the optical transitions. In this contribution we evidence the possible distinction of two different sublevel coherence effects whenever the velocity distribution of atoms in gas phase is taken into account.

2. THEORETICAL BACKGROUND

We consider the backward DFWM configuration of Fig.1. The gas medium is composed of Λ-type three-level atoms with two optical transitions of equal dipole moments d . We admit equal dipole coherence decay rates $\Gamma = \frac{1}{2}(\gamma_0 + \gamma)$, being the lower to upper level population decay ratio $r = \gamma/\gamma_0 \ll 1$.The field-free population is equally distribu-

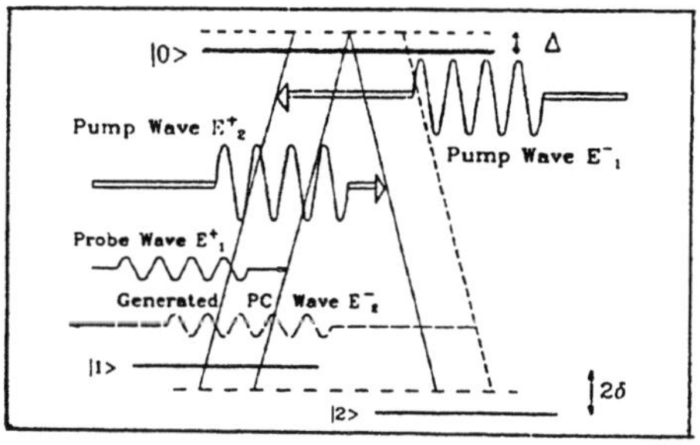

Figure1.- Geometry and optical transition scheme of the considered DFWM interaction.

buted among the closely spaced lower levels ($P_1 = P_2$), and their velocity distribution is assumed to be Maxwellian, with most probable velocity u. The generated signal and incident waves are described by

$$E_j^{\mu} = 1/2\, A_j^{\mu}\, \vec{e}_j\, exp[-i(wt - \mu kz)] + c.c. \qquad \mu = \pm 1; \quad j = 1,2\;, \tag{1}$$

and have the adequate polarization $\vec{e}_j$ to drive only the $0-j$ transition. Their detuning, lying within the one-photon Doppler profile, is defined by

$$\Delta_1 = \Delta - \delta \qquad \Delta_2 = \Delta + \delta, \tag{2}$$

where $|\Delta| > \Gamma$ is a fixed and common optical detuning of the fields, while 2δ accounts for the variable Zeeman sublevel splitting.

The nonlinear mixing is evaluated within the semiclassical framework of the density matrix ρ, exactly in the arbitrarily intense forward A_2^+ and backward A_1^- pump amplitudes, and up to first order in the weak probe amplitude A_1^+ ($dA_1^+/2\hbar \ll \Gamma, \gamma$). For an optically thin sample and in the slowly varying envelope approximation, the intensity of the PC signal wave is proportional to the phase-matched Fourier component $|\rho_{02(-k)}^{(1)}|^2$

The medium response requires in general a numerical evaluation when Doppler-broadening and moreover saturation of two arbitrarily intense pump beams comes into play [3]. However, when the relaxation ratio $r \ll 1$ and TOP dominates over saturation of the optical transitions, the mixing response $\rho_{02(-k)}^{(1)}$ is basically determined by sublevel coherence effects. Owing to motional washout effects, coherent population trapping induced by the nonlinear coherent coupling of the intense counterpropagating pump waves has however a negligible incidence [5]. Signal generation is domminated by a Zeeman coherence grating induced by the crosspolarized forward pump and probe beams, on which backward pump is diffracted along the path of the phase-conjugated wave. The mixing response admits thus a closed analytical expression

$$\rho^{(1)}_{02(-k)} \simeq \frac{2i\sqrt{\pi}\, d\, A_1^{+*}}{ku\ 2\hbar\Gamma} \sqrt{I_1^- I_2^+} \left(C_1 + C_2\right), \tag{3}$$

where the pump intensity parameters

$$I_j^\mu = \frac{|A_j^\mu|^2}{I_{SAT}^{TOP}(\Delta=0)} = \frac{(dA_j^\mu)^2}{4\hbar^2 r\gamma_0\Gamma}, \tag{4}$$

express the effective saturation of the three-level system in the TOP regime [4], and

$$C_1 = \frac{P_1 I_1^-}{J_1(1+J_1+2i\Delta_1/\Gamma)(1+J_1-2i\delta/\Gamma)Z_1} \tag{5}$$

$$C_2 = \frac{-P_2(1+J_2)}{2J_2(1+J_2-2i\Delta_2/\Gamma)Z_2} \tag{6}$$

The saturation parameters $J_j = \sqrt{2I_j^\mu + 1}$ and the sublevel coherence resonance denominators are given by

$$Z_j = 1 - \frac{2i\delta}{\gamma} + \frac{I_j^\mu}{1+J_j-2i\delta/\Gamma} + \frac{I_{3-j}^{-\mu}}{1+J_j-\mu\, 2i\Delta_j/\Gamma} \tag{7}$$

The two C_j terms describe respectively the share of P_j ground-level population participating in the sublevel coherence generation mechanisms, in accordance with forward and backward pump saturation effects. Note, however, that the P_1 contribution has a counterphase incidence and only participates effectively in the mixing process in case of an intense backward pump.

The resonance behaviour of signal generation is basically described by the Z_j denominators. In particular, for an $\Gamma < \Delta << ku$, and

$\delta < \Gamma$, the resonance position of the C_j contribution is approximately given by

$$\left.\frac{2\delta}{\Gamma}\right|_{Cj} \simeq \frac{2r\, I_{3-j}^{-\mu}\, \mu\, \Delta/\Gamma}{1+\sqrt{2I_j^{\mu}+1}+I_j^{\mu}+2\Delta^2/\Gamma^2} \tag{8}$$

being the effective halfwidth of the narrow C_j resonance

$$\left.\gamma^{ef}\right|_{C_j} \simeq 2r\Gamma\left(1+\frac{I_j^{\mu}}{1+\sqrt{2I_j^{\mu}+1}}+\frac{I_{3-j}^{-\mu}\left(1+\sqrt{2I_j^{\mu}+1}\right)}{2\left(1+\sqrt{2I_j^{\mu}+1}+I_j^{\mu}+2\Delta^2/\Gamma^2\right)}\right). \tag{9}$$

3. DFWM RESONANCES

The type of mechanisms that participate in PC depend markedly on the predominance of backward or forward saturation. Figure 2 illustrates typical spectra for $r = 0.01$ in the considered Doppler limit of $ku = 400\Gamma$. If only forward pump E_2^+ saturates the medium (Fig.2a), emission line-shapes are always centered at $\delta = 0$, while only the C_2 term contributes to PC. This means that a Raman transition transferring P_1 population to the most depleted level $|2>$ does not effectively participate in the sublevel grating mechanism. An enhancement of PC emission results from a resonant mechanism of P_2-population trapping in a sublevel superposition state in which the atoms are not perturbed by unequal light-shifts of the ground levels.

Higher generation efficiencies are actually obtained in the complementary case of backward pump saturation, while PC spectra are characterized by an intense laterally power-shifted peak (Fig.2b). It results from the resonant enhancement of a light-shift sensitive $2 \rightarrow 1$ Raman transition, induced by the grating waves whenever $I_1^- > I_2^+$ (see also Fig.3). But backward pump interference with the weak grating fields results also in P_1 coherent population trapping mechanisms of counterphase incidence, and a narrow antiresonance may be observed at $\delta = 0$ in the TOP regime. Since unilateral coherent population trapping effects decrease rapidly with an increase of the optical detuning Δ/Γ, their observation besides the intense Raman-line requires a minimum spectral separation, i.e. $\delta^{shift}|_{C2} \geq 2\gamma^{ef}|_{C1}$.

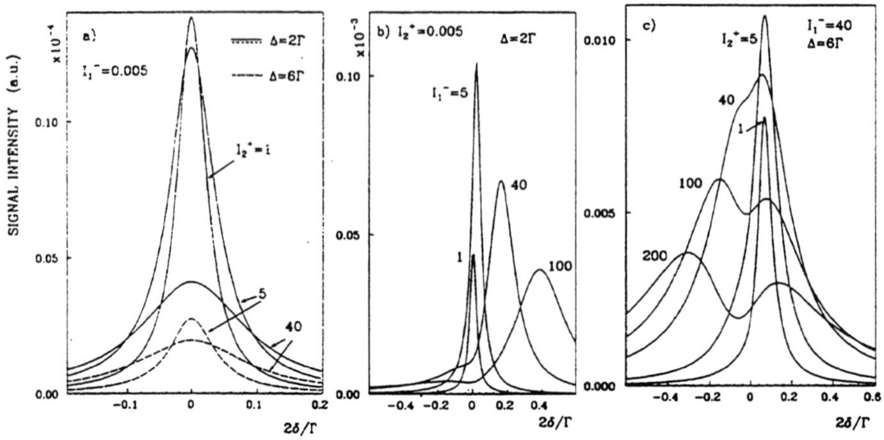

Figure **2**.- Characteristic PC emission lineshapes in case of a) forward, b) backward, c) bilateral saturation, being $r = 0.01$ and $ku = 400\Gamma$.

If however $I_2^+ > I_1^- > 1$, Raman processes transfering P_1- population to the most depleted $|2>$ level are effective, and backward pump induces counterphase effects of increased incidence. As both pump fields are intense, power-shift effects on the two ground levels become equiparable and the two intense counterphase PC resonances cannot be spectrally resolved. The complicated dispersion-like curves of Fig.2c arise thus from the competition of two counteracting P_1- and P_2-population transferring Raman resonances, where the position of the hole follows closely $2\delta/\Gamma|_{C1}$.

Population transfer and trapping effects may be certainly somewhat mixed and are hardly distinguishable when level-shift effects almost compensate. But for $\Delta = 0$, light-shift sensitive Raman processes are absent [4], while the unilateral coherent population trapping mechanisms are greatly enhanced and merge into a single intense PC line (see Eq.8), exhibiting for $I_1^- > I_2^+$ a narrow saturation dip that never annuls PC emission at line center owing to the dominating P_2-trapping contribution sustained by the grating fields [5].

4. CONCLUSIONS

All the most interesting features pointed here out arise in the inhomogeneously broadened regime and can be interpreted in terms of two counteracting sublevel coherence contributions, induced by either pump field, where the choice of pump intensity imbalance not only allows to vary the degree of incidence of the two contributions but also selects the type of competing sublevel coherence mechanisms. In particular, two different sublevel coherence mechanisms, as unilateral coherent population trapping mechanisms and two-photon processes transferring population to the most depleted ground level may be distinguished on the basis of their characteristic level-crossing tuning behaviour, whenever the laser frequency is sligthly detuned from the degenerate one-photon transitions. In contrast to fluorescence spectroscopy [1], both types of sublevel coherence mechanisms may have an inphase or counterphase incidence in PC. However, counterphase resonance lines may be only observed in case of predominant backward pump saturation.

ACKNOWLEDGEMENT

This work was partially supported by the Dirección General de Investigación Científica y Tecnológica (Spain), Project PB89-0319-C03-01.

REFERENCES

1.- G.P. Agrawal, 'Saturation effects in degenerate four-wave mixing on homogeneously broadened coupled transitions', *Phys. Rev. A*, Vol 28, No 4, 1983, pp. 2286-2295.
2.- E. Köster, J. Mlynek, and W. Lange, 'Zeeman coherence effects in degenerate backward four-wave mixing: saturation studies on coupled transitions', *Opt. Commun.*, Vol 53, No 1, 1985, pp. 53-58.
3.- C. Schmidt-Iglesias, G. Orriols and F. Pi, 'Phase- conjugate amplification through transverse optical Zeeman pumping in resonant Doppler-broadened degenerate four-wave mixing', *Appl. Phys. B*, Vol 47, 1988, pp. 27-33.
4.- G. Orriols, 'Nonabsorption resonances by nonlinear coherent effects in a three-level system', *Nuovo Cimento*, Vol 53B, No 1, 1979, pp.1-24.
5.- C. Schmidt-Iglesias, 'Nonlinear sublevel-coherence effects in the phase conjugate emission of Doppler-broadened dilute media', *Phys. Rev. A*, to be published end 1992.

Studies in Classical
and Quantum Nonlinear Optics

ISBN 1-56072-168-5

GENERATION OF LIGHT BULLETS BY USING THE MODULATION INSTABILITY OF THE GROUND STATE IN A SATURABLE SELF-FOCUSING MEDIUM

José María Soto-Crespo[1] *and Nail Akhmediev*[2]
[1]Instituto de Optica, C.S.I.C., Serrano 121, 28006 Madrid, Spain
[2]Optical Sciences Centre, Institute of Advanced Studies,
The Australian National University, Canberra, ACT 2601, Australia

Abstract. We study the propagation of the cylindrically symmetric ground state in a saturable self-focusing medium under the combined effects of diffraction, group velocity dispersion and nonlinear self-phase modulation. We have numerically solved the full wave equation and showed that a self-trapped beam in a nonlinear medium produces, through propagation, a train of three dimensionally optical solitons or light bullets.

1. INTRODUCTION

There is currently great interest in 3-D localized pulses or 'optical bullets' [1]-[3] in nonlinear media with anomalous group-velocity dispersion. It has been shown [3] that these 3-D self-focused pulses are robust in the sense that its shape is preserved even if it collides with another similar pulse. This behaviour resembles that of solitons in 1-D systems, so we shall call them '3-D solitons' sometimes in this paper. The question arises as to whether it is possible to generate these light bullets using the process of modulation instability of self-focused beams as suggested in Ref. [4]. The effect of modulation instability of a cw wave in a single mode optical fiber has been predicted [5], theoretically investigated [6] and experimentally observed [7],[8].

In this paper, we present an investigation of the modulation instability of the cylindrically symmetric ground state that exists in a self focusing medium with saturation. For these media only the ground state is stable relative to angular perturbations [10],[11] and this will be the only one we shall consider. We shall study the process of generation of a train of 3-D solitons as a result of this instability.

The remainder of this paper is organized as follows. Section 2 describes the nonlinear wave equation and notation we have employed and discusses the 3-D self-trapped solutions. The results of the numerical simulations are given in Sec.3, and Sec.4 contains our summary and conclusions.

2. BASIC THEORY

We consider the propagation of a linearly polarized wave in a homogeneous saturable self-focusing medium. In the slowly varying envelope approximation, in a reference frame traveling at the group velocity and adopting the scaling and notation of Ref.[11], the nonlinear wave equation for the electric-field envelope ψ can be written in the following form:

$$i\frac{\partial\psi}{\partial z} = -\nabla_T^2\psi - \frac{\partial^2\psi}{\partial t^2} - \frac{|\psi|^2\psi}{1+|\psi|^2}, \tag{1}$$

where the transverse Laplacian describing beam diffraction in cylindrical coordinates is:

$$\nabla_T^2 = \frac{\partial^2}{\partial r^2} + \frac{1}{r}\frac{\partial}{\partial r} + \frac{1}{r^2}\frac{\partial^2}{\partial\theta^2}. \tag{2}$$

We assume as in [1],[3] that the group velocity dispersion is negative.

Neglecting the dispersive term, that is for continuous wave, Eq.(1) has solutions with cylindrical symmetry which have a constant profile on propagation [11] : $\psi(r,z) = A(r)\exp(iq^2z)$, where q plays the role of a scaled effective index for the nonlinear wave. Introducing this expression in the Eq.(1) we obtain for A(r) the following equation:

$$\frac{d^2A}{dr^2} + \frac{1}{r}\frac{dA}{dr} - q^2A + \frac{|A|^2A}{1+|A|^2} = 0. \tag{3}$$

For each allowed value of q^2 there exist an infinite number of solutions of Eq.(3), each one characterized by the number n of field nodes . The fundamental solution or ground state being nodeless. This is the only one we shall consider. The others are unstable relative to angular perturbations [10]-[11], and therefore their behavior can be quite complicated.

2.1 Optical Bullets

It is also possible to find solutions of Eq.(1) corresponding to pulses that propagate without changing its temporal or spatial shape:

$$\psi(r,t,z) = U(\rho)\ \exp(i\gamma z), \tag{4}$$

where $\rho = \sqrt{r^2+t^2}$ and γ is the propagation constant. Following to Silberberg [1], we can call them 'optical bullets'. Now, Eq.(1) reduces to:

$$\frac{d^2U}{d\rho^2} + \frac{2}{\rho}\frac{dU}{d\rho} - \gamma U + \frac{|U|^2U}{1+|U|^2} = 0. \tag{5}$$

We have obtained numerical solutions of Eq.(5) using a shooting method. For a given value of γ the corresponding nodeless fundamental solution U(r) was obtained. From these numerical solutions we then calculated the self-trapped energy of each optical bullet as:

$$I = 4\pi \int_0^\infty |U(\rho)|^2 \rho^2 \, d\rho. \tag{6}$$

Figure 1.a shows this energy I as a function of γ. Note that there exist two regions with different curvature. We can expect [9] that the branch of this curve with positive slope is stable and the branch with negative slope is unstable. Figure 1.b shows the corresponding field profiles for γ=0.1, 0.2, and 0.5.

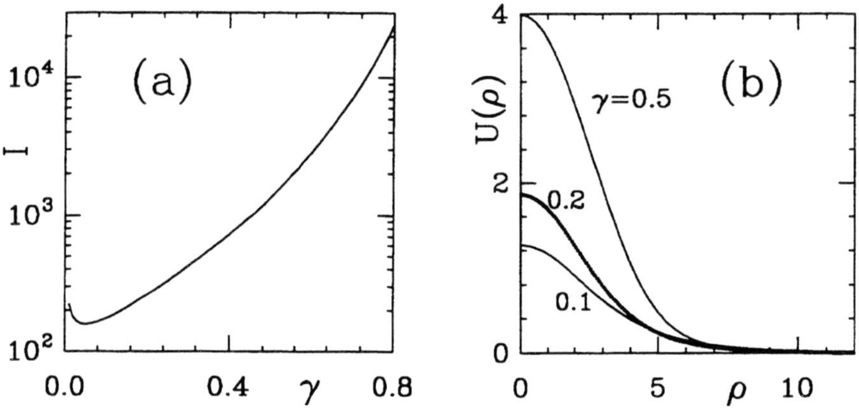

Figure 1.- a) Energy versus propagation constant for optical bullets. b) Profiles of the optical bullets corresponding to γ=0.1, 0.2 and 0.5

3. NUMERICAL SIMULATIONS

It was shown in Ref.[4] that the fundamental solution of Eq.(3), stable relative to angular perturbations, is unstable with respect to time periodic perturbations: for a certain range of frequencies there exist perturbations which grow exponentially with exponent equals to δz. This perturbative study is valid only when the perturbation is much smaller than the stationary ground state. To know what happens when the perturbation grows and becomes not much smaller than the stationary solution is necessary to solve numerically the nonlinear wave equation.

We have solved Eq.(1) with the initial condition:

$$\psi(r,t,0) = A_o(r) + \mu\, f(r) \cos(\nu t), \tag{7}$$

where μ is a small parameter, $\nu = 2\pi/T$ is the frequency of a periodic modulation, and $f(r)$ is the perturbation function. This function is complex in general. The function $f(r)$ has been calculated in [4] and is normalized so that $|Re(f(0))| + |Im(f(0))| = A_0(0)$. Figure 2.a shows the dependence of the instability growth rate (δ) on the modulation frequency (ν) for different values of the parameter q^2. The point A ($\nu = 0.395, \delta = 0.168$) in this figure corresponds to the frequency of maximum growth rate for $q^2 = 0.1$ that we shall study in detail in the rest of the paper. The corresponding perturbative function $f(r)$ is shown in Fig.2.b together with the stationary c.w. solution A_0 for $q^2 = 0.1$.

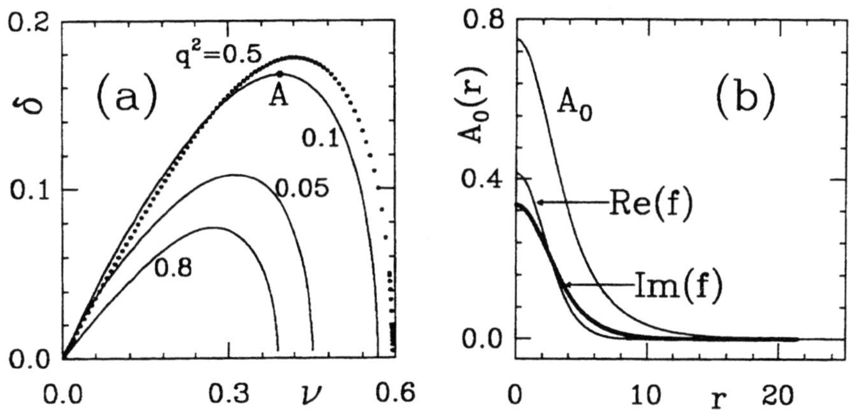

Figure 2. a) Dependence of the growth rate δ on the modulation frequency ν for q^2=0.05, 0.1, 0.5, and 0.8. b) Real and Imaginary parts of the perturbation along with the shape of the initial self-focused beam corresponding to q^2=0.1.

We have solved Eq.(1) numerically, and assuming that the cylindrical symmetry is conserved, using a split step method. We divide the equation into a dispersive part and a diffractive and nonlinear part. At each step we solve the dispersive part ($i\psi_z = -\psi_{tt}$) using the Fast fourier transform which automatically imposes time periodic boundary conditions. The diffractive plus the nonlinear part ($i\psi_z = -\psi_{rr} - \frac{1}{r}\psi_r - \frac{|\psi|^2\psi}{1+|\psi|^2}$) is solved using a Crank-Nicholson scheme with zero boundary condition at the maximum value of r considered. The grid size was chosen to have up to 2000 points in r-direction and up to 256 (usually 128) points in the t-direction. The step size in the z direction was typically taken to be $\Delta z = 0.1$. Numerical accuracy was checked by repeating the simulations for different grid sizes and step sizes. To ensure the reliability of our code we verified that 'optical bullets' with propagation constant α on the branch of energy dispersion curve (Fig.1a) with positive slope propagate long distances without experimenting any change in both their temporal and spatial profiles. These runs serve to confirm that optical bullets are stable on the positive slope branch of this curve. We also were able of confirming all the results found perturbatively.

Fig.3a shows an example of the evolution of the field intensity $|\psi|^2$ corresponding to the initial condition given by Eq.(7) in the point of its maximum amplitude ($r = 0, t = 0$). The modulation period was T=15.9 and $A_0(r)$ and $f(r)$ are those shown in Fig.2b. Initially the field stays as stationary solution until $\exp(\delta z)$ becomes comparable with μ^{-1}. Then the field intensity increases rapidly reaching a maximum value (point (a) in Fig.3a) and later on, the behaviour of the field intensity becomes oscillatory with an amplitude that decays smoothly. At the point (a), the field in each period is highly compressed in three dimensions (3-D) and the whole solution loks like a train of 3-D self-compressed pulses. After reaching the point (a) the degree of compression decreases up to the point of the first minimum of the curve in Fig.3a, where pulses become more spread out but still well separated. Then the pulses in each period are compressed again up to the point of the next maximum (lower than the first one). The process is repeated periodically but the amplitude decreases, converging to some fixed field intensity value ≈ 2.95. The same happens with the temporal and spatial widths of the wave (which become equal after certain propagation distance): they tend to the value ≈ 1.9. Figure 3.b shows the spatial and temporal half width at half maximum, as defined by the following expression:

$$|\psi(hwhm_r, 0)|^2 = \frac{1}{2}|\psi(0,0)|^2 = |\psi(0, hwhm_t)|^2. \tag{8}$$

If we look at the phase at which the field at the origin changes, we find that it oscillates periodically with z, the period being equals to 35.9 (frequency=0.175). The solution of Eq.(5) corresponding to a propagation constant α=0.175 has exactly $|U(0)|^2$=2.95 and $hwhm_\rho = 1.9$. In fact the whole profile of this solution coincides with the profile to which our numerical solution converges. That is, what we have obtained is a train of 'optical bullets'.

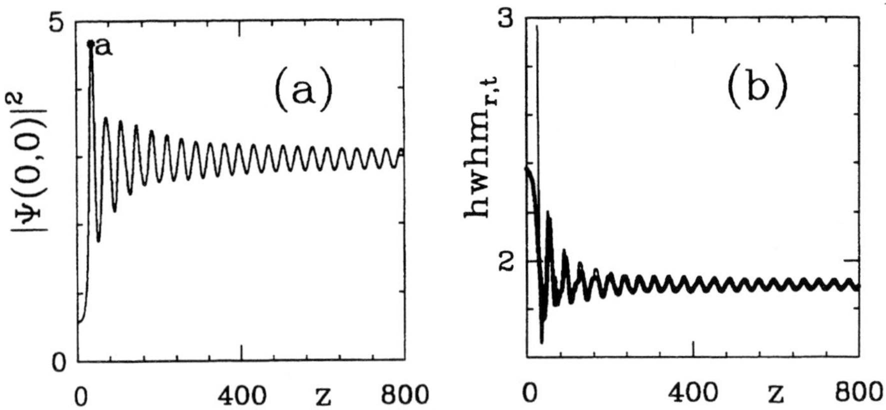

Figure 3.- a) Field intensity at the origin versus the propagation distance z for the initial condition given by Eq.(7). b) Spatial and temporal half widths at half maximum versus z.

Similar behaviour was observed for different values of the effective scaled index q^2. In addition, the formation of a train of 'optical bullets' was also observed for other kind of initial conditions such as:

$$\psi(r,t,z=0) = \sqrt{2}A_0(r)\cos(\frac{2\pi t}{T}), \tag{9}$$

and also with Gaussian profiles, which indicates that optical bullets could be excited using a wide spectra of initial conditions.

4. SUMMARY AND CONCLUSIONS

We have shown that an initially homogeneous self-focused beam breaks up into separate optical "clumps" as result of the modulation instability, and that each of these "clumps" converges to a 3-D soltion (optical bullet). Other initial configurations can also produce a periodic train of light bullets upon propagation. This shows that optical bullets are attractors in the Hilbert space of the solutions of the 3-D nonlinear Schrődinger equation.

References

[1] Y. Silberberg, 'Collapse of optical pulses',*Opt. Lett.*, **15**, No 22, pp.1282-1284 (1990).

[2] A.B.Blagoeva, S.G.Dinev, A.A.Dreischuh, and A.Naidenov. 'Light bullets formation in a bulk media', *IEEE J.Quant.Electron*, **QE-27**, No 8, pp.2060-2065 (1991).

[3] D.E.Edmundson and Enns R.H. 'Robust bistable light bullets', *Opt. Lett.*, **17**, No 8, p.586-588 (1992).

[4] N.N.Akhmediev, V.I.Korneev, and R.F.Nabiev, 'Modulation instability of the ground state of nonlinear wave equation. Optical machine-gun', *Opt.Lett.*, **17**, No 6, pp.393-395 (1992).

[5] A.Hasegawa and W.F.Brinkman, *IEEE J.Quantum Electron*, **QE-16**, 694 (1980).

[6] N.N.Akhmediev, V.M.Eleonskii, and N.E.Kulagin, "Generation of periodic trains of picosecond pulses in an optical fiber: Exact solutions," *Sov. Phys. JETP*, **62**, No 5, pp.894-899 (1985).

[7] 7. K.Tai, A.Hasegawa, and A.Tomita," Observation of modulational instability in optical fibers," *Phys.Rev.Lett.*, **56**, No 2, pp.135-138 (1986).

[8] K.Tai, A.Tomita, J.L.Jewell, and A.Hasegawa, "Generation of subpicosecond solitonlike optical pulses at 0.3 THz repetition rate by induced modulational instability," *Appl.Phys.Lett.*, **49**, No 5, pp.236-238 (1986).

[9] A.A.Kolokolov. 'Stability of the dominant mode of the nonlinear wave equation in a cubic medium', *Zh.Prikl.Mekh.Tekh.Fiz.*, No3, pp.152-155, (1973). [Engl.Transl.pp.426-428].

[10] A.A.Kolokov and A.I.Skykov, "Instability in the higher modes of a nonlinear equation", *J.Appl.Mech.Tech Phys.*, No.4, 1975, pp.519-522.

[11] J.M.Soto-Crespo, D.R.Heatley, E.M.Wright and N.N.Akhmediev, "Stability of the higher-bound states in a saturable self-focusing medium", *Phys.Rev.A*, Vol.44, No.1, 1991, pp.636-644.

Studies in Classical
and Quantum Nonlinear Optics

ISBN 1-56072-168-5

THE PERFORMANCE CHARACTERISTICS AND COMPARISON OF THREE KINDS OF LIQUID CRYSTAL LIGHT VALVES

W. Xue, N. Collings, and K.J. Weible
Institute of Microtechnology, University of Neuchâtel,
CH-2000 Neuchâtel, Switzerland

1. INTRODUCTION

Liquid Crystal Light Valves (LCLVs) have found application in many proposed and demonstrated optical computing systems and have been considered the key component for the systems. We have investigated three different types of LCLVs: based on twisted nematic, aligned nematic and ferroelectric (smectic C) liquid crystals.[1-3] The initial motivation of doing the characterizations was to find the optimal drive condition and some basic characteristic parameters, such as the sensitivity, contrast ratio, and gain, in order to use the LCLVs for a neural network application. But soon afterwards we became aware that the characterization itself is of significance because usually the characteristics of different types of LCLV are not always available, and in particular, some important characteristic parameters, such as spatial uniformity and read-out efficiency, are always ignored.

2. DESCRIPTION OF THE LCLVS

A LCLV transfers information from one beam of light (called the write light), which can be incoherent, to a second beam of light (called the read light), which is usually coherent. A typical device structure is shown in Figure.1. It consists of three main parts, the photoconductor, the liquid crystal layer and the light blocking layer and mirror assembly, which are sandwiched between two transparent electrodes. The light block layer acts to separate the read and write beams, reducing the effect of the read beam on the photoconducor.

There are two types of liquid crystalline phases employed in light valves, the nematic and the smectic C. The nematic is configured by appropriately preparing the bounding plates of the cell. For example, when the surfaces of the plates are treated so as to promote almost perpendicular alignment of the molecules, we refer to the homeotropic configuration.

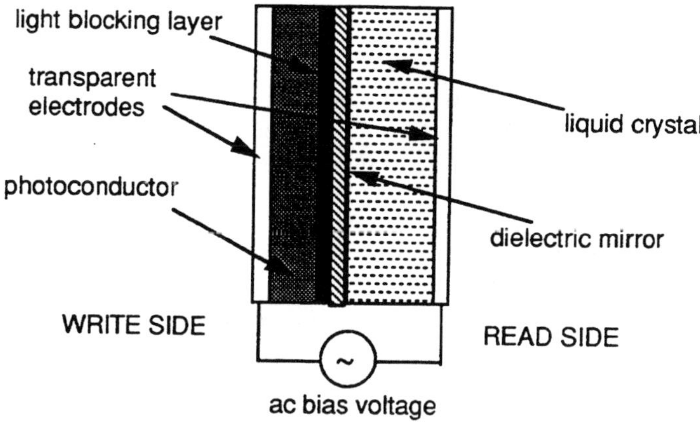

Figure 1. Structure of a LCLV

The homeotropic configuration of the nematic is used with liquid crystals of negative dielectric anisotropy when we wish to employ the electrically controlled birefringence (ECB) effect uniquely. The negative dielectric anisotropy implies that the liquid crystal director rotates perpendicular to the field, i.e. parallel to the plane of the cell, when a voltage is applied across the cell. If the device is oriented so that the plane of rotation of the liquid crystal director is at 45° to the plane of polarization of the incident light beam, then a crossed analyzer will give a dark OFF state and a light ON state. The reflectance of the valve in the ON state is a function of the retardation of the liquid crystal layer. When this is a quarter wavelength, the reflectance is a maximum. The LCLV1 had this configuration. The photoconductor is, in this case, sputtered cadmium sulphide CdS, and the diameter of the working aperture is 37 mm.

In a twisted nematic cell the walls are coated with an alignment layer which forces the liquid crystal molecules to lie parallel to the cell walls with their optical axes along the alignment direction in the "off" state. In the case of the 45° twisted nematic cell the two alignment directions are at 45° to each other, and the planes of molecules twist between the front and back surface of the cell. The LCLV2 had this configuration. In this state light incident with polarization along the front alignment direction is guided by the molecules and emerges with the same polarization after reflection from the dielectric mirror. Therefore, there will be a good dark state when the valve is placed between a crossed polarizer/analyzer pair with the director orientation on the front plate parallel to the polarization of the incident light field. When a write beam is incident on the photoconductor, the resulting voltage across the liquid crystal breaks the twist with the consequence that the polarization vector no longer follows the twist of the director. The light now only experiences a birefringent region near each wall of the cell. These regions rotate the polarization of the light so it no longer emerges with the original polarization. An analyzer is then used to convert the polarization to amplitude modulation. This provides the modulation between the "on" and "off" states of the device. The photoconductor in LCLV2 is amorphous silicon (a-Si:H), and the working aperture is 40 x 40 mm^2.

The LCLV3 is a prototype device using a chiral smectic C layer with a-Si:H photoconductor and a 30x30 mm^2 aperture. The dielectric mirror is replaced by a pixellated metal mirror in this device. In the FLC smectic C phase, the molecular directors **n** are arranged in layers. The director **n** is tilted by angle θ_t with respect to the layer normal **z**. A thin layer of FLC material was placed between two glass plates. Surface interactions between the material and glass plates can suppress the cone so that only two energy states exist at angles $\pm\theta_t$ with respect to the layer normal. Application of an electric field **E** produces a torque which switches the orientation of the FLC between the two stable states. One polarity of **E** results in **n** at angle $+\theta_t$. Reversing **E** rotates **n** to $-\theta_t$. When the tilt angle $2\theta_t = 45^\circ$ and the interaction length or thickness of the material is a quarter wavelength, a 180° phase shift occurs between the ordinary and the extraordinary eigenmodes of propagation, and the FLC LCLV operates as a switchable halfwave plate.

3. CHARACTERISTICS OF THE LCLVS

3.1 Read-out Efficiency Curve

The read-out efficiency is defined as the ratio of read-out light intensity to read-in light intensity, which directly indicates the power loss due to the LCLV in the optical system. Optical systems exploiting the threshold performance of the LCLV and employing feedback to achieve different processing operations need a high efficiency read-out.

The read-out and write light intensities were detected simultaneously as the write light intensity was changed. Then the measured read-out light intensity was normalized by the read-in light intensity and plotted as a function of the write light intensity (see Figure 2). The ordinate in these plots is the read-out efficiency. The read-out light intensity was detected behind the analyzer and corrected for the absorption in the analyzer.

The read-out efficiency of LCLV3 is relatively lower, it is partly because of its special drive mechanism.[4] We may also obtain the write sensitivity and the contrast ratio from the efficiency curve. The write sensitivity is represented by the minimum intensity of write light which switches the valve ON (90% of full ON-state). The contrast ratio is defined as the ratio of the read-out light intensity in the ON-state to the intensity in the OFF-state and measured after the analyzer which is crossed with respect to the polarizer.

The efficiency curve for LCLV1 was obtained by using a dye laser at 555 nm for both read and write beam. For LCLV2 and LCLV3 both read and write light were Ar^+ laser at 515 nm. We should keep in mind that the transfer curves are valid only for a certain read-in light intensity and on a certain drive conditions.

3.2 Effective Gain

The gain means the amplifying ability of a LCLV. With an ideal light blocking layer, write light at the threshold intensity would be capable of activating the liquid crystal layer and modulating a read beam of infinite intensity giving an LCLV with infinitely high gain. However this is not possible in real devices, since in a LCLV a small percentage of read light passes through the mirror and affects the photoconductor.

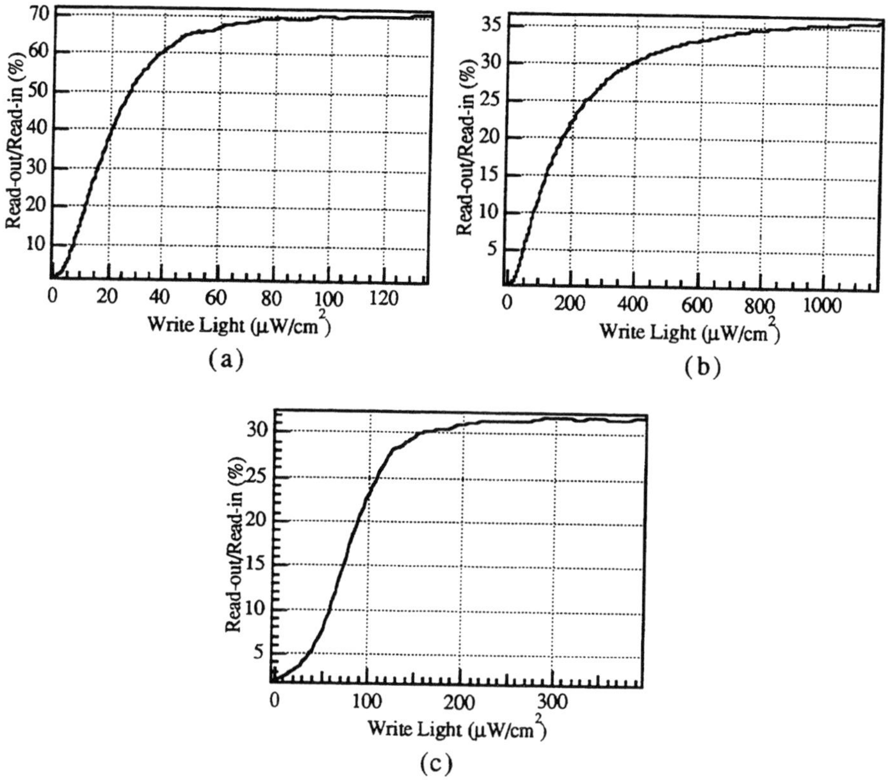

Figure 2. Efficiency curves: (a) for LCLV1; (b) for LCLV2; (c) for LCLV3.

In [4], we introduced a definition of effective gain which takes account of the read-out efficiency. The effective gain is defined as the ratio of the read-out light intensity to the write light intensity for 90% of ON-state. The maximum effective gain which is obtained when both read-out efficiency and read-in light intensity reach their maximum value, is the usable amplifying power of a valve.

Obviously the maximum read-in light intensity which does not affect the efficiency curve is the key to obtain high gain. We measured the effect of read-in light intensity on the read-out without write light. The results are shown in Figure 3. We see from the curves that LCLV1 was quite tolerable to the read light, but LCLV2 was almost switched on when the read-in light was over 1 mW/cm^2.

3.3 Spatial Uniformity

The uniformity of spatial light modulators is very important. Each parameter of a LCLV may change spatially across the valve. All parameters above were measured on a small area of each LCLV, because the optical performance was not homogeneous on the whole aperture of the LCLVs. The uniformity is represented by a uniformity area, for which the variation of read-out light intensity

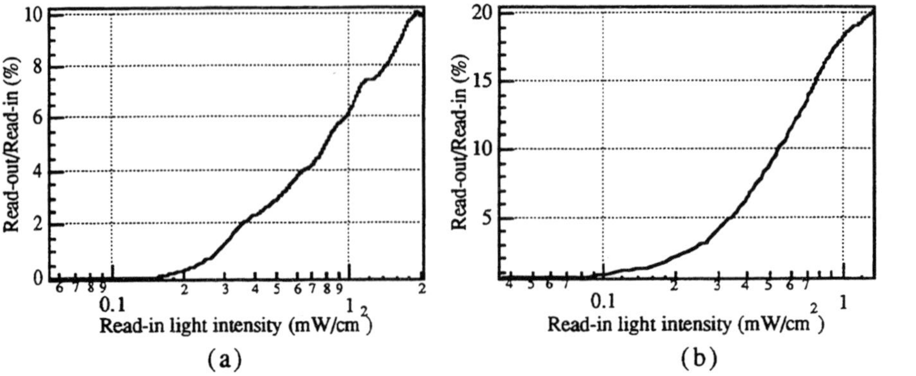

Figure 3. The effect of read-in light on read-out: (a) for LCLV1; (b) for LCLV2.

is less than ±10% of the ON-state light intensity. The results are given in Figure 4. Usually we measure each valve on its center, that means the result is probably not the best one. From Figure 4(b), e.g., we see the read-out efficiency is very low at the center, but it is nearly doubled at the edge of the valve.

3.4 Polarization Rotation and Response Time

The polarization rotation versus write intensity was measured. We found that only LCLV1 generates a linearly polarized read-out light, and the read-out light from LCLV2 or LCLV3 is no longer linearly polarized, but elliptically polarized. We use the principal axis of the ellipse to represent its orientation.

The response time of the three valves were measured, respectively. The rise time (τ_{on}) is from 10% to 90% of full ON-state, and the fall time (τ_{off}) is from 90% to 10% of full ON-state.

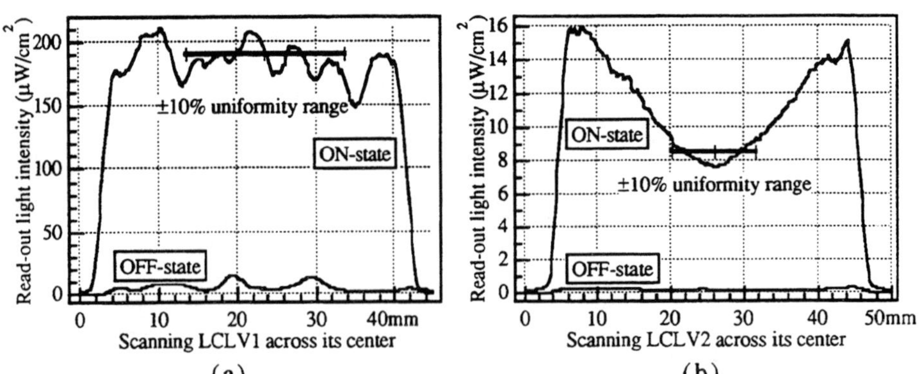

Figure 4. ON- & OFF-state reflectivities as a function of position: (a) on LCLV1; (b) on LCLV2.

Device		LCLV1	LCLV2	LCLV3
Spectral Sensitivity (nm)		540~600	450~650	450~650
Write Sensitivity		$45\mu W/cm^2$ (at 555 nm)	$514\mu W/cm^2$ (at 515 nm)	$133\mu W/cm^2$ (at 515 nm)
Contrast Ratio		41 : 1	50 : 1	16 : 1
Polarization rotation (state of polarization) in fully ON-state		90° (linear)	74° (elliptical)	82° (elliptical)
Response Time (ms)	rise	120	10	1.4
	fall	540	14	0.6
Uniformity Area (cm^2)		4 (±10%)	1.5 (±10%)	3.2 (±10%)
Read-out Efficiency		70%	36%	32%
Effective Gain		17 (at 555 nm)	0.21	0.34

Table 1. The characteristic parameters for the three LCLVs.

4. CONCLUSIONS

The characterization of three kinds of liquid crystal light valves has been performed and the results are summarized in Table 1. We find that the spatial nonuniformity is a common and severe problem for the three valves, and gain is only realized in LCLV1. However, these problems are technological rather than fundamental. A more fundamental problem is the polarization rotation generated by the valve. This depends on the configuration of the liquid crystal. Among our present devices, only the ECB effect in LCLV1 generates a full 90° rotation to a linearly polarized reflected beam.

REFERENCES

1. The LCLV1 was fabricated for our institute by ASULAB S.A., pass Max-Meuron 6, Neuchâtel, Switzerland.
2. The LCLV2 was fabricated commercially by GEC-Marconi Research Centre, Chelmsford, Essex CM2 8HN, England (*Techbrief* No M342, 'Amorphous Silicon Nematic Spatial Light Modulator').
3. D. Williams, S. G. Latham, C. M. J. Powles, M. A. Powell, R. C. Chittick, A. P. Sparks, and N. Collings, "An amorphous silicon/chiral smectic spatial light modulator", *J.Phys. D*, Vol.21, 1988, pp. S156-S159.
4. W. Xue, N. Collings, and K. J. Weible, "The performance characteristics of a ferroelectric liquid crystal light valve with pixellated metal reflector", *Optical Computing and processing*, Vol. 2, No. 2, 1992, pp. 69-76.

Studies in Classical
and Quantum Nonlinear Optics

ISBN 1-56072-168-5

A RESONANT MEDIUM IN A MODULATED OPTICAL FIELD

Zigmund Yankauskas
Technical University, Vilnius, Lithuania

The behaviour of a resonant medium in a modulated optical field can be described by the help of the Bloch equations

$$\dot{P} = EN + \Delta\omega Q \ , \tag{1}$$

$$\dot{N} = -EP \ , \tag{2}$$

$$\dot{Q} = -\Delta\omega P, \qquad \Delta\omega = \omega - \omega_0 \tag{3}$$

where N is the density of active atoms, P, Q the real and imaginary parts of the polarization, ω the optical frequency of the field E, and ω_0 the resonance frequency of the medium. The dipole moment of the transition, d_0, and Planck's constant, h, are assumed to be equal to unity so that the Rabbi frequency, d_0E/h, coincides with E.

In the case of AM modulation

$$E = E_0 + E_1 \cos \Omega t, \qquad \Omega \ll \omega \ . \tag{4}$$

If $\Delta\omega = 0$ the system of Eqs. (1)-(3) can be solved exactly. For $\Delta\omega \neq 0$ only various numerical methods are in use. Therefore from a methodical and practical point of view the cases where the Bloch equations may be solved analytically are of considerable importance.

From Eqs. (1)-(3) we have

$$\dot{P} = (E_0 + E_1 \cos \Omega t)N - \Delta\omega^2 \int P dt \ , \tag{5}$$

$$\dot{N} = -(E_0 + E_1 \cos \Omega t)P \ , \tag{6}$$

or in dimensionless form

$$\left(t \rightarrow \Omega t, m \rightarrow \frac{E_0}{\Omega}, 2q \rightarrow \frac{E_1}{\Omega}, \alpha \rightarrow \frac{\Delta\omega}{\Omega}\right)$$

$$\dot{P} = (m + 2q \cos t)N - \alpha^2 \int P dt \ , \tag{7}$$

$$\dot{N} = -(m + 2q \cos t)P \ . \tag{8}$$

Now, we seek solutions for P and N of the form

$$P = exp(i\beta t) \sum_{n=-\infty}^{\infty} a_n \; exp(int) \; , \tag{9}$$

$$N = i \; exp(i\beta t) \sum_{n=-\infty}^{\infty} b_n \; exp(int) \; . \tag{10}$$

In order to determine the expansion coefficients a_n and b_n, we substitute the two series into Eqs. (7) and (8) and compare the expressions for the same exponents. In this way a set of recurrent equations for the coefficients a_n and b_n is obtained, i.e.

$$\left(\beta + n\,\Omega - \frac{\alpha^2}{\beta + n\Omega}\right) a_n = q(b_{n-1} + b_{n+1}) + mb_n \; , \tag{11}$$

$$(\beta + n\Omega)b_n = q(a_{n-1} + a_{n+1}) + m\; a_n \; . \tag{12}$$

The condition

$$0 \le \beta \le 1 \tag{13}$$

can always be realized by a change of the summation index n. For the set of linear homogenous equations (11) and (12) to have a solution different from zero the system determinant must vanish, i.e.

$$| D |= 0 \; . \tag{14}$$

The determinant D is written out explicitly in Fig. 1.

The zeros of this determinant were sought for taking various real values of the parameters m, q, and α, and from this the corresponding real characteristic frequency β was determined.

For example, in Figs. 2 and 3 the curves with $\beta =$ const are given in a $\alpha - q$ plane for $m = 0$ and $m = 1$, respectively.

When β is determined the coefficients a_n and b_n can be calculated by the help of the recurrent equations in (11) and (12) taking into account that for $n > 7\alpha$

$$a_n \approx b_n \approx J_n(2q) \; ,$$

where $J_n(2q)$ are n^{th} order Bessel function of the first kind (Points A and B in Figs. 2 and 3 are the first zeros of $J_0(2q)$ and $J_1(2q)$ respectively).

In the case of FM modulation

$$\Delta\omega = \Delta\omega_0 + \omega^I cos\; \Omega t \; , \tag{15}$$

$$E = E_0 \; . \tag{16}$$

... b_{-3}	a_{-2}	b_{-2}	a_{-1}	b_{-1}	a_0	b_0	a_1	b_1	a_2	b_2	a_3 ...
...											
q	$(\beta-2)-\frac{\alpha^2}{(\beta-2)}$	m		q							
	m	$(\beta-2)$	q								
		q	$(\beta-1)-\frac{\alpha^2}{(\beta-1)}$	m		q					
	q		m	$(\beta-1)$	q						
				q	$\beta-\frac{\alpha^2}{\beta}$	m		q			
			q		m	β	q				
						q	$(\beta+1)-\frac{\alpha^2}{(\beta+1)}$	m		q	
					q		m	$(\beta+1)$	q		
								q	$(\beta+2)-\frac{\alpha^2}{(\beta+2)}$	m	
							q		m	$(\beta+2)$	q
											...

Figure 1. *System Determinant* $|D|$ *of Eqs.(11) and (12).*

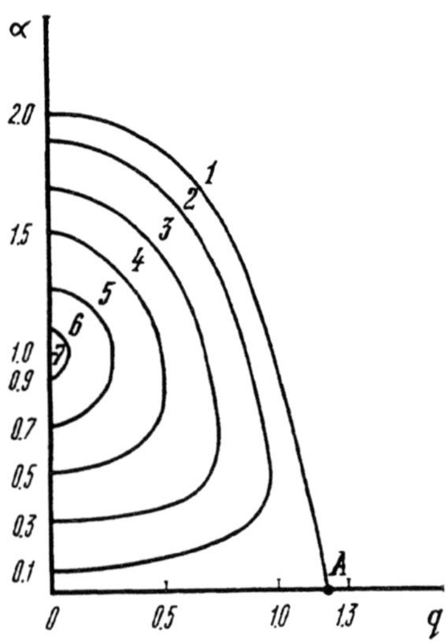

Figure 2. $m = 0$; β : (1) $\pm$ 0.0, (2) $\pm$ 0.1, (3) $\pm$ 0.3, (4) $\pm$ 0.5, (5) $\pm$ 0.7, (6) $\pm$ 0.9, (7) $\pm$ 1.0.

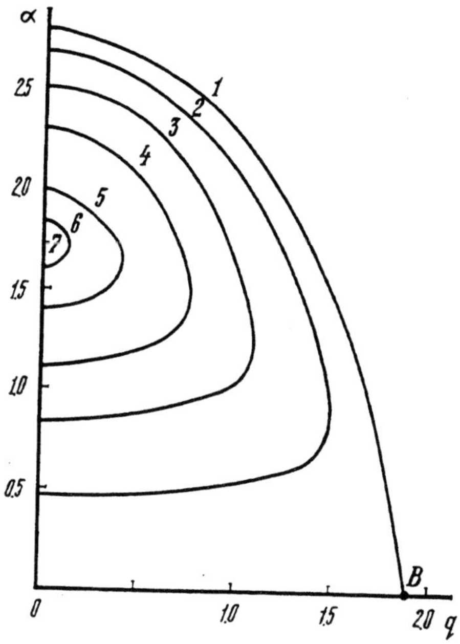

Figure 3. $m = 1$; β : (1) $\pm$ 0.0, (2) $\pm$ 0.1, (3) $\pm$ 0.3, (4) $\pm$ 0.5, (5) $\pm$ 0.7, (6) $\pm$ 0.9, (7) $\pm$ 1.0

In this case Eqs. (1)-(3) can be written as follows

$$\dot{P} = (\Delta\omega_0 + \omega^I \cos \Omega t)Q - E_0^2 \int P dt \ , \tag{17}$$

$$\dot{Q} = -(\Delta\omega_0 + \omega^I \cos \Omega t)P \ . \tag{18}$$

Eqs. (17) and (18) are equivalent[1] to Eqs. (5) and (6) and the aforementioned calculations can be applied to the FM as well as to the AM modulated field case.

[1] Z. Yankauskas, Optics and Spectroscopy, 1989, Vol. 67, p. 149-151 (Sov.Phys.)

Studies in Classical
and Quantum Nonlinear Optics

ISBN 1-56072-168-5

PHASE LOOP AM MODE LOCKING OF FLASH LAMP PUMPED Nd: GLASS LASER WITH NEGATIVE FEEDBACK CONTROL: THIRD HARMONIC GENERATION FROM Si

P. Yankov and G. Petrov
Sofia University, Institute for Laser Technology
"Galitchiza", 33A, Sofia 1126, Bulgaria

In this work we report on the construction of a Nd:glass laser with negative feedback control to be used in harmonic generation measurements from surfaces and interfaces.

The large bandwidth of the Nd:glass active media allows shorter pulses to be generated. In comparison to the cw systems the pulsed lasers offer the advantages to obtain much higher pulse energies and greater mechanical stability. But due to the larger amplification there is a tendency to simultaneous Q-switching. In this case the application of a real negative feedback overcomes this tendency and the laser generates a long, smooth pulse train in the whole range of pumping. The large number of round trips of the pulse leads to the formation of ps pulses with duration defined by the way of mode locking. Due to the large bandwidth of the gain material (Nd:glass) shorter pulses could be obtained, but the instabilities between the modulation and the cavity round-trip frequencies should be kept below 5.10^{-7}. We applied a combined scheme suitable for operation in pulsed system. It incorporates the advantages of both systems: stable ps pulses as for the cw systems and high output energy as for the pulsed systems. On

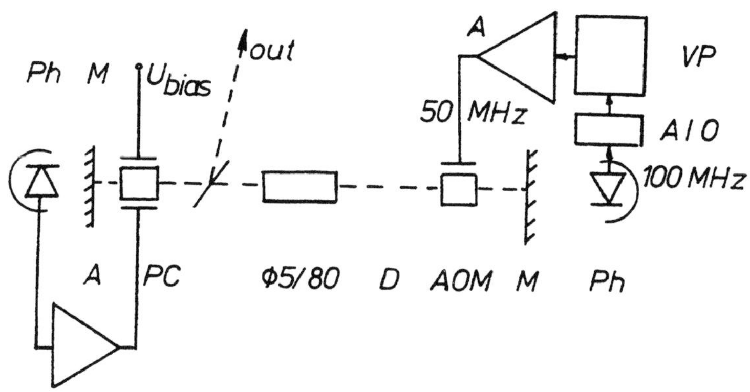

Figure 1. Laser set-up.

Fig.1 the laser is shown. The active media is a Nd:phosphate glass ϕ 5/80 mm, with luminescence bandwidth of 22nm. The photodiode Ph1, the amplifier A and the Pockels cell PC form a negative feedback that controls the cavity losses, so smooth pulse trains are generated. The other feedback consisting Ph2, the amplifier /oscillator A/O, the manually variable phase circuit VP and amplifier A drives the acoustoptic modulator AOM . The 50 MHz amplifier - oscillator is close to the threshold of self oscillation and is synchronized by the subharmonic of the 100 MHz repetition rate of the ps pulses from the resonator.This oscillator has a +/- 2 MHz locking range. After suitable amplification and an adjustable phase shift this signal drives the acoustooptic mode-locker. The manually driven phase compensates the constant phase shift from the electronic circuit and the AOM. The overall delay of the whole feedback is 15-20 ns, i.e. the phase and frequency are corrected after the third or forth pulse. Thus the phase instabilities between the pulse train and the mode locking frequency are determined by the dispersion of the high

Q-factor of the acoustoptic modulator, which in our case allows a cavity length mismatch of +/- 40 μm. The ps pulse durations are measured by a multiple shots background free autocorrelator with accuracy of 0.5 ps. A 100 ns gate allows us to measure the ps pulses duriation at different delays from the beginning of the pulse train. We compare the pulse durations in the middle of the train (20 μs delay) without and with the phase loop. We were not able to obtain pulses shorter than 35 ps with frequency synthesizer driven mode locker ($\delta\nu/\nu = 10^{-7}$), while with the phase locked oscillator the duration of the pulses is less than 10-12 ps in the middle of the train. This value is close to the pulse duration predicted by the theory for our acoustooptic modulator. The maximum energy of the 50 μs smooth pulse train is 20 mJ and is determined by our power supply.The single pulse energy is kept below 10 μJ by the negative feedback to avoid the self phase modulation in the active medium. Nevertheless a slight broadening of the spectra is observed in the pulses at the end of the train giving a value $\delta\nu \cdot \tau_p = 0.7$. The spectrum is measured with a Fabry-Perrot interferometer and a gated optical multichanel analyzer. The pulse train energy stability is +/- 2,5%, i.e. the same as in the free running regime. We should note that the 1.4 m resonator is not temperature stabilized and the parameters of the laser are kept in several days work.

So an AM mode locked Nd:phosphate glass laser is reported.Two independent feed backs are applied: a negative one giving a long and smooth pulse train and a positive one synchronizing the phase and frequency of the mode locking with the repetition rate of the traveling ps pulse in the cavity. We have measured the parameters of the system with a Nd:YAG laser rod as active media as well. In this case 65 ps, 10-15 μJ single pulses with repetition rate of up to 30 Hz are obtained. In this case an unstable resonator is applied and the energy of the whole 100-120 μs long train reaches 80-90 mJ.

Harmonic generation from surfaces of samples, the bulk of which possess inversion symmetry has evoked much interest during the last decade. It was soon realized that it is the upper surface layers with broken symmetry that contribute with their dipole allowed transitions to the second harmonic signal together with the bulk quadrupole contribution. More attention has been paid to band structure calculations of the nonlinear response functions of tetrahedral semiconductors that reveal the despersion of that response.

Our attention has been drawn to the possibility to study third harmonic generation (THG) from silicon surface at a wavelength close to the bandgap and slightly vary this gap with temperature. Since nonlinear response is expected to exhibit resonant enhancement under these conditions, such measurements would yield an experimental ground for comparison with band structure calculations of the nonlinear response functions.

THG is a bulk dipole allowed process in centrosymmetric silicon and nonlinear polarization at 3ω is given by:

$$P_i^{(3)D}(3\omega)=(\chi_{1122}+\chi_{1212}+\chi_{1221})E_i(E.E)+\xi_D E_i^3 \qquad (1)$$

where $\xi_D=\chi_{1111}-(\chi_{1122}+\chi_{1212}+\chi_{1221})\neq 0$ for the cubic symmetry. D.J.Moss et al. [1,2] have carried band structure calculations and measurements that revealed the dispersion of THG from silicon in the wavelength range from 0,7 to 2,0 μm. They have shown that the anisotropy parameter ξ_D is a more sensitive quantity to electronic structure variations let alone χ_{1111} quantity. They have observed a weak structure of the anisotropy parameter near the wavelength of YAG generation that has been attributed to 3ω resonance of the E_1 critical point (direect bandgap) of silicon.

We have studied Si samples of (111) orientation placed at 45 degrees of incidence to the pump beam of 1,064 μm of our picosecond system. Input polarizer and output analyzer where set to a parallel position and rotated synchronously from p-p to s-s position. We note that the s-s relationship for THG from Si(111) is an isotropic one independent of sample orientation. This ralationship of the intensity of THG reads:

$$I_{TH}^{s-s} = F_{\perp}^{R}(\theta,\varepsilon)\,|\chi_{1111} - 1/2\,\xi_D|^2\,|E(\omega)|^6 \qquad (2)$$

where $F_{\perp}^{R}(\theta,\varepsilon)$ plays the role of a Fresnel coefficient, that depends on the angle of incidence and linear dielectric functions at ω and 3ω. $E(\omega)$ is the amplitude of the fundamental frequency. Temperature has been varied from 25°C to 520°C and measured with a thermocouple. Calculations of $F_{\perp}^{R}(\theta,\varepsilon)$ have been performed on the basis of the known temperature dependence of $\varepsilon(\omega)$ and $\varepsilon(3\omega)$ in the cited temperature range. These quantities are given in Table 1.

T,K	$\varepsilon'(\omega)$	$\varepsilon''(\omega)$	$\varepsilon'(3\omega)$	$\varepsilon''(3\omega)$
294	12,46	$1,7.10^{-3}$	24,0	35,0
465	12,74	$1,9.10^{-2}$	21,0	30,0
646	13,10	$3,7.10^{-2}$	18,0	28,5
972	13,51	$1,2.10^{-1}$	15,0	28,0

Table 1. Values of real and imaginary parts of dielectric function of silicon versus temperature.

It has been found that nonlinear Fresnel coefficient varies insignificantly within that temperature range and is equal to 5.10^{-4}. We have observed a monotonic decrease of THG signal with temperature rise (see Fig. 2) and deduced the following relationship between the elements of the nonliner polarizability tensor:

$$|\chi_{1111} - 1/2\,\xi_D|_{25^\circ C} \,/\, |\chi_{1111} - 1/2\,\xi_D|_{520^\circ C} = 1.7 \qquad (3)$$

This result is consistent with the observations of D.J.Moss et al. on the despersion of the anisotropy parameter. In our case we have a somewhat similar situation - temperature rise will shrink the gap and detune the 3ω

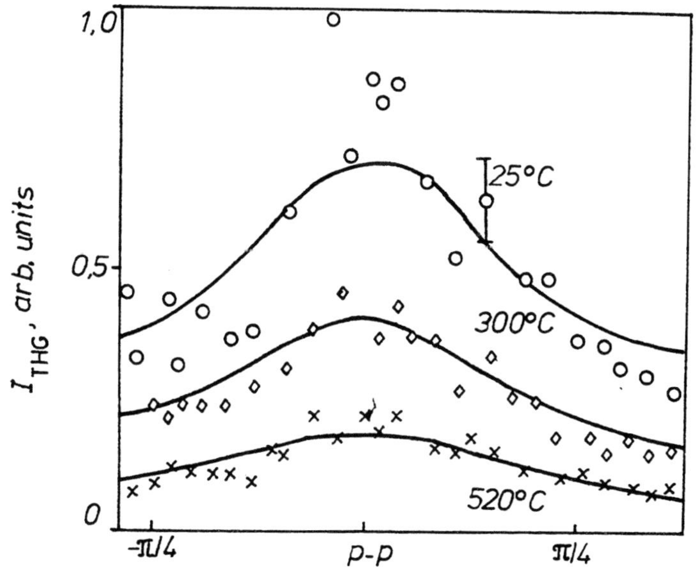

Figure 2. Temperature dependence of THG from Si (111) surface

resonance that is the overwhelming factor leading to THG from silicon when pump wavelength is 1,064 μm.

REFERENCES

1. D. J. Moss, H. M. van Driel and J. E. Sipe, "Dispersion of the Anisotropy of Third Harmonic Generation in Silicon", *Optics Letts.*, Vol. 14, No1, 1989, pp. 57-59.

2. D. J. Moss, E. Ghahramani, J. E. Sipe, H. M. van Driel, "Band Structure Calculations of Dispersion and Anisotropy in (3) for Third Harmonic Generation in Si, Ge and GaAs", *Phys. Rev. B*, Vol. 41, No 53, pp. 1542-1560.

Studies in Classical
and Quantum Nonlinear Optics

ISBN 1-56072-168-5

LASER DOPPLER VELOCIMETRY MEASUREMENTS ON INTERMITTENT PIPE FLOWS

Jun Zhang, Mogens T. Levinsen and Preben Alstrøm
Physics Laboratory, H. C. Ørsted Institute,
Universitetsparken 5, DK-2100 Copenhagen Ø, Denmark

ABSTRACT

Laser Doppler velocimetry is playing a key role for noncontact velocity measurements in various flows. It measures velocities of particles suspended in the flow. The measuring mechanism is discussed in this work. The results show interesting properties of a pipe flow in the intermittent regime.

Pipe flows are common in daily life as well as in scientific applications, the technical studies and scientific discussions have been around for centuries. The intermittency transition to turbulence in pipe flows is one of the oldest studied problems in turbulence [1]. There are more and more carefully designed measurements carried out by use of optical methods or by other means [2,3]. Many aspects of the behavior of pipe flows can be unveiled by utilizing these tools.

In this work, we are going to present an experiment on pipe flows, where we will focus our attention on the onset of intermittency, i.e. where the pipe flow shows turbulent spots but still is far from the fully turbulent state. In this regime, the system changes temporarily from laminar to turbulent flow under the same imposed conditions. Laser Doppler velocimetry (LDV for short) is used to read out the velocity signals coming from particles suspended in (and following) the flow.

I. THE EXPERIMENT

A circular glass pipe, 1 cm in inner diameter and 150 cm in length, was placed horizontally and connected via plastic tubes (4 cm in inner diameter) to reservoirs containing deionized water. Keeping the water levels inside the reservoirs at different heights, a pressure drop exists between the inlet and the outlet of the pipe, and a flow is observed running through the pipe. The plastic tubes were connected to short glass tubes that were welded to the pipe by smooth contraction (Fig. 1). Throughout the glass contraction regions the angle between the wall and the axial direction is less than 9°. Upstream from the pipe, tightly packed straws, 9 cm long and 0.5 cm in diameter, were fitted inside the plastic tube. A fly-screen (average aperture size = 2.0 mm) was placed 5 cm downstream from the straws and 30.5 cm upstream from the pipe entrance. The difference in height level, and consequently in pressure drop, between the two reservoirs was maintained by means of a pump. The value of the pressure drop was controlled by adjusting the pumping rate.

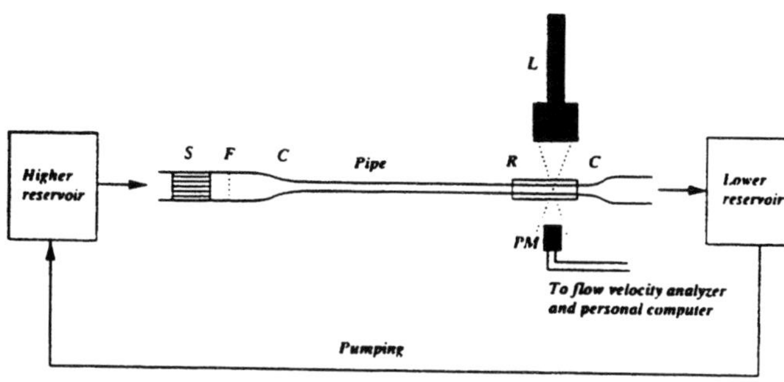

Figure 1. Schematic illustration of the experimental setup. S: straws; F: fly-screen; C: smooth contractions; L: He-Ne laser; R: rectangular cell; PM: photomultiplier.

The water was seeded with 1 to 2 ml homogenized milk (concentration ratio $\sim 10^{-6}$). LDV was used to measure the central axial velocity near the end of the pipe. Two parallel linearly polarized laser beams with 60 mm separation splitted from a He-Ne laser, were focused into the measuring volume by a lens with focal length 120 mm. The laser was operated in the TEM_{00} mode, with a 15 mW output. The size of the measuring volume in the axial direction of the pipe was about ~ 0.1 mm. To avoid the undesirable refraction from the curved pipe wall, a rectangular glass cell (filled with water) was added around the pipe where the velocity was measured.

One of the laser beams was shifted by about 40 MHz with respect to the other by use of a Bragg cell. The scattered light from particles suspended in the flow was detected by a photomultiplier (PM). The output from the PM went to a DANTEC 58N20 flow velocity analyzer (with internal frequency = 40 MHz). This was operated so that it rejected Doppler signals with less than 60 cycles, to avoid erroneous measurements.[1] The arriving time and the velocity of each particle were transmitted to a personal computer. The average sampling rate was about 100 Hz.[2] The exact Bragg cell frequency was set to optimize the velocity resolution.[3]

[1] Setting the number of cycles is called validation setting.
[2] Controlled by means of seeding concentration, PM bias voltage, and validation setting.
[3] See section II.

As we have observed from the experiment, when the pressure drop is kept below a critical value Δp_0, the flow stays laminar. The quality of the laminar flow was examined. The laminar velocity profiles were measured both near the entrance and at the end of the pipe and found to be in good agreement with descriptions of Newtonian fluids: very flat at the entrance and almost parabolic at the end.

For our concern, the pressure drop Δp was controlled to be just above Δp_0, where the system changes temporarily from a laminar to a turbulent flow. A spreading turbulent flow between laminar flows is called a turbulent *slug* [2,4]. They have seemingly random and sudden appearance. The slugs are mostly formed near the entrance due to the departure from the parabolic profile and due to the finite-size disturbances occurring there. A slug is convected by the flow downstream while at the same time it spreads - laminar fluid at both ends of the slug becomes turbulent. In pipe flows driven by gravity (constant pressure difference) an additional complication arises because the velocity constantly varies. The reason for this is that the friction is larger for a turbulent flow than for a laminar flow; therefore the bulk velocity decreases as a slug spreads. Near the onset of turbulence the formation of one slug thus prevents the formation of new slugs until the slug has left the pipe and the flow again is laminar.

When the pressure drop is increased, the time the flow spends in the turbulence phase gets longer and longer, and finally the flow will be fully turbulent. A measure of how much time the flow spends in each of the two states is given by the intermittency factor γ, which is defined as the fraction of the time the flow is turbulent. Fig 2 shows us a typical velocity time series over 20 seconds. The intermittency factor γ was measured to be 0.11 in a 1800 seconds run. It is convenient to introduce an important dimensionless quantity, known as the Reynolds number. The Reynolds number Re is defined as $Re \equiv Ud/\nu$, where U is the bulk velocity, d the pipe diameter, and ν the viscosity of water. For the whole flow system, the Reynolds number can be different from place to place. For our current experiment, the Reynolds number in the pipe is around 20 000 (the highest value in the system), and inside the plastic tube it is in the order of 5 000. The smallest Reynolds number is found to be less than 650 inside the straws.

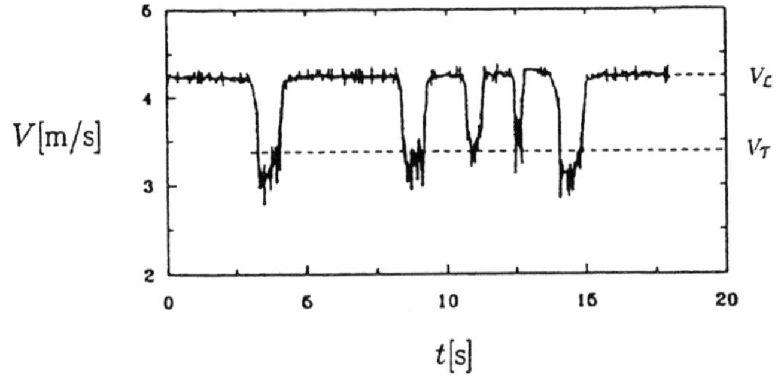

Figure 2. Axial velocity V versus time t, measured at the center and near the exit of the pipe ($\Delta p = 7120$ Pa). A sharp transition between a laminar state with $V = V_{\mathcal{L}} \simeq 4.24$ m/s and a turbulent state with $V = V_{\mathcal{T}} \simeq 3.34$ m/s is observed.

II. MEASURING MECHANISM

The mechanism of LDV measurement can be explained in terms of the Doppler effect. But for many LDV users, the *fringe model* is favored because it makes sense intuitively.

Two highly coherent beams coming from the same light source, like the two parallel beams splitted from a laser as in our case, are focused by a common lens such that the beams intersect where both beams have their minimum cross section (waist). Highly visible interference pattern are formed within the intersection. Since in a short range inside the waist of each laser beam, wavefronts of a Gaussian beam are very much like plan waves, a set of regular *fringe planes* equally distributed in the intersection region are formed. Denote the spacing between two adjacent fringes by Λ. When a moving particle crosses the bright and dark fringe planes, it modulates the light intensity received by a detector (the PM section), giving rise to a periodic signal[4] with a Gaussian envelope. The Doppler frequency f_D is proportional to the particle velocity component V normal to the fringe planes,

$$\Lambda = \frac{\lambda}{2\sin(\theta/2)} \quad , \qquad f_D = \frac{V}{\Lambda} = \frac{2V}{\lambda}\sin\frac{\theta}{2}. \tag{1}$$

Here λ is the laser wavelength, and θ is the angle between the two incident beams.

The accuracy[5] for the velocity measurement was designed to be about 1%, and this was checked in our experiment. For our present concerns, we pay attention in order to distinguish flow states between laminar and turbulent. 1% is enough to serve this purpose, since we are measuring in a central point of the pipe, near the end, where we have the maximum axial velocity in the entire flow system when the flow is laminar. If the flow goes turbulent, the bulk velocity has a certain drop, and furthermore, the average velocity profile in this case tends to be more flat compared to the nearly parabolic laminar flow. It is therefore advantageous to measure in the center where we have a comparatively high velocity change in the axial direction while, at the same time, the transverse velocity gradient is at its minimum. This furthermore makes it easy to align the optical components.

For later studies, better accuracy will be demanded when we need to look into the fine structures of turbulence. We found that it is possible to improve the accuracy considerably from the original design by electronic means. We will describe this part in few words since it is not crucial for the current experiment.

The original system measures a velocity range that spans over a short negative region and a rather wide positive region at any frequency bandwidth. Hence, for velocities well above zero, the number of velocity channels used becomes small, i.e. the resolution becomes poor. To improve the resolution, we therefore send a frequency f_m to the Bragg cell from a synthesizer which is different from the reference frequency of the signal analyzer (40 MHz). The frequency and thus the velocity range is then shifted to a new range centered around the velocity to be measured. The size of the shift is the difference between the frequency f_m to the Bragg cell and the reference

[4]Assuming that the flow velocity does not change much over 0.1 mm.

[5]Defined as $\Delta V/\overline{V}$, where ΔV denotes the measured velocity uncertainty and $\overline{V}$ the mean velocity for steady laminar flow.

frequency. By selecting the frequency bandwidth properly, the accuracy can be improved to be $0.1 - 0.4\%$, depending on the nature of the flow. In some extreme cases, say, very high laminar velocities, the accuracy can be made even better than 0.1%.

Going back to eq.(1), by changing the optical arrangement, one can optimize the frequency range for fixed flow rate. However, in practice, one should keep in mind that the detected frequency $f_m + f_D$ has to be controlled within certain limits. Consider our He-Ne laser, which has a typical natural gain linewidth 1.5×10^9 Hz. From its 47 cm long resonator, the frequency spacing between longitudinal modes is about 320 MHz. Thus it is possible that a few longitudinal modes may survive and get involved in the scattered signals. That will limit us to make $f_m + f_D$ less than 320 MHz.[6] Otherwise the beating between different modes can make the result of the measurements ambiguous. Also, the channel width is limited by the electronic noise from the flow analyzer, the stability of the frequency f_m, etc. It is advisable to set the channel width to be higher than about 1 kHz.

In some circumstances, when the data rate is too high, phase noise will appear caused by the dense and random appearance of particles in conjunction with the finite measuring volume which may cause ambiguous measurements [5]. In our case, we keep a very low data rate ($\sim$ 100 Hz); in the velocity range considerd ($\sim$ 4 m/s), we get a ratio between the mean free path for a particle and the measuring size (0.1 mm) that is 400, far greater than 1. Hence, the chance for two particles to appear in the measuring volume at the same time is very small. We conclude that our data rate is far below the order of magnitude that may cause phase noise.

III. RESULTS AND DISCUSSION

In our experiment, the pressure drop Δp is the control parameter and this was changed in steps. At each different Δp, we performed a 30 min measurement; about 180 000 particles were recorded. As shown in Fig. 2, the sharp transition between a laminar velocity $V_{\mathcal{L}}$ and a turbulent velocity $V_{\mathcal{T}}$ suggests a well-defined interface between the laminar and the turbulent flow and allows a clear distinction between these two states. The intermittency factor is found to increase monotonically with pressure drop.[7]

Further quantitative characterization of the dynamics of the transition intermittency can be obtained through a statistical analysis of the turbulent and laminar times. The turbulent time $t_{\mathcal{T}}$ is the time it takes for a slug to pass the measuring point, and the laminar time $t_{\mathcal{L}}$ is the time elapsing between the end of a slug (the trailing edge) and the beginning (the leading edge) of the next. We consider here the mean turbulent time $\overline{t_{\mathcal{T}}}$,[8] the mean laminar time $\overline{t_{\mathcal{L}}}$,[9] and the inverse cumulative distribution of laminar times $P(t_{\mathcal{L}})$, defined as the fraction of laminar times that are larger than $t_{\mathcal{L}}$. Figure 3(a) shows the relation between γ and the mean turbulent

[6] If there exist more than one transverse modes, the frequency spacing will be even smaller.

[7] At sufficiently high values of Δp, above the range considered here, the flow becomes fully turbulent with $\gamma = 1$.

[8] The total time the flow at the measuring point (near the outlet) is turbulent divided by the number of slugs observed at the outlet during the elapsed period.

[9] Defined similar to the mean turbulent time $\overline{t_{\mathcal{T}}}$.

time (rescaled by a constant[10] f_0). Notice that at the onset of intermittency the mean turbulent time remains finite. In Fig. 3(b) a log-log plot of the rescaled mean laminar time versus γ is shown. The data show that $\overline{t_{\mathcal{L}}} \sim \gamma^{-1}$. For obvious reasons the data lie on a curve with negative slope, but that they align on a rather straight line with a slope close to -1 calls for further theoretical studies.

In Fig. 4 we show the (normalized) inverse cumulative distributions $P(t_{\mathcal{L}})$ for different values of γ close to the onset of intermittency. The data show that the distributions can be well described as exponentially decaying, $P(t_{\mathcal{L}}) \sim \exp(-\sigma f_0 t_{\mathcal{L}})$, with a decay rate σ that is proportional to γ [cf. inset in Fig. 4]. At larger values of γ, deviations from an exponential decay are observed.

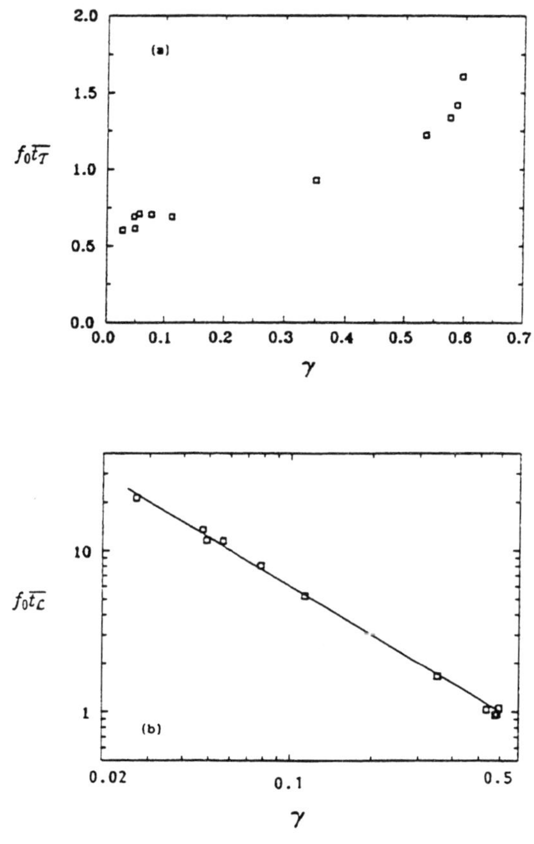

Figure 3. (a) Rescaled mean turbulent time $f_0\overline{t_{\mathcal{T}}}$, plotted as a function of the intermittency factor γ. (b) Double-logarithmic plot of the rescaled mean laminar time $f_0\overline{t_{\mathcal{L}}}$ versus γ; solid reference line in the plot has a slope -1.

[10] $f_0 \equiv V_0/2L$, where V_0 is the axial velocity at the onset of intermittency, and L is the pipe length.

We have already mentioned that the two essential mechanisms for interpreting the above observations are the generation of slugs in the entrance region of the pipe and the growth process which takes place as the slugs are convected by the flow. The generation of slugs is due to the level of disturbance and amplification effects which both are related to the Reynolds number. The exponential decay of $P(t_{\mathcal{L}})$ suggests that the disturbance to flows can be treated as white noise at low intermittency factors: once there is a laminar flow, it may change into a slug with a certain *probability*[11] at any fixed time unit. The departure from the exponential decay for large γ may be due to the feedback dynamics involved in the formation of slugs (or laminar flows).

Further detailed understanding of the growth process from a certain level of local *disturbance* to a well-developed *slug* needs more theoretical and experimental research. A phenomenological model based on experiments and using stochastic arguments has been explored in [6].

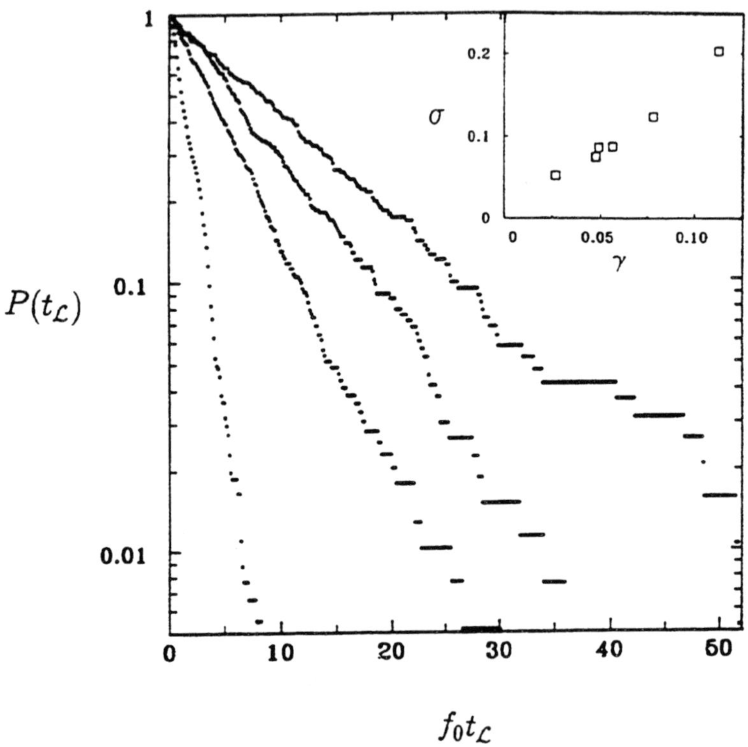

Figure 4. Inverse cumulative distributions $P(t_{\mathcal{L}})$ of laminar times obtained experimentally at different values of the intermittency factor γ. From right to left: γ=0.05, 0.075, 0.11, 0.35. *Inset:* Decay rate σ versus γ.

[11] Which is a function of the Reynolds number.

In conclusion, based on LDV measurements, statistical information for intermittent pipe flows have been unveiled, giving us a picture for understanding the aspects of intermittency. LDV is a convenient tool which may bring us to a better stage in understanding the transition to turbulence in pipe flows.

ACKNOWLEDGEMENT

We are greatful for helpful discussions with D. Stassinopoulos, B. Christiansen and K. R. Sreenivasan. We also thank S. Pantman and J. Brinchmann for their skillful mechanical assistance. JZ was supported by the Danish Research Academy and the Daloon Foundation.

REFERENCES

1. O. Reynolds, "On the dynamical theory of incompressible viscous fluids and the determination of the criterion", *Philos. Trans.*, Vol. 186, 1895, pp. 123-164
2. I. J. Wygnanski and F. H. Champagne, "On transition in a pipe. Part 1. The origin of puffs and slugs and the flow in a turbulent slug", *J. Fluid Mech.*, Vol. 59, part 2, 1973, pp. 281-335
3. K. R. Sreenivasan and R. Ramshankar, "Transition intermittency in open flows, and intermittency routes to chaos", *Physica D*, Vol. 23, 1986, pp. 246-258
4. D. J. Tritton, *Physical Fluid Dynamics*, Oxford: Oxford Science Publications, 1988, pp. 12-19 and pp. 281-288
5. L. Lading, "The fundamental limits to turbulence measurements - revived", in *Laser Anemometry - Advances and Applications*, Ed. A. Dybbs and B. Ghorashi, New York: ASME, 1991, pp. 58-62
6. J. Zhang, D. Stassinopoulos, P. Alstrøm, and M. T. Levinsen, "Stochastic transition intermittency in pipe flows: experiment and model", submitted to *Phys. Fluids A*, 1992

SUBJECT INDEX